Physical Methods in Advanced Inorganic Chemistry

Physical Methods in Advanced Inorganic Chemistry

Edited by

H.A.O. Hill

Fellow of The Queen's College, Oxford

P. Day

Fellow of St. John's College, Oxford

1968

INTERSCIENCE PUBLISHERS

a division of John Wiley and Sons

London New York Sydney

First Published 1968 by John Wiley & Sons Ltd.

Library of Congress catalog card No. 68–19106
SBN 470 39610 5

Made and printed in Great Britain by William Clowes and Sons, Limited
London and Beccles

FOREWORD

Inorganic chemistry is traditionally the study of the comparative chemistry of the elements. Until recently, this merely meant listing the properties of series of compounds against the background of the periodic table. With advances in physicochemical studies, a more sophisticated approach has developed in which the properties of each compound are discussed against the background of a knowledge of the molecular (or crystal) and electronic structure of that compound. It is therefore essential that research workers in inorganic chemistry should have a thorough understanding not so much of the detailed theory of the different physical methods as of the manner in which any one technique can be applied to a given problem and the type of information it will provide. This book meets that need. Thus, in each of the eleven chapters, an author has been chosen who is extremely knowledgeable about the application of a given technique to inorganic chemistry. For the most part, neither the technique nor its theory are described in any depth—for this purpose other texts exist—but the authors skilfully show what the method can reveal about inorganic compounds. It is left for the editors to illustrate how a combination of the methods can be utilized. Rather than attempting an exhaustive survey, they have carefully chosen a few examples to show that the methods in combination are powerful indeed. Only when a great number of compounds have been subjected to similar study will the new comparative inorganic chemistry be securely founded.

R. J. P. Williams

PREFACE

In recent years, physical methods have become a feature of most research projects in inorganic chemistry. Indeed, it could be argued that the much acclaimed renaissance in inorganic chemistry has been due to the increasing application of spectroscopic and diffraction techniques. Before the extensive use of these methods, there was little an inorganic chemist could do to rationalize the results of his preparative ingenuity. Their development has made available the means of arriving at a structural and electronic picture of the extraordinary variety of compounds now known.

Though many reviews of the better known techniques have appeared, they are widely scattered and few have attempted critical assessments of their value and limitations. We have therefore collected in a single volume a number of accounts whose intention is to illustrate the character of the information obtained from each method rather than the details of the methods of obtaining it. We hope that they will indicate what can and cannot be deduced in each case and, by means of carefully chosen examples, show how the methods have been used to solve problems in many branches of inorganic chemistry. The articles, therefore, are not intended to be reviews, although we hope that they will serve as points of reference.

Because the methods are at different stages of development, the approach to the subject matter used in the various chapters differs considerably but, in our opinion, legitimately. For example, it would be impracticable and indeed superfluous, to include in the chapter on nuclear magnetic resonance spectroscopy a detailed description of the theory, since this is readily available to chemists. By comparison, it is necessary to give a fairly extensive introduction to the theory of nuclear quadrupole resonance spectroscopy. In addition to the better known physical methods, we have sought accounts of some which, though not widely used at present, appear to have considerable potential. Among the various spectroscopic and diffraction methods, a chapter on thermochemical measurements has been included because, as the authors point out, it is important that inorganic chemists understand the origin and relevance of thermochemical data, since the explanation of such data is the aim of much theoretical work. We have omitted discussions of some physical properties, such as dipole moments or bulk paramagnetism, because either they have been extensively reviewed for chemists or else the application of the method is suffering a decline.

We hope that this book will be helpful to those already active in inorganic research and to those about to start. It is important that a graduate student beginning research should have some idea of the various methods used to tackle problems in inorganic chemistry and that this knowledge should encourage him to adopt a critical attitude to the relevance of a particular technique to the problem in hand. In short, our hope is that the book will encourage the design of good experiments.

A knowledge of inorganic chemistry to the level of Phillips and Williams' *Inorganic Chemistry* (Oxford University Press, London, 1965) or Cotton and Wilkinson's *Advanced Inorganic Chemistry*, 2nd edition (Interscience, New York, 1966) is assumed, and also an acquaintance with group theory and quantum chemistry as given in such books as Cotton's *Chemical Applications of Group Theory* (John Wiley and Sons, New York, 1964) and Murrell, Kettle, and Tedder's *Valence Theory* (John Wiley and Sons, London, 1965).

Finally, we wish to thank our collaborators for their help and enthusiasm, Mr. C. F. Hague for the translation of Dr. Bonnelle's contribution, and Drs. P. R. Edwards, C. K. Jørgensen and R. J. P. Williams for assistance with a number of points.

August, 1967

H. A. O. Hill
P. Day

Contributing Authors

BONNELLE, C. — *Laboratoire de Chimie-Physique de la Faculté des Sciences de Paris, Paris, France*

CHRISTENSEN, J. J. — *Brigham Young University, Provo, Utah, U.S.A.*

DANON, J. — *Centro Brasileiro de Pesquisas Físicas, Rio de Janeiro, Brazil*

DAY, P. — *Inorganic Chemistry Laboratory, University of Oxford, Oxford, England*

EATON, D. R. — *Central Research Department, E. I. du Pont de Nemours & Co., Wilmington, Delaware, U.S.A.*

GILLARD, R. D. — *University of Kent, Canterbury, England*

HILL, H. A. O. — *Inorganic Chemistry Laboratory, University of Oxford, Oxford, England*

IZATT, R. M. — *Brigham Young University, Provo, Utah, U.S.A.*

KÖNIG, E. — *Institut für Physikalische Chemie der Universität Erlangen-Nürnberg, Germany*

PROUT, C. K. — *Chemical Crystallography Laboratory, University of Oxford, Oxford, England*

SCHMIDTKE, H.-H. — *Cyanamid European Research Institute, Geneva, Switzerland*

SILLESCU, H. — *Institut für Physikalische Chemie der Universität Frankfurt-am-Main, Germany*

TURNER, D. W. — *Physical Chemistry Laboratory, University of Oxford, Oxford, England*

WARE, M. J. — *University of Manchester, Manchester, England*

Contents

1

Diffraction Methods

C. K. PROUT

I. INTRODUCTION

It is neither feasible nor desirable to describe in detail the complex techniques of the diffraction methods of structure analysis within the narrow confines of a single chapter. These techniques have already been described with great clarity by many authors. The object of the present chapter is to provide the reader with such information as to allow him to make a reasonable assessment of the quality and value of the structure analyst's work. Even this limited objective is too extensive unless some previous acquaintance with the field is assumed. It is recommended that the reader should be familiar with the basic principles of the method as described in one or more of the texts in Section A of the Bibliography.

The diffraction methods of structure analysis cover vapour-phase, thin-film, and single-crystal electron diffraction, liquid and amorphous and crystalline solid-state x-ray diffraction, and solid-state neutron diffraction. Each technique has its particular value but single-crystal x-ray diffraction is at present the simplest and most powerful. Electron-diffraction measurements in the gas phase are only applicable to relatively simple stable molecules but in the solid state find considerable application in the study of thin films. The necessary high-vacuum conditions restrict solid-state work to involatile phases. Cost is the greatest single restriction to the use and application of neutron diffraction; a recent estimate prices thermal neutrons at one dollar a hundred. The number and availability of high-flux sources is also limited.

X-ray diffraction structure analysis is usually regarded as one of the most tedious of physicochemical techniques. This impression, although originally true, is now one of the major misconceptions of structural chemistry. In 1962, Wheatley (ref. A.2 of the Bibliography) stated that the structure determination of vitamin B_{12} employed ten workers for ten years. A more realistic evaluation of the situation is the fact that the x-ray crystallographic investigation of the 109-atom 5,6-dimethylbenzimidazolecobamide coenzyme[1] by an experienced crystallographer took one year, and that a totally inexperienced graduate student was able to determine the structure of copper glycollate[2] by three-dimensional methods in four weeks from start to finish.

The relative importance of the three techniques for solid-state problems can be demonstrated from *Acta Crystallographica*, Vol. **18**, Jan.–June 1965. There are 111 major works in the x-ray field, two in the field of neutron diffraction, and no electron-diffraction determinations. This preponderance of x-ray diffraction work may be attributed to cost, convenience, experimental simplicity, and the heavy-atom method. As the predominant technique, the major part of this chapter will be devoted to the x-ray method.

II. CHOICE OF RADIATION

Many analyses fail, from the outset, to have any real hope of reaching their objective because the wrong radiation is used for the structure under consideration. The choice of radiation at first seems large—electrons, neutrons, or x-rays, and a wide choice of wavelength for each type; yet, returning again to Vol. **18** of *Acta Crystallographica*, of the 113 major determinations not only did 111 use x-rays but, of these 111, 90 used the Cu $K\alpha$ lines and 17 the Mo $K\alpha$ lines. The routine structure analysis will tend to employ Cu $K\alpha$ radiation or possibly Mo $K\alpha$ independent of the physicochemical problem. There is a dual convenience in using these radiations. Firstly, they are very satisfactory target materials and allow x-ray tubes to be made with two to three times the flux obtainable from comparable x-ray sources using targets of neighbouring metals in the periodic table. Secondly, the wavelengths of the Cu $K\alpha$ lines (1.5418 Å) and Mo $K\alpha$ lines (0.717 Å) allow the recording of more or less the total available data on cylindrical and flatplate cameras, respectively, for the average molecular crystal. Only on a few occasions are other x-ray sources used. It is, however, undesirable to use a radiation that has either an energy fractionally less than the absorption edge of a major constituent atom of the crystal or an energy which will excite intense fluorescence from a major constituent atom. Both these circumstances will lead to inaccurate intensity data, although the anomalous dispersion which is maximized before an absorption edge may be turned to advantage in the process of phase determination.

X-rays are scattered by the electrons of the atoms in the crystal and, for scattering with the Bragg angle equal to zero degrees, the scattering amplitude of a given atom is directly proportional to the number of electrons in the atom. The wavelength of x-rays used is comparable to the size of the atom and as the Bragg angle increases the scattering amplitude decreases due to interference between radiation scattered by different parts of the atom. The percentage decrease in scattering amplitude is greater for the atoms with fewer electrons. From a given crystal, the scattering from the heavier atoms will tend to mask that from the lighter atoms and this effect will be more marked at high Bragg angles. It is commonly recognized that it is difficult to detect and accurately locate hydrogen atoms by x-ray diffraction, particularly in the presence of elements other than those in the first row of the periodic table. It is not always realized that it is almost equally difficult to detect the first-row elements in the presence of the heavier elements such as uranium. The use of x-ray diffraction in the structure analysis of crystals containing these elements, in particular binary compounds, e.g. UO_2[3] or TeO_2[4], has led to incorrect conclusions. Neutrons, however, are scattered principally by nuclei. It is not possible to calculate the scattering amplitude for neutrons but experience shows that, although there is an underlying slow increase in scattering amplitude with atomic weight, the effect is outweighed by that of resonance scattering and the resulting scattering amplitudes show a random variation throughout the periodic table. The variation between the smallest and the largest scattering amplitude is less than one order of magnitude and the majority lie within a factor of two or three of each other. In addition, there is no fall-off of neutron-scattering amplitude with Bragg angle. As a rough guide, it may be said that when the atoms of chemical interest in a structure constitute less than 10% of the total scattering matter and have about one-fifth or less of the atomic weight of the major constituents of the crystal, then it is not possible to get highly accurate atomic locations for these atoms from x-ray data. It may be necessary to query not only interatomic distances but in certain cases general stereochemical features. There will be no similar disadvantages to a neutron determination. Neutron diffraction has the further value in that the magnetic dipole of the neutron will interact with an atom with a permanent magnetic dipole to give the so-called magnetic scattering from which the magnetic superstructure of ferro-, ferri-, and antiferro-magnetic crystals may be determined[5] and which, in certain cases, will give information about the magnetic properties of paramagnetic crystals[6].

The chief scientific disadvantages of neutron diffraction are (*a*) that crystals ten to one hundred times the volume of those used in x-ray diffraction are required, (*b*) that even with large crystals and a high flux reactor as source the time required for the intensity measurement is at least ten times that required per reflection for x-ray work on a similar compound with an analogous data-collection technique, and (*c*) that the high incoherent neutron scattering from hydrogen atoms makes it difficult to determine structures with large numbers of hydrogen atoms, i.e. large organic molecules. The effects of (*b*) have been largely eliminated by the introduction of automatic computer-controlled data-collection equipment and the fallacy of (*c*) has been demonstrated by the determination of the structure of a vitamin B_{12} derivative by Hodgkin, Moore, and Willis[7].

Electrons which interact with the electric potential field of a crystal have the distinct advantage that the scattering amplitude falls off less rapidly with the atomic number of the scattering atom than a radiation which interacts only with the electrons of the atom. Making use of the Thomas–Fermi statistical theory for the atom, it can be inferred that at zero Bragg angle the scattering amplitude for electrons approximates to the one-third power of the atomic number[8]. If the Hartree–Fock method is used, finer variations in the scattering amplitude are observed. Within the first two periods of the periodic table ($Z = 1$–2 and 3–10), the scattering amplitudes *decrease*, not increase, with increasing atomic number. There is a sudden and considerable increase in scattering amplitude in passing from one period to another (i.e. from $Z = 2$ to $Z = 3$, or from $Z = 10$ to $Z = 11$). This is because of the size dependence of the potential. In general, electrons are better at locating light atoms than x-rays but not as good as neutrons. The relative values of scattering amplitudes of x-rays, neutrons, and electrons are summarized in Figure 1. Single-crystal electron

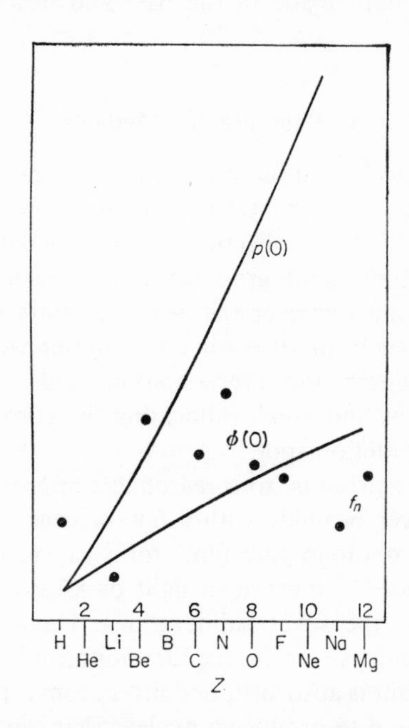

Figure 1. Curves of the detectability of atoms: in x-ray analysis $p(0)$, f_n in electron diffraction ϕ in neutron diffraction (0) (points). [Reproduced, by permission, from *Quart. Rev.* (*London*), **14**, 109 (1960), without change in nomenclature.]

diffraction has been developed over the last 20 years by Z. G. Pinsker and B. K. Vainshtein in the U.S.S.R. and by J. M. Cowley in Australia but it has not yet achieved its full impact as a structure analysis technique.

III. RECORDING AND MEASUREMENT OF INTENSITY DATA

Before the common use of large digital computers in structure analysis, the quality of the intensity measurement, providing it reached a certain minimum standard, was relatively unimportant because the numerical analysis of the measurements was never continued beyond the early stages. Large computers have allowed the analysis to be taken to the best possible agreement between observed and calculated structure amplitudes, e.g. by using the method of least squares. Today a major limiting factor to the accuracy of structure determination is the quality of the experimental measurement. It is therefore of great interest to examine in some detail how these measurements have been made in the past and how they may be made in the future.

A. Photographic Methods

Neutrons do not affect photographic emulsions but may be recorded photographically by using neutron-sensitive fluorescent screens. The technique is not used for the collection of intensity data.

Until the present time, photographic methods have been normal practice in x-ray diffraction but over the next few years the method should be replaced by counter techniques for most routine work. In general, the moving-film Weissenberg and precession methods have predominated because of the ease of 'indexing'—allocating the correct set of reflecting planes to a given diffraction spot.

A photographic emulsion is, to a reasonable approximation, a suspension of grains of silver bromide with a few percent of silver iodide in a gelatine base. Most photographic films for x-ray work are coated with emulsion on both sides. Exposure to light or x-rays produces nuclei of metallic silver within the silver halide grains. The sensitized grains are converted into metallic silver in the development process. When one quantum of visible light is absorbed, one silver atom is produced but about 300 quanta are required to produce a nucleus that can be developed. It is believed that a nucleus capable of development is produced for each quantum of x-rays absorbed. The total number of nuclei formed is proportional to the number of quanta reaching the film and to the total amount of silver present after development; consequently, the blackening of the film at any point is proportional to the intensity of x-rays falling at that point and to the length of time that the film was exposed to x-rays. Commonly, the intensity of an x-ray reflection is measured in terms of the blackening of the x-ray film. To measure exactly the integrated intensity of blackening of a photographic film is, in principle, a simple process but in practice requires elaborate and expensive apparatus. Machines to scan the whole film on a small grid, i.e. in regions from 10 μ square up to as large as 40 μ square, have been constructed. The information, i.e. the photocell output, is fed via an analogue-digital convertor to a computer and the integrated intensities of reflections are calculated. Some excellent results have been obtained by Abrahamsson in Göteborg with apparatus of this type but they are as yet unpublished.

By far the most common method of measuring intensities is by visual comparison with a standard set of spots made by exposing a strip of film to x-rays reflected from a chosen set of crystal planes, for known intervals of time. A superficial examination suggests that the quantity measured by this process is the maximum blackening for a given reflection. This will only approximate to a measure of the integrated intensity if the peak profile of the measured reflection is very similar to that of the reference reflection. However, if the reference reflection is roughly the same shape as the measured reflection, then the eye tends to act as an integrating mechanism and the result is a measure of something between that of peak density and integrated density. Providing the eye examines shades of grey rather than the very black reflections, the results are quite good and the final structure analysis based on this type of measurement will yield general stereochemical information and show gross differences in interatomic distances. If more accurate results are required, more care must be taken in measurement.

The optical density of the film may be measured photometrically. Single-beam microdensitometers measure peak density only, in terms of photocell output. Double-beam instruments compare peak density with that of a density 'wedge', e.g. a film of linearly graded blackening. The method is well suited to the production of something akin to a peak profile on a chart recorder. A third type of microdensitometer, the 'flyspot' instruments, scan the spot with a very small beam of light after the fashion of the electron beam scanning a television cathode-ray tube screen. A photocell records transmission for each point of the scan and an integrated intensity is produced. Experience of the first two types of microdensitometer in Oxford has suggested that the measurements produced are only a little better than visual measurements unless 'integrating' cameras are used and are very tedious. The third type of instrument has only appeared on the market recently and there is little information as to the quality of measurements. It is equally tedious to use.

The 'integrating' Weissenberg and precession cameras, devised by Wiebenga, Cochran, and others, use auxillary movements of the film cassette to superpose a reflection upon itself with a small lateral displacement to produce a diffraction spot with a flat-topped or plateau form. The maximum peak density of plateau is then a much improved measure of the integrated intensity. The use of integrating cameras and microdensitometers gives about a five-fold reduction in the percentage error in intensity measurement.

B. Counter Methods

Ionization, Geiger Müller, proportional, and scintillation counters have all been used at various times for measuring x-ray intensity. For a given intensity, the error in measurement by counter methods is generally less than for photographic measurements and excellent estimates of the probable errors can be made. The major obstacle to the use of counters has been the placing of the counter and crystal. There are two approaches to overcoming these difficulties. The first is to use an analogue setting device, the linear diffractometer. There are a number of designs but essentially they are all mechanical systems to convert linear movements corresponding to reciprocal lattice translations into the necessary shifts to bring crystal and counter from one reflecting position to another. Calculation of crystal settings is eliminated and the control of the linear motion requires only simple servosystems. Movement from reflection to reflection is automatic and intensity data output is on paper tape or direct to a computer.

The second system employs a Eulerian cradle crystal support (Figure 2). The counter rotates about a vertical axis and incident and diffracted rays always lie in a horizontal plane. The crystal rotates about each of the Eulerian axes. Given a known starting point, the rotations about each Eulerian axis and the counter position can be computed. This information is fed from the computer either directly or via paper or magnetic tape systems and the necessary servosystems perform each of the required four setting rotations. Output of intensity data is again either on to paper or magnetic tape or directly to the computer.

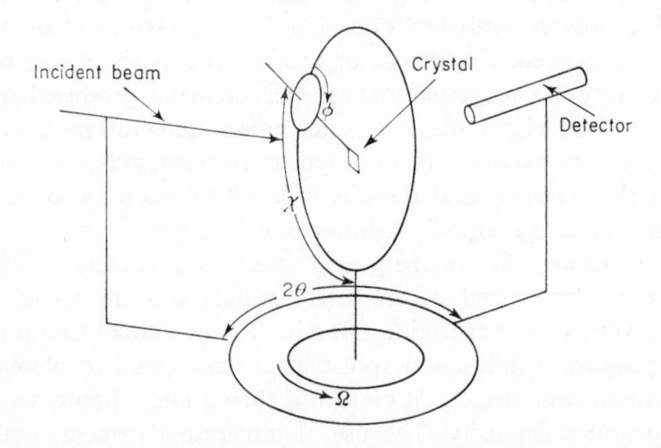

Figure 2. The Eulerian cradle system of a four-circle diffractometer. The detector and crystal move independently.

A typical operating routine of an automatic diffractometer is (1) read setting information, (2) move to background, (3) count background intensity n_1 for a time T, (4) scan through reflection by rotation of the crystal (Ω scan), either in equal steps or at constant velocity with total counting time $2T$ and total count n, (5) count background, n_2 beyond reflection for time T. The intensity of the reflection is

$$I = n - (n_1 + n_2)$$

The x-ray quanta (or neutrons) emitted in the direction of a diffraction vector are randomly distributed in time and have a Poisson distribution. The observation of n independent events, e.g. emitted quanta, is therefore subject to a variance $(\sigma)^2$ equal to n. The variance of a linear combination of independent variates is given by the sum of the respective variances. The variance $(\sigma_I)^2$ in the intensity I is

$$(\sigma_I)^2 = n + n_1 + n_2$$

If it is assumed that there are only independent small random errors in the physical and geometrical factors affecting I, then the structure amplitude F may be considered as a function of I.

$$|F| = f(I)$$

$$\sigma_{|F|} = \frac{F}{I}\sigma_I$$

$$\sigma_{|F|} = \frac{n + n_1 + n_2}{k(n - n_1 - n_2)} \tag{1}$$

where k is the product of a constant for the experiment and known correction terms, e.g. Lorentz, polarization, and absorption corrections. If the value of n is made large compared with the background, then the accuracy can be very high. Neutron diffraction uses a monochromatic beam and in the absence of hydrogen and certain other elements the incoherent scatter is low and the background is small. X-ray diffraction tends to use filtered rather than monochromatic radiation and a balanced-filter technique may be used to reduce the background. The $K\alpha$ lines of the target material, atomic number Z, provide the normal radiation used in diffraction. The absorption edge for x-rays of the element, $Z - 1$, is at a wavelength slightly shorter than that of the $K\alpha$ lines of the element Z but at longer wavelength than that the $K\beta$ lines of Z. The absorption edge of $Z - 2$ (or $Z - 3$) is at longer wavelengths than the $K\alpha$ lines of Z. Otherwise the absorption profiles of $Z - 1$ and $Z - 2$ (or $Z - 3$) are very similar (Figure 3) and can be made almost equivalent by adjusting the filter thickness.

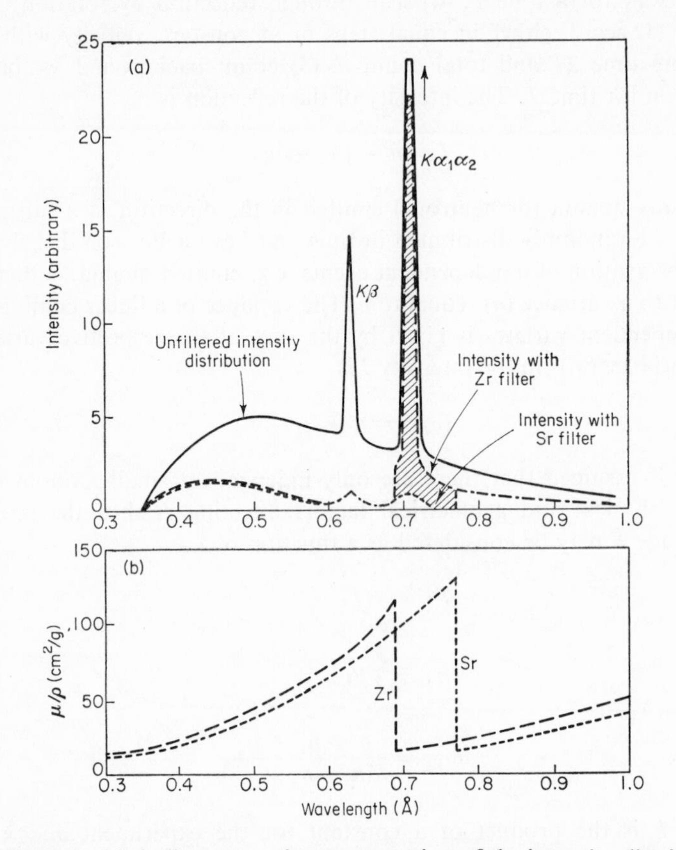

Figure 3. (a) A diagrammatic representation of the intensity distribution of x-rays from a Mo target (i) unfiltered, (ii) with Zr filter, (iii) with Sr filter. (b) Mass absorption coefficients of Zr and Sr as a function of wavelength. [Reproduced, by permission, from *Intern. Tables X-ray Crystallog.*, Vol. 3, p. 78.]

The counting routine is: count background, reflection, and background with filter $(Z - 1)$ in the beam then repeat with filter $(Z - 2)$:

$$I = n' - n_{11} - n_{21} - n'' + n_{12} + n_{22}$$

If the apparatus is correctly set up, n_{11}, n_{21}, n_{12}, and n_{22}, the backgrounds, will be nearly equal to each other and I will represent the reflection of a narrow band of wavelengths between the two absorptions edges. Good x-ray counter measurements require either balanced filters or, better, a monochromator if Mo $K\alpha$ radiation is used. With copper radiation, proportional or scintillation counters with pulse-height discrimination circuitry may be good enough.

All neutron-diffraction data are collected by the counter method. The apparatus is of the Euler cradle type with the heavily shielded BF_3 counter moving in the horizontal plane. The heavy shielding of the counter requires that the apparatus be of more massive construction.

IV. DEDUCTION OF STRUCTURE AMPLITUDES FROM INTENSITY MEASUREMENTS

The intensity $I_{\mathbf{h}}$ of x-rays reflected from a set of planes, for which the Miller index triple is $\mathbf{h}(hkl)$, is related to the structure amplitude $|F_{\mathbf{h}}|$ by an expression of the form

$$I_{\mathbf{h}} = KL_{\mathbf{h}}P_{\mathbf{h}}A_{\mathbf{h}}|F_{\mathbf{h}}|^2$$

where K is a constant for the experiment, $P_{\mathbf{h}}$ is the polarization factor

$$P_{\mathbf{h}} = \frac{1 + \cos^2 \theta}{2}$$

i.e. it is independent of the experiment and is due to the polarization of x-rays on reflection, and $L_{\mathbf{h}}$ is the Lorentz factor and is specific to the experiment, dependent on the motion performed by the crystal during the recording process. For uniform rotational motion of the crystal, $L_{\mathbf{h}}$ is given by

$$1/L_{\mathbf{h}} = \cos \mu \cos \nu \cos \gamma$$

where the rotation axis makes an angle of $(\pi/2) - \mu$ with the incident beam, $(\pi/2) - \nu$ is the semiangle of the cone of diffraction, and γ is the projection of the angle 2θ between the incident and reflected beams on to the zero layer of the reciprocal lattice. For non-uniform motion, as in the precession camera with universal joint coupling, the expression is more complex. In all cases, its value can be computed exactly for an accurately set crystal. If there are large errors in values of μ, ν, and θ attributed to a particular reflection, there may be very large errors in the computed correction. $A_{\mathbf{h}}$ represents the physical factors affecting the intensity $I_{\mathbf{h}}$. A number of these factors are important and may be corrected for with varying difficulty.

A. Extinction

In diffraction techniques, the word 'extinction' is used to describe the attenuation of the direct beam by the crystal. In 'primary extinction' the direct beam is reflected by a set of planes (**h**) with the usual phase change of $\pi/2$. The same set of planes (**h**) then reflect the diffracted beam back into the primary beam again with a phase change of $\pi/2$. This twice-reflected beam is now out of phase with the primary beam by an amount π and diminishes the intensity of the direct beam by destructive interference. Therefore, as the direct beam penetrates the crystal it is progressively attenuated. If a real crystal could be considered as a single block of the same orientation, then extinction predicts that $I_{\mathbf{h}}$ would be proportional to $|F_{\mathbf{h}}|$ but experience shows that $I_{\mathbf{h}}$ is usually proportional to $|F_{\mathbf{h}}|^2$, a relation predicted for a small volume element or for an 'ideally imperfect' crystal formed from a 'mosaic' of small blocks each so small that extinction does not occur. Fortunately the 'ideally imperfect' crystal is a very good approximation to a real crystal and primary extinction affects only a few strong reflections from a given crystal. In some highly perfect crystals, e.g. diamond, it is very important and the relation $I_{\mathbf{h}} \propto |F_{\mathbf{h}}|^2$ must be replaced by $I_{\mathbf{h}} \propto |F_{\mathbf{h}}|$. For intermediate cases, there is no useful correction although various processes, e.g. dipping in liquid air, have been suggested to reduce the effect. Primary extinctions affect reflections with large $|F_{\mathbf{h}}|$ and small $\theta_{\mathbf{h}}$, giving observed values of $|F_{\mathbf{h}}|$ smaller than the calculated values. These reflections are often omitted from the calculation of R (equation 6) and observed values are replaced by calculated values in F_{obs} synthesis.

Secondary extinction behaves much like absorption. If the incident beam fails to satisfy the Bragg condition for any set of planes, then it is only attenuated by absorption. If, however, it satisfies the Bragg condition, it is further attenuated by the diversion of energy from the incident beam into the diffracted beam. The earlier work on the theoretical treatment of this problem by Darwin[9] and others has been shown to be in error and is replaced by a treatment by Zachariasen[10]. It is an important correction in neutron diffraction but it is not usually significant in x-ray diffraction. Empirical corrections[11] have been suggested.

B. Absorption

X-rays (and neutrons) are absorbed by crystals according to Lambert's law, i.e.

$$-\frac{\mathrm{d}I}{I} = \mu\, \mathrm{d}t \qquad I = I_0\, \mathrm{e}^{-\mu t}$$

where μ is the linear absorption coefficient and t is the thickness of crystal that the beam has passed through. The linear absorption coefficient may be calculated from the chemical composition, density, and mass absorption coefficients of the elements present, for a given material. The mass absorption coefficient is a function of the element and of the wavelength and type of radiation use (*International Tables for X-ray Crystallography*, Vol. 3).

The calculation of the transmission factor for a real crystal requires a detailed knowledge of the size and shape of the crystal. Consider a small volume element dV. Within a polyhedral crystal of volume V (Figure 4)

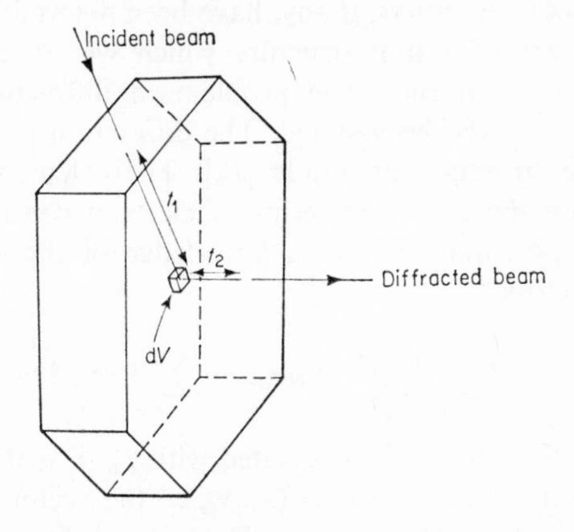

Figure 4.

the transmission factor T is

$$T = \int^V \frac{e^{-\mu t}\, dV}{V}$$

where t, the total path, comprises t_1, the incident beam path, and t_2, the diffracted beam path. The values of t_1 and t_2 depend on the directions of the incident and diffracted beam and on the crystal shape. The integration over the volume of the crystal shape is not difficult but is very time consuming. When performed for a set of three-dimensional diffraction measurements, the function is integrated numerically by the gaussian integration technique using a large digital computer.

Various attempts have been made to simplify the problem by grinding the crystal to a sphere or cylinder. These methods work well for hard crystals but are impractical for most molecular crystals. The most usual procedure is to choose a small crystal so as to minimize absorption effects. Absorption corrections should always be applied when there are heavy atoms present and when an analysis of the thermal-vibration parameters is required. Experience shows that, at the level of accuracy of the measurement of intensity data currently used, failure to apply absorption corrections to measurements made on crystals for which the product of average path and linear absorption coefficients is less than 2, has little effect on the space parameters and a very considerable effect on the magnitude of the thermal parameters but much less on their direction.

V. DETERMINATION OF A TRIAL STRUCTURE

Very few crystal structures, if any, have been determined without preliminary knowledge of a trial structure which was determined from x-ray diffraction measurements. The problem in diffraction analysis is well known and has often been stated. The diffraction pattern yields a scalar quantity, the structure amplitude $|F_{\mathbf{h}}|$. The calculation of the periodic pattern within the crystal in terms of electron density, nuclear density, or electrical potential requires a knowledge of the structure factors $F_{\mathbf{h}}$ which are vectors.

$$F_{\mathbf{h}} = |F_{\mathbf{h}}| \exp(i\alpha_{\mathbf{h}}) = \sum_{j=1}^{n} f_j \exp(2\pi i \mathbf{h}\mathbf{r}_j) \tag{2}$$

where $\alpha_{\mathbf{h}}$ is the phase angle associated with $F_{\mathbf{h}}$, f_j is the atomic scattering factor of the jth atom, and $\mathbf{r}_j = (x_j, y_j, z_j)$ the vector whose components are the coordinates of the jth atom. The integral (3) expresses the structure

$$F_{\mathbf{h}} = \int \rho(\mathbf{r}) \exp(2\pi i \mathbf{h}\mathbf{r})\, d\mathbf{r} \tag{3}$$

factor $F_{\mathbf{h}}$ in terms of the electron density $\rho(\mathbf{r})$, and $\rho(\mathbf{r})$ may be expressed in terms of $F_{\mathbf{h}}$ by use of the Fourier inversion theorem to give

$$\rho(\mathbf{r}) = \frac{1}{V} \sum_{\mathbf{h}} F_{\mathbf{h}} \exp(-2\pi i \mathbf{h}\mathbf{r}) \tag{4}$$

Atoms are represented by maxima in $\rho(\mathbf{r})$, the electron density, and V is the unit-cell volume.

It appears from equation (2) that there are, associated with any given set of $F_{\mathbf{h}}$, an arbitrary set of phases $\alpha_{\mathbf{h}}$. This is not the case, since there are severe restrictions on the values of $\rho(\mathbf{r})$. These are that $\rho(\mathbf{r})$ is non-negative throughout space and may be represented by the superposition of a given number of accurately known atomic electron distributions. In principle, this information together with the known magnitudes of $|F_{\mathbf{h}}|$ is sufficient to determine uniquely the values of the phases $\alpha_{\mathbf{h}}$.

In equations (2) to (4), the known quantities are $|F_{\mathbf{h}}|$, f_j, and $\mathbf{h}$, and the unknowns are $\alpha_{\mathbf{h}}$ and $\mathbf{r}_j$, the phases and atomic coordinates. Of these unknowns, $\mathbf{r}_j$ is the ultimate object of the analysis and $\alpha_{\mathbf{h}}$ acts as an intermediary in the determination of $\mathbf{r}_j$.

There are two possible approaches to the determination of crystal structure. The first is to determine the values of $\mathbf{r}$ directly, based on the use of the Patterson function. The second is to determine the values of $\alpha_{\mathbf{h}}$ and hence, by equation (4), the values of $\mathbf{r}_j$. The second group includes isomorphous replacement and anomalous dispersion techniques and the so-called direct methods.

A. The Patterson Function

In 1934, Patterson introduced the Fourier series generally known by his name:

$$P(\mathbf{r}) = \sum_{\mathbf{h}} |F_{\mathbf{h}}|^2 \exp(-2\pi i \mathbf{hr})$$

This function has maxima at the ends of the vectors between atom locations. Thus for every distinct pair of maxima in the electron density there is a maximum in the 'Patterson density'. The three-dimensional $F_{\mathbf{h}}$ or electron-density synthesis for a system with j atoms in the unit cell has j maxima at about 1.0–2.0 Å separation. In the corresponding $|F_{\mathbf{h}}|^2$ synthesis, the j atoms give j^2 vectors and hence j^2 maxima, many of which will be exactly superposed by the symmetry of the system and many more which may be found very close together in space by accident. The Patterson function is much less well resolved than the electron-density function. It is common practice to modify or to 'sharpen' the $|F_{\mathbf{h}}|^2$ values so that they approximate to values which might be expected from 'point atoms at rest'. The modification function is of the form

$$|F_{\mathbf{h}}|^2 \text{ (modified)} = |F_{\mathbf{h}}|^2 \frac{\exp(2B \sin^2 \theta/\lambda^2)}{\hat{f}^2}$$

where B is the estimated overall isotropic temperature factor and $\hat{f}$ is the weighted mean unitary scattering factor for the atoms in the structure. If B is correctly chosen, the sharpening should decrease the volume occupied by each peak and increase its maximum height, thus improving the resolution of the peaks.

The 'Patterson space' or 'vector space' has a unit cell no larger than that in the real crystal space, has at least the symmetry elements of the real space, and is always centrosymmetric. At the origin of the vector space is a large density peak formed by the superposition of the vectors between atoms and their equivalents in neighbouring unit cells. The integrated peak density of the origin peak is proportional to $\sum |F_{\mathbf{h}}|^2$ and to $\sum (Z_j)^2$. The integrated peak density of a maximum corresponding to the vector between the jth and ith atom has a magnitude proportional to nZ_iZ_j, where n is the multiplicity of the vector. In general, peak height may be taken as a measure of integrated peak density. The most prominent peaks on a Patterson function will tend to be those between the atoms of largest atomic number.

Vectors between atoms and their symmetry-related partners are known as Harker vectors. Any one- or two-dimensional section in a Patterson distribution, with a high concentration of Harker vectors, is referred to as a Harker line or Harker section and may be identified by the proportionally large number of maxima along the line or in the section. Harker sections at $z = 0$ are the consequence of symmetrical vectors for atoms related by a pure rotation axis parallel to the c axis. Similarly, any screw axis parallel to c relates atoms of a certain level difference s and all symmetrical vectors have z components equal to s and terminate in the section of the three-dimensional Patterson function with $z = s$. Harker lines are produced by atoms related by reflection symmetry. The presence of a concentration of vectors in the Patterson function may be used to identify elements of space-group symmetry not involving translation and not identified by systematic absences in the $|F_{\mathbf{h}}|$ values. Further, Harker sections provide information which permits the location of atoms with respect to the axes of rotation. Implication theory, the theory of the detailed interpretation of Harker sections, is due to Buerger[12]. Buerger describes the construction of an 'implication diagram' from the Harker section by taking each point on the Harker section, shrinking its radial polar co-ordinate, and rotating its azimuth. The shrinkage and rotation are constant for a given axial rotation symmetry. The resulting implication diagram is a projection of the crystal structure on to the plane normal to the rotation axis. Implication diagrams are subject to numerous ambiguities.

If the crystal structure contains one or more heavy atoms, the most prominent features of the Harker lines and sections will be the heavy-atom vectors and it is these Harker vectors that are commonly used in the location of the heavy atoms in real space. When the heavy atoms have been located, they may be used to give approximately correct phases to the structure amplitudes and analysis proceeds by $F_{\mathbf{h}}$ synthesis (3), each cycle locating more light atoms and leading to improved phases until the full structure is revealed.

Frequently, Patterson distributions calculated for crystals containing heavy atoms contain sufficient information themselves to permit the location of the majority of non-hydrogen light atoms. The simplest method of attempting to extract this information from the Patterson function is 'vector convergence' due to Beevers and Robertson[13]. They describe the method as applicable if there is a heavy atom present and the crystal contains moderately high symmetry so that there are at least four equivalent atoms available. The origin of a separate Patterson distribution is placed on each heavy-atom location. The locations of light atoms are then taken to be where high peaks on several Patterson functions coincide. After the heavy-atom peaks, the next most prominent in the distribution are the heavy–light vectors which originate from each of the heavy atoms. If the crystal has i heavy-atom locations, the Patterson distribution should contain i images of the structure. When the Patterson origins are placed at heavy-atom sites, the i images should coincide to form the common image of the light atoms. In this method, the 'heavy' atom need not be heavier than phosphorus if the other atoms present are carbon, nitrogen, oxygen, and hydrogen. It has been found that one phosphorus, sulphur, or chlorine atom to up to 50 light atoms is all that is required if there are about 10 or more observations per parameter to be determined.

Buerger's image-seeking functions are an important development on the vector-convergence idea. Of these image-seeking functions, the minimum function is the most valuable[14]. In its simplest form, the minimum function is derived by superposing the Patterson maps as in the vector-convergence method and then drawing a map that takes the minimum contour from the superposition. The minimum function has a number of advantages as an image-seeking function. The density levels of the minimum function are ideally proportional to electron density and the background is always the minimum of the superposed Patterson functions. The superposition points need only be major features of the function and not necessarily heavy-atom locations. Heavy atoms are not a necessary prerequisite of the method. Image-seeking theory and its use are considered in great detail in Buerger's book *Vector Space* (ref. B.14 of Bibliography).

B. Isomorphous Replacement

This method requires that there should be available a pair, or preferably a series, of crystals of essentially the same structure with a majority of light atoms in a fixed configuration and a few heavy atoms which may be added or replaced. The simplest example is a pair of isomorphous centrosymmetric crystals, one containing the invariant group of light atoms PA, and the other containing PA with a heavy atom HA. Then

$$F_{\mathrm{PA}} + F_{\mathrm{HA}} = F_{\mathrm{PA+HA}}$$

If the location of the heavy atom is known, for example from a Patterson function, the phase of F_{HA} is known. If F_{HA} is not near zero, it is possible to determine the phase of the larger structure factor from F_{PA} and $F_{\mathrm{PA+HA}}$[15]. For the non-centrosymmetric crystal, at least three isomorphous crystals PA, PA + HA_1, and PA + HA_2 are required to give an unambiguous phase for any reflection, and the greater the number of derivatives available the better the determination of the phases. The method has been applied with great success by Perutz, Kendrew, and their school to the determination of protein structures.

C. Anomalous Dispersion

If a scattering atom is irradiated with radiation of wavelength near the absorption edge for that atom, the atomic scattering factor changes both with regard to amplitude and phase. Its atomic scattering factor is of the form

$$f = f_0 + \Delta f' + \mathrm{i}\,\Delta f''$$

where f_0 is the normal scattering factor and $\Delta f'$ and $\mathrm{i}\,\Delta f''$ are the real and imaginary anomalous corrections. For neutrons, the atomic scattering factor has a similar form but the values of $\Delta f'/f_0$ and $\Delta f''/f_0$ are about two orders of magnitude larger than for x-ray scattering. In certain non-centrosymmetric space groups, this leads to a breakdown of Friedel's law and the consequent inequality of the value of $|F_{\mathbf{h}}|$ for the planes hkl and $\bar{h}\bar{k}\bar{l}$[16]. Figure 5 shows the relationship between the phases of $F_{\mathbf{h}}$ and $F_{\bar{\mathbf{h}}}$ for a non-centrosymmetric crystal containing a centrosymmetric arrangement of atoms A giving anomalous dispersion effects and sited at known locations, and the remaining atoms R.

The real and imaginary contributions to the scattering of the anomalous scatterers are F_A and F''_A, respectively. The phase of $F_{\mathbf{h}}$ is determined, but not unambiguously, if the contribution of A to the total scattering is large.

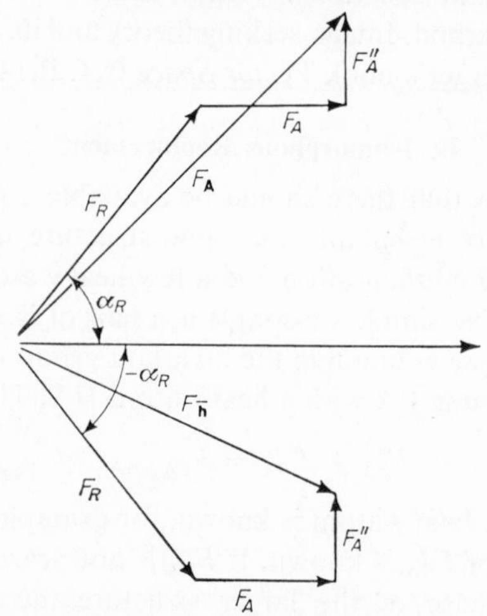

Figure 5. The relationship between the phases of $F_{\mathbf{h}}$ and $F_{\bar{\mathbf{h}}}$ for a non-centrosymmetrical crystal with a centrosymmetrical arrangement of atoms A showing anomalous dispersion and sited at known locations.

The technique has been applied to the determination of the structures of the vitamin B_{12} derivatives, factor V_{Ia}[17], and the DBC coenzyme[18] using the anomalous scattering of Cu $K\alpha$ radiation by the cobalt atom. The technique can be applied to neutron-diffraction measurements. Four nuclei show anomalous dispersion in the thermal neutron range: ^{151}U, ^{113}Cd, ^{149}Sm, and ^{157}Gd. The work on the vitamin B_{12} derivatives suggests that, by using x-rays and compounds with one cobalt atom ($\Delta f''/f_0 \simeq 0.15$), it is possible to solve structures containing 100 atoms. On this basis, using neutrons and compounds with one ^{113}Cd per asymmetric unit ($\Delta f''/f_0 = 10$–12), it must be possible to phase reflections from a structure containing 3000 atoms per asymmetric unit[19].

D. Direct Method

At the present time, there are a few examples in the literature of x-ray structure analysis where the phases of the structure factors were computed directly from their magnitudes. In all the examples, the structures were centrosymmetric; therefore, the phase angles have had the only possible values of 0 and π. The processes employed are often referred to as sign-determination techniques

The earliest attempts to derive a set of 'signs' from structure amplitudes are to be found in the work of Ott[20] in the early 1930s but it was not pursued for some 20 years. Harker and Kasper[21] in 1948 proposed a set of inequality relationships between structure factors. These were expanded by Karle and Hauptman[22] and were shown to be based on the requirement of positive electron density. The complete set of inequalities can be expressed as determinants of higher and higher order. The simplest are

$$F_{000} \geqslant 0$$

$$|F_{\mathbf{h}}| \leqslant 0$$

$$\left| F_{\mathbf{h}} - \frac{F_{\mathbf{k}}F_{\mathbf{h}-\mathbf{k}}}{F_{000}} \right| \leqslant \frac{\left| \dfrac{F_{000}F_{-\mathbf{k}}}{F_{\mathbf{k}}F_{000}} \right|^{\frac{1}{2}} \left| \dfrac{F_{000}F_{-\mathbf{h}+\mathbf{k}}}{F_{\mathbf{h}-\mathbf{k}}F_{000}} \right|^{\frac{1}{2}}}{F_{000}} \tag{5}$$

The application of these, together with the higher-order members of the infinite set, could lead to the determination of the sign of $|F_{\mathbf{h}}|$ but as the order of the determinants increases the increasing computational complexity, together with a requirement of data of extraordinary accuracy, precludes any general practical application. However, a few structures, including the important orthorhombic decaborane, have been solved by this method.

It was shown by Sayre in 1952[23] that, for a centrosymmetric structure, the sign of the product $F_{\mathbf{h}}F_{\mathbf{k}}F_{\mathbf{h}-\mathbf{k}}$ was most probably positive. If the magnitudes of $F_{\mathbf{k}}$ and $F_{\mathbf{h}-\mathbf{k}}$ are large and comparable to the value of F_{000}, then the right-hand side of the inequality (5) is very small and the magnitude of $F_{\mathbf{h}}$ must be similar to the magnitude of $F_{\mathbf{h}}F_{\mathbf{h}-\mathbf{k}}$. As $F_{\mathbf{k}}$ and $F_{\mathbf{h}-\mathbf{k}}$ become smaller, then although the magnitudes of $F_{\mathbf{h}}$ and $F_{\mathbf{k}}F_{\mathbf{h}-\mathbf{k}}$ may deviate, it is probable that they will still have the same sign; hence the Sayre equation. Probability considerations have long been used for the determination of the absolute scale of x-ray intensities (Wilson plot)[24] and in the detection of symmetry centres ($N(Z)$ plot)[25]. Karle and Hauptman[26] applied probability considerations to the determination of phases. All the derived relationships of Karle and Hauptman are in terms of the normalized structure factor $E_{\mathbf{h}}$.

$$E_{\mathbf{h}} = \frac{F_{\mathbf{h}}}{\left(n\sum_{j=1}^{N} f_j^2\right)^{\frac{1}{2}}}$$

where n is a symmetry number for the appropriate space group. The Karle and Hauptman approach starts from the idea that the probability of a structure factor having a positive sign is one-half so long as no other structure amplitudes are known, but once the full set of structure amplitudes are given then this probability deviates from one-half. The manner in which the theoretical treatment proceeds yields the probability of the structure factor having a positive sign in terms of the fraction of all possible atomic arrangements that will yield a positive sign for the structure factor under consideration, given the magnitudes and signs of the structure factors involved in the particular phase-determining relationship. Many of these 'joint probability functions' and derived phase-determining formulae are given in the Karle and Hauptman monograph[26]. They may be illustrated by the simpler phase-determining formulae for the space group $P\bar{1}$

$$\textstyle\sum_1: \quad \mathrm{s}E_{2\mathbf{h}} \sim \mathrm{s}\sum_{\mathbf{h}} (E_{\mathbf{h}} - 1)$$

$$\textstyle\sum_2: \quad \mathrm{s}E_{\mathbf{h}} \sim \mathrm{s}\sum_{\mathbf{k}} E_{\mathbf{k}}E_{\mathbf{h}-\mathbf{k}}$$

$$\textstyle\sum_3: \quad \mathrm{s}E_{\mathbf{h}} \sim \mathrm{s}\sum_{\mathbf{k}} E_{\mathbf{k}}(E^2_{(\mathbf{h}+\mathbf{k})/2} - 1)$$

where $\sim$ implies 'probably equal to' and $\mathrm{s}E_{\mathbf{h}}$ is used to represent 'sign of' $E_{\mathbf{h}}$. These will be modified for other space groups. The $\sum_1$ formulae for $P2_1/c$, for example, has a special form

$$\mathrm{s}E_{2h,0,2l} \sim \mathrm{s}\sum (-1)^{k+1}(E^2_{hkl} - 1)$$

Each phase-determining formula is derived from an expression giving the probability of $E_{\mathbf{h}}$ having a positive sign. For example, the probability $P + (E_{2h,0,2l})$ of $E_{2h,0,2l}$ having a positive sign in space group $P2_1/c$ is

$$P + (E_{2h,0,2l}) = \frac{1}{2} + \frac{1}{2} \tanh \frac{\sigma_3}{2\sigma_2} \frac{3}{2} |E_{2h,0,2l}| \sum_{\mathbf{k}} (-1)^{k+1}(E_{hkl}^2 - 1)$$

where σ_n is $\sum_{j+1}^{N} Z_j^n$.

The phase of $E_{\mathbf{h}}$ can be determined together with the probability of the answer being the correct one. In general, the probability of the answer being correct will only approach unity if $E_{\mathbf{h}}$ is about unity, i.e. if $I_{\mathbf{h}}$ is strong.

A systematic application of the formulae of the type $\sum_1$, $\sum_2$, and $\sum_3$, modified according to space group, either with or without the help of inequalities and other techniques, can lead to the correct determination of the phases of a sufficient fraction of the $E_{\mathbf{h}}$ values, the largest few hundred of a three-dimensional set, to compute a Fourier map with $E_{\mathbf{h}}$ as coefficients which is capable of interpretation in terms of the crystal structure. In *p,p'*-dimethoxybenzoquinone, a structure with two molecules, 36 atoms, in the asymmetric unit, Karle and coworkers found that 6% of the non-zero data gave an $E_{\mathbf{h}}$ map with 37 maxima, 1 spurious, and 36 at what were subsequently proved to be atomic locations. The systematic application of sign-determining formulae appears one of the best methods of fully computerized structure analysis. Techniques of interpreting Fourier figure fields are well developed and at no stage need there be any intervention from the operator.

VI. REFINEMENT

Given a satisfactory trial structure, it is necessary to refine the structure to get the best possible structure from the intensity measurements. The usual criterion of the agreement between the observations and the proposed structure is the residual (R value). The residual R is usually of the form

$$R = \frac{\sum \| F_{\text{obs}}| - |F_{\text{calc}}\|}{\sum |F_{\text{obs}}|} \tag{6}$$

Wilson has shown that for entirely wrong structures R will be 0.828 for centrosymmetric and 0.586 for non-centrosymmetric cases. There is an approximate relationship between R and $\sigma(x_i)$, the mean standard deviation of the position coordinates of the ith atom

$$\sigma(x_i) \sim \frac{R}{\bar{S}} \left[\frac{(N_i)}{8p} \right]^{\frac{1}{2}} \tag{7}$$

where $\bar{S}$ is the r.m.s. reciprocal radius for the observed planes, p is the number of observations less the number of parameters, and N_i is the number of atoms of the same type as the ith atom needed to give the scattering power at $\bar{S}$ equal to that of the asymmetric unit of the whole structure. It is essential that R should be less than 0.20 in any zone of the measurements and desirable for it to be less than 0.15. Very accurate analyses will have R values of about 0.05.

Refinement was the major bottle-neck in x-ray structure analysis until large digital computers came into general use, that is before about 1958. In relatively few structures determined before this date is the refinement pushed to the limit of the measurements. Before the late 1950s, a great deal of effort went into devising methods to maximize the amount of refinement gained for a given amount of computational effort and numerous refinement methods were produced. Today, the majority of refinement is by the method of least squares, with difference Fourier techniques and differential synthesis having importance in certain specialized situations. All three processes are iterative and the choice of method must depend to some extent on the expected rate of convergence. The difference Fourier synthesis ($F_{obs} - F_{calc}$ maps) is computationally the simplest but the interpretation must be more or less subjective according to the problem. The least-squares method is least subject to the intuition of the operator and is often regarded as less useful in abnormal situations, e.g. disorder, twinning, etc.

A. The Difference Fourier Map ($F_{obs} - F_{calc}$ Synthesis)

From equation (4),

$$\rho_{obs} - \rho_{calc} = 1/V \sum_{\mathbf{h}} (F_{\mathbf{h},obs} - F_{\mathbf{h},calc}) \exp(-2\pi i \mathbf{h}\mathbf{r})$$

This is the difference synthesis of Booth and Cochran[27]. The synthesis may only be used at such a stage in the analysis that the vast majority of the phases of $F_{\mathbf{h}}$ are known and in general it is much more effective in the centrosymmetric case. It is a fundamental property of the difference synthesis that when the proposed model is identical with the true structure the synthesis is featureless except for minor random fluctuations. The nature of any deviation of the model from the true structure is revealed by topographical features characteristic of the deviation. The required modifications to the model may be deduced and the trial model corrected if certain rules are followed. The difference synthesis can be calculated relatively rapidly and is free from series termination errors. Figure 6 shows three types of topographical features that may be found in a difference synthesis. In Figure 6a, the atom in the model is displaced from the

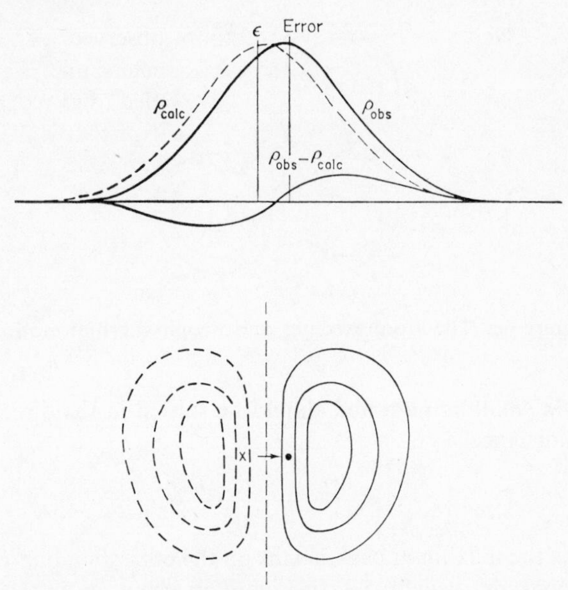

Figure 6a. The atom displaced from its true site by a small amount ϵ.

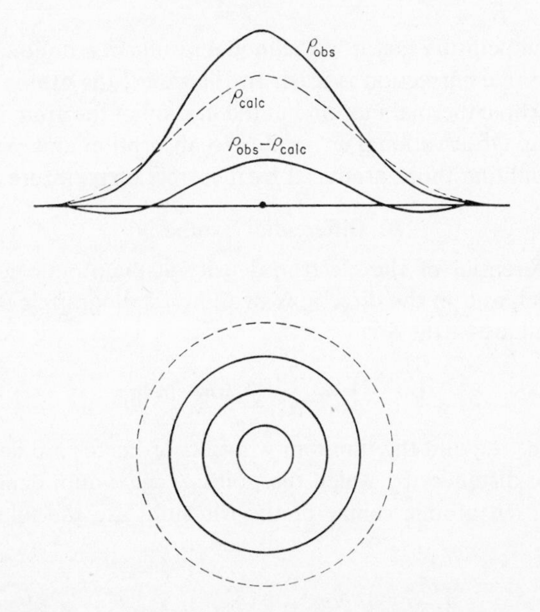

Figure 6b. The atom has too high a temperature factor.

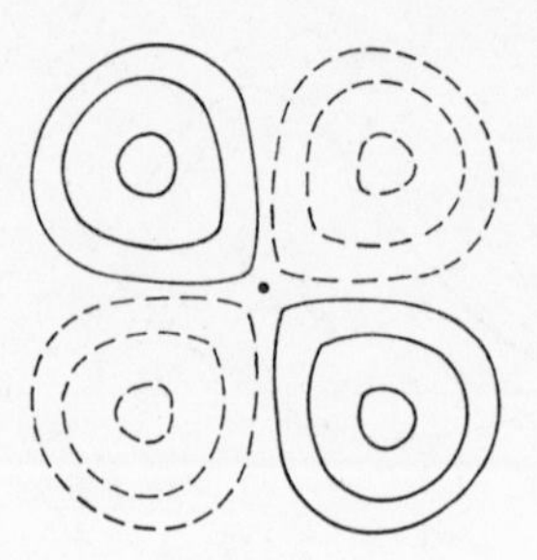

Figure 6c. The atom executes anisotropic thermal motion.

true site by a small amount and should be shifted in the direction of the arrow by a distance

$$r = \frac{\mathrm{d}(\rho_{\mathrm{obs}} - \rho_{\mathrm{calc}})/\mathrm{d}r}{2\rho_{\mathrm{obs}}p} \tag{8}$$

where ρ_{obs} is the maximum peak height on the corresponding $F_{\mathbf{h}}$ synthesis and p is a constant, usually 5.0. In Figure 6b, the atom in the model has too high a temperature factor. The temperature factor used in the model must be reduced by an amount ΔB, given by

$$\Delta B = \frac{-2B(\rho_{\mathrm{obs}} - \rho_{\mathrm{calc}})}{\rho_{\mathrm{obs}}}$$

Too low a temperature factor in the model results in a hollow instead of a hill and the same correction is used. In Figure 6c, the atom in the model executes isotropic thermal motion but the motion of the atom in the crystal is anisotropic. Observational errors due to absorption and extinction give features resembling those produced by incorrect temperature factors.

B. Differential Synthesis

The first differential of the electron density at the atomic centres of the current model, and in the directions of the crystallographic axes, may be evaluated and are of the form

$$\left(\frac{\partial \rho_{\mathrm{obs}}}{\partial x}\right)_j = \frac{2\pi}{aV} \sum \mathbf{h} F_{\mathbf{h}} \sin \mathbf{hr} \tag{9}$$

for the x direction and the jth atom whose coordinates are defined by the vector $\mathbf{r}$. The distances by which the point of maximum density deviates from the chosen atomic centre of the jth atom are the solutions to the equations

$$\left(\frac{\partial^2 \rho_{\mathrm{obs}}}{\partial x^2}\right)_j \Delta x_j + \left(\frac{\partial^2 \rho_{\mathrm{obs}}}{\partial y^2}\right)_j \Delta y_j + \left(\frac{\partial^2 \rho_{\mathrm{obs}}}{\partial z^2}\right)_j \Delta z_j + \left(\frac{\partial \rho_{\mathrm{obs}}}{\partial x}\right)_j = 0, \text{ etc.}$$

For orthogonal axes and centrosymmetric atoms,

$$x_j = \frac{(-\partial\rho_{obs}/\partial x_j)}{(\partial^2\rho_{obs}/\partial x_j^2)}, \text{ etc.} \tag{10}$$

The second differential is approximated by the equation

$$\frac{\partial^2\rho_{obs}}{\partial x_j^2} = 2pz_j(p/\pi)^{\frac{3}{2}}$$

where p is the same constant as is used in determining parameter shifts from a difference synthesis (8)[28]. The calculation is fast and the arithmetic simple. The method has been used very much with smaller computers such as the IBM 1620 and with very great success. The principles may be extended to include refinement of isotropic and anisotropic temperature factors.

C. Method of Least Squares

This method is quickly becoming the most used refinement technique. The theory of errors predicts that if the errors in the measured $F_{\mathbf{h}}$ are random and follow the gaussian law then the best atomic and thermal parameters will be those giving the minimum value of

$$\sum w_{\mathbf{h}}(|F_{\mathbf{h},obs}| - |F_{\mathbf{h},calc}|)^2$$

where $w_{\mathbf{h}}$ is the weight of a particular observation and is proportional to the inverse of the square of the probable error of that observation. The mathematical details of the minimization process of a function of this type to give the least squares best set of parameters for a linear system is given by Whittaker and Robinson[29] and this treatment may be expanded to cover certain non-linear situations if a reasonable set of trial parameters are available. The process is iterative in the non-linear case. Each observation $F_{\mathbf{h}}$ gives an observational equation of the form

$$\sum_0^j \sqrt{w_{\mathbf{h}}}\,\frac{\partial(F_{\mathbf{h},calc})/K}{\partial x_j}\,\partial x_j = \sqrt{w_{\mathbf{h}}}\,(|F_{\mathbf{h},obs}| - 1/K\,|F_{\mathbf{h},calc}|)$$

where x_j is the jth parameter of the n parameters to be refined and $1/K$ is the scale factor required to place the values of $F_{\mathbf{h},calc}$ on the scale of $F_{\mathbf{h},obs}$. The function minimized is

$$\sum_{\mathbf{h}} w_{\mathbf{h}}(|F_{\mathbf{h},obs}| - 1/K\,|F_{\mathbf{h},calc}|)^2$$

If in matrix notation the observational equations are written in the form $Ax = b$, then there are n normal equations of the form $A^TAx = A^Tb$, the solutions of which will shift the parameters x towards the values required to satisfy the minimization condition. In this case $(A'A)_{ij}$ is

$$\sum_{\mathbf{h}} w_{\mathbf{h}} \frac{\partial |F_{\mathbf{h},\mathrm{calc}}|}{\partial x_i} \frac{\partial (F_{\mathbf{h},\mathrm{obs}})}{\partial x_j}$$

and $(A^T b)_j$ is

$$\sum_{\mathbf{h}} w(\mathbf{h})(K|F_{\mathbf{h},\mathrm{obs}}| - |F_{\mathbf{h},\mathrm{calc}}|) \frac{\partial (F_{\mathbf{h},\mathrm{calc}})}{\partial x_j}$$

For a structure analysis with j atoms defined by n parameters and with M observations, there will be n elements of the vector $A^T b$ and $(n + 3)n/2$ independent elements of the matrix $A^T A$, each element of each involving a sum over all M observations. It is desirable that M should be greater than n by a factor of 10. The computation involved for even small structures is massive but feasible. For large structures, e.g. 40 or more independent atoms, it is often necessary to ignore terms in the normal matrix that are off-diagonal except those terms between x, y, z and B_{ij} of the same atom. The method is efficient and converges rapidly if the full matrix is used. Accelerating devices may be necessary if the block-diagonal approximation has to be used. The least-squares method is particularly effective with a full set of three-dimensional data and has the advantage over the difference synthesis that subjective judgement of figure fields is not required.

VII. ESTIMATION OF ERRORS

The theoretical treatment of the errors in structure analysis has been worked out largely by Cruikshank[30]. Following his notation, the electron density at a given point

$$\rho(\mathbf{r}) = 1/V \sum_{\mathrm{indep}} \left[|F_{\mathbf{h}}| \left(\sum_{\mathrm{form}} \cos 2\pi \mathbf{hr} \right) \right]$$

where $\sum_{\mathrm{indep}}$ is the summation over all symmetry-independent planes and $\sum_{\mathrm{form}}$ is the sum over a particular plane and its symmetry equivalents. If each independent $|F_{\mathbf{h},\mathrm{obs}}|$ is subject to an error $\sigma(F_{\mathbf{h},\mathrm{obs}})$ and these errors are random, then the error in the electron density is simply the variance of the sum of a set of random variables, that is

$$\sigma^2(\rho(\mathbf{r})) = 1/V^2 \left[\sum_{\mathrm{indep}} \left\{ \sigma^2 F_{\mathbf{h}} \left(\sum_{\mathrm{form}} \cos 2\pi \mathbf{h}x \right)^2 \right\} \right]$$

In this formula, $\sigma^2(\rho(\mathbf{r}))$ is the variance in the electron density by the summation over a particular finite series that approximates to the electron density and as a consequence shows the apparent anomaly that as the number of observations increases so does the error in the electron density determined by the observations. An approximate form of this equation giving average error in electron density

$$\sigma(\rho) = (1/V)^2 \sum \sigma^2(F_{\mathbf{h}})$$

may be used, but it has the particular disadvantage of not taking into account the increase in the error of the electron density that usually occurs at positions of specialized symmetry within the unit cell. Coordinate errors may be estimated from electron-density errors. In equation (10), $\partial^2\rho/\partial x^2$ is much larger than $\partial\rho/\partial x$, so that the major errors in Δx_j are due to errors in $\partial\rho/\partial x$. The standard deviation in the parameter x_j is

$$\sigma(x_j) = \sigma\left(\frac{\partial\rho}{\partial x}\right) \bigg/ \left(\frac{\partial^2\rho}{\partial x^2}\right)$$

$\partial\rho/\partial x$ is given by (9), so that $\sigma(\partial\rho/\partial x)$ is given by

$$\sigma\left(\frac{\partial\rho}{\partial x}\right) = \frac{2\pi}{aV}\left\{\sum_{\text{indep}} \sigma^2(F_{\mathbf{h}})\left(\sum_{\text{form}} h \sin 2\pi\mathbf{hr}\right)^2\right\}^{\frac{1}{2}}$$

The parameters must have been corrected for finite series effects by the back-shift method of Booth.

Variances may be derived from the least-squares normal matrix. If the normal matrix is B, then the variances are the diagonal elements $(B^{-1})_{jj}$ of the inverse of B, if the weights are on an absolute scale. For relative weights

$$\sigma_{jj}^2 = (B^{-1})_{jj} \frac{\sum w(|F_{\text{obs}}| - |F_{\text{calc}}|)^2}{M - n}$$

where there are M observations and n parameters. The off-diagonal terms of the inverse of B, $(B^{-1})_{ij}$ are the covariances of the parameters i and j. For the variances in the parameters to be valid, the correct weights for the observations should be used. Ideally, the weight for a particular observation is the inverse of its variance. If the observations are made by any counter method, then the variance is known (1). In least-squares refinement, it is more usual, and for photographic measurements essential, to use some analytical weighting scheme. Such a scheme must reflect the accuracy of the measurements and for most analyses the most accurate measurements are made at the centre of the spread of values of $F_{\mathbf{h},\text{obs}}$ and schemes of the type (1) $w = 1$ if $F_{\mathbf{h}} > F^*$; $w = 1/(F_{\mathbf{h}})^2$ if $F_{\mathbf{h}} < F^*$ or of the type (2) $w = 1/1 + ([F_{\mathbf{h}} - b]/a)^2$, where a and b are chosen to give the desired shape and maximum value to the function with respect to the values of $(F_{\mathbf{h}})$. If an inappropriate weighting scheme is used, it can lead to large errors in scale and thermal parameters but the errors caused in space parameters will be less important. It will also lead to large errors in the estimated standard deviations, up to 15%. The use of a block-diagonal approximation may lead to gross underestimation of the errors in all parameters if the off-diagonal terms of the normal matrix are not small.

Estimation of the errors in space coordinates from electron-density determinations and from the least-squares analysis will not give the same results. It has been shown that there is a close formal relationship[30] between refinement of coordinates by $F_{obs} - F_{calc}$ syntheses with the criterion of zero slope at atomic sites and the least-squares method. The $F_{obs} - F_{calc}$ synthesis is the equivalent of minimizing $\sum 1/f_j(F_{obs} - F_{calc})^2$. This is equivalent to the least-squares method with each reflection weighted inversely as the atomic-scattering factor. Consequently, a different weighting function for the normal equations for each atom must be used and the off-diagonal terms governing the interactions between parameters of different atoms eliminated. It is fortunate that all the f_j functions are very similar except for scale. If this were not the case, the $F_{obs} - F_{calc}$ synthesis would have little value. Since the Fourier methods are the equivalent of the least-squares methods with incorrect weighting schemes, it might be expected that they would yield higher estimated standard deviations (e.s.d.s) than the least-squares method. Often the converse is true in that the Fourier e.s.d.s are lower. These underestimates arise from (1) the neglect of the off-diagonal terms, (2) the proper diagonal terms are often numerically smaller than the $\partial^2\rho/\partial x^2$ terms of the simple differential synthesis, and (3) the misuse of the Fourier methods in two-dimensional analyses with insufficiently resolved projections. This latter is the equivalent of neglecting the off-diagonal terms under conditions which these terms assume great importance. The estimated standard deviations of bond length and inter-bond angle are related to the e.s.d.s of the atom parameters by equations (11)[31] and (12)[32] if the covariances are neglected.

$$\sigma^2(1) = \sigma^2(x_1) + \sigma^2(x_2)\cos^2\alpha + (\sigma^2 y_1 + \sigma^2 y_2)\cos^2\beta + (\sigma^2(z_1) + \sigma^2(x_2))\cos^2\gamma \qquad (11)$$

where x_1, y_1, z_1 and x_2, y_2, z_2 are the coordinates of the two atoms and cos α, cos β, and cos γ are the direction cosines of the atom. If the atoms 1 and 3 subtend the angle θ at atom 2 and the bond between atoms 1 and 2 has a length l_1 and between 1 and 3, l_3 then

$$\sigma^2(\theta) = \left(\frac{1}{l_1 l_3}\sin\theta\right)^2 \times [A_3^2\sigma^2(x_1) + (A_1 + A_3)\sigma^2 x_2 + A_1^2\sigma^2(x_3) + \text{similar terms in } y \text{ and } z] \qquad (12)$$

$$A_1 = (x_2 - x_1) - l_1/l_3 \cos\theta(x_2 - x_3)$$

$$A_3 = (x_2 - x_3) - l_3/l_1 \cos\theta(x_2 - x_1)$$

etc.

VIII. RESULTS*

The results obtained by the structure analyst are used in every branch of chemistry. The most spectacular structural work is without doubt with proteins, and the constant drive to determine the structures of more and more complex crystals has led to a partial neglect of much of interest in inorganic structural chemistry. The structures of many binary compounds have not been redetermined since the 1930s when any small differences in structure would have been neglected. There are, for example, more than 40 structures said to be of the cadmium hydroxide type, but perhaps only in one or two of these structures has there been any real attempt to determine the only variable space parameter required to define the crystal structure. It is generally assumed that this parameter, the height of the hydroxyl group up the *c* axis, has the value of $\frac{1}{4}$. The wide range of values of the ratio of the *a* axis length to that of the *c* axis indicate that in many cases large deviations from the value $\frac{1}{4}$ might be expected. In spite of this number of only approximately known $Cd(OH)_2$ structures, they were used quite recently in the estimation of a set of transition metal radii. Reexamination of the structures of the fluorides MF_2 where M is Fe, Co, or Ni[33] has shown that the metal coordination spheres are tetragonally distorted octahedra, a fact so far without satisfactory explanation. Detailed work on these and similar metal binary compounds could yield very interesting information about metals, but in the present climate of opinion as to what is worthwhile in crystallographic research it is being neglected.

The value of, and the questions answered by, a structure analysis depend very much on the details of the analyst's approach to the problem. The techniques employed will depend very much on the quality and stability of the sample, the size of the problem, the investigator's estimate of the possible value of the results, and the technology of the time when the work was executed. Structure analysis has always been, and will remain, one of the most expensive of the physical methods both in terms of effort and finance and there is often a tendency to strike a balance between the scientific worth of the problem and its cost in these terms. It is important to review the merits of the results obtained from the various approaches.

A. Powder Method

The powder diffraction method with either x-rays or neutrons is widely used in inorganic structure analysis and in guessing possible structural

* EDITORS' NOTE: To illustrate his critical appraisal of the merits and drawbacks of the various methods of x-ray structure analysis, the author has chosen examples mainly taken from his own work, although some of these, in fact, refer to organic molecules.

relationships. The major drawbacks are the relatively small amount of intensity data usually obtained and the difficulty of indexing powder patterns of low symmetry. De Wolff and his school[34] have shown that, given a fairly sophisticated analysis of a set of accurately known line positions, a powder pattern of a low symmetry crystal can usually be indexed if the unit cell is not excessively large (say in excess of 1000 Å^3 in volume). The method is advantageous in only one case, namely when single crystals of sufficient size cannot be obtained, although it might be possible to justify its use when there are very few atomic parameters to determine. Using the best-quality crystalline powder under ideal experimental conditions, it is rarely possible to observe more than 100 powder lines and very often the number is much less than this. In single-crystal work, the number of observable reflections in the powder diffraction pattern tend to form a continuous background, this limiting the useful observations. Each atom in a triclinic structure is defined by three variable space parameters and it is desirable to have three or four independent observations for each parameter to be determined. To achieve this with the powder method the structure must be limited to five or six atoms to the asymmetric unit. Perhaps one of the best examples of refined powder techniques is the determination of the structure of $Mg_2(OH)_3Cl \cdot 4H_2O$ by de Wolff and Walter-Lévy[35]. In this triclinic structure, the space parameters of the eight independent atoms of the asymmetric unit were determined. A readily interpretable electron-density (010) projection calculated from 20 powder lines with $\theta \leqslant 35°$ gave the atomic positions, and interatomic distances were estimated with an error of about $\pm$ 0.1 Å.

Such analyses are rare and the more general use of powder patterns is to demonstrate that certain solids are similar in structure. To demonstrate that two substances have the same unit-cell dimensions and space group is not evidence for similarity of structure unless it is supported by independent information indicating the same conclusions. The zinc(II) and copper(II) dithiocarbamates[36], for example, are isomorphous in that they have the same unit cell and space group but there are considerable and important differences in structure between these two complexes. It could be suggested that an examination of the line intensities as well as positions might have revealed this but the subjectiveness of the interpretation of intensity measurements in this type of situation may have led to the wrong conclusion. In simpler cases, the failure to observe weak lines, or the presence of impurities in the sample giving extra lines, may lead to false conclusions.

The powder method has very important applications in qualitative and quantitative analysis and this will remain its main value as a physicochemical tool.

B. Two-dimensional X-ray Structure Analyses

In assessing any analysis, two questions may be asked: are the general features outlined in the analysis correct and, if so, how accurately determined are the details of the structure? In two-dimensional analyses, it is common experience that interatomic distances and angles are not well determined in the majority of cases and it is found in a few cases that the analysis is completely wrong. There are a variety of ways of arriving at an incorrect but plausible solution.

Two-dimensional analyses have relied very heavily on the use of electron-density projections in refinement and for confirmation of the conclusions. However, in most electron-density Fourier syntheses the phases are supplied by the experimenter according to some model he believes to be correct, and the resulting synthesis will always have some resemblance to the model even if the model is wrong. Furberg and Hassel[37] demonstrated this effect in a report on the structure of bi-1,3-dioxacyclopentyl. The

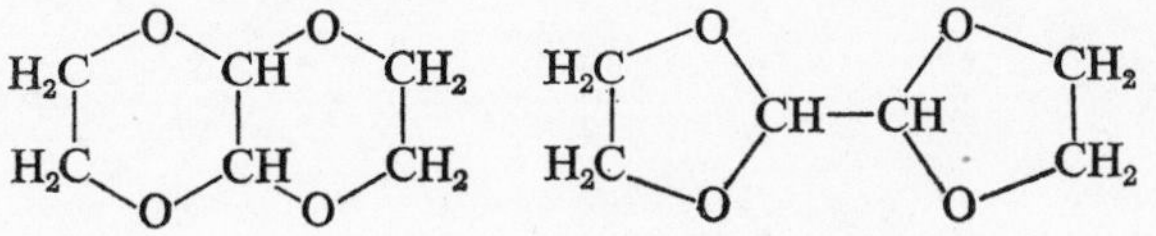

1 **2**

molecule, which may be **1** or **2**, is required by space-group considerations to have a centre of symmetry. The incorrect model, Figure 7a, gives a reasonable fit to a projection F_{obs} synthesis but will not refine to an R value better than 38%. The correct model, Figure 7b, gives a much cleaner F_{obs} synthesis and refinement yields an R value of 13%. Refinement demonstrates, without doubt, which is the correct structure. It is unfortunate that in a number of two-dimensional analyses, particularly involving heavy-atom coordination complexes, the structures are based on a single F_{obs} synthesis without refinement. In these cases, there must always be some small element of doubt in the validity of the structure. The correct and incorrect models in Figure 7 have certain characteristics. With the correct model, the atom peaks are nearly circular, the contour lines are evenly spaced, the peak heights are in the ratio of the atomic numbers of the scattering atoms, and the background is uniformly low. With the incorrect model, the contrary is true. The difficulty is greater if the projection is without a centre of symmetry. Experiments performed by Ramachandran[38] have shown that, if the phases given by a certain model are taken with a random set of values of structure amplitudes or even the structure amplitudes derived from an entirely different structure, then the resulting F_{obs} projection synthesis will be a reasonable representation

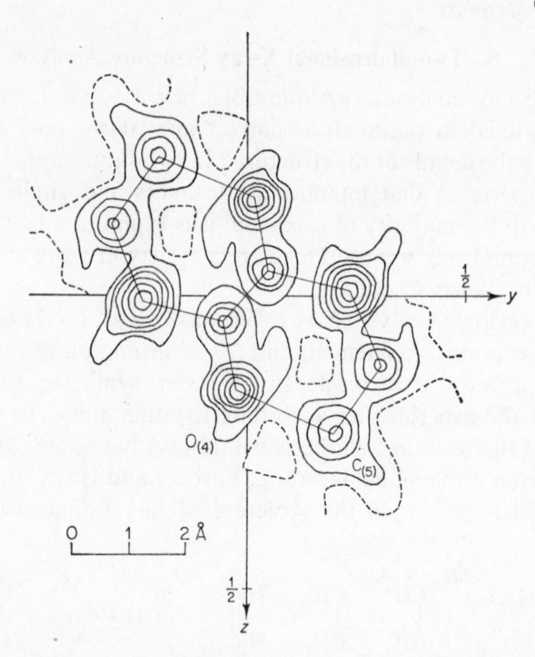

Figure 7a. Electron density map obtained by using the naphthodioxane formula. Contours at intervals of 1 eÅ^{-2}. The unit contour is dotted.

of the original model whose phases were used. Indeed, in non-centrosymmetric structures the tendency is always to get back from an F_{obs} synthesis whatever model the experimenter is using. Whether the structure has or has not a centre of symmetry, a totally incorrect model will not refine to give a low R value. Projections with R less than about 20% tend to have the correct model unless the F_h values are dominated by a correctly placed heavy atom with incorrectly placed light atoms.

A model using a molecule of the correct shape but in the wrong place in the unit cell will give better agreement between observed and calculated structure factors. This may arise from the interpretation of the Patterson function either if the wrong vectors are chosen as between heavy atoms or if a general vector lies close to a Harker section and is interpreted as a Harker vector. Analyses involving only one heavy atom which is not at some special position in the unit cell such as a symmetry axis, and is in the middle of an array of light atoms, are particularly prone to this type of error. This may be illustrated by a recent redetermination of the structure of trisacetylacetonatochromium(III) crystals[39], in which the chromium atom is displaced a quarter of a unit cell from the position used in the original

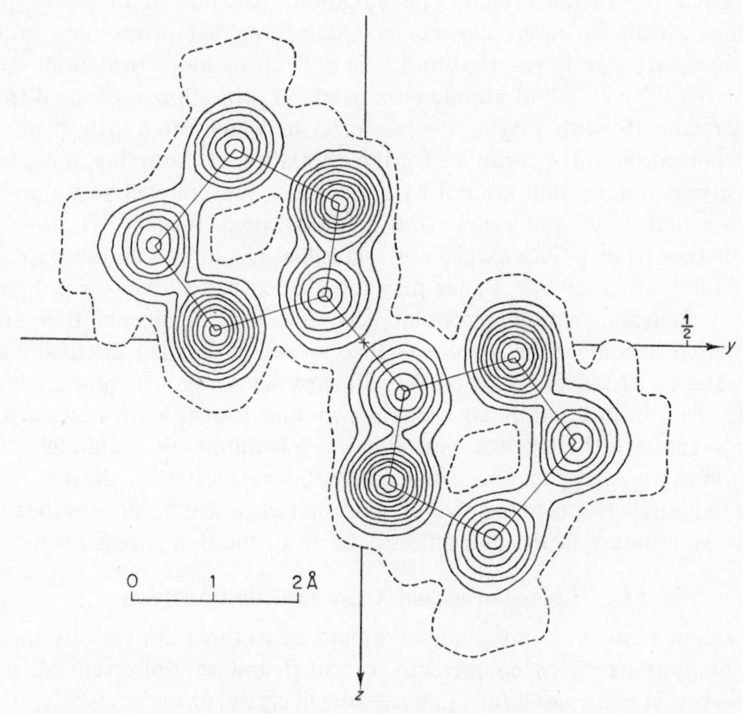

Figure 7b. Electron-density map obtained by using the bidioxacyclopentenyl formula **2**. Contours at intervals of 1 eÅ^{-2}. The unit contour line is dotted. [Reproduced, by permission, from *Acta Chem. Scand.*, **4**, 1887, 1588 (1950).]

determination[40]. The shape and size of the complex in the earlier incorrect structure is remarkably similar to those found for the correctly determined structure, but the molecular packing is quite wrong.

Even if the projections are individually correct they may be put together wrongly to give an incorrect crystal structure. This is quite easily done in the more complex orthorhombic and tetragonal space groups and if generalized projections, projections down a generalized zone axis [*hkl*], are used.

The interatomic distances and angles from projection work are in general unreliable, as are observations of small deviations from planarity in supposedly planar groups of atoms and other similar types of observation. Interatomic distances are fairly good for planar molecule stacked in a crystal of low symmetry with the molecular plane perpendicular to a principal axis of the crystal. This situation gives one well-resolved projection down the molecular plane. Other principal projections of the molecule are poorly resolved and it is difficult to judge deviations from planarity. The e.s.d.s of atomic parameters in projection work tend to be underestimates with varying degrees of error which often arise from the misapplication of the formulae for the estimation of the errors in electron density to systems that are not well resolved. This, coupled with the subjective nature of the refinement methods used (usually $F_{obs} - F_{calc}$ syntheses), tends to give exceedingly plausible structures with low e.s.d.s—structures which at first glance may look much better than three-dimensional analyses of data of similar accuracy. It is significant that most reputable scientific journals now refuse to accept detailed discussions of interatomic distances based on two-dimensional x-ray structure analysis. With neutron diffraction, cost makes two-dimensional work necessary in many cases, although the introduction of automatic data-collection equipment is changing this situation. Solid-state electron diffraction is almost entirely confined to a two-dimensional treatment because of the two-dimension nature of the diffraction patterns of the thin specimens used.

C. Three-dimensional X-ray Structure Analyses

All recent x-ray structure analyses should be in three dimensions and, if the analysis has been competently executed and carefully refined, it is unlikely that there will be any major errors in crystal or molecular structure if the crystal structure has little or no disorder. Crystals in which the molecules do not form an ordered array are quite common and are very difficult to deal with if the molecular structure is not known in advance. Crystals of the bis-8-hydroxyquinolinatocopper(II) benzotrifuroxan molecular complex[41] give sharp diffraction patterns with no spurious reflections, diffuse reflections, or streaking. From the Patterson function, it is easy to deduce the shape and position of the copper chelate but the region in which copper to benzotrifuroxan vectors are expected is diffuse and without significant features. An F_{obs} synthesis calculated using the phases from the atomic locations of the copper chelate gave a region of electron density of the form shown in Figure 8. If an incorrect dinitrosobenzodifuroxan model is fitted to the electron density (Figure 8a), the structure will refine to give $R = 0.148$, a reasonable value for a correct structure. The high-temperature factors of the nitrogen and oxygen atoms of the benzotrifuroxan and some unacceptable bond-length distances necessitate the rejection of the model in favour of some disordered model such as Figures 8b and 8c. The criteria of chemically reasonable bond length, van der Waal contacts, and temperature factors are as important as the statistical estimates of error.

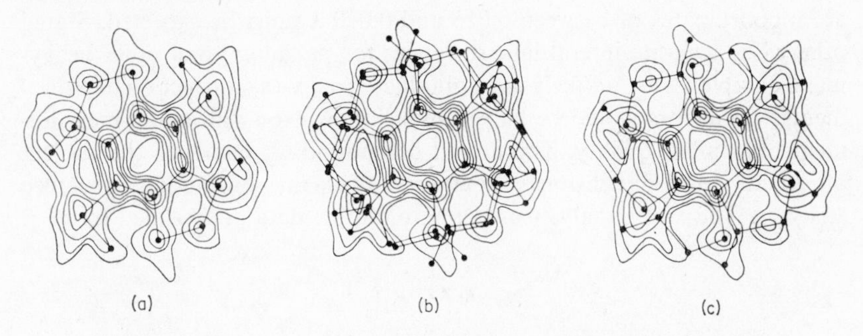

Figure 8. The electron density in a plane approximately parallel to the plane of the metal complex in the bis-8-hydroxyquinolinato-copper(II) benzotrifuroxan molecular compound. In (a) a centrosymmetric dinitrosobenzodifuroxan molecule is superposed on the electron density. In (b) four and in (c) only two equally occupied sites are assigned to benzotrifuroxan molecules. [Reproduced, by permission, from *J. Chem. Soc.*, **1965**, 4883.]

The typical three-dimensional analysis based on intensities measured by eye from Weissenberg photographs will, on average, refine to convergence at an R value of between 10 and 15%. If the value for a refined routine three-dimensional analysis is greater than 20%, the analysis must be suspect. Counter intensity measurements will, under favourable conditions, give R values of around 5%. Examination of Cruickshank's formula (7), relating R value to expected e.s.d. of atomic coordinate, shows the relationship to be linear for a given set of conditions, crystal, temperature, etc. The formula suggests two ways of improving the e.s.d.s: one is to reduce R by increasing the accuracy of the measurements, and the second is to increase S and P by reducing the temperature and hence the vibration amplitudes for the atoms of the crystal and then the temperature factors B.

For many structures, the determination is carried out using a heavy-atom derivative in which the heavy atom is irrelevant to the structure—a brominated organic molecule, for example. If we consider a 20-carbon atom molecule and neglect the hydrogen atoms, then Cruickshank's formula suggests that, if the carbon atoms have a temperature factor B equal to 4.0 and 2000 observations are made out to a maximum Bragg angle of 75°, then the mean e.s.d. of atomic coordinate will be about 0.02 Å when $R = 13\%$, reducing to 0.008 Å when $R = 5\%$. If a heavy atom, say iodine, is introduced and all else remains the same, the e.s.d.s of the carbon atom positions will more than double, although the e.s.d.s of the heavy-atom coordinates will be much less. If we imagine a similar situation but with 20 carbon atoms in a palladium organometallic compound, then with good counter data and an R value of 5% e.s.d.s of the carbon atom coordinates of between 0.015 and 0.020 Å must be expected. Stated otherwise, accurate interatomic distances for palladium and other heavy-metal complexes are unlikely to be obtained from x-ray diffraction measurements. The errors would be halved by using neutron-diffraction measurements of similar quality. The uranyl compound (**3**) has been the subject of recent three-dimensional neutron and x-ray diffraction structure analyses. The x-ray analysis used photographic data and refined to an *R*

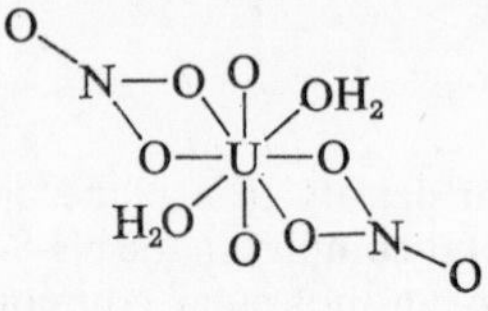

3

value of 10%. However, the e.s.d.s in the U—O and N—O bond lengths were 0.06 and 0.09 Å, respectively. The neutron determination gave an *R* value of 3.9% and e.s.d.s in all bond lengths of about 0.005 Å. Comparison with the following example shows the improvement in accuracy to be much greater than that to be expected from improved data alone. Both determinations gave nitrato groups with the N—O bonds non-equivalent. The N—O bond not engaged in coordination with the metal atom was observed to be shorter than the other two. Only in the neutron determination could the shortening be shown to be significant by statistical tests.

Agreement between observed and 'calculated' (or estimated) bond distances is often better than might be expected from the observed e.s.d.s in atomic coordinates. This can often be explained in terms of possible errors in the positioning of the molecule as a whole. Further, the calculation from which the e.s.d.s are derived rests on the assumption that all errors in measurement are random. This is rarely, if ever, the real case.

The effect of increasing the accuracy of the data with the same observation to parameter ratio is demonstrated by two determinations of benzotrifuroxan. One is of the molecular complex[42] from benzotrifuroxan (**4**) and the molecule **5**. There are about 3.9 observations per parameter and the visually estimated photographic data gives an *R* value of 13.7%. The

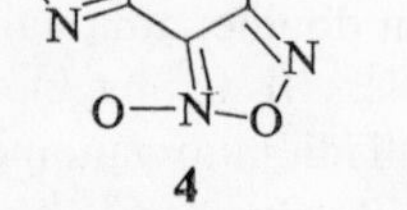

4 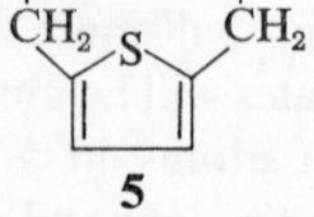 **5**

second is of benzotrifuroxan (**4**) itself[43]. The observation to parameter ratio is 3.8 and the counter data gives an R value of 4.0%. In both determinations, the refinements were performed by the least-squares method, were not biased towards a particular model, and assumed anisotropic thermal motion. The final interatomic distances in the benzotrifuroxan molecule are shown in Figure 9. In the more accurate analysis, the e.s.d.s in bond

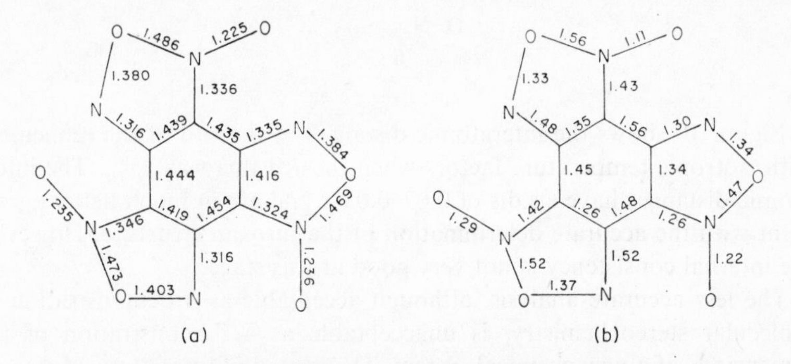

Figure 9. Interatomic distances (a) in benzotrifuroxan and (b) in the benzotrifuroxan polarization-bonded complex.

length are about 0.005–0.008 Å and in the less accurate 0.03–0.05 Å. The latter may be underestimated due to the block-diagonal approximation in the refinement but are reasonable and expected by equation (7). In the more accurate analysis, there is internal consistency between the furoxan rings and although it is not entirely clear if there is an alternation of bond length in the carbon ring it appears very probable. The less accurate model is inconsistent with the more accurate and the situation is not improved either by averaging bond lengths expected to be equal or by accepting the majority agreements for likely equalities. The less accurate model is of little use in any theoretical discussion of the molecule. It does, however, show some similar trends to those found in the better model in an exaggerated form, e.g. alternation of bond lengths in the carbon ring. It is possible that the interatomic distances in the molecule in the charge-transfer complex have been considerably modified by complex formation but other evidence, e.g. infrared spectra, is contrary to this idea and the x-ray analysis is not sufficiently good to reach a decision. It is disturbing that in the less accurate model there are discrepancies between equivalent bonds of more than three standard deviations.

If the parameter to observation ratio is increased, the situation is improved. In an investigation of the product (**6**) of the reaction between benzotrifuroxan and triphenylphosphine[44], the observation to parameter ratio is about 8 for anisotropic thermal motion.

6

Figure 10 shows the interatomic distances at the end of the refinement with isotropic temperature factors when the R value was 15%. The interatomic distances have e.s.d.s of 0.02–0.03 Å and are in much better agreement with the accurate determination of the furoxan structure. However, the internal consistency is not very good at this stage.

The less accurate analysis, although acceptable as a demonstration of molecular stereochemistry, is unacceptable as a demonstration of the dimensions of a new chemical system. The spread of interatomic distances in routine analysis may be demonstrated by the available information on

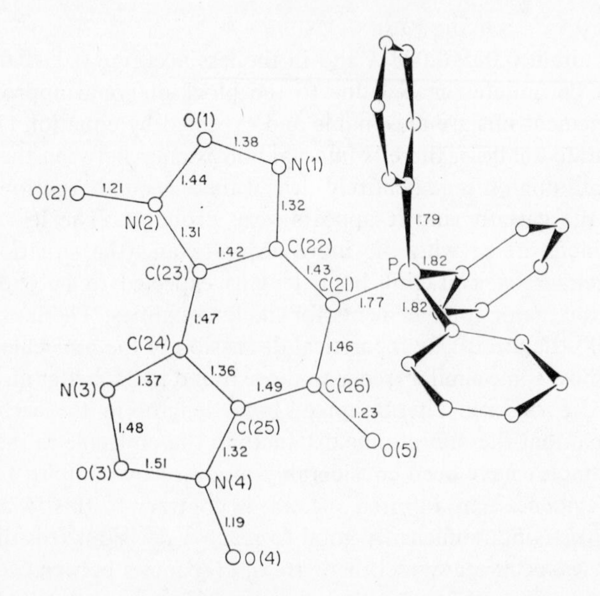

Figure 10. The molecule **6** projected on to the least-squares best plane of benzyltrifuroxan residue. [Reproduced, by permission, from *Chem. Commun.*, **1966**, 665.]

the 8-hydroxyquinoline ligand (7). There are now at least ten determinations of the structure of this system in coordination complexes with an

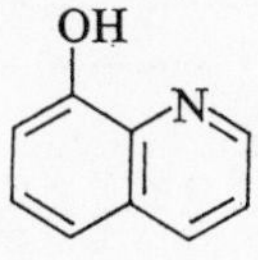

OH N

7

aromatic electron-acceptor molecule. The results are given in Table 1. The most outstandingly different are those for compound (2), the previously discussed complex of 8-hydroxyquinalinatocopper(II) and benzotrifuroxan. This reflects the effect of a disordered benzotrifuroxan molecule on the observed bond distances in the apparently well-ordered copper chelate. Apart from this, agreement is fairly good, although in some cases the differences in observed bond length for the same bond in different systems may be as much as 0.1 Å and in many cases differences are between two and three standard deviations and are on the verge of statistical significance. For example it would appear that the C—O bond in 8-hydroxyquinoline itself is longer than in the metal chelate complexes but there are similar differences in bonds in the 8-hydroxyquinoline system that can be presumed to be the same for all complexes. The effect of the heavy palladium atoms in 8-hydroxyquinolinatopalladium(II), with rather few light atoms, is clearly visible. The spread of the light-atom bond lengths is disproportionately large for the relatively low R value of 9%.

This discussion presents a rather depressing picture of the discrepancies found in the measurement of bond length by diffraction methods. Higher accuracy can be achieved and in a number of specialized cases is achieved after very much time and effort. In most of the examples in Table 1, corrections have not been applied to take into account the molecular vibrations which produce small shifts of electron-density maxima, but no absorption corrections were applied either, so that the temperature factors from which the molecular vibration is estimated would be wrong and the vibration corrections pointless. To increase the absolute accuracy of bond-length measurement not only are highly accurate counter-intensity data required, but the systematic errors of absorption and extinction must be corrected for. Even then, the dominance of the scattering of the heavy atom may preclude accurate measurements by x-ray methods in many inorganic systems. Present-day x-ray diffraction measurements are invaluable to the chemist as long as the limitations are realized. X-ray diffraction is a remarkably reliable method for determining the stereochemistry of systems

Table 1. The interatomic distances in 8-hydroxyquinoline charge-transfer complexes and metal 8-hydroxyquinolinates. (1) Bis-8-hydroxyquinolinatocopper(II)[45], (2) bis-8-hydroxyquinolinatocopper(II) benzotrifuroxan[41], (3) bis-8-hydroxyquinolinato-copper(II) tetracyanoquinodimenthane[46], (4) bis-8-hydroxyquinolinatocopper(II) picryl azide[47], (5) bis-8-hydroxyquinolinato-zinc(II) dihydrate[48], (6) bis-8-hydroxyquinolinatopalladium(II) 1,2,4,5-tetracyanobenzene[49], (7) bis-8-hydroxyquinolinato-palladium(II) chloranil[50], (8) bis-8-hydroxyquinolinatopalladium(II)[51], (9) bis-8-hydroxyquinolinatocopper(II) 1,2,4,5-tetra-cyanobenzene[52], (10) 8-hydroxyquinoline chloranil[53], (11) calculated from the bond orders for 8-hydroxyquinoline from a Hückel LCAO model.

Bond	(1)	(2)	(3)	(4)	(5)	(6)	(7)	(8)	(9)	(10)	(11)
M—N	1.973(Cu)	1.97(Cu)	1.946(Cu)	1.96(Cu)	2.099(Zn)	2.00(Pd)	1.97(Pd)	2.02(Pd)	1.954(Pd)	—	—
M—O	1.930	1.89	1.920	1.95	2.066	2.00	1.98	2.02	1.900	—	—
N—$C_{(1)}$	1.324	1.39	1.328	1.31	1.328	1.33	1.31	1.33	1.338	1.33	1.32
$C_{(1)}$—$C_{(2)}$	1.403	1.32	1.395	1.44	1.394	1.37	1.42	1.42	1.42	1.40	1.40
$C_{(2)}$—$C_{(3)}$	1.355	1.48	1.368	1.37	1.361	1.34	1.35	1.44	1.339	1.33	1.38
$C_{(3)}$—$C_{(4)}$	1.415	1.36	1.415	1.44	1.420	1.44	1.46	1.41	1.433	1.40	1.41
$C_{(4)}$—$C_{(5)}$	1.399	1.44	1.413	1.39	1.407	1.42	1.33	1.44	1.418	1.40	1.41
$C_{(5)}$—$C_{(6)}$	1.363	1.24	1.372	1.41	1.359	1.39	1.39	1.46	1.335	1.38	1.38
$C_{(6)}$—$C_{(7)}$	1.435	1.38	1.404	1.42	1.388	1.37	1.45	1.39	1.404	1.44	1.41
$C_{(7)}$—$C_{(8)}$	1.375	1.48	1.388	1.40	1.386	1.38	1.38	1.44	1.383	1.32	1.38
$C_{(8)}$—$C_{(9)}$	1.427	1.34	1.424	1.38	1.448	1.39	1.45	1.48	1.436	1.42	1.41
$C_{(9)}$—$C_{(4)}$	1.417	1.42	1.414	1.43	1.423	1.41	1.42	1.45	1.395	1.44	1.42
$C_{(8)}$—$O_{(1)}$	1.320	1.31	1.321	1.35	1.314	1.33	1.31	1.29	1.315	1.38	—
$C_{(9)}$—N	1.355	1.31	1.352	1.39	1.342	1.38	1.32	1.39	1.335	1.35	1.35
R value (%)	7.1	—	~6	13.4	13.1	10.2	10.7	9.1	9.8	11.4	—

when the hydrogen atom locations are not essential. It is a much less reliable method of measuring bond lengths unless used with great care.

It is now well known that hydrogen atoms can be found by three-dimensional x-ray structure analysis in a molecule containing only the lighter atoms, say the elements of the first and second short periods, if the measured intensities are sufficiently accurate. However, the accuracy with which the hydrogen atoms are located is low. The inorganic chemist is often interested in the much more difficult problem of locating hydrogen in the presence of heavy metals in transition metal hydrido complexes, for example. In these cases, recourse to neutron or electron diffraction is not always possible or desirable on the grounds of crystal size, volatility, chemical instability, and so on.

The difficulties of using x-ray structure analysis for hydrogen atom location in heavy-metal compounds have been demonstrated by Abrahams and Ginsberg[54] in a note on a recent structure determination of dihydridodicyclopentadienylmolybdenum[55]. Abrahams and Ginsberg claim to have demonstrated that it was not possible to locate the hydrogen atoms from the rather incomplete but quite accurate three-dimensional data used in the original determination, in which hydrogen atom locations were claimed. Ibers and La Placa[56] have put forward conditions for the location of hydrogen atoms after what they claim to be the successful location of hydrido hydrogen in hydridocarbonyl-tris-triphenylphosphine rhodium. These conditions are (*a*) good intensity data, (*b*) a large unit cell with many intensity observations of low Bragg angle θ, and hence proportionally larger hydrogen atom contribution, (*c*) low thermal motion, (*d*) a heavy-atom scattering content that is not overwhelming, i.e. big organic ligands present, and (*e*) the use of group refinement. In the group-refinement method, the analyst accepts standard forms for well-known atom groupings such as benzene rings, and defines the position of the group as a whole and refines this position with such restraints as to maintain the standard form.

IX. CONCLUSIONS

The vast majority of diffraction structure analyses in the past have been done with x-rays. This will continue to be the situation in the future. Solid-state electron diffraction will always be restricted to solids of very low volatility able to survive in a high vacuum, and the cost in terms of money, time, and scarcity value will restrict the application of neutron diffraction for many years to come. X-ray diffraction, in spite of limitations, will remain the major tool of structure analysts in the foreseeable future. Much of the tedium is being taken out of the analyses. Three-dimension

Fourier syntheses which at Oxford in 1956 took one to three months to compute on electromechanical Hollerith equipment now take ten minutes or less. Automatic counter diffractometers are replacing visually measured photographs for routine work, decreasing the labour involved and increasing accuracy. The analyst is able to think in terms of weeks and months for the duration of analyses; soon it may be days and weeks. Direct methods and improved quality of data will eliminate the need for the often chemically redundant heavy atom, leading to a further increase in accuracy of atomic locations in these crystals. Inorganic chemists are beginning to realize that they can be their own crystallographers, and crystallographers are beginning to look for fine details such as the distribution of bonding electrons.

X. BIBLIOGRAPHY

A. Introductory Texts

1. J. C. D. Brand and J. C. Speakman, *Molecular Structure, the Physical Approach*, Arnold, London, 1961.
2. P. J. Wheatley, *The Determination of Molecular Structure*, revised ed., Oxford University Press, London, 1962.
3. C. W. Bunn, *Chemical Crystallography*, 2nd ed., Oxford University Press, London, 1961.
4. G. E. Bacon, *Neutron Diffraction*, 2nd ed., Oxford University Press, London, 1962.
5. B. K. Vainshtein, *Structure Analysis by Electron Diffraction*, Pergamon, Oxford, 1964. Solid state only.

B. A Selection of Useful Texts

1. F. C. Philips, *An Introduction to Crystallography*, 3rd ed., Longmans, London, 1963. Classical crystallography.
2. M. A. Jaswon, *An Introduction to Mathematical Crystallography*, Longmans, London, 1965. Crystal symmetry.
3. R. W. James, *The Optical Principles of the Diffraction of X-rays*, Bell, London, 1948. Very useful, but watch out for recent developments which tend to make the book rather dated in places.
4. M. J. Buerger, *Elementary Crystallography*, Wiley, New York, 1956. Crystal symmetry.
5. B. N. H. Hartshorre and A. Stuart, *Practical Optical Crystallography*, Arnold, London, 1964. The polarizing microscope.
6. N. F. M. Henry, H. Lipson, and W. A. Wooster. *The Interpretation of X-ray Diffraction Photographs*, Macmillan, London, 1960. Good introduction.
7. M. J. Buerger, *X-ray Crystallography*, Wiley, New York, 1942. Detailed treatment of x-ray photography. Rather difficult for the beginner but very useful later.
8. M. J. Buerger, *The Precession Method*, Wiley, New York, 1964. Use of the Buerger precession camera.

9. U. W. Arndt and B. T. M. Willis, *Single Crystal Diffractometry*, Cambridge University Press, Cambridge, 1966. Counter diffractometers.
10. H. Lipson and W. Cochran, *The Determination of Crystal Structures*, 3rd ed., Bell, London, 1966.
11. M. J. Buerger, *Crystal Structure Analysis*, Wiley, New York, 1960. A gigantic work; a must for the researcher.
12. A. I. Kitaygorodsky, *The Theory of Crystal Structure Analysis*, English ed., Consultants Bureau, New York, 1961. Direct method and intensity statistics. Also see the Karle and Hauptman monograph[26].
13. H. Lipson and C. A. Taylor, *Fourier Transform and X-ray Diffraction*, Bell, London, 1958. Also valuable in teaching.
14. M. J. Buerger, *Vector Space*, Wiley, New York, 1959. The Patterson function.
15. J. S. Rollett (Ed.), *Computing Methods in Crystallography*, Pergamon, Oxford, 1965. Computer programming.
16. *The International Tables for X-ray Crystallography* (Vol. I, 'Symmetry Groups'; Vol. 2, 'Mathematical Tables'; Vol. 3, 'Physical and Chemical Tables'), Kynoch Press, Birmingham, 1952–1962.
17. J. D. H. Donnay (Ed.), *Crystal Data*, A.C.A. Monograph No. 5, A.C.A., New York, 1963.
18. H. D. Megaw, *Crystallographic Book List*, International Union of Crystallography. Available for $3.00 through the Polycrystal Book Service, G.P.O. Box No. 620, Brooklyn 1, New York, U.S.A. Unesco coupons accepted for payment.

XI. REFERENCES

1. D. C. Hodgkin and G. Lenhert in *Vitamin B_{12} and Intrinsic Factor* (Ed. H. C. Heinrich), Ferdinand Anke, Stuttgart, 1962, p. 105.
2. J. G. Forrest and C. K. Prout, *Chem. Commun.*, **1966**, 685.
3. B. T. M. Willis, *Proc. Roy. Soc. (London), Ser. A*, **274**, 122, 124 (1963).
4. J. Laciejeewicz, *Z. Krist.*, 116, **345** (1961).
5. C. G. Schull and M. K. Wilkinson, *Rev. Mod. Phys.*, **25**, 100 (1953).
6. C. G. Schull, W. A. Strauser, and E. O. Wollan, *Phys. Rev.*, **83**, 298 (1951).
7. D. C. Hodgkin, F. Moore, and B. T. M. Willis, *Acta Cryst.*, in press.
8. B. K. Vainshtein in *Advances in Structure Research by the Diffraction Method*, Interscience, New York, 1964, p. 27.
9. C. G. Darwin, *Phil. Mag.*, **43**, 800 (1922).
10. W. H. Zachariasen, *Acta Cryst.*, **16**, 1139 (1963).
11. P. R. Pinnock, J. Lipson, and C. A. Taylor, *Acta Cryst.*, **4**, 289 (1951).
12. M. J. Buerger, *J. Appl. Phys.*, **17**, 576 (1946).
13. C. A. Beevers and J. H. Robertson, *Acta Cryst.*, **3**, 164 (1950).
14. M. J. Buerger, *Proc. Natl. Acad. Sci. U.S.*, **39**, 674 (1953).
15. H. Lipson and W. Cochran, *The Determination of Crystal Structures*, Bell, London, 1953, p. 212.
16. J. M. Bijvoet, *Nature*, **173**, 888 (1954).
17. D. Dale, D. C. Hodgkin, and K. Venkateson in *Crystallography and Crystal Perfection* (Ed. G. N. Ramachandran), Academic Press, New York, 1963, p. 237.

18. C. Nockolds and D. C. Hodgkin, unpublished work.
19. D. Dale, *United Kingdom Atom Energy Research Establishment Report*, AERE–R5195, 1966.
20. H. Ott, *Z. Krist.*, **66**, 136 (1928).
21. J. S. Kasper and D. Harker, *Acta Cryst.*, **1**, 70 (1948).
22. J. Karle and H. Hauptman, *Acta Cryst.*, **3**, 181 (1950).
23. D. Sayre, *Acta Cryst.*, **5** 60 (1952).
24. A. J. C. Wilson, *Nature*, **150**, 152 (1942).
25. E. R. Howells, D. C. Philips, and D. Rogers, *Acta Cryst.*, **3**, 210 (1950).
26. J. Karle and H. Hauptman, 'Solution of the phase problem I', *The Centrosymmetric Crystal*, A. C. A. Monograph No. 3, A.C.A., New York, 1953.
27. A. D. Booth, *Nature*, **161**, 765 (1948); W. Cochran, *Acta Cryst.*, **4**, 408 (1951).
28. A. D. Booth, *Trans. Faraday Soc.*, **42**, 444 (1946).
29. E. T. Whittaker and C. J. Robinson, *The Calculus of Observations*, 2nd ed., Van Nostrand, New York, 1930.
30. D. W. J. Cruickshank, *Acta Cryst.*, (a) **2**, 65, 154 (1949), (b) **3**, 72 (1950), (c) **5**, 511 (1952).
31. D. W. J. Cruickshank and F. R. Ahmed, *Acta Cryst.*, **6**, 385 (1953).
32. S. F. Darlow, *Acta Cryst.*, **13**, 683 (1960).
33. A. F. Wells, *Structural Inorganic Chemistry*, 3rd ed., Oxford University Press, London, 1961, p. 338.
34. P. M. de Wolff, *Acta Cryst.*, **10**, 590 (1957); **11**, 664 (1958).
35. P. M. de Wolff and L. Walter-Lévy, *Acta Cryst.*, **6**, 40 (1953).
36. M. Bonamico, G. Dessy, A. Magnoli, A. Vacciago, and L. Zambonelli, *Acta Cryst.*, **19**, 887 (1965).
37. O. Hassel and S. Furberg, *Acta Chem. Scand.*, **4**, 1584 (1950).
38. G. N. Ramachandran, Lecture at Oxford University, unpublished.
39. B. Morosin, *Acta Cryst.*, **17**, 705 (1964).
40. L. H. Shkolnikova and E. A. Shagan, *Soviet Phys.–Cryst.* (*English Transl.*), **5**, 32 (1960).
41. C. K. Prout and H. M. Powell, *J. Chem. Soc.*, **1965**, 4882.
42. B. Kamenar and C. K. Prout, *J. Chem. Soc.*, **1965**, 4838.
43. H. H. Cady, A. C. Larson, and D. T. Cromer, *Acta Cryst.*, **20**, 336 (1966).
44. A. S. Bailey, T. S. Cameron, J. M. Evans, and C. K. Prout, *Chem. Commun.* **1966**, 664.
45. G. J. Palenik, *Acta Cryst.*, **17**, 687 (1964).
46. S. C. Wallwork and R. M. Williams, to be published.
47. A. S. Bailey and C. K. Prout, *J. Chem. Soc.*, **1965**, 4867.
48. G. J. Palenik, *Acta Cryst.*, **17**, 696 (1964).
49. B. Kamenar, C. K. Prout, and J. D. Wright, *J. Chem. Soc.*, *A*, **1966**, 661.
50. B. Kamenar, C. K. Prout, and J. D. Wright, *J. Chem. Soc.*, *A*, **1966,** 661.
51. C. K. Prout and A. G. Wheeler, *J. Chem. Soc.*, *A*, **1966**, 1286.
52. P. Murray-Rust and J. D. Wright, *J. Chem. Soc.*, *A*, in press.
53. C. K. Prout and A. G. Wheeler, *J. Chem. Soc.*, *A*, **1967**, 469
54. S. C. Abrahams and A. P. Ginsberg, *Inorg. Chem.*, **5**, 500 (1966).
55. M. Gerloch and R. Mason, *J. Chem. Soc.*, **1965**, 296.
56. S. T. La Placa and J. A. Ibers, *Acta Cryst.*, **18**, 511 (1965).

2

X-ray Spectroscopy

C. BONNELLE

I. INTRODUCTION

X-ray spectroscopy has contributed in many ways to our understanding of the properties of matter. Moseley's empirical law, defining the relation between the characteristic x-ray frequencies of an element and its atomic number, confirmed by later experimental measurements, was of considerable help in establishing the periodic table. Precise measurements of the energy levels in atoms were obtained by this technique long before magnetic spectral analysis of photoelectrons removed from the corresponding inner shells provided a suitable alternative[1]. It has contributed a great deal to the field of atomic physics and the study of the interaction of radiation with matter.

For many years, x-ray spectroscopy has also been used to study the behaviour of electrons in solids. An analysis of x-ray emission and absorption spectra, due to transitions between a discrete inner atomic level of known characteristics and the various levels in the energy bands, usually provides the most straightforward means of determining the distribution of the energy levels making up the band. These levels are weakly bound to the atom and describe the distribution of the occupied and unoccupied electron states in the solid[2]. The physical and chemical states of an element thus modify its x-ray absorption edges and the neighbouring emission lines or bands, and information about the element's electronic orbitals in various chemical states can be obtained from x-ray spectra.

We shall give a brief summary of the mechanisms which result in x-ray emission and absorption spectra for the ideal case of a free atom as well as for metallic and other solids. We shall describe the methods offered by x-ray spectroscopy for the measurement of electronic energy levels and the steps to be followed for determining true electron distributions from experimental data, but point out the difficulties which can sometimes prevent rigorous conclusions. After a brief look at experimental techniques, we shall give a few typical results as examples.

II. BASIC THEORY

A. Atoms

1. Emission

When an inner shell, e.g. the K shell, of an atom is ionized, either by electron collision, photoelectric effect, or some other process, the atom acquires an excess potential energy with respect to its normal state. The atomic electrons immediately tend to reorganize themselves so as to restore the stable state, i.e. the state with the lowest energy. An electron in the outermost shell, say the L shell, can be transferred to the hole in the K shell and the energy difference $E_K - E_L$ between the two states may be emitted as radiation of frequency $(E_K - E_L)/h$. A series of spectral lines will be observed with discrete frequencies characteristic of the target atom. However, the energy $E_K - E_L$ may be transferred to an electron of an outermost level which will therefore be ejected from the atom. This is called the Auger effect.

The various transitions have different probabilities and some are practically absent in the emission spectrum. The probability P_e that a spontaneous transition between two states i and f shall take place in unit time with the emission of one photon having frequency ν is, according to quantum electrodynamics,

$$P_e \propto \nu |\mathbf{M}_{if}|^2$$

where $\mathbf{M}_{if}$ represents the transition matrix element. If the non-relativistic Schrödinger equation is used, and the 'electric dipole' approximation is made, then $\mathbf{M}_{if}$ is written

$$\mathbf{M}_{if} = \int \psi_f^* \sum_n \text{grad}_n \psi_i \, d\tau$$

where ψ_i and ψ_f are the wave functions which characterize the two states and the summation is taken over the n electrons. The emitted intensity is proportional to $h\nu P_e$. The transition probabilities have been calculated for hydrogen-like atoms, i.e. for the case where the matrix element of the momentum $\mathbf{M}_{if}$ is proportional to ν^2 times the matrix element of the coordinate (or dipole moment). Then

$$P_e \propto \nu^3 \left| \int \psi_f^* \sum_n x_n \psi_i \, d\tau \right|^2$$

Simple selection rules have been established which are similar to those in use for atomic optical spectra involving one electron, and can be formulated as $\Delta l = \pm 1$ and $\Delta j = 0, \pm 1$ where l and j are the quantum numbers of the appropriate levels.

The shape and width of the spectral emission depend, as do its frequency and intensity, on the energy levels participating in the transition. A mean lifetime τ_X can be associated with a discrete state X, which has energy E_X. This lifetime τ_X is inversely proportional to the sum of all the transition probabilities (between the state X and any other state), whether these transitions give rise directly to radiation or not. According to Heisenberg's uncertainty principle, the width of the emission measured in energy units is

$$\Gamma_X = \frac{h}{2\pi\tau_X} \quad \left(\text{or } \frac{1}{2\pi\tau_X} \text{ if measured in frequency units}\right)$$

According to the Weisskopf and Wigner quantum theory[3], a transition between two atomic states A and B has a frequency distribution of the Lorentz type

$$\mathscr{L}(\nu)\, d\nu \propto \frac{\Gamma/2\pi}{(\nu_{AB} - \nu)^2 + (\Gamma/2)^2} \, d\nu$$

symmetrical about the frequency maximum $\nu_{AB} = (E_A - E_B)/h$, with a width Γ at half maximum. A frequency distribution of the Lorentz type of width Γ_A (or Γ_B) can be associated with either state A or B such that $\Gamma = \Gamma_A + \Gamma_B$. It follows that the energy distribution of a transition AB can generally be represented by the convolution integral of the distribution of levels A and B.

2. Absorption

During a photoelectric process involving electrons of the inner atomic shell, an electron can either be removed completely from the atom or temporarily occupy one of the optical excitation levels. A series of absorption lines are formed by a mechanism similar to that which gives atomic spectral lines in the optical region, but their width is such that the lines overlap each other. They form the Kossel structures which have been resolved for monatomic gases [4], i.e. for what may be considered as the 'free' atom case. The limit of the series coincides with the ionization energy (Figure 1).

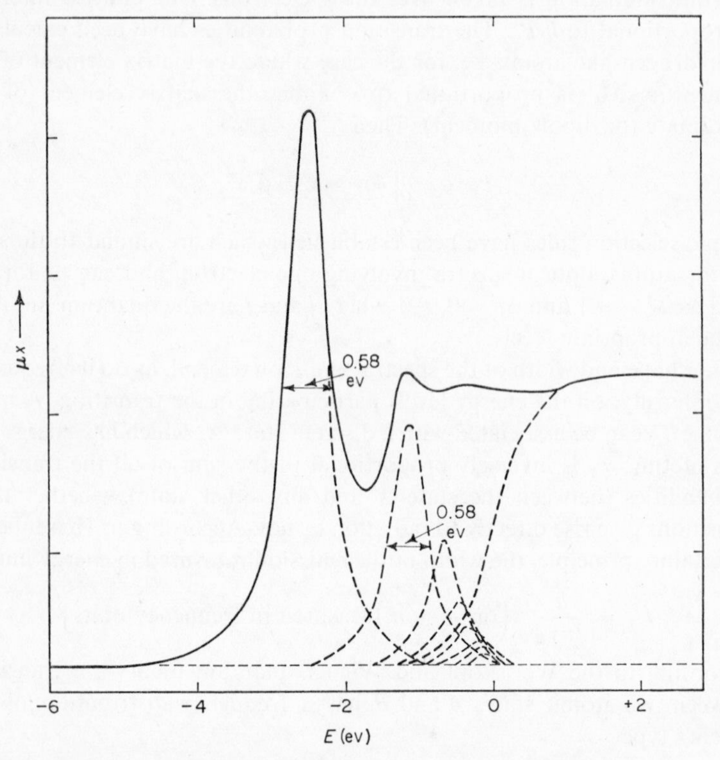

Figure 1. *K* absorption in argon [according to Parratt [4]].

Conventionally, the position of an x-ray absorption edge is defined as the point of inflexion on the low-frequency side of the first absorption line. The width of the absorption line, measured at half maximum, will give the width of the excited state and therefore its lifetime. If the atomic emission widths are also determined, the width of the other energy levels can be deduced.

The probability P_a that radiation of frequency ν and incident intensity $I_0(\nu)$ shall be absorbed in unit time by a photoelectric effect is given by

$$P_a \propto \frac{I_0(\nu)}{\nu^2} |\mathbf{M}_{fi}|^2$$

The absorption coefficient $\mu(\nu)$ is then

$$\mu(\nu) \propto \frac{h\nu}{I_0(\nu)} P_a \propto \frac{1}{\nu} |\mathbf{M}_{fi}|^2$$

The selection rules are shown schematically below for dipolar transitions from or towards *s*, *p*, or *d* states.

E.g. for absorptions $K, L_I, M_I, \ldots$ $s \rightarrow p$

$L_{II,III}, M_{II,III}, \ldots$ $p \Rightarrow d$ ↘ s

$M_{IV,V}, \ldots$ $d \Rightarrow f$ ↘ p

The energies of atomic levels are determined from measurements of conveniently chosen x-ray absorption and emission frequencies. Tables of singly ionized energy levels have been drawn up for the elements of atomic number 3 to 92[5]. These energies are measured with respect to a reference level which in each case corresponds to the first unoccupied level associated with the absorption on which the calculation was based. An approximate theoretical value for the ionization energy of each level may be obtained from the Hartree–Fock one-electron parameter for the electron which has been removed, if Koopmans' theorem is applied (adiabatic approximation).

So far, we have assumed that the x-ray absorptions and emissions correspond to one-electron transitions between singly ionized states. However, lines are also found in emission spectra which cannot be explained by such transitions[6]. They are called 'satellite lines' and have been thought to correspond to transitions in multiply ionized atoms following either an Auger effect or an initial double ionization, i.e. the simultaneous ejection of two electrons. This may arise, for example, from an 'electron shake-off'[7,8], the 'sudden perturbation' due to the ionization of the inner shell causing a bound electron from an outer shell to undergo a transition to an excited state.

X-ray absorption transitions corresponding to a double excitation were detected for the first time in the argon K spectrum[9,10] and from these measurements the double ionization energies could be calculated. In argon, they are the KM_I and $KM_{II,III}$ transitions.

B. Metals

While the distribution of the outermost energy levels, whether occupied or unoccupied, is well defined in a free atom, in a solid it is deformed by atomic interactions. An emission much broader and, in general, weaker than the atomic lines will appear at the end of each series in the x-ray spectrum of a solid, and is termed an emission band. The shape of this x-ray band is determined by the way in which a hole can be distributed over the unperturbed and occupied outer energy levels when allowance has been made for the transition probabilities and lifetimes.

The observed emission or absorption curve $\sigma(\nu)$ is, in fact, a convolution integral of the distribution of these levels, occupied or unoccupied, respectively, and the $\mathscr{L}$ distribution of the inner level. Let us first consider the case of a metal (Figure 2). If the distribution of its electron states is

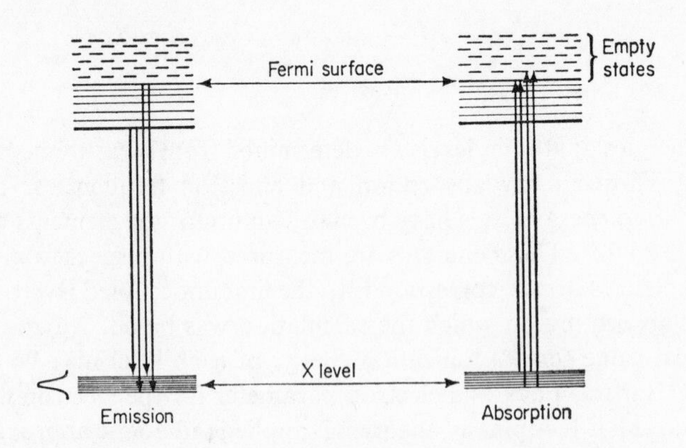

Figure 2. Schematic diagram of emission band and absorption in a metal.

given by $N(\varepsilon)$, where the energy ε is measured relative to an average electron potential in the solid, and if the variations in transition probabilities are neglected along the $\sigma(\nu)$ curve, the latter is given by

$$\sigma(\nu) = \int_{-\infty}^{+\infty} N(\varepsilon)\mathscr{L}(\varepsilon - \nu)\,\mathrm{d}\varepsilon \tag{1}$$

The emission band of a metal should have a discontinuity with a sharp cut-off towards the higher frequencies characterizing the upper limit of the Fermi distribution ε_F. A marked drop in intensity is indeed observed for certain metals though it is broadened by the smearing effect of the $\mathscr{L}$ distribution and the finite resolving power of the measuring instrument[11].

Absorption limits also show a sharp drop in intensity. Their shape has been studied in the general case where an internal electron is transferred towards a series of normally unoccupied levels, which are assumed to have a uniform energy distribution, forming a continuum of unoccupied states[12]. This is quite a good approximation to what happens in a metal which satisfies the Fermi distribution law, i.e. where

$$N(\varepsilon)\,\mathrm{d}\varepsilon = F(\varepsilon)\varepsilon^{\frac{1}{2}}\,\mathrm{d}\varepsilon$$

since the Fermi function $F(\varepsilon)$ is almost constant at normal temperatures and $F(\varepsilon)\varepsilon^{\frac{1}{2}}$ varies very little, at least over the limited range of ε values in which measurements are carried out.

If Weisskopf and Wigner's theory[3] is applied to the electron transfer between an internal level with a Lorentz distribution and the continuum of states, cut off at ε_F, the variation of the absorption coefficient $\mu(\nu)$ can be obtained by rewriting equation (1) as

$$\mu(\nu) \propto \int_{\nu_{XE_F}}^{\infty} \frac{\Gamma_X/2\pi}{(\nu_{XE}-\nu)^2+(\Gamma_X/2)^2}\,\mathrm{d}\nu_{XE}$$

$$\propto \frac{1}{2} - \frac{1}{\pi}\arctan\frac{\nu_{XE_F}-\nu}{\Gamma_X/2} \qquad (2)$$

where $\nu_{XE} = (E_X - E_i + \varepsilon)/h$, and E_i is the mean potential of the crystal. This assumes that the width of the ground state is negligible compared with that of the excited state Γ_X.

The distance between the two points C and D situated on either side of the inflexion point on the arctangent curve, at quarter and three-quarter height, is Γ_X, which by definition is the width of the internal level (Figure 3).

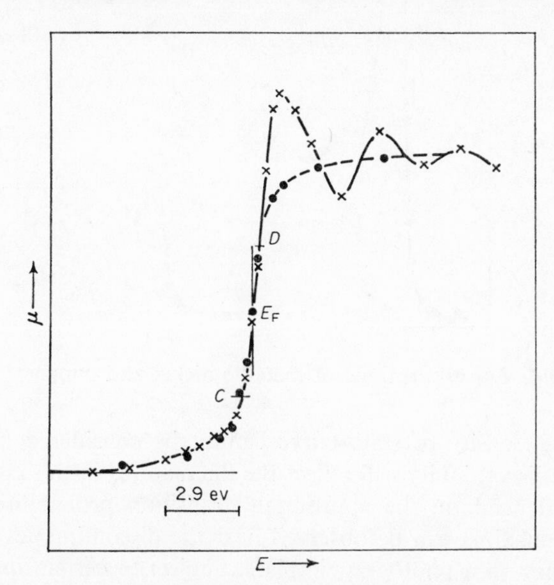

Figure 3. L_{III} absorption of metallic copper (——) and the arctangent curve (---).

The point of inflexion gives the position of the Fermi limit. Experimentally, metals such as the alkali metals, magnesium, aluminium, and the noble metals have absorption edges which can be described by arc-tangent curves.

The emission discontinuities which have been observed coincide with the absorption edges to within the limits of experimental error. For instance, they coincide to within 0.1 ev for the aluminium K spectrum[13], confirming that the two phenomena are reversed with respect to each other at the Fermi level.

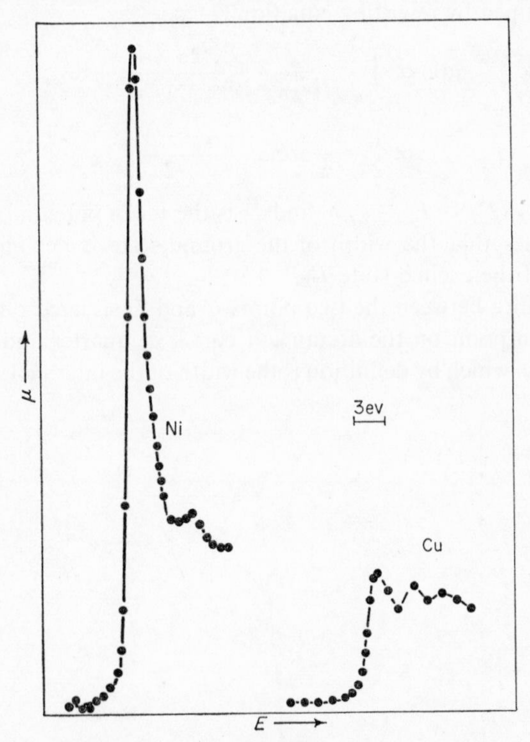

Figure 4. L_{III} absorptions of metallic nickel and copper[14] to the same scale.

When the density of state curve cannot be considered uniform, the absorption discontinuity will reflect the fluctuations in this curve, but the effect will depend on the appropriate transition probabilities. Marked absorption maxima can be observed near the discontinuities and appear as white lines on a photographic plate. Subject to certain qualifications, such an absorption line indicates a high density of states whose spectral characteristics correspond to the most probable transition. Thus L_{II} and

L_{III} absorption discontinuities, arising from internal *p*-state transitions, show, for the transition elements, marked absorption lines which indicate the presence of unoccupied *d* states at the Fermi level in the metal. The striking difference in shape between the nickel and copper L_{III} discontinuities illustrates this point[14] (Figure 4). Similarly, intense absorption lines can be expected for the M_{IV} and M_V discontinuities in the rare earths and transuranic elements which have a high density of unoccupied *f* states localized near the Fermi surface in the metal. This has been observed experimentally and will be discussed later.

The form and width of absorption lines, which are normally asymmetrical, provide information on the distribution of unoccupied levels. For instance, the number of holes in the 3*d* levels of cobalt and nickel were determined qualitatively from the *L* absorption lines by comparing their widths to those of the corresponding emission bands[15].

Fine structure is observed in the absorption spectra beyond the principal discontinuity towards higher frequencies, sometimes several hundred electron volts away. A number of theories have been put forward to interpret this structure. Kronig[16] postulated that the structure arose from forbidden energy bands in a crystal—forbidden, that is, for electrons with a given momentum. Their position would thus depend on the crystalline structure, i.e. on a long range order. However, from certain experimental results and later calculations carried out by Kronig[17], Petersen[18], and other workers[19,20], it would seem that the structure depends mainly on the arrangement of nearest-neighbour atoms, i.e. on a short range order.

C. Compounds

It was at first thought that only the outer levels were influenced by chemical bonding, since the changes in binding state influenced mainly the position and shape of the x-ray emissions and the absorption edges concerned with these levels. In fact, any modification to the outer-electron configuration also changes the energy of the deeper atomic levels because the ionization energy of an electron is altered by the screening effect of the other electrons, whatever their wave functions. This has been examined by Cauchois[21] in connexion with calculations on the ionization energy carried out with the Hartree–Fock self-consistent field method for an ion in various electron configurations[22,23]. The ionization energy change should be of about the same magnitude for each of the inner levels. The corresponding displacements of the atomic spectral lines are therefore very small and difficult to observe. High-resolution spectrographic studies have revealed displacements of the most intense lines as a function of the degree of oxidation, but only for the lighter elements. More generally, the energy of the inner levels depends on the spatial distribution of the electron cloud surrounding the emitting atom. It follows that the position of the atomic lines also depends on the hybridization of the valence orbitals, the degree of covalence, and the type of coordination. A few experimental examples will be given later.

An x-ray emission band of a compound, whether a semiconductor or insulator, is usually of lower energy than the corresponding emission from the metal itself. On the other hand, the absorption discontinuities are generally displaced towards higher energies. According to Kunzl's empirical law[24], this displacement is proportional to the degree of oxidation of the element in the compound. Experiments have been carried out to determine the oxidation state of an element in a compound from x-ray absorption-edge measurements. To our knowledge, there are only a few anomalies, the $L_{\mathrm{II}}L_{\mathrm{III}}$ absorptions of cupric oxide, which we shall mention later[25] and, the $M_{\mathrm{IV}}M_{\mathrm{V}}$ absorptions of the few elements having unfilled *f* levels.

The shifts in emission and absorption spectra confirm that the occupied and unoccupied orbitals of the metal ion in a compound are normally separate whether it is a semiconductor or an insulator. For a compound which can be described to a good approximation by band theory, the energy separation between the high-frequency side of the emission and the inflexion point of the absorption discontinuity must correspond to the forbidden energy range termed the 'gap'. Subject to certain qualifications, x-ray spectra will therefore provide a direct measurement of the gap. This has been confirmed by the author for cuprous oxide[15]. Indeed, in this compound, the absorption edge can be fitted to an arctangent curve which is typical of a transition to an unoccupied conduction band with a virtually uniform distribution.

However, for many compounds, the distance between the limit of the occupied states and the inflexion point of the absorption edge has no simple physical meaning. This is especially true when an intense absorption line is superimposed on the discontinuity. If the line is symmetrical and has a form closely resembling a Lorentz distribution, it can be attributed to a transition towards a localized orbital with an almost atomic character. For example, the L_{II} and L_{III} absorption edges of compounds of the elements at the end of each transition series have very marked absorption lines, characteristic of transitions to the unfilled *d* orbitals. For each atom or ion in the crystal lattice, the more the *d* wave functions are localized and the weaker the hybridization with *s* and *p* states, the narrower will be the absorption lines, if we assume that all other factors remain the same.

The absorption edges sometimes have a complex form, i.e. there are steps in the absorption discontinuity, or occasionally secondary absorption maxima near the main edge, though these may not always be well resolved. They arise because of the possible existence of transitions to other orbitals which have different probabilities, or sometimes because of interactions with holes in the same or some other level in the ion[26].

The electron distribution and the relative positions of orbitals of various symmetries of an element in a given physical and chemical state can be determined from the appearance of the spectrum in the region of each ionization frequency, and computation of the energies from the emission and absorption transitions. Thus, because selection rules exist, data obtained from K and L spectra or the five M absorption edges of heavy elements provide complementary information for determining the outer-electron configurations.

The absorption edge in a metal generally corresponds to transfer of an inner electron to the Fermi surface, whatever the l quantum number. Indeed, within experimental limits the energy of each absorption discontinuity is equal to that of its corresponding levels, calculated from other absorption transition and emission lines. Anomalies do exist, however. Thus, at the M_{IV} and M_V absorption edges in the elements from about 70 to 83, transitions occur to unoccupied $5f$ states, well above the Fermi limit for these elements[27]. We shall show that absorption edges suffer energy displacements in a compound with respect to the metal and that these are usually quite different, according to whether the spectrum arises from an s-, p-, or d-electron transition. The inner levels are probably displaced by a similar amount, but each characteristic unoccupied orbital is separated one from the other because of the ligand-field effect.

D. Energy Distribution of Electron States

Let us come back to some simple considerations on the subject of x-ray emission and absorption processes. The intensity emitted as a function of the frequency, $I(\nu)$, is dependent on the density of the occupied states, while the absorption coefficient $\mu(\nu)$ is related to the density of unoccupied states. The relations, to a first approximation, are as follows:

$$I(\nu) \propto \int \nu P_e(\varepsilon) N_{occ}(\varepsilon) \mathscr{L}(\varepsilon - \nu)\, d\varepsilon$$

$$\propto \nu^2 \int \overline{|\mathbf{M}|^2} N_{occ}(\varepsilon) \mathscr{L}(\varepsilon - \nu)\, d\varepsilon$$

$$\mu(\nu) \propto \int \nu P_a(\varepsilon) N_{unocc}(\varepsilon) \mathscr{L}(\varepsilon - \nu)\, d\varepsilon$$

$$\propto \frac{1}{\nu} \int \overline{|\mathbf{M}_a|^2} N_{unocc}(\varepsilon) \mathscr{L}(\varepsilon - \nu)\, d\varepsilon$$

Both the I and μ expressions contain a factor which is a more or less complex function of frequency depending on the approximations made in calculating the wave functions and estimating the matrix elements. Accordingly, the power of ν varies between 2 and 4 for I and -1 and $+1$ for μ. However, the frequency variations across the emission band and absorption edge are almost negligible unless the wavelength is greater than about 100 Å, so that this factor can be neglected in most cases.

It is not possible to evaluate the matrix elements rigorously, but if we suppose that the electrons are almost free, it can be shown that for a metal at a low value of ε, i.e. at the lower limit of the conduction band, $\overline{|\mathbf{M}|^2}$ is proportional to ε for transitions to s-electron states, while $\overline{|\mathbf{M}|^2}$ is equal to a constant for transitions to p states[28]. The intensity of the s and p emission bands increases, for such an approximation, as $\varepsilon^{\frac{3}{2}}$ and $\varepsilon^{\frac{1}{2}}$, respectively, at the long-wavelength end of the band. This is observed for a large part of the $K\beta$ band in aluminium[13]. When the hybridization of the occupied and unoccupied levels is not complete, $N(\varepsilon)$ reveals the state symmetries, and allowance has to be made for the transition probabilities by applying the inner-level selection rules to electron states (or to the distribution of states $N(\varepsilon)$) from which or towards which the transitions take place.

We have seen that a continuous distribution with a sharp edge combined with an inner-level Lorentz distribution gives an arctangent curve, while two Lorentz distributions produce a Lorentz curve. The former applies to transitions to free electron states in solids and the latter to transitions between the levels of a free atom. All intermediate cases are possible, so that experimental emission and absorption curves ought to be adjusted to take account of the convolution with the $\mathscr{L}$ distribution. Simplified calculations give the order of magnitude of the correction in particular cases[15]. The $N_{\text{occ}}(\varepsilon)$ and $N_{\text{unocc}}(\varepsilon)$ curves are not appreciably disturbed if the width Γ_X of the internal level is small compared to that of the electron distributions to be determined. It follows that such experiments must be carried out in a region where the characteristic Γ_X widths are small.

We have already seen that an x-ray emission band describes the distribution of the final state, i.e. of a hole in the unperturbed electron band, provided that the possible spreading of the lower energy levels in the band, arising from its shorter lifetime, is neglected. In the case of an absorption, a hole is formed in an inner level. The conduction electrons in a metal tend to neutralize the charge of the crystal in the neighbourhood of the hole, but the perturbation produced by the ionized atom most probably remains small, and the Fermi distribution in the unperturbed metal is generally observed. It takes the form of a discontinuity with a sharp edge whose finite width, once the various factors mentioned have been corrected for, is attributable to the width of the Fermi distribution at the experimental temperature. It could be expected that semiconductors or insulators show discrete excitation states below the conduction band[29], but these levels would lie close to the band and would be unresolved. We can, therefore, consider that the distribution of the occupied or unoccupied states is practically unchanged when the inner electron is removed. Then any effect due to an inner hole can be neglected, especially when comparing the spectrum of the element in the metal and in its compounds. An example of this is that the widths of the discontinuities in cuprous oxide are nearly equal to those of copper itself.

III. EXPERIMENTAL TECHNIQUES

Many laboratories now possess standard equipment such as spectrometers and electron probes for the qualitative and quantitative analysis of all but the lightest elements by x-ray emission techniques. Observations are made of the intense atomic lines obtained either by electron excitation or secondary excitation (fluorescence excited by primary x-rays). Generally, the resolving power of these instruments is not good enough to detect the small displacements of the emissions arising from variations in the chemical state.

The study of the effect of physical and chemical changes on x-ray spectra and the determination of energy levels in the atoms requires instruments with high dispersion and very good energy-resolving power. Dispersion, in energy units, becomes greater as the wavelength increases, even if the dispersion measured in wavelength units stays about the same. Moreover, the width of an x-ray level, which varies approximately as the square of its energy, decreases as the wavelength increases. It follows that all such studies should, as far as possible, take place in the soft x-ray region. However, radiationless transitions predominate in this spectral region and so observations are often very difficult. Also, the experimental conditions sometimes limit the type of work which can be carried out; the radiation is easily absorbed so that the instrument must be evacuated, only thin foils can be placed in the radiation path, and only superficial emissions can be studied.

The absorption coefficient μ can be determined for a monochromatic radiation from the intensity measurements of the incident radiation I_0 and transmitted radiation I, respectively, before and after absorption by a homogeneous film of thickness x. Then

$$I/I_0 = \exp(-\mu x)$$

For short wavelengths, the practical problems involved in the preparation of thin films can usually be solved quite readily; but for wavelengths above about 8 Å, the films must be but a few thousand ångströms thick if the absorption edges are to be observed. Films of metals can be prepared by vacuum deposition, but it is not always possible to make sufficiently thin films of compounds. They can sometimes be obtained from the metal film by chemical transformation. Thus, up till now, x-ray absorption studies of compounds have been limited by these difficulties.

The type of instrument used depends on the wavelength region to be studied. For short wavelengths (less than 2 Å), where the radiation is not easily absorbed, a Cauchois bent-crystal transmission spectrometer of high luminous efficiency can be used or, alternatively, a double plane-crystal spectrometer (two crystals are placed in antiparallel positions) (Figures 5 and 6). Both these types operate at atmospheric pressure. Wavelengths in the 2 to 25 Å range are studied with bent-crystal reflection spectrometers of the Johann type (Figure 5); here again a double-crystal spectrometer can be used. In this region, the x-ray tube, spectrometer, and detector must all be placed under vacuum. Yet, in the range up to 5 Å it is sometimes convenient to fill the airtight tank with helium to just below atmospheric pressure, especially when studying radioactive elements[30]. The dispersion of all these instruments is a function of the spectrometer dimensions and the crystal-lattice spacing producing the Bragg reflections. Their resolving power depends largely on the 'quality' of the crystals. Above 25 Å, concave gratings are used, either singly or in pairs[31], at grazing incidence.

In all these cases, the spectra are recorded by photographic or electron

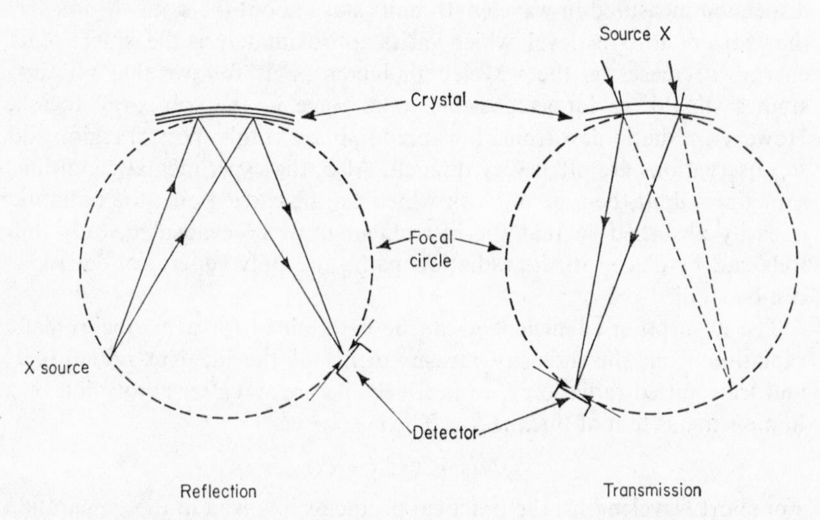

Figure 5. The principle of a bent-crystal spectrograph.

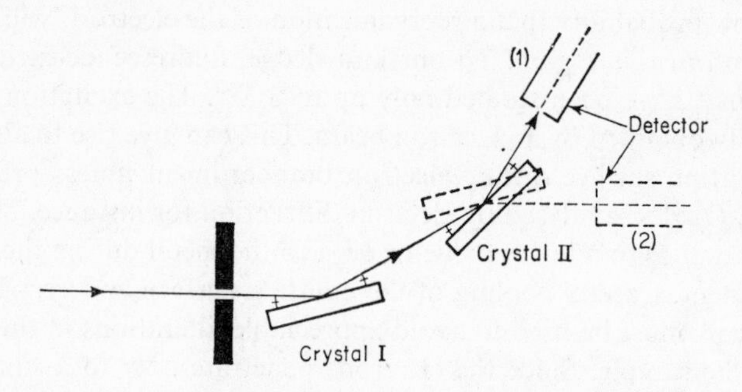

Figure 6. The principle of a double spectrometer. (1) Antiparallel position, (2) parallel position.

detectors. Further details of the many types of instruments in use will be found in the specialized literature[32,33] and in the recent papers already referred to.

It should be pointed out that 'true' absorption and emission intensity curves can only be obtained if the instrumental errors can be corrected for, using an instrument function $F(\lambda)$. For a given instrument, this is defined as the distribution of an infinitely narrow spectral line seen by its detector. The experimental curves $I(\lambda)$ and $\mu(\lambda)$ appear as the convolution integral of the real curves and the $F(\lambda)$ function. F cannot be known precisely, but in the case of a double-crystal spectrometer a good approximation is obtained by observing the distribution in the zero dispersion position where the two crystals are parallel. The width and shape of the correction factor can be assessed for a single bent-crystal spectrometer. These problems are discussed elsewhere for several experimental conditions encountered for this type of apparatus[34].

Most x-ray spectroscopic studies have been on solids, though very thorough verifications of the physicochemical state have not been made. Little work has been done on gases because of the technical difficulties involved, with the exception of some work on the rare gases. Because of the low intensity of x-ray emission in the long-wavelength region and its low penetration, absorption edges of only a few metals have been observed beyond 20 Å. In the bent-crystal vacuum spectrometer region, most of the studies have been carried out on metals, alloys, and a few oxides. The experimental measurements on more or less complex compounds have been largely performed in the region below 2 Å, i.e. in the region of the K spectra of chromium and elements of higher atomic numbers and the L spectra of the rare earths and above.

Fluorescence excitation of characteristic x-ray emissions is virtually impossible in the long-wavelength region because of its very poor efficiency and the low probability that a reorganization of the electrons will give rise to radiation in this region. To our knowledge, fluorescence-excited weak spectral lines have been studied only up to 8 Å[13]. The excitation is therefore usually obtained by an electron beam. This can give rise to alterations in the emitting surface during electron bombardment unless precautions are taken. Oxides, analysed by electron diffraction for instance, have been known to reduce to a lower oxide or even to the metal during the electron bombardment. Careful cooling of the emitting surface and very low electron currents must be used to avoid appreciable alterations in the composition of the sample. Since the electrons penetrate only to a small depth in the material, the excitation is limited to a superficial layer. The condition of the emitting surface is less critical for secondary excitation since the useful x-ray excitation usually takes place to a greater depth.

It is impossible to give a complete summary of all the work carried out, but further information can be obtained from the specialized bibliography already mentioned[2,33,35] and recent international conferences on x-ray spectroscopy[36,37]. We have picked out a few examples which will each illustrate a type of result which can be obtained with x-ray spectroscopy.

IV. PRACTICAL EXAMPLES

A. $K\alpha_1$ and $K\alpha_2$ Lines

As an example of the effect of physical and chemical modifications on x-ray spectra, we will first consider the experiments carried out on the $K\alpha$ doublet of sulphur, situated at 5.4 Å. This work has shown the effect which the degree of oxidation has on the wavelength of atomic lines.

According to Faessler[38], the doublet is displaced towards longer wavelengths with respect to that of the free element when the sulphur is in a negative oxidation state, e.g. in a sulphide where it is -2. On the other hand, if the oxidation state is positive, as in the thiosulphates ($+2$), sulphites ($+4$), and sulphates ($+6$), the shift occurs towards a shorter wavelength and increases with the degree of oxidation. This displacement of $K\alpha_1$ varies between -0.14 and $+1.19$ ev. For a given oxidation number, the position of the doublet $K\alpha$ also depends on the electronegativity of the atoms surrounding the emitting atom, in this case sulphur. The more electronegative they are, the more $K\alpha$ is displaced towards shorter wavelengths. Since the positive charge of the central atom may increase with the increase in negative charge of the surrounding atoms, Faessler concluded that the position of $K\alpha$ depends on the charge of the emitting atom and conversely that the measurement of the wavelength of $K\alpha_1\alpha_2$ can generally be used to deduce unambiguously the value of this charge. He was able to determine the charge of several sulphur ions in compounds of various degrees of oxidation such as thiosulphates, polythionates, etc. He was also able to determine the ionic or covalent bond characteristics of various compounds such as ZnS from the charge of the sulphur[39]. A similar study of phosphorus and silicon led to the same conclusions.

A more recent experiment was carried out by Meisel[40], who studied the width, shape, and position of the $K\alpha_1$ lines from the first series of transition elements, both in the metals and in many of their compounds. For low oxidation states, the bond affects mainly the $4s$ electrons in the metal ion. The $K\alpha$ line of this ion in the compound then shifts towards shorter wavelengths with respect to its position in the metal, i.e. in the same direction as in the previous case. When the oxidation state is greater than or equal to $+3$, as in the case of some V and Cr compounds[41], the displacement occurs in the opposite direction, towards longer wavelengths, and the shift increases with the degree of oxidation. It is often of the same order of magnitude as the experimental errors.

Theoretical calculations bearing on the energy of x-ray levels have been carried out by the self-consistent field method for some atoms in various ionization states[22,23,42,43]. According to calculations made by Watson on the free atoms and ions in the iron group assuming a $3d^n$ configuration, the wavelength of $K\alpha_1$ should increase with the degree of oxidation, but the theoretical displacement is always considerably greater than that observed experimentally. Before calculations on the free atom or ion can be applied to a metal or compound in solid form, the radial and angular distributions of (in the present example) the $3d$ wave functions would have to be determined and allowance made for bond effects, directional effects, and the degree of hybridization of the electron function in the solid. Generally speaking, any change in the hybridization of the wave functions modifies the relative positions of the $3d$ and $4s$ states and the shape of the $3d$-state distribution, which alters the energy distribution of the inner levels.

The dependence of the $K\alpha$ shift in transition metal compounds on the degree of oxidation may therefore arise from the presence of $3d$ states more or less perturbed by interactions between neighbouring ions.

B. Absorption Curves of an Element in a Series of Compounds

A large number of experiments in x-ray spectroscopy have been carried out on the absorption spectra of a given metal ion in series of its compounds. Normally, the object of the work is to detect any differences which may occur between the absorption curve of a compound whose electron configuration is to be found and a compound of known electron configuration. As an example, we shall mention experiments carried out on hydrated nickel bisacetylacetonate[44]. Collet was able to suggest that it had an octahedral structure by comparing its absorption curves with those of other octahedral complex compounds (Figure 7). The figure shows the difference in shape and the shift of the absorption curve of the metal in various compounds with respect to the metal alone.

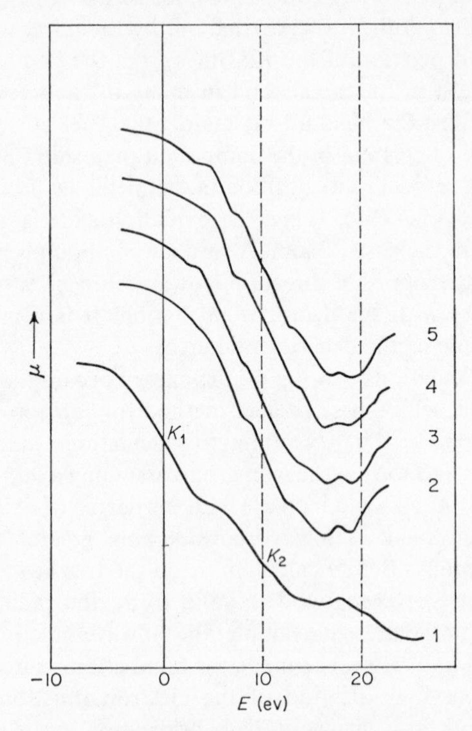

Figure 7. *K* absorption of nickel in the metal (1) and various compounds: (2) [Ni(dipyridyl)$_3$](ClO$_4$)$_2$, (3) [Ni(C$_5$H$_7$O$_2$)$_2$]·2H$_2$O, (4) [Ni(ethylenediamine)$_3$](ClO$_4$)$_2$, (5) [Ni(NH$_3$)$_6$](ClO$_4$)$_2$ [according to Collet[44]].

Many other experiments of this type can be mentioned; for example, the compounds of the first transition series can be classified according to the changes in shape of the *K* absorption curves, in three categories: octahedral ionic hydrates, oxides, etc., or covalent cyanides and carbonyls, or tetrahedral compounds[45] (Figure 8). This classification has been applied to the determination of the coordination number of cobalt in several compounds[46]. Sinha and Mande[47] have suggested that the differences observed between octahedral or tetrahedral compounds could be explained by different *p*-orbital overlap. According to these authors, the *p* orbitals have a more pronounced atomic character in tetrahedral compounds whose *K* absorption edge is characterized by marked steps.

Much work has been done on the structures observed on the high-energy side of absorption edges. Systematic experiments were performed to establish their positions as a function of the crystalline structure and in many cases a qualitative agreement was obtained with Kronig's predictions[48]. However, absorption structures were also found in liquids and

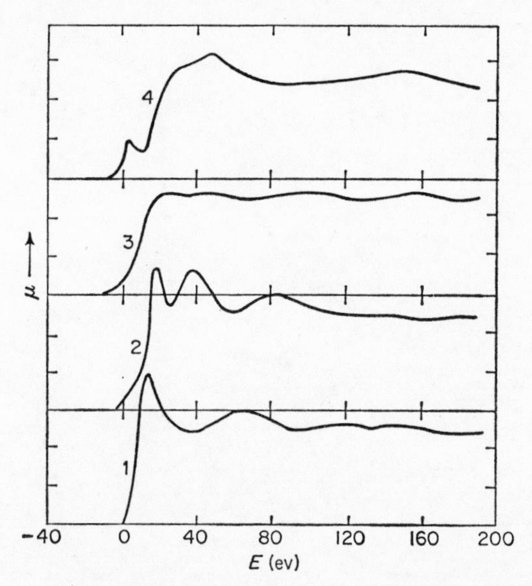

Figure 8. *K* absorption of manganese in various chemical states: (1) $MnCl_2 \cdot 4H_2O$, (2) $K_3Mn(CN)_6$, (3) Mn metal, (4) $KMnO_4$ [according to Von Nordstrand[45]].

amorphous materials, similar to those in the corresponding crystalline solid. For *K* discontinuities, we cite the following cases: divalent nickel either in aqueous solution or in a crystalline salt[49]; liquid or solid gallium[50]; hexagonal or amorphous germanium oxide (GeO_2)[50a]. This suggests that the fine structure is more strongly influenced by the arrangement of neighbouring atoms. Moreover, the structures which accompany *s* and *p* discontinuities do not correspond, whether in a single metal or between a number of metals, with the same crystalline structure. The theoretical work of Hayashi should also be mentioned[51]. According to this author, the structure of the absorptions would be due to transition towards 'quasi-stationary' states which depend on the appropriate hole.

C. Aluminium and Aluminium Oxide

Aluminium is an example of a 'good' metal. Its *K* spectrum at about 8 Å is shown in Figure 9[13]. The emission and absorption discontinuities can be described by arctangent curves and have coincident inflexion points at the Fermi limit. The $K\beta$ band has been observed in fluorescence[13,52]. The excitation conditions have a marked effect on the shape of the x-ray band of aluminium because the influence of surface contamination varies considerably with the manner of excitation (cf. p. 60) and because of the

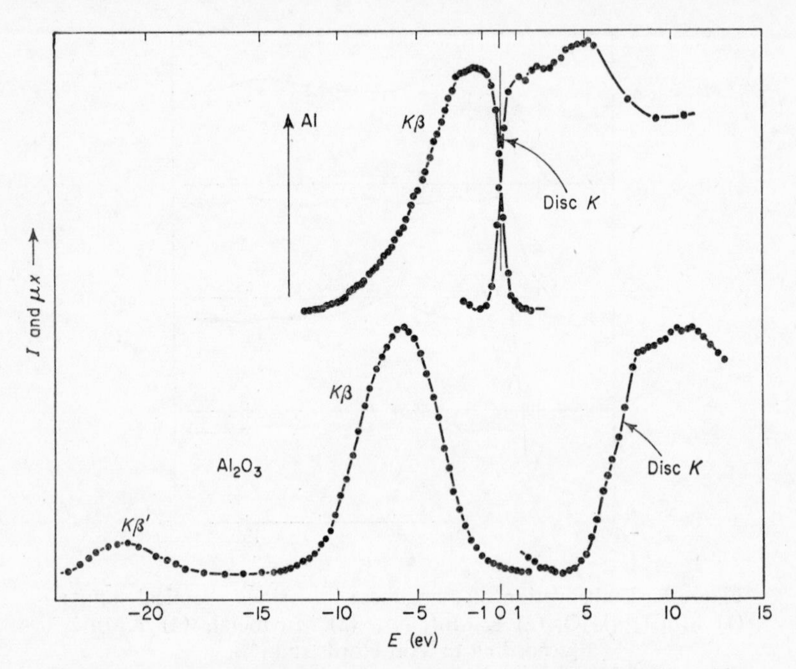

Figure 9. The K spectra of aluminium and aluminium oxide [according to Senemaud[13]].

'getter' properties of this metal. Thus the curves obtained by electron excitation are composed of the emission distribution of the pure metal and that of the superficial contamination probably produced by electron bombardment. The width of the K level, measured by the method described on page 51, is equal to 0.5 ev and is negligible compared with the width of $K\beta$. The shape of the latter describes the distribution of occupied levels in the conduction band of the metal, if the influence of the transition probabilities can be neglected.

The shape of the $K\beta$ emission of the aluminium in Al_2O_3 is quite different from that of the metal. The emission no longer shows a discontinuity and becomes nearly symmetrical. It describes the electron distribution of the filled bonding orbitals and is strongly displaced towards lower frequencies. The K absorption discontinuity is displaced in the opposite direction. Its form is similar to that of the metal but the absorption structures are different. The discontinuity must correspond to transitions towards relatively delocalized empty antibonding orbitals.

Finally, another emission, called $K\beta'$, is observed at about -15 ev from the maximum of the $K\beta$ band in the oxide. It also appears, but more faintly, in the spectrum of the metal when its surface is contaminated. An emission similarly positioned with respect to the metal emission band has been observed in the L spectra [53]; it has been interpreted as due to satellite emission from metallic aluminium [54].

In a compound such as Al_2O_3, the intensity of the $K\beta$ emission can be attributed partly to electronic transitions from the oxygen $2p$ orbital towards the aluminium ion $1s$ hole. The $K\beta'$ band might correspond to oxygen $2s$ electronic transitions also towards the $1s$ hole. Such transitions are termed 'cross transitions' and were already suggested by Skinner[55] as an interpretation of the structures in aluminium $L_{II}L_{III}$ bands appearing in Al_2O_3. The K absorption in the compound arises from a K-electron transition towards the empty $3p$ levels in the metal ion.

D. Copper Oxides

We have already mentioned (p. 54) that the L_{III} and L_{II} absorption edges of copper in CuO are displaced towards lower frequencies, while Kunzl's law [24] suggests that the shift should occur in the opposite direction. These shifts are -1.7 and -1.3 ev, respectively. The shape of the absorption curves is quite different from that observed in the metal. The K absorption edge also has a markedly different shape but shifts by $+5$ ev in the oxide. We must bear in mind that copper in the metal has arctangent-type discontinuities representing transitions towards an unoccupied $4sp$ conduction band (see Figure 3). As for L emission bands which, subject to certain assumptions, describe the distribution of the d-type levels, they are nearly symmetrical and therefore are characteristic of a totally filled d band. It is clear that the population of the copper $4s$ and $3d$ orbitals, the type of wave function describing the inner level concerned in the transition, and the charge distribution it imposes are the main factors affecting the shape and position of discontinuities in cuprous and cupric oxide, when the instrumental functions have been taken into account. Data are available on the K and L emission and absorption spectra for copper and its two oxides, so we shall examine the distribution of the various orbitals and their respective positions [15] (Figures 10 and 11).

The L_{III} and L_{II} absorption edges in cuprous oxide are quite similar in form to those observed in the metal and are displaced by $+0.4$ ev. These edges can then be attributed to transitions towards the copper empty $4s$ states in the conduction band, and their inflexion point positions the bottom of the band. The width of the forbidden band can be obtained from the absorption and emission spectra, since the latter give the distribution of the filled 3*d* levels. It is found to have, after correction, a width of 2.1 ev which is in good agreement with other determinations obtained from optical spectra and conductivity measurements (Figure 12, p. 68).

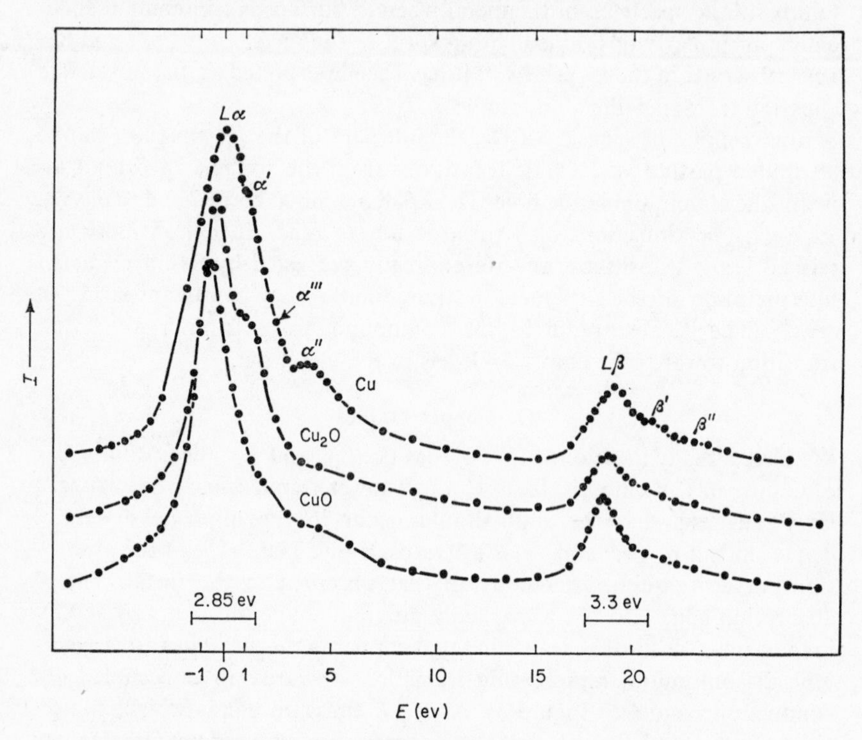

Figure 10. *L*α and *L*β emissions of copper in the metal, Cu_2O, and CuO[15].

The L_{III} and L_{II} discontinuities of copper in CuO, on the other hand, have intense symmetrical lines which can be attributed to a 2*p*-electron transfer towards an empty, almost atomic, 3*d* state whose width is about 1 ev. In a compound of this type, the inflexion point of the absorption line has no simple physical significance; neither, indeed, does the distance between the inflexion point and the side of the emission band. The *K* edge represents a 1*s*-electron transition to an empty 4*p* antibonding orbital.

The emission bands of Cu_2O shift to a lower energy with respect to the metal and those from CuO are displaced still further. Similarly, there is a reduction in the band width, more pronounced in the case of CuO, suggesting a contraction of the 3*d*-electron shell of the copper, which becomes more and more marked as the copper atoms become more widely separated and the overlap of the wave functions is reduced.

The frequency of an absorption or emission depends on two energy states so that the frequency modification of the 1*s* or 2*p* levels must be taken into account before the results can be analysed. No displacement

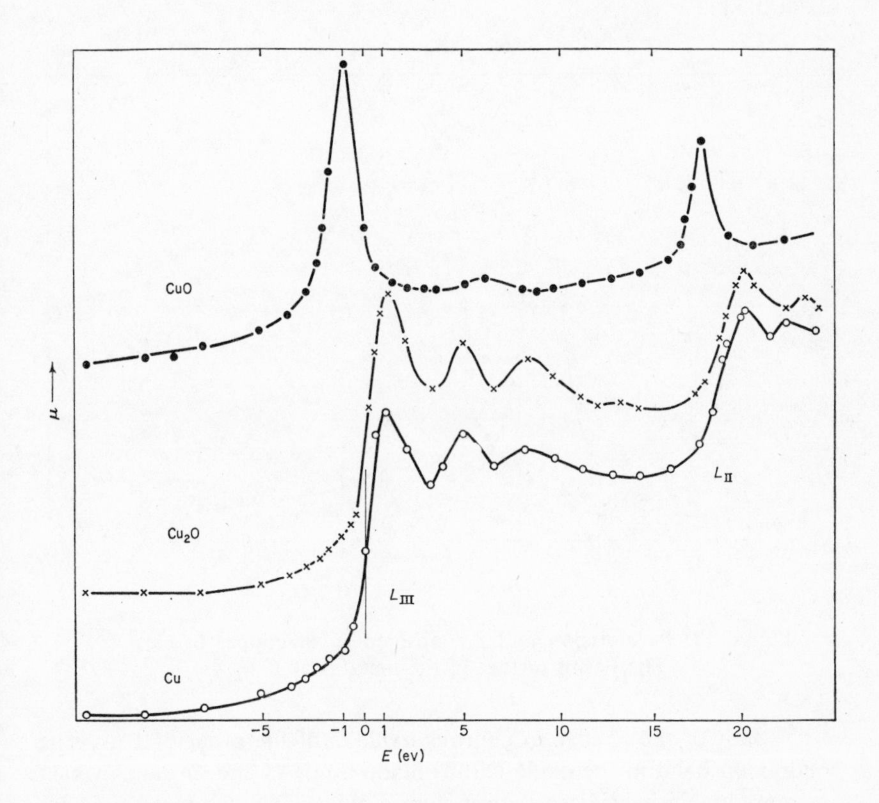

Figure 11. L_{III} and L_{II} absorptions of copper in the metal, Cu_2O, and CuO[15].

of $K\alpha_1$ and $K\alpha_2$ lines has been observed to within 0.2 ev (the experimental precision) in cuprous or cupric oxides. From this we may conclude either that the 1*s* and 2*p* levels shift only very slightly or that they both shift in the same direction by about equal amounts. We have already pointed out that

K and L discontinuities in a metal are transitions of inner s and p electrons, respectively, towards the Fermi surface.

The difference in displacement between the K absorption and L_{III} absorption in CuO, for instance, can thus give the distance between the empty $3d$ and $4p$ orbitals in this compound. This difference is about 6.5 ev. We attribute the presence of the absorption structure on the high-energy side of the L_{III} absorption line to a transition towards $4s$ orbitals. The energy difference between the empty $3d$ and $4s$ orbitals is, therefore, about 3.5 ev.

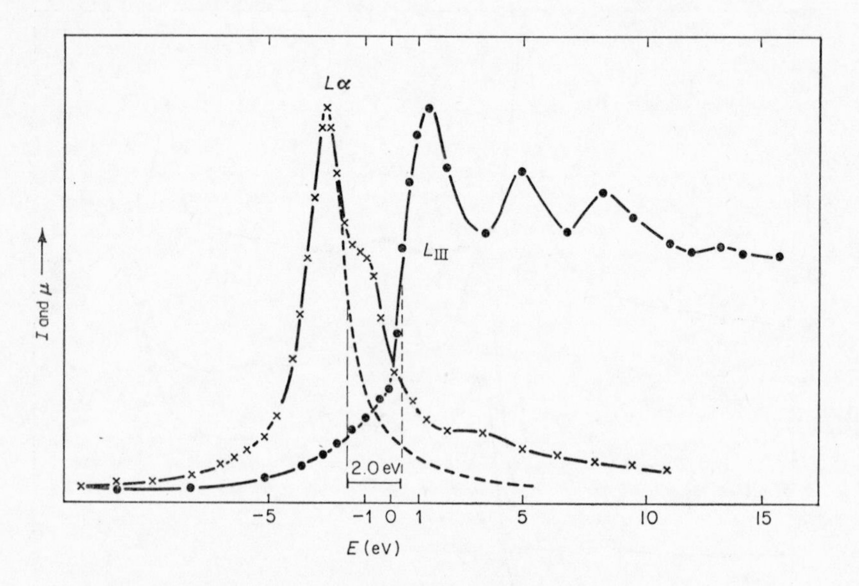

Figure 12. $L\alpha$ emission and L_{III} absorption of copper in Cu_2O[15]. The Fermi surface of the metal is at $E = 0$.

The shift of the K edge in cuprous oxide is of the order of 1 ev. The conduction band in the oxide is thus made up of $4s$ and $4p$ states which are more highly hybridized than in cupric oxide, since the metal ions are less far apart and the bond is more nearly covalent.

In order to compare the respective positions of either absorption edges or emission bands in a metal and in its oxide, they must be measured from the same reference level; for a free atom, it would be logical to choose the level whose energy is zero *in vacuo*. From simple considerations connected with photoelectric measurements and the shift of the discontinuities mentioned above, the energy of the L_{III} level must be lower in CuO than in copper. This is the opposite of what is usually suggested by free-atom calculations. But we have already mentioned that the energy of an x-ray level is related to the spatial distribution of the electron cloud. The marked contraction of the 3*d* wave functions in cupric oxide should account for the variation in energy of the L_{III} and L_{II} levels.

E. Plutonium Metal and Oxide

From work carried out on the *L* and *M* emission and absorption spectra of the heavy elements and their oxides, we now have fairly complete data on the plutonium and plutonium oxide spectra [56]. The experimental results cannot be given in detail here, but we shall give a brief account of the most important conclusions which can be drawn from the experiments.

The respective positions of the empty energy states were obtained from the five *M* absorption edges, if the transition probabilities are taken into account. Let us first consider the metal. A comparison of its *M* absorption edge energies with the energy levels calculated from the L_{III} edge and *L* emission lines shows that empty *f* states are present at the Fermi level overlapped with *d* and *s* states. Conversely, the *M* discontinuities are electron transitions towards the Fermi limit whatever the state of the inner level concerned.

The displacements in the oxide with respect to the metal are different for each of the *M* edges. They are as follows:

M edge	E_{Pu-PuO_2} (ev)
M_V	−1.2
M_{IV}	−1.5
M_{III}	+5.6
M_{II}	+4.1
M_I	+20

The shift in the inner levels due to chemical bonding probably remains small so that the respective positions of the *f*, *d*, or *p* states in the compound are deduced from the above data. Depending on the transitions involved, the shifts are accompanied by a more or less marked change in shape of the discontinuities. The M_{III} and M_{II} discontinuities have more pronounced maxima in the oxide than in the metal which suggests the presence of localized *d* states in the compound (Figure 13). The M_V and M_{IV} discontinuities have very similar shapes (Figure 14). An absorption structure, at about 20 ev from M_V on the high-frequency side in the PuO_2 spectrum, can be attributed to transitions towards *p* states.

Research in the field of x-ray spectroscopy is becoming more and more concerned with the study of gases and simple molecules for which theoretical work can most easily be carried out. Best[57] and Schnopper[58] have recently described experiments on the K emission and absorption spectra of Mn, Cr, and V in molecules of the $KMnO_4$ type and of Cl in $KClO_3$

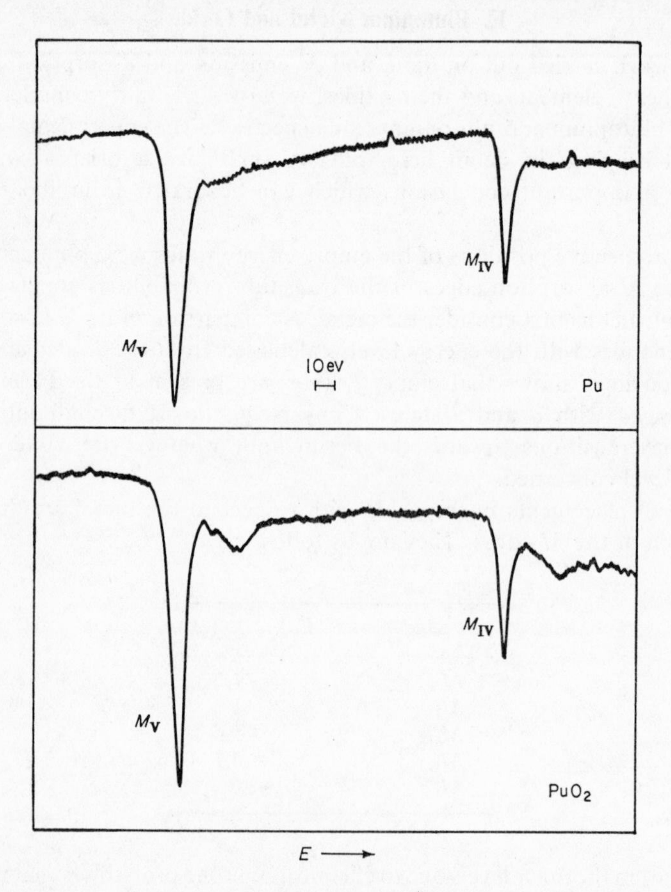

Figure 13. M_V and M_{IV} absorptions of plutonium in the metal and PuO_2[56].

and $KClO_4$ with such comparisons in mind. Even here, disagreement exists between the experimental and theoretical values of the ionization energies which cannot be accounted for by experimental errors. The monoionization energies of the rare gases alone can, at present, be calculated with precision (argon, for instance[59]) and the agreement with experimental results improves as deeper electron shells are considered. It remains

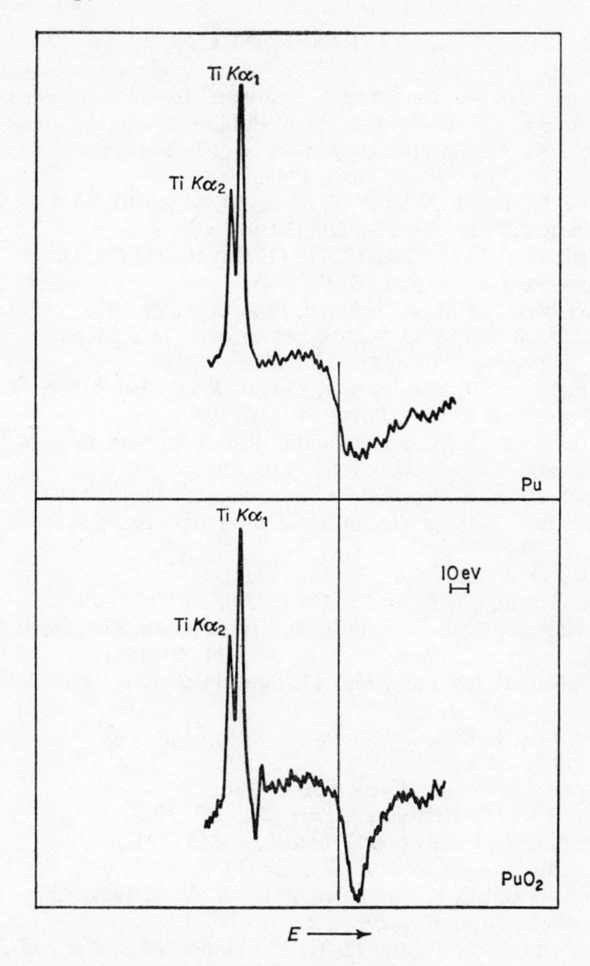

Figure 14. M_{III} absorptions of plutonium in the metal and PuO_2[56].

to be hoped that it will become possible to compare accurate experimental results with theoretical values less subject to restrictions than the present free-atom model.

V. ACKNOWLEDGEMENTS

It is a pleasure to acknowledge the encouragement and invaluable advice given me by Professor Y. Cauchois, Director of the Laboratoire de Chimie Physique de Paris. Thanks are also due to Dr. C. Hague, who translated this chapter into English.

VI. REFERENCES

1. S. Hagström, C. Nordling, and K. Siegbahn, *Alpha-, Beta-, and Gamma-ray Spectroscopy*, Vol. 1, North-Holland Publishing Co., Amsterdam, 1965.
2. Y. Cauchois, *Les Spectres de Rayons X et la Structure électronique de la Matière*, Gauthier Villars, Paris, 1948.
3. V. Weisskopf and E. Wigner, *Z. Phys.*, **63**, 54 (1930); **65**, 18 (1930).
4. L. G. Parratt, *Phys. Rev.*, **56**, 295 (1939).
5. Y. Cauchois, *J. Phys. Rad.*, **13**, 113 (1952); **16**, 253 (1955).
6. Y. Cauchois, *J. Phys. Rad.*, **5**, 1 (1944).
7. M. Wolfsberg and M. L. Perlman, *Phys. Rev.*, **99**, 1833 (1955).
8. T. A. Carlson and M. O. Krause, *Phys. Rev.*, **137**, A1655 (1965).
9. H. W. Schnopper, *Phys. Rev.*, **131**, 2558 (1963).
10. C. Bonnelle and F. Wuilleumier, *Compt. Rend.*, **256**, 5106 (1963).
11. L. G. Parratt, *Rev. Mod. Phys.*, **31**, 616 (1959).
12. F. K. Richtmyer, S. W. Barnes, and E. Ramberg, *Phys. Rev.*, **46**, 843 (1934).
13. C. Senemaud, *Thèse 3ème Cycle*, Paris, 1965.
14. Y. Cauchois and C. Bonnelle, *Compt. Rend.*, **245**, 1230 (1957).
15. C. Bonnelle, *Thèse de Doctorat d'Etat*, Paris, 1964; *Ann. Phys.*, **1**, 439 (1966).
16. R. de L. Kronig, *Z. Phys.*, **75**, 191 (1932).
17. R. de L. Kronig, *Z. Phys.*, **75**, 468 (1932).
18. D. R. Hartree, R. de L. Kronig, and H. Petersen, *Physica*, **1**, 895 (1934).
19. A. I. Kostarev, *Zh. Eksperim. Teor. Fiz.*, **11**, 60 (1941).
20. T. Shiraiwa, T. Ishimura, and M. Sawada, *J. Phys. Soc. Japan*, **13**, 847 (1958).
21. Y. Cauchois, *Colloq. Spectr. Intern., 9th, Lyon, 1961*, G.A.M.S.-Muray-Print, Paris, 1962.
22. A. Sureau, *Thèse 3ème Cycle*, Paris, 1961.
23. A. Sureau and G. Berthier, *J. Phys.*, **24**, 672 (1963).
24. V. Kunzl, *Coll. Trav. Chim. Tchécosl.*, **4**, 213 (1932).
25. C. Bonnelle, *Compt. Rend.*, **248**, 2324 (1959).
26. C. Bonnelle and C. K. Jørgensen, *J. Chim. Phys.*, **1964**, 826.
27. Y. Cauchois, *Disq. Mat. Phys.*, **2**, 319 (1942).
28. H. Jones, N. F. Mott, and H. W. B. Skinner, *Phys. Rev.*, **45**, 370 (1934).
29. Y. Cauchois and N. F. Mott, *Phil. Mag.*, **40**, 1260 (1949).
30. Y. Cauchois, *Proc. Colloq. Spectr. Intern., 10th, Washington, 1962*, Spartan Books, Washington, 1963.
31. P. Jaegle, *Thèse de Doctorat d'Etat*, Paris, 1965.
32. A. H. Compton and S. K. Allison, *X-rays in Theory and Experiment*, McMillan, London, 1935.
33. A. E. Sandstrom, *Handbuch der Physik*, Band 30, Springer Verlag, Berlin, 1957, p. 246.
34. Y. Cauchois, *Symp. Crystal Diffr. Nuclear Gamma Rays, Athens, 1964.*
35. M. A. Blokhin, *The Physics of X-rays*, State Publishing House of Technical Theoretical Literature, Moscow, 1957.
36. *Conf. Phys. X-ray Spectra, Ithaca, New York, 1965*; *Bull. Am. Phys. Soc., Ser. II*, **10**, 1218 (1965).
37. *Intern. Symp. X-ray Spectra Chem. Binding, Leipzig, 1965*, Institut der Karl Marx Universität, Leipzig, 1966.

38. A. Faessler and M. Goehring, *Naturwissenschaften*, **39**, 169 (1952).
39. A. Faessler and M. Goehring, *Z. Phys.*, **142**, 558 (1952).
40. A. Meisel, *Bull. Am. Phys. Soc., Ser. II*, **10**, 1225 (1965).
41. A. Meisel, *Z. Anorg. Allgem. Chem.*, **339**, 1 (1965); *J. Prakt. Chem.*, **29**, 192 (1965).
42. A. Feillet and G. Berthier, *J. Phys.*, **62**, 11 (1965).
43. R. E. Watson, *Tech. Rept.*, No. **12**, Institute of Technology, Cambridge, Massachusetts, 1959; *Phys. Rev.*, **118**, 1036 (1960).
44. V. Collet, *Thèse de Doctorat d'Etat*, Paris, 1959.
45. R. A. Von Nordstrand, *Noncrystalline Solids*, Wiley, New York, 1960.
46. C. Mande and A. R. Chetal, *Intern. Symp. X-ray Spectra Chem. Binding, Leipzig, 1965*, Institut der Karl Marx Universität, Leipzig, 1966, p. 194.
47. K. P. Sinha and C. Mande, *Indian J. Phys.*, **37**, 257 (1963).
48. Y. Cauchois, *Acta Cryst.*, **5**, 351 (1952).
49. Y. Cauchois, *Compt. Rend.*, **224**, 1556 (1947).
50. M. Vidal, *Diplôme d'Etudes Supérieures*, Paris, 1949.
50a. W. F. Nelson, I. Siegel, and R. W. Wagner, *Phys. Rev.*, **127**, 2025 (1962).
51. T. Hayashi, *Sci. Rept. Tohoku Univ.*, **33**, 123, 183 (1949); **34**, 185 (1950).
52. Y. Cauchois, C. Bonnelle, and C. Senemaud, *Compt. Rend.*, **257**, 1051 (1963).
53. G. A. Rooke, *Phys. Letters*, **3**, 234 (1962).
54. R. A. Ferrell, *Rev. Mod. Phys.*, **28**, 308 (1956).
55. H. W. B. Skinner and H. M. O'Bryan, *Proc. Roy. Soc., Ser. A*, **176**, 229 (1940).
56. Y. Cauchois, C. Bonnelle, and L. de Bersuder, *Compt. Rend.*, **256**, 112 (1963); **257**, 2980 (1963).
57. P. E. Best, *Bull. Am. Phys. Soc.*, **10**, 29 (1965).
58. H. W. Schnopper, *Intern. Symp. X-ray Spectra Chem. Binding, Leipzig, 1965*, Institut der Karl Marx Universität, Leipzig, 1966.
59. P. S. Bagus, *Phys. Rev.*, **139**, A 619 (1965).

3

Molecular Photoelectron Spectroscopy

D. W. TURNER

I INTRODUCTION AND HISTORICAL SURVEY

Vacuum ultraviolet radiation ($\lambda < 1750$ Å) has enough energy to cause excitation, bond rupture, and ionization in a molecule; frequently all three occur together in complex molecules. By focusing attention only upon the electrons ejected, we can use the Einstein relation $E = h\nu - I$ to determine the ionization energies of the various electrons within a molecule by measuring E by electron spectrometry, if $h\nu$ is fixed[1-3]. In addition, for isolated molecules the Einstein relation can be extended to include vibrational and rotational excitation so that

$$E = h\nu - I_{\text{electronic}} - E_{\text{vib}} - E_{\text{rot}}$$

where $I_{\text{electronic}}$ takes the values of the various ionization potentials of the different electronic levels and E_{vib} and E_{rot} are the quantized energies of vibration and rotation of the various electronic states of the ion which result from the loss of the different electrons. We will suppose for the present that the molecule is in its ground vibronic state. In order to see why this attractively simple concept has suddenly led to a fresh insight into molecular electronic structure, it is of interest to survey briefly the development of the subject.

The geometrical structure of molecules results from the interplay of the coulombic electrical forces between the positively charged nuclei on the one hand and the negatively charged electrons on the other. Since not only the attractive forces (nucleus–electron) but also repulsive forces (electron–electron and nucleus–nucleus) are subject to subtle variations as the geometrical arrangement is varied, the most stable structure of the molecule is a compromise involving many interparticle dimensional parameters which can be known only in principle.

It is well recognized, however, that whilst the momentary positions and momenta of all the particles may be expressed theoretically in the form of a molecular wave function, which is a linear combination of one-electron wave functions, the most easily accessible experimental data relate to the energy values associated with this wave function. Thus the visible and ultraviolet absorption spectrum of a vapour or a dilute solution gives an indication of the energy difference between the molecular orbitals occupied by an electron when promoted from one orbital to another. Whilst such energy differences leave open the question of the absolute energy scale, it is often possible to form a good idea of the order and spacing of the energy levels which are occupied in the ground state of the molecule if sufficient interrelated energy differences can be obtained. This, however, requires the detailed analysis of an absorption spectrum, a task which is formidable for any but the simplest molecular spectra.

For polyatomic molecules, it has been necessary first to form some expectation of the molecular electronic energy levels from a consideration of the more precisely known electronic structures of the constituent atoms, a procedure pioneered by Hund[4] and Mulliken[5] in the years following

1926. In formulating the rules by which molecular orbitals were formed from combinations of atomic orbitals, Mulliken made extensive use of the rather few molecular ionization potentials which were then available experimentally. A period of vigorous growth in the understanding of molecular electronic structure followed in which molecular orbitals were calculated from linear combinations of atomic orbitals, a procedure associated initially with the names of Pauling, Morse, and Stückelburg. This has now reached the point where entirely theoretical calculation of the energy-level structure of simple molecules has been achieved. In examples of more practical interest, which concern more complex molecules or atoms of high nuclear charge, theoretical procedures are of necessity empirical and, in addition, reliable experimental data on any but the highest occupied orbital levels are scarce.

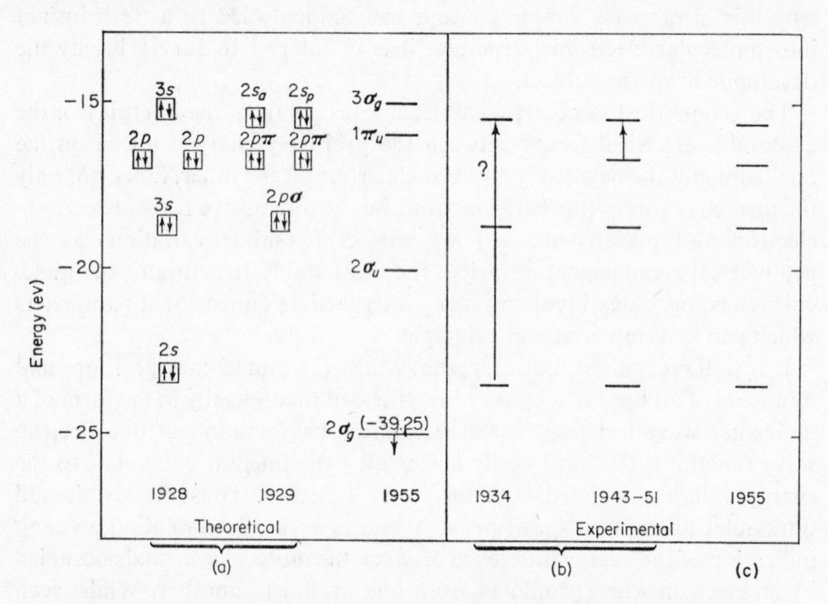

Figure 1. The valence-shell electronic energy levels of the nitrogen molecule according to the ideas of Lennard-Jones and Mulliken, the more recent calculations by Scherr[26], and experimental spectroscopic studies.

We may illustrate the manner in which certainty fades as complexity increases by reference first to nitrogen and then to acetylene.

a. Nitrogen Hund and Mulliken were able to classify the electronic states of the diatomic molecules in terms of their symmetry elements, and to show how the constituent molecular orbitals correlated with the atomic orbitals in the separated atoms and in the hypothetical united atom in which the internuclear distance was reduced to zero. On semitheoretical grounds, the molecular orbitals were put in order of energy by Mulliken (1928) and by Lennard-Jones (1929) as shown in Figure 1a. At about this period, the ionization energies of only two of these orbital levels were known precisely (by Rydberg series) corresponding to the ground (X) state of the ion (Mulliken, 1934) and the second excited state (B) (Hopfield, 1930). Much more recently, Rydberg series have been identified leading to the A (Worley, 1943) and C states (removal of an electron from the $2p$ and $2s$ orbital levels) (Figure 1b). Corresponding breaks have been detected in the electron-impact ionization cross-section curves (Frost and McDowell, 1955) which have been associated with higher ionization potentials and, by way of illustration, these electron-impact values are shown in Figure 1c. Apart from the direct Rydberg series data, emission spectroscopy had early suggested the location of the A (Meinel, 1951) and C (Watson and Koontz, 1934) states of the ion.

Thus there is a high degree of certainty about the experimental data concerning these four levels. The semiempirical background provided by the Mulliken description of the structure accounts for these observations in a general way but a completely non-empirical calculation using the self-consistent field treatment for all the electrons (Scherr, 1955) predicts orbital energies which are in somewhat better agreement with observation. It also confirms that the bonding properties to be expected of electrons in these orbitals differ somewhat from those originally inferred by Mulliken[6]. In particular, it would appear that the highest occupied level ($2p\sigma_g$) is nearly non-bonding instead of being responsible for about one-third of the triple bond in Mulliken's original description.

b. Acetylene In contrast to the harmony between theory and experiment for the nitrogen molecule, in the case of acetylene, which has the same number of electrons, only one ionization potential, the first (11.41 ev), was quickly discovered by Price (1935). Of the three other values expected[7] for L-shell electrons, two (13.25, 16.99 ev) were reported by Collin[8] (in 1962) using electron-impact methods but the first of these is almost certainly due to an autoionization process and does not relate to an ionization energy. No Rydberg series other than the first have been detected. Several theoretical calculations of the orbital energies have been carried out with varying degrees of refinement[9,10] and although little variation in the highest level results, the lower values vary widely. For example, the value for the $\sigma_g 2p$ orbital level ranges from 13.14 to 23.5 ev. The results of observations and calculations on acetylene are illustrated in Figure 2.

It will be apparent, therefore, that while it is extremely informative to discuss molecular structure in terms of molecular orbitals, to relate the latter to atomic orbitals it is necessary to know something of the relative order and energy spacing of the molecular orbitals, about which information becomes increasingly vague as molecular complexity increases.

Before 1961, all methods for determining molecular ionization potentials could be described as threshold measurements. Thus, whether one measured the convergence of a Rydberg series in absorption spectroscopy, the change

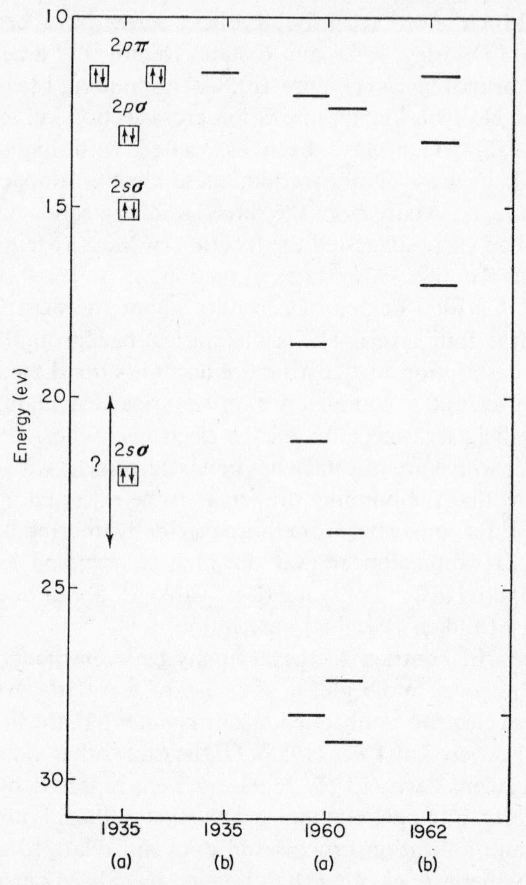

Figure 2. The energy levels of valence-shell electrons of acetylene from (a) theoretical considerations and (b) as determined experimentally.

of electron-impact ionization cross-section, or the change of ionization mode following the impact of positive ions of different energies, the essential feature was the association of an alteration in the state of the system as the energy supplied was varied, with certain critical energy values—the convergence limit of the Rydberg series, or a discontinuity in the electron-impact cross-section curve. Unhappily, processes other than direct ionization give rise to such discontinuities. Autoionization occurs when an outer electron is ejected after initial excitation of an inner electron and photodissociation commences at the critical energy for such a process as:

$$AB \xrightarrow{h\nu} A^+ + B^-$$

Furthermore, the vacuum ultraviolet absorption spectra of complex molecules consist for the most part of intense overlapping band spectra associated with photoexcitation:

$$AB \xrightarrow{h\nu} AB^*$$

which makes the detection of Rydberg series difficult.

II. PHOTOELECTRON SPECTROSCOPY

A. History

The technique of photoelectron spectroscopy in vapours was applied to molecular electronic structure determination as a way of overcoming these fundamental restrictions. It appears to have been developed independently at about the same time in Leningrad and in London[1-3]. At Imperial College, London, experiments with a helium resonance light source showed a most convenient and copious source of essentially monochromatic ionizing radiation ($h\nu = 21.21$ ev). Whilst performing basically the same experiment (retardation of photoelectrons by coaxial cylindrical grids), the Russian workers employed as a source of ionizing radiation the output of a hydrogen discharge tube, a many-lined spectrum from which a narrow band was selected by a vacuum monochromator incorporating LiF windows. This arrangement denied them access to photon energies greater than about 11.7 ev and also much reduced the photon flux available.

B. Experimental Methods

1. General arrangement

The two earlier experimental arrangements are illustrated in Figures 3 and 4. The distinctive feature of the apparatus described by Al-Joboury and Turner[2] (Figure 3) is a helium resonance light source L, separated from the ionization chamber RC by two aligned sections of precision bore (0.5 mm) Pyrex capillary tube, which prevents the target vapour from penetrating into the light source without the necessity for a window and also provided a narrow well-collimated photon beam which could be aligned with the axis of the electron-retarding grid structure G_1, G_2. It will be shown below that this arrangement contributes markedly to the energy resolution attainable.

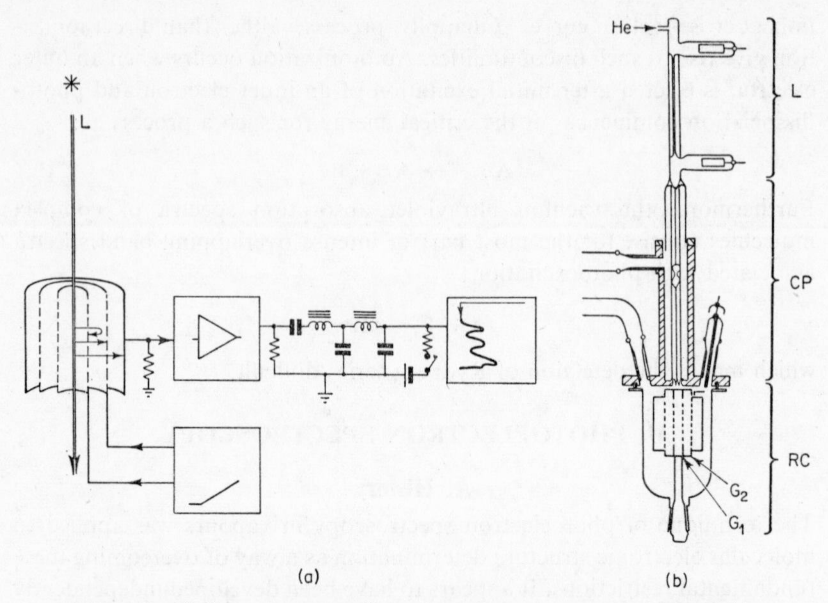

Figure 3. Helium resonance line photoelectron spectrometer of Al-Joboury and Turner[2]. (a) General arrangement; (b) cross-section of light source L, collimating section CP, and target chamber RC attached to the exit slit of (a).

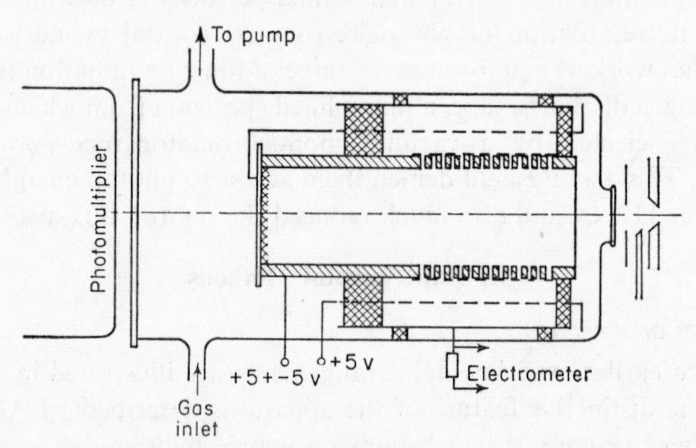

Figure 4. Photoelectron spectrometer for vacuum monochromator [after Vilesov and colleagues[3]].

2. Light-source monochromaticity

Under the appropriate conditions, the helium emission spectrum emitted by a d.c. discharge consists of the series leading to the ionization limit (24.47 ev) of which by far the strongest line is He 584 Å (21.21 ev). This resonance line then accounts for at least 99% of the emission in this spectral region. To longer wavelengths, we have the $^3P \rightarrow {}^3S$ series, of which the shortest wavelength emission is 3000 Å ($\sim$4 ev), followed by a number of lines in the visible, notably $\lambda = 5875$ Å (yellow). It is clear, therefore, that in the vast majority of substances, which have an ionization potential (I.P.) of 5 ev or greater, essentially the only ionization caused is due to the resonance line He 584 Å. Traces of hydrogen, which are only removed with difficulty, cause emission of the Lyman α line 1215 Å (10.20 ev) and both oxygen and nitrogen, readily released on outgassing of the lamp structure, have a many-line spectrum of shorter wavelengths than 1000 Å. Such emissions are also capable of exciting ionization in most substances and can confuse the resultant electron energy spectra. It is essential, therefore, that the gas in the discharge tube is extremely pure. Fortunately, this is not too difficult using a copper oxide furnace and a liquid nitrogen cooled charcoal trap to purify the helium. The state of the discharge is readily assessed by examining the visible spectrum for absence of the H Balmer series and O and N visible emission lines. When operating correctly, the light source is a yellowish peach colour, free from blue around the electrodes. Features in electron energy spectra which arise from ionization by those lines due to impurities in the discharge tube are easily recognized, however, as they increase in intensity relative to the remainder of the spectrum when the spectral purity of the light source deteriorates. For example, a weak but sharp line in the CS_2 photoelectron spectrum (see below), earlier assigned tentatively[11] to the sixth I.P., has now been proved to arise from photoionization of electrons from the highest occupied level (I.P. = 10.11 ev*) by hydrogen Lyman β photons ($h\nu$ = 12.07 ev), since on adding hydrogen to the helium stream, the peak feature increases very much in intensity while the remainder of the spectrum is quenched.

It has also been found possible[12] to make use of the resonance lines of elements other than helium by adding a small proportion to the helium flow. Thus, adding hydrogen in small amount results in the main light output becoming the Lyman α line of atomic hydrogen ($h\nu$ = 10.20 ev). Argon yields mainly the resonance line at 1067 Å ($h\nu$ = 11.62 ev), but in both cases quite large intensities reside in other lines than the desired one, the Lyman β and γ lines in hydrogen and the other resonance line in argon. Helium remains the most generally satisfactory radiation source.

3. Photoelectron energy analysis

The methods for determining the kinetic energy spectrum of a 'polychromatic' beam of electrons are many and varied, ranging from the simplest, which employ merely electrostatic retarding fields, through deflexion

* Mean of doublet, spectroscopic values 10.08, 10.13 ev.

analysis by means of radial electrostatic or magnetic fields and combinations of these, to highly chromatic lens systems. The subject has been recently reviewed by Klemperer[13] who describes many examples in detail. Some of these designs are, however, inherently unsuited to the present task owing to their low collection efficiency for an electron flux emerging radially from a line source.

Both of the early photoelectron spectrometers employed electrostatic retarding fields to analyse the spectrum of kinetic energies obtained on photoionization. For an electron to pass through two grids, it must have an initial kinetic energy at least equal to the potential difference between them. This is exactly sufficient provided that the electron is travelling along a

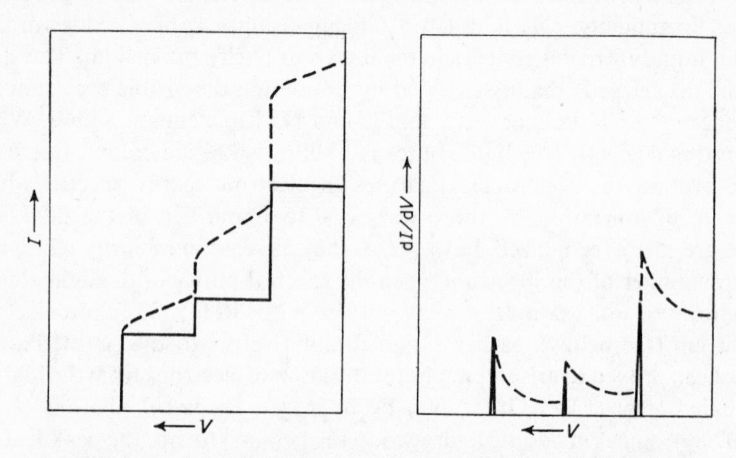

Figure 5. Idealized forms of the curve of collector current *I* versus retarding potential *V*, and its first derivative, the photoelectron energy distribution function. The effect of the deviations of the electron paths from the electric vector direction is indicated by dotted lines.

line of force, i.e. is always normally incident upon the equipotential surfaces. To make use of as much of the photoelectron flux as possible, cylindrical grids were used for which this condition of normal incidence applies for electrons emitted radially from the axis. The electron energy spectrum is obtained as the first derivative with respect to retarding field of the electron current emerging from this second grid to a suitably biased collector electrode (Figure 5). The differentiation can be done manually by point-by-point plotting of the current changes following a succession of small potential increments, the technique used by Vilesov and Kurbatov and subsequently by Schoen[14]. Alternatively, a continuous spectrum can be recorded by applying a potential sweep which is linear with time and simultaneously recording the first derivative of the output current, as was described by Turner and Al-Joboury. An elegant variation on this method has been described by Comes[15] who divided both retarding grid structures into two parts with a small difference dV in the retarding field between them. Using two collector electrodes to receive the separate currents I_1 and I_2, the difference in these currents I_1 and I_2 was recorded as proportional to dI/dV. Some such continuous recording technique is to be preferred, especially when fine structure is present in the spectrum but it introduces some problems in obtaining an adequate signal-to-noise ratio. Probably it cannot be applied with the very small signals obtained when a monochromator is used as a light source.

4. Energy resolution and the form of the spectrum from grid spectrometers

The resolving power is readily tested by examining the electron energy spectra of the rare gases, which give simple doublets[2,16] because spin–orbit splitting distinguishes the two states of the ion $^2P_{\frac{3}{2}}$ and $^2P_{\frac{1}{2}}$ (Figure 6).

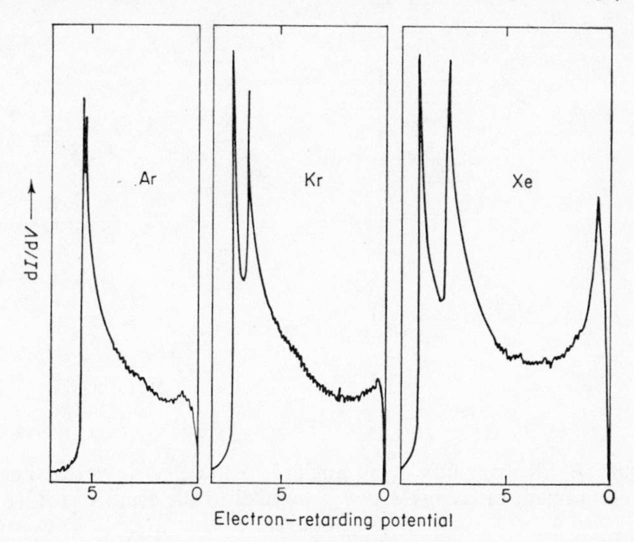

Figure 6. Photoelectron spectra for argon, krypton, and xenon, excited by helium resonance radiation (584 Å, 21.21 ev) in the apparatus of Figure 3.

The ultimate sharpness of the lines in the photoelectron energy spectrum will be determined by the sum of the Doppler broadening of the helium atoms and target gas molecules (about 0.2 and 0.05 mv, respectively, at room temperature) and by the contribution to the photoelectron velocity from thermal motion of the target molecules (about 2 mv in cases of practical interest).

By careful choice of mesh spacing in relation to grid separation, by gold coating the electrodes, and by excluding all insulating materials (which can acquire surface charges and consequently cause field distortion) from between the grids, it has been possible[2] to achieve a resolving power $E/\Delta E$ of about 50 judged by performance with the rare gases. This is sufficient to distinguish the doublet components of Ar^+ when using He 584 Å radiation, a difference of 0.18 ev for ~5 ev electrons.

The peaks are seen to be broadened on the low-energy side and it is this marked asymmetry (cf. Figure 5) which is a major limitation to attaining high resolution. It seems to arise partly from electron scatter at the first grid but is partly an inherent defect of using cylindrical grids. Electron photoemission at low excess energies follows a $\cos^2 \theta$ law where θ is the angle between the electric vector of the incident light and the electron trajectory (Figure 7)*. Electrons for which $\theta \neq 0$ enter the retarding field

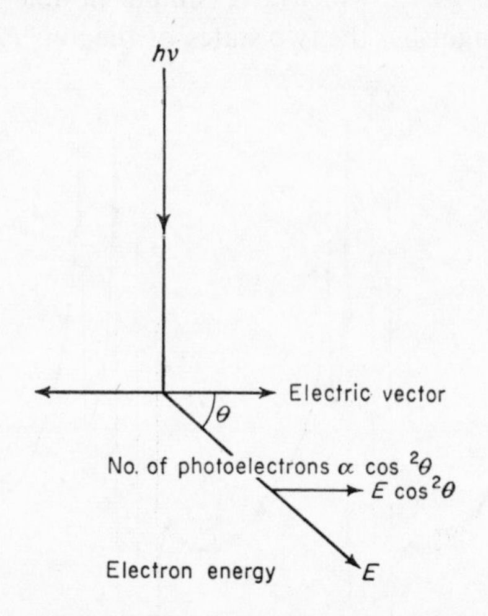

Figure 7. Electron flux at an angle θ to the electric vector has a component of energy $E \cos^2 \theta$ parallel to the electric vector.

with only a part of their kinetic energy ($E \cos^2 \theta$) available for penetration of this field, and are thus rejected at less than the maximum retarding field. In addition, electrons originating at points remote from the grid axis suffer similar loss of effective kinetic energy (Figure 8). It can be shown that to avoid peak broadening due to this latter phenomenon, the photon beam must be within 0.5 mm of the axis of a 7 mm diameter grid.

* There is, however, some evidence that in a few cases the distribution is isotropic, or may even contain an additional small $\sin^2\theta$ component.

The low energy tail due to the $\cos^2 \theta$ nature of the emission could be eliminated in principle by the use of spherical geometry. McDowell and colleagues[18] have obtained considerable improvement in this respect using a spherical grid structure surrounding a photoionization region small compared with the grid diameter (see also reference 18a).

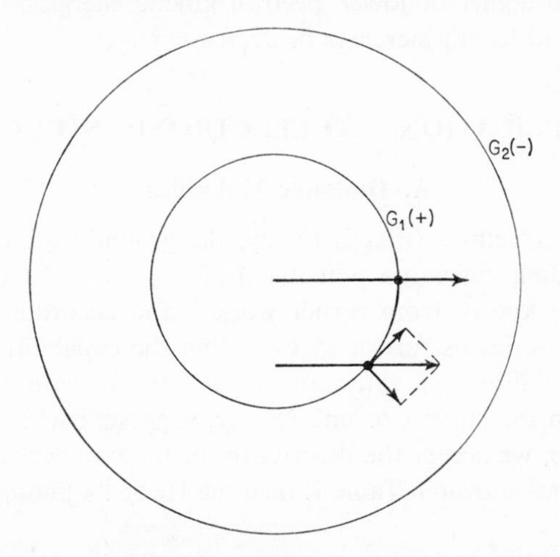

Figure 8. Loss of effective kinetic energy for electrons emitted away from the axis of a cylindrical (or spherical) retarding field grid system.

5. Deflexion-type photoelectron spectrometers

The use of a focusing deflecting field, either magnetic (180°) or electrostatic (127°), offers the promise of much greater energy resolution and an approach to the ideal spectrum with symmetrical peaks, since electron scatter can be minimized and only those electrons moving parallel to the electric vector need be used, though at the loss of some sensitivity. Both of these spectrometer types have been investigated in the author's laboratory. A small 180° magnetic deflexion analyser, having fixed slits and low resolving power[16], demonstrated that the improvement in peak shape alone greatly extended the usefulness of the method for estimating vertical ionization potentials and Franck–Condon factors in ionization (see below). A larger 127° electrostatic velocity selector with adjustable slits is still being developed. It is hoped to reach a resolving power $E/\Delta E$ of the order of 1000 with this instrument. If this can be attained, a resolution in the He 584 Å photoelectron spectra of fine structure with about 5 mv spacing should be possible. This energy is of the order of some rotational quanta so that a vibration–rotation analysis for ions may eventually be feasible. The best that has been achieved at the time of writing is a full-width at half-height of 15 mv for the peaks in the argon spectrum. Since this is less than kT at room temperature (~24 mv), we can obtain some indication of the shape of the rotational envelope. In a diatomic molecule which undergoes a large dimensional change on ionization, this reveals the 'shading' to higher or lower electron kinetic energies, depending on whether the bond length increases or decreases [27].

III. APPLICATIONS TO ELECTRONIC STRUCTURE

A. Diatomic Molecules

The electronic structures (that is to say, the ground-state configurations and corresponding ionization potentials) of O_2, N_2, and CO are, for the most part, well known from earlier work [17] and accordingly these substances form a series useful for investigating the capabilities of the new technique. In addition, an extension to NO, the structure of which has previously been the subject of uncertainty, suggests itself.

If, as a guide, we accept the description of these molecules in the configurational terms shown in Table 1, then the He 584 Å photoelectron spectra should have three bands in the case of N_2 and CO and four in the case of NO and O_2, since both the K shell and the $2s\sigma_g$ orbital levels are known to be below -21.21 ev.

Table 1. The *a priori* electronic configurations of the nitrogen, carbon monoxide, and oxygen molecules.

Molecular orbital	Nitrogen, carbon monoxide	Oxygen
$2p\pi_g$		↑ ↑
$2p\sigma_g$	↑↓	↑↓
$2p\pi_u$	↑↓ ↑↓	↑↓ ↑↓
$2s\sigma_u$	↑↓	↑↓
21.21 ev		
$2s\sigma_g$	↑↓	↑↓
K	↑↓	↑↓
K	↑↓	↑↓

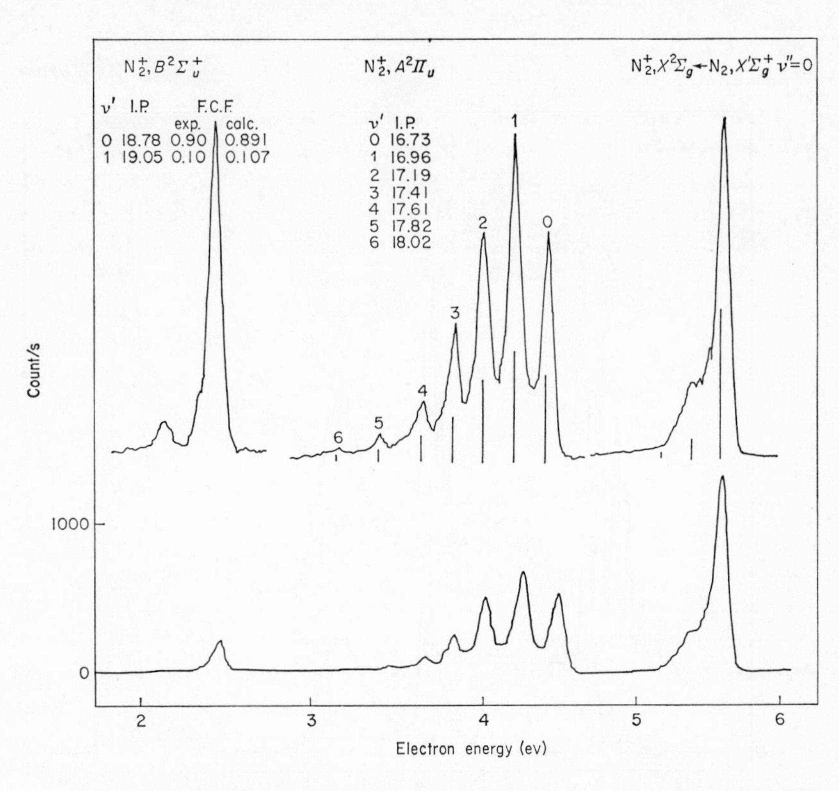

$N_2^+, B^2\Sigma_u^+$
ν' I.P. F.C.F.
exp. calc.
0 18.78 0.90 0.891
1 19.05 0.10 0.107
$N_2^+, A^2\Pi_u$
ν' I.P.
0 16.73
1 16.96
2 17.19
3 17.41
4 17.61
5 17.82
6 18.02
$N_2^+, X^2\Sigma_g \leftarrow N_2, X'\Sigma_g^+ \, v''=0$
Count/s
1000
0
2
3
4
5
6
Electron energy (ev)

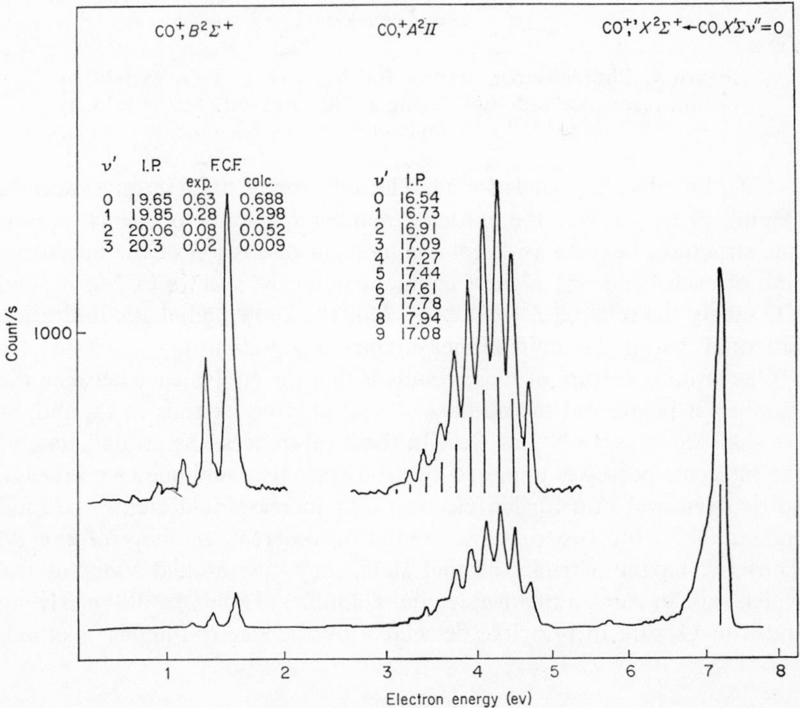

$CO^+, B^2\Sigma^+$
$CO^+, A^2\Pi$
$CO^+, X^2\Sigma^+ \leftarrow CO, X'\Sigma \, v''=0$
ν' I.P. F.C.F.
exp. calc.
0 19.65 0.63 0.688
1 19.85 0.28 0.298
2 20.06 0.08 0.052
3 20.3 0.02 0.009
ν' I.P.
0 16.54
1 16.73
2 16.91
3 17.09
4 17.27
5 17.44
6 17.61
7 17.78
8 17.94
9 17.08
Count/s
1000
1
2
3
4
5
6
7
8
Electron energy (ev)

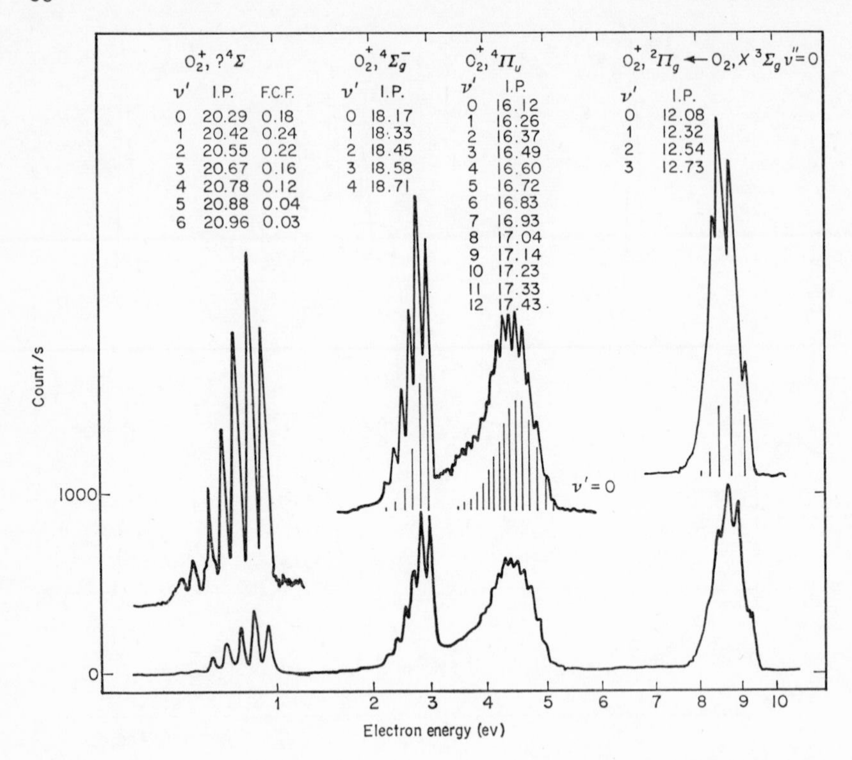

Figure 9. Photoelectron spectra for N_2, CO, and O_2 excited by helium resonance radiation, using a 180° magnetic sector velocity analyser[16].

This, broadly, is what is found. The 180° sector analyser gives spectra (Figure 9) which have the expected number of bands, on which appear fine structure. Leaving aside for the moment discussion of the fine structure of each band, the highest components in the spectra of N_2, O_2, and CO satisfy the relation $E = 21.21 - I$ for the known adiabatic ionization potentials within the limits of measurement ($\sim \pm 20$ mv).

One striking feature of these results is that the equivalence between the number of bands and the number of orbital levels extends to O_2 and, as we shall see later, to NO as well. In these substances, the ground state of the molecule possesses unpaired electron spins (two and one respectively), so that removal of a further electron may increase or decrease the total spin number, the two processes requiring different amounts of energy. Thus O_2, having a triplet ground state, may give excited states of O_2^+ which exist in pairs, a quadruplet and a doublet (Table 2). All the existing states of O_2^+ are in principle detectable by the electron-impact method,

Table 2. The possible states of O_2^+ resulting from direct electron removal.

Electron removed from O_2, ${}^3\Sigma_g^-$	Ion produced
$2p\pi_g$	O_2^+, ${}^2\Pi_g$ ← (b)
$2p\pi_u$	${}^4\Pi_u$ ← (b); ${}^2\Pi_u$ → (a)
$2p\sigma_g$	${}^4\Sigma_g$ → (b); ${}^2\Sigma_g$
$2s\sigma_u$	${}^4\Sigma_u$; ${}^2\Sigma_u$
$2s\sigma_g$	${}^4\Sigma_g$; ${}^2\Sigma_g$

Transitions: (*a*) produces the first negative band system
(*b*) produces the second negative band system

which does not distinguish between states of different multiplicity. The He 584 Å photoelectron spectrum presents a simpler picture. Each possible configuration of the ion of O_2^+ is represented by a single band and this corresponds in each case to the quartet states; the doublet does not contribute. A one-to-one relationship can also be traced between the bands of the NO photoelectron spectrum (Figure 11) and the configurations corresponding to the loss of one electron in turn from each orbital level, even though each excited configuration may give rise to either singlet or triplet states.

The close agreement between the photoelectron spectral data and the known energies of the various ionic configurations of O_2, N_2, and CO[8,12] is the main justification for using the spectra in more obscure cases as the basis of an orbital energy level diagram with a one-to-one correspondence between photoelectron and occupied orbital levels.

B. Vibrational Fine Structure; Franck–Condon Factors in Ionization

Each band in the spectra given in the previous section shows fine structure, the resolution of which into separate lines merely requires an improvement in energy-resolving power. The best that has yet been attained in the author's laboratory is shown in Figure 10 for O_2. The calculated Franck–Condon factors for ionization to the ground state of O_2^+ are indicated by the vertical lines and the measurements made by Frost and coworkers[18] by the arrows. The various lines are the consequence of ionization leaving the molecular ion in different vibrational states. Since at the temperature of the experiment none of the excited vibrational states of the molecule are highly populated (5.74×10^{-4} in $\nu' = 1$ for O_2 at n.t.p.), the line at highest electron energy derives from ionization to the molecular ion in its lowest vibrational state, i.e. to the 'adiabatic' ionization process. The most intense peak then defines the most probable (vertical) ionization process and gives the vertical ionization potential.

In many cases, the relative intensities of the different lines within a band approximate quite closely to the Franck–Condon (F.C.) factors, which define the relative probabilities for the excitation of the different vibrational levels accompanying the ionization. It is so unusual to find quantum-mechanical matrix elements appearing directly in primary experimental data that it is worth examining the limitations and implications of this observation more closely.

That the strength of vibrational components may be represented by numerical (F.C.) factors is due to the well-known separability of the probabilities for vibrational and electronic excitation in vibronic transitions. In the present case, therefore, the different electron fluxes are proportional to the F.C. factors to the extent that σ_i, the 'electronic' part of the ionization cross-section, is invariant with changes in the excess energy

$$[h\nu - (I_{\text{adiabatic}} + h\nu_{\text{vibration}})]$$

In fact, σ_i decreases slowly with excess energy in those substances for which data exist. Hydrogen shows a change of 10% per ev near 21 ev, which would enhance the $v',0 \leftarrow v,0$ component (I.P. 15.45 ev) by 10% relative to the $v',4 \leftarrow v,0$ component (I.P. 16.43 ev). The individual corrections are thus small and insufficient to change the relative order of the F.C. factors which would be deduced directly from the photoelectron spectrum[16].

Another small correction may be required, originating in the spectrometer itself. In sector spectrometers in which, as usual, line contours are determined by electron optical aberrations and by stray electrostatic fields rather than by the slit widths, it is strictly the band areas, i.e. the number of electrons belonging to each vibrational component, which form the primary experimental data. When the spectra are scanned by varying the momentum (180° magnetic, H varying) or kinetic energy (127° electrostatic, V varying) selected by fixed slits, the spectral slit width increases with increasing focusing field H or V. The peak electron

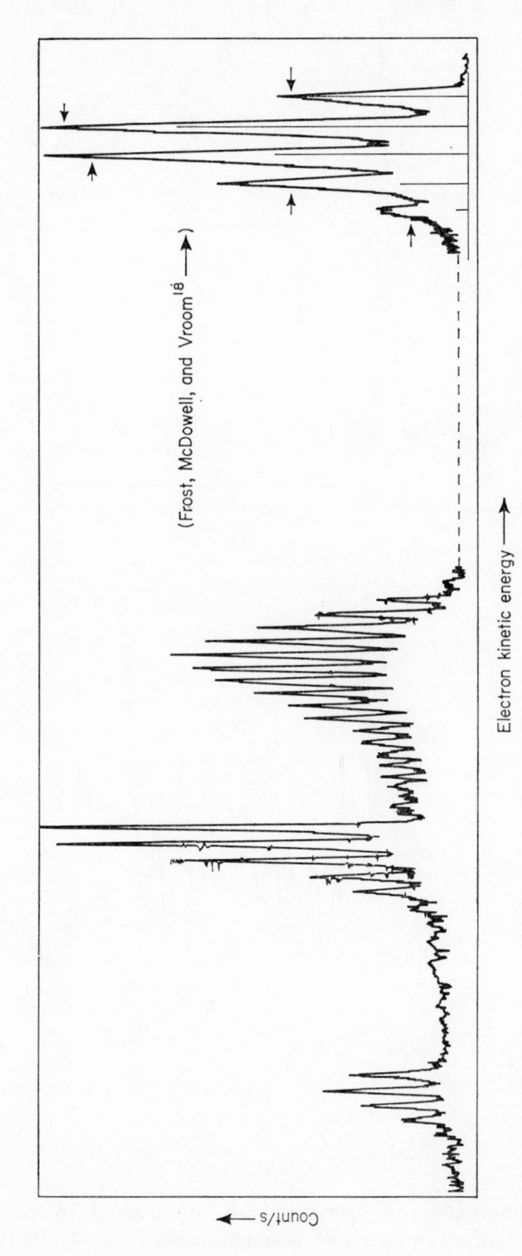

Figure 10. Photoelectron spectrum for O_2 excited by helium resonance radiation, using a 10 cm radius 127° electrostatic velocity analyser.

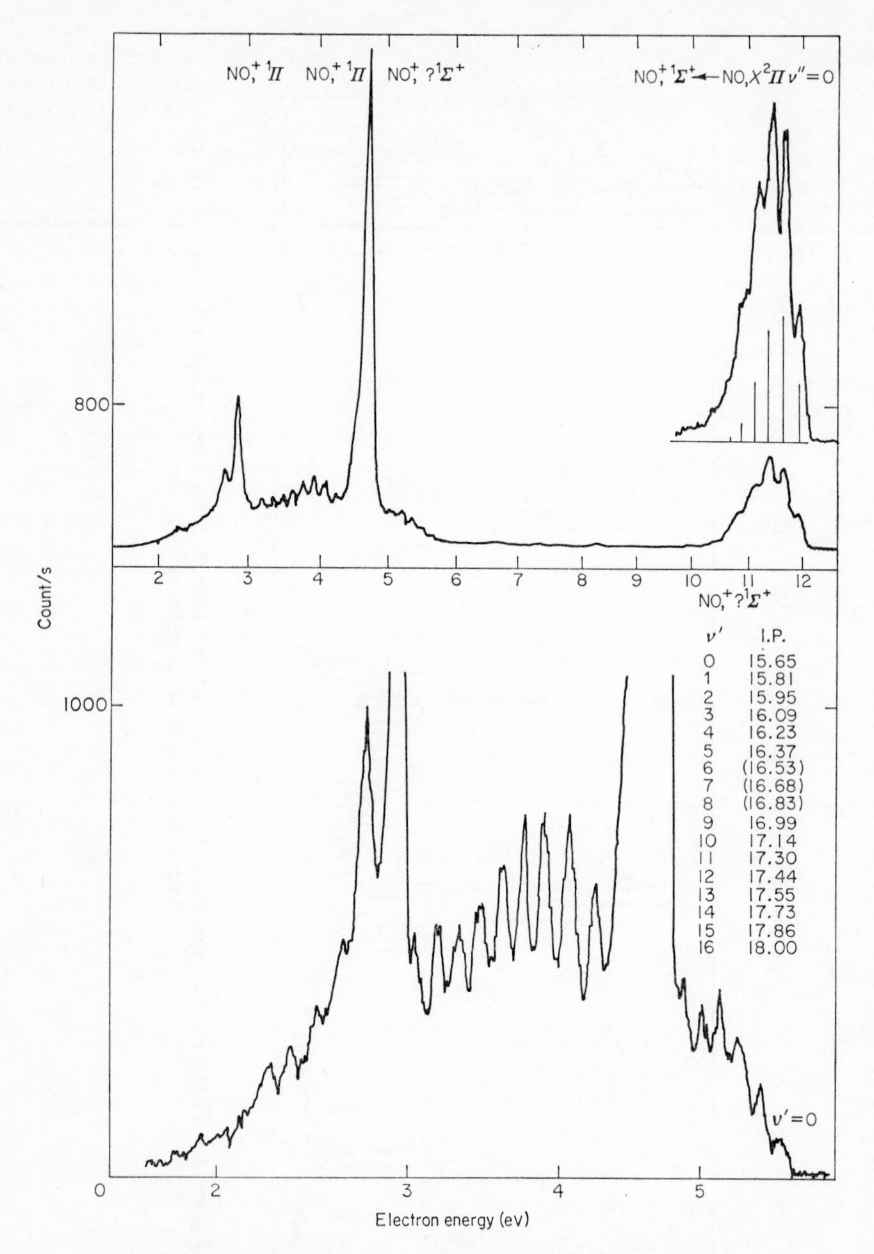

Figure 11. Photoelectron spectrum for NO excited by helium resonance radiation, using a 180° magnetic sector velocity analyser (same conditions as for Figure 9).

count rate, or collected current, needs to be divided by the particular H or V value to give a figure proportional to the true associated electron flux. This correction operates in the opposite direction from that for σ_i variations (above) and over small ranges (say 1 ev in 10) they effectively cancel, so that the spectra may very closely represent plots of F.C. factors.

Thus the photoelectron spectrum from a sector spectrometer shows in a clear manner both adiabatic and vertical ionization potentials for each orbital level. Even when vibrational structure is not resolved, the same arguments apply to the interpretation of the position of the high electron energy edge and the peak point in the envelope. Retarding field instruments cannot give reliable vertical I.P. values directly, since there is an inherent asymmetry in the band shapes, giving rise to a 'building up' of overlapping bands towards lower electron energy (cf. Figure 7). In either case, however, the difference between adiabatic and vertical I.P.s quantitatively or qualitatively indicates the amount of dimensional change undergone when each electron is ejected. Alteration of dimensions, especially of bond length, is recognized as being the major factor controlling the extent of a Franck–Condon envelope but usually these reflect changes in force constant. The adiabatic–vertical I.P. difference is thus a measure of the bonding character of the ejected electron. The loss of a completely non-bonding electron leaves the ion with the same dimensions and vibrational frequencies as the molecule and only the $0 \leftarrow 0$ transition appears in the photoelectron spectrum. An increase in bonding character is reflected by an increasingly broad band in the photoelectron spectrum. Note that only when the vibrational structure is resolved can a distinction be made between bonding and antibonding electrons (increase of r, decrease of V; and vice versa) both of which can give an extensive Franck–Condon envelope. Nitric oxide affords examples of electrons with similar I.P.s having extremes of bonding character. Compare (Figure 11) the appearance of the single strong peak at 4.69 ev (I.P. 16.52 ev) with the long series starting with the $0 \leftarrow 0$ component at 5.56 ev (I.P. 15.65 ev). The former corresponds to removal of a $2p\sigma_g$ electron which, in the other diatomic molecules mentioned, has little bonding character and here is apparently completely non-bonding, a conclusion which agrees well with theoretical calculations. The latter, however, corresponds to removal of a $2p\pi_u$ electron which is strongly bonding in all the other diatomic molecules but in nitric oxide gives the most extensive Franck–Condon envelope yet resolved. $\nu = 11$ is the strongest component and the $\nu = 23$ can just be seen at 2.2 ev (I.P. = 19 ev). The vertical–adiabatic I.P. difference is about 1.65 ev, which can be compared with 0.23, 0.37, and 0.50 for the corresponding levels in N_2, CO, and O_2, respectively. These figures confirm what comparison of the different spectra indicates: that what can be superficially described as bonding character becomes concentrated in the π_u electrons along the series N_2, CO, O_2, NO.

C. Some Triatomic Molecules

1. Carbon dioxide, carbon oxysulphide, and carbon disulphide

The energy levels of these three molecules have been less surely based on ionization potential data than those of some of the diatomic molecules described above. There should be six ionization potentials corresponding to the configuration

$$(\sigma_g)^2_{\mathrm{X}} \quad (\sigma_u)^2_{\mathrm{X}} \quad (\sigma_g)^2_{\mathrm{C-X}} \quad (\sigma_u)^2_{\mathrm{C-X}} \quad (\pi_u)^4_{\mathrm{C-X}} \quad (\pi_g)^4_{\mathrm{C-X}}$$

We have here an isoelectronic series with a regular increase in nuclear shielding which should be reflected in an increase in all the energy levels. This close relationship is clear in the photoelectron spectra (Figure 12), which also show the expected regular displacement of all the bands to higher energies. The similarities between the spectra are traced on an energy-level diagram in Figure 13 with assignments which derive from detailed arguments possible only for CO_2 which has been the most fully investigated[19]. It is apparent that the first, third, and fourth bands are fairly sharp throughout and this is consistent with their assignments to

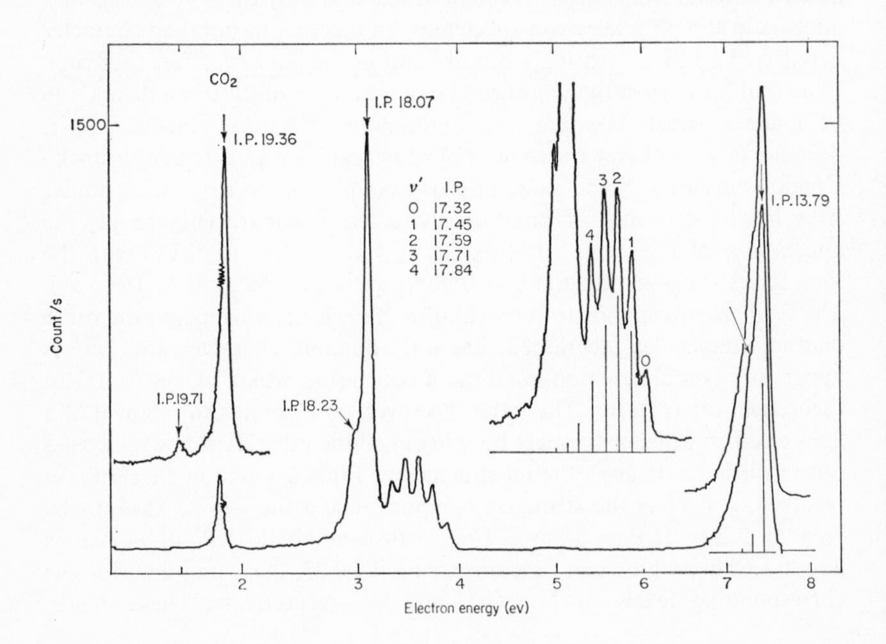

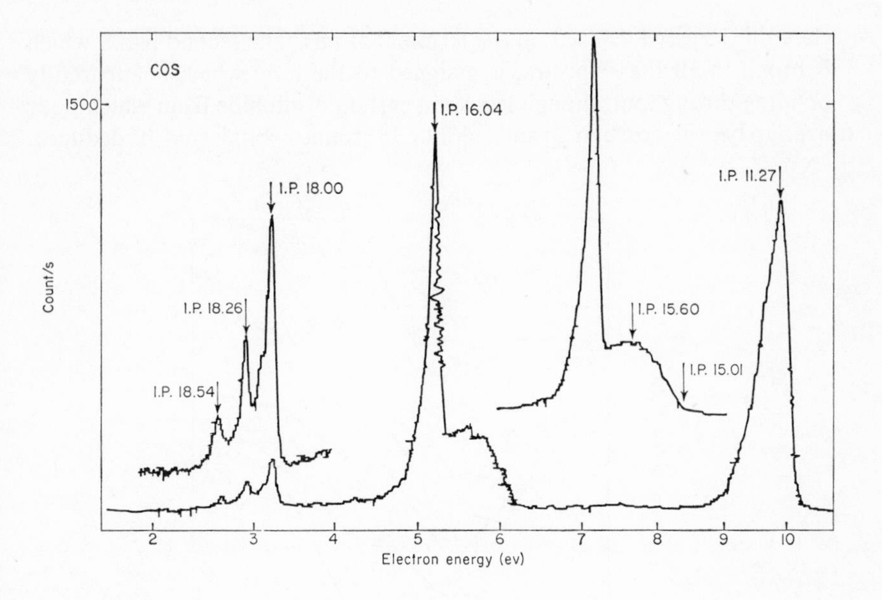

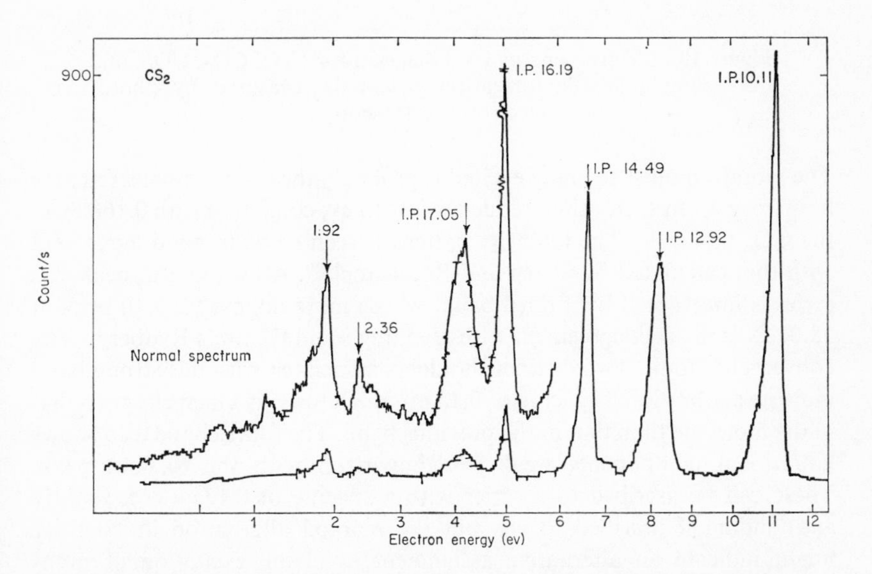

Figure 12. Photoelectron spectra for CO_2, COS, and CS_2 excited by helium resonance radiation using a 180° magnetic sector velocity analyser [D. P. May and D. W. Turner, unpublished work].

electrons largely localized on the terminal atoms. The second band, which is broad in all these spectra, is assigned to the level which is apparently bonding throughout, though less so in carbon disulphide from which poor overlap between carbon $2p$ and sulphur $3p$ atomic orbitals may be deduced.

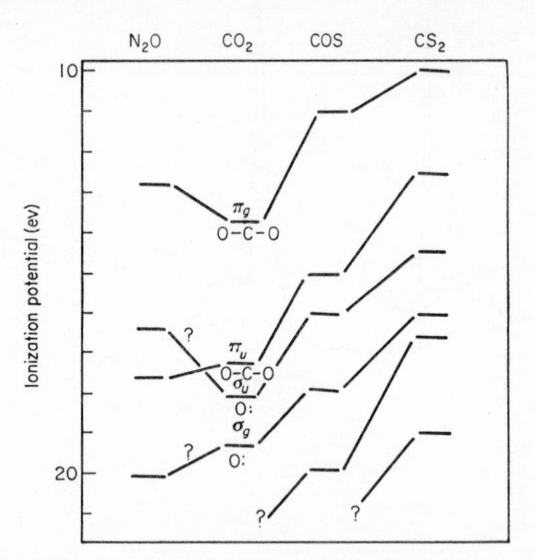

Figure 13. Electron energy level diagram for N_2O, CO_2, COS, and CS_2 using adiabatic ionization potentials measured by photo-electron spectroscopy.

The vibration most strongly excited is probably the symmetrical stretching frequency ν_1. In CO_2^+, this frequency is 0.13 ev, compared with 0.166 ev in the CO_2 molecule. The intensity pattern is seen to be in good agreement with that calculated by Sharp and Rosenstock[20]. After the fifth peak this series is interrupted by a third band, whose more intense (0, 0, 0) peak at 18.07 ev is in exact agreement with Henning's and Ogawa's Rydberg series convergence limit. Two distinct shoulders associated with this strong band indicate a vibrational spacing of 0.15 ev which for ν_1 is much closer to that in the molecule than that in the previous band. The fourth band also shows vibrational components weak in comparison with the (0, 0, 0) peak. These can be ascribed to a series with a spacing of 0.17 ev (i.e. slightly antibonding if ascribed to ν_1), but the marked alternation in intensity might indicate an alternative assignment involving excitation of even-numbered quanta of ν_3, the antisymmetric vibration. This is expected on the basis of the vibronic selection rules but such an assignment requires the postulation a very large change in the energy of this quantum (from 0.295 to 0.16 ev). It is noteworthy that a similar but less marked alternation of intensity is to be seen in this region of the N_2O spectrum (see below).

In carbon oxysulphide, a regular series of vibrational quanta are observed in the fourth I.P. band with a spacing (0.27 ± 0.02 ev) which is close to the value for ν_3 in the molecule (0.258 ev), indicating the removal of a very weak antibonding electron.

2. *Nitrous oxide*

As in carbon dioxide, only four of the six expected ionization potentials are smaller than 21.21 ev, but the arrangement of orbitals here clearly differs as the photoelectron spectrum (Figure 14) is quite different from

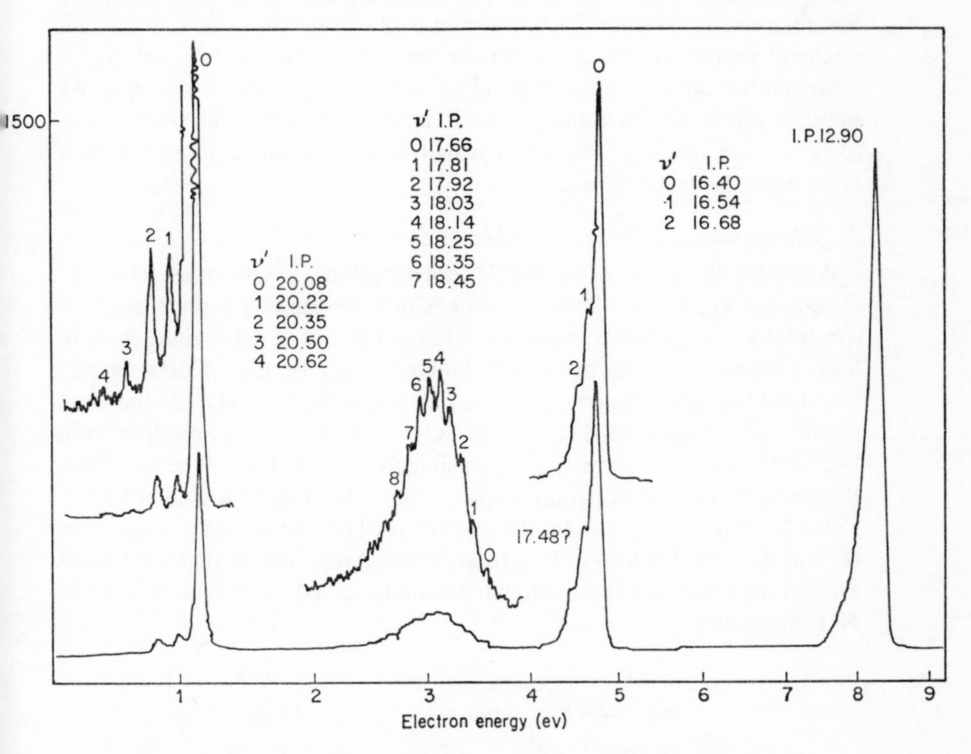

Figure 14. Photoelectron spectrum for N_2O excited by helium resonance radiation, using a 180° magnetic sector velocity analyser [D. P. May and D. W. Turner, unpublished work].

that of carbon dioxide. The main bonding level is now the third, and the fine structure is due to the stretching frequency 0.11 ev, markedly less than ν_1 in the molecule (0.159 ev). The first, second, and fourth bands apparently correspond to levels which are almost entirely non-bonding. Accordingly, the configuration of N_2O can be described by

$$(\sigma)^2\text{N—O} \quad (\sigma)^2\text{N—N} \quad (\sigma)^2\text{O:} \quad (\pi_u)^4 \quad (\sigma)^2\text{N:} \quad (\pi_g)^4$$

Comparing the spectrum of N_2O with that of CO_2 suggests that the difference in ordering is mainly due to an increase in the energy of the higher lone-pair orbital in CO_2 to give one mainly localized on the terminal atom, now N. The bonding orbital which is non-localized would not be expected to be so sensitive to the nuclear charge redistribution and is seen to change only slightly in the direction of lower energy.

The same vibrational spacing can be found in the second as in the fourth band, and since this is similar to ν_1 in the molecule, the removal of weakly bonding electrons might be inferred in each case. However, the unusual intensity alternation remarked on for the fourth band of CO_2 (above) is again noticeable in the corresponding band in N_2O and once more we have to admit the possibility that it is the antisymmetric vibration ν_3 (0.278 ev for the N_2O molecule) which is being excited, though with a large decrease in frequency.

3. Sulphur dioxide, nitrogen dioxide, and ozone

A similar discussion of electronic configurations[21,22] for other triatomic molecules—H_2O, SO_2, NO_2, O_3—for which the He 584 Å photoelectron spectra have now been measured (Figure 15) will not be given here in detail. However, some points of similarity may be traced between the spectra of the three latter compounds. This is expected since O_3 has the same number of valence-shell electrons as SO_2 and just one more electron than NO_2 in the 4σ, orbital which is responsible for the bonding of the molecule. Removal of the one electron present in this orbital in NO_2 leads to NO_2^+ which is linear in its ground state, and removal of one of the two present in O_3 may be expected also to lead to an increase in dihedral angle. O_3^+ in its ground state may be expected to have a similar dihedral angle to that of the NO_2 molecule:

		$\text{A–}\hat{\text{B}}\text{–A}$
O_3,	KKK --- $(4a_1)^2$	116.8°
	$\downarrow$	
O_3^+,	KKK --- $(4a_1)^1$	
	$\approx$	
NO_2,	KKK --- $(4a_1)^1$	134°
	$\downarrow$	
NO_2^+,	KKK --- $(4a_1)^0$	180°

In fact, for both O_3 and NO_2, the highest energy band is broad, consistent with a large change in dimensions, but it seems probable from the spectra that the first I.P. of NO_2 is no lower than 10.97 ev in spite of earlier evidence to the contrary. This difference between O_3 and NO_2 ($\Delta I = 2.3$ ev) is reasonable, since the change in electronegativity on replacing nitrogen by oxygen would most affect the highest occupied orbital in NO_2 because it

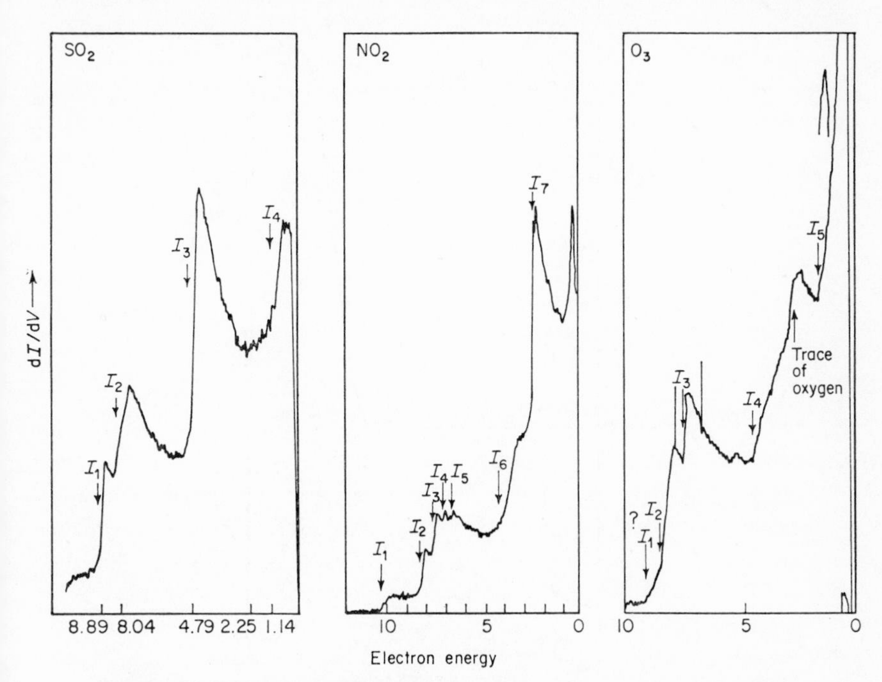

Figure 15. Photoelectron spectra for SO_2, NO_2, and O_3 excited by helium resonance radiation, using a cylindrical grid energy analyser (cf. refs. 21 and 22).

is largely localized on the central atom. The change can be compared with the first I.P. difference between NO and O_2 ($\Delta I = 3.09$ ev) where a similar change occurs when a single electron largely localized on nitrogen is replaced by two largely localized on (different) oxygen atoms.

4. *Water*

From the fact that its He 584 Å photoelectron spectrum shows only three bands[22] (Figure 16), we conclude that water has only three orbital levels of energy greater than 21.21 ev, and not four as deduced from electron-impact studies[23].

There seems little doubt from its relative sharpness and also from theoretical considerations that the highest electron-energy band relates to the removal of an oxygen 2*p* 'lone-pair' electron. Since the oxygen 2*s* electrons are almost certainly at a much lower energy than 21.21 ev, the remaining two bands (both of which are broad and thus represent strongly bonding electrons) can be attributed to the O—H bonding molecular

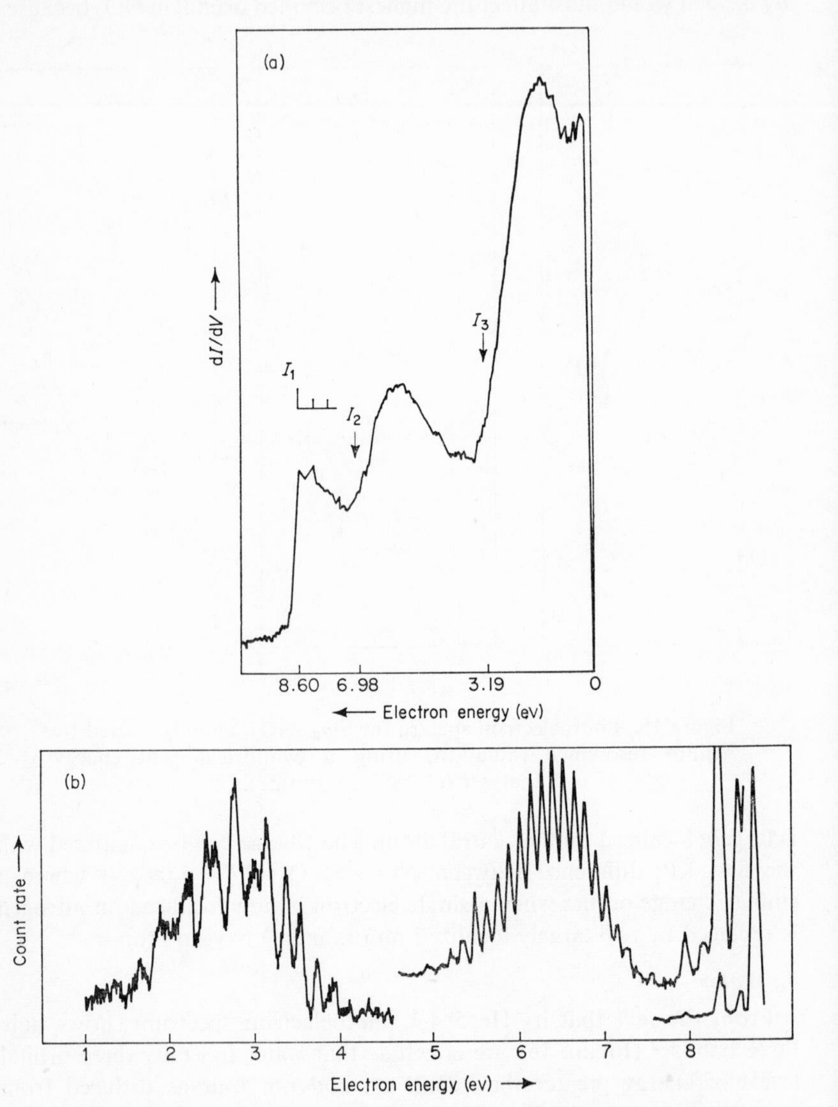

Figure 16. Photoelectron spectrum for H_2O excited by helium resonance radiation, using (a) a cylindrical grid analyser (b) a 127° electrostatic sector analyser. (NOTE. The electron energy scale is from right to left in (a) but from left to right in (b).)

orbitals. These can be regarded, from the equivalent-orbital viewpoint, as in-phase and out-of-phase combinations of the two localized O—H bond orbitals. Using the much higher resolving power of the 127° electrostatic analyser (Figure 16b), the main vibrational structure of the two highest energy bands becomes clear and helps to confirm the assignment given above. Only the highest energy band shows the presence of a large vibrational spacing (0.41 ev) comparable with the symmetrical stretching frequency of the molecule:

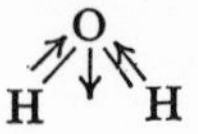

ν_1 3651.7 cm^{-1} (0.453 ev)

There appears in combination, but much more weakly, a lower energy quantum (0.17 ev) which is similar in magnitude to the bending mode:

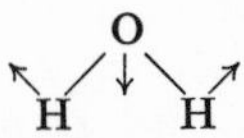

ν_2 1595.0 cm^{-1} (0.198 ev)

The similarity between the magnitudes of the vibrational quanta in molecule and ion respectively, the strength of the $0 \rightarrow 0$ transition and the favouring of the stretching mode over that involving deformation all support the assignment of the first band to removal of a (nearly) non-bonding electron from the oxygen atom.

This contrasts with the very different bonding character indicated by the structure of the second band, which exhibits a long vibrational series with an interval of 0.125 ev, comparable only with the bending mode in H_2O (0.198 ev). In other words, the main dimensional change has been of bond angle, not length. Thus we conclude that the orbital concerned is strongly bonding between the hydrogen atoms, in good agreement with theoretical predictions by Ellison and Shull[24]. The structure of the deepest lying band in the H_2O spectrum is more complex and must contain at least two long series of different vibrations. These cannot yet be analysed but clearly the orbital concerned must contribute strongly to the determination of both H–$\hat{O}$–H bond angle and O—H bond length.

D. Some Polyatomic Molecules

1. Ammonia, hydrazine, methane, and ethane

Water provided an example of strong interaction between geminal bond orbitals. Rather weaker resonance interaction can also be detected in hydrazine between the pairs of vicinal N—H bond orbitals and each of the nitrogen lone pairs. The two bands in the ammonia spectrum[21] (Figure 17) correspond to the antisymmetric N—H bonding orbital level (cf. methane) and the nitrogen lone pair, respectively.

Both bands appear at about the same energy in hydrazine[22] but both are broadened to the point at which a split into two pairs is only just evident. An analogous symmetrical splitting (to higher and lower energies) is found when ethane is compared with methane[21,25] (Figure 18), and it is

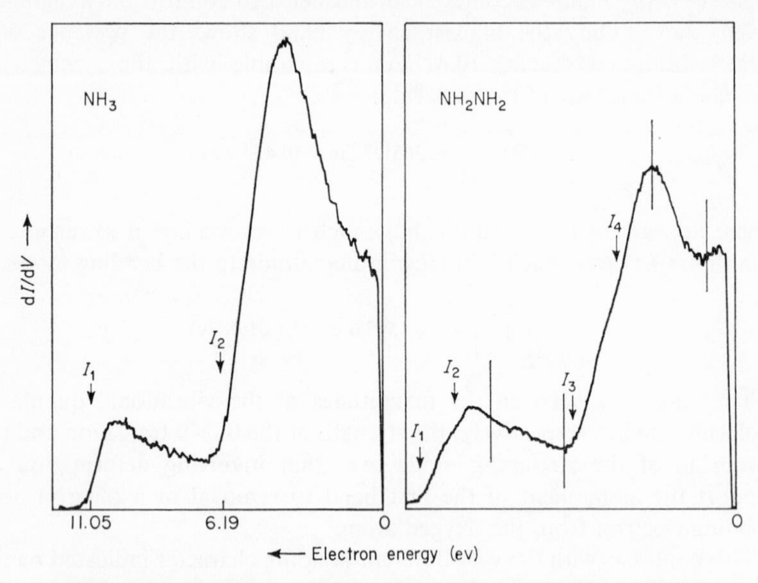

Figure 17. Photoelectron spectra for NH_3 and N_2H_4 excited by helium resonance radiation, using a cylindrical grid analyser.

interesting that in propane, though the bands are broad (and incidentally involve a number of overlapping levels), a division into three is apparent.

2. Ferrocene

It is often found that the electron flux from each orbital level reflects the degeneracy of the level. This can be seen at a glance in the spectra of the diatomic molecules for which the greatest electron flux is associated with ionization from a π orbital, which is doubly degenerate. A similar example from the organometallic field is provided by ferrocene. In ferrocene ($\mathbf{D}_{5d}$ symmetry), the five initially degenerate metal d orbitals must be split by the presence of the cyclopentadienyl groups. Only the three highest in energy remain localized on the iron atom and of these the highest pair are still degenerate in energy[30], a situation clearly revealed by the photoelectron spectrum (Figure 19). The 2:1 intensity ratio for the two components of the highest energy band nicely reflects the expected molecular orbital description $\ldots(a_{1g})^2(e_{2g})^4$. In addition, the next deeper level, at an energy appropriate for ligand orbitals, is also split, presumably

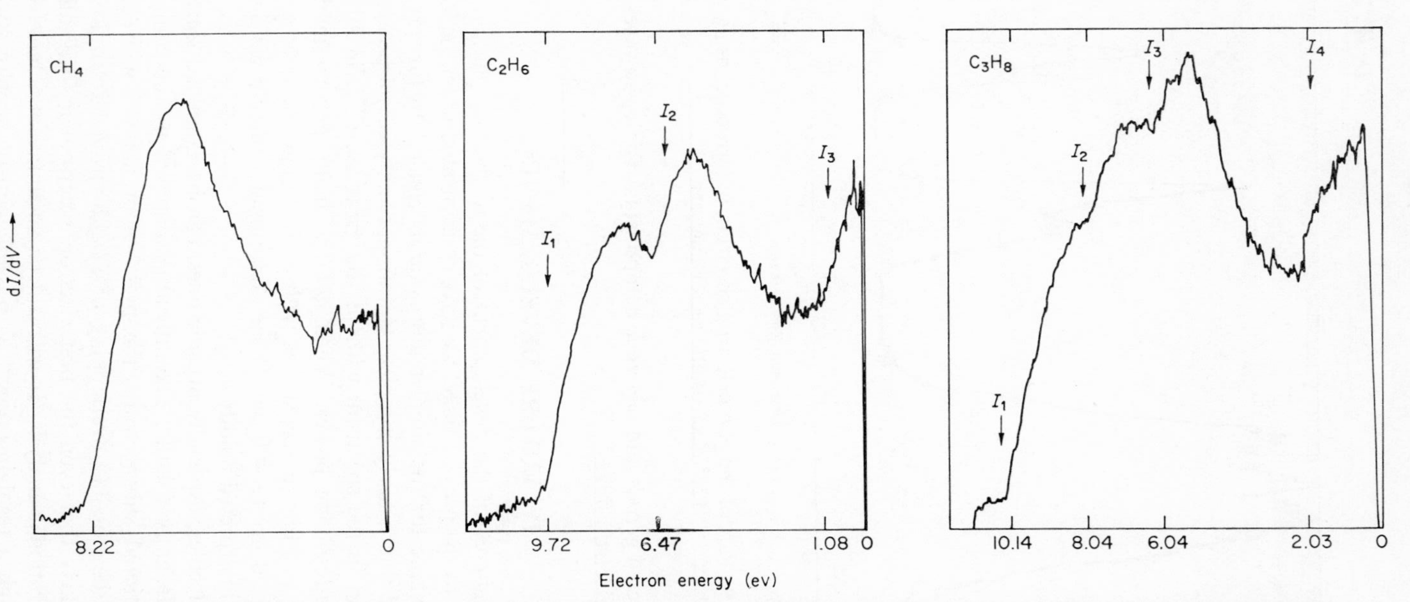

Figure 18. Photoelectron spectra for CH_4, C_2H_6, and C_3H_8 excited by helium resonance radiation, using a cylindrical grid analyser.

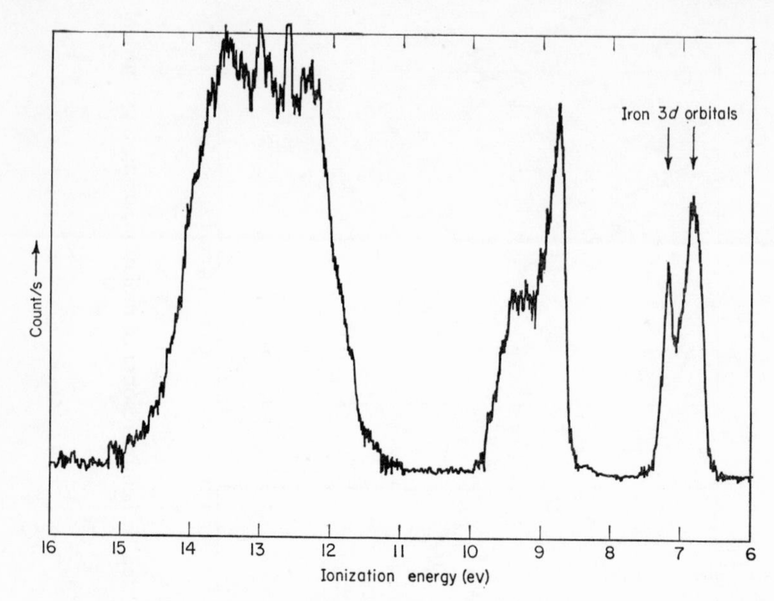

Figure 19. Part of the photoelectron spectrum of ferrocene, using a 127° electrostatic sector analyser.

into the expected *gerade* and *ungerade* components. These, however, have comparable electron fluxes.

IV. FUTURE DEVELOPMENTS

In this brief survey of the development of the method of photoelectron spectroscopy to the present stage, we have concentrated upon the kind of results obtainable for the simpler inorganic molecules. In this group, it can be hoped that a fairly complete vibrational analysis will be achieved for each band in the spectrum without too great an effort in attaining better energy-resolving power. When not too many normal modes of vibration are possible, it seems that with even a quite modest resolving power of say 50 mv ($\sim$400 cm^{-1}) the vibrational structure within each band can be interpreted readily.

Larger molecules, especially organic ones, can be examined since they are often sufficiently volatile. The difficulties imposed by the complexity of the vibrational subspectrum, the presence of many low frequency modes, and the possible overlapping of neighbouring electronic levels make it unlikely that even the best electron spectrometry will allow a complete vibrational analysis. It will still be advantageous, however, to employ as high a resolving power as possible consistent with adequate sensitivity, since it may still be possible to identify the larger vibrational modes, particularly in molecules of high symmetry. In benzene, for example[28] (Figure 20), even a 50 mv resolution allows the ring-breathing mode ν_2 to be identified in three of the bands, one being associated with a deep σ molecular orbital of high symmetry, since ν_2 is apparently the only mode excited. If ever the practical limit of ~ 2 mv can be reached, we may expect to obtain vibration–rotation data for simple molecular ions in all their electronic configurations and quite detailed vibrational data for the various configurations of complex molecular ions.

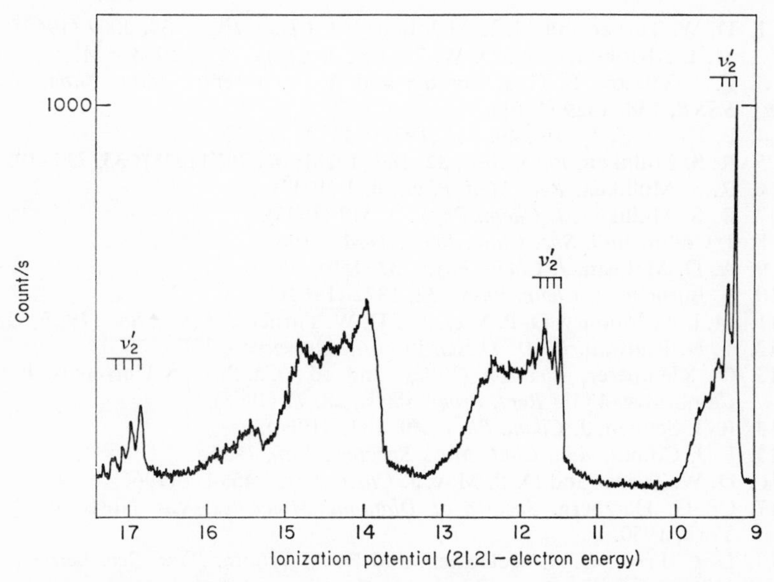

Figure 20. Photoelectron spectrum for benzene excited by helium resonance radiation, using a 127° electrostatic sector analyser.

Detailed analysis by conventional spectroscopic refinement is not the only use of photoelectron spectroscopy. By exploiting instead the conceptual simplicity of the primary experimental data, where each occupied orbital is represented by a band in the spectrum, a very rewarding discussion of substituent effects can be had upon examination of the spectra of related groups of compounds without using any vibrational analysis. This is perhaps easier in the field of organic chemistry, where ultraviolet and visible absorption spectroscopy have already been applied in this way by very many workers.

From these remarks, it will be evident that in all probability this form of spectroscopy will be developed, as have other forms of spectroscopy, by attracting both 'fundamental' and 'empirical' studies. We can add to such uses the probable employment of the data to test quantum-mechanical computational methods by equating the calculated eigenvalues with the observed ionization potentials for the various levels, an approach which has not generally been attempted by theoretical chemists because of the lack of experimental data. Eventually, however, it may even become convenient to make use of a large body of such fundamental data to help frame alternative theoretical approaches.

V. REFERENCES

1. D. W. Turner and M. I. Al-Joboury, *J. Chem. Phys.*, **37**, 3007 (1962).
2. M. I. Al-Joboury and D. W. Turner, *J. Chem. Soc.*, **1963**, 5141.
3. F. I. Vilesov, B. C. Kurbatov, and A. N. Terenin, *Dokl. Akad. Nauk SSSR*, **138**, 1329 (1961).
4. F. Hund, *Z. Physik*, **40**, 742 (1927); **42**, 93 (1927).
5. R. S. Mulliken, *Phys. Rev.*, **32**, 186 (1928); **32**, 761 (1928); **33**, 730 (1929).
6. R. S. Mulliken, *Rev. Mod. Phys.*, **4**, 1 (1932).
7. R. S. Mulliken, *J. Chem. Phys.*, **3**, 517 (1935).
8. J. Collin, *Bull. Soc. Chim. Belg.*, **71**, 15 (1962).
9. A. D. McLean, *J. Chem. Phys.*, **32**, 1595 (1960).
10. L. Burnelle, *J. Chem. Phys.*, **32**, 1872 (1960).
11. M. I. Al-Joboury, D. P. May, and D. W. Turner, *J. Chem. Soc.*, **1965**, 6350.
12. T. N. Radwan, *Ph.D. Thesis*, London University, 1966.
13. O. Klemperer, *Electron Optics*, 2nd ed., Cambridge University Press, Cambridge, 1953; *Rept. Progr. Phys.*, **28**, 77 (1965).
14. R. I. Schoen, *J. Chem. Phys.*, **40**, 1830 (1964).
15. F. J. Comes, *Ann. Conf. Mass Spectry., 13th, 1965.*
16. D. W. Turner and D. P. May, *J. Chem. Phys.*, **45**, 471 (1966).
17. Cf. G. Herzberg, *Spectra of Diatomic Molecules*, Van Nostrand, New York, 1950.
18. D. C. Frost, C. A. McDowell, and D. A. Vroom, *Phys. Rev. Letters*, **15**, 612 (1965).

18a. D. C. Frost, J. S. Sandhu, and D. A. Vroom, *Nature*, **212**, 604 (1966).

19. Y. Tanaka, A. S. Jursa and F. J. LeBlam, *J. Chem. Phys.*, **32**, 1199 (1960).
20. T. E. Sharp and H. M. Rosenstock, *J. Chem. Phys.*, **41**, 3453 (1964).
21. T. N. Radwan and D. W. Turner, *J. Chem. Soc.*, **1966**, 85.
22. M. I. Al-Joboury, *Ph.D. Thesis*, London University, 1964.
23. W. C. Price and T. M. Sugden, *Trans. Faraday Soc.*, **44**, 108 (1948).
24. F. O. Ellison and H. Shull, *J. Chem. Phys.*, **23**, 2348 (1955).
25. D. W. Turner and M. I. Al-Joboury, *Bull. Soc. Chim. Belge*, **73**, 428 (1964).
26. D. W. Turner, unpublished results.
27. D. W. Turner, *Nature*, **213**, 795 (1967).
28. D. W. Turner, *Tetrahedron Letters*, **1967**, No. 35, 3419.
29. J. Berkowitz and H. Erhardt, *Phys. Letters*, **21**, 531 (1966).
30. J. N. Murrell, S. F. A. Kettle, and J. M. Tedder, *Valence Theory*, Wiley, London, 1965, pp. 342, 343.

4

Electronic Absorption Spectroscopy

H.-H. SCHMIDTKE

I. INTRODUCTION

A. Preliminaries

In the past 15 years, there has been a marked revival of interest in inorganic chemistry. In particular, our knowledge of transition metal complexes, with which we are primarily concerned in this article, has grown exceptionally. There are essentially two reasons for this, one practical, the other theoretical. The practical reason for the development of modern coordination chemistry is the commercial availability of the complicated instruments which are necessary for different kinds of spectroscopic investigation and which had previously to be built by the scientist himself. Instead of devoting himself to the questions actually of interest to him, he had to spend time doing electronics! For visible and ultraviolet spectroscopy, there are now excellent spectrophotometers on the market, which are easily operated even by scientists who have no detailed knowledge about their mechanism. The theoretical part, however, which ought to explain the experimental results in detail, remains a more difficult task. Here considerable effort is necessary in every new case and no routine work is possible. A careful knowledge of the theoretical foundations is also advisable in order to avoid any misinterpretation of the experimental findings which may arise from an ignorance of the limitations on which a theory is based. However, the development and application of wave mechanics in general, and of crystal- and ligand-field theory in particular, has been an important stimulus to the chemistry of coordination compounds. It is only by applying these theoretical methods that an understanding of the physical properties of the substances can be approached.

We shall describe instrumental matters only briefly in this chapter, and shall devote ourselves mainly to theoretical questions, in particular the interpretation of electronic spectra. Since the molecules which we shall be discussing show certain symmetries in their nuclear frameworks, it is convenient to use the corresponding point symmetries when describing the electronic structure of the systems. The mathematical tool for handling symmetry properties is group theory and the symbols of irreducible representations characterize the one- or many-electron terms. It may be mentioned in this connexion that spectroscopists classified energy levels for atoms and ions in terms of quantum numbers long before quantum mechanics was developed and before group theory was introduced in this field. Today we are in the happy position that theory can at least predict the number of terms in a molecule that may arise from a certain configuration; it can also give degeneracy numbers and symmetry properties of these states. Usually not all of these energy levels are actually observed, but theory nevertheless is able to supply us with information about the possibilities that have to be taken into account.

B. Limitation and Goal

We shall essentially confine ourselves to a discussion of complex compounds of the transition elements, i.e. open-shell atoms and ions in solution, in single crystals, and as ions in host crystals. These molecules, of course, do not represent all the inorganic compounds one can think of, but the restriction is imposed for theoretical and practical reasons. For reasons we will discuss later, it appears that theories like ligand-field theory are mainly useful only for open-shell central atoms, and the transitions occurring between the electronic levels determined by this theory are found in spectral regions covered by normally available instruments. The electronic absorption of these compounds is found in the near-infrared, visible, and ultraviolet regions. Here one is usually limited at 200 mμ wavelength by the Schumann vacuum ultraviolet region. Figure 1 shows an example of

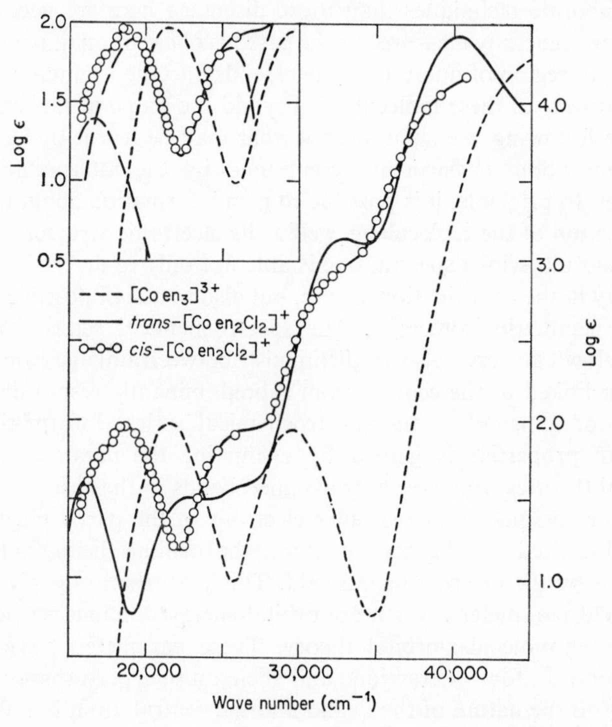

Figure 1. Absorption spectrum of $[Co\,en_3]^{3+}$, *cis*- and *trans*-$[Co\,en_2Cl_2]^+$ in aqueous solution. In the upper corner, a curve analysis is made for the *cis* isomer. [Reproduced, by permission, from *Z. Anorg. Allgem. Chem.*, **271**, 109 (1953).]

the type of experimental information obtained from visible and ultraviolet spectroscopy and which we wish to rationalize in this chapter. Several bands with very different extinction coefficients are found in this spectral region, and these can be explained only by different transition mechanisms with different transition probabilities. The spectra are discussed in detail below. For closed-shell coordination compounds, the interpretation of electronic absorption spectra is by no means as developed as it is for open-shell complexes. To make definite assignments of intramolecular charge-transfer bands, internal ligand bands, and also bands from a closed central atom shell, a rigorous molecular-orbital treatment of the molecular unit would be necessary. Such a calculation in the strict sense has been performed up to now only for the open-shell case $KNiF_3$[1,2]. Another large field is the electronic absorption spectroscopy of small inorganic molecules. However, in order to record the spectra of these compounds, more elaborate techniques than those discussed here are necessary. The spectra are generally measured in the gaseous phase or on thin films and the wavelength region of interest mostly extends into the vacuum ultraviolet[3]. The treatment of these molecules is beyond the scope of this chapter.

In the following, we shall discuss what may be learnt under optimum conditions about a particular compound by the interpretation of its spectrum. In principle, it is possible to gain information about the nuclear configuration of the molecule as well as its electronic structure. From the visible and ultraviolet spectra, one is able not only to determine the point symmetry in the coordination sphere, but also to assign possible coordination or geometric isomers[4,5]. The spectrochemical series discussed in detail below can serve also to distinguish between linkage isomers, since the atom linked to the central atom is predominantly responsible for the position of a ligand in the spectrochemical series. Information about electronic properties is gained by evaluating the parameters of semi-empirical theories from correctly assigned bands in the spectrum. Ligand-field theory assumes that the outer electrons of the partly filled shell are mainly localized on the central atom, the orbitals being only slightly perturbed by the external ligand field. This perturbation depends on the ligand-field parameters, which are orbital-energy differences when we talk in terms of molecular-orbital theory. These parameters vary with the ligand–central atom interactions and their mutual perturbation depends not only on the nature of the ligand and the central atom but also on the form of the interacting orbitals. Therefore, it is possible from these parameters to draw conclusions about the chemical bonding and the electron distribution in a complex compound and compare the results to the properties of its constituents.

Another set of parameters originating from the theory are the inter-electron-repulsion integrals which can be determined from the assigned bands of the spectra. When compared to the free-ion parameters, they serve as a measure of electron delocalization in the orbitals of the partly filled shell of the compound. A series of ligands ordered according to experimental electron-repulsion parameters is called the nephelauxetic series[6]. It has been used to describe in a simple way the covalency effects of the central metal–ligand bonds.

Other localized electronic systems in the complex compound can be found on the ligands. Some of them, particularly coordinated atoms or ions, do not absorb at all in the region in which we are interested. Among those ligands which absorb at wavelengths greater than 200 mμ, we may distinguish between two sets, one where the ligand spectrum is hardly changed in comparison to the free molecule, e.g. pyridine, and the other where we observe a dramatic change in the molecular absorbance, e.g. thiocyanate. We conclude that in the first group of ligands the electronic structure of the coordinated species is subject to only a slight change, which theoretically could be handled by a perturbation[7]. This approach does not seem to be valid for the second group of ligands in which the orbital system is obviously so different from the free molecule that a molecular-orbital treatment of the whole complex entity appears to be necessary.

Furthermore, bands can be found due to electron transitions between the central metal ion and the two delocalized systems just described. The classification of these bands as due to charge transfer is only possible when weakly interacting systems are assumed. The electron-jump from one molecular part to another is energetically favoured if the electronegativities are not very different. The positions of the electron-transfer bands of the compounds might therefore give some idea of the electronegativity difference of the central atom and the ligand system[8]. To treat this subject in a fundamental way, a more complete theory is necessary than ligand-field theory is able to supply. Attempts to order and understand these phenomena are more or less empirical.

Another sort of band which can be detected in the spectral region of interest here results from transitions between ions, and was first observed by Linhard and Weigel[9]. It is explained by an intermolecular charge transfer from the anion to the empty antibonding *d* orbitals of the central metal atom and has been found in ion pairs in solution and in crystals[10,11]. These intermolecular charge-transfer processes, which are common in organic complexes, have been studied theoretically even less than those occurring within one complex molecule which we mentioned before.

C. Instruments

Recently, several books on chemical spectroscopy have appeared[12–15] which give a detailed description of the nature of light, experimental devices, and the practical instruction necessary for recording spectra. We will therefore only make some brief remarks on these subjects.

According to Bohr's condition

$$E_f - E_i = h\nu \tag{1}$$

light of frequency ν is absorbed when the energy of an atomic or molecular system is changed from an initial to a final state. The mechanism of absorption generally goes from the initial to the final state; the energy being absorbed by the system is supplied from the incident light of frequency ν and later is radiated again through molecular vibrations or some other energy-transfer path. The unit of frequency is usually oscillations per second with the dimension $\sec^{-1}$, but since these are very large numbers for the visible and ultraviolet region, spectroscopists also use the fresnel unit f ($1\ \mathrm{f} = 10^{12}\ \sec^{-1}$). Instead of classifying in frequencies, electromagnetic waves are also characterized by their wavelength λ determined by

$$\nu\lambda = c \tag{2}$$

where c is the velocity of light *in vacuo*. λ is measured in the visible region in ångströms ($1\ \text{Å} = 10^{-8}$ cm) or in millimicrons ($1\ \mathrm{m}\mu = 10^{-7}$ cm). When plotting absorption spectra, the wave number unit $\bar{\nu} = 1/\lambda$ is preferentially used, since in this case the scale varies linearly with the absorbed energy

$$\varepsilon = E_f - E_i = h\nu = h\frac{c}{\lambda} = hc\bar{\nu} \tag{3}$$

The units of wave numbers are cm^{-1} or kayser (K) with 1000 kayser = 1 kilokayser (kK).

The energy of a molecule is quantized in electronic, in vibrational, and in rotational motion. Translational quantization is negligibly small. As seen from Figure 2, the quantum units of the three possible motions are very different. The heavy lines *b* and *a* represent the electronic levels, say the bonding and antibonding $1s\sigma$ orbital states of H_2^+, between which an electronic transition could take place. Superimposed on each of these states, the equidistant vibrational term scheme is drawn in normal lines including the zero-point energy. The rotational states have even smaller quantum units as illustrated in Figure 2 by dotted lines. Since electronic transitions are usually accompanied by simultaneous vibrational and rotational transitions, we do not observe in a spectrum taken at normal temperature a single line with a certain frequency ν as (1) would suggest, but rather a whole series of transitions forming a band with a maximum value of absorbance nearly at a position that would correspond to the pure electronic transition.

Absorption spectra in the visible and ultraviolet region are recorded either by photographic plates or by photocells as detecting devices. It proves to be rather difficult to determine exact extinction coefficients by photographic spectra, since the sensitivity of the photographic material varies rather widely with the frequency of light. Therefore, the preferred

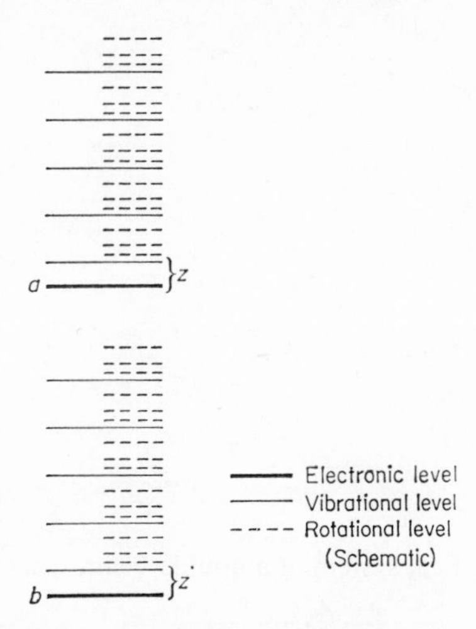

Figure 2. Order of electronic, vibrational, and rotational level separation (schematic).

spectrophotometers measure the intensity of light electrically. Such a spectrophotometer consists essentially of four parts:

1. light source (hydrogen and tungsten lamp for the ultraviolet and visible range, respectively),
2. monochromator (a system of prisms which select the desired frequency),
3. absorption cell assembly, and
4. detector (e.g. PbS phototube).

A schematic diagram of a recording double-beam spectrophotometer is shown in Figure 3. In this instrument, the monochromatic beam is separated before it enters the absorption cells. The intensities of the sample and reference cell are measured continuously against one another. By this arrangement, the possible absorption or reflection of the solvent is automatically subtracted from that of a solution so that only the absorption of the solute is recorded by the instrument. After passing through the absorption cells, the two beams enter the photocell compartment where they either hit two photomultipliers (Cary 15) whose wavelength sensitivity is accurately tuned, or the two beams impinge on the same detector and are separated again by an electronic commutator. Sample and reference currents are compared and transferred to a pen which records the extinction or transmission. In common instruments, the plot on the abscissa is linear in wavelength.

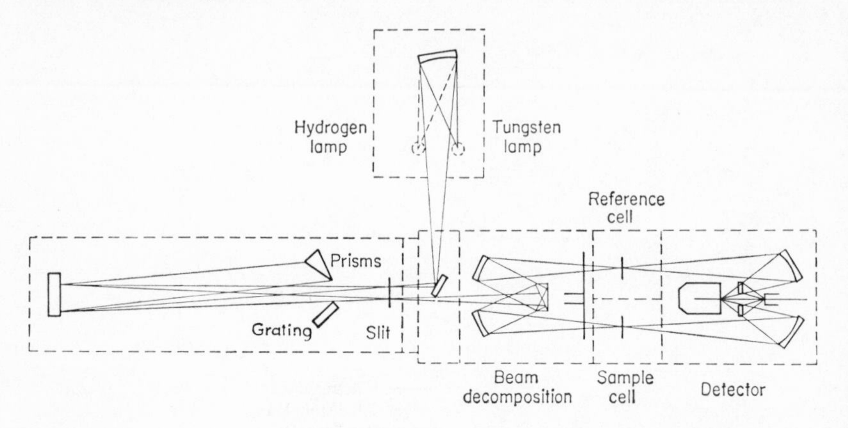

Figure 3. The optical system of a double-beam spectrophotometer.

Figure 4 shows a list of optical materials used as prisms, lenses, and windows in different spectral regions. Common spectrophotometers cover an effective range up to 50 kK in the ultraviolet, but in nitrogen or rare gas atmospheres this could be extended by a good quartz optic to 59 kK. Photomultipliers or other photocells (PbS photoconductors) serve as light detectors and cover the range from the far ultraviolet to near infrared. In Figure 4 are also listed the various excitations which a molecule can experience and which are observed by absorption spectroscopy. For further descriptions of spectrophotometric instrumentation, the reader must consult the literature[16-20].

D. Band Characteristics

In 1925, quantitative methods were introduced in absorption spectroscopy. These enabled both the wave number and the intensity of an absorption to be determined. In the 1940s, the introduction of a spectrophotometer made possible a faster and more exact quantitative measurement of absorbed or transmitted light.

Light passing through an absorbing solution of thickness d suffers a loss of intensity. If this loss $-\delta I$ in the slice δd is proportional to its intensity, i.e.

$$\frac{-\delta I}{\delta d} = kI \tag{4}$$

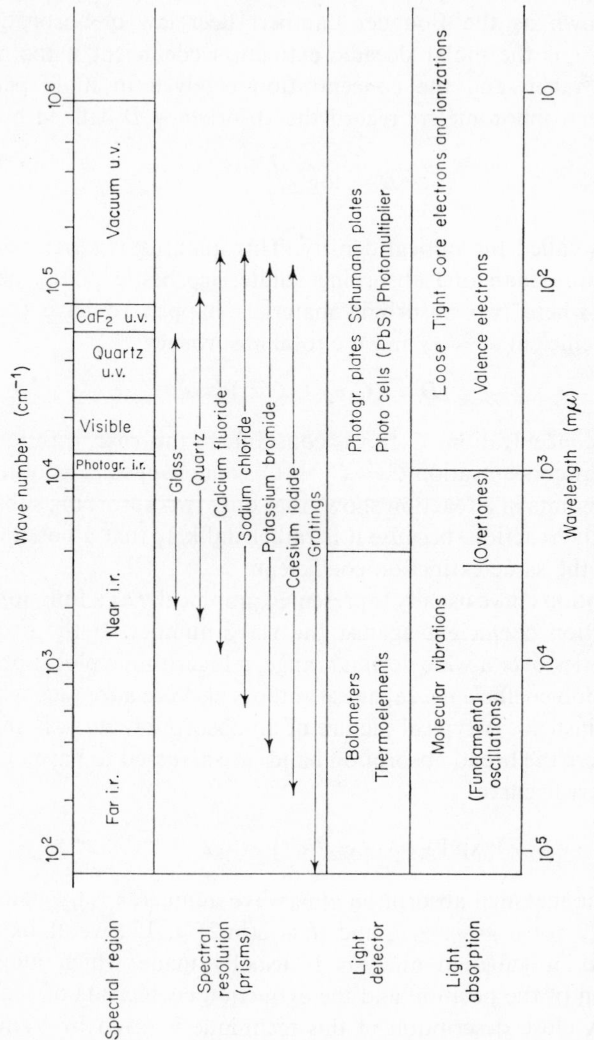

Figure 4. **Infrared, visible, and ultraviolet spectral regions, prism materials, and light detectors for absorption and reflection spectroscopy.**

and $k = \varepsilon C$ is expressed by the concentration C, the absorption equation is given by

$$\log_{10} \frac{I}{I_0} = -\varepsilon C d \tag{5}$$

which is known as the Bouguer–Lambert–Beer law of absorption. In equation (5), ε is the molar decadic extinction coefficient if the decadic logarithm is taken and the concentration is given in mole per litre. Common spectrophotometers record the absorbency D defined by

$$D = \log_{10} \frac{I_0}{I} \tag{6}$$

which also is called the optical density. This quantity is strictly additive in case of more than one absorbing solute. Isosbestic points occur at wavelengths where two absorbing materials happen to have the same extinction coefficient $\varepsilon_1 = \varepsilon_2$ and the total absorbency

$$D = (C_1\varepsilon_1 + C_2\varepsilon_2)d \tag{7}$$

for different concentrations C_1, C_2 is constant. In this case, it follows that the total molar concentration $C = C_1 + C_2$ is unchanged. The occurrence of isosbestic points in a reaction shows that only two absorbing substances take part in the reaction, because it is rather unlikely that a possible third one also has the same extinction coefficient.

The absorption curve usually is presented graphically as a function of the molar extinction coefficient against the wave number. If the extinction coefficient varies over a wide decimal range, a logarithmic plot is preferred in the extinction coefficients. Japanese authors also use a frequency plot on the abscissa instead. A typical picture of an absorption curve is shown in Figure 5, where the broad absorption bands are assumed to have the shape of gaussian error curves

$$\varepsilon = \varepsilon_0 \exp\left[-(\sigma - \sigma_0)^2/d^2\right] = \varepsilon_0 2^{-(\sigma-\sigma_0)^2/\delta^2} \tag{8}$$

where ε_0 is the maximal absorption at σ_0 wave number. δ is the half-width, since $\varepsilon = \varepsilon_0/2$ for $\sigma = \sigma_0 \pm \delta$, and d is $\delta(\ln 2)^{-\frac{1}{2}}$. If several bands are superimposed, a gaussian analysis is usually made which allows the determination of the position and the extinction coefficients of each band separately. A close description of this technique is given by Vandenbelt and Henrich[21]. For bands not symmetrically shaped, gaussian curves with different half-widths towards smaller and larger wave numbers have been suggested[22]. In a logarithmic ε plot, the gaussian curves take on a parabolic shape.

As mentioned above, electronic transitions are usually accompanied by vibrational and rotational excitations of the nuclear framework in the molecule. Since the quantum units of the two latter motions are small, these effects are seen at normal temperature only as an appreciable broadening of the absorption lines in the spectrum. The width of each band in the spectrum can differ very much. This is explained by the Franck–Condon principle. We assume that most of the vibrational excitation is due to the totally symmetric stretching mode so that the potential hypersurface of the complex molecule can be represented in a simple two-dimensional plot which is known from diatomic molecules. Since electronic transitions are generally fast compared with rearrangements of the nuclear

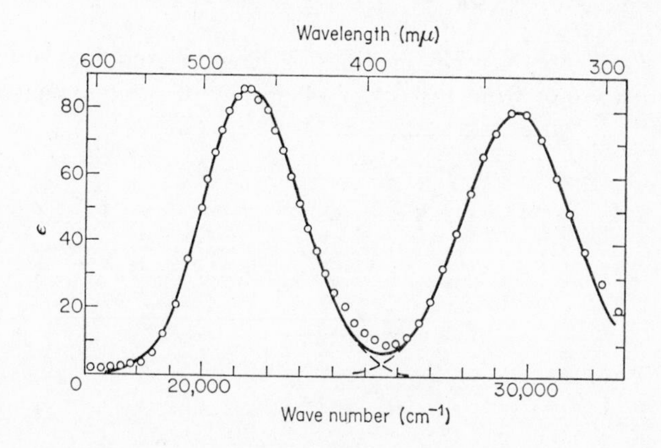

Figure 5. Absorption spectrum of [Co en_3]$^{3+}$. The curves are drawn as gaussian type. [Reproduced, by permission, from *Acta Chem. Scand.*, **8**, 1495 (1954).]

system, transitions should take place keeping the nuclear configuration unchanged (Franck–Condon). Thus all possible transitions between potential energy curves have to be drawn in Figure 6 as vertical lines. If the equilibrium distances of the ground state and the excited state are different, a larger number of excited vibrational states can be reached from the electronic ground state than is the case for equal distances. An example of this situation is shown in Figure 6. The spin-forbidden bands in octahedral d^3 complexes turn out to be narrow and the spin-allowed bands are broadened. The different equilibrium distances for the excited states result from changes in subshell configurations.

The intensity of a band is connected with its extinction coefficient by the relation

$$I = \int \varepsilon(\nu)\, d\nu \tag{9}$$

The integration extends over the whole absorption band. Theoretically, the intensity or the oscillator strength of a dipole transition can be calculated from the wave functions. The probability of absorption between an initial (i) and a final (f) electronic state is given by

$$P_{\text{if}} = \frac{8\pi^3 e^2}{3h^2 c} s_{\text{f}} \mathbf{D}_{\text{if}} \tag{10}$$

where e is the charge of an electron, h is the Planck constant, c is the velocity of light, s_{f} is the degeneracy number of the excited state, and $\mathbf{D}_{\text{if}}$ is the dipole strength

$$\mathbf{D}_{\text{if}} = \left[\int \Psi_{\text{i}}^* \sum_n \mathbf{r}_n \Psi_{\text{f}} \, \mathrm{d}\tau \right]^2 \tag{11}$$

in which Ψ_{i}, Ψ_{f} are the wave functions of the initial and final states and $\mathbf{r}_n$ is the radius vector from the centre of gravity of positive charge to the

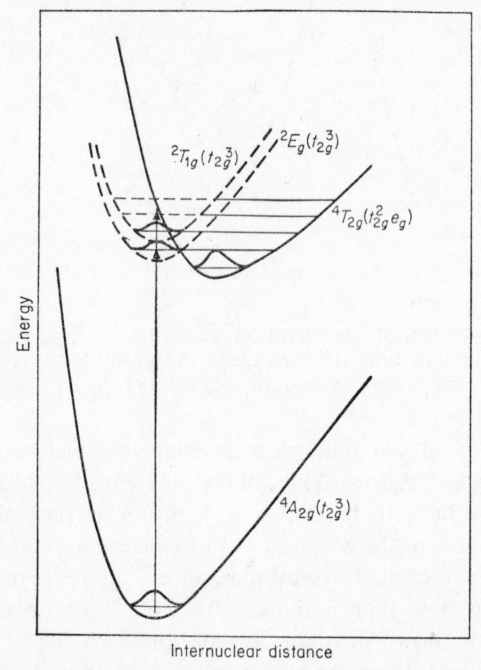

Figure 6. Potential curves for an octahedral d^3 complex.

nth electron. The integral in (11) is also called the transition moment. Instead of using the probability of absorption P_{if}, the oscillator strength f is introduced as measure of intensity by

$$f = \frac{8\pi^2 mc}{3h} \nu s_{\text{f}} \mathbf{D}_{\text{if}} = 1.096 \times 10^{11} \, \nu s_{\text{f}} \mathbf{D}_{\text{if}} \tag{12}$$

where ν is the frequency of absorption. The oscillator strength is related to the intensity defined in (9) by

$$f = 4.315 \times 10^{-9} I \tag{13}$$

In practice, extinction coefficients vary over a range of some millions. Generally, their magnitudes are determined by various selection rules, which are derived from the symmetry properties of the wave functions Ψ in (11).

One property is the inversion symmetry, which is only applicable to molecules which have a centre of symmetry. For this case, all wave functions are either symmetric (g) or antisymmetric (u) under symmetry inversion of the system. Since the radius vector of the dipole radiation has odd (u) symmetry, the integral in (11) is only different from zero if the two wave functions Ψ_i, Ψ_f have opposite inversion symmetry (parity). Therefore, transitions between states having equal inversion symmetry are parity forbidden or, as it is frequently called, Laporte-forbidden.

Another severe restriction which is put on transition probabilities is the multiplicity of the participating states. As the dipole moment vector does not depend on the spin coordinates, it is obvious from orthogonality reasons in the spin functions that only those transition moments are finite which connect functions with equal multiplicity.

A third selection rule is concerned with the symmetry of the space part of the wave functions. As is well known from group theory, an integral is only different from zero if the representation of the integrand contains the identity representation of the certain group in question in the reduction of its direct product. If the wave functions Ψ_i and Ψ_f transform according to the irreducible representations Γ_i and Γ_f, respectively, and the electric dipole moment vector according to Γ, then the product $\Gamma_i \times \Gamma \times \Gamma_f$, representing the integrand of the transition moment, must contain the totally symmetric representation if the transition is symmetry allowed. The electric dipole moment transforms in spherical symmetry according to P, in lower symmetry according to its equivalent irreducible representation, e.g. in octahedral symmetry $\mathbf{O}_h$ to T_{1u}. Using the multiplication table of octahedral symmetry, we find for $\mathbf{O}_h$ the following symmetry-allowed transitions (neglecting inversion symmetry):

$$A_1 \rightarrow T_1 \qquad T_1 \rightarrow A_1, E, T_1, T_2$$
$$A_2 \rightarrow T_2 \qquad T_2 \rightarrow A_2, E, T_1, T_2$$
$$E \rightarrow T_1, T_2$$

At this point the use of polarized crystal spectra as an aid to band assignment must be emphasized. In this case, the transition moment (11) contains the appropriate component of the vector $\mathbf{r}$, so that the oscillator strength f (12) may become anisotropic for certain electronic states Ψ_f. The crystal spectrum then has different band intensities for polarized light which is oriented in different directions with respect to the crystal.

In practice, it appears that these selection rules are not fulfilled in their strict sense. We always find bands which according to the selection rules are forbidden and actually should not be there at all. These forbidden bands nevertheless appear as weak absorptions in the spectra. In particular, more complicated molecules do not seem to obey the selection rules; there are several reasons for this. We have to consider not only electric dipole radiation but also magnetic dipole and electric quadrupole radiation, whose irreducible representations Γ are different from that of the dipole moment. Moreover, the vibrational part of nuclear motion has to be in principle included in the wave functions, thus changing the parity character of the total function. By an intermixing of odd vibrations into the electronic states, a gain of intensity is achieved which gives rise to higher absorption. Also the multiplicity rule could partly break down when spin–orbit coupling effects are appreciable. Spin-forbidden transitions may 'borrow' intensity by intermixing different multiplets by virtue of spin–orbit interaction. Furthermore, second-order effects, to which one may attribute physical significance, are also likely to operate, but it proves to be difficult to decide and distinguish between their possible causes.

II. THEORY

A. Ligand-field Bands

1. Crystal-field theory

The development and application of crystal-field theory, together with its different extensions, has been in recent years the subject of extensive research in the field of inorganic chemistry and solid-state physics. Several books[23-27] and review articles[28-30] have appeared describing in detail the original ideas of Bethe[31,32], Kramers[33], and Van Vleck[34,35] and the stimulating development after the Second World War. For this reason, we present only the basic ideas and formalisms which are currently accepted by inorganic chemists.

Crystal-field theory deals with the electronic structure of the outer-shell electrons of transition metal and rare earth salts in solutions or in the crystalline state.

The theory can be handled semiempirically where its parameters are determined experimentally or it can also be interpreted, with considerably less success, purely theoretically; in this case its parameters are calculated using atomic functions of the central atom. Crystal-field theory as a model helps us to gain insight into the physical properties of sets of inorganic compounds and should be appreciated as such instead of considering it as a tool for the accurate reproduction of experimental numbers.

The idea is that only nearest atomic neighbours of a central atom or ion are considered, as also proposed by the usual bonding scheme. For coordination compounds, only the interaction between the central ion and its ligands is taken into account and other effects, like ligand–ligand and next-nearest neighbour interactions, are neglected. Furthermore, this interaction is assumed to be weak and essentially electrostatic in character so that it can be mathematically described as a perturbation of the central atom electronic levels by a point-charge field having the symmetry of the nuclear framework of the ligand. The restriction to purely electrostatic forces does not imply that crystal-field theory can only describe ionic bonding effects. Partial delocalization of bonded electrons is also contained in this simple model when second-order perturbation effects are considered. As we will see later, however, it is the inclusion of symmetry properties that constitutes the success of crystal-field theory and which is in common for both the model and the real system.

The total hamiltonian for the electrons of the metal ion is given by

$$\mathscr{H} = -\frac{\hbar^2}{2m}\sum_i \nabla_i^2 - \sum_i \frac{Ze^2}{r_i} + \sum_i \xi(r_i)(\mathbf{l}_i \cdot \mathbf{s}_i) + \sum_{i>j} \frac{e^2}{r_{ij}} + V \qquad (14)$$

where the different terms describe the kinetic energy, the electron-core potential, the spin–orbit coupling, the electron–electron repulsion, and the crystal-field potential, respectively. The summation i goes over the electron number. Except for the electron-repulsion term, the Schrödinger equation with its hamiltonian $\mathscr{H} = \sum_i \mathscr{H}_i$ is separable with respect to the coordinates of different electrons, in which case the system can be handled as a one-electron problem. Comparing the magnitude of the expectation values for the last three terms of (14), we distinguish between three cases

1. $\xi(r)(\mathbf{l}\cdot\mathbf{s}) < V < e^2/r_{ij}$ Russell–Saunders coupling, weak-field case
2. $\xi(r)(\mathbf{l}\cdot\mathbf{s}) < e^2/r_{ij} < V$ Russell–Saunders coupling, strong-field case
3. $V < \xi(r)(\mathbf{l}\cdot\mathbf{s})$ intermediate and (j, j) coupling, weak or medium field

High-spin transition metal complexes belong in particular to the first group, the second group is exemplified by low-spin complexes, and the third by rare earth and actinide complexes.

We shall now formulate the model using octahedral symmetry as an example. Suppose we put six charges ϵ at a distance d from the origin on the axes of a coordinate system as shown in Figure 7. The point charges occupy the respective vertices of an octahedron. An electron with charge e interacts with this electric field via the potential energy

$$V = \sum_{n=1}^{6} V_n = \sum_{n=1}^{6} \frac{e\epsilon}{r_n} \tag{15}$$

where r_n is the distance from the nth ligand to the electron. Expanding each of the partial potential energies V_n from the origin of the coordinate system, where the central atom is located, we get for $r < d$, e.g.

$$V_3 = \frac{e\epsilon}{r_3} \frac{e\epsilon}{\sqrt{r^2 + d^2 - 2dr\cos\vartheta}} = e\epsilon \sum_{l=0}^{\infty} \frac{r^l}{d^{l+1}} P_l(\cos\vartheta) \tag{16}$$

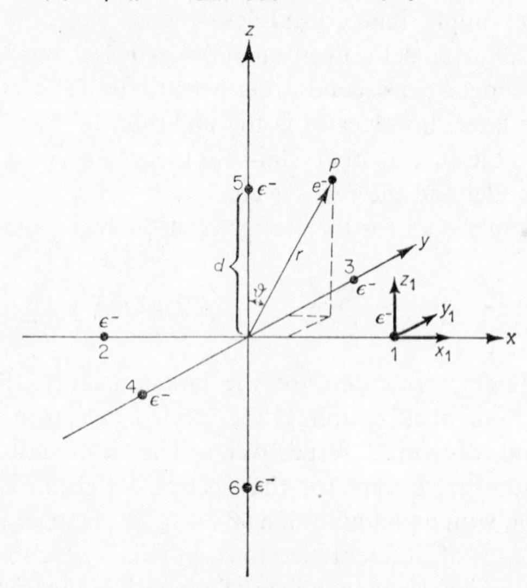

Figure 7. Choice of the coordination system for an octahedral point-charge model.

with the Legendre polynomials P_l. Adding all contributions to (15) and neglecting terms higher than $l = 4$ in the summation (16), the octahedral potential energy is given by

$$V = e\epsilon\left[\frac{6}{d} + \frac{35}{4d^5}(x^4 + y^4 + z^4) - \frac{21}{4}\frac{r^4}{d^5}\right] \tag{17}$$

The calculation of higher terms in the expansion series is not necessary when the perturbation is applied to d electrons. It can be shown by group theory that the integrals arising from higher contributions vanish. The expansion of the coordinate system from the origin of the central atom is useful because of the reduction of the molecular system to a one-centre problem. The octahedral term of the expansion (17) is given in the literature[36] also in the form

$$V_{\text{oct}} = D(x^4 + y^4 + z^4 - \tfrac{3}{5}r^4) \tag{18}$$

with the Schlapp and Penney[36] parameter

$$D = \frac{35}{4}\frac{e\epsilon}{d^5} \tag{19}$$

The same potential expressed in normalized spherical harmonics Y_l^m is

$$V_{\text{oct}} = \frac{4\sqrt{\pi}}{15} Dr^4 \left\{ Y_4^0 + \sqrt{\frac{5}{14}}\,[Y_4^4 + Y_4^{-4}] \right\} \tag{20}$$

This form is particularly useful when evaluating the integrals, since the hydrogenic wave functions are also given in terms of Y_l^m. By far the largest energy contribution to the octahedral field potential results from the constant term in (17). This is positive and shifts all energy levels equally in the term scheme. Therefore, it is negligible for spectroscopic and certain other applications but it is nevertheless important for other properties (e.g. thermodynamic heat of formation). The energy levels in the complex are now calculated by performing a first-order perturbation calculation using (18) or (20) as perturbation operator. We know from group theory that an atomic d level splits into a two-fold e_g and a three-fold t_{2g} degenerate level when the symmetry of the field is reduced to octahedral symmetry

$$d \rightarrow e_g + t_{2g} \tag{21}$$

When d functions are adapted to the symmetry $\mathbf{O}_h$ (symmetry orbitals), they transform as

$$e_g\colon\ d_{z^2}, d_{x^2-y^2} \qquad t_{2_g}\colon\ d_{xy}, d_{xz}, d_{yz} \tag{22}$$

Since off-diagonal elements of the perturbation matrix between different d orbitals vanish for symmetry reasons, we get immediately the energy shifts due to the octahedral field from the integrals

$$\begin{aligned} \Delta E_{e_g} &= \int (d_{x^2-y^2})^2 V_{\text{oct}} r^2 \sin\vartheta \,\mathrm{d}r\,\mathrm{d}\vartheta\,\mathrm{d}\varphi \\ \Delta E_{t_{2g}} &= \int (d_{xy})^2 V_{\text{oct}} r^2 \sin\vartheta \,\mathrm{d}r\,\mathrm{d}\vartheta\,\mathrm{d}\varphi \end{aligned} \tag{23}$$

These integrals are separable into a radial and an angular part, so that we can write

$$\Delta E_{e_g} = 6Dq \quad \text{and} \quad \Delta E_{t_{2g}} = -4Dq \tag{24}$$

where the radial integral is abbreviated by

$$q = \frac{2}{105} \int R_{nd}^2(r) r^4 r^2 \, dr \tag{25}$$

and R_{nd} is the normalized radial part of the central atom function, being equal for all *nd* orbitals. Since q always appears multiplied by D, the product Dq is used as a ligand-field parameter, which in the semiempirical theory can be determined experimentally. Instead of Dq, the separation $\Delta = E_1 - E_2$ of the e_g and t_{2g} levels is introduced by different authors[23,37,38] as a parameter.

$$\Delta = \Delta E_{e_g} - \Delta E_{t_{2g}} = E_1 - E_2 = 10Dq = \frac{5}{3} \frac{e\epsilon}{d^5} \int R_{nd}^2(r) r^4 r^2 \, dr \tag{26}$$

For positive D, i.e. negative ligand point charges ϵ, we see from (24) that the e_g orbital energy is higher than the t_{2g} energy. This is also found for non-ionic ligands where electric dipole charges are at the positions of the ligands[37,38]. For a detailed analysis the reader must refer to the literature.

For the general case, we develop the crystal-field energy in normalized spherical harmonics

$$V = e\epsilon \sum_i \sum_l \sum_{m=-l}^{+l} \frac{4\pi}{2l+1} \frac{r_<^l}{r_>^{l+1}} Y_l^{m*}(\vartheta, \varphi) Y_l^m(\vartheta_i, \varphi_i) \tag{27}$$

where the summation i goes over the positions ϑ_i, φ_i of the ligands and $r_<$ is the smaller and $r_>$ the larger of the two distances d and r defined in Figure 7.

For tetrahedral symmetry, we expect from group theory the same splitting scheme

$$d \rightarrow e + t_2 \tag{28}$$

The splitting parameter can be determined easily as follows. Let us insert in the matrix element (23) the function d_{z^2} having $m_l = 0$, when its angular part $\mathbf{a}_{00}$ becomes

$$\mathbf{a}_{00} = \sum_i \sum_l \sum_{m=-l}^{l} \int [Y_2^0(\vartheta, \varphi)]^2 Y_l^{m*}(\vartheta, \varphi) Y_l^m(\vartheta_i, \varphi_i) \sin \vartheta \, d\vartheta \, d\varphi \tag{29}$$

For matrix elements diagonal in l, m_l only those terms in the sum over m which have $m_l = 0$ do not vanish. Furthermore, we know that only fourth-order terms appear in cubic symmetry so that (29) is reduced to

$$\mathbf{a}_{00} = \sum_i \int [Y_2^0(\vartheta, \varphi)]^2 Y_4^0(\vartheta, \varphi) Y_4^0(\vartheta_i, \varphi_i) \sin \vartheta \, d\vartheta \, d\varphi \tag{30}$$

When relating the energy of the d_{z^2} orbital in a tetrahedron to that in an octahedron, we get for the case of equal radial parts $R_{nl}(r)$ of the functions and equal charges ϵ on the ligand positions:

$$\frac{\mathbf{a}_{00}^{\text{tetr}}}{\mathbf{a}_{00}^{\text{oct}}} = \frac{\sum_{i=1}^{4} Y_4^0(\vartheta_i)}{\sum_{i=1}^{6} Y_4^0(\vartheta_i)} \tag{31}$$

For an octahedron, the ϑ_i position of the ligands are

$$\cos \vartheta_5 = 1, \quad \cos \vartheta_6 = -1, \quad \text{and} \quad \cos \vartheta_i = 0$$

for $i = 1, 2, 3, 4$, using the numbering in Figure 7, and for a tetrahedron

$$\cos \vartheta_j = \pm \tfrac{1}{3}$$

for $j = 1, 2, 3, 4$.

The φ_i coordinate positions are irrelevant, since Y_4^0 does not depend on φ_i. With these numbers, (31) becomes

$$\frac{\mathbf{a}_{00}^{\text{tetr}}}{\mathbf{a}_{00}^{\text{oct}}} = -\frac{4}{9} \tag{32}$$

Because of the baricentre rule of perturbation, each t_{2g} orbital energy will also follow the same property. Consequently the corresponding relation is also valid for their crystal-field parameters

$$\frac{\Delta_{\text{tetr}}}{\Delta_{\text{oct}}} = -\frac{4}{9} \tag{33}$$

In this way, corresponding fractions can be derived for other cubic symmetries.

Tetragonal and trigonal ligand fields are derived by superimposing an axial potential along the four-fold axis or the three-fold axis of an octahedral potential

$$\begin{aligned} \mathbf{D}_{4h}&: \quad V_{\text{tetr}} = V_{\text{oct}} + V_{\infty} \\ \mathbf{D}_{3d}&: \quad V_{\text{tri}} = {}^{\text{tri}}V_{\text{oct}} + V_{\infty} \end{aligned} \tag{34}$$

In ${}^{\text{tri}}V_{\text{oct}}$, we choose the trigonal axis of the octahedron to coincide with the z axis of the coordinate system[39]

$${}^{\text{tri}}V_{\text{oct}} = \frac{8}{45}\sqrt{\pi}\, Dr^4 \left\{ Y_4^0 + \sqrt{\frac{10}{7}}\, [Y_4^3 - Y_4^{-3}] \right\} \tag{35}$$

The axial potential in $\mathbf{D}_{\infty h}$ symmetry is

$$V_{\infty} = 4\sqrt{\pi} \left\{ \frac{1}{\sqrt{5}} B_2 r^2 Y_2^0 + \frac{4}{3} B_4 r^4 Y_4^0 \right\} \tag{36}$$

For a square planar complex ($\mathbf{D}_{4h}$ symmetry), the ligand-field potential energy is

$$V_{\rm pl} = -4\sqrt{\frac{\pi}{5}}\,B_2 r^2 Y_2^0 + 4\sqrt{\pi}\,B_4 r^4 \left\{ Y_4^0 + \frac{1}{3}\sqrt{\frac{35}{2}}\,[Y_4^4 - Y_4^{-4}] \right\} \quad (37)$$

In a point-charge model, the B_2 and B_4 parameters in (36) and (37) are

$$B_2 = \frac{e\epsilon_i}{d_i^3}; \qquad B_4 = \frac{e\epsilon_i}{4d_i^5} \quad (38)$$

where d_i is the distance from the central atom to the ith ligand of charge ϵ_i.

In order to calculate the orbital energies in $\mathbf{D}_{4h}$ symmetry, we again separate the integrals of the perturbation matrix elements into a radial and an angular part. When reducing the symmetry from octahedral to tetragonal, we expect a further splitting of the orbital levels, viz.

$$\mathbf{O}_h \rightarrow \mathbf{D}_{4h} \begin{cases} e_g \rightarrow a_{1g} + b_{1g} \\ t_{2g} \rightarrow b_{2g} + e_g \end{cases} \quad (39)$$

When evaluating the integrals over the angular parts, the energies perturbing the octahedral levels are given by

$$\begin{aligned} \Delta E_{a_{1g}} &= -2Ds - 6Dt \\ \Delta E_{b_{1g}} &= 2Ds - Dt \\ \Delta E_{b_{2g}} &= 2Ds - Dt \\ \Delta E_{e_g} &= -Ds + 4Dt \end{aligned} \quad (40)$$

In these relations, the integrals over the radial part are denoted by s and t corresponding to those of the octahedral case (25)[25,29]. These parameters, multiplied by D, are observables and can be used together with Dq as parameters to determine the orbital-energy differences in tetragonal symmetry.

In Figure 8, various symmetry cases are compiled to show the splitting of d-orbital energy levels in a semiquantitative way. Although the exact distances in the term scheme are determined by the actual values of the crystal-field parameters, the order of the levels and their relative splittings is usually that represented in the figure. For small deviations from octahedral symmetry, it is important to know in which sense the distortion operates. In the figure, an example is shown for $\mathbf{C}_{4v}$ and $\mathbf{D}_{4h}$ where the Y ligands interact more weakly with the central atom than the X ligands. This effect is described in the electrostatic picture by a larger distance $d_{\rm Y}$ or a smaller ligand charge $\epsilon_{\rm Y}$ of the B parameters (38) of the ligand Y.

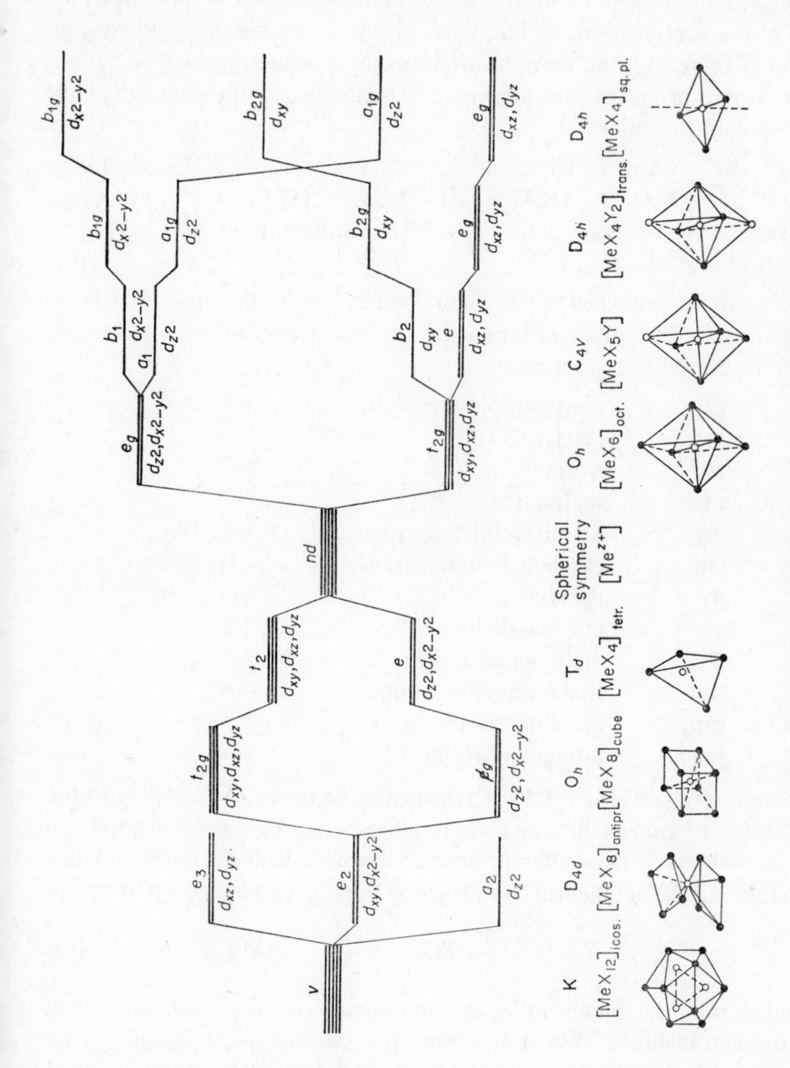

Figure 8. *d*-Orbital splitting scheme for different symmetries.

The influence of the ligand on spectroscopic properties of the complex had been investigated before the development of ligand-field theory. Fajans[40] and Tsuchida[41] observed that the replacement of I^- by Br^-, Cl^- by H_2O, etc., in a coordination compound of a certain symmetry shifts the band system to shorter wavelength. This effect is quite independent of the central atom or ion, particularly in octahedral symmetry, no matter if there are one or more *d* electrons present. A series of ligands giving rise to progressively shorter wavelength absorption is known as the spectrochemical series:

I^-, Br^-, $OCrO_3^-$, Cl^-, SCN^-, N_3^-, dtp^-, F^-, SSO_3^{2-}, urea, OCO_2^{2-}, OCO_2R^-, ONO^-, OH^-, OSO_3^{2-}, ONO_2^-, ox^{2-}, H_2O, mal^{2-}, NCS^-, gly^-, $enta^{4-}$, py, NH_3, en, dien, tren, SO_3^{2-}, dip, phen, NO_2^-, $C_5H_5^-$, CN^- (41)

The ligands are attached to the central metal ion by the atom first listed in each case. The abbreviations used and the atoms by which they are linked to the central atom are:

dtp^-	diethyldithiophosphate	2-S
urea	$(NH_2)_2CO$	1-O
ox^{2-}	oxalate	2-O
mal^{2-}	malonate	2-O
gly^-	aminoacetate, glycinate	1-O, 1-N
$enta^{4-}$	ethylenediaminetetraacetate	4-O, 2-N
py	pyridine	1-N
en	ethylenediamine	2-N
dien	diethylenediamine	3-N
tren	tris(aminoethyl)amine	4-N
dip	α,α'-dipyridyl	2-N
phen	*o*-phenanthroline	2-N

(42)

The same series is also valid for tetrahedral or lower symmetries. In the latter case, it happens that the wave numbers σ of the absorption spectrum of a complex with two different ligands can be calculated approximately from those of the octahedral complexes by the rule of average environment

$$\sigma(MX_nY_{6-n}) \simeq \frac{n}{6}\sigma(MX_6) + \frac{6-n}{6}\sigma(MY_6) \tag{43}$$

provided that the bands in each compound arise from corresponding electronic transitions. We notice that the position of a ligand in the spectrochemical series is mainly determined by the atom that is linked to the central ion. A corresponding series for these atoms binding the ligands to the central metal is approximately

$$\text{I, Br, Cl, S, F, O, N, C} \tag{44}$$

If we neglect electron interaction in central atoms with more than one d electron, the first transition σ observed in the spectrum is due to a transition between the orbital states split by the crystal field. For octahedral symmetry, this is the transition from the t_{2g} to the e_g subshell; its energy in wave numbers is equal to the crystal-field parameter Δ. Instead of using the wave number of the first band when establishing the spectrochemical series (41) or the rule of average environment (43), their crystal-field parameters also serve as spectrochemical classification of the ligands. As a matter of fact, ordering according to crystal-field parameters is preferred because it sorts out various effects of electron interaction.

Using the expression Δ for a point-charge model of octahedral symmetry (26), the spectrochemical series (44) for atoms linked to the central ion is rationalized if we bear in mind that for a given metal this series runs parallel with decreasing metal–ligand distance. Deviations from this general rule are explained by steric hindrance or π-bonding effects, which is treated in more detail in the molecular-orbital section. These effects, however, can only be understood by a more general theory, in which the ligands do not consist of simple point charges but also have more elaborate electronic structure.

One use of the spectrochemical series is for assigning linkage isomers[41,42]. Ambidentate ligands like SCN^- and NO_2^- are found on different positions in the spectrochemical series, depending on the atom by which they are linked to the central ion. Therefore they give rise to different ligand-field transitions which can be used for the determination of the possible isomers.

If we compare the crystal-field parameters of different central ions which are to some extent ligand dependent, some regularities are observed. A corresponding spectrochemical series of increasing crystal-field parameters for the central ions is given after Jørgensen[26] by

$$\mathrm{Mn^{II}, Ni^{II}, Co^{II}, Fe^{II}, V^{II}, Fe^{III}, Cr^{III}, V^{III}, Co^{III}, Mn^{IV}, Mo^{III},}\\ \mathrm{Rh^{III}, Ru^{III}, Pd^{IV}, Ir^{III}, Re^{IV}, Pt^{IV}} \qquad (45)$$

We notice that higher oxidation numbers give rise to larger crystal-field parameters as is predicted from a simple point-charge model. Elements of higher transition series also tend to have larger ligand-field parameter values. This is not so easy to understand from the simple model. Spectrochemical series of ligands in compounds of lower symmetry can be constructed when we use the octahedral crystal-field parameter as an approximation. In fact, Tsuchida has not only compared octahedral MX_6 complexes in constructing his spectrochemical series but he also used the spectra of pentaammines MX_5Y or tetraammines MX_4Y_2 together with the rule of average environment (43).

Deviations from octahedral symmetry may sometimes disturb the order of the ligands in the spectrochemical series. This is especially the case for Cr^{II} or Cu^{II} complexes. The reduction of symmetry may be due to the Jahn–Teller[43–45] effect. According to this theorem, the nuclear configuration of a non-linear molecule is not stable if its electronic ground state is orbitally degenerate. The stable form is one of lower symmetry in which all orbital degeneracy is abolished. A complex compound with a d^1 central ion electron configuration, for instance, should not exist either in octahedral or in tetrahedral environment. The symmetry is rather reduced to trigonal, tetragonal, or even rhombic symmetry in which orbital degeneracy of the ground level is removed. It is also possible for this distortion to oscillate within a molecule in different directions (dynamical Jahn–Teller effect), so that statistically no deviation from the higher symmetry is observed. Also static distortions due to the normal Jahn–Teller effect are sometimes so small that they can hardly be detected. Experience shows that distortions are nearly negligible for T_{1g} and T_{2g} octahedral states while they are important for E_g ground states (e.g. d^9 configurations).

2. Spin–orbit coupling

An additional interaction between the electron and the nucleus is due to spin forces. Its energy, due to relativistic effects, is more pronounced in heavy metal ions, where the interaction between the orbital and the spin angular momentum is larger. The operator of the spin–orbit interaction for atoms (spherical symmetry) is

$$\mathscr{U} = -\frac{e}{2m^2c^2}\,\frac{1}{r}\,\frac{\mathrm{d}V}{\mathrm{d}r}\,(\mathbf{l}\cdot\mathbf{s}) \tag{46}$$

where c is the velocity of light, m is the electron mass, and V is the scalar potential of the central field. The integral over the radial parts of the operator (46) and the wave functions $R(r)$ is usually taken as the coupling constant (Landé parameter)

$$\zeta_{nl} = -\frac{e}{2m^2c^2}\int\frac{1}{r}\,\frac{\mathrm{d}V}{\mathrm{d}r}\,R_{nl}^2(r)r^2\,\mathrm{d}r \tag{47}$$

General calculation procedures for matrix elements of the energy matrix including the operator (46) are given in the literature[25,46,47]. Some experimental ζ_{nl} parameters for transition group elements are listed by Dunn[48]. According to the magnitude of this parameter, one distinguishes between three different coupling cases. In Russell–Saunders coupling, the electron-repulsion energy is large compared with the spin–orbit interaction. On the other hand, spin–orbit forces are more pronounced in (j, j) coupling than electron repulsions. If the two effects are comparable in magnitude, an intermediate coupling case exists where matrix elements have to be considered mixing different multiplets (L, S), so that L and S lose their property of being good quantum numbers.

For molecules, the spin–orbit coupling operator consists of a spherical part (46) and an additional one which has the symmetry of the molecule but was shown[39] to be in general negligibly small. It is therefore sufficient to use the operator $\zeta_{nl}(\mathbf{l}\cdot\mathbf{s})$ or its equivalent $\zeta'_{nl}(\mathbf{t}\cdot\mathbf{s})$ in lower symmetry transformations when calculating the energy levels of a molecule including spin–orbit coupling. In Figure 9, a term-splitting scheme is shown which

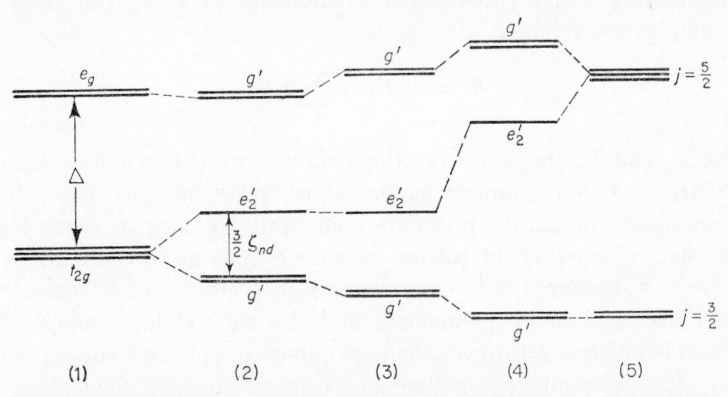

Figure 9. Schematic d-orbital splitting in octahedral symmetry for different coupling cases: (1) strong octahedral fields, (2) weak spin–orbit coupling, (3) intermediate spin–orbit coupling, (4) (j,j) coupling and octahedral field, (5) free-ion (j,j) coupling.

presents schematically all coupling cases for a single d electron in octahedral symmetry.

3. *Molecular-orbital theory*

The molecular orbital (MO) theory today is considered the more appropriate tool to describe the electronic structure of coordination compounds. This arises from the fact that in this theory the orbital structure of the ligands is also taken into consideration. The atomic overlap in metal–ligand bonds allows the d electrons to penetrate from the central atom to the ligand. The direct evidence for this electron delocalization is given by electron spin resonance data. The hyperfine structure of the absorption band due to the unpaired electron in the spectrum of $[IrCl_6]^{2-}$ has been explained by assuming that this electron is only about 70% a $5d$ electron and 30% of the time it is localized on the ligands[49]. Other evidence is found by nuclear magnetic resonance studies. The ^{19}F resonance in a crystal of $KNiF_3$ indicates[1] that, even for compounds in which ionic bonds are predominant, d-electron delocalization is appreciable, so that a molecular-orbital model is necessary for a satisfactory description of the experiments. Indirect evidence for significant covalent bonding in complex compounds is given by the relatively large intensities of the ligand-field absorption bands. The gain in oscillator strength cannot be fully explained by an interaction of the d functions with vibrational wave functions alone. A mixing of ligand functions into the central-ion functional set can raise the intensity of an absorption band by one order of magnitude[50].

The molecular orbitals are generally approximated by a linear combination of atomic orbitals (MO–LCAO) as originally proposed by Van Vleck[51,52] for complex compounds:

$$\Psi = a_A \Psi_A + \sum_L b_L \Psi_L \tag{48}$$

where Ψ_A and Ψ_L are atomic functions of the central atom and the ligands, respectively. For a complete mathematical treatment, this theory has a serious disadvantage: the necessary computational effort is much higher than that of crystal-field theory because of the many-centre integrals involved. Considered as a semiempirical method, molecular-orbital theory needs so many parameters that the method loses much of its physical significance. However, there are numerous approximations possible which are physically acceptable and help to simplify the calculation considerably.

The simplified molecular-orbital theory, as it is generally applied to coordination compounds, has much in common with ordinary Hückel theory which also neglects all interatomic interactions of non-bonded atoms and is based on an effective one-electron hamiltonian. While Hückel's π-electron theory only considers the bonding of similar atoms (carbon) with one orbital (p_π orbital) per atom, the MO theory of coordination compounds is more complicated, since it has to deal with the interaction of different atoms with several orbitals per atom and the orbitals also could be degenerate. As in Hückel theory, we can use the symmetry of the molecule to find suitable orbitals which reduce the secular equation to lower degrees and we may consider or neglect the orbital overlap.

The basic concepts of MO theory can be mentioned here only briefly; for a detailed description of the method and its application to coordination compounds, we will refer to the literature[25,26,30,53], give only some introductory remarks, and present some new approaches which are not found in the textbooks.

We shall again exemplify the situation by looking at a transition group central ion in octahedral coordination as we did above when discussing crystal-field theory. The basis of interacting orbitals is limited to the outer-shell atomic functions of the central atom and the ligands. From the central atom, not only the five *nd* functions as in crystal-field theory but also the $(n + 1)s$ and the three $(n + 1)p$ functions are taken into account if they are sufficiently stable. From the ligands, only σ- and π-bonding *p* functions are added to the orbital basis. The correct linear combinations adapted to octahedral symmetry are, for σ- and π-bonding types:

σ bonds:

$$a_{1g} \quad \frac{1}{\sqrt{6}}(\sigma_1 + \sigma_2 + \sigma_3 + \sigma_4 + \sigma_5 + \sigma_6)$$

$$e_g \quad \frac{1}{2}(\sigma_1 + \sigma_2 - \sigma_3 - \sigma_4)$$

$$\frac{1}{\sqrt{12}}(\sigma_1 + \sigma_2 + \sigma_3 + \sigma_4 - 2\sigma_5 - 2\sigma_6)$$

$$t_{1u} \quad \frac{1}{\sqrt{2}}(\sigma_1 - \sigma_2); \quad \frac{1}{\sqrt{2}}(\sigma_3 - \sigma_4); \quad \frac{1}{\sqrt{2}}(\sigma_5 - \sigma_6)$$

π bonds:

$$t_{1g} \quad \tfrac{1}{2}(z_3 - z_4 - y_5 + y_6); \quad \tfrac{1}{2}(z_1 - z_2 - x_5 + x_6); \quad \tfrac{1}{2}(y_1 - y_2 - x_3 + x_4)$$

$$t_{1u} \quad \tfrac{1}{2}(x_3 + x_4 + x_5 + x_6); \quad \tfrac{1}{2}(y_1 + y_2 + y_5 + y_6); \quad \tfrac{1}{2}(z_1 + z_2 + z_3 + z_4)$$

$$t_{2g} \quad \tfrac{1}{2}(z_3 - z_4 + y_5 - y_6); \quad \tfrac{1}{2}(z_1 - z_2 + x_5 - x_6); \quad \tfrac{1}{2}(y_1 - y_2 + x_3 - x_4)$$

$$t_{2u} \quad \tfrac{1}{2}(x_3 + x_4 - x_5 - x_6); \quad \tfrac{1}{2}(y_1 + y_2 - y_5 - y_6); \quad \tfrac{1}{2}(z_1 + z_2 - z_3 - z_4) \qquad (49)$$

Here the numbering and *p*-orbital positions of Figure 7 are used. The normalization factors hold for zero ligand–ligand orbital overlap. The symmetry orbitals can be derived from group theory. The character tables of the groups also allow us to determine the irreducible representations for atomic orbitals of lower symmetry and their σ- and π-bonding type. They are for the central atom:

σ bonds:	π bonds:
$s(a_{1g})$	$p(t_{1u})$
$p(t_{1u})$	$d(t_{2g})$
$d(e_g)$	

(50)

Corresponding tables that are valid for other symmetries with various coordination numbers are given by Eisenstein[54]. Figure 10 shows the possible combinations of symmetry functions of the ligands on one side and the central atom on the other. This molecular-orbital term scheme is a qualitative presentation and gives only the approximate order of the energy levels. We notice that the lowest unoccupied orbitals t_{2g}^* and e_g^* are pure antibonding π and σ orbitals, respectively, which correspond in crystal-field theory to the separated t_{2g} and e_g levels originating from the central-ion d orbitals.

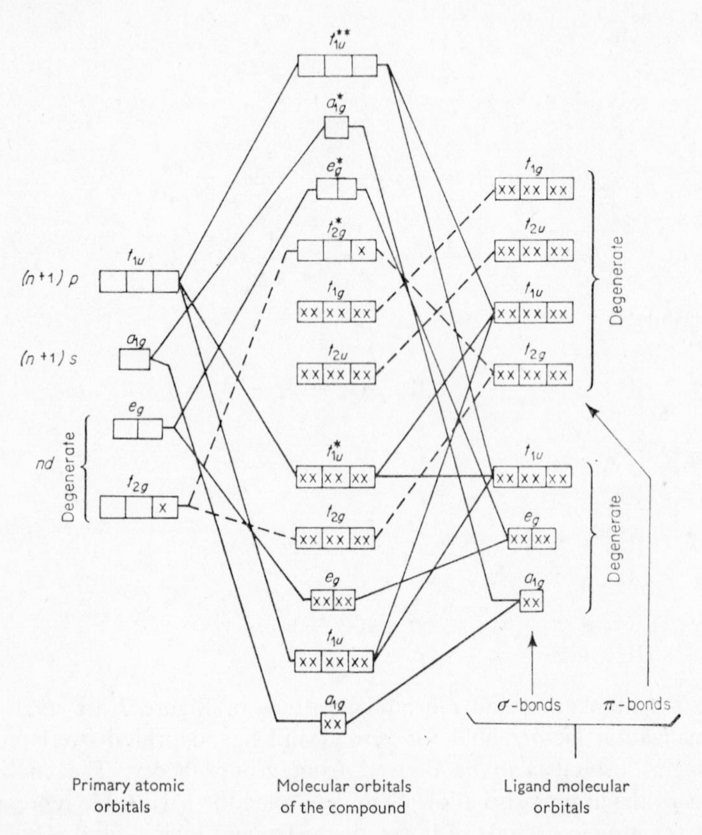

Figure 10. Molecular-orbital combination diagram for $\mathbf{O}_h$ symmetry.

The orbital-energy levels are calculated by a variational procedure, which also determines those coefficients of the linear combination (48) that are not given by symmetry. Following the usual course, the method leads to a secular equation

$$|H_{ij} - S_{ij}E| = 0 \tag{51}$$

with the integrals

$$H_{ij} = \int \Psi_i^* \mathscr{H} \Psi_j \, d\tau, \qquad S_{ij} = \int \Psi_i^* \Psi_j \, d\tau$$

For interacting atoms or ions with very different orbitals energies, the secular equation (51) can be simplified. We will show this for a simple example of two orbitals Ψ_A, Ψ_L belonging to the central atom A and the ligand L whose atomic orbital energies E_A, E_L are very different. Substituting the functions Ψ_A and another Ψ_L^0 which is orthogonal to Ψ_A in the secular equation where

$$\Psi_L^0 = (1 - S_{AL}^2)^{-\frac{1}{2}}(\Psi_L - S_{AL}\Psi_A); \; S_{AL} = \int \Psi_A^* \Psi_L \, d\tau$$

and solving the corresponding determinant, we find that one of its roots is given by

$$E_1^A = \frac{H_{AA}}{1 - S_{AL}^2} + \frac{H_{AL}^2 - H_{AL}^2 S_{AL}^2 - 2H_{AA}H_{AL}S_{AL} + H_{AA}H_{LL}S_{AL}^2}{(1 - S_{AL}^2)(H_{AA} - H_{LL} + 2S_{AL}H_{AL})} \tag{52}$$

if the second quotient is small compared with one. This condition is fulfilled when assuming weakly interacting orbitals

$$H_{AL} \ll H_{AA} - H_{LL} \tag{53}$$

In (52), several terms can be neglected since the following order of magnitude is generally obeyed

$$H_{AA}, H_{LL}, 1 > H_{AL}, S_{AL} > H_{AL}^2, H_{AL}S_{AL}, S_{AL}^2 \tag{54}$$

To a good approximation, E_1^A then simplifies to

$$E_1^A = H_{AA} + \frac{H_{AL}^2 - 2H_{AA}H_{AL}S_{AL} + H_{AA}H_{LL}S_{AL}^2}{H_{AA} - H_{LL}} \tag{55}$$

Moreover, we take advantage of the possibility to resolve the molecular hamiltonian into an atomic part $\mathscr{H}_A$ and the crystal-field potential part $\mathscr{V}$

$$\mathscr{H} = \mathscr{H}_A + \mathscr{V}$$

so that the orbital-energy change with respect to E_A is given by

$$\Delta E_A = E_1^A - E_A = \int \Psi_A^* \mathscr{V} \Psi_A \, d\tau + \frac{H_{AL}^2 - 2E_A H_{AL} S_{AL} + E_A E_L S_{AL}^2}{E_A - E_L} \tag{56}$$

where the perturbation $\int \Psi_A^* \mathscr{V} \Psi_A \, d\tau$ can be neglected in comparison with the atomic energies E_A and E_L which are at least an order of magnitude larger. For more than one ligand, the following formula results

$$\Delta E_A = \int \Psi_A^* \mathscr{V} \Psi_A \, d\tau + \sum_{L \neq A} \frac{H_{AL}^2 - 2E_A H_{AL} S_{AL} + E_A E_L S_{AL}^2}{E_A - E_L} \tag{57}$$

This expression, although it has some formal similarity, is not equivalent with the normal Rayleigh–Schrödinger perturbation theory, because it contains also orbitals of the perturbing atom L. It can, however, be derived from the Brillouin–Wigner perturbation formula if overlap effects are included. The first term in (57) describes crystal-field bonding effects and the second-order terms represent the weak covalent part of the bond. The first integral corresponds in Hückel theory to the Coulomb integral α, in the second term we find H_{AL} to be equivalent to the Hückel resonance integral β. E_A or E_L would be, in the simple π-electron theory, atomic p-orbital energies in the trigonal valence state.

The crystal-field parameter Δ, which in MO theory is an orbital-energy difference now called the ligand-field parameter, is given for octahedral symmetry by

$$\Delta = E_{e_g^*} - E_{t_{2g}^*} \tag{58}$$

where the orbital energies are determined by the use of (55) or (57). For tetrahedral symmetry, a relation for the ligand-field parameters can be derived as in (33) where now weak covalent bonding is included by means of (57). From the reduction Table 1 for the tetrahedral group $\mathbf{T}_d$

Table 1. Reduction table of group $\mathbf{T}_d$ with four tetrahedral bonds (degeneracy numbers in parentheses).

		a_1 (1)	a_2 (1)	e (2)	t_1 (3)	t_2 (3)
	s (1)	1	0	0	0	0
Central	p (3)	0	0	0	0	1
atom	d (5)	0	0	1	0	1
	f (7)	1	0	0	1	1
Ligands	σ (1)	1	0	0	0	1
	π (2)	0	0	1	1	1

with four tetrahedral bonds, we see that the e orbital has pure π-antibonding character while the other split d orbital t_2 has σ- and π-antibonding character. When resolving a t_2 orbital with respect to a corner of a tetrahedron (y' axis pointing to a corner)

$$d_{yz} = -\tfrac{2}{3}d'_{x^2-y^2} - \sqrt{\tfrac{1}{18}}d'_{xy} + \sqrt{\tfrac{1}{3}}d'_{xz} - \sqrt{\tfrac{1}{6}}d'_{yz} \tag{59}$$

it was shown[55] that this function has $\frac{4}{9}$ σ- and $(\frac{1}{18} + \frac{1}{6} =)$ $\frac{2}{9}$ π-antibonding character. For a representative e orbital, this resolution becomes

$$d_{x^2-y^2} = \sqrt{\tfrac{1}{3}}d_{xz} + \sqrt{\tfrac{2}{3}}d'_{yz} \tag{60}$$

with a π-antibonding part of $\frac{2}{3}$. Putting the resolved orbitals in the energy expressions (57) and (58), we see that the relation

$$\frac{\Delta_{\text{tetr}}}{\Delta_{\text{oct}}} = -\frac{4}{9}$$

is also fulfilled for weak covalent σ and π bonding. The expressions H_{AL}^2, $H_{\text{AL}}S_{\text{AL}}$, S_{AL}^2 transform equivalently under any transformation of the central-atom orbital A when all ligand–ligand interactions are neglected. It is therefore not necessary to approximate the off-diagonal element H_{AL} in the secular equation by the overlap integral S_{AL} as is done in the angular-overlap model proposed by Schäffer and Jørgensen[56].

In principle, other symmetries can be handled by the above technique, but the method proves to be rather complex when applied to low symmetry coordination compounds, for which the Ξ method[57] is more applicable. The latter is based on the Helmholz–Wolfsberg[58] model and approximates the ligand wave functions by Kronecker δ functions. The method of Helmholz and Wolfsberg has recently been employed on numerous complex systems. It should, however, be accepted only with strong reservations[59,60] since the choice of the diagonal elements and the constant k in the off-diagonal element

$$H_{\text{AL}} = kS_{\text{AL}}(H_{\text{AA}} + H_{\text{LL}})/2 \tag{61}$$

is rather arbitrary and lacking in physical significance. If we do not try to reproduce experimental results in detail but are interested only in qualitative understanding of the electronic structure in the molecule, the Helmholz–Wolfsberg procedure is useful by virtue of its dependence on the group-overlap integral G_{AL}, in which ligand orbitals enter as symmetry functions. By this, full advantage is taken of the known symmetry of the molecule, and the order of the energy levels in the orbital term scheme is mainly governed by the overlap integrals. So, a lot of structures can be rationalized by using very few assumptions other than the relative size of group-overlap integrals[59]. Having recognized the key position of the group-overlap integral in this theory, it is sufficient to consider its behaviour for each electron level and leave the other terms in (61) untouched. Such a procedure is realized in the Ξ method[57] where the overlap integral is separated into an angular part Ξ and a radial part S_{AL}^*

$$S_{\text{AL}} = \Xi S_{\text{AL}}^* \tag{62}$$

where Ξ depends only on the ligand positions in space

$$\Xi = N_{\text{A}} \sum_{i=1}^{N} A(x_i, y_i, z_i)N_i \tag{63}$$

This expression results from the functions that are chosen such that

$$\Psi_A = N_A A(x, y, z) R(r)$$
$$\Psi_L = \sum_{i=1}^{N} N_i \,\delta(x - x_i)\,\delta(y - y_i)\,\delta(z - z_i) \tag{64}$$

when the ligand symmetry functions are approximated by Kronecker δ symbols. This simplified method was recently applied on several systems with significant success[61–63].

For lower symmetry complexes, certain splittings of the octahedral levels are expected. Deviations from cubic symmetry are more distinct for complexes having different ligands at different position in the spectrochemical series because for these ligands the electronic structure of the complex is highly unsymmetrical. Yamatera[4] and McClure[5] independently showed a way to calculate band splittings due to substitution in octahedral complexes. From different band splittings, it is possible by this method to assign the geometric isomers. It is beyond the scope of this chapter to present the procedure of Yamatera and McClure but we will later use their results which are given by coefficients describing σ- and π-bonding energy shifts[63].

4. *Electron interaction*

Until now the effect of electron repulsion has not been considered in the theory. The inclusion of electron–electron interaction is, however, necessary because its energetic effects are large enough to affect visible and ultraviolet spectra. The wave-mechanical treatment of the electron repulsion described by the term $1/r_{ij}$ in the hamiltonian is outlined in many textbooks. The wave functions of a many-electron system are chosen antisymmetrically with respect to an exchange of two electrons. The many-electron energy levels (denoted by capital letters) for a certain electron configuration l^n depend on the coulomb and exchange integrals $J(\mathrm{a, b})$ and $K(\mathrm{a, b})$, respectively. In crystal-field theory, we distinguish between weak-field and strong-field calculations. In the first case, the crystal field is weak compared with the electron interaction, so that the quantum numbers L, S of a multiplet retain their validity and the levels are weakly perturbed without mixing different multiplets. The strong-field case starts with a certain subshell configuration, with its functions adapted to the symmetry, and considers the electron interaction as a perturbation. For molecular-orbital theory, the second procedure is more appropriate since the orbitals transform like the strong-field functions. In the following, we therefore pursue the molecular-orbital approach.

The lowest subshell configuration for two d electrons in octahedral symmetry is t_{2g}^2. From group theory and by the restriction given by the Pauli principle, four molecular levels arise from this configuration which have the energy in terms of J and K integrals

$$\begin{aligned} {}^3T_{1g}&: \; J(xy, xz) - K(xy, xz) \\ {}^1T_{2g}&: \; J(xy, xz) + K(xy, xz) \\ {}^1E_g&: \; J(xy, xy) - K(xy, xz) \\ {}^1A_{1g}&: \; J(xy, xy) + 2K(xy, xz) \end{aligned} \tag{65}$$

where xy, xz mean the functions d_{xy}, d_{xz} or other functions transforming equivalently. The integral expressions for other important subconfigurations are found in the literature[8,24,26]. For our case, we must evaluate nine different repulsion integrals, which in crystal-field theory (pure d functions) can be expressed in Racah parameters[64] A, B, C by the relations[26]

$$\begin{aligned} J(z^2, x^2 - y^2) &= J(z^2, xy) = A - 4B + C \\ J(x^2 - y^2, xy) &= A + 4B + C \\ J(xy, xy) &= A + 4B + 3C \\ J(xy, xz) &= A - 2B + C \\ K(z^2, x^2 - y^2) &= K(z^2, xy) = 4B + C \\ K(x^2 - y^2, xy) &= C \\ K(xy, xz) &= 3B + C \end{aligned} \tag{66}$$

Table 2 compiles the perturbation energies for certain subshell configurations in the strong-field case, neglecting all configuration interactions between different subshells. Tanabe and Sugano give in their fundamental paper[65] the energy matrix elements of all possible configurations also including configuration intermixing. Energy-level schemes that depend on the crystal-field parameter Dq in units of Racah parameter B (and assuming $C = 4B$) are called Tanabe–Sugano diagrams[65]. A corresponding plot used by Orgel[23] is a level diagram as a function of $\Delta = 10Dq$ with fixed Racah parameters B, C. The A values are unimportant because they vanish when subtracting any two energy differences.

As seen from Table 2 and also from the Tanabe–Sugano diagram, the magnetic behaviour can be different for different subshell configurations. One distinguishes between low- and high-spin complexes according to which energy level of certain spin multiplicity is the ground state. The different behaviour is governed by the competition between the orbital-energy difference Δ and the interelectron-repulsion energy parameters.

Table 2. Energies of pure subshell configurations in octahedral d^n complexes given by $a\Delta + bB + cC$. The atomic ground states are chosen as zero energies[26].

			a	b	c
d^2	t_2^2	$^3T_1$	-0.8	3	0
		^1E	-0.8	9	2
		$^1T_2$	-0.8	9	2
		$^1A_1$	-0.8	18	5
	t_2e	$^3T_2$	0.2	0	0
		$^3T_1$	0.2	12	0
		$^1T_2$	0.2	8	2
		$^1T_1$	0.2	12	2
	e^2	$^3A_2$	1.2	0	0
		^1E	1.2	8	2
		$^1A_1$	1.2	16	4
d^3	t_2^3	$^4A_2$	-1.2	0	0
		^2E	-1.2	9	3
		$^2T_1$	-1.2	9	3
		$^2T_2$	-1.2	15	5
	t_2^2e	$^4T_2$	-0.2	0	0
		$^4T_1$	-0.2	12	0
. . .	. . .	. . .	. . .	. . .	. . .
	t_2e^2	$^4T_1$	0.8	3	0
. . .	. . .	. . .	. . .	. . .	. . .
d^4	t_2^4	$^3T_1$	-1.6	6	5
		^1E	-1.6	12	7
		$^1T_2$	-1.6	12	7
		$^1A_1$	-1.6	21	10
	t_2^3e	^5E	-0.6	0	0
. . .	. . .	. . .	. . .	. . .	. . .
	$t_2^2e^2$	$^5T_2$	0.4	0	0
. . .	. . .	. . .	. . .	. . .	. . .
d^5	t_2^5	$^2T_2$	-2.0	15	10
	t_2^4e	$^4T_1$	-1.0	10	6
		$^4T_2$	-1.0	18	6
		$^2A_2$	-1.0	12	9
. . .	. . .	. . .	. . .	. . .	. . .
	$t_2^3e^2$	$^6A_1$	0.0	0	0
		$^4A_1$	0.0	10	5
		^4E	0.0	10	5
d^5		$^4T_2$	0.0	13	5
		^4E	0.0	17	5
		$^4T_1$	0.0	19	7
		$^4A_2$	0.0	22	7
. . .	. . .	. . .	. . .	. . .	. . .
	$t_2^2e^3$	$^4T_1$	1.0	10	6
		$^4T_2$	1.0	18	6
d^6	t_2^6	$^1A_1$	-2.4	5	8
	t_2^5e	$^3T_1$	-1.4	5	5
		$^3T_2$	-1.4	13	5
		$^1T_1$	-1.4	5	7
		$^1T_2$	-1.4	21	7
	$t_2^4e^2$	$^5T_2$	-0.4	0	0
. . .	. . .	. . .	. . .	. . .	. . .
	$t_2^3e^3$	^5E	0.6	0	0
. . .	. . .	. . .	. . .	. . .	. . .
d^7	t_2^6e	^2E	-1.8	7	4
	$t_2^5e^2$	$^4T_1$	-0.8	3	0
. . .	. . .	. . .	. . .	. . .	. . .
	$t_2^4e^3$	$^4T_2$	0.2	0	0
		$^4T_1$	0.2	12	0
. . .	. . .	. . .	. . .	. . .	. . .
	$t_2^3e^4$	$^4A_2$	1.2	0	0
. . .	. . .	. . .	. . .	. . .	. . .
d^8	$t_2^6e^2$	$^3A_2$	-1.2	0	0
		^1E	-1.2	8	2
		$^1A_1$	-1.2	16	4
	$t_2^5e^3$	$^3T_2$	-0.2	0	0
		$^3T_1$	-0.2	12	0
		$^1T_2$	-0.2	8	2
		$^1T_1$	-0.2	12	2
	$t_2^4e^4$	$^3T_1$	0.8	3	0
		^1E	0.8	9	2
		$^1T_2$	0.8	9	2
		$^1A_1$	0.8	18	5

According to Table 2 (comparing maximum and minimum multiplicity), low-spin behaviour occurs for

$$
\begin{aligned}
d^4&: \quad \Delta > 6B + 5C \\
d^5&: \quad 2\Delta > 15B + 10C \\
d^6&: \quad 2\Delta > 5B + 8C \\
d^7&: \quad \Delta > 4B + 4C
\end{aligned}
\tag{67}
$$

if second-order effects are neglected.

In the semiempirical theory, the ligand-field strength and the electron-repulsion integrals are considered as parameters. When comparing the experimentally determined Racah parameters for the complex molecule with those of the free central ion, different values are expected due to bonding effects in the wave functions. Several authors [65–67] have suggested that these parameters in octahedral complexes are smaller than those of the corresponding free ions, which explains the generally observed red shift of the bands in the coordinated species. Schäffer and Jørgensen [6,68] proposed a nephelauxetic effect on the antibonding orbitals of the central ion when they are bonded to the ligand system. The expansion of the central-atom electron clouds causes a decrease in all interelectron-repulsion integrals (Slater–Condon parameters F_k and Racah parameters). A series of ligands with progressively decreasing parameter B is called the nephelauxetic series:

F^-, H_2O, urea, NH_3 en, ox^{2-}, NCS^-, Cl^-, CN^-, N_3^-, Br^-, I^-, dtp^- (68)

A nephelauxetic ratio

$$\beta = \frac{B_{\text{complex}}}{B_{\text{free ion}}} \tag{69}$$

is defined, which describes the parameter decrease quantitatively. Obviously the nephelauxetic effect is an approximate way to describe covalent bonding. From smaller interelectron-repulsion parameters and consequently expanded central-atom wave function, more strongly interacting wave functions are expected, since the central-atom ligand orbital overlap is then increased. When inserting larger overlap integrals in (61), the energetic expression (57) gives a higher covalent contribution to the metal–ligand bond.

The change of the atomic Landé parameters ζ_{nl} with respect to those in the molecule is less pronounced but it nevertheless shows the same trend to lower parameter values. The corresponding relation

$$\beta^* = \frac{\zeta_{nl}^{\text{complex}}}{\zeta_{nl}^{\text{free ion}}} \tag{70}$$

has been called the relativistic nephelauxetic quotient [8].

B. Charge-transfer Bands

The intramolecular electron-transfer bands in coordination compounds, which in general have much higher oscillator strengths than the ligand-field bands (*d–d* transition), have been extensively investigated in recent years [62,69,70]. Transitions that can be assigned to a charge-transfer process are observed only for systems with localized electronic structures. In coordination compounds, the central atom and the ligands are generally separated systems interacting only weakly with each other so that electron transfer between the two parts of the molecule can be observed in both directions. The transfer from the ligands to the central atom is preferred energetically when, in the ground state, the ligand has a higher electronegativity than the central ion. Charge transfer in the reverse direction is most probable when the ligand possesses easily available empty orbitals, e.g. π^* orbitals in coordinated π-electron systems. The occurrence of an electron transfer is energetically favoured for small orbital electronegativity differences. One empirical approach to electron-transfer band positions is that of optical electronegativity, which includes corrections due to electron repulsion. According to Jørgensen [69,70], the charge-transfer band of a transition ion or rare earth complex has its maximum wave number at

$$\bar{\nu}_{\sigma,\pi}[\mathrm{k\textsc{k}}] = 30(x_{\mathrm{L}} - x_{\mathrm{A}}) + \Delta + \delta SP + aE + b\zeta_{nl} \qquad (71)$$

The factor 30 converts the usual electronegativity scale into kilokayser and x_{L}, x_{A} are the optical electronegativities of the ligand and the central atom, respectively. For the halogens, they follow exactly the numbers of Pauling's electronegativity. Different values for x_{L} have to be used according to whether the transfer originates from σ- or π-bonded ligand orbitals. Δ, the orbital-energy difference due to the ligand-field splitting, applies only to transition group complexes for which the lowest subshell is filled so that the electron transition reaches a higher sublevel. *SP* stands for the spin-pairing energy which for *q* electrons is given to good approximation by

$$SP = \Pi\left[\frac{3}{4}q - \frac{3q(q-1)}{16l+4} - S(S+1)\right] \qquad (72)$$

Π is the spin-pairing parameter depending on the electron-repulsion integrals; for *d* electrons it is $\Pi = \frac{7}{6}(\frac{5}{2}B + C)$, i.e. some 5–6 kK for the first transition series and about 3 kK for 4*d* and 5*d* elements. Rare earth complexes have $\Pi = \frac{9}{8}E^1$ (Racah parameter for *f*-electron interactions) with $\Pi = 6.5$ kK. In (71), δSP means the difference of spin-pairing energy for the $(q+1)$- and the q-electron case. In this, only the change of electron-repulsion energy close to the central atom is considered while the change of spin-pairing energy near the ligands is felt to be negligible because of their larger spatial extent. The E term enters in (71) only when higher terms with the multiplicity of the ground term exist (^3F, ^4F for d^2, d^3, d^7, d^8 and ^3H, ^4I, ^5I, ^6H for f^2, f^3, f^4, f^5, etc.). This arises from the fact that the formula of spin-pairing energy is only valid for the baricentre of a given multiplicity. E for transition ions (d electrons) is equal to the Racah parameter B; for rare earth ions $E = E^3$. The last term in (71) is only significant for rare earth complexes where the spin–orbit coupling constant is sufficiently large. The coefficients a and b in rare earth ions are listed in ref. 70. Values for optical electronegativities are found in refs. 8 and 71, and an example of their use is given later.

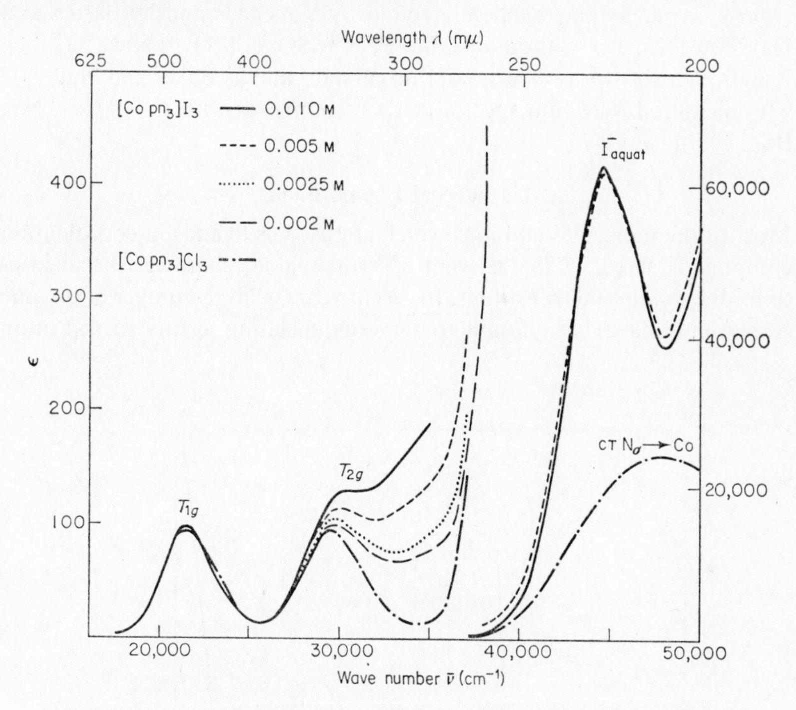

Figure 11. Absorption spectrum of $[Co\ pn_3]^{3+}$ as chloride or iodide for different concentrations in aqueous solution.

A charge transfer in the reverse direction, i.e. from the central atom to the ligand, is likely to happen in complexes with central atoms having small ionization potentials and ligands with easily available empty π^* orbitals. The dipyridyl or phenanthroline complexes of Fe^{II} represent a good example[72]. In the spectrum of these compounds, the bands about 510 mμ ($\varepsilon \sim 10{,}000$) are assigned to such a charge transfer from the central ion.

Another type of charge transfer is due to ion-pair formation in solution. The complex cation associates with polarizable anions (like iodide) forming ion pairs, in which a charge-transfer transition from the reducing anion to the cation can take place. The strength of these Linhard–Weigel bands[9,73] is dependent on the formation constant of the ion pair and consequently on the concentration of the complex. Figure 11 shows typical spectra of this kind where, in the range of 34,000 cm^{-1}, a concentration-dependent band appears which is due to intermolecular charge transfer. The same complex exhibits at 47.6 kK an intramolecular electron-transfer transition from the σ-bonded N atom of propylenediamine to the central-atom e_g level. Similar bands are found by Yoneda[74] and by Barnes and Day[75] in ion pairs with anions like SCN^-, SO_3^{2-}, $S_2O_3^{2-}$, and CO_3^{2-}, by Kondo[10] in absorption spectra of thin crystals, and by Baker and Phillips[76] who measured reflection spectra of Cr^{III} complexes with I^-, N_3^-, SCN^-, Br^-, Cl^- as anions.

C. Internal Ligand Bands

Most of the molecules and ions which are used as ligands in coordination compounds show, in the relevant spectral region, characteristic absorptions due to internal transitions. In recent years, a large number of organic compounds have been found to serve as chelating agents to transition

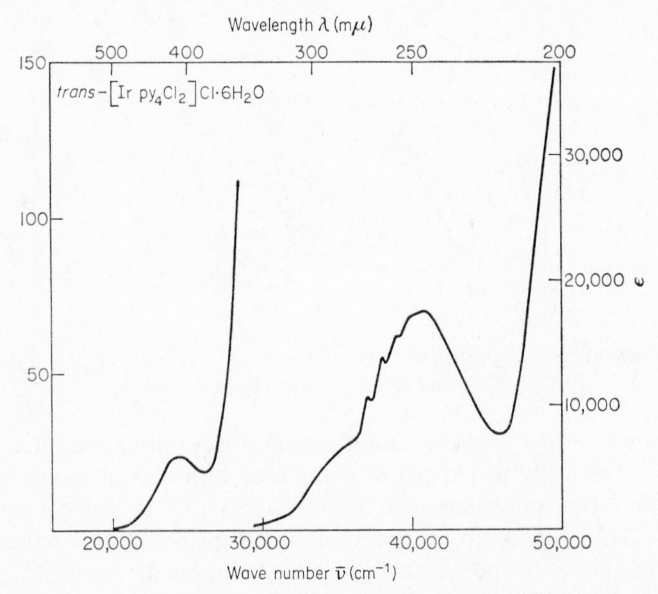

Figure 12. Absorption spectrum of *trans*-[Ir py_4Cl_2]Cl·$6H_2O$ in aqueous solution.

group and rare earth ions. These ligands, which are sometimes very large and complicated, are attributed specific absorptions which in most cases are so strong that their bands may cover a part or even all ligand-field and charge-transfer transitions. This is a great disadvantage because in this case very little can be said about the properties of the central atom–ligand bond. For ligands with closed π-electron systems, in general only a slight change of the internal bonds is found in comparison to the free molecules. As seen from Figure 12, even the vibrational structure of the π–π^* transition band in coordinated pyridine can be discovered. In the complex, the whole band is only shifted by 5–7 mμ to longer wavelength.

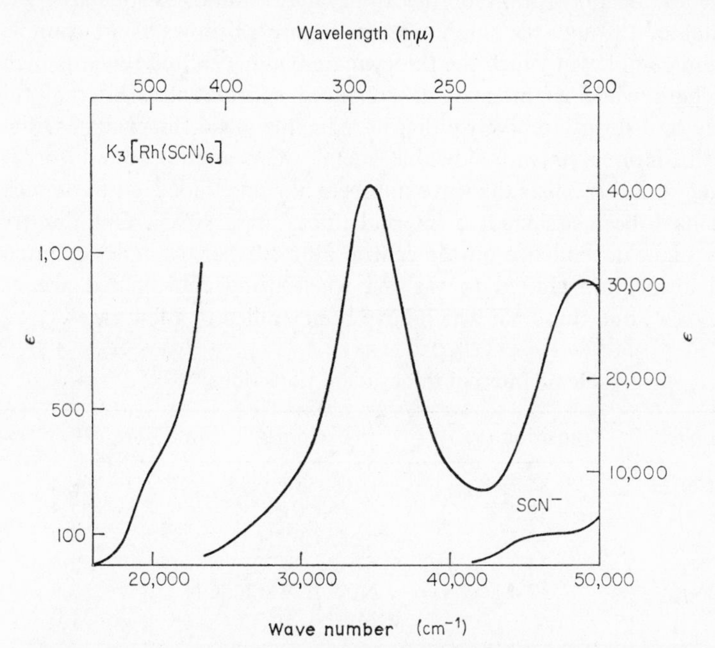

Figure 13. Absorption spectrum of $K_3[Rh(SCN)_6]$ and KSCN in aqueous solution.

Similar slight shifts are detected in dipyridyl, *o*-phenanthroline, and other ligands of this kind [7]. In general, one can say that the influence of the central atom on the ligands has been by no means so carefully studied as the reverse effect of the ligands on the central atom. There are several reasons for this. The most important one is that for the interpretation of the ligand spectrum we do not have such a simple method available today as ligand-field theory supplies for *d*–*d* transitions.

There are also some small molecules which in their free-ion states absorb in the near-ultraviolet spectral region. From the simple ligands listed in the spectrochemical series (41), the following ions show strong absorption in aqueous solution:

I^-	44.2 kK	SO_3^{2-}	50 kK	
SCN^-	47	NO_3^-	49	
$SeCN^-$	42.5	NO_2^-	47.6	(73)
N_3^-	(42.5)	$(COS)_2^{2-}$	39	
$S_2O_3^{2-}$	46	ox^{2-}	(40)	

The internal transitions of some of these ions are shifted to lower wave numbers when they are coordinated to a metal. A striking example is the change in the absorption of the thiocyanate anion, particularly when coordinated through the sulphur atom. Figure 13 shows as an example a rhodium complex in which the thiocyanate band is shifted towards higher wavelength while its intensity is enhanced substantially. Internal thiocyanate transitions are likely either in reducible metal thiocyanates bound through sulphur or in oxidizable metal thiocyanates bound through nitrogen[75]. Table 3 lists the wave numbers of some thiocyanate complexes which have been assigned to internal thiocyanate[26,75,77]. One discovers only a slight dependence on the central atom. Other examples of intense ligand absorption shifted to the red when being coordinated are N_3^-, NO_2^-, ox^{2-}, and thiourea, which have been studied in a few cases.

Table 3. Internal thiocyanate transitions[26,75,77].

Complex	Transition (kK)	Complex	Transition (kK)
$Rh(SCN)_6^{3-}$	34.7	$Cr(NCS)_6^{3-}$	32.4
$Pd(SCN)_4^{2-}$	32.5	$Co(NCS)_4^{2-}$	32.8
$Os(SCN)_6^{2-}$	30.5	$Co(NCS)(H_2O)_5^+$	36.7
$Pt(SCN)_6^{2-}$	34.8	$Co\ en_2(NCS)_2^+$	32.5
$Pt(SCN)_4^{2-}$	37.4	Ni^{2+} in 9 M KSCN	33.9
		$Mo(NCS)_6^{3-}$	31.0

D. Limitations of the Band Classification

It is obvious that for a larger delocalization of the central atom and ligand electrons the classification into *d–d* transitions, internal ligand, and electron-transfer transitions is no longer valid. The main reason for the stronger interaction of the ligand and central-atom orbitals lies in their similar orbital energies which in the extreme case can even be degenerate. The highest possible interaction is then attained when the orbitals have identical symmetry and their resonance integral

$$H_{AL} = \int \Psi_A^* \mathscr{H} \Psi_L \, d\tau$$

is large. The molecular orbitals are then highly delocalized and the observed band classification into *d–d* and charge-transfer transitions is not relevant.

An example where this effect is manifested is the series of iridiumpentaammine compounds. While the hexaammine and aquopentaammine have two weak absorptions undoubtedly assigned to *d–d* transitions, the classification of the second band for the chloro and bromo complex is not possible. The extinction coefficient of this band (see Table 4) is too large to be due to a pure *d–d* transition; this assignment also yields too large a distance between the two ligand-field transitions $^1T_{1g}$, $^1T_{2g}$, as is seen from the evaluation of the parameters in the semiempirical theory, which give nephelauxetic β ratios which are too high.

Table 4. Ligand-field and charge-transfer bands of $Ir(NH_3)_5X^{n+}$ in kK. Molar extinction coefficients are given in parentheses.

	X = H_2O	NH_3	Cl^-	Br^-	I^-	
$^1T_{1g}$	38.8 (86)	39.8 (92)	35.0 (73)	33.0 (100)	29.7 (372)	
$^1T_{2g}$	47 (128)	46.8 (160)				
$^1T_{2g}$ / CT X → Ir			44.1 (333)	43.5 (800)		
CT X → Ir					42.7 (4600)	46.5 (6700)
β	0.84	0.71	[0.95]	[1.12]	—	

According to Englman[78], the gain in intensity of the second band is explained by a mixing of odd-bonding states into the even antibonding ones through the vibrations of the molecule. This mechanism is described by the formula

$$f = \alpha \frac{E_f - E_i}{(E_c - E_f)^2} \tag{74}$$

where E_i and E_f are the initial and final ligand-field states and α is a factor which is nearly constant for different compounds. It is seen that the transition $E_i \rightarrow E_f$ gains oscillator strength f when the charge-transfer state E_c approaches the final ligand-field state. That also explains why in Table 4 the second transitions are more affected than the first.

Other groups of coordination compounds for which a band classification is also difficult are π-electron systems (dipyridyl, phenanthroline, etc.) coordinated to transition ions with unusually low oxidation numbers (Cr^+, Cr^{2+}, V^0, Ti^-)[79–81]. These central ions are extremely 'soft' (polarizable) in the Chatt–Ahrland–Pearson classification[82,83] so that their bonding electrons are strongly delocalized. A series of planar dithiolate compounds which has recently been investigated in detail[84–86] also belongs to this group. These complexes have maleonitriledithiol, toluene-3,4-dithiol, or similar compounds as ligands which are bound through the sulphur atom. A Helmholz–Wolfsberg[58] calculation shows[87] that the higher antibonding orbitals of the *d* shell are very much delocalized and therefore a classification of the terms in the molecular-orbital scheme according to *d* levels loses its validity. Consequently, no distinction can be made between *d–d* and electron-transfer transitions. In the rhodium maleonitriledithiolate complex, for instance, rhodium carries a formal charge of two[88,89]. However, the calculation[87] shows that the last occupied orbital $4a_{1g}$ has its maximum electronic density on the ligands giving rise to a net charge on Rh higher than two. It is even more probable that the last *d* electron of Rh^{II} occupies a non-bonding π orbital of the ligand, in which case the common oxidation state Rh^{III} would result.

For these complex compounds, which have a large number of low-lying empty orbitals, another peculiarity arises from the electron correlation. If more than one electron configuration exists with similar energy, the many-electron wave function of the molecular ground state is composed of a linear combination of Slater–Heisenberg determinants, each of which describes a low-energy electron configuration. When those configurations belonging to different occupation numbers for ligand and central atom orbitals are intermixed, it happens that the oxidation number of the central ion in the molecular ground state is not clearly determined any more as in complexes with normal (innocent)[90] ligands. These conditions allow the formation of compounds with unusual formal oxidation numbers for the central ions which are stabilized by their ligand systems.

III. PRACTICAL EXAMPLES

Before embarking on specific examples it should be emphasized that, as was already indicated in the preceding sections, there are no golden rules in the application of theory to the interpretation of spectra. Though the results obtained from more or less empirical calculations can help, one needs to apply a large amount of chemical intuition, aided by the comparison of a series of related compounds and the use of intensity or polarization arguments. For definite decisions on band assignments, more elaborate techniques like optical rotation or magnetic optical rotation dispersion (Faraday effect) measurements may prove to be necessary.

A. Six-coordinated d^1 and d^9 Complexes: Ti^{III}, Cu^{II}

We now turn to some selected examples for which the theoretical methods outlined in the preceding paragraphs are demonstrated. Simple cases are represented by the one-electron complexes d^1 or, because of the hole-equivalency principle, by d^9 compounds, since the valance-shell energies of these compounds do not contain electron-repulsion terms. Typical representatives for octahedral d^1 complexes are $[Ti(H_2O)_6]Cl_3$, $[Ti(urea)_6]I_3$, and the titanium alum $CsTi(SO_4)_2 \cdot 12H_2O$ which also has a Ti^{III} ion surrounded by six H_2O molecules[91]. The visible and near-ultraviolet spectrum of a dilute HCl solution of different $TiCl_3$ concentrations is given in Figure 14. It is seen that, particularly for low concentrations, the

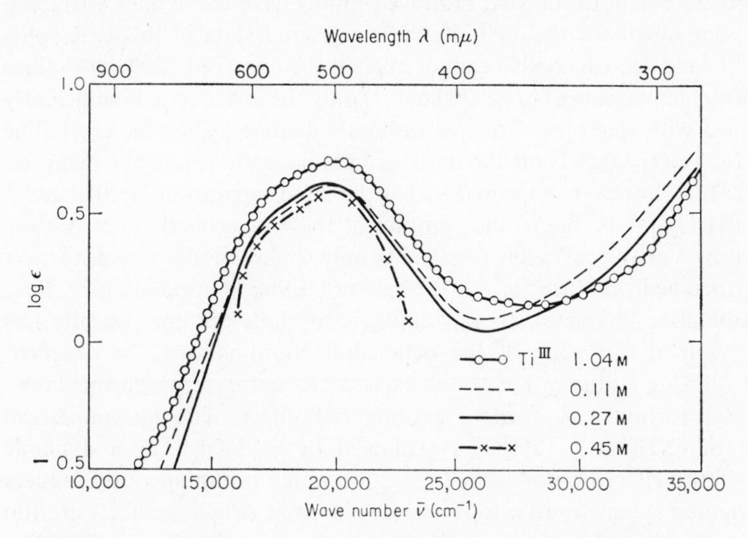

Figure 14. Absorption spectrum of $TiCl_3$ for different molarity M in dilute hydrochloric acid solution. [Reproduced, by permission, from *Jahrbuch der Akademie der Wissenschaften*, Göttingen, 1962, p. 19.]

Bouguer–Lambert–Beer law (5) is fulfilled, and therefore the presence of only one compound for different concentrations can be inferred. If we assume an environment for the Ti^{III} central ion in which the six water molecules occupy the vertices of a regular octahedron, it follows from group theory that only one ligand-field transition should occur (see Figure 8). Ligand-field and simple MO theory, moreover, states (24 and 60) that a weak band is expected arising from the transition $^2T_{2g} \rightarrow {}^2E_g$ at a wavelength which is equal to the ligand-field parameter Δ. This transition is spin and symmetry allowed but parity forbidden. As seen from Figure 14, a weak but distinct band ($\varepsilon = 4$) shows up at 500 mμ wavelength. However, this band is broad and unsymmetrical; it looks rather as if two bands are superimposed. A curve analysis yields a band separation of approximately

2000 cm^{-1}. Other Ti^{III} compounds mentioned above show[91] similar band structures, and the reflection spectra even have two separated bands. The occurrence of more than one transition is in agreement with the theorem of Jahn and Teller according to which the nuclear framework of a molecule with an electronic ground state $^2T_{2g}$ should be unstable. The necessary distortion of the octahedron can go either to trigonal, tetragonal, or even rhombic symmetry. The trigonal deviation does not seem probable because, for this symmetry, the orbital degeneracy of the excited state is not removed. However, alums have commonly a trigonal field component for the Ti^{III} site[92]. The ground state in this case splits into $^2A_1$ and ^2E, which also cannot explain the observed 2000 cm^{-1} since at normal temperature ($kT \approx 200$ cm^{-1}) only the lower state is sufficiently occupied with electrons. Russian authors[93] assume a rhombic crystalline field for $[Ti(H_2O)_6]Cl_3$ on the basis of paramagnetic resonance measurements. It is, however, improbable that the band separation of 2000 cm^{-1} in $[Ti(H_2O)_6]^{3+}$ is due to the splitting of the $^2T_{2g}$ ground state. As was shown by Van Vleck[94], this term splits only slightly under any distortion of an octahedron since the t_{2g} orbitals are either non-bonding or have π-antibonding character. In this case, the orbitals are energetically less sensitive to a distortion of the octahedral coordination. The observed band splitting is due rather to the excited ^2E_g state, on which the Jahn–Teller distortion has a more pronounced effect. The unsymmetrical band for $CsTi(SO_4)_2 \cdot 12H_2O$ is explained by Schläfer[91] by a dynamic Jahn–Teller effect on the excited ^2E_g state which instantaneously reduces the trigonal symmetry to a lower one. Spin–orbit coupling effects are too small to explain the experimental findings because the free-ion coupling constant of Ti^{III} has only a value of $\zeta_{3d} = 154$ cm^{-1}.

According to the hole-equivalency principle, the electronic levels in Cu^{II} complexes are inverted. For these complexes, the ^2E_g state is the ground state and an octahedral configuration should not be stable but rather strongly distorted. In fact, most of the six-coordinated Cu^{II} complexes have the axial ligands of the octahedron farther away than the equatorial ligands[95]. The symmetry in this case is tetragonal $\mathbf{D}_{4h}$. An analysis of the polarized absorption spectrum of $CuSO_4 \cdot 5H_2O$ shows three transitions[96]. The crystal structure of this compound was investigated by Beevers and Lipson[97]. The environment of each Cu^{II} consists of four water molecules arranged in an approximate square with two axial sulphate oxygens at slightly greater distance from the central ion. For this symmetry, the bands at 10.5, 13.0, and 14.5 kK from the polarizations can be assigned to $^2B_{1g} \rightarrow {}^2A_{1g}$, $^2B_{2g}$, and ^2E_g (Figure 15). Even for Cu^{II} compounds with six identical ligands, a distorted octahedral environment has to be assumed. The absorption spectrum of $[Cu(H_2O)_6]^{2+}$ in solution has an unsymmetrical band which originates from at least two different absorptions. The corresponding ammine complex is not sufficiently characterized to decide from the absorption spectrum if a hexaammine or pentaammine complex is formed in liquid ammonia[98]. Cu^{II}, surrounded by six chloride or bromide ions, also shows two transitions, which are found close to each other

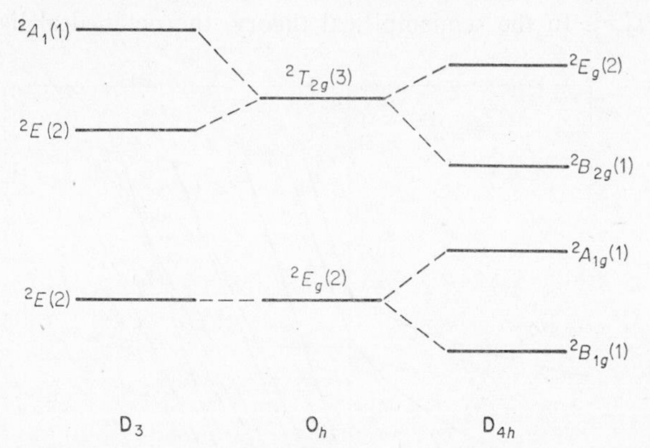

Figure 15. Energy level diagram of Cu^{II} for different symmetries (schematic).

in the spectrum[99]. The electronic structures of these compounds were interpreted[62] on the basis of simple molecular-orbital theory also assuming $\mathbf{D}_{4h}$ symmetry. Unsymmetrical bands are observed[100] also in spectra of Cu^{II} substituted in MgO crystals. From this spectrum, a distortion of the cubic MgO host lattice occurs, originating from the Cu^{II} ion which in this symmetry has a Jahn–Teller unstable environment. The theory of hexacoordinated Cu^{II} was thoroughly discussed by Ballhausen[101] and by Belford[102].

B. Low-spin d^6 Complexes: Co^{III}, Rh^{III}, Ir^{III}

As a typical example for complexes where the *d*-electron repulsion plays an important role for the interpretation of the visible spectra, we shall discuss some diamagnetic d^6 complexes for which ligand-field effects exceed the spin-pairing energy. As seen from the relevant Tanabe–Sugano diagram in Figure 16, the multiplicity of the molecular ground term changes at a certain Dq/B value. High spin d^6 ions are found in Fe^{II} complexes, but $[CoF_6]^{3-}$ is the only high-spin Co^{III} complex. The ground state is a $^5T_{2g}$ arising from the ^5D ground term of the free ion. However, the majority of Co^{III} complexes and their homologues are diamagnetic with all their spins coupled, forming a filled octahedral t_{2g} subshell. The ground state of the t_{2g}^6 configuration is $^1A_{1g}$ and the states belonging to the lowest excited configuration $t_{2g}^5 e_g$ are $^3T_{1g}$, $^3T_{2g}$, $^1T_{1g}$, and $^1T_{2g}$ (see Table 2). We consequently expect two spin-forbidden and two spin-allowed transitions, which are actually found, for example, in the reflection spectrum of $K_3[RhF_6]$[103]. Figure 17 shows two weak transitions on the long wavelength side of the spectrum and two higher energy bands which are spin-allowed transitions. The shoulder near 37 kK might be due to a double excitation $t_{2g}^4 e_g^2$. In the semiempirical theory, the octahedral ligand-field

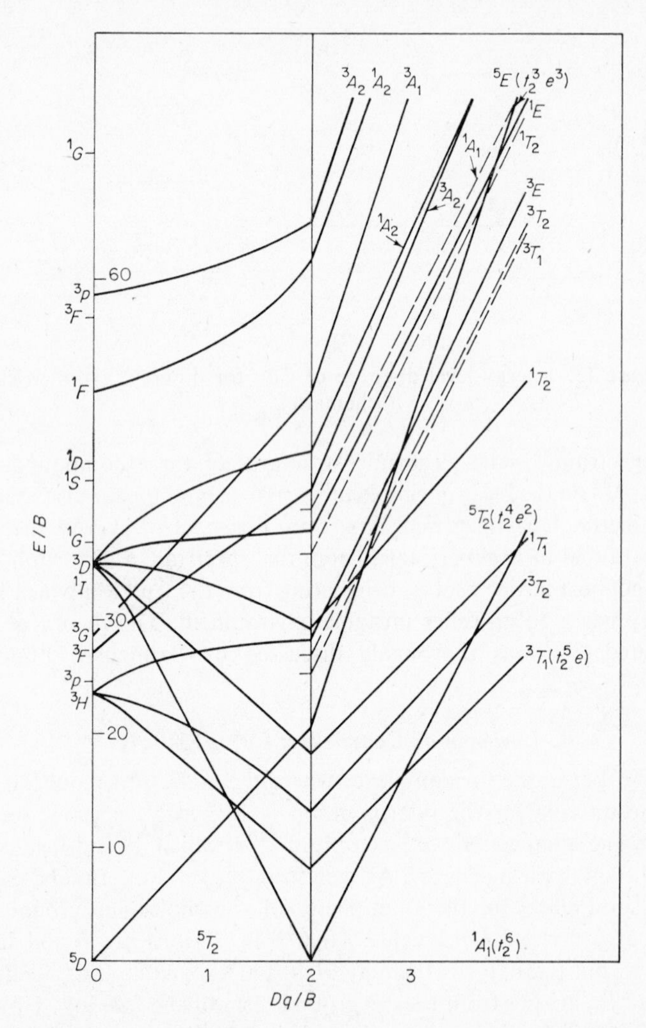

Figure 16. Tanabe–Sugano diagram for octahedral d^6 complexes. [Reproduced, by permission, from *J. Phys. Soc. Japan*, **9**, 766 (1954).]

parameters and the Racah parameter B are calculated as follows. As usual we assume $C = 4B$; then the two spin-allowed transitions σ_1 and σ_2 are expressed in terms of parameters Δ and B by the equations

$$\sigma_1 = \Delta - 4B + 86\frac{B^2}{\Delta}$$
$$\sigma_2 = \Delta + 12B + 2\frac{B^2}{\Delta} \qquad (75)$$

The linear terms in (75) are given by the diagonal terms in the energy matrix (see Table 2); the quadratic contributions are correction terms due to the subshell configuration interaction. They are listed by Jørgensen[68] and

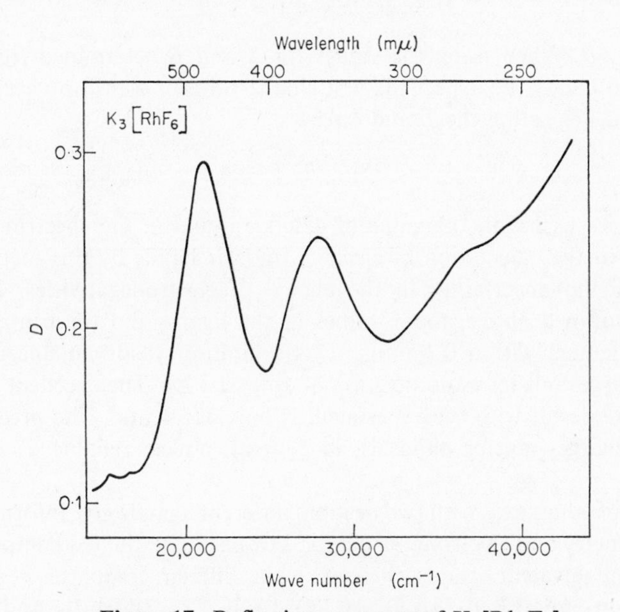

Figure 17. Reflection spectrum of $K_3[Rh\ F_6]$.

serve in the most important cases to calculate the Δ and B values more accurately. From the two singlet transitions in $K_3[RhF_6]$ which occur at 21.3 and 27.8 kK, the parameters Δ and B are calculated by (75) as $\Delta = 22.3$ kK and $B = 0.46$ kK. The latter value leads to a nephelauxetic ratio of $\beta = 0.64$.

We shall now demonstrate the empirical calculation of the wavelength of an electron-transfer band, using $[Co\ pn_3]^{3+}$ (Figure 11) as an example. The chloride shows a strong band at 47.6 kK with an extinction coefficient of $\varepsilon = 24{,}400$. The optical electronegativities of Co^{III} and aliphatic amines, like propylenediamine, are given approximately by $x_{Co^{III}} = 2.35$ and $x_{pn} = 3.3$[8,71]. The ligand-field parameter of the octahedral Co^{III} complex[73] is $\Delta = 22.5$ and the Racah parameter $B = 0.58$. The spin-pairing energy is calculated by the use of (72) as

$$q = 6,\ S = 0: \quad SP(q = 6) = 2\Pi$$
$$q = 7,\ S = \tfrac{1}{2}: \quad SP(q = 7) = \Pi$$

With this, the change of spin-pairing energy when adding another electron is

$$\delta SP = SP(q = 7) - SP(q = 6) = -\Pi \tag{76}$$

and the spin-pairing parameter is

$$\Pi = \tfrac{7}{6}(\tfrac{5}{2}B + C) = \tfrac{91}{12}B \tag{77}$$

with $C = 4B$. When using the values for Δ and B determined from the ligand-field bands, we expect the first charge-transfer transition according to (71) and neglecting the E and ζ part

$$\bar{\nu}_{CT} = 30(3.3 - 2.35) + 22.5 - 4.4 = 46.6 \text{ kK} \tag{78}$$

which agrees well with the value of 47.6 kK found in the spectrum. The accuracy for the calculation of electron-transfer bands by this method is limited by the uncertainty in the choice of electronegativities. In the example outlined above, the x values of the ligand and the central ion are only defined within 0.1 units. This limitation yields an uncertainty for the charge-transfer band position of at least 3 kK. The excellent agreement of our result with the experiment is only accidental. The prediction of other charge-transfer bands is, in general, not so reliable as in our example.

Spectra of complexes with two or more different ligands give information about geometry and electronic structure. For hexacoordinated compounds with two inequivalent ligands, there are eight different complexes possible: MA_5B, *trans*-MA_4B_2, *cis*-MA_4B_2, (1,2,6)MA_3B_3, (1,2,3)MA_3B_3, and those where A and B are exchanged. We now discuss the spectra of *cis*- and *trans*-$[Co\ en_2Cl_2]^+$ compared to octahedral $[Co\ en_3]^{3+}$, as mentioned in Section I.B of this chapter. Figure 1 shows that both ligand-field spectra of the *cis* and *trans* complex are shifted to longer wavelengths[104]. This is due to the fact that a ligand is substituted by another one appearing earlier in the spectrochemical series (41) than ethylenediamine. Besides this, we notice a pronounced splitting of the two spin-allowed transitions which is due to the reduction in symmetry. Most important, however, is the observation that the ligand-field absorption intensity of the stereoisomers is so different that a distinction, and perhaps also an assignment, on the

Table 5. Predicted shifts and splittings of the two spin-allowed transitions of octahedral d^6 complexes due to a replacement of ligands according to Yamatera[4]. δ_σ and δ_π are the orbital shifts of σ and π interactions, respectively.

Transition	MA_5B $\mathbf{C}_{4v}$	*trans*-MA_4B_2 $\mathbf{D}_{4h}$	*cis*-MA_4B_2 $\mathbf{C}_{2v}$	(1,2,6)MA_3B_3 $\mathbf{C}_{2v}$	(1,2,3)MA_3B_3 $\mathbf{C}_{3v}$
	0 (A_2)	0 (A_{2g})	$\frac{1}{2}\delta_\sigma + \frac{1}{2}\delta_\pi$ (B_1)	$\frac{1}{2}\delta_\sigma + \frac{1}{2}\delta_\pi$ (A_2)	$\frac{1}{2}\delta_\sigma + \frac{1}{2}\delta_\pi$
T_{1g}:	$\frac{1}{4}\delta_\sigma + \frac{1}{4}\delta_\pi$	$\frac{1}{2}\delta_\sigma + \frac{1}{2}\delta_\pi$	$\frac{1}{4}\delta_\sigma + \frac{1}{4}\delta_\pi$	$\frac{1}{4}\delta_\sigma + \frac{1}{4}\delta_\pi$ (B_2)	$\frac{1}{2}\delta_\sigma + \frac{1}{2}\delta_\pi$ (A_2, E)
	(E)	(E_g)	(A_2, B_2)		
	$\frac{1}{4}\delta_\sigma + \frac{1}{4}\delta_\pi$	$\frac{1}{2}\delta_\sigma + \frac{1}{2}\delta_\pi$	$\frac{1}{4}\delta_\sigma + \frac{1}{4}\delta_\pi$	$\frac{3}{4}\delta_\sigma + \frac{3}{4}\delta_\pi$ (B_1)	$\frac{1}{2}\delta_\sigma + \frac{1}{2}\delta_\pi$
	$\frac{1}{3}\delta_\sigma$ (B_2)	$\frac{2}{3}\delta_\sigma$ (B_{2g})	$\frac{1}{6}\delta_\sigma + \frac{1}{2}\delta_\pi$ (A_1)	$\frac{1}{2}\delta_\sigma + \frac{1}{2}\delta_\pi$ (A_2)	$\frac{1}{2}\delta_\sigma + \frac{1}{2}\delta_\pi$
T_{2g}:	$\frac{1}{12}\delta_\sigma + \frac{1}{4}\delta_\pi$	$\frac{1}{6}\delta_\sigma + \frac{1}{2}\delta_\pi$	$\frac{5}{12}\delta_\sigma + \frac{1}{4}\delta_\pi$	$\frac{3}{4}\delta_\sigma + \frac{1}{4}\delta_\pi$ (B_2)	$\frac{1}{2}\delta_\sigma + \frac{1}{2}\delta_\pi$ (A_1, E)
	(E)	(E_g)	(A_2, B_2)		
	$\frac{1}{12}\delta_\sigma + \frac{1}{4}\delta_\pi$	$\frac{1}{6}\delta_\sigma + \frac{1}{2}\delta_\pi$	$\frac{5}{12}\delta_\sigma + \frac{1}{4}\delta_\pi$	$\frac{1}{4}\delta_\sigma + \frac{3}{4}\delta_\pi$ (B_1)	$\frac{1}{2}\delta_\sigma + \frac{1}{2}\delta_\pi$

basis of their spectra is possible. From the coefficients given by Yamatera[4] and McClure[5] (see Table 5) for the band shifts of the two spin-allowed transitions, we see that the first band of the *trans* component undergoes a shift with a larger transition probability compared to the *cis* isomer. The $\frac{1}{2}\delta_\sigma + \frac{1}{2}\delta_\pi$ shift belongs in the *trans* compound to a two-fold degenerate state (E_g) while in the *cis* component it describes the shift of a non-degenerate state (B_1). The shift of the other component (A_2, B_2) of the first spin-allowed band is predicted for the *cis* complex to be only half as large, namely $\frac{1}{4}\delta_\sigma + \frac{1}{4}\delta_\pi$. The theory is nicely confirmed in practice. A curve analysis of the first band in the spectrum of the *cis* complex (Figure 1) shows that the (A_2, B_2) component lies in the centre between the first component (E_g) of the *trans* complex and the first band (T_{1g}) of the unsubstituted trisethylenediamine complex. With this observation, we are able to make a safe assignment of the two isomers: the complex absorbing at longer wavelengths is the *trans* component, the other one, which has also a higher extinction coefficient because it lacks a centre of inversion, is the *cis* isomer. The other bands are assigned to the following transitions (only the excited states are given)[105]:

cis:	16.5 kK	$B_1(T_{1g})$	*trans*:	16.2 kK	$E_g(T_{1g})$
	18.9	$A_2, B_2(T_{1g})$		22.0	$A_{2g}(T_{1g})$
	25.6	$A_1, A_2, B_2(T_{2g})$		25.8	$B_{2g}, E_g(T_{2g})$
	32.0	CT $\pi_{Cl} \rightarrow e_g$		32.5	CT $\pi_{Cl} \rightarrow e_g$
	~42	CT $\sigma_{Cl} \rightarrow e_g$		40.3	CT $\sigma_{Cl} \rightarrow e_g$

Charge-transfer bands are also found in the spectra which do not depend very much on the geometric structure of the complex.

C. Tetrahedral and Square Planar d^7 and d^8 Complexes: Co^{II}, Pd^{II}, Pt^{II}

Tetrahedral complexes are far less frequently encountered than octahedral complexes because their crystal-field stabilization for most d^n configurations is not so large compared to the octahedral coordination[95]. However, there are quite a number of examples, such as Co^{II}, Ni^{II}, Mn^{II}, Fe^{III}, and Cu^{II}, where tetrahedral coordination is definite.

In combination with the halogens and thiocyanate, Co^{II} occurs in tetrahedral coordination but octahedral Co^{II} complexes are known, so that this ion serves as an excellent example to test the validity of the relation between the cubic ligand-field parameters (33) derived from crystal-field and simple MO theory. If we compare the ligand-field parameter of $Co(NH_3)_6^{2+}$ and $Co(NCS)_4^{2-}$, i.e. $\Delta_{oct} = 10.1$ kK, $\Delta_{tetr} = -4.9$ kK[26], we find that the relation $\Delta_{tetr}/\Delta_{oct} = -\frac{4}{9}$ is rather well obeyed. This suggests that we might be able to compare complexes which have different ligands in the tetrahedral and octahedral species if these ligands are at a similar position in the spectrochemical series. Recently, Reinen[106] measured the light absorption of Co^{II} and Ni^{II} in solid oxides with garnet structure where these two ions occupy octahedral and tetrahedral interstices in the lattice. It was found that a semiempirical calculation fits the bands in the spectrum if one assumes for Co^{II} a parameter $\Delta_{oct} = 9.2$ kK in the octahedral and $\Delta_{tetr} = -4.2$ kK in the tetrahedral environment, so that these values also reproduce the ratio $-\frac{4}{9}$ very well. For Ni^{II} the relation is not quite as well fulfilled but the result agrees with the spectral properties of tetrahedral Ni^{II} halogen complexes in fused salts of LiCl and KCl. Several other workers[107,108] performed a closer analysis of the spectra of these compounds and came to the conclusion that an interpretation of the results is only possible by a complete calculation, as was done on the d^2 and d^8 configuration by Liehr and Ballhausen[109]. The ligand-field parameter derived from the fused salt spectra is $\Delta_{tetr} = -3.8$ kK which is compared to $\Delta_{oct} = 7.2$ kK of $NiCl_6^{4-}$ giving the relation $\Delta_{tetr}/\Delta_{oct} = -0.53$ which is in agreement with Reinens[106] results (i.e. -0.54) for four- and six-coordinated Ni^{II} in the oxide lattice.

Complexes with the coordination number four but having another symmetry are realized in square planar d^8 compounds of Pd^{II} and Pt^{II}. The nuclear configuration has the symmetry of the point group $\mathbf{D}_{4h}$. As seen from Figure 8, a d-orbital state is split in this field into $d \rightarrow a_{1g}, b_{1g}, b_{2g}, e_g$. The symmetry functions are also given in the figure. Extensive calculations on planar d^8 complexes have been performed on the basis of ligand-field theory by Fenske, Martin, and Ruedenberg[110] and by means of simple MO theory by Gray and Ballhausen[111]. They report the ordering of d-orbital energies as shown in Figure 8 which disagrees with the earlier results of Chatt, Gamlen, and Orgel[112] who proposed the order $a_{1g} < e_g < b_{2g} < b_{1g}$. Another sequence $e_g < b_{2g} < a_{1g}$ for occupied orbitals was given in the case of cyanide complexes by Perumareddi, Liehr, and Adamson[113]. All authors agree that the b_{1g} orbital state is the highest. From this situation the following molecular states arise if the eight d electrons are placed in the orbitals of $\mathbf{D}_{4h}$ symmetry

$e_g^4 a_{1g}^2 b_{2g}^2$	${}^1A_{1g}$	ground state
$e_g^4 a_{1g}^2 b_{2g} b_{1g}$	${}^{1,3}A_{2g}$	
$e_g^4 a_{1g} b_{2g}^2 b_{1g}$	${}^{1,3}B_{1g}$	
$e_g^3 a_{1g}^2 b_{2g}^2 b_{1g}$	${}^{1,3}E_g$	

Nothing can be said about the ordering of molecular states in the energy scheme if the correct orbital sequence is not established. In each case, therefore, the authors give different band assignments for the spectrum of

Table 6. Band assignments in the spectrum of aqueous $PtCl_4^{2-}$ (as taken from ref. 114).

Band (kK)	Extinction coeff.	Assignment to excited state			
		Ref. 112	Ref. 110	Ref. 111	Ref. 114
17.2	(2)	$^3A_{2g}$		$^3A_{2g}$	
21.0	15	^3E_g	$(^3A_{2g}, {}^3B_{1g})$	$^1A_{2g}$	$^1B_{1g}$
25.4	57	$^1A_{2g}$	$^1A_{2g}$	$^1B_{1g}$	$^1A_{2g}$
30.3	62	^1E_g	$^1B_{1g}$	^1E_g	^1E_g

aqueous $PtCl_4^{2-}$ (see Table 6). The absorption spectrum of single crystals with light polarized parallel to and perpendicular to the four-fold symmetry axis, together with the results of the Faraday effect, now seem to clarify these questions. Day and coworkers[115] came to the conclusion that the assignments given by Fenske and coworkers[110] and Chatt and coworkers[112] are the only ones compatible with their polarized light spectra. Martin and collaborators[116] agreed in their second paper with the Chatt, Gamlen, and Orgel[112] orbital sequence. Their MO calculations, which included also spin–orbit coupling effects, also indicate that the band assignments given by the latter workers[112] are the most probable ones. This was finally confirmed by the results obtained from the magnetic circular dichroism of the $PtCl_4^{2-}$ ion. These measurements[117] allowed a definite assignment of the 30.3 kK band to the ^1E_g excited state.

For other ligands, e.g. cyanide or ammines, these arguments need not be relevant. It is rather probable that out-of-plane π-bonding effects in ammines are so weak that for these ligands the e_g-orbital state is the lowest.

D. Low-symmetry Rare Earth Complexes MX_9

We shall now discuss a series of rare earth complex molecules with a low point symmetry which is rather unusual for transition group central ions but quite common for rare earth complexes. The enneaaquo complexes of M^{III} rare earth ions MX_9 have six ligands arranged in a trigonal prism and three others in an equilateral triangle parallel to the triangles of the prism but rotated by 60°. This type of coordination belongs to the point symmetry group $\mathbf{D}_{3h}$. We will now use this example to demonstrate briefly the application of the Ξ method outlined in the discussion of MO theory.

From symmetry considerations, we know that an f-orbital state is split in $\mathbf{D}_{3h}$ symmetry into

$$f \rightarrow a_1' + a_2' + e' + a_2'' + e'' \quad (79)$$

The symmetry orbitals for σ ligand–central field interactions are given[57] in Table 7. The orbitals belonging to a_2' describe pure π-bonding effects which in general can be neglected for water ligands.

Table 7. Some symmetry orbitals for MX_9 chromophores in $\mathbf{D}_{3h}$.

	f-orbital angular part	Ligand orbitals[a]
a_1'	$\frac{1}{4}(70)^{\frac{1}{2}}x(x^2 - 3y^2)/r^3$	$(\sigma_1 + \sigma_2 + \sigma_3)/(3)^{\frac{1}{2}}$
		$(\sigma_4 + \sigma_5 + \sigma_6 + \sigma_7 + \sigma_8 + \sigma_9)/(6)^{\frac{1}{2}}$
a_2''	$\frac{1}{2}(7)^{\frac{1}{2}}z(5z^2 - 3r^2)/r^3$	$(\sigma_4 + \sigma_5 + \sigma_6 - \sigma_7 - \sigma_8 - \sigma_9)/(6)^{\frac{1}{2}}$
e'	$\frac{1}{4}(42)^{\frac{1}{2}}y(5z^2 - r^2)/r^3$	$(\sigma_2 - \sigma_3)/(2)^{\frac{1}{2}}$
		$(\sigma_5 - \sigma_6 + \sigma_8 - \sigma_9)/2$
e''	$(105)^{\frac{1}{2}}(xyz)/r^3$	$(\sigma_5 - \sigma_6 - \sigma_8 + \sigma_9)/2$

[a] The numbers 1–3 denote the ligands of the equilateral triangle, the numbers 4–9 those of the prism.

When inserting these functions for every symmetry state in (63), we get the following Ξ values:

$$\begin{aligned} a_1' &\quad \Xi^2 = \tfrac{105}{8} + \tfrac{105}{4}\chi^6 \\ a_2'' &\quad \Xi^2 = \tfrac{189}{2}\chi^4\xi^2 - 126\chi^2\xi^4 + 42\xi^6 \\ e' &\quad \Xi^2 = \tfrac{63}{16} + \tfrac{63}{8}\chi^6 - 63\chi^4\xi^2 + 126\chi^2\xi^4 \\ e'' &\quad \Xi^2 = \tfrac{315}{4}\chi^4\xi^2 \end{aligned} \tag{80}$$

where χ and ξ denotes the position of a ligand in the prism. These co-ordinates represent the parameters which are given when choosing the relative dimensions of the trigonal prism and the equatorial triangle. If equal distances are assumed for all nine ligands, these parameters are connected by $\chi^2 + \xi^2 = 1$. Using (62) and assuming, with Helmholz and Wolfsberg[58], that the orbital energy of the antibonding states is proportional to S_{AL}^2, we get the following orbital-energy shifts relative to the atomic system

$$a_1' \quad 16.41\sigma^* \qquad e' \quad 12.80\sigma^*$$
$$a_2'' \quad 1.31\sigma^* \qquad e'' \quad 9.84\sigma^*$$

where σ^* is essentially a two-centre integral over the radial parts of the function which can be considered as a scaling parameter. One may also calculate it using Hartree–Fock functions, which are known[118] for the central ions. Figure 18 shows the orbital energies derived from the absorption spectra of lanthanide enneaaquo ethylsulphates compared to the predicted levels when using two parameter sets for χ, ξ in the present

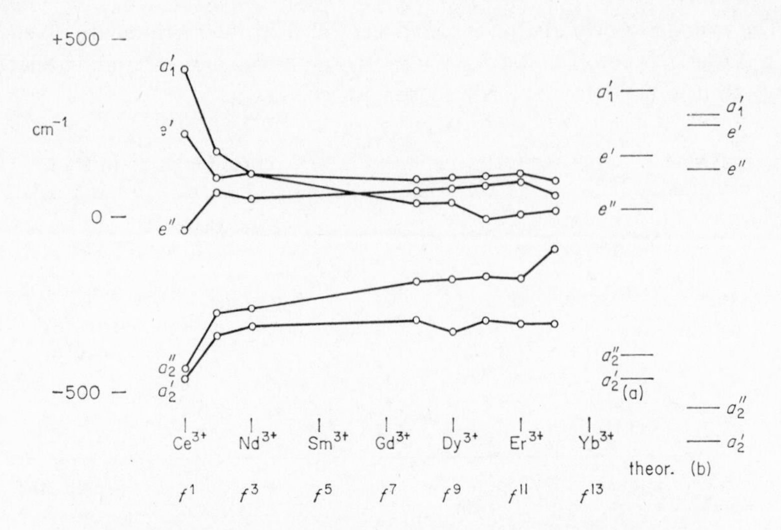

Figure 18. Orbital-energy diagram for enneaaquo ethylsulphates of different lanthanides together with the predicted results from different parameter sets. [Reproduced, by permission, from *J. Chem. Phys.*, **39**, 1422 (1963).]

theory. The agreement of theory and experiment, particularly for the f^1 and f^2 complexes, is remarkably good if one bears in mind the number of simplifications which have been made in this theory.

E. Closed-shell Compounds

Complex formation is also possible for central atoms or ions with empty or completely filled *d*-shell configurations. Although the electron distribution in closed-shell ions is spherically symmetrical, these ions are nevertheless able to form low-symmetry complexes. As we have seen from the preceding paragraphs, ions with incomplete *d*-electron shells, which in general have an unsymmetrical electron distribution, also form highly symmetric (cubic) compounds. Unfortunately, the value of the information obtainable from visible and ultraviolet spectroscopy is far less for closed-shell than it is for partly filled shell complex compounds. Since a spherically symmetrical electronic term is not affected by a ligand-field component of lower symmetry, we cannot expect, when applying symmetry arguments, that a splitting of the atomic terms will occur from which ligand-field bands in the spectrum will originate. It is, however, this part of the spectrum which is of predominant importance for drawing conclusions about the symmetry or electronic structure of metal complexes.

In spite of this drawback, other types of bands still remain to be interpreted which belong to other transition mechanisms. There are charge-transfer bands due to electron-jumps from the ligands to the empty d-orbital shell or for post-transition ions to higher atomic s orbitals. More frequent, however, are $nl \rightarrow n'l'$ central-ion transitions, which are energetically favoured for post-transition elements which have low-lying ($\sim$10 ev) excited states. Table 8 gives an account of possible transi-

Table 8. Possible transitions for closed-shell complex ions.

Ground-state configuration	Excited-state configuration	Possible examples
A. *Central-ion transitions*		
nd^{10}, ^1S	$nd^9(n+1)s$, $^{1,3}D$	Cu^+, Ag^+, Au^+ Zn^{2+}, Cd^{2+}, Hg^{2+}
$nd^{10}(n+1)s^2$, ^1S	$nd^{10}(n+1)s(n+1)p$, $^{1,3}P$	In^+, Tl^+, Sn^{2+}, Pb^{2+} As^{3+}, Sb^{3+}, Bi^{3+}
np^6, ^1S	$np^5(n+1)s$, $^{1,3}P$	halogenides
B. *Charge-transfer transitions*		
nd^0	$\sigma^{-1}d$ or $\pi^{-1}d$	V^{5+}, Cr^{6+}, Mn^{7+} Mo^{6+}, Tc^{7+}, Ru^{8+} W^{6+}, Re^{7+}, Os^{8+}
nd^{10}	$\sigma^{-1}s$ or $\pi^{-1}s$	Cu^+, Ag^+, Au^+ Hg^{2+}, Sn^{4+}, Pb^{4+} Sb^{5+}
nd^{10}	$nd^9\pi^*$	Cu^+, Ag^+

σ^{-1}, π^{-1} denotes a hole in a σ- or π-bonded ligand orbital, respectively.

tions for complexes of closed-shell ions. Usually, the absorption arising mainly within the metal ion occurs on the long-wavelength side of the spectrum for both the transition and post-transition ions. For open-shell transition element complexes, a $d^q \rightarrow d^{q-1}s$ transition is very rare. For instance, the 40.3 kK band of $[Fe(H_2O)_6]^{2+}$ is believed[105] to be a $t_{2g}(3d) \rightarrow a_{1g}(4s)$ transition. On the other hand, $nf \rightarrow (n+1)d$ transitions are quite common in lanthanides and actinides[30].

Typical d closed-shell complexes with $nd \rightarrow (n+1)s$ transitions are $Cu(CN)_4^{3-}$, $Ag(H_2O)_2^+$, $Ag(NH_3)_2^+$, or $HgCl_4^{2-}$ which have moderately high absorption coefficients, except perhaps for the mercury halide solid-state spectra where a charge-transfer mechanism must be involved[119,121]. The assignment of the 24 kK band of tetrahedral complexes of solid CuCl and CuI also is not clear; it may be due either to a charge transfer π (ligands) $\rightarrow a_1$(Cu), a $3d$(Cu) $\rightarrow a_1$(Cu), or to a transition where both mechanisms are mixed. McMahon and Forro[120,121] found a temperature dependence for the absorption strength of Cu^+ and Ag^+ in NaCl or KCl which indicates the participation of a $d \rightarrow s$ transition because of differing vibrational perturbations of the forbidden transitions.

Examples of $s^2 \rightarrow sp$ transitions in crystals are frequent. Seitz[122] assigned the 40.8 and 51 kK bands of Tl^{I} in KCl to transitions from the $6s^2$, $^1S_0$ ground state to the $6s6p$, $^3P_1$ and $^1P_1$ excited levels. McClure[30] discusses this spectrum at length together with other alkali halides doped with Tl^{I} for which a large quantity of experimental results is available. Jørgensen[123] remarks that for a series of complexes the wave numbers of these transitions decrease in the same order as the nephelauxetic series (68). The spectra of Pb^{2+} in alkali halides were measured by Merritt and co-workers[119] and by Yuster and Delbecq[124]. The shape of the chromophore has not been definitely determined for these compounds. In KCl, a hexa-coordinated species seems to exist[125] while in the bromides and iodide a PbX_4^{2-} chromophore is assumed. However, the symmetry of the environment does not seem to be important for these transitions. In certain cases of lead and tin halides, the J components of the excited ^3P state are also detected in the spectra. In Table 9, the wave numbers of the $s \rightarrow p$ transitions in these compounds are given.

Table 9. Wave numbers (kK) of $s \rightarrow p$ transitions in solid tin and lead halide compounds.

		$^3P_1$	$^1P_1$			$^3P_1$	$^1P_1$
Sn^{II}	free ion	55.2	80	Pb^{II}	free ion	64.5	95
	Cl^-	34–35.5	40–44		Cl^-	36.5	51
	Br^-	~32	35.5–40		Br^-	33.2	45
	I^-	~27	31.5–35		I^-	27.5	41

Electron-transfer spectra to empty d shells are found in highly ionized transition elements which are stable as oxides. The deep colour of the permanganate ion is due to a series of charge-transfer processes from the ligand σ- or π-bonding orbitals to the empty e and t_2 central-ion orbitals. In the spectrum, a distinct vibrational structure is observed[126]. The peaks at 19.0 and 32.3 kK are assigned[58] to $^1A_1 \rightarrow {}^1T_2$ transitions which are essentially charge-transfer processes from π-ligand orbitals to the lower lying e subshell. Other representatives of this category of charge-transfer transitions are CrO_4^{2-}, MoO_4^{2-}, TcO_4^-, RuO_4(gas), WO_4^{2-}, ReO_4^-, and OsO_4 which all show the same type of bands in their spectra.

Other electron-transfer processes are found in the spectra of post-transition elements. In complexes of these ions, the electron jumps to a higher s shell, giving rise to very strong absorption. The 44.9 kK band

($\varepsilon \sim 10{,}000$) of $SnCl_6^{2-}$ and the absorption at 36.9 kK ($\varepsilon \sim 8000$) in $SbCl_6^-$ is due to a parity-allowed transition from the $t_{1u}(\pi)$ ligand to the $a_{1g}(5s)$ central-ion orbital. Another example is given by $PbCl_6^{2-}$ which has two strong bands in the ultraviolet spectrum due to a parity-allowed transition from σ- and π-bonded ligands to the $a_{1g}(6s)$ central-ion orbital[69]. Also the 35.7 kK band of $Hg(SCN)_4^{2-}$ seems to be an electron-transfer transition of the same type.

Reverse electron-transfer transitions in coordination compounds of closed-shell central ions have so far been observed only in a few cases. The dipyridyl and phenanthroline complexes of Cu^I have strong bands at 23.0 kK[127] which are assigned to a transition from the t_2 central-atom orbital to an empty antibonding π^* molecular orbital of the organic ring system[105].

The list of central-ion, charge-transfer, and internal ligand transitions in the spectra of coordination compounds can be extended at great length and its number grows every day, since new unexpected compounds are found or old ones are investigated more carefully using new techniques in visible and ultraviolet spectroscopy. In any case, the approach outlined in this article may lead to a closer understanding of physical and chemical properties of coordination compounds.

IV. REFERENCES

1. R. G. Shulman and S. Sugano, *Phys. Rev.*, **130**, 506 (1963).
2. R. E. Watson and A. J. Freeman, *Phys. Rev.*, **134**, A1526 (1964).
3. A. D. Walsh, *J. Chem. Soc.*, **1953**, 2260.
4. H. Yamatera, *Bull. Chem. Soc. Japan*, **31**, 95 (1958).
5. D. S. McClure, 'Advances in the chemistry of coordination compounds', *Proc. Intern. Conf. Coord. Chem., 6th* (Ed. S. Kirschner), Macmillan, New York, 1961, p. 498.
6. C. E. Schäffer and C. K. Jørgensen, *J. Inorg. Nucl. Chem.*, **8**, 143 (1958).
7. H. L. Schläfer, *Z. Phys. Chem., Neue Folge*, **8**, 373 (1956).
8. C. K. Jørgensen, *Orbitals in Atoms and Molecules*, Academic Press, London, 1962, p. 94.
9. M. Linhard and M. Weigel, *Z. Anorg. Allgem. Chem.*, **266**, 73 (1951).
10. Y. Kondo, *Bull. Chem. Soc. Japan*, **28**, 497 (1955).
11. G. Briegleb, *Elektronen-Donator-Acceptor-Komplexe*, Springer, Berlin, 1961.
12. R. E. Dodd, *Chemical Spectroscopy*, Elsevier, Amsterdam, 1962.
13. H. H. Jaffé and M. Orchin, *Theory and Application of Ultraviolet Spectroscopy*, Wiley, New York, 1962.
14. M. Pestemer, *Anleitung zum Messen von Absorptionsspektren im Ultraviolett und Sichtbaren*, Georg Thieme, Stuttgart, 1964.
15. R. N. Dixon, *Spectroscopy and Structure*, Methuen, London, 1965.
16. H. H. Cary and A. O. Beckman, *J. Opt. Soc. Am.*, **31**, 628 (1941).
17. H. H. Cary, *Rev. Sci. Instr.*, **17**, 588 (1946).

18. W. R. Brode, *Chemical Spectroscopy*, Wiley, New York, 1943.
19. G. R. Harrison, R. C. Lord, and J. R. Loofbourow, *Practical Spectroscopy*, Prentice-Hall, Englewood Cliffs, N. J., 1948.
20. F. X. Mayer and A. Luszczak, *Absorptionsspektralanalyse*, De Gruyter, Berlin, 1951.
21. J. M. Vandenbelt and C. Henrich, *Appl. Spectry.*, **7**, 173 (1953).
22. C. K. Jørgensen, *Acta Chem. Scand.*, **8**, 1495 (1954).
23. L. E. Orgel, *An Introduction to Transition-metal Chemistry*, Methuen, London, 1960.
24. J. S. Griffith, *The Theory of Transition-metal Ions*, Cambridge University Press, Cambridge, 1961.
25. C. J. Ballhausen, *Introduction to Ligand Field Theory*, McGraw-Hill, New York, 1962.
26. C. K. Jørgensen, *Absorption Spectra and Chemical Bonding in Complexes*, Pergamon Press, Oxford, 1962.
27. H. L. Schläfer and G. Gliemann, *Einfuhrung in der Ligandenfeldtheorie*, Akademische Verlagsgesellschaft, Frankfurt-am-Main, 1967.
28. H. Hartmann and H. L. Schläfer, *Angew. Chem.*, **66**, 768 (1954).
29. W. Moffitt and C. J. Ballhausen, *Ann. Rev. Phys. Chem.*, **7**, 107 (1956).
30. D. S. McClure, *Solid State Phys.*, **9**, 399 (1959).
31. H. Bethe, *Ann. Phys.*, [5], **3**, 133 (1929).
32. H. Bethe, *Z. Phys.*, **60**, 218 (1930).
33. H. A. Kramers, *Proc. Acad. Sci.* (*Amsterdam*), **32**, 1176 (1929); **33**, 959 (1930).
34. J. H. Van Vleck, *Electric and Magnetic Susceptibilities*, Oxford University Press, London, 1932, p. 384.
35. J. H. Van Vleck, *Phys. Rev.*, **41**, 208 (1932).
36. R. G. Schlapp and W. G. Penney, *Phys. Rev.*, **41**, 194 (1932); **42**, 666 (1932).
37. F. E. Ilse and H. Hartmann, *Z. Phys. Chem.*, **197**, 239 (1951).
38. H. Hartmann and F. E. Ilse, *Z. Naturforsch.*, **6a**, 751 (1951).
39. H.-H. Schmidtke, *Z. Naturforsch.*, **18a**, 276 (1963).
40. K. Fajans, *Naturwissenschaften*, **11**, 165 (1923).
41. R. Tsuchida, *Bull. Chem. Soc. Japan*, **13**, 388, 436 (1938).
42. H.-H. Schmidtke, *J. Am. Chem. Soc.*, **87**, 2522 (1965).
43. H. A. Jahn and E. Teller, *Proc. Roy. Soc.* (*London*), *Ser. A*, **161**, 220 (1937).
44. H. A. Jahn, *Proc. Roy. Soc.* (*London*), *Ser. A*, **164**, 117 (1938).
45. E. Ruch and A. Schönhofer, *Theor. Chim. Acta* (*Berlin*), **3**, 291 (1965).
46. E. U. Condon and G. H. Shortley, *The Theory of Atomic Spectra*, Cambridge University Press, Cambridge, 1959.
47. H.-H. Schmidtke, *Fortschr. Phys.*, **13**, 211 (1965).
48. T. M. Dunn, *Trans. Faraday Soc.*, **57**, 1441 (1961).
49. J. H. E. Griffiths and J. Owen, *Proc. Roy. Soc.* (*London*), *Ser. A*, **226**, 96 (1954).
50. C. J. Ballhausen and A. D. Liehr, *J. Mol. Spectry.*, **2**, 342 (1958).
51. J. H. Van Vleck, *J. Chem. Phys.*, **3**, 803, 807 (1935).
52. J. H. Van Vleck and A. Sherman, *Rev. Mod. Phys.*, **7**, 167 (1935).
53. C. J. Ballhausen and H. B. Gray, *Molecular Orbital Theory*, Benjamin, New York, 1964.

54. J. C. Eisenstein, *J. Chem. Phys.*, **25**, 142 (1956).
55. H.-H. Schmidtke, *Z. Naturforsch.*, **19a**, 1502 (1964).
56. C. E. Schäffer and C. K. Jørgensen, *Mol. Phys.*, **9**, 401 (1965).
57. C. K. Jørgensen, R. Pappalardo, and H.-H. Schmidtke, *J. Chem. Phys.*, **39**, 1422 (1963).
58. M. Wolfsberg and L. Helmholz, *J. Chem. Phys.*, **20**, 837 (1952).
59. F. A. Cotton and T. E. Haas, *Inorg. Chem.*, **3**, 1004 (1964).
60. R. F. Fenske, *Inorg. Chem.*, **4**, 33 (1965).
61. W. G. Perkins and G. A. Crosby, *J. Chem. Phys.*, **42**, 407 (1965).
62. P. Day and C. K. Jørgensen, *J. Chem. Soc.*, **1964**, 6226.
63. C. E. Schäffer and C. K. Jørgensen, *Kgl. Danske Vidensk. Selsk., Math.-fys. Medd.*, **34**, No. 13 (1965).
64. C. Racah, *Phys. Rev.*, **62**, 438 (1942); **63**, 367 (1943).
65. Y. Tanabe and S. Sugano, *J. Phys. Soc. Japan*, **9**, 753, 766 (1954).
66. L. E. Orgel, *J. Chem. Soc.*, **1952**, 4756.
67. J. Owen, *Proc. Roy. Soc. (London), Ser. A*, **227**, 183 (1955).
68. C. K. Jørgensen, *Progr. Inorg. Chem.*, **4**, 73 (1962).
69. C. K. Jørgensen, *Mol. Phys.*, **2**, 309 (1959).
70. C. K. Jørgensen, *Mol. Phys.*, **5**, 271 (1962).
71. C. K. Jørgensen, 'Inorganic chromosphores', in *Essays in Coordination Chemistry* (dedicated to G. Schwarzenbach), Birkhäuser, Basel, 1964.
72. R. J. P. Williams, *J. Chem. Soc.*, **1955**, 137.
73. H.-H. Schmidtke, *Z. Phys. Chem., Neue Folge*, **38**, 170 (1963).
74. H. Yoneda, *Bull. Chem. Soc. Japan*, **28**, 125 (1955).
75. J. C. Barnes and P. Day, *J. Chem. Soc.*, **1964**, 3886.
76. W. A. Baker, Jr., and M. G. Phillips, *Inorg. Chem.*, **4**, 915 (1965).
77. C. K. Jørgensen, *Absorption Spectra of Complexes of Heavy Metals*, research report to European Research Office, U.S. Department of the Army, Frankfurt/Main, contract No. DA-91-508-EUC-247, 1958.
78. R. Englman, *Mol. Phys.*, **3**, 48 (1960).
79. S. Herzog and R. Taube, *Z. Chem.*, **2**, 208, 225 (1962).
80. E. König, H. L. Schläfer, and S. Herzog, *Z. Chem.*, **4**, 95 (1964).
81. E. König, *Z. Naturforsch.*, **19a**, 1139 (1964).
82. S. Ahrland, J. Chatt, and N. R. Davies, *Quart. Rev. (London)*, **12**, 265 (1958).
83. R. G. Pearson, *J. Am. Chem. Soc.*, **85**, 3533 (1963).
84. A. Davison, N. Edelstein, R. H. Holm, and A. H. Maki, *Inorg. Chem.*, **2**, 1227 (1963).
85. H. B. Gray and E. Billig, *J. Am. Chem. Soc.*, **85**, 2019 (1963).
86. J. H. Waters, R. Williams, H. B. Gray, G. N. Schrauzer, and H. W. Finck, *J. Am. Chem. Soc.*, **86**, 4198 (1964).
87. S. I. Shupack, E. Billig, R. J. H. Clark, R. Williams, and H. B. Gray, *J. Am. Chem. Soc.*, **86**, 4594 (1964).
88. E. Billig, S. I. Shupack, J. H. Waters, R. Williams, and H. B. Gray, *J. Am. Chem. Soc.*, **86**, 926 (1964).
89. A. H. Maki, N. Edelstein, A. Davison, and R. H. Holm, *J. Am. Chem. Soc.*, **86**, 4580 (1964).
90. C. K. Jørgensen, *Z. Naturwiss.-Med. Grundlagenforsch.*, **2**, 248 (1965).
91. H. L. Schläfer, *Proc. Symp. Theory Structure Complex Compounds, Wroclaw, Poland, 1962*, Pergamon Press, Oxford, 1964.

92. See B. W. Low, *Paramagnetic Resonance in Solids*, Academic Press, New York, 1960, p. 80.
93. V. J. Avvakumov, N. S. Garif'Yanov, S. G. Salikhov, and E. J. Semenova, *Fiz. Tverdogo Tela*, **3**, 2111 (1961).
94. J. H. Van Vleck, *J. Chem. Phys.*, **7**, 72 (1939).
95. J. D. Dunitz and L. E. Orgel, *Advan. Inorg. Chem. Radiochem.*, **2**, 1 (1960).
96. O. G. Holmes and D. S. McClure, *J. Chem. Phys.*, **26**, 1686 (1957).
97. C. E. Beevers and L. Lipson, *Proc. Roy. Soc.* (*London*), *Ser. A*, **146**, 570 (1934).
98. J. Bjerrum, C. J. Ballhausen, and C. K. Jørgensen, *Acta Chem. Scand.*, **8**, 1275 (1954).
99. P. Day, *Proc. Chem. Soc.*, **1964**, 18.
100. O. Schmitz-Dumont and H. Fendel, *Monatsh.*, **96**, 495 (1965).
101. C. J. Ballhausen, *Kgl. Danske Vidensk. Selsk., Math.-fys. Medd.*, **29**, No. 4 (1954).
102. R. L. Belford, M. Calvin, and G. Belford, *J. Chem. Phys.*, **26**, 1165 (1957).
103. H.-H. Schmidtke, *Z. Phys. Chem., Neue Folge*, **40**, 96 (1964).
104. M. Linhard and M. Weigel, *Z. Anorg. Allgem. Chem.*, **271**, 101 (1952).
105. C. K. Jørgensen, *Advan. Chem. Phys.*, **5**, 33 (1963).
106. D. Reinen, *Z. Anorg. Allg. Chem.*, **327**, 238 (1964).
107. D. M. L. Goodgame, M. Goodgame, and F. A. Cotton, *J. Am. Chem. Soc.*, **83**, 4161 (1961).
108. C. Furlani and G. Morpurgo, *Z. Phys. Chem., Neue Folge*, **28**, 93 (1961).
109. A. D. Liehr and C. J. Ballhausen, *Ann. Phys.* (*N.Y.*), **6**, 134 (1959).
110. R. F. Fenske, D. S. Martin, and K. Ruedenberg, *Inorg. Chem.*, **1**, 441 (1962).
111. H. B. Gray and C. J. Ballhausen, *J. Am. Chem. Soc.*, **85**, 260 (1963).
112. J. Chatt, G. A. Gamlen, and L. E. Orgel, *J. Chem. Soc.*, **1958**, 486.
113. J. R. Perumareddi, A. D. Liehr, and A. W. Adamson, *J. Am. Chem. Soc.*, **85**, 249 (1963).
114. D. S. Martin, Jr., and C. A. Lenhardt, *Inorg. Chem.*, **3**, 1368 (1964).
115. P. Day, A. F. Orchard, A. J. Thomson, and R. J. P. Williams, *J. Chem. Phys.*, **42**, 1973 (1964).
116. D. S. Martin, Jr., M. A. Tucker, and A. J. Kassman, *Inorg. Chem.*, **4**, 1682 (1965).
117. D. S. Martin, Jr., J. G. Foss, M. E. McCarville, M. A. Tucker, and A. J. Kassman, *Inorg. Chem.*, **5**, 491 (1966).
118. A. J. Freeman and R. E. Watson, *Phys. Rev.*, **127**, 2058 (1962).
119. C. Merritt, H. M. Hershenson, and L. B. Rogers, *Anal. Chem.*, **25**, 572 (1953).
120. A. M. McMahon, *Z. Phys.*, **52**, 336 (1928).
121. M. Forro, *Z. Phys.*, **56**, 235 (1929); **58**, 613 (1929).
122. F. Seitz, *J. Chem. Phys.*, **6**, 150 (1938).
123. C. K. Jørgensen, *Solid State Phys.*, **13**, 375 (1962).
124. P. Yuster and C. Delbecq, *J. Chem. Phys.*, **21**, 892 (1953).
125. H. Fromherz and Kun-Hou Lih, *Z. Phys. Chem.* (*Leipzig*), A**153**, 321 (1931).
126. G. Den Boef, H. J. Van der Beek, and T. Braaf, *Rec. Trav. Chim.*, **77**, 1064 (1958).
127. R. J. P. Williams, *J. Chem. Soc.*, **1955**, 137.

5

Optical Rotatory Dispersion and Circular Dichroism

R. D. GILLARD

I. INTRODUCTION

The subject of optical activity has a long history even in its application to coordination chemistry. The Cotton effect was first discovered in tartrato complexes of transition metals, 70 years ago, before any transition metal complex had been resolved. This convenient term, 'Cotton effect', refers[1] to all the behaviour of electromagnetic radiation in the neighbourhood of an absorption band in an enantiomeric molecule.

In the inorganic field, steady advances have been made in the area of optical activity since the days of Cotton, and the present position draws deeply on the work of the pioneers Werner, Jaeger, Lifschitz, Kuhn, and Mathieu. The chief developments of the last 20 years have been in three directions; firstly, in instrumentation; secondly, in the possibility of establishing absolute configurations by experiment; and thirdly, in the interrelationship of optical configuration and spectroscopy.

The aim of the present chapter is to cover in detail a few examples of applications of the techniques of optical activity and to compare the results with theoretical predictions. We have been selective in the sense that many applications to chemical problems are not mentioned. However, in the more general area of inorganic applications of optical activity, recent reviews[2-5] are more comprehensive. Instrumental methods are not described in detail, although relevant matters are discussed.

There is a slight element of repetition in this chapter; this is intentional, on the grounds that optical activity is a subtle subject and that recent work incorporates several rather unfamiliar areas of inorganic research, ranging from the conformational analysis of chelate rings[6] to the vibrational fine structure of the circular dichroism of oriented single crystals. The whole picture of inorganic optical activity is made up of strands from these several apparently unrelated fields; we hope that the present account may remove some of the obscurity surrounding this extremely valuable method. There is the further difficulty that accepted spectroscopic ideas in the field have recently been challenged and an attempt has been made to describe the difficulties which have arisen, in the hope that this will stimulate critical experimentation.

Further, a number of points of difficulty are raised here which have often been ignored. The subject is now at the stage where rigorous arguments are quite essential and some of the more subtle features of coordination chemistry must be considered in any application of optical activity.

II. NOMENCLATURE AND CONFIGURATION

A. Nomenclature

The simplest observable property of an optically active compound is the rotation of the plane of polarization of monochromatic plane-polarized light of wavelength λ. The observed rotation α is measured in degrees;

dextrorotation (α positive) is clockwise rotation from the observer's standpoint. The specific rotation, for the wavelength λ, is defined as

$$[\alpha]_\lambda = \frac{100\alpha}{lc\rho} \tag{1}$$

where l is the path length of the light in the optically active medium, usually a solution, in *decimetres* (the use of this rather odd unit is a historical accident), c is the weight of the sample in 100 g of solution, and ρ is the density of the solution. The molecular rotation $[\phi]$ (in accordance with standard practice in organic chemistry) is given by

$$[\phi] = \frac{M[\alpha]}{100} \tag{2}$$

where M is the molecular weight of the optically active compound.

The yellow light of a sodium flame or lamp ($\lambda \sim 589$ mμ for the so-called D line, in fact a doublet) has commonly been used for measuring rotations, which are then called $[\alpha]_D$. Most organic compounds are called dextro- or laevorotatory, according to their effect on plane-polarized sodium yellow light. In the case of coordination compounds, the wavelength of observation should be appended (or at least clearly understood) after signs of rotation, since rotatory power changes, both in sign and in magnitude, with the wavelength of measurement. The sign (+) or (−) without a wavelength suffixed *always* refers to sodium yellow light. We can denote a particular enantiomer by such nomenclature as (+)-$[Co\,en_3]^{3+}$ or as $(-)_{5461}$-$[Rh\,en_3]^{3+}$. It should also be borne in mind that the nature of the medium may affect sizes (and occasionally signs) of rotation; for example, a large number of ion-pairing effects have been observed[7-10] which can alter the rotatory power of cations rather dramatically. Presumably, similar effects are to be expected for optically active anions forming ion pairs with optically inactive cations, though there is only one indication[7] that such effects occur.

The effect of the solvent on optical activity is commonly ignored. This has led to errors; one example of this is described later. In general, theoretical treatments of the solvent dependence of both electric dipole absorption and of circular dichroism are based on the traditionally employed Lorentz field in a dielectric medium. In the commonly used equations of the quantum theory of optical activity, which are discussed later, terms in $(n^2 + 2)/3$, where n is the refractive index, are mainly determined by the solvent and influence the observed rotations. However, there are also effects on *rotational strengths* of individual bands which are extremely difficult to treat theoretically, although notable efforts have been made[11].

Such effects arise from solvation or distortion of the molecules. In one of the very few cases where the effect of solvent variation on the rotatory power of a coordination compound was studied, the rotation of (+)-1,2,6-tris-(+)-hydroxymethylenecamphoratocobalt(III) at 500 mμ plotted against $\epsilon - 1$ for the solvent gave[12] a straight line. In any case, it is useful practice to describe the exact conditions of measurement, including solvent, temperature, and concentration.

B. Configuration

A complex molecule which lacks certain symmetry elements may be resolved[13] (kinetics permitting) into enantiomers, one enantiomer showing dextrorotation at a given wavelength, the other showing equal laevorotation. The problem of determining absolute configuration is to decide which absolute stereochemistry corresponds to a given sign of rotation for a given wavelength.

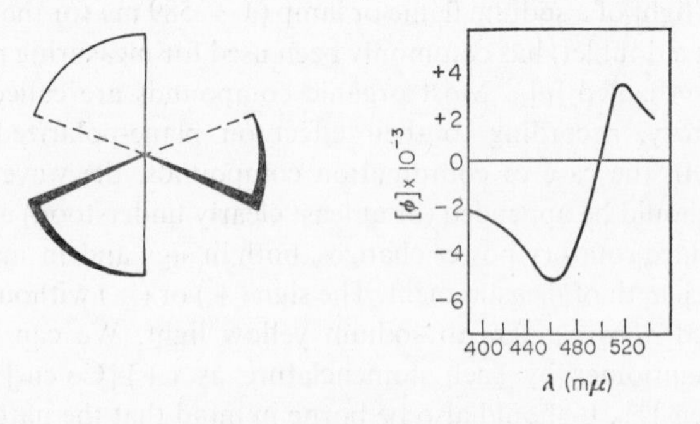

Figure 1. The absolute configuration and the rotatory dispersion spectrum of (+)-$[Co\ en_3]^{3+}$.

Three optically active complexes of known absolute configuration are shown in Figures 1–3, with their optical rotatory properties. The complex (+)-$[Co\ en_3]^{3+}$ of Figure 1 has been given several configurational labels (see Figure 4 below). The first of these[14] was D. This convention reflects perhaps the feeling that sign of rotation and configurational label should be similar where possible. Indeed, it happens to be true[15] that, for many complexes $[Co\ en_2XY]^{n+}$, the D enantiomer is dextrorotatory at the sodium *D* line. Subsequent conventions for the configurational label attached to (+)-$[Co\ en_3]^{3+}$ have all used the symmetry properties of the ion in its ground state. For example, the *Λ* nomenclature of Piper[16] indicates the left-handed helicity about the *principal* axis of the complex. This has the disadvantage that the relationship between complexes of the bis- and trisbidentate chelate series is obscured, since D-$[Co\ en_2Cl_2]^+$ would be called Δ in this convention, whereas D-[Co en$_3$]$^{3+}$ is called *Λ*. Later schemes take account of this by using all the axes of rotation of the complexes. (+)-[Co en$_3$]$^{3+}$ was called[17] $S(C_3)R(C_2)$, meaning *s*inistral helicity about the three-fold axis (Schoenflies' symbol $\mathbf{C}_3$) and *r*ight-handed (rectus) helicity about the two-fold axis (Schoenflies' symbol $\mathbf{C}_2$). This use of *S* and *R* was adapted from organic practice current[18] at that time. An alternative form[19] is $M(C_3)P(C_2)$, *M* standing for *m*inus and *P* for *p*lus. As shown in Figure 4, both these conventions preserve the natural relationship between D-[Co en$_2$X$_2$]$^+$ species, where X = Cl or $\frac{1}{2}$en.

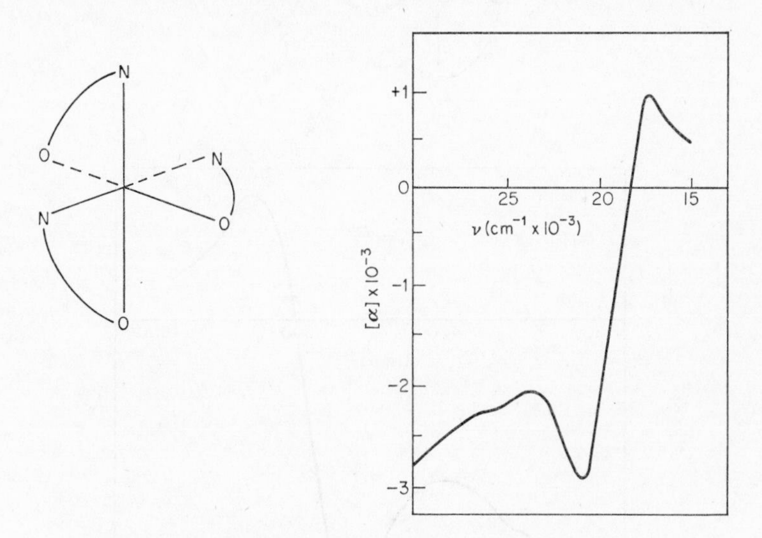

Figure 2. The absolute configuration and the optical rotatory dispersion spectrum of purple α-(+)-[Co(L-ala)$_3$].

The simplest method of naming complexes (D or L) is adequate in most of the more commonly encountered systems. However, the ideal configurational nomenclature is one where ambiguity is impossible, i.e. where the name or symbol given cannot possibly refer to any other configuration than that intended. For this reason, Cahn, Ingold, and Prelog produced[20] a rigorous configurational nomenclature designed to cover all cases. This is best regarded as a reference method; for convenience, the following simple D and L system for chelated compounds is used in this chapter.

It often transpires that the dominant sign of the Cotton effect in the longest wavelength absorption band of complexes with the d^3 or d^6 (spin-paired) electronic configuration reflects, irrespective of their nature, the helicity of the chelate rings around the metal ion. This is the basis of the

DL nomenclature. The configurational nomenclature for bis- and tris-chelated complexes used here is based on that for axially symmetric complexes. Figure 4 shows D configurations for complexes with axial symmetry, with the alternative nomenclature based on earlier proposals. To complete the nomenclature based on helicity for complexes with no axes of symmetry,

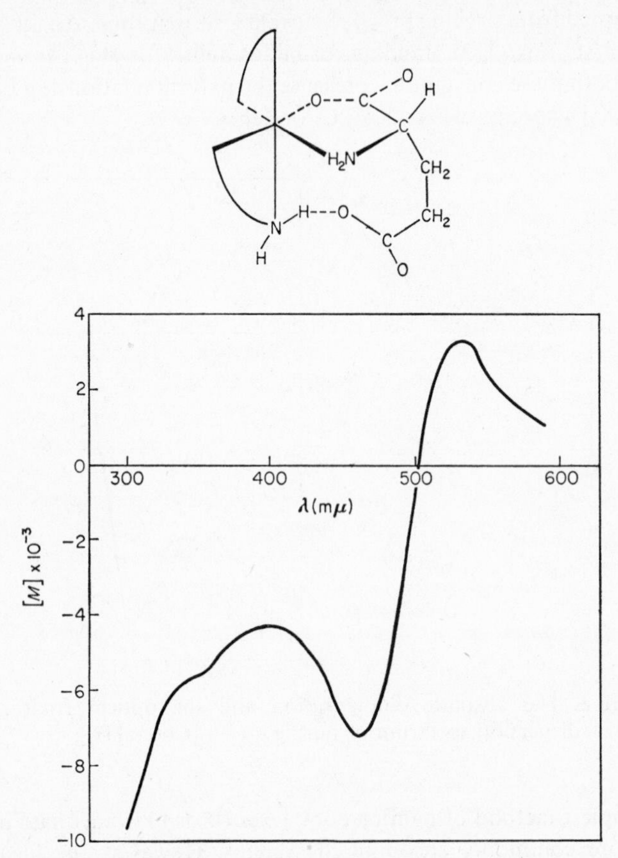

Figure 3. The absolute configuration and the optical rotatory dispersion spectrum of (+)-[Co en_2(glu)]ClO_4, where (glu) is the dianion of L-glutamic acid.

the following steps are taken (the treatment effectively neglects the nature of the bonding atoms and considers only the relative disposition of the chelate rings):

1. Replace each monodentate ligand by the letter X.
2. Replace any unsymmetrical chelating ligand AB or CD by AA.

The axial symmetry of the figure thus obtained is used to decide the configurational nomenclature, since the figure will either belong to the $\mathbf{D}_3$ or $\mathbf{C}_2$ point group and can be named D or L as in Figure 4. Examples are shown in Figure 5.

The absolute configuration of molecules may be determined by a number of methods, including the use of x-ray analysis. The fundamental advance in this field was made[21] by Bijvoet, who used the fact that Friedel's law

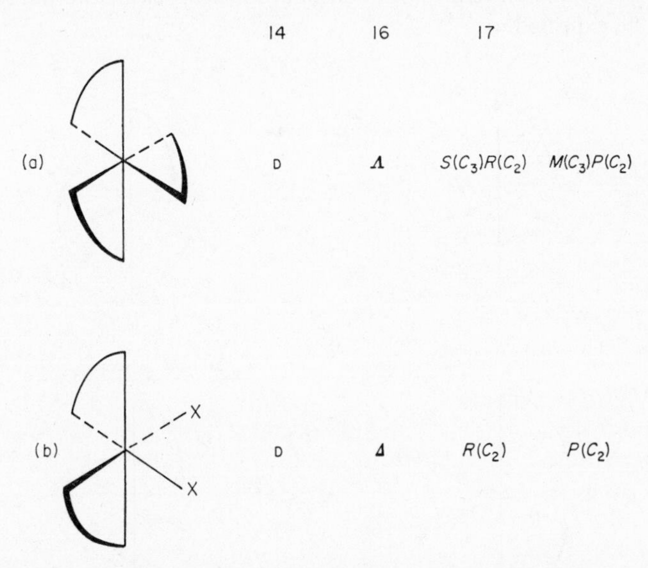

Figure 4. Configurational nomenclatures for (a) $[M(AA)_3]$ and (b) $[M(AA)_2X_2]$, where AA is a symmetrical chelating ligand, such as oxalate. The numbers are references.

$F_{hkl} = F_{\bar{h}\bar{k}\bar{l}}$ breaks down when a constituent of the crystal absorbs x-rays of the wavelength used. This method of anomalous dispersion of x-rays was used both for (+)-$[Co\ en_3]^{3+}$ [14] and for (−)-$[Co(-pn)_3]^{3+}$ [22], which had the D and L configurations, respectively. Very recently, the configuration of the stereospecifically formed complex (−)-sarcosinatobisethylenediaminecobalt(III) ion has been determined[23] by Bijvoet analysis, as has that[24] of (−)-$[Fe\ phen_3]^{2+}$.

A more easily applicable x-ray method relies on the presence in the ligand of an asymmetric atom of absolute configuration already known from another method. Normal x-ray methods give the relative configurations of all the sources of asymmetry; the known centre is then used to give absolute configurations. This method was used[25,26] for the complexes of

Figures 2 and 3. An advantage of this method is that the absolute configurations of a very large number of organic molecules are reliably known[18] from the classical work on relative configurations, combined with the more recent Bijvoet studies of the absolute configurations of a number of important reference compounds. Among the organic molecules of known configuration are a number of important ligands, including hydroxy acids, 1,2-diamines, α-amino acids, and peptides. Many more applications of this use of an asymmetric carbon atom as an internal reference are to be expected.

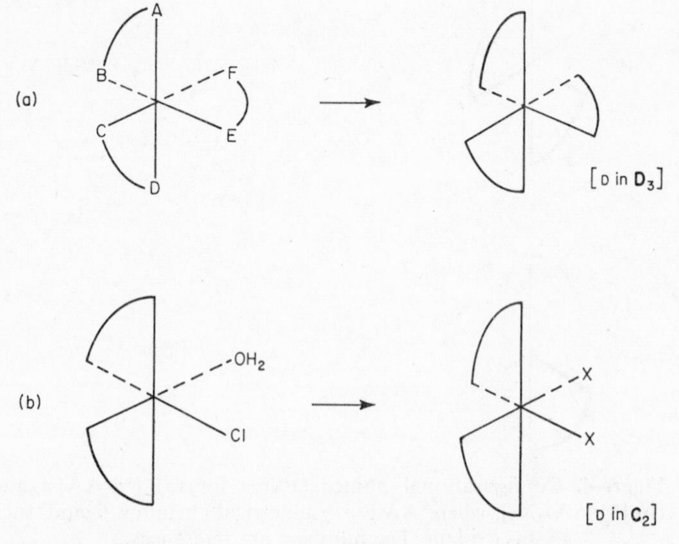

Figure 5. Applications of simplified D and L nomenclature to (a) a tris-chelated molecule of C_1 symmetry, and (b) a molecule with two chelate rings of C_1 symmetry.

Several other methods are available, particularly for relative configurations of coordination compounds. These include chemical correlations, in which[15] one optically active compound is converted into another without breaking any of the metal–ligand bonds. (This method avoids the difficulty of establishing the optical mechanism of reactions where metal–ligand bonds are broken.) Also useful has been the rule of less soluble diastereoisomers, which has been discussed in detail and restated[27] as 'If the less soluble diastereoisomers formed by certain enantiomers of two ions with a given resolving counter ion are isomorphous, then those enantiomers have related configurations'. For example, the less soluble diastereoisomers with (+)-bromocamphorsulphonate are formed by (+)-$[Co\,en_2Cl_2]^+$, (+)-$[Cr\,en_2Cl_2]^+$, and (−)-$[Rh\,en_2Cl_2]^+$; these enantiomers therefore have related configurations.

III. THE COTTON EFFECT

A. Nature of the Cotton Effect

The rotation of plane-polarized light may be accounted for by considering the nature of the light in classical terms. Light is regarded as an electromagnetic radiation with vibrations transverse to the direction of propagation. Plane-polarized light[28], where the electric vector vibrates only in one plane, may be decomposed (physically) into two circularly polarized components with equal intensities, in which the electric vector rotates with respect to the axis of propagation, the frequency of rotation being a function of the frequency of the light. This is described more fully in equation

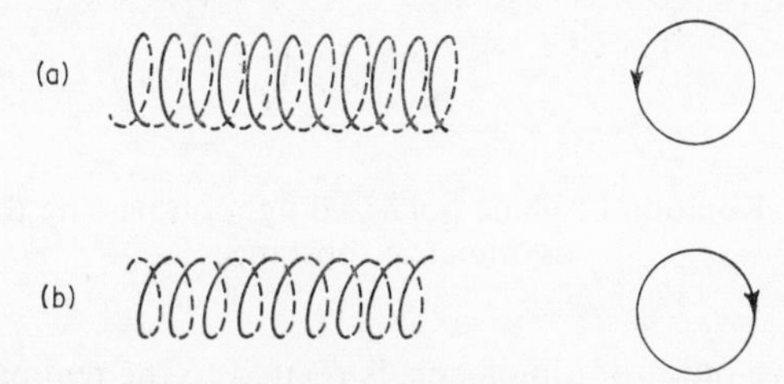

Figure 6. Envelope of amplitudes and sectional patterns for (a) left circularly polarized light and (b) right circularly polarized light.

(3) for the amplitude a of circularly polarized light, propagated in the x direction, through a medium of refractive index n with a frequency ν.

$$a = a_0[\mathbf{y} \cos 2\pi\nu(t - nx/c) \pm \mathbf{z} \sin 2\pi\nu(t - nx/c)] \qquad (3)$$

where t is the time, c is the velocity of light *in vacuo*, and **y** and **z** are unit vectors perpendicular to x, forming a right-handed system.

The + sign in equation (3) gives the equation for left circularly polarized light, while the − sign refers to right circularly polarized light.

At time $t = 0$, the envelope of the amplitudes has the form shown in Figure 6. To an observer looking towards the light source, the sectional patterns (the traces of the amplitudes) at $x = 0$ appear to be rotating clockwise for right and anticlockwise for left circularly polarized light, as also shown in Figure 6.

After a plane-polarized beam has been resolved into its two circularly polarized components, it is possible to recombine them to give the original plane-polarized light. However, if the refractive indices of the medium for the left and right circularly polarized components differ, Fresnel pointed out that their recombination will give plane-polarized light but with a rotated plane of polarization, as shown in Figure 7. Fresnel attributed optical rotation to such a difference of refractive indices.

$$\alpha = \frac{n_l - n_r}{\lambda} \tag{4}$$

where λ is the wavelength of incident light, n_l and n_r are the refractive indices of the medium for left and right circularly polarized light, respectively, and α is the rotation in radians per unit length (the length is measured in the same units as λ).

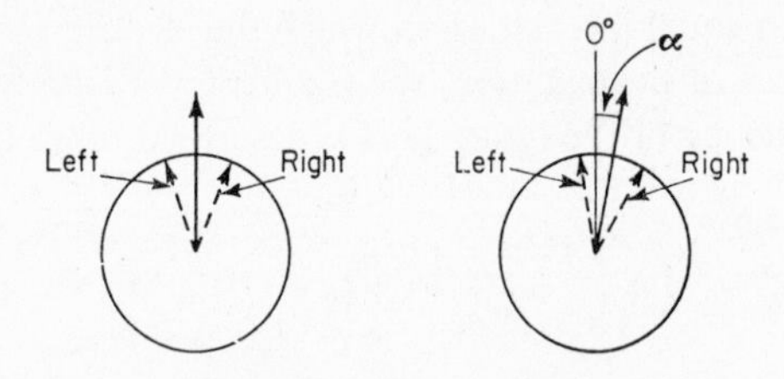

Figure 7. Rotation of plane-polarized light on passing through an asymmetric substance.

The refractive index of a molecule is related to the real part *a* of a complex variable, the polarizability ($a + ib$). The electronic absorption of a molecule is related to the imaginary part *b* of the same complex variable. Pairs of parameters related in this way (one proportional to the real part, and the other to the imaginary part of the complex variable) are themselves related by the Kronig–Kramers transforms, so that it is possible to derive the refractive index from absorption spectrum and, conversely, to derive the absorption spectrum from the dispersion of the refractive index. Hence, the refractive index shows dispersion with the frequency of the light used to measure it and has a marked anomaly in the region of absorption bands.

The optical rotation $[\alpha]$ is a function of the difference in refractive indices for left and right circularly polarized light. The property obtained by Kronig–Kramers integration of rotatory dispersion is the differential absorption of left and right circularly polarized light, i.e. the circular dichroism. Circular dichroism is defined as the difference in molar decadic extinction coefficients for left and right circularly polarized light, $\epsilon_l - \epsilon_r$. The rotatory dispersion and circular dichroism for an isolated chromophore are shown in Figure 8. If the refractive index n_l is larger than n_r at wavelengths where no absorption occurs, then at absorption wavelengths ϵ_l should be larger than ϵ_r.

The methods of calculating circular dichroism (CD) from optical rotatory dispersion (ORD) using Kronig–Kramers relations have been set out[29] by Moscowitz. There are a number of more simple semiempirical conversions between ORD and CD, set out in detail by Kuhn[30] and summarized here as

$$[\phi]_{\max} - [\phi]_{\min} = [A] = 4028\,(\epsilon_l - \epsilon_r) \qquad (5)$$

and

$$\Delta\nu = 0.925(\nu_{\max} - \nu_{\min}) \qquad (6)$$

where $\Delta\nu$ is the CD bandwidth at half maximum height. The quantities in these equations are shown in Figure 8.

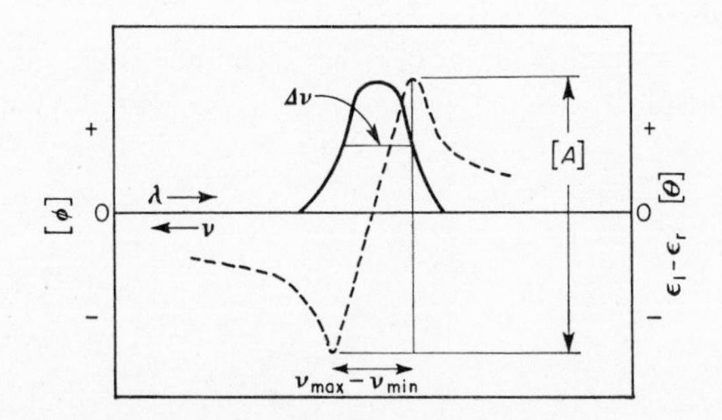

Figure 8. Optical rotatory dispersion (------) and circular dichroism (——) for an isolated asymmetric chromophore. $[A]$ is the amplitude, $\Delta\nu$ is the width at half-height.

The unit (Ψ), the molar ellipticity, was used by Mathieu in his classical work[31] on complex compounds, and it has been recommended recently[32] for use in preparing diagrams (since the numerical size of units of ellipticity is about the same as those of molecular rotations). The ellipticity is a third phenomenon associated with the Cotton effect; it arises because not only is the plane of polarization rotated by an optically active sample but, in the region of an absorption band, the right and left components are differentially absorbed, so that, on recombination, the light obtained will not be truly plane polarized but elliptically polarized. The ellipticity of the emergent light will obviously depend directly on the difference in absorption, i.e. the circular dichroism, and molar ellipticity is indeed related to molar circular dichroism by

$$(\Psi) = 3300(\epsilon_l - \epsilon_r) \qquad (7)$$

Mathieu's results[31] can be compared with more recent work after this equation has been applied.

All three facets of the behaviour of polarized light in the region of an absorption band in an optically active molecule were known[1] to Cotton, and the term 'Cotton effect' is a generic term including the anomalous dispersion of optical rotation, the circular dichroism, or the ellipticity.

B. Comparison of the ORD and CD Techniques

The operational relationship between these techniques will now be considered briefly.

Rotatory dispersion is only occasionally the preferred technique, and usually then for instrumental reasons which may be grouped under the headings (1) optical density, (2) accuracy, and (3) spectral range covered.

1. Optical density

A limiting instrumental factor in either technique is the ratio of optical activity to optical density, i.e. $[\phi]/\epsilon$ or $(\epsilon_l - \epsilon_r)/\epsilon$. Rotatory dispersion is

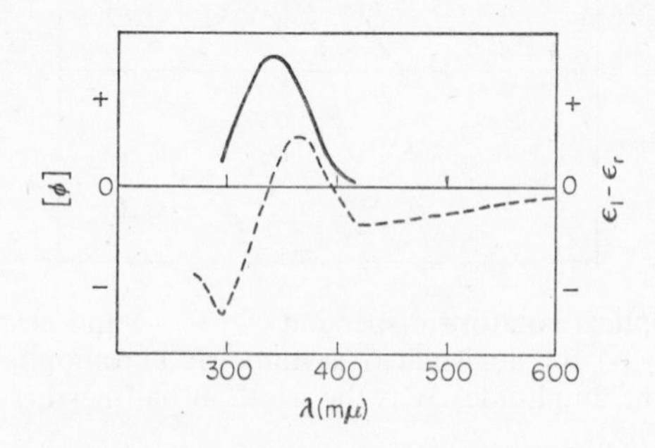

Figure 9. The optical rotatory dispersion (------) and circular dichroism (———) of D-(−)-$[Rh\ en_3]^{3+}$.

measureable outside absorption bands, while circular dichroism is not, and hence rotatory dispersion is the best method for detecting weak optical activity in a chemically pure material. However, if it is necessary to study the optical activity of a particular chromophore in a medium of high background optical activity, circular dichroism is preferable, since only the asymmetry in the particular chromophore is detected. For example, in concentrated solutions of (+)-diethyl tartrate, the hexamminecobalt(III) ion associates with the asymmetric polar molecules, and circular dichroism shows that the ligand-field transitions of the cobalt ion become optically active[33]. The circular dichroism bands of the (+)-diethyl tartrate background medium do not interfere, being in the further ultraviolet region, whereas the contribution of the (+)-diethyl tartrate to rotatory dispersion in the visible region is dominant.

The longest wavelength *d–d* transition in most optically active complexes shows strong Cotton effects whereas the Cotton effects for the *d–d* bands at shorter wavelength are normally weak. In circular dichroism, the latter weak Cotton effects are commonly observable[34], whereas the rotatory dispersion spectrum shows little or no inflexion in the same region.

All the chromophores in a molecule contribute to the rotatory power $[\alpha]_\lambda$ at a given wavelength, and even those in the far ultraviolet may give a significant rotation in the regions where Cotton effects arise from *d–d* transitions. This is the reason for the apparently 'plain negative' rotatory dispersion spectrum[31b] of (–)-$[Rh\ en_3]^{3+}$ through the visible region, shown in Figure 9. Although the Cotton effect of the lowest energy *d–d* band at about 300 mμ is, in fact, positive (as shown quite clearly in the circular dichroism spectrum), this is superimposed on a much stronger negative Cotton effect in the ultraviolet.

2. Accuracy

Commercial spectropolarimeters are some 10–50 times more sensitive than commercial circular dichroism spectrophotometers, although recent modifications to the Bendix–Bellingham and Stanley–N.P.L. 'Polarmatic' spectropolarimeter are highly promising in this respect. This very sensitive instrument utilizes a Faraday magnetic compensation of rotation and it

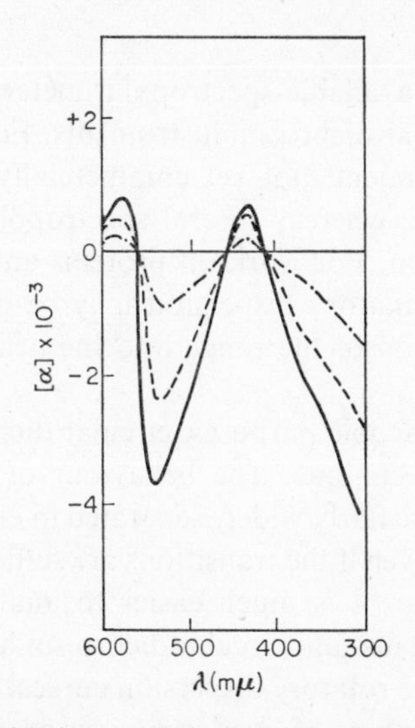

Figure 10. Racemization of $(-)_{500}$-$[Fe\ phen_3]^{2+}$ followed by optical rotatory dispersion. Rate of traverse of spectrum 600–300 mμ, 10 min. Starting at 0 min (———), 20 min (------) and 50 min (–·–·–·–·–).

has recently proved possible to obtain circular dichroism spectra with either optical or electronic modification of the present instrument. It seems likely that very accurate circular dichroism spectra will be obtained with such equipment.

At present, for analytical purposes, rotatory dispersion is more useful than circular dichroism. For example, in work[35] on stereoselectivity involving the relative amounts of diastereoisomers in mixtures, measurements of $[\alpha]_\lambda$ have been applied. More precise results[36] are similarly obtainable from rotatory dispersion than from circular dichroism in kinetic studies. An example is given in Figure 10, which shows the curves obtained during the racemization of $[Fe(o\text{-phen})_3]^{2+}$.

The observation of isorotatory points (common points of $[\alpha],\lambda$) on the base line proves that the reaction studied is a genuine racemization and that no other optically active compound of significant lifetime is produced. An analysis of the occurrence of isorotatory points has been given[37]. Like isosbestic points[38], their use requires caution, since they prove that the concentrations of *all* optically active species present are linearly related, not that there are only two species present. Further, to assume that the observation of isorotatory points during a reaction* proves retention of configuration is unjustified. However, in conjunction with isosbestic points and other information, the observation of isorotatory points can be of great value.

3. Spectral range covered

The spectral range of available spectropolarimeters is significantly different from that of circular dichroism instruments. For example, only one circular dichroism instrument, not yet commercially available, is useful from 210 mμ to 186 mμ, whereas several spectropolarimeters cover this range with some precision. For work on proteins and other biopolymers in this spectral region, rotatory dispersion may be preferred. In the inorganic field, the extension of the range into the near infrared would be more useful.

In general, for spectroscopic purposes, circular dichroism is, for several reasons, the preferred technique. The behaviour of individual chromophores, so long as they are fairly widely separated in energy, can be studied more conveniently and, even if the transitions are sufficiently close for their Cotton effects to overlap, it is much easier to analyse the overlapping integral (approximately gaussian) circular dichroism bands than to analyse the overlapping derivative rotatory dispersion curves. An extra factor contributing to the easier analysis of overlapping circular dichroism bands is that they are narrower than the corresponding rotatory dispersion curves.

* Such points were first observed by Woldbye[39], and Kling[40] proposed to call them isodinetic (*isodinetische*) points. The self-evident name is preferred here.

One of the most obvious applications of circular dichroism to complex compounds was to reveal[31,34] the existence of more transitions in certain regions of the spectrum than the absorption spectrum showed. The superior resolution of circular dichroism arises principally from the extra parameter of sign, as shown in Figure 11, though the bandwidth of circular

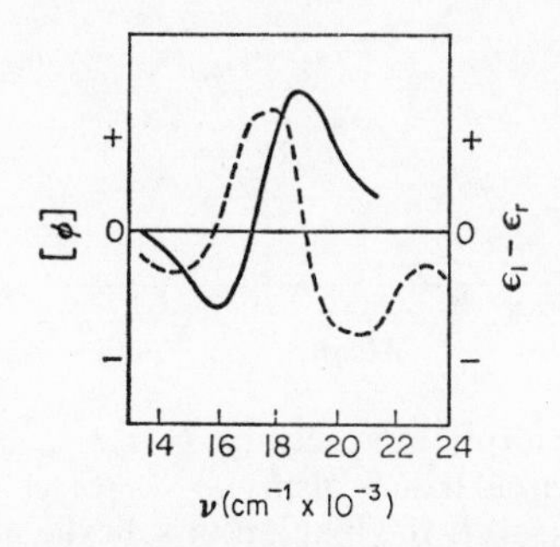

Figure 11. The circular dichroism (———) and optical rotatory dispersion (------) of (+)-[Co en_2Cl_2]$^+$.

dichroism bands is usually smaller than those of corresponding electronic absorption bands (perhaps because certain vibrations which contribute to vibronic electric dipole intensity are ineffective in generating optical activity, as discussed later). In this comparatively favourable case, where the separation between the two components is rather large, the presence of two components is increasingly obvious in the order: absorption spectrum < rotatory dispersion < circular dichroism. Indeed, the analysis of the rotatory dispersion curve had given[41] the same result as the direct measurement of circular dichroism.

However, in more complex cases, the rotatory dispersion curve is useful only to indicate the larger components, as shown in Figures 12–14 (unless a

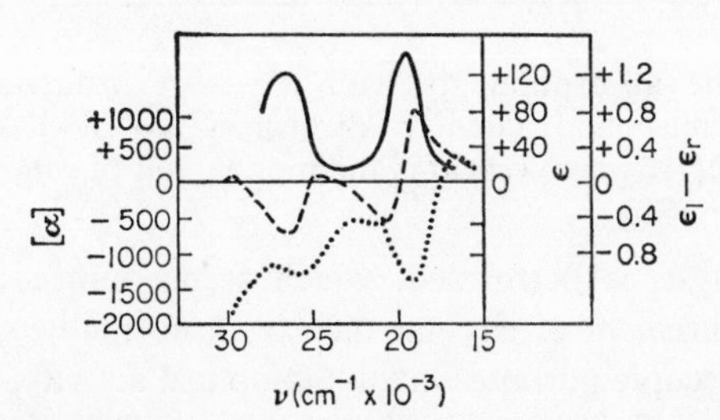

Figure 12. The absorption spectrum (———), circular dichroism (------), and optical rotatory dispersion (······) of red β-(+)-[Co(L-ala)$_3$].

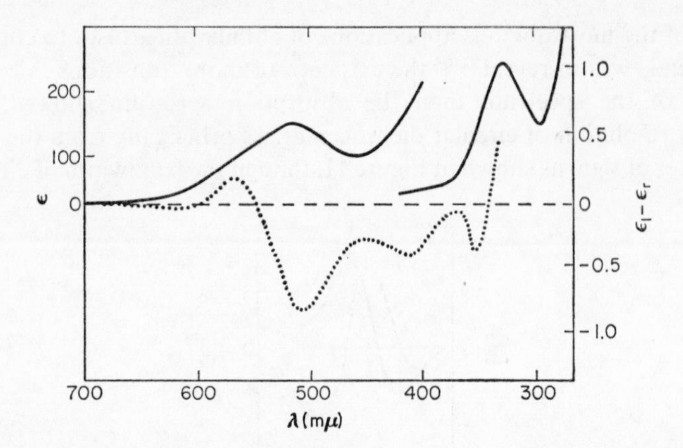

Figure 13. The absorption spectrum (———, ordinate × 10^{-1} below 400 mμ) and circular dichroism (······) of (+)-[Co en_2(sal)]$^+$, where (sal) is the dianion of salicylic acid.

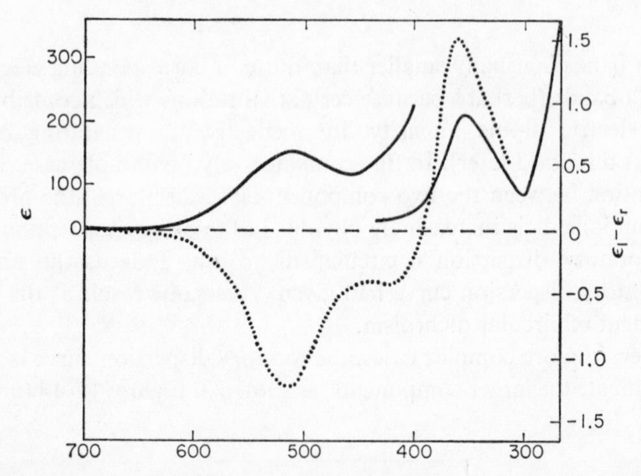

Figure 14. The absorption spectrum (———, ordinate × 10^{-1} below 400 mμ) and circular dichroism (······) of (+)-[Co en_2(Hsal)]$^{2+}$, where (Hsal) is the monoanion of salicylic acid.

very lengthy analysis is performed, which seems unnecessary, since a single rapid measurement of circular dichroism normally gives the same data). For spectroscopic purposes involving optical activity, circular dichroism is the method of choice and most of the remainder of this review is concerned with it.

Several claims have been made[42] that circular dichroism spectra have been analysed into their components by gaussian analyses, though no details of the method employed have been given. These analyses have usually given the number of components required by symmetry arguments (e.g. three for complexes of cobalt(III) with $\mathbf{C}_2$ or $\mathbf{C}_1$ symmetry). However, it now appears likely that the number of optically active bands observed is not necessarily related to splitting of spectroscopic levels in low symmetry fields but may be more intimately related to Jahn–Teller effects in the

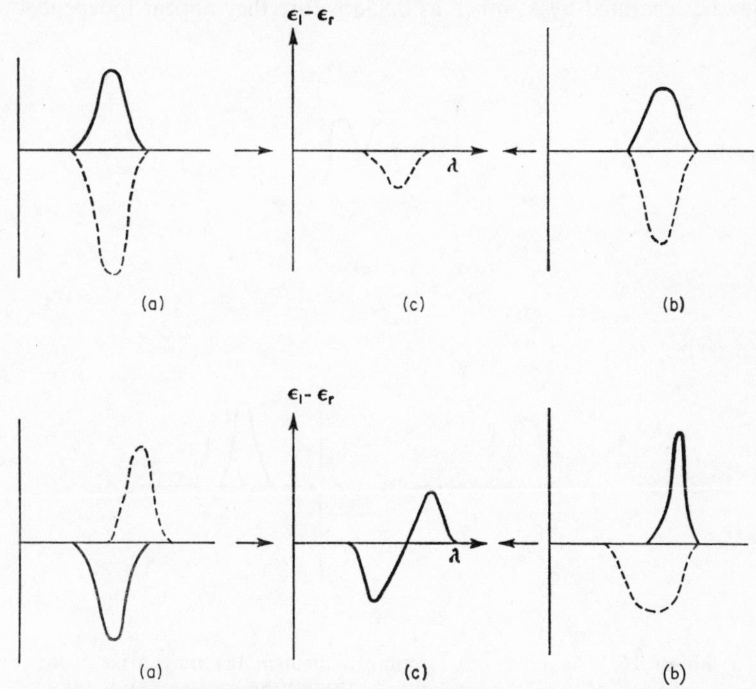

Figure 15. The entirely different situations (a) and (b) may give the same resultant experimental spectrum (c).

excited state. Some comments on such analyses are required, since complications are possible which have not been generally realized. In analysing overlapping electronic spectral bands through gaussian analysis, whether using a computor programme or not, the requirement is a reasonable guess of at least three parameters for each component: the energy, intensity, and width at half height. Successive approximations of these parameters then bring the calculated curve into coincidence with that observed. In the case of the circular dichroism of complex compounds, the individual optically active components usually derive from a single parent, and the splitting is commonly small (about 1000 cm^{-1} or less). The overlap of two components of opposite sign is presented in Figure 15. By varying either the relative positions, or relative intensities, or both, of the two components, it is possible to obtain the same resultant curves in several ways, as shown in Figure 16.

In a discussion of the implications of band shape, Schäffer used[43] an empirical representation of a gaussian absorption band, similar to that given by Schmidtke in Chapter 4. Among Schäffer's conclusions were that (*a*) two gaussian curves with the same height and the same half-width may be separated by as much as 0.85δ before they appear independently

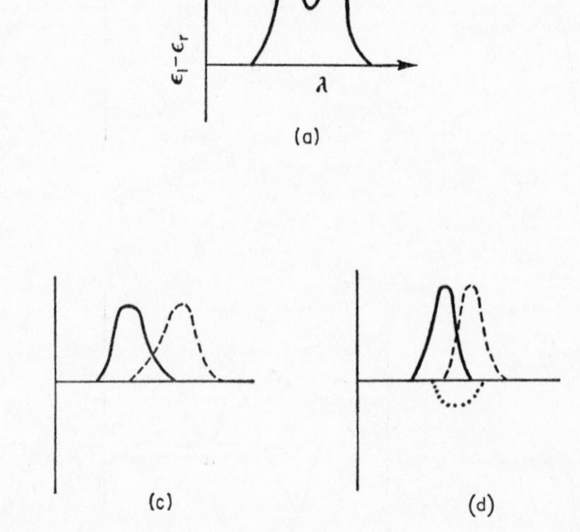

Figure 16. The observed circular dichroism (a) may arise from overlap of two components, e.g. (b), or three components (c).

with a minimum between them, and that (*b*) if a negative gaussian is superimposed on a positive one, in general three band peaks appear. He pointed out that in circular dichroism, the number of observed peaks will in general be larger than the number of energy levels. There is also the difficulty that the situation shown in Figure 17 is not impossible.

Clearly, even for circular dichroism spectra formed by only two overlapping components, analysis is extremely difficult and if the parent spectroscopic state were three-fold degenerate, so that the optically active complex may have three overlapping Cotton effects, then analysis is probably not yet possible. A useful adjunct to spectroscopic studies, at present in its infancy, is using a change in the ionic strength or some other property of the solution to alter the relative strength of components. In one or two cases[44], this procedure, perhaps by stabilizing one conformer in a conformational equilibrium, has revealed extra components.

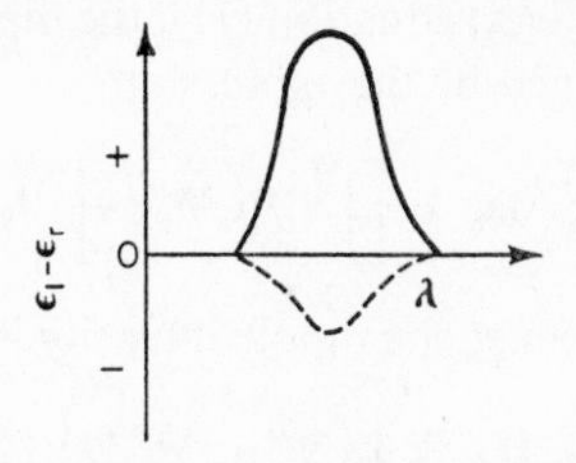

Figure 17. A further source of complication in analysing circular dichroism spectra arising from a highly degenerate transition in octahedral fields. The simplest gaussian analysis of the resultant would suggest wrongly that it arose from two overlapping *positive* Cotton effects.

Finally, it should be mentioned that the differences in selection rules discussed later between optical activity and unpolarized absorption can in certain cases be particularly advantageous. For example, spin-forbidden *d–d* transitions often have rather large circular dichroisms; thus, the Cotton effects observed[45] in the region of the so called 'chrome-doublet' lines, arising from quartet → doublet absorption in complexes of chromium(III) may enable assignments to be made. In a similar way, the rather strong Cotton effect at[31b] about 315 mμ in $[Ir\ en_3]^{3+}$ almost certainly arises from a singlet–triplet absorption band. Thus, comparisons of configuration should not be made using this Cotton effect and the longest wavelength Cotton effect at about 310 mμ for $[Rh\ en_3]^{3+}$, since the latter arises from a singlet–singlet absorption.

IV. SELECTION RULES

The unequal absorption of left and right circularly polarized light by an optically active chromophore depends on the mechanism of absorption. For electronic absorption to occur, several mechanisms are possible; these, in order of their importance in electronic spectroscopy, are (*a*) the electric dipole mechanism, (*b*) the magnetic dipole mechanism, and (*c*) the electric quadrupole mechanism.

Drude showed[46] that when a charge is moved along a helical pathway, the translational motion produces an electric dipole moment, whereas the rotational motion gives a magnetic dipole moment. Depending on the right or left handedness of the helical pathway, the electric and magnetic moments are parallel or antiparallel. The quantum theory of optical activity extends these ideas to account for the origin of optical rotatory power in electronic transitions which have parallel or antiparallel electronic and magnetic transition moments.

As described in Chapter 4, equation (11), the dipole strength for electric dipole intensity is governed by the expression

$$\mathbf{D}_{\mathrm{e}} = \left[\int \Psi_{\mathrm{i}}^{*} \mu_{\mathrm{e}} \Psi_{\mathrm{f}} \, \mathrm{d}\tau\right]^{2} \tag{8}$$

The dipole strength for magnetic dipole intensity is given by

$$\mathbf{D}_{\mathrm{m}} = \left[\int \Psi_{\mathrm{i}}^{*} \mu_{\mathrm{m}} \Psi_{\mathrm{f}} \, \mathrm{d}\tau\right]^{2} \tag{9}$$

μ_e and μ_m are the electric and magnetic dipole moment operators, and Ψ_i and Ψ_f are the wave functions of the initial and final states, respectively. $\mathbf{D}_m$ is much smaller than $\mathbf{D}_e$. Further, μ_e transforms as a translation (i.e. electric dipole intensity may be regarded as arising from translation of electric charge during the transition from the initial to the final state). μ_m transforms as a rotation (i.e. a magnetic dipole intensity may be regarded as arising from the rotation of electron density). Both translation and rotation are two-dimensional processes and are identical in each enantiomer.

However, the combination of electric and magnetic dipole transition moments is a three-dimensional helical process. The displacement of electric density is helical for a transition with both magnetic and electric dipole character. This is the basis of the modern theory of optical activity —namely that, for an electronic transition to give rise to optical activity, the transition is required to be *both* electric and magnetic dipole allowed. The rotational strength is defined as

$$\mathbf{R} = Im\left\{\left[\int \Psi_{\mathrm{i}}^{*} \mu_{\mathrm{e}} \Psi_{\mathrm{f}} \, \mathrm{d}\tau\right]\left[\int \Psi_{\mathrm{i}}^{*} \mu_{\mathrm{m}} \Psi_{\mathrm{f}} \mathrm{d}\tau\right]\right\} \tag{10}$$

We now consider the application of this general theory* to transition metal complexes.

As is well known, *d–d* transitions in centric symmetry are orbitally forbidden (so-called Laporte rule). However, even in centric molecules, the vibronic relaxation of the strict *gerade* or *ungerade* character of orbitals leads to appreciable electric dipole intensity. In any transition, the more critical requirement for optical activity to arise is the magnetic dipole condition. The magnetic dipole selection rules are such that no change in

* For an extension of this theory, see Condon[47].

parity is required (i.e. the rules are $g \leftrightarrow g$, or $u \leftrightarrow u$, but $g \not\leftrightarrow u$). Thus Laporte-forbidden transitions within the *d–d* manifold may be magnetic dipole allowed, and indeed, in general, the lowest energy spin-allowed transition between the Stark levels of an octahedral complex is allowed in magnetic dipole radiation. This magnetic dipole character has been useful in an experimental verification of spectroscopic assignments; some examples are given below.

Equation (10) may be rewritten

$$\mathbf{R} = \mu_e \mu_m \cos\theta \tag{11}$$

where θ is the angle between the two transition moments. Experimentally, from the area of the circular dichroism band,

$$\mathbf{R} = \left(\frac{3hc10^3 \log_e 10}{32\pi^3 N}\right) \int \left[\frac{\epsilon_l - \epsilon_r}{\nu}\right] d\nu \tag{12}$$

$$= 22.9 \times 10^{-40} \int \left[\frac{\epsilon_l - \epsilon_r}{\nu}\right] d\nu \tag{13}$$

and, from the area of the absorption band, the total dipole strength $\mathbf{D}$ is given by

$$\mathbf{D} = \left(\frac{3hc\ 10^3 \log_e 10}{8\pi^3 N}\right) \int \frac{\epsilon}{\nu}\, d\nu \tag{14}$$

$$= 91.8 \times 10^{-40} \int \frac{\epsilon}{\nu}\, d\nu \tag{15}$$

Kuhn introduced the so-called anisotropy factor[30] g, and a criterion[48] for assigning magnetic dipole transitions has been derived from it:

$$g = \frac{\epsilon_l - \epsilon_r}{\epsilon} \tag{16}$$

$$\approx \frac{4\mathbf{R}}{\mathbf{D}} \tag{17}$$

$$= \frac{4\mu_e \mu_m \cos\theta}{\mu_m^2} \tag{18}$$

By inserting realistic values of the moments μ_m and μ_e, it was found[48] that values of g which were greater than 0.01 could arise only from magnetic dipole transitions. In agreement with Moffitt's prediction[49], observed Cotton effects in complexes of any symmetry show that magnetic dipole transitions dominate the optical rotatory spectra. Typical g factors for a number of transition metal compounds are shown in Table 1. It is clearly the case that the magnetic dipole allowed absorption of the octahedral parent compounds continues to dominate the optical activity even in complexes of symmetry much lower than octahedral.

Table 1. *g* factors for transition metal complexes.

Complex[a]	$^1A_{1g} \rightarrow {}^1T_{1g}$[b]			$^1A_{1g} \rightarrow {}^1T_{2g}$[b]		
	ν_{abs}	ν_{CD}	g ($\times 10^2$)	ν_{abs}	ν_{CD}	g ($\times 10^2$)
$[Co(C_2O_4)_3]^{3-}$	16,500	16,100	1.4	23,800	23,800	0.1
$[Co\ en_3]^{3+}$	21,400	20,300	2.2	29,400	27,500	0.33
$[Rh(C_2O_4)_3]^{3-}$	25,000	24,700	2.02	30,000	—	—

[a] The formula $[M(C_2O_4)_3]^{3-}$ is employed for trisoxalato complexes throughout this chapter without prejudice to their actual structures.
[b] Bands in octahedral parents.

The rather naive view taken by this application of the *g* factor must, of course, be refined in accordance with the actual symmetry of the ligand field. The most usually studied optically active complexes have trigonal symmetry ($\mathbf{D}_3$) in the ground state, and the calculation of formal selection rules reduces to the usual procedure of establishing, from the character table for the appropriate group (say $\mathbf{D}_3$), whether the representation of the direct product of the ground and excited states contains the representation of the appropriate transition moment operator. The assumption made in any such calculation of selection rules is that the symmetry elements of the excited state are the same as those of the ground state. There is the complication (not included in current treatments) that, for any d^n electronic configuration, either the ground state or the excited state is subject to Jahn–Teller distortion, and the simple spectroscopic approach outlined here is in the course of modification as described later.

As an application of this procedure and to show the potential usefulness of optical activity measurements on oriented crystals, some little-known data for single crystals of nickel sulphate hexahydrate will now be analysed in terms of trigonal symmetry. The salt happens to crystallize in an enantiomorphic space group, so that any single crystal will contain nickel(II) ions in an asymmetric environment. There is, of course, no observable asymmetry associated with the hexaaquonickel(II) species in solution.

The optical activity of the *d–d* bands in complexes of nickel(II) has received little attention, presumably because tris-chelated nickel(II) complexes, such as trisethylenediaminenickel(II), are not normally resolvable for kinetic reasons. However, asymmetry may be induced at the nickel atom in several ways, including crystallization in an asymmetric lattice, as in α-nickel sulphate hexahydrate, or coordination with an optically active ligand, as in the tris-(+)-propylenediaminenickel(II) cation.

The magnetic dipole moment operator belongs to T_{1g} in the point group $\mathbf{O}_h$, transforming as a rotation. In spin-free octahedral complexes of nickel(II), transitions are assigned as $^3A_{2g} \rightarrow {}^3T_{2g}$ (at lowest energy), $^3A_{2g} \rightarrow {}^3T_{1g}(F)$ and $^3A_{2g} \rightarrow {}^3T_{1g}(P)$ (at highest energy). Only the lowest energy transition $^3A_{2g} \rightarrow {}^3T_{2g}$ is magnetic dipole allowed and, as pointed out[49] by Moffitt, such magnetic transitions should have rotational strengths an order of magnitude lower than magnetic dipole forbidden transitions. This rule can be verified experimentally; as shown in Table 2, $^3A_{2g} \rightarrow {}^3T_{2g}$ dominates the optical activity. (Assignments are given in terms of $\mathbf{O}_h$ symmetry, this approximation being justified by the strength of the selection rules for $\mathbf{O}_h$ in other complexes where the symmetry is, in fact, no higher than $\mathbf{C}_3$).

Table 2. Rotational strengths for Ni^{2+} in asymmetric environments.

Compound	Transition	Absorption		Circular dichroism		
		λ_{max}	ϵ_{max}	λ_{max}	$(\epsilon_l - \epsilon_r)_{max}$	g
$NiSO_4 \cdot 6H_2O$[a]	$^3A_{2g} \rightarrow {}^3T_{2g}$	1160	—	1155	—	0.24
	$^3A_{2g} \rightarrow {}^3T_{1g}(F)$	675	—	690	—	0.03
	$^3A_{2g} \rightarrow {}^3T_{1g}(P)$	385	—	—	—	0.01
$[Ni(+pn)_3]^{2+}$[b]	$^3A_{2g} \rightarrow {}^3T_{2g}$	990	8	—	0.18[c]	0.23
	$^3A_{2g} \rightarrow {}^3T_{1g}(F)$	550	8	550	−0.036[d]	0.004
	$^3A_{2g} \rightarrow {}^3T_{1g}(P)$	350	15	—	—	0.01[e]

[a] Data from Rudnick and Ingersoll[54].
[b] Tris-(+)-propylenediaminenickel(II) ion.
[c] At 900 mμ, the highest wavelength used, circular dichroism was still rising. The value at the absorption maximum would be $\geqslant 0.2$, so $g \geqslant 0.23$.
[d] The circular dichroism of $[Ni(+pn)_3]^{2+}$ was previously studied[31a] by Mathieu; his results (430–680 mμ) are qualitatively similar to those reported here. However, his propylenediamine had $[\alpha]_D = 13°$, whereas that used here had $[\alpha]_D = 33°$.
[e] In rotatory dispersion, the absorption at 350 mμ shows an extremely weak Cotton effect; g is certainly $\ll 0.01$.

Recently, the optical absorption, magnetic anisotropy, and paramagnetic resonance of several hexahydrated salts of the Ni^{2+} ion were correlated by assuming[50] an orthorhombic perturbation of the energy levels for a cubic field. However, such a treatment is inappropriate for nickel sulphate hexahydrate at least, since the transitions within the *d–d* manifold for this substance show marked Cotton effects. The Ni^{2+} ion must therefore be situated in an asymmetric environment. It was assumed by Bose and Chatterjee that the orthorhombic field found[51] in paramagnetic resonance studies of Tutton salts could be extended to nickel sulphate hexahydrate.

The naive approach using $\mathbf{O}_h$ symmetry gives a reasonable account of the rotational strengths of the transitions for both complexes in Table 2, because the selection rules for $\mathbf{O}_h$ are strong. In fact, the symmetry of tris-(+)-propylenediaminenickel(II) ion is no higher than $\mathbf{C}_3$. The nickel ion in nickel sulphate hexahydrate may be treated in a similar way, using $\mathbf{D}_3$ symmetry, a trigonal perturbation being the simplest means of introducing optical asymmetry about a metal ion. A_2, T_2, and T_1 in $\mathbf{O}_h$ go over, respectively, into A_2, $A_1 + E$, and $A_2 + E$ in $\mathbf{D}_3$. The circular dichroism studies[52] of crystalline nickel sulphate hexahydrate were carried out with radiation propagated parallel to the optic axis; under such conditions, only those transitions $A_2 \rightarrow E$ are allowed. One component only appeared for each region of absorption; these are the E components, leading to the assignments for bands I and II (using the nomenclature[50] of Bose and Chatterjee) given in Table 3.

Table 3. Assignment of strong peaks in the spectrum of $NiSO_4 \cdot 6H_2O$.

Notation of ref. 50	Assignment in $\mathbf{O}_h$ (see Table 2)	Experimental values (cm^{-1} at 291°K)	Assignment in $\mathbf{D}_3$
A	$^3A_{2g} \rightarrow {}^3T_{2g}$	8,430	$^3A_2 \rightarrow {}^3E$
B		8,900	$^3A_2 \rightarrow {}^3A_1$
C		13,974	$(^3A_2)_1 \rightarrow (^3E)_0$
D	$^3A_{2g} \rightarrow {}^3T_{1g}(F)$	14,110	$(^3A_2)_0 \rightarrow (^3E)_0$
E		15,450	$^3A_2 \rightarrow {}^3A_1$
F	$^3A_{2g} \rightarrow {}^3T_{1g}(P)$	25,720	$^3A_2 \rightarrow {}^3E$
G		26,060	$^3A_2 \rightarrow {}^3A_1$

Support for the assignment $^3A_2 \rightarrow {}^3E$ for band I arises from the variation in circular dichroism with temperature. The general significance of such variations is first considered. Providing that the direct product of U and G contains the representation of the magnetic dipole operator in the appropriate point group, the transition $G_0 \rightarrow U_0$ (where U represents the upper state, G the ground state, and the subscripts refer to vibrational levels) is a pure magnetic dipole transition, for which no change of parity is necessary. Any other transitions like $G_1 \rightarrow U_0$ or $G_0 \rightarrow U_1$ have also an electric dipole component, achieved by the vibronic mechanism. The rotational strength $\mathbf{R}$ of a transition involves the magnetic dipole strength, whereas the dipole strength $\mathbf{D}$ involves very largely only the electric dipole strength. The g factor is a function of $\mathbf{R}/\mathbf{D}$, and will therefore increase with the magnetic dipole character of the transition. Now, on cooling, the occupancy of G_0 relative to $G_1, G_2, \ldots$ increases, so that the magnetic dipole character of $(G \rightarrow U)$, due largely to $G_0 \rightarrow U_0$, will increase with decreasing temperature. The g factor of an isolated magnetic dipole allowed transition is therefore expected to increase at lower temperature, as argued by Moffitt and Moscowitz[53] for $n \rightarrow \pi^*$ transitions of ketones.

For nickel sulphate hexahydrate, with the radiation parallel to the optic axis, only transitions $A_2 \rightarrow E$ are allowed. The assignment of the circular dichroism band at 1170 mμ to a single electronic transition ${}^3A_2 \rightarrow {}^3E$ is supported by the observation[54] that the maximal wavelength of the circular dichroism band does not change when the temperature is lowered from 25° to −188°c. As the generalization above would indicate, for a band arising from transitions from several vibrational levels of the lower state to several levels of the upper state, the g factor should increase. At room temperature, the g factor is 0.24, and at −188°c it is 0.4. This may be attributed to the increasing population of $({}^3A_2)_0$ and $({}^3E)_0$, and to the consequently greater contribution of $({}^3A_2)_0 \rightarrow ({}^3E)_0$ to the circular dichroism at the lower temperature.

The assignment of bands C, D, and E has given rise to a good deal of discussion. Band E has been said[55,56] to occur through L–S coupling, but it has also been assigned[57] to a transition to a singlet upper state. Calculations of both band positions[58] and intensities[59] appear to rule out coupling with a singlet level and the double peak observed in solution is therefore most likely due to spin–orbit fine structure. However, in the crystal, the situation is different. Bose and Chatterjee pointed out[50] that the observed magnetic anisotropy[60] of Ni^{2+} in nickel sulphate hexahydrate required the components of band II to be separated by 1450 cm^{-1}. It appears that E may be due to one component, while C and D (separated from E by about 1400 cm^{-1}) may represent transitions to the other component $({}^3A_2 \rightarrow {}^3E)$ from neighbouring vibrational levels of the ground state. The assignment of C and D to $({}^3A_2 \rightarrow {}^3E)$ is supported by the fact that both bands contribute to circular dichroism with radiation propagated parallel to the optic axis, whereas band E (assigned as ${}^3A_2 \rightarrow {}^3A$ and forbidden under such circumstances) does not contribute to circular dichroism. Distinction between C and D is based on the variation of circular dichroism with temperature; both bands contribute about equally to the Cotton effect at room temperature, while at −188°c the circular dichroism is dominated by band C which may therefore be assigned as $({}^3A_2)_0 \rightarrow ({}^3E)_0$. As pointed out[61] by Piper and Koertge, fine structure in this region for other salts containing the hexaaquonickel(II) ion may be attributed to differing occupancies of the vibrational states arising from the asymmetric Ni—O stretching mode. In the case of hexaaquonickel(II) sulphate, a similar vibrational quantum could lead to the observed separation between C and D. Since band C at lower frequencies becomes relatively more intense at low temperature, it can only arise from a non-vibrating ground state if the excited states for C and D differ, as in

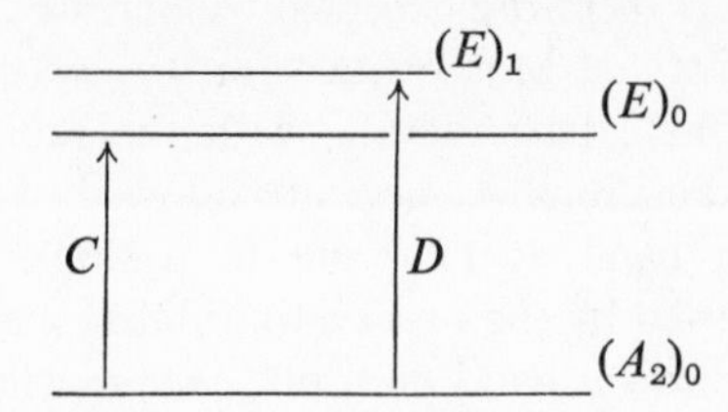

So far as bands F and G are concerned, while $\mathbf{D}_3$ symmetry could account for the splitting observed, the assignments given in Table 3 may be reversed; experimental data on the absorption of polarized light by oriented crystals are lacking in the region around 25000 cm^{-1}, although a weak Cotton effect was observed in rotatory dispersion, consistent with one component being an $A_2 \rightarrow E$ transition in $\mathbf{D}_3$ symmetry.

The suggestion that the effective ligand field of the hexaaquonickel(II) ion has $\mathbf{D}_3$ symmetry has been made previously[62] for nickel fluorosilicate hexahydrate; on the basis of paramagnetic resonance experiments, a trigonal field superimposed on the cubic crystalline field was indicated. No paramagnetic resonance data are yet available for nickel sulphate hexahydrate. It is certainly increasingly clear that the detailed structure of the $[Ni(H_2O)_6]^{2+}$ ion varies from salt to salt and that it is dangerous to assume that data obtained from one compound can be applied to other compounds. It is noteworthy that nickel sulphate heptahydrate shows[63] Cotton effects in its *d–d* bands, suggesting that treatment by an acentric field will be more appropriate than the orthorhombic field most recently suggested[50].

A treatment based on circular dichroism of the ordering of spectroscopic levels in solutions of the blue and yellow Lifschitz salts of optically active diamines has been given[64]. Otherwise, the optical activity of nickel(II) complexes has attracted little spectroscopic attention.

V. SPECTROSCOPY AND CONFIGURATION

Pasteur's rule, that an optically active molecule is not superposable on its mirror image, has long been a guide in stereochemical relationships and deductions. It was applied with huge success by Werner in establishing the basic rules of octahedral stereochemistry. The rule has been expanded[65] to state that the molecule must have no centre, plane, or alternating axis of symmetry, a statement which contracts in the symmetry operation terms of modern coordination chemistry (since $S_2 \equiv i$, and $S_1 \equiv \sigma$) to the form: The molecule must have no improper axis.

This at once correlates resolvability with spectroscopic transitions giving rise to Cotton effects: the helical combination of electric and magnetic dipole character, in a molecule with axial symmetry only, must be non-identical with its mirror image. In one enantiomer of the molecule, a particular transition will have one helicity and, in the enantiomeric molecule, the corresponding transition between the same two states will have the opposite helicity. Attempts to relate configuration with spectroscopy have all had this link as their basis. In both classical and quantum theories of optical activity, an electron promoted in a right-handed helix will give a positive Cotton effect (i.e. $\epsilon_l - \epsilon_r$ positive). The situation for a tris- and a bis-chelated molecule of the D configuration is shown in Figure 18. In an

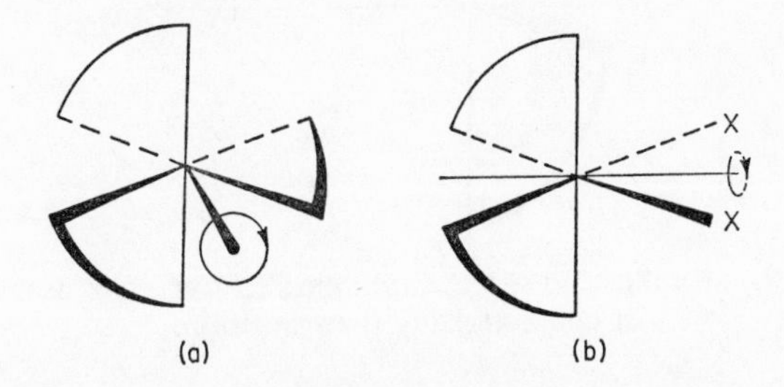

Figure 18. Helicities parallel to (a) C_3 axis of $\mathbf{D}_3$ complex (*S*), and (b) C_2 axis of $\mathbf{C}_2$ complex (*R*).

experiment described in detail later, Mason showed[66] that tris-chelated molecules with the D configuration gave positive Cotton effects for the *E* component and negative Cotton effects for the *A* component. This led to a spectroscopic means of establishing configuration: a tris-chelated molecule with a positive rotational strength for the *E* transition has the D configuration.

A good deal of attention has been given to extending this rule to less symmetrical complexes—a particularly important prerequisite for work involving the asymmetric reactivity of complexes, since most of the interesting stereoselective effects[2] arise in molecules with $\mathbf{C}_2$ or $\mathbf{C}_1$ symmetry. A subsidiary spectroscopic rule would state that complexes with $\mathbf{C}_2$ symmetry and positive rotational strength for the *A* transition have the D or $R(C_2)$ configuration. The difficulty lies in characterizing the *A* transition, since no work has yet been done with oriented crystals of known structure in which the complex has $\mathbf{C}_2$ symmetry. Indeed, only in one case[15] has the *A* transition been identified. The *cis*-(+)-diisothiocyanatobisethylenediaminecobalt(III) ion forms a 1:1 adduct with mercuric ion in solution. As shown in Figure 19, the mercuric ion lies on the C_2 axis, and will therefore modify the *A* transition which is polarized parallel with this axis. For the (+) isomer, the Cotton effect which was strongly shifted by complexing with mercuric ion was positive, so that the D configuration may be assigned. The (+)-$[Co\ en_2(NCS)_2]^+$ ion was converted without breaking bonds at the cobalt (and therefore with retention of configuration), through the (+)-$[Co\ en_2(NH_3)(NCS)]^{2+}$ ion, into the (+)-$[Co\ en_2(NH_3)_2]^{2+}$ ion, to which was assigned the D configuration.

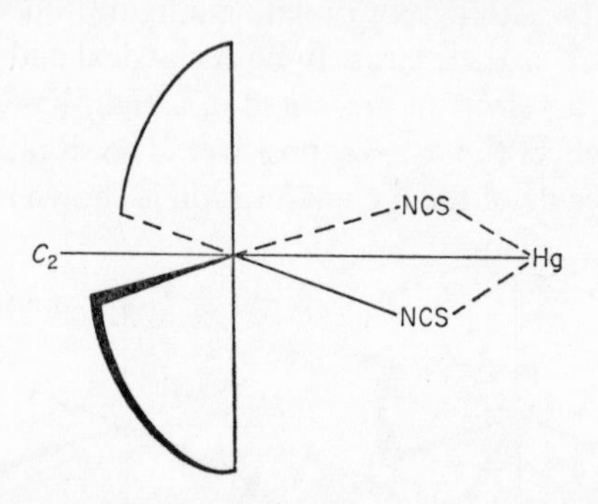

Figure 19. Modification of transition parallel to C_2 axis as a means of characterizing the transition.

The Cotton effects of this D diammine complex are extremely similar to those of the D-(+)-$[Co\ en_3]^{3+}$ ion, as might be expected. There is at present a marked shortage of chemical correlations of this kind, and further work in this area is awaited with interest.

VI. TRISDIAMINE COMPLEXES

The basic principles and major applications of circular dichroism may be stated as follows:

1. Each electronic transition of a resolved molecule may show a Cotton effect, the sign of which, for an individual transition, is related to the absolute configuration of the molecule.
2. The selection rules for optical activity (as for any other electronic excitation) are controlled by the symmetry of the molecule in the electronic states giving rise to the transition. This may be useful in structural deductions, since the number of allowed Cotton effects is, in principle, predictable.
3. The rotational strength of the transition depends on *selection rules*, as does the oscillator strength. This itself may be useful as a spectroscopic method.

4. The differences in selection rules for optical activity and electronic absorption may give weak transitions strength, so that, for example, spin-forbidden *d–d* bands may show up strongly in circular dichroism.
5. Optically active transitions are polarized and may be used in spectroscopic work with oriented crystals.

The present situation with respect to the trisdiamine complexes of kinetically inert metal ions (especially $[Co\ en_3]^{3+}$) demonstrates the range of information accessible from measurements of circular dichroism.

A. Theoretical Considerations

Moffitt[49] applied the quantum theory of optical rotatory power to the tris-chelated coordination compounds of the $\mathbf{D}_3$ point group. He first established that, in octahedral complexes of point group $\mathbf{O}_h$, the spin-allowed T_{1g} transition of lowest energy (T_{1g} denoting the direct product of the irreducible representations of the ground and excited states) has a magnetic moment while the T_{2g} transition at higher energy has not. The total magnetic moment of transitions from the ground state of a metal ion with orbital angular moment L is $[L(L+1)]^{\frac{1}{2}}\beta_e$. Of course, if there is only one magnetic dipole allowed transition from this state, *all* the moment is in this single transition. For complexes with the d^3 electronic configuration, the magnetic moment of ${}^4A_{2g} \rightarrow {}^4T_{2g}$ is $\sqrt{12}\beta_e$; for cobalt(III) and other spin-paired d^6 complexes, the transition ${}^1A_{1g} \rightarrow {}^1T_{1g}$ has a magnetic moment of $\sqrt{24}\beta_e$.

The accuracy of the magnetic dipole criterion is clear. In the whole range of transition metal complexes studied, the longest wavelength spin-allowed absorption band has always dominated the optical activity. Further, the detailed predictions of the magnetic transition moment are rather satisfactorily confirmed by experiments. The magnetic moment concentrated in ${}^1A_{1g} \rightarrow {}^1T_{1g}$ of cobalt(III) ($\sqrt{24}\beta_e$) is isotropic and, in the ions containing cobalt(III) with $\mathbf{D}_3$ symmetry, the moments[67] are $2\sqrt{2}\beta_e$ for the A_2 and $4\beta_e$ for the E_a transitions. The dipole strength of ${}^1E_a$ in $[Co\ en_3]^{3+}$ is 1500×10^{-40} cgs units; the corresponding absorption (${}^1T_{1g}$) of $[Co(NH_3)_6]^{3+}$ has a dipole strength of 880×10^{-40} cgs units. It was assumed that the difference in these dipole strengths arises from the 'static' intensity of the E_a transition because the trigonal field has made it formally electric dipole allowed. This extra 'static' dipole strength has been taken to give the square of the electric moment for the transition. The equation relating rotational strength, electric moment, magnetic moment, and the angle θ (taken as zero) gave a magnetic moment of 3.4 β_e from the circular dichroism, compared with the theoretical value of 4.0 β_e. This agreement, in view of the assumptions made, is probably illusory, particularly since the static intensity of $[Co\ en_3]^{3+}$ is in fact only about 10% of the vibronic intensity. Similar arguments have been given[68] for complexes of chromium(III).

The major points of debate in the field of optical activity are connected with the origin of rotational strength. Moffitt employed a crystal-field model for the tris chelates, in which the *d–d* transitions derived their electric dipole allowedness from an admixture of the 4*p* character through the operation of a trigonal field (with *ungerade* character). Although Sugano[69] used symmetry arguments to show that the *d–p* mixing proposed by Moffitt could not account for optical activity, numerous calculations[70–72] on the tris-chelated compounds have employed arguments based on a trigonal perturbation of the octahedral levels, varying from Moffitt's in detail but not in approach.

A crystal-field calculation showed[72] that neither *d–p* nor *d–f* mixing could account for optical activity but that a net rotational strength is obtained by taking trigonal splittings of energy levels into account. In all subsequent theories, the larger the trigonal splitting parameters (*K* hereafter), the larger the optical activity.

B. Results

The models using the crystal field theory give results which may be summarized as follows[73]:

$$\mathbf{R}(^1A_1 \rightarrow {}^1A_2) = -\mathbf{R}(^1A_1 \rightarrow {}^1E_a)\ [d^6] \tag{19}$$

$$\mathbf{R}(^4A_2 \rightarrow {}^4A_1) = -\mathbf{R}(^4A_2 \rightarrow {}^4E_b)\ [d^3] \tag{20}$$

(i.e. the rotational strengths of the first band should disappear when *K*, the trigonal splitting parameter, is equal to zero).

$$\mathbf{R}(^1A_1 \rightarrow {}^1E_b) \ll \mathbf{R}(^1A_1 \rightarrow {}^1E_a)\ [d^6] \tag{21}$$

$$\mathbf{R}(^4A_2 \rightarrow {}^4E_a) < \mathbf{R}(^4A_2 \rightarrow {}^4E_b)\ [d^3] \tag{22}$$

(i.e. the rotational strength of the formally magnetic dipole allowed *E* component of the second band should be much smaller than that of the *E* component of the first band).

$$\mathbf{R}(^1A_1 \rightarrow {}^1A_1) = 0\ [d^6] \tag{23}$$

$$\mathbf{R}(^4A_2 \rightarrow {}^4A_2) = 0\ [d^3] \tag{24}$$

A number of elegant experiments have utilized oriented single crystals to attempt to establish the size and sign of the trigonal splitting parameters, and the degree of correlation with the various theoretical predictions. The experimental approach may be divided into two periods—up to 1965, and the subsequent work (largely at low temperatures) which has thrown the whole subject back into the melting pot.

One rather disturbing result which was apparent even from the earliest work (although it has not previously been commented on) was that the predicted similarity[3] of sign of Cotton effects for the two E transitions was not observed. Two clear-cut cases where the bands assigned to $^1A_1 \rightarrow {}^1E_a$ and $^1A_1 \rightarrow {}^1E_b$ had rotational strengths of opposite sign are given in Table 4. Further cases of contradiction emerge if the following commonly

Table 4. Signs of Cotton effects for transitions of E symmetry.

Complex	Band I				Band II	
	$^1A_1 \rightarrow {}^1E_a$		$^1A_1 \rightarrow {}^1A_2$		$^1A_1 \rightarrow {}^1E_b$	
	λ	$\epsilon_l - \epsilon_r$	λ	$\epsilon_l - \epsilon_r$	λ	$\epsilon_l - \epsilon_r$
$[Co(+hmc)_3]$[a]	6030	+6.2	7100	−0.5	4520	−8.0
$[Cr(C_2O_4)_3]^{3-}$	5550	+2.9	6300	−0.6	4150	−1.0

[a] hmc = hydroxymethylenecamphorate ion.

made statement is accepted: if under band I only one Cotton effect is seen, then this arises from the E component. No explanation in terms of symmetry of the transitions seems likely to help, since even accepting that borrowing of rotational intensity occurs from transitions of E symmetry further into the ultraviolet, it seems unlikely that the sign of rotational strength could be changed in this way.

A brief account of results in satisfactory agreement with the present theory is now given. Using plane-polarized light and optically inactive crystals, earlier work established the results shown in Table 5. The appropriate selection rules are discussed in detail in the references.

Table 5. Trigonal splitting of band I.

Complex	K[a] (cm^{-1})	$(\epsilon_l - \epsilon_r)_{max}$[b]	Ref.
$[Cr(C_2O_4)_3]^{3-}$	+270	2.8	74
$[Cr\ acac_3]$	+600	1.1[c]	75
$[Co\ acac_3]$	+600	1.3[c]	75
$[Co\ en_3]^{3+}$	+45	1.9	76
$[Cr\ en_3]^{3+}$	+67	1.4	76

[a] Defined as in ref. 16. Note that Schäffer[101] defines K with the opposite sign.
[b] No significance is attached here to the sign of the circular dichroism.
[c] Probably not optically pure.

An extension of these earlier experiments with plane-polarized light to work with optically active molecules was implicit in the observations[77] of Drouard and Mathieu. They studied the optical rotatory dispersion and circular dichroism of crystals of (+)-[Co en_3]Br_3·$2H_2$0 with radiation propagated in the direction of the optic axis. The solution circular dichroism showed bands at 480 mμ (+1.7) and 440 mμ (weak negative), whereas in the crystal only the positive component at 480 mμ (+18) appeared. These workers pointed out that the symmetry of the ion in the crystal was probably not so high as that in solution. A similar experiment[66] using a crystal of accurately known structure was performed by Ballard, McCaffery, and Mason, with 2-(+)-[Co en_3]Cl_3·NaCl·$6H_2$0; the C_3 axes

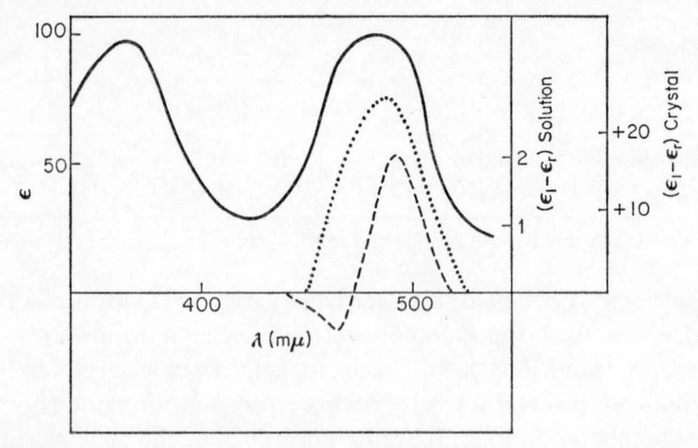

Figure 20. The isotropic circular dichroism (------) and absorption (———) spectra of [Co en_3]$^{3+}$ in solution, and the circular dichroism (······) of a single crystal (see text) with radiation parallel to the optic axis.

of the complex ions were known[14] to be parallel to the optic axis of the crystal. With circularly polarized radiation along the optic axis, only the transitions of E symmetry are allowed. The results (Figure 20) were taken to show that the E transition has a positive rotational strength for the (known) D configuration of the complex (+)-[Co en_3] and, by comparison with the solution circular dichroism, that energetically $E < A$. The results of Drouard and Mathieu, of course, agree.

A number of apparently satisfactory correlations between trigonal splittings and optical activity emerged from the earlier work, and several effects may be explained in a reasonable way. As examples, we will take the effect of ring size on optical activity, the configurations of oxalato- and malonatochromate(III) species, and the effect of ion pairing on the observed circular dichroism of [Co en_3]$^{3+}$. The estimate of magnetic moments for isolated transitions has already been mentioned; from the prediction[32] $\mathbf{R}(^4A_2 \rightarrow {}^4E_a) = -\mathbf{R}(^4\mathrm{A}_2 \rightarrow {}^4A_1)$, the optical activity should disappear when the trigonal distortion is vanishingly small. This was used to explain the difference[45] between the large observed rotatory dispersion of $[\mathrm{Cr\ en_3}]^{3+}$, which contains five-membered chelate rings, and the small Cotton effects of $[\mathrm{Cr\ tn_3}]^{3+}$, which contain six-membered rings. In the latter case, the flexibility of the rings will probably lead to a smaller trigonal distortion at the metal ion. While this explanation may be correct, there is a general effect[78] of the chelate ring size on optical activity. The larger the chelate rings, the smaller the observed optical activity, even in such cases as $[\mathrm{Co\ en_2(ox)}]^+$ compared with $[\mathrm{Co\ en_2(mal)}]^+$. There is, further, the possibility that the larger number of available conformations of the six-membered rings may be connected with the fall in optical activity.

The case of the ions (+)-$[\mathrm{Co\ ox_3}]^{3-}$ and (−)-$[\mathrm{Co\ mal_3}]^{3-}$ is perhaps a little less straightforward. The trigonal splittings[74,79] derived from plane-polarized crystal spectra are shown in Table 6.

Table 6. Trigonal splittings in tris-chelated complexes of chromium(III).

Complex	$^4A_2 \rightarrow {}^4E_a$	Predicted sign of Cotton effect	$^4A_2 \rightarrow {}^4A_1$	Predicted sign of Cotton effect
$[\mathrm{Cr\ ox_3}]^{3-}$	shorter λ	+	longer λ	−
$[\mathrm{Cr\ mal_3}]^{3-}$	longer λ	+	shorter λ	−

Thus, the complexes with related helicity, say $S(C_3)R(C_2)$, should have the signs shown and their circular dichroism spectra should be enantiomeric. The ions giving the less soluble diastereoisomers with (−)-strychnine were taken to be (+)-$[\mathrm{Cr\ ox_3}]^{3-}$ and (−)-$[\mathrm{Cr\ mal_3}]^{3-}$ and these ions certainly[80] have enantiomeric CD curves. However, it is not by any means certain that they have related configurations, since the rule relating the configuration of ions forming the less soluble diastereoisomers with the same resolving agent can only be applied rigorously where the diastereoisomers are isomorphous. The oxalato complexes are particularly dangerous in this respect, since, depending[81] on conditions, either diastereoisomer with (−)-strychnine may crystallize first!

There is a danger in any comparison of the absorption of plane-polarized light by crystalline racemic complexes with the circular dichroism of crystals of the same complex when resolved. This arises from our ignorance of the detailed structures of the two types of crystal, which may contain complex ions which are either different or located in different environments. As an example, let us consider the case of the trisoxalato complex of rhodium. There is good evidence from broad-line proton magnetic resonance experiments[82] that when racemic potassium trisoxalatorhodate (III) is crystallized from water, the crystals are properly formulated $K_6[Rh(C_2O_4)_3][Rh(C_2O_4)_2(HC_2O_4)(OH)]\cdot 8H_2O$ (i.e. in half of the complex ions in the solid state there is a constitutive molecule of water). There is no necessary connexion between this structure and that of the same salt when resolved, since the racemic crystal will have possible symmetry elements unavailable in the optically active form. Indeed, it appears that the composition of the optically active crystal is different (a monohydrate) from that of the racemic crystal.

The one case where the symmetry of the racemic complex in crystals is undoubtedly exactly that of the resolved complex is where spontaneous resolution occurs, but such cases are extremely rare among the complexes of interest.

As an illustration of some difficulties that have not altogether been appreciated in this field, let us consider a claim[83] by Werner that potassium rhodioxalate resolved spontaneously, i.e. that single crystals obtained from a racemic solution contained only one enantiomer of the complex ion. Against this claim is the crystallographic work of Herpin[84], who found different compositions, space groups, and structures for resolved and racemic potassium rhodioxalate. These apparently contradictory findings may yet be reunited by a study of the effect of temperature on the possibility of spontaneous resolution, since Jaeger reported[85] that crystallization of trisoxalate complexes, like Pasteur's original sodium potassium tartrate, showed a transition temperature. On one side of this temperature, there was spontaneous resolution (i.e. crystals with non-centric space groups) and, on the other side, the unit cell of the crystals contained both D and L complex ions. A start has been made[86] on determining the exact nature of the complex ions in crystals of the trisoxalate complexes, and related work in solution is valuable. For example, the observation by Kuhn and Bein[87] that resolved trisoxalatocobaltate(III) gave two circular dichroism components under band II ($^1A_{1g} \rightarrow {}^1T_{2g}$ of $\mathbf{O}_h$) was confirmed by McCaffery and Mason[88] who pointed out that, since in complexes with $\mathbf{D}_3$ symmetry only one component ($^1A_1 \rightarrow {}^1E_b$) should appear in this region, the likely explanation was that optically active cobalt species of lower symmetry such as *cis*-$[Co(C_2O_4)_2(HC_2O_4)(OH)]^{3-}$ were also present in solution. Later, however, it was found[89] that circular dichroism in the band II ($^1A_{1g} \rightarrow {}^1T_{2g}$ of $\mathbf{O}_h$) is not reliable for structural assignments.

Finally, while the unpolarized *d–d* absorption spectra of trisdiamine-cobalt(III) complexes in the visible region are essentially unchanged in the presence of polarizable anions, e.g. phosphate, the circular dichroism is modified. For example[44], with (+)-[Co en_3]$^{3+}$ and with (+)-[Co(+pn)$_3$]$^{3+}$, there is a decrease in $\mathbf{R}(E_a)$ and a corresponding increase in $\mathbf{R}(A_2)$ under band I, and an increase in $\mathbf{R}(E_b)$ under band II. This is thought to be due to the formation of an ion pair in which the phosphate ion lies on the three-fold axis of the complex ion, so that the ion pair retains a three-fold axis. The charge-transfer transition from anion to metal, which has E symmetry, leads to mixing of the magnetic moment of E_a with that of E_b. The rotational strength of E_b increases, that of E_a decreases, and (since $^1A_2$ and ^1E_a overlap) that of $^1A_2$ appears to increase.

These, and similar observations on tris-chelated compounds, are all in accord with the theoretical picture that the optical activity arises from the trigonal splitting. However, very recent work[90–91] suggests that the trigonal splitting parameter is a great deal smaller than had been assumed (and it will be remembered that one prediction of all the trigonal theories, whether they refer to trigonal distortions of the ligand atoms or of their orbitals, is that the splitting must be finite to give rise to rotational strengths).

The conclusions drawn from the experiments of Mason[66] and Mathieu[77] have been questioned[90]. They had said that the E transition lay at longer wavelength than the A and that results could be carried over from crystal to solution, circular dichroism being a property of the molecule. Both conclusions have been criticized; we shall take the points in turn.

The validity of comparing crystal and solution results was questioned because of the ion-pairing effects described earlier. It was felt[90] that the anions in the crystal might perturb the cations so strongly that results could not be carried over to solution. Indeed, Drouard and Mathieu[77] had already commented that the symmetry of the environment of [Co en_3]$^{3+}$ in solution and in crystals were probably different. However, while the detailed lattice position of the optically active cations may be important in certain spheres, McCaffery, Mason, and Norman[92] defended their original assumption that they were concerned with a property of the [Co en_3]$^{3+}$ ion and not of ionic complexes by measuring the circular dichroism of (+)-[Co en_3]$^{3+}$ in a variety of site environments, using potassium halide discs. The measured circular dichroism was, in each case, the same, lending support to the idea that the experiments showing that the positive component of circular dichroism of (+)-[Co en_3] had E character also applied in solution.

A far more serious objection to any theory based on a trigonal splitting in $\mathbf{D}_3$ of the parent triply degenerate levels of $\mathbf{O}_h$ arises from some extremely elegant work[90] by Dingle on the identification of the electronic transitions of [Co en_3]$^{3+}$. His data refer to the crystal DL-2-[Co en_3]Cl_3·NaCl·$6H_2O$ but the findings summarized here are also applicable to the optically resolved crystals. This has been directly confirmed by Denning[91] in a study of almost colourless D-[(Rh,Co)en_3]Cl_3·NaCl·$6H_2O$ with about 1 mole percent of cobalt.

Firstly, in the region of band I [$^1T_{1g}(\mathbf{O}_h)$], the ^1E state is at higher energy than the $^1A_2$ state. This opposes directly the conclusion from the circular dichroism experiments. Secondly, the vibrational fine structure in the spectra of 5°K coincide for π, σ, and axial spectra; therefore the splitting of $^1T_{1g}(\mathbf{O}_h)$ by the trigonal field is 0 ± 2 cm^{-1}. (Denning found[91] 3.5 cm^{-1} for the resolved complex.) The intensity, based on the non-totally symmetric vibrations 185 ± 5 cm^{-1}, 345 ± 5 cm^{-1}, and about 400 ± 20 cm^{-1} which are reasonably assigned as arising from t_{1u} and t_{2u} modes of the octahedron, is about 90% vibronic. There is independent evidence from the observed magnetic circular dichroism that the trigonal splitting in this important case is extremely small.

It has been suggested[92] that the difficulty lies in the different properties measured by plane-polarized and circularly polarized light. The plane-polarized technique emphasizes the electric dipole intensity (largely vibronic), whereas the magnetic dipole intensity is centred at the band origins (0–0) of the transitions. However, it is difficult to see that the discrepancies can at present be unified by considerations of this kind and perhaps the most hopeful approach is that recently discussed[91] by Denning. On the basis of measurements of the magnetic circular dichroism, he concluded that the two bands observed in circular dichroism were, in fact, components of a Jahn–Teller distortion of the excited state.

No vibronic analysis has yet appeared of the optical activity of transition metal compounds. However, there is now available some spectroscopic evidence on the vibrations responsible for oscillator and rotational strength, and the underlying considerations are presented here.

Moffitt and Moscowitz analysed[53] the situation for the rotational strength of a 'magnetic-allowed, electric-forbidden transition' and came to the qualitative conclusion that circular dichroism would arise for totally symmetric modes, while absorption would occur in asymmetric modes and their totally symmetric progressions. This, of course, predicts a shift in the CD maxima relative to the absorption maxima (and indeed, such a shift is often found for the non-degenerate vibronically allowed $n \rightarrow \pi^*$ transitions of organic ketones). Weigang added[93] a term in which both magnetic and electric moments are borrowed.

In his very recent study, Denning measured[91] the vibrational fine structure at the low-energy side of the first ligand-field band for both absorption and circular dichroism. The same vibronic structure and hence the same intensity mechanism was found for both properties. The vibrational components had energies agreeing with the six skeletal vibrations of the CoN_6

framework determined[94] by Mathieu and his coworkers but there was no obvious relationship of intensity to the *g* or *u* character of the vibration. Denning pointed out that, if the first excited state of the $[Co\ en_3]^{3+}$ ion is Jahn–Teller distorted, then there will be static contributions to magnetic moment in non-totally symmetric modes. The work of both Moffitt and Moscowitz and of Weigang requires identical equilibrium symmetries of the nuclei in both ground and excited states.

It therefore seems likely that a full solution of the problem of the nature of the spectroscopic levels giving rise to the observed circular dichroism will depend on characterizing Jahn–Teller components, rather than the components split from an octahedral parent by a trigonal field. Further results will be keenly awaited, since the whole basis of the subject is about to undergo drastic reappraisal.

C. Conformational Factors

Conformational aspects of chelation have been reviewed recently[6] and the present treatment refers only to those matters which have been important in the relationship of spectroscopy and configuration in optically active complexes. The puckered nature of ethylenediamine rings giving enantiomeric forms *k* and *k′* has been known[95] for many years. Corey and Bailar showed[96] that there would be several ways of combining three puckered ethylenediamine rings with a metal ion; these were named *kkk*, *kkk′*, *kk′k′*, and *k′k′k′*. For the D configuration of the metal, the *kkk* structure is favoured; for the L configuration, it is the *k′k′k′* structure. It was further shown that optically resolved diamines, employed as ligands, would dictate a preferred configuration at the central metal ion by virtue of the equatorial substituents on the chelate rings. For example, (+)-propylenediamine will give D-$[M(+pn)_3]^{3+}$ rather than L-$[M(+pn)_3]^{3+}$. This is a thermodynamic rather than a kinetic effect. The deduction of configurations of metal complexes from the known absolute configurations of asymmetric diamine ligands has been extremely useful[17,96,97], since this method is independent of spectroscopic assumptions.

One suggestion[5] was that, instead of being due to the split trigonal components of ${}^1T_{1g}(\mathbf{O}_h)$, the large circular dichroism band of (+)-$[Co\ en_3]^{3+}$ in this region arose from complex ions in the predominant *kkk* conformation, and the small negative band from some other (*kkk′*, *kk′k′*, or *k′k′k′*) conformation. However, the conformation of D-(+)-$[Co(+pn)_3]^{3+}$ is fixed by virtue of the three equatorial methyl groups and this complex has a circular dichroism spectrum extremely similar to that of (+)-$[Co\ en_3]^{3+}$ with two Cotton effects[97a] of opposite sign in the visible region. These cannot be due to differing conformations of the complex ion; they were thought to arise from the two electronic transitions of the ion. A closely similar argument was used[98] in a later paper, which also extended the measurements to the further ultraviolet and suggested that the sign of the charge-transfer Cotton effect at 220mμ depends on the conformation (k or k') of the diamine.

This in turn suggested that, since the sign of the circular dichroism in the charge-transfer region depends on the helicity of the diamine ligand without regard to the configuration of the whole complex, theories ascribing optical activity to trigonal perturbations were unsatisfactory—a criticism which is probably justifiable but with the wrong reasons. If the signs of Cotton effects for the *d–d* transitions were related to those of the charge-transfer bands, reflecting the conformations of the chelate rings, then clearly the orbitals or spectroscopic levels which are responsible for the visible optical activity would need to be modified to take account of the carbon atoms of the ligands. However, the *reverse* is found: the sign of Cotton effects for the *d–d* bands reflects the configuration about the metal and only in a few cases does it reflect the conformations of the individual chelate rings. The conformational possibilities will be important in the cases of saturated ligands but not for such complexes as the trisoxalates, for which, at present, the theory using trigonal splitting* is unchallenged.

Related to this criticism[98] of the theories relying on distortions of the ligand atoms, or their orbitals[99], or their charges, from octahedral symmetry is an attempt[100] to demonstrate the effects of isotopic substitution in the ligands remote from the metal. The electronic and circular dichroism spectra of (+)-$[Co\ en_3]^{3+}$ were measured in water and in heavy water. The observed differences were attributed to participation by the N–H or N–D bands in the electronic wave functions of the states giving rise to the optical activity and the isotropic absorption. However, Dingle showed[90] for the isotropic absorption that vibrational progressions in the N—H mode were not involved in giving vibronic intensity, and the careful work[101] of Boudreaux, Weigang, and Turner explained the anomaly. They measured the isotropic absorption spectrum of $[Co\ en_3]Cl_3$ in D_2O and in H_2O, confirming the difference found in the earlier results. However, the spectra of deuterated $[Co\ en_3]^{3+}$ in perdeuterodimethyl sulphoxide and of ordinary $[Co\ en_3]^{3+}$ in ordinary dimethyl sulphoxide are virtually identical, so that the difference between H_2O and D_2O is a solvent effect. The suggestion[100] that the *d*-orbital wave functions extend to the carbon and hydrogen as well as to the nitrogen atoms of the ethylenediamine rings is so far not proved.

* Known[74] to be appreciable in these complexes.

VII. CONCLUSION

A. The State of Empirical Knowledge

Whatever the origin of the optical activity in tris-chelated compounds, there is certainly a mass of empirical information on configurations which will most likely prove extremely useful in checking spectroscopic theories. Whether the two components in optical activity observed for $[Co\ en_3]^{3+}$ arise from Jahn–Teller splitting of the excited level or not, it is clear that comparisons of optical configuration may legitimately be made using Cotton effect curves *provided that no change in the spectroscopic situation has occurred.* It seems reasonable, therefore, to compare the series of, say, tris-1,2-diamine complexes of cobalt(III), or of chromium(III), or of rhodium(III), or the series 1,2,6-tris-α-amino acid complexes of a given metal. Though such empirical comparisons have been criticized [80], it seems likely that they are generally reliable. While such comparisons are based on intuition, they are readily defensible. Series of complexes may be compared which are spectroscopically similar, without knowing the detailed energy levels involved. For a comparison to be legitimate, the order of energy levels must stay the same in the compared complexes and clearly, for the series just mentioned, e.g. $[Co(1,2\text{-diamine})_3]^{3+}$, this is the case. The criticism [80] of empirical approaches was perhaps rather sweeping, since it was based on two examples; firstly, $[Cr(C_2O_4)_3]^{3-}$ and $[Cr\ mal_3]^{3-}$, which have chelate rings of different sizes and therefore should not be compared empirically and, secondly, complexes (−)-$[Fe\ phen_3]^{2+}$ and (−)-$[Fe\ bipy_3]^{2+}$ with parallel circular dichroisms, which might be thought to have the same absolute configurations. It was claimed [80] on spectroscopic grounds and on the solubility properties of the diastereoisomers that the complexes (−)-$[Fe\ phen_3]^{2+}$ and (+)-$[Fe\ bipy_3]^{2+}$ with quasi-enantiomeric circular dichroism spectra had the same configurations. Intuitively, this seems unlikely, and the recent Bijvoet analysis [24] of (−)-$[Fe\ phen_3]^{2+}$ is a useful step in establishing the origin of the discrepancy.

There have been several attempts to establish empirical rules which relate optical configurations of complexes to the observed Cotton effects for the longest wavelength absorption band in solution. These attempts have, in general, been directed in two independent ways. Firstly, the nature of spectroscopic transitions has been established and only Cotton effects arising from transitions of the same symmetry type compared. Secondly, absolute or relative configurations have been established in a variety of ways and the Cotton effects then compared to try to establish configurational rules independent of detailed spectroscopic knowledge. Some discussion of the present position in our knowledge of configuration is given, with particular attention to complexes of amino acids, $RCH(NH_2)COOH$.

The first, spectroscopic, approach has been highly successful with tris-chelated complexes having high symmetry ($\mathbf{D}_3$) but is less effective for complexes with only one axis of symmetry (C_2 or C_3) and is, of course, inappropriate to the case of C_1 symmetry, where electronic transitions are not symmetry determined.

As the examples of the second approach, the configurations now known with certainty are listed:

1. From Bijvoet analysis: D-(+)-$[Co\ en_3]^{3+}$, L-(−)-$[Co\ pn_3]^{3+}$, L-(−)-[Co en_2(sarcosine)], and (−)-$[Fe\ phen_3]^{2+}$.
2. From an internal configurational reference of known configuration: 1,2,6-D-(+)-[Co(L-ala)$_3$] and D-(+)-[Co en_2(L-glutamate)]$^+$.

A large number of relative configurations is known (usually through conversion of one kinetically inert, thermodynamically unstable complex into another chemically distinct species without bond breakage at the metal).

Other methods of configurational determination have been employed, notably stereoselectivity, and the so-called method of less soluble diastereoisomers. There is now a considerable body of consistent evidence on configurations of complex compounds, although a few errors have occurred. These will now be discussed in order to illustrate the pitfalls of such methods.

Arguments concerning configurations of complexes of low symmetry (e.g. $\mathbf{C}_1$) based on parentage* are often inconclusive. For example, an apparently reasonable argument of this kind led[103] to the assignment of (+)-1,2,6-[Co(L-ala)$_3$] to the L configuration, whereas a direct crystallographic study[25] showed it to be of the D configuration. Similarly parentage arguments in the series [Co en_2(α-amino acid)] led to the assignment[104] L-(+)-[Co en_2($\alpha\alpha$)]. (All enantiomers, for a series of R, had dominantly positive Cotton effects for their absorption bands near 492 mμ.) However, this is in error, since it was subsequently shown[26] crystallographically that the kinetically stable diastereoisomer (+)-[Co en_2(L-glutamate)]ClO_4, with a large positive Cotton effect for the 492 mμ band, had the D configuration.

There is a difficulty here, which is, as usual, spectroscopic in origin. How do we know that the Cotton effects of the complexes $[Co\ en_2(NH_2CHRCOO)]^{n+}$ are comparable, when R varies from CH_3 in alanine to $CH_2CH_2COO^-$ or $CH_2CH_2CO_2H$ in the glutamate complexes of known configuration? The intuitive idea that the spectroscopic components will appear in comparable orders for the whole series seems

* A detailed and extended discussion of parentage arguments in relating absolute configurations has been given by Schäffer[102].

reasonable but is not sufficient. Fortunately, there is very strong chemical support[105] for the idea. (−)-[Co en_2(glycine)] (R = H) has been converted chemically, through the Knoevenagel reaction, into (−)-[Co en_2 (serine)] (R = CH_2OH) and (−)-[Co en_2(threonine)] (R = $CH(OH)CH_3$) with no great change in the observed Cotton effects*. No bonds to cobalt are broken in the reaction, so that products and factors have the same helicity about the cobalt. In this series, at least, complexes with similar Cotton effect curves for the *d–d* bands have the same optical configuration about the metal ions.

A further point arises from the latter error in the series $[Co\ en_2(\alpha\alpha)]^{2+}$: the solubilities[104] of the various less soluble diastereoisomers *within the series* fitted the rule put forward by Werner, that for two complexes resolved with the same agent, the less soluble diastereoisomers contain the complexes of the same absolute configuration. This is not an unreasonable application of the rule, since only very small changes occur in passing from one amino acid or ligand to another (though it is still by no means rigorous). However, any comparison with $[Co\ en_3]^{3+}$, which might give diastereoisomers of completely different charge type, is completely inappropriate. A discussion of situations where the rule of less soluble diastereoisomers might be rigorous has been given[27]. Since solubility is such a complex function of lattice and ionic parameters, the conclusion was that the only safe application to deducing relative configurations was for isomorphous diastereoisomers; this can be strikingly useful for a series in which only the metal ion varies, such as $[M\ en_2Cl_2]$ with M = Cr, Co, Rh, where the less soluble diastereoisomers with the (+)-bromocamphorsulphonate ion are (−)-$[Co\ en_2Cl_2]$. (+)-$[Cr\ en_2Cl_2]$, and (−)-$[Rh\ en_2\ Cl_2]^+$.

There have been several attempts to establish a general rule relating Cotton effects and configurations for complexes, rather like the octant rule[18] for organic ketones, where a measurement of Cotton effects in solution is now usually sufficient to establish absolute configuration. However, the problems arising from the degenerate nature of transitions in the octahedral complexes are perhaps less serious than had been thought. For example, empirical methods have been criticized[80] on the ground that only analogous transactions should be compared, since the effect of small changes in ligands on the ordering of split components is unknown. (Indeed, it is not even clear that the electric dipole components are those of relevance.) It is certainly true that comparisons of complexes of polydentate ligands with the analogous bidentate compounds have been dangerous. However, the empirical observation, first made by Mathieu for the series $[Co\ en_2XY]^{n+}$, may be extended. He suggested[31], intuitively,

* The same was true starting from (+)-$[Co\ en_2(glycinate)]^{2+}$.

that those enantiomers in this series having dominant Cotton effects for band I of the same sign had the same configuration. On the basis of a variety of studies, this rule has been extended[105a,106] to state that 'If for a monomeric complex of cobalt(III) with two or more bidentate ligands, the Cotton effect for the longest wavelength absorption band is positive, then the complex has the D configuration'.

As final examples of the difficulty inherent in this field, some problems of apparent contradictions in configurational assignments will be discussed. Difficulties have arisen in complexes of polydentate ligands. The stereoselective coordination to metal ions of both *R*-(−)-mepenten[107] *N,N,N′,N′*-tetrakis-(2′-aminoethyl)-1,2-(−)-diaminopropane and of *R*-(−)-PDTA[108] (−)-1,2-diaminopropane-*N,N,N′,N′*-tetraacetate ion gives complexes of known [$R(C_2)$] absolute configuration (both ligands are made from *R*-(−)-propylenediamine) and, by empirical comparison of Cotton effect curves, gives the configurations $R(C_2)$-(−)-[Co penten] and $R(C_2)$-$(+)_{5461}$-[Co(EDTA)]. However, under band I ($^1A \rightarrow {}^1A + {}^1B + {}^1B$) both enantiomers have Cotton effects as follows: larger negative (at lower energy), smaller positive (at higher energy). This is in contrast to the situation in complexes of bidentate ligands obeying the simple rule above, where the order of circular dichroism bands for the $R(C_2)$ configuration is, in general: larger positive (at lower energy), smaller negative (at higher energy).

The apparent discrepancy may possibly be resolved on the following lines. For complexes with $\mathbf{C}_2$ symmetry, the transition $^1A \rightarrow {}^1A$ (parallel to the C_2 axis) involves the promotion of an electron in a helical path. In either classical or quantum theories of optical activity, an electron constrained to move in a right-handed helix gives a positive Cotton effect. Thus, for the $R(C_2)$ configuration, $^1A \rightarrow {}^1A$ will show positive rotational strength and, for the $S(C_2)$ configuration, $^1A \rightarrow {}^1A$ will show a negative rotational strength.

The order and rotational strengths of the *B* transitions are not deducible, since they are not symmetry determined. Empirical assignments of configuration using Cotton effect curves are unreliable in cases where the electronic transitions of the chromophore considered are degenerate in the zero order and where splittings in energy levels may vary between the compounds compared. This is certainly true for attempted comparisons of polydentate with bidentate complexes, since the effects of small changes of molecular geometry on the ordering of components of band I are not yet predictable.

The *A* component of band I can usually be assigned, since it is symmetry determined. For $(+)_{5461}$-[Co{*R*-(−)-PDTA)}]$^-$ and $(+)_{5461}$-[Co(EDTA)]$^-$, and for (−)-[Co penten]$^{3+}$ and (−)-[Co{*R*-(−)-mepenten}]$^{3+}$, all known to have the $R(C_2)$ configuration, the positive Cotton effect may be assigned to $^1A \rightarrow {}^1A$. The order of the spectroscopic levels in the polydentate complexes is then[107] $A > B$, in contrast to the complexes covered by rule I, where it is $B > A$.

A very similar example arises in comparisons of complexes of cobalt(III) and chromium(III). For the trisdiamine chelates [Co en$_3$]$^{3+}$ and [Cr en$_3$]$^{3+}$, the splitting of the spectroscopic levels is the same in each case ($A > E$). For complexes of polydentate ligands, this is often not true; for example, with the ligand (+)-PDTA, the circular dichroism curves for the complexes of cobalt(III) and chromium(III) are approximately enantiomeric[108], although the complexes have the same configuration. This is better explained by a spectroscopic effect, e.g. changes in the order (or relative rotational strengths) of energy levels of chromium(III) relative to cobalt(III), than by an inversion[107] of octant sign.

In the case of (−)-[Co penten], it appears that the order of levels is $A > B$. The virtual disappearance of the Cotton effect due to the A transition arises because the A and B levels are so nearly degenerate. The total negative rotational strength of the B levels is (in this case) greater and $^1A \rightarrow {}^1A$ appears only as a residual wing absorption. The similar situation for [Co(EDTA)]$^-$ is more clear cut, since the components $^1A + {}^1B + {}^1B$ are more distinctly split.

It is likely that assignment[109] of configurations to complexes of $\mathbf{C}_2$ or lower symmetry on the basis of alterations in Cotton effects due to ion pairing with anions such as selenite may be unreliable. (It was on this basis that (−)-[Co penten]$^{3+}$ was incorrectly assigned to the same configuration as (−)-[Co en$_2$(NH$_3$)$_2$]$^{3+}$.)

B. Conclusion

A number of questions arise in forming a theory of the optical activity of asymmetric molecules.

1. Which molecules are capable of exhibiting optical activity?
2. How does the radiation interact with the asymmetric molecule to give rise to optical activity?
3. How are optical activity and absolute configuration related?
4. How big is the optical activity of a given molecule? What does $[\alpha]_\lambda$ depend on?

The conditions for asymmetry are known. The requirement (that the molecule is not superposable on its image in a mirror) is expressed in the rule that the molecule should not have an improper axis of symmetry (centres and planes of symmetry are special cases of improper axes). The third question has been partly answered experimentally; active research in this field has been a feature of organic chemistry for the past 10 years. The configurations of coordination compounds are now being established by a variety of means. Calculations on this subject appear not to have been successful. The fourth question is discussed at length in this chapter. It will be clear from the statement of the present position that the surface has merely been skimmed and that a great deal of experimentation will be required to set up rules in answer to the questions formulated above. In particular, the next few years should see a great deal of work on the difficult problem of the spectroscopic levels responsible for optical activity.

The optical activity of inorganic complexes has once more become a field of intense activity, largely because of the instrumental advances and because of the increasingly sophisticated experimental elucidation of the spectroscopic facts underlying optical activity, as summarized in this chapter. An aspect of great chemical and biological importance, not discussed here but certainly attracting increasing attention, is the interaction of a number of types of asymmetry within the same molecule or in the transition state of a reaction, giving rise to the effects commonly called stereoselectivity. It is perhaps reasonable to expect that the ability to relate optical configurations to observable properties will lead rapidly to biological applications of increasing complexity.

VIII. REFERENCES

1. S. Mitchell, *The Cotton Effect*, Bell, London, 1933.
2. J. H. Dunlop and R. D. Gillard, *Advan. Inorg. Chem. Radiochem.*, **9**, 185 (1966).
3. S. F. Mason, *Quart. Rev.* (*London*), **17**, 20 (1963).
4. R. D. Gillard, *Progr. Inorg. Chem.*, **7**, 215 (1966).
5. F. Woldbye, *Recent Chem. Progr.*, **24**, 197 (1963).
6. R. D. Gillard and H. M. Irving, *Chem. Rev.*, **65**, 616 (1965).
7. M. J. Albinak, D. C. Bhatnagar, S. Kirschner, and A. J. Sonnessa, *Can. J. Chem.*, **39**, 2360 (1961).
8. H. L. Smith and B. E. Douglas, *J. Am. Chem. Soc.*, **86**, 3885 (1964).
9. R. Larsson, *Acta Chem. Scand.*, **16**, 2267 (1962).
10. S. F. Mason and B. J. Norman, (a) *Chem. Commun.*, **1965**, 73; (b) *Proc. Chem. Soc.*, **1964**, 339.
11. O. E. Weigang, *J. Chem. Phys.*, **41**, 1435 (1964).
12. J. H. Dunlop, R. D. Gillard, and R. Ugo, *J. Chem. Soc.*, *A*, **1966**, 1540.
13. F. Basolo and R. G. Pearson, *Mechanisms of Inorganic Reactions*, Wiley, New York, 1958.
14. K. Nakatsu, M. Shiro, Y. Saito, and H. Kuroya, *Bull. Chem. Soc. Japan*, **30**, 158 (1957).
15. K. Garbett and R. D. Gillard, *J. Chem. Soc.*, **1965**, 6084.
16. T. S. Piper, *J. Am. Chem. Soc.*, **83**, 3908 (1961).
17. J. H. Dunlop and R. D. Gillard, *J. Inorg. Nuclear Chem.*, 1965, **27**, 361.

18. W. Klyne, 'Stereochemical correlations', *Roy. Inst. Chem. Lecture Ser.*, **1962**, No. 4.
19. A. J. McCaffery, S. F. Mason, and R. E. Ballard, *J. Chem. Soc.*, **1965**, 2883.
20. R. S. Cahn, C. K. Ingold, and V. Prelog, *Angew. Chem. (Intern. Ed. Engl.)*, **5**, 385 (1966).
21. J. M. Bijvoet, *Endeavour*, **14**, 71 (1955).
22. Y. Saito, H. Iwasaki, and H. Ota, *Bull. Chem. Soc. Japan*, **30**, 1543 (1963).
23. J. F. Blount, H. C. Freeman, A. M. Sargeson, and K. R. Turnbull, *Chem. Commun.*, **1967**, 324.
24. D. H. Templeton, A. Zalkin, and T. Ucki, *Intern. Congr. Symp. Crystal Growth, 7th Moscow, 1966*, p. A163.
25. M. G. B. Drew, J. H. Dunlop, R. D. Gillard, and D. Rogers, *Chem. Commun.*, **1966**, 42.
26. J. H. Dunlop, R. D. Gillard, N. C. Payne, and G. B. Robertson, *Chem. Commun.*, **1966**, 874.
27. K. Garbett and R. D. Gillard, *J. Chem. Soc., A*, **1966**, 802.
28. W. A. Shurcliff, *Polarized Light*, Oxford University Press, London, 1961.
29. A. Moscowitz in *Optical Rotatory Dispersion* (Ed. C. Djerassi), McGraw-Hill, New York, 1960.
30. W. Kuhn, *Ann. Rev. Phys. Chem.*, **9**, 417 (1958).
31. J. P. Mathieu, (a) *Ann. Phys. (Paris)*, **19**, 335 (1944); (b) *J. Chim. Phys.*, **33**, 78 (1936); (c) *Bull. Soc. Chim. France*, **3**, 476 (1936); (d) *Bull. Soc. Chim. France*, **5**, 105 (1938); (e) *Bull. Soc. Chim. France*, **5**, 773 (1938); (f) *Bull. Soc. Chim. France*, **6**, 873 (1939); (g) *Contributions to the Study of Molecular Structure* (Victor Henri Commem. Vol.), Desoer, Liège, 1947, p. 111.
32. E. Bunnenberg and C. Djerassi, *Proc. Chem. Soc.*, **1963**, 299.
33. S. F. Mason and B. J. Norman, *Chem. Commun.*, **1965**, 335.
34. R. D. Gillard, *Nature*, **198**, 580 (1963).
35. J. H. Dunlop and R. D. Gillard, *J. Chem. Soc.*, **1965**, 6531.
36. P. V. Dowley, K. Garbett, and R. D. Gillard, *Inorg. Chim. Acta*, (1967) in press.
37. K. Garbett and R. D. Gillard, *J. Chem. Soc., A*, **1966**, 204.
38. M. D. Cohen and E. Fischer, *J. Chem. Soc.*, **1962**, 3044.
39. F. Woldbye, *Optical Rotatory Dispersion of Transition Metal Complexes*, European Research Office, U.S. Dept. of the Army, Frankfurt-am-Main, 1959.
40. O. Kling, *Acta Chem. Scand.*, **15**, 229 (1961).
41. Y. Shimura, *Bull. Chem. Soc. Japan*, **31**, 315 (1958).
42. B. E. Douglas, R. A. Haines, and J. G. Brushmiller, *Inorg. Chem.*, **2**, 1194 (1963).
43. C. Schäffer, *Symp. Structure Coord. Comp. Bratislava, 1964*, p. 66.
44. (a) R. Larsson, S. F. Mason and B. J. Norman, *J. Chem. Soc., A*, **1966**, 301; (b) S. F. Mason and B. J. Norman, *J. Chem. Soc., A*, **1966**, 307.
45. O. Kling and F. Woldbye, *Acta. Chem. Scand.*, **15**, 704 (1961).
46. P. Drude, *Lehrbuch der Optik*, Leipzig, 1900. English version published by Dover, New York, 1959.
47. E. U. Condon, *Rev. Mod. Phys.*, **9**, 432 (1937).
48. S. F. Mason, *Proc. Chem. Soc.*, **1962**, 137.

49. W. Moffitt, *J. Chem. Phys.*, **25**, 1189 (1956).
50. A. Bose and R. Chatterjee, *Proc. Phys. Soc.* (*London*), **82**, 23 (1963).
51. J. H. E. Griffiths and J. Owen, *Proc. Roy. Soc.* (*London*), *Ser. A*, **213**, 459 (1952).
52. L. R. Ingersoll, P. Rudnick, F. G. Slack, and N. Underwood, *Phys. Rev.*, **57**, 1145 (1940).
53. W. Moffitt and A. Moscowitz, *J. Chem. Phys.*, **30**, 648 (1959).
54. P. Rudnick and L. R. Ingersoll, *J. Opt. Soc. Am.*, **32**, 622 (1942).
55. C. J. Ballhausen, *Kgl. Danske Vidensk. Selsk., Mat.-fys. Medd.*, **8**, 29 (1955).
56. H. Hartmann and H. Muller, *Discussions Faraday Soc.*, **26**, 49 (1958).
57. C. K. Jorgensen, *Acta Chem. Scand.*, **9**, 1362 (1955).
58. A. D. Liehr and C. J. Ballhausen, *Ann. Phys.* (*N.Y.*), **6**, 134 (1959).
59. C. J. Ballhausen and A. D. Liehr, *Mol. Phys*,, **2**, 123 (1959).
60. A. Mookherji and N. S. Chhonkar, *Indian J. Phys.*, **34**, 363 (1960).
61. T. S. Piper and N. Koertge, *J. Chem. Phys.*, **32**, 559 (1960).
62. R. P. Penrose and K. W. H. Stephens, *Proc. Phys. Soc.* (*London*), **63**, 29 (1950).
63. L. Longchambon, *Compt. Rend.*, **173**, 89 (1921).
64. B. Bosnich, J. H. Dunlop, and R. D. Gillard, *Chem. Commun.*, **1965**, 274.
65. W. Voigt, *Ann. Phys.*, **18**, 649 (1905).
66. R. E. Ballard, A. J. McCaffery and S. F. Mason, *Proc. Chem. Soc.*, **1962**, 331.
67. A. J. McCaffery and S. F. Mason, *Mol. Phys.*, **6**, 359 (1963).
68. A. J. McCaffery and S. F. Mason, *Trans. Faraday Soc.*, **59**, 1 (1963).
69. S. Sugano, *J. Chem. Phys.*, **33**, 1883 (1960).
70. H. Poulet, *J. Chim. Phys.*, **59**, 584 (1962).
71. N. K. Hamer, *Mol. Phys.*, **5**, 339 (1962).
72. (a) T. S. Piper and A. G. Karipides, *Mol. Phys.*, **5**, 475 (1962); (b) A. G. Karipides and T. S. Piper, *J. Chem. Phys.*, **40**, 674 (1964).
73. T. Burer, *Helv. Chim. Acta*, **46**, 242 (1963).
74. T. S. Piper and R. L. Carlin, *J. Chem. Phys.*, **35**, 1809 (1961).
75. T. S. Piper, *J. Chem. Phys.*, **36**, 3330 (1962).
76. S. Yamada and R. Tsuchida, *Bull. Chem. Soc. Japan*, **33**, 98 (1960).
77. E. Drouard and J. P. Mathieu, *Compt. Rend.*, **236**, 2395 (1953).
78. R. D. Gillard, *J. Inorg. Nuclear Chem.*, **26**, 657 (1964).
79. W. E. Hatfield, *Inorg. Chem.*, **3**, 605 (1964).
80. A. J. McCaffery, S. F. Mason, and B. J. Norman, *Proc. Chem. Soc.*, **1964**, 259.
81. C. H. Johnson and A. Mead, *Trans. Faraday Soc.*, **31**, 1621 (1935).
82. A. L. Porte, H. S. Gutowsky, and G. M. Harris, *J. Chem. Phys.*, **34**, 66 (1961).
83. A. Werner, *Chem. Ber.*, **47**, 2955 (1914).
84. P. Herpin, *Bull. Soc. Franc. Mineral. Crist.*, **81**, 245 (1958).
85. F. M. Jaeger, *Spatial Arrangements of Atomic Systems and Optical Activity*, McGraw-Hill, New York, 1930.
86. R. H. Fenn, A. J. Graham and R. D. Gillard, *Nature*, **213**, 1012 (1967).
87. W. Kuhn and K. Bein, *Z. Anorg. Chem.*, **216**, 321 (1934).
88. A. J. McCaffery and S. F. Mason, *Proc. Chem. Soc.*, **1962**, 388.
89. E. Larsen and S. F. Mason, *J. Chem. Soc.*, *A*, **1966**, 313.

90. R. Dingle, *Chem. Commun.*, **1965**, 304.
91. R. G. Denning, *Chem. Commun.*, **1967**, 120.
92. A. J. McCaffery, S. F. Mason, and B. J. Norman, *Chem. Commun.*, **1966**, 661.
93. O. E. Weigang, *J. Chem. Phys.*, **43**, 3609 (1965).
94. J. P. Mathieu, J. M. Terrasse, and H. Paules, *Spectrochim. Acta*, **20**, 305 (1964).
95. J. V. Quagliano and S. I. Mizushima, *J. Am. Chem. Soc.*, **75**, 6084 (1953).
96. E. J. Corey and J. C. Bailar, *J. Am. Chem. Soc.*, **81**, 2620 (1959).
97. (a) R. D. Gillard and G. Wilkinson, *J. Chem. Soc.*, **1964**, 1368; (b) J. H. Dunlop, R. D. Gillard, and G. Wilkinson, *J. Chem. Soc.*, **1964**, 3160.
98. A. J. McCaffery, S. F. Mason, and B. J. Norman, *Chem. Commun.*, **1965**, 49.
99. A. D. Liehr, *J. Phys. Chem.*, **68**, 665 (1964).
100. S. F. Mason and B. J. Norman, *Chem. Commun.*, **1965**, 48.
101. E. A. Boudreaux, O. E. Weigang and W. Turner, *Chem. Commun.*, 1966, 378.
102. C. Schäffer, *Proc. Roy. Soc.* (*London*), *Ser. A*, **297**, 96 (1967).
103. B. E. Douglas and S. Yamada, *Inorg. Chem.*, **4**, 1561 (1965).
104. C. T. Liu and B. E. Douglas, *Inorg. Chem.*, **3**, 1356 (1964).
105. (a) R. D. Gillard, *Chemistry in Britain*, 1967; (b) R. D. Gillard and P. M. Harrison, *J. Chem. Soc.*, *A*, **1967**, in press.
106. R. D. Gillard, *Proc. Roy. Soc.* (*London*), *Ser. A*, **297**, 134 (1967).
107. J. R. Gollogly and C. J. Hawkins, *Chem. Commun.*, **1966**, 833.
108. R. D. Gillard, *Spectrochim. Acta*, **20**, 1431 (1964).
109. S. F. Mason and B. J. Norman, *Chem. Commun.*, **1965**, 73.

6

Vibrational Spectroscopy

M. J. WARE

I. INTRODUCTION

At a time when inorganic chemistry is enjoying a period of renaissance, it is not surprising to find that parallel advances have occurred in vibrational spectroscopy, which, with a few notable exceptions, has for so long been the domain of the organic and the physical chemist. Indeed, the developments in the two fields are related; the demand of the inorganic chemist to know more about vibrations involving atoms other than those in the first row of the periodic table has resulted in the extension of the long-wavelength limit of commercial infrared spectrometers. Moreover, the widespread interest in transition metal compounds, which are for the most part highly coloured, has been answered in the Raman field by the development of excitation at the long-wavelength end of the visible spectrum, where the customary method of using blue radiation is quite inapplicable. Inorganic chemistry owes much to the pioneers in these fields.

Although infrared and Raman spectroscopy are the established and predominant methods of investigating the vibrations of a molecule in its electronic ground state, it cannot go unremarked that vibrational transitions have also recently been observed by the methods of slow-neutron scattering[1], circular dichroism[2], visible absorption spectroscopy[3], photoionization[4], and the so-called 'stimulated'[5], 'inverse'[6], and 'hyper'[7] Raman effects. These observations hold promise for the future. However, it is the purpose of this present chapter to indicate the way in which infrared and Raman studies can provide both a means of determining molecular structure and an insight into molecular dynamics. The former, as will be seen, arises from the close relationship between the spectroscopic activity of the vibrations and the symmetry of the molecule. Knowledge of the latter may be deduced from the vibration frequencies and leads to the restoring forces which operate when a molecule is slightly deformed from its equilibrium configuration. These forces are related to the potential field of the molecule and their evaluation can make a useful contribution to the understanding of the chemical bond.

The demands of space have necessarily limited the scope of this account; the application of vibrational spectroscopy for the purposes of qualitative[8–13] and quantitative[14] analysis, and the determination of equilibria[14,15], important as they are to inorganic chemistry, will not be discussed. As a necessary preliminary, the theory of molecular vibrations and of Raman and infrared spectra is outlined in Section II. No attempt has been made to achieve mathematical rigour (for which the reader should consult the authoritative works by Herzberg[16] and Wilson, Decius, and Cross[17]), but rather it has been the object to explain the methods of the practising inorganic spectroscopist. The brief description of experimental techniques in Section III is confined to the more recent and less widely documented developments and may be greatly amplified by reference elsewhere[18,19]. Following an account of the methods of vibrational assignment[20] in Section IV, the determination of molecular structure is illustrated by selected examples in Section V. Surveys[9,14,21,22] more comprehensive than this will provide valuable further reading.

Regarding the order of molecular complexity which can reasonably be resolved by a method of structure determination, vibrational spectroscopy may be judged to occupy a position intermediate between microwave spectroscopy and electron diffraction on the one hand, and x-ray and neutron diffraction on the other. But, unlike these methods, it may also be described as 'sporting'[23], in that the results obtained are not always accurate or unambiguous. In compensation, vibrational spectroscopy may claim the advantage that it is a rapid method which often demands no great amount of interpretative effort and can be applied, though not with equal ease, to all states of matter. Finally, Section VI deals with the effects of electronic structure on vibrational spectra. Foremost among these is the calculation of force constants from vibration frequencies[16,17] and their interpretation in terms of chemical bonding[24,25], but the intensities of

Raman[26] and infrared[27] bands have also yielded information useful to inorganic chemistry. Despite the approximations which are almost inevitable in force-constant calculations and the lack of a completely satisfactory intensity theory, this aspect of vibrational spectroscopy holds much interest and provides many of the current points of growth in the field.

II. THEORETICAL BASIS

A. Molecular Vibrations

It is convenient to approach the problem of molecular vibrations from the viewpoint of classical mechanics and then proceed to the quantum-mechanical treatment. The simplicity of the diatomic molecule makes it a useful starting point, since it exemplifies many of the vibrational properties of more complex molecules.

1. Diatomic molecules

We adopt as a model (Figure 1) two point masses m_1 and m_2 (the nuclei) joined by a weightless spring (the bond) which obeys Hooke's law; that is,

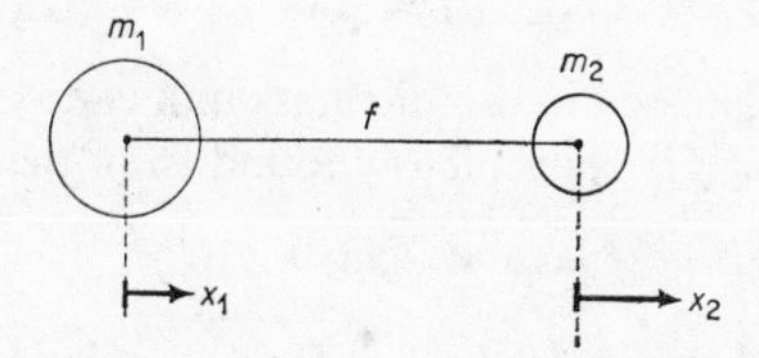

Figure 1. Model for diatomic molecules.

the potential function is quadratic for small displacements x_1 and x_2 along the internuclear axis. The potential energy V is then given by

$$2V = f(x_1 - x_2)^2$$

where f is the bond force constant. The kinetic energy T is

$$2T = m_1\dot{x}_1^2 + m_2\dot{x}_2^2$$

where

$$\dot{x}_1 = \left(\frac{\partial x_1}{\partial t}\right)$$

By substituting these expressions into the general equations of motion[28]

$$\frac{\mathrm{d}}{\mathrm{d}t}\left(\frac{\partial T}{\partial \dot{x}_i}\right) + \frac{\partial V}{\partial x_i} = 0 \tag{1}$$

it is readily shown that the internal motion of this system is the same as that of a single particle of mass μ, executing linear harmonic oscillation with a frequency ν (in cm^{-1}) given by

$$\nu = \frac{1}{2\pi c}\sqrt{\frac{f}{\mu}}$$

where

$$\frac{1}{\mu} = \frac{1}{m_1} + \frac{1}{m_2}$$

Thus both nuclei perform simple harmonic oscillation along their line of centres with the same phase and frequency but in general with differing amplitudes. This is the *normal mode* of vibration for a diatomic molecule; a single coordinate, $(x_1 - x_2) = s$, serves to define it:

$$s = A \cos (\lambda^{\frac{1}{2}}t + \epsilon)$$

where A is an amplitude constant, ϵ a phase constant, and λ is defined by

$$\lambda = 4\pi^2c^2\nu^2 \tag{2}$$

Thus

$$f = \mu\lambda$$

If m_1 and m_2 are expressed in atomic mass units and ν in cm^{-1}, the value of the force constant f is conveniently found from the expression

$$f = \mu \times 0.5889(\nu/10^3)^2$$

in units of millidynes/ångström (mdyn/Å) (see Table 1).

2. Polyatomic molecules

If we visualize a polyatomic molecule as an assembly of point masses and springs, as above, then it is apparent that an arbitrary 'blow' to the system will produce a periodic, but not necessarily simple, harmonic motion in each of its constituent atoms. This complex and varied motion may be resolved into a limited number of simple independent modes of vibration, known as the *normal modes*, each being described by one *normal coordinate*. The number of normal modes is equal to the number of internal degrees of freedom possessed by the N-atomic system, i.e. $3N - 6$ for non-linear molecules and $3N - 5$ for linear molecules.

For any nucleus i, a small displacement from its equilibrium position is defined by three cartesian displacement coordinates Δx_i, Δy_i, and Δz_i. These are conveniently replaced by mass-weighted displacement coordinates, with a consistent notation q_i:

$$q_1 = m_1^{\frac{1}{2}}\,\Delta x_1;\quad q_2 = m_1^{\frac{1}{2}}\,\Delta v_1;\quad q_3 = m_1^{\frac{1}{2}}\,\Delta z_1;\quad q_4 = m_2^{\frac{1}{2}}\,\Delta x_2;\ \ldots\text{etc.}$$

The kinetic energy is then simply written as

$$2T = \sum_{i=1}^{3N} \dot{q}_i^2 \qquad (3)$$

The potential energy is a function of all the q_i coordinates, and may be expanded as a Taylor series:

$$2V = 2V_0 + 2\sum_{i=1}^{3N}\left(\frac{\partial V}{\partial q_i}\right)_0 q_i + \sum_{i,j=1}^{3N}\left(\frac{\partial^2 V}{\partial q_i\,\partial q_j}\right)_0 q_i q_j + \cdots$$

We ignore terms higher than quadratic by applying the harmonic oscillator approximation for small displacements. Moreover, when $q_i = 0$ the molecule is in its equilibrium configuration, for which V is a minimum. Therefore

$$\left(\frac{\partial V}{\partial q_i}\right)_0 = 0$$

and we choose the arbitrary zero of energy so that $V_0 = 0$. The potential energy expression is then

$$2V = \sum_{i,j=1}^{3N} f_{ij} q_i q_j \qquad (4)$$

where $f_{ij} = (\partial^2 V/\partial q_i\,\partial q_j)_0$ is the force constant relating to the simultaneous displacements q_i and q_j. These generalized expressions for the energy may be compared with those obtained earlier for the specific case of a diatomic molecule. The potential energy V is not yet in a tractable form for solution of the equations of motion, owing to the presence of cross-products which involve different displacement coordinates in equation (4), i.e. $f_{ij} \neq 0$ when $i \neq j$. To overcome this difficulty, the coordinate system is changed from the q_i to a new set Q_k which are defined in such a way as to be orthogonal. The Q_k coordinates are generated from the q_i by a linear transformation

$$Q_k = \sum_{i=1}^{3N} B_{ik} q_i \qquad k = 1, 2, 3, \ldots, 3N$$

in which the coefficients B_{ik} are chosen so that the energy expressions contain no cross-terms:

$$2T = \sum_{k=1}^{3N} \dot{Q}_k^2 \qquad \text{and} \qquad 2V = \sum_{k=1}^{3N} \lambda_k Q_k^2$$

On substitution of these expressions for T and V, the equations of motion (1) are resolved into $3N$ independent equations of the linear harmonic oscillator type:

$$\ddot{Q}_k + \lambda_k Q_k = 0 \qquad k = 1, 2, 3, \ldots, 3N$$

having solutions

$$Q_k = Q_k^0 \cos(\lambda_k^{\frac{1}{2}} t + \epsilon_k) \qquad k = 1, 2, 3, \ldots, 3N \qquad (5)$$

where $\lambda_k = 4\pi^2c^2\nu_k^2$ and ϵ_k is a phase constant. These oscillations are the normal modes of vibration*, each having a characteristic frequency ν_k and being defined by a single normal coordinate Q_k which embodies all the displacements that the nuclei suffer during the execution of the normal mode. Thus, a normal coordinate is a time-dependent variable, comprising a linear combination of the nuclear displacements, suitably mass-weighted and normalized. During a normal mode of vibration, the relative magnitudes and directions of these displacements remain fixed but their absolute values fluctuate periodically with time. Any vibrational motion of the

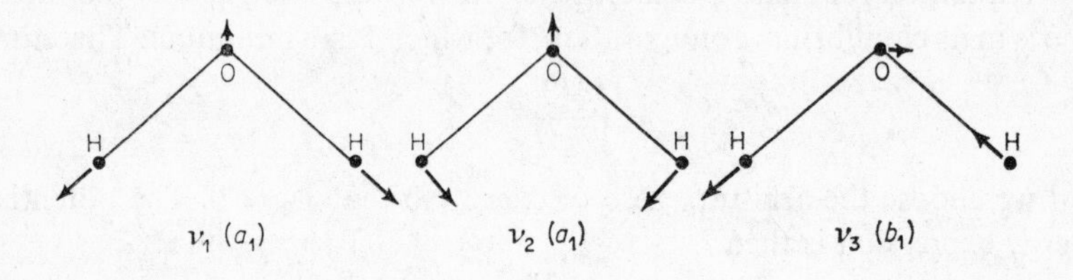

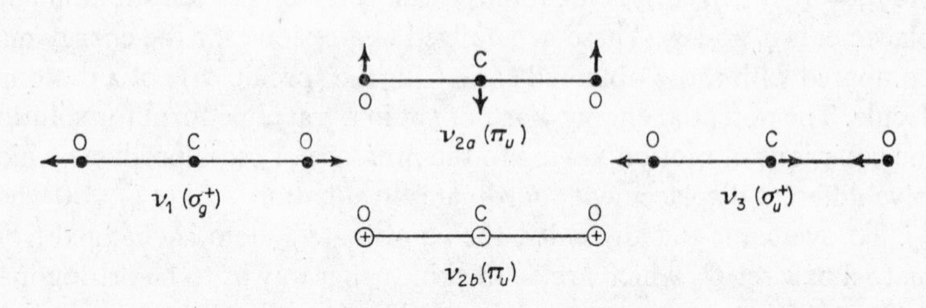

Figure 2. Normal modes of H_2O and CO_2. The arrows indicate relative nuclear displacements; ⊕ and ⊖ signify motion perpendicular to the page.

molecule whatsoever may then be represented by a superposition of one or more normal modes. As an example, the normal modes of H_2O and CO_2 are illustrated schematically in Figure 2. The example of CO_2 introduces the important idea of *degeneracy* in normal modes: although CO_2 has $3N - 5 = 4$ normal modes, ν_{2a} and ν_{2b} obviously have the same frequency, as a direct consequence of the molecular symmetry. It is usual, therefore, to refer to a single, doubly degenerate normal mode ν_2. In a

* Although $3N$ equations (5) are obtained, there are only $3N - 6$ solutions having non-zero frequencies. It is a consequence of the use of cartesian coordinates that the solutions include those for 'non-genuine' vibrations, i.e. translation and rotation.

degenerate normal mode, the nuclei still oscillate with the same frequency but, owing to the arbitrary phase with which the components may be combined, their motion is no longer necessarily rectilinear, as it is in non-degenerate normal modes.

Although the existence of normal modes may be demonstrated mathematically and their number and symmetry properties may be readily derived as we shall see shortly, it is not generally possible to calculate normal coordinates *ab initio* because of our ignorance of the molecular force field. Indeed, it is one of the objects of vibrational spectroscopy to derive from the observed frequencies ν_k satisfactory values of the f_{ij}—the molecular force constants. The relationship between ν_k and f_{ij} is not simple, however, and no knowledge of it is necessary for the purposes of structure determination. Discussion of this aspect will therefore be deferred until Section VI.

3. *Quantization*

Referring to the results for a diatomic molecule obtained previously, the potential energy is

$$V = \tfrac{1}{2}fs^2$$

and the kinetic energy is

$$T = \tfrac{1}{2}\mu\dot{s}^2$$

The momentum p conjugate to s is $p = \mu\dot{s}$, so the kinetic energy becomes

$$T = \frac{p^2}{2\mu}$$

The total vibrational energy E_v is therefore

$$E_v = V + T = \frac{p^2}{2\mu} + \tfrac{1}{2}fs^2$$

and the vibrational hamiltonian becomes

$$\mathscr{H}_v = -\frac{h^2}{2\mu}\frac{\partial^2}{\partial s^2} + \tfrac{1}{2}fs^2$$

The Schrödinger wave equation for vibration

$$\mathscr{H}_v\psi_v = E_v\psi_v$$

takes the form of a one-dimensional harmonic oscillator wave equation, having eigenvalues

$$E_v = \text{h}c\nu(v + \tfrac{1}{2})$$

where $v = 0, 1, 2, \ldots$, the vibrational quantum number. Quantization therefore gives rise to a set of equally spaced terms $G(v)$, given (in cm^{-1}) by

$$G(v) = \nu(v + \tfrac{1}{2})$$

whose population is governed by the Boltzmann distribution

$$n_v = n_0 \exp(-vhc\nu/kT)$$

Transitions between these terms may be induced by electromagnetic radiation, the frequency of light absorbed or emitted being given by the Bohr frequency rule

$$\nu_{\text{observed}} = \frac{E_{v'} - E_{v''}}{hc} = G(v') - G(v'')$$

It follows that for a transition $v \rightarrow v \pm 1$ the observed spectroscopic frequency is equal to the classical vibrational frequency of the normal mode. It must be emphasized that this is not a trivial result but arises because the quantization of vibrational energy is linear in the harmonic oscillator approximation. Departures from this approximation may be caused by mechanical anharmonicity* and lead to a discrepancy between the observed and the harmonic vibration frequencies. Anharmonicity corrections can only be made for a limited number of simple molecules, in particular diatomics; fortunately, the corrections are small except for systems involving very light atoms, as shown in Table 1.

Table 1. Vibration frequencies and force constants of diatomic species.

Molecule	ν_0 (cm^{-1})	ν_e (cm^{-1})	f (mdyn/Å)
H_2	4161	4395	5.15
HF	3962	4139	8.85
HCl	2886	2990	4.82
HBr	2558	2649	3.85
HI	2230	2310	2.93
N_2	2331	2360	22.4
O_2	1555	1580	11.4
NO	1876	1904	15.5
CO	2143	2170	18.6
F_2	892	—	4.46
Cl_2	557	565	3.19
Br_2	317	323	2.45
I_2	213	215	1.70
CN^-	ca. 2080	—	15.3
OH^-	ca. 3640	—	8.4
NO^+	ca. 2220	—	21.7

ν_0 is the observed fundamental frequency.
ν_e is the harmonic vibration frequency (see text below).
The f values are calculated from ν_0.

* If a more realistic potential function is used (e.g. the Morse function), the term values become $G(v) = \nu_e(v + \frac{1}{2}) - x_e\nu_e(v + \frac{1}{2})^2 \ldots$, where ν_e is the harmonic vibration frequency.

On proceeding to polyatomic molecules, the particular advantages of employing normal coordinates become apparent. Since the normal coordinates are orthogonal by construction, the vibrational hamiltonian may be written

$$\mathscr{H}_v = -\frac{\hbar^2}{2}\sum_{k=1}^{3N}\frac{\partial^2}{\partial Q_k^2} + \frac{1}{2}\sum_{k=1}^{3N}\lambda_k Q_k^2$$

and the Schrödinger wave equation for nuclear motion factors into $3N$ wave equations of the linear harmonic oscillator type. The equations are analogous to those for the diatomic molecule and are obtained by replacing the simple change in internuclear distance s by the linear combination of nuclear displacements Q_k. The eigenvalues for the $3N - 6$ 'genuine' normal modes are

$$E_k = hc\nu_k(v_k + \tfrac{1}{2}) \qquad k = 1, 2, 3, \ldots, 3N - 6$$

Thus, v_k generates a set of terms for each normal mode

$$G(v_k) = \nu_k(v_k + \tfrac{1}{2})$$

and the total vibrational energy for a given state of the molecule is

$$G(v_1, v_2, \ldots, v_{3N-6}) = \sum_{k=1}^{3N-6} \nu_k(v_k + \tfrac{1}{2})$$

As an example of a term scheme, which also conveniently introduces the nomenclature associated with vibrational transitions, we may consider a molecule having three normal modes, ν_1, ν_2, and ν_3. Each term in the manifold $G(v_1, v_2, v_3)$ is designated by values of the three quantum numbers; a few of the lower-lying terms are illustrated in Figure 3. The accompanying Table 2 summarizes the types of transition possible and the approximate relation between the observed spectroscopic frequency and the harmonic vibration frequencies of the normal modes.

It is a direct consequence of the quantum-mechanical treatment of the linear harmonic oscillator that a selection rule restricts the 'permitted' transitions to those in which only one quantum number changes, and that by one unit only, i.e.

$$\Delta v_k = \pm 1, \qquad \text{all other} \quad \Delta v = 0$$

This very stringent condition allows only fundamentals and hot bands in the spectra, and the probability of the latter will be small except at elevated temperatures or for low frequencies. In practice, the spectra prove to be richer than this selection rule would predict, because combination, difference, and overtones may be permitted to an extent which depends on the breakdown of the linear harmonic oscillator selection rule through anharmonicity of the normal modes. Anharmonicity therefore assumes importance as the mechanism by which forbidden transitions may gain sufficient intensity to be observed but its quantitative effect on the frequencies must usually be ignored. The phenomenon known as Fermi resonance[21] is another indirect consequence of anharmonicity. It may

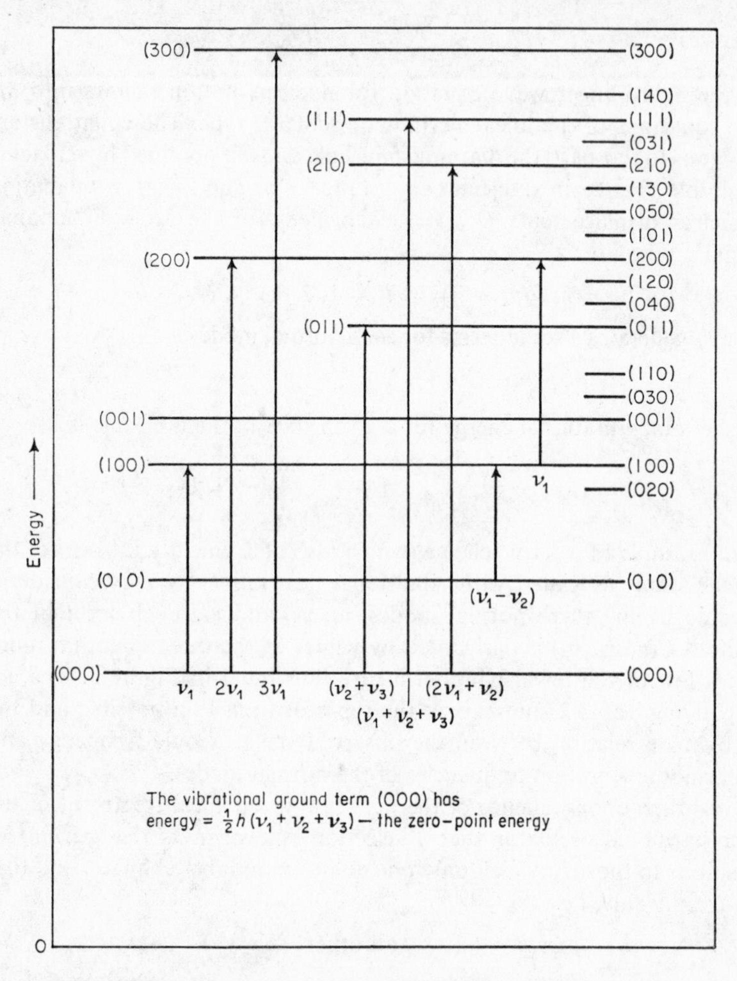

Figure 3. Simplified vibrational-term scheme for a molecule having three normal modes.

happen that two vibrational terms of the same symmetry are accidentally degenerate. Quantum-mechanical resonance then causes the terms to 'repel' one another, taking up values respectively greater and less than the energy of the unperturbed system. By this mechanism, combination or

Table 2. Nomenclature of vibrational transitions.

Transition	Name	Approximate observed frequency
$(000) \rightarrow (100)$	Fundamental	ν_1
$(000) \rightarrow (200)$	First overtone	$2\nu_1$
$(000) \rightarrow (300)$	Second overtone	$3\nu_1$
$(000) \rightarrow (011)$	Binary combination	$\nu_2 + \nu_3$
$(000) \rightarrow (111)$	Ternary combination	$\nu_1 + \nu_2 + \nu_3$
$(000) \rightarrow (210)$	Ternary combination	$2\nu_1 + \nu_2$
$(010) \rightarrow (100)$	Difference tone	$\nu_1 - \nu_2$
$(100) \rightarrow (200)$	Hot band	ν_1

overtones, which would normally be weak, may 'borrow' intensity by mixing with fundamentals and appear strongly in the spectra. This phenomenon is encountered quite commonly, for example in CO_2, CS_2, CCl_4, CH_4, and C_6H_6.

Having resolved the vibrations of a molecule into normal modes and determined the manifold of terms which arises from their quantization, we may now examine the means by which transitions between these terms may be induced in an observable manner.

B. Infrared Spectra

1. Classical view

Infrared spectra result from the direct absorption of electromagnetic radiation. If the oscillating electric vector $\mathbf{E}$ of incident light with frequency ν_k

$$\mathbf{E} = \mathbf{E}^0 \cos 2\pi c\nu_k t \tag{6}$$

can fall in step with an oscillating electric dipole in the molecule, then transfer of energy may occur with absorption (or emission) of radiation. It has already been pointed out that the only *simple* periodic motions of the molecule are the normal modes, so the only frequencies absorbed are the normal frequencies. Thus the infrared spectrum effectively performs a harmonic analysis of the molecular motion.

The displacement of charge which occurs during vibration causes an oscillating dipole moment $\boldsymbol{\mu}$ in the molecule which may be expanded as a function of the normal coordinate:

$$\boldsymbol{\mu} = \boldsymbol{\mu}^0 + \left(\frac{\partial \boldsymbol{\mu}}{\partial Q_k}\right)_0 Q_k + \text{higher anharmonic terms}$$

Remembering that*

$$Q_k = Q_k^0 \cos 2\pi c\nu_k t$$

* Since incoherent light is used, the phase constant is irrelevant and has been omitted.

it is clear that μ will only contain a component fluctuating periodically with frequency ν_k, provided that

$$\left(\frac{\partial \mu}{\partial Q_k}\right)_0 \neq 0$$

This is the simple selection rule for infrared activity: there must be a change in dipole moment during vibration. It should be noted that the presence, or absence, of a permanent dipole moment is irrelevant to the vibrational infrared selection rule.

In the above remarks, the directional properties of **E** and μ, which are vector quantities, have been ignored. In the main, this is justified because most infrared spectra are obtained using natural (unpolarized) light and randomly oriented samples. If polarized light and oriented single crystals are used, a further restriction must be imposed that the oscillating dipole should have a component *in the same direction* as the incident electric vector.

2. Quantum view

The probability P of a transition between two vibrational terms $v'' - v'$ is shown by quantum mechanics[28–30] to be proportional to the squares of the transition moments **M**

$$P \propto (\mathbf{M}_x)^2_{v'v''} + (\mathbf{M}_y)^2_{v'v''} + (\mathbf{M}_z)^2_{v'v''}$$

where

$$(\mathbf{M}_g)_{v'v''} = \int \psi^*_{v'} \mu_g \psi_{v''} \, \mathrm{d}\tau \qquad (7)$$

in which $\psi_{v''}$ is the time-independent vibrational eigenfunction of the lower term v''; $\psi^*_{v'}$ is the (complex conjugate) vibrational eigenfunction of the upper term v', and μ_g is the g component of the electric dipole moment vector. If the probability of the transition is to be non-zero, one or more of the integrals over all spatial coordinates in equation (7) must also be non-zero. This can only be so if one or more of the triple products $\psi^*_{v'}\mu_g\psi_{v''}$ is *totally symmetric* with respect to all the symmetry operations of the molecular point group. As an illustration, let us suppose that $\psi^*_{v'}\mu_g\psi_{v''}$ is antisymmetric (i.e. changes sign) with respect to a plane of symmetry in the molecule. The integral over all space would then obviously be zero, due to the mutual cancellation of terms with opposite sign. By considering the symmetry properties of the individual components of the triple products, the infrared selection rule may be derived in its most general and useful form. Consider a fundamental transition ($v'' = 0 \rightarrow v' = 1$); inspection of the mathematical form of the vibrational eigenfunctions[28] shows that $\psi_{v''}$, the ground state, is always totally symmetric and $\psi^*_{v'}$ has the same symmetry properties as the normal mode itself. The symmetry of the triple product is therefore unaffected by $\psi_{v''}$ and depends on $\psi_{v'}^{*}\mu_g$. The result may be stated thus: for infrared activity, the normal mode must have the same symmetry as one or more of the electric dipole components μ_g because the product of two species of the same symmetry always contains the totally symmetric representation. The relevance and use of this result will become apparent when we consider symmetry properties in more detail later.

C. Raman Spectra

The Raman effect, named after its discoverer[31] in 1928, is a second-order phenomenon observed in the scattering of monochromatic light by a homogeneous medium. If the incident light has frequency ν_0 (usually in the visible-ultraviolet region) and is not absorbed by the sample, the scattered light is found to contain not only the original frequency ν_0 (Rayleigh scattering), but also weaker lines of frequency ν' (the Raman lines) displaced from the exciting line ν_0 in a symmetrical manner:

$$\nu' = \nu_0 \pm \nu_k$$

where the displacements ν_k are equal to vibration frequencies of the scattering species.

1. Classical theory of the Raman effect

The electric vector **E** of the incident radiation perturbs the molecule by inducing an electric dipole $\boldsymbol{\mu}$.

$$\boldsymbol{\mu} = \boldsymbol{\alpha}\mathbf{E} \tag{8}$$

where $\boldsymbol{\alpha}$ is the polarizability of the molecule. **E** is, of course, fluctuating with time according to the frequency ν_0 of the incident light (equation 6); whence

$$\boldsymbol{\mu} = \boldsymbol{\alpha}\mathbf{E}^0 \cos 2\pi\nu_0 ct \tag{9}$$

The polarizability $\boldsymbol{\alpha}$ depends on the disposition of electrons in the molecule and may vary as the molecule vibrates. If so, $\boldsymbol{\alpha}$ is a function of the normal coordinate Q_k and may be expanded as a Taylor series. If electrical harmonicity is assumed, the expansion approximates to

$$\boldsymbol{\alpha} = \boldsymbol{\alpha}^0 + \left(\frac{\partial \boldsymbol{\alpha}}{\partial Q_k}\right)_0 Q_k$$

From equation (5), it follows that

$$\boldsymbol{\alpha} = \boldsymbol{\alpha}^0 + \left(\frac{\partial \boldsymbol{\alpha}}{\partial Q_k}\right)_0 Q_k{}^0 \cos 2\pi\nu_k ct \tag{10}$$

On substituting this expression for $\boldsymbol{\alpha}$ into equation (9), it may be seen that the induced dipole moment $\boldsymbol{\mu}$ has two components: the first,

$$\boldsymbol{\alpha}^0 \mathbf{E}^0 \cos 2\pi\nu_0 ct$$

is always present (a molecule always has a finite polarizability) and oscillates with frequency ν_0; it therefore reemits light of the same frequency—the Rayleigh scattering. The second term may be rearranged to the form

$$\mathbf{E}^0 \left(\frac{\partial \boldsymbol{\alpha}}{\partial Q_k}\right)_0 Q_k^0 [\cos 2\pi(\nu_0 + \nu_k)t + \cos 2\pi(\nu_0 - \nu_k)t]$$

It contains oscillations of frequency $(\nu_0 + \nu_k)$ and $(\nu_0 - \nu_k)$, and may give rise to scattered light of these frequencies—the Raman lines. To draw an analogy with electronics, the molecular vibrations may be said to 'frequency modulate' the scattered light. Raman lines will only appear, however, if

$$\left(\frac{\partial \boldsymbol{\alpha}}{\partial Q_k}\right)_0 \neq 0$$

The simple selection rule for Raman spectra therefore requires a change of polarizability during vibration. Comparison of this rule with that governing infrared spectra, $(\partial \boldsymbol{\mu}/\partial Q_k)_0 \neq 0$, serves to emphasize the difference between the two methods.

In order to understand the Raman selection rule based on symmetry arguments and the behaviour of polarized light in the Raman effect, it is desirable to examine more closely the nature of the molecular polarizability $\boldsymbol{\alpha}$. In equation (8), it relates two vectors $\boldsymbol{\mu}$ and $\mathbf{E}$. If the molecule is perfectly isotropic, $\boldsymbol{\mu}$ and $\mathbf{E}$ have the same direction and $\boldsymbol{\alpha}$ is then a simple scalar quantity (Figure 4a). If the molecule is anisotropic there will exist in it certain preferred directions of polarizability, along which the molecule is most readily 'deformed' electronically. For the moment, we shall regard the anisotropic molecule as space-fixed with respect to a set of cartesian axes, x, y, z; $\boldsymbol{\mu}$ will no longer have the same direction as $\mathbf{E}$, but will tend to be 'slewed round' toward the direction of maximum polarizability (Figure 4b). This confers on $\boldsymbol{\alpha}$ the properties of a quantity which can transform two vectors, namely a *tensor*. Equation (8) is then an equation in matrix notation, which may be written at length:

$$\begin{aligned}
\mu_x &= \alpha_{xx}E_x + \alpha_{xy}E_y + \alpha_{xz}E_z \\
\mu_y &= \alpha_{yx}E_x + \alpha_{yy}E_y + \alpha_{yz}E_z \\
\mu_z &= \alpha_{zx}E_x + \alpha_{zy}E_y + \alpha_{zz}E_z
\end{aligned}$$

whereby the components of $\boldsymbol{\mu}$ along the cartesian axes are related to the components of $\mathbf{E}$ by the six polarizability tensor elements $\alpha_{gg'}$ (for it may be shown that the matrix $\boldsymbol{\alpha}$ is symmetric, i.e. $\alpha_{gg'} = \alpha_{g'g}$). The polarizability tensor $\boldsymbol{\alpha}$ determines the properties of the Rayleigh scattering; but it is the derivative of $\boldsymbol{\alpha}$ with respect to the normal coordinate, $(\partial\boldsymbol{\alpha}/\partial Q_k)_0 = \boldsymbol{\alpha}'$, the *derived polarizability tensor*, that interests us as the factor controlling Raman scattering. We may reformulate the selection rule thus: for Raman activity, at least one of the matrix elements $\alpha'_{gg'}$ of the derived polarizability tensor must be non-zero. It is possible to decide the Raman activity of a normal mode by examining all the elements of $\boldsymbol{\alpha}'$. To this end, the polarizability $\boldsymbol{\alpha}$ is represented by an ellipsoid, whose motion is visualized as the normal mode is performed. Although this approach provides a valuable insight, the selection rules are much more conveniently derived by symmetry arguments (Section II.D); the reader is referred elsewhere[14,16] for a discussion of the polarizability ellipsoid model.

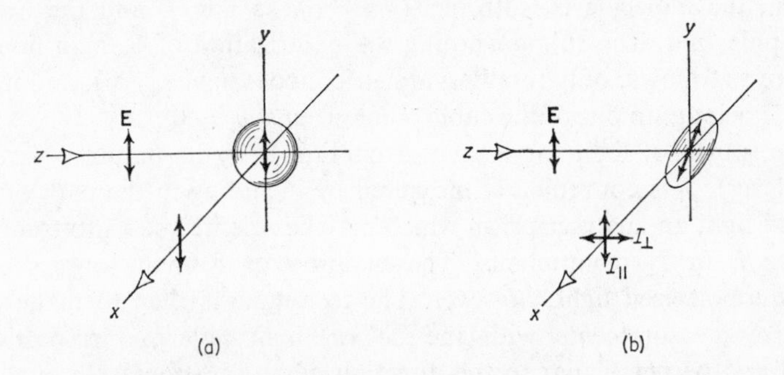

Figure 4. The effect of molecular polarizability on scattered light. (a) Isotropic molecule, (b) anisotropic molecule.

2. Polarization of Raman lines

Referring to Figure 4, we consider plane-polarized light incident on the molecule along the z axis, with its electric vector in the y direction. The Raman scattering is viewed along the x axis and the intensities of its components respectively parallel and perpendicular to the incident vector are designated as $I_{\parallel}$ and $I_\perp$. The depolarization ratio ρ_l of a Raman line is defined by

$$\rho_l = \frac{I_\perp}{I_{\parallel}}$$

$I_\perp$ and $I_{\parallel}$ are related to the elements of the derived polarizability tensor. Since the majority of Raman measurements are obtained from fluids, it is necessary to remove the restriction of a space-fixed molecule and make an average of the induced dipole moment over all possible orientations of the molecule. The results of this averaging process are conveniently expressed in terms of two *invariants* of the derived polarizability tensor (i.e. properties which are independent of orientation), the mean value $\bar{\alpha}'$ and the anisotropy γ' which are defined by

$$\bar{\alpha}' = \tfrac{1}{3}(\alpha'_{xx} + \alpha'_{yy} + \alpha'_{zz})$$

$$2(\gamma')^2 = (\alpha'_{xx} - \alpha'_{yy})^2 + (\alpha'_{yy} - \alpha'_{zz})^2 + (\alpha'_{zz} - \alpha'_{xx})^2 + 6(\alpha'^2_{xy} + \alpha'^2_{yz} + \alpha'^2_{zx})$$

The depolarization ratio is then found to be

$$\rho_l = \frac{3(\gamma')^2}{45(\bar{\alpha}')^2 + 4(\gamma')^2}$$

$\bar{\alpha}'$ is the totally symmetric component of the derived polarizability tensor and will only be non-zero if the normal mode is also totally symmetric; in which case $0 < \rho_l < \frac{3}{4}$, and the Raman line is said to be polarized. If, in addition, the molecule is isotropic, $(\gamma')^2 = 0$, so $\rho_l = 0$ and the line is totally polarized. The rule governing the polarization of Raman lines is therefore as follows: only totally symmetric modes have $\rho_l < \frac{3}{4}$, and if the molecule belongs to one of the cubic point groups, $\rho_l = 0$.

If the source of light for Raman excitation is by its nature polarized (e.g. a laser), ρ_l is conveniently measured by including in the path of the scattered light an analyser prism which may be oriented so as to transmit only the $I_\parallel$ or $I_\perp$ components. The majority of Raman sources emit natural unpolarized light, however. The technique is then to make two successive measurements with the *incident* light polarized respectively parallel and perpendicular to the direction of observation*. No analyser is used, but the total intensities scattered parallel $I_{\text{obs}\parallel}$ and perpendicular $I_{\text{obs}\perp}$ to the incident vector are measured. The depolarization ratio for natural unpolarized light ρ_n is defined as

$$\rho_n = \frac{I_{\text{obs}\parallel}}{I_{\text{obs}\perp}}$$

and is given by[16,17]

$$\rho_n = \frac{6(\gamma')^2}{45(\bar{\alpha}')^2 + 7(\gamma')^2}$$

The polarization rule is obtained as above but with the limit $\frac{3}{4}$ replaced by $\frac{6}{7}$.

3. *Quantum view of the Raman effect*

For the appearance of a fundamental in the Raman spectrum, an incident photon $h\nu_0$ must induce a vibrational transition, $v_k = 0 \rightarrow v_k = 1$. The inelastically scattered photon is then depleted in energy by a vibrational quantum $h\nu_k$ and has a frequency $\nu_0 - \nu_k$. This will appear in the spectrum as a 'stokes' line, on the long-wavelength side of ν_0. Conversely,

* This is accomplished by surrounding the sample with sheets of Polaroid.

if the transition $v_k = 1 \rightarrow v_k = 0$ is induced, the resulting Raman line has frequency $\nu_0 + \nu_k$ and is observed as an 'antistokes' line on the short-wavelength side of ν_0. The Raman spectrum appears as a pattern of lines which are symmetrically disposed about the exciting line in frequency, but not in intensity: the antistokes lines are weaker than their stokes counterparts owing to the smaller Boltzmann population of the excited vibrational states (e.g. see Figure 7).

Although the terms 'stokes' and 'antistokes' are taken over from fluorescence, the Raman effect should not be confused with this phenomenon. In the Raman effect, the molecule does not 'absorb' the incident photon in the sense of being excited to a well-defined electronic state. The electronic wave function is perturbed by the photon for a period of time in the order of 10^{-15} s, a value which may be compared with the half-life of about 10^{-8} s for the fluorescence effect. The difference is clearly demonstrated by the fact that fluorescence may be 'quenched' (i.e. the excited state thermally robbed of its energy) but the Raman effect cannot. The time scale of the effect becomes particularly significant when structural studies of non-rigid or labile species by different physical methods are compared. The period of molecular vibrations is in the order of 10^{-12}–10^{-14} s. Raman and infrared spectra therefore give a picture of the molecule which is more nearly 'instantaneous' than that found by techniques such as n.m.r., where at room temperature the time average may span a period long enough for the interconversion of different conformational isomers or species.

To obtain the selection rules for Raman activity on a quantum-mechanical basis, we proceed in a manner analogous to the infrared, but the components of the dipole moment are replaced by the elements of the polarizability tensor. Raman activity therefore demands a non-zero value for one or more of the six triple products $\int \psi_{v'}^* \alpha_{gg'} \psi_{v''} d\tau$. The argument runs parallel to that given in Section II.B and results in the requirement that the normal mode must have the same symmetry as one or more of the polarizability tensor elements $\alpha_{gg'}$.

D. Application of Symmetry Theory

The power of symmetry arguments is now widely appreciated by inorganic chemists. In what follows, some acquaintance with the theory of groups and molecular symmetry[32–34] has been assumed in order to present only the essential steps in its application to vibrational spectroscopy. For a given molecular geometry, we may readily learn:

1. The number, symmetry properties, and degeneracies of the normal modes.
2. Their spectroscopic activity.

3. Their behaviour to polarized light in both Raman and infrared spectra.
4. Which normal modes are likely to involve chiefly the deformation of certain bonds or interbond angles.
5. The probable contour shapes conferred on infrared bands of gases by unresolved rotational fine structure.
6. How to set up symmetry coordinates, as a first step in simplifying the solution of the molecular force field (normal coordinate analysis).

A molecule in its equilibrium configuration possesses certain symmetry elements which may be determined by inspection. These can include (in the Schönflies notation): mirror planes (σ), axes of n-fold rotation (C_n), an inversion centre (i), and axes of n-fold rotation–reflection (S_n). Each symmetry *element* gives rise to one or more symmetry *operations* which, when performed on the molecule, carry it into a configuration indistinguishable from that in which it started. The collection of distinct symmetry operations constitutes the symmetry point group of the molecule, whose properties are embodied in the group character table.

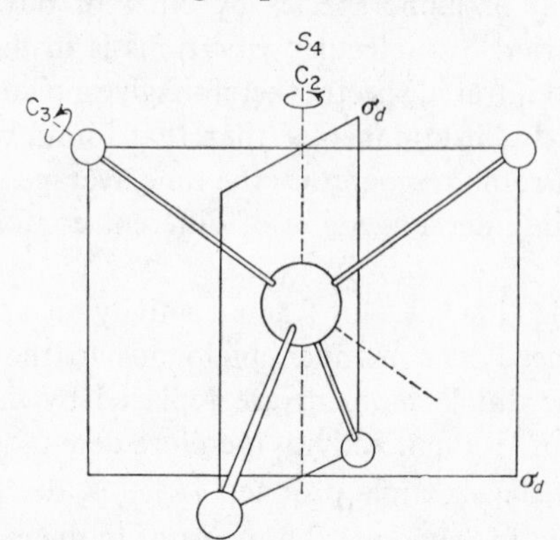

Figure 5. Typical symmetry elements of a regular tetrahedral molecule (T_d).

As an example, we take a regular tetrahedral molecule of type MX_4 (Figure 5). The point group is T_d and its character table is displayed at the top of Table 3. The symmetry operations (including the identity operation E) are listed in the first row of the table; their total number g is 24 and the number of operations of each respective class g_j is shown by the coefficient appearing before the symbol for that operation. The numerical entries in the table are the *characters* χ_j for each type of symmetry operation; and a row of characters, which is labelled by a species symbol in the left-hand column, constitutes an *irreducible representation* of the group. The symbols

Table 3. Character table and derivation of selection rules for MX_4 (point group T_d).

T_d	E	$8C_3$	$3C_2$	$6S_4$	$6\sigma_d$	Translation and rotation	Polarizability
A_1	1	1	1	1	1		$(x^2 + y^2 + z^2)$
A_2	1	1	1	−1	−1		$(2z^2 - x^2 - y^2)$
E	2	−1	2	0	0		$(x^2 - y^2)$
T_1	3	0	−1	1	−1	(R_x,R_y,R_z)	
T_2	3	0	−1	−1	1	(x,y,z)	(xy,xz,yz)

Cartesian displacement coordinates

n_j	5	2	1	1	3	(no. of unshifted atoms)
$n_jC_j = \chi_j^R$	15	0	−1	−1	3	(reducible representation)
$g_j\chi_j^R$	15	0	−3	−6	18	

$$\left.\begin{aligned}
A_1 \quad p &= (15 + 0 - 3 - 6 + 18)\ /\ 24 = 1\\
A_2 \quad p &= (15 + 0 - 3 + 6 - 18)\ /\ 24 = 0\\
E \quad p &= (30 + 0 - 6 + 0 + 0)\ /\ 24 = 1\\
T_1 \quad p &= (45 + 0 + 3 - 6 - 18)\ /\ 24 = 1\\
T_2 \quad p &= (45 + 0 + 3 + 6 + 18)\ /\ 24 = 3
\end{aligned}\right\}\quad \begin{aligned}&A_1 + E + T_1 + 3T_2\\ &\text{less } (T_1 + T_2)\\ &= A_1 + E + 2T_2\end{aligned}$$

Internal coordinates

Δr *set*

χ_j^R	4	1	0	0	2	(reducible representation)
$g_j\chi_j^R$	4	8	0	0	12	

$$\left.\begin{aligned}
A_1 \quad p &= (4 + 8 + 0 + 0 + 12)\ /\ 24 = 1\\
E \quad p &= (8 - 8 + 0 + 0 + 0)\ /\ 24 = 0\\
T_2 \quad p &= (12 + 0 + 0 + 0 + 12)\ /\ 24 = 1
\end{aligned}\right\}\quad A_1 + T_2$$

$\Delta\theta$ *set*

χ_j^R	6	0	2	0	2	(reducible representation)
$g_j\chi_j^R$	6	0	6	0	12	

$$\left.\begin{aligned}
E \quad p &= (12 + 0 + 12 + 0 + 0)\ /\ 24 = 1\\
T_2 \quad p &= (18 + 0 - 6 + 0 + 12)\ /\ 24 = 1
\end{aligned}\right\}\quad E + T_2$$

(in Mulliken's notation) for the various symmetry species, or classes, express in shorthand form the transformation properties with respect to the symmetry operations, for example:

- *A* Symmetric with respect to principal axis } non-degenerate
- *B* Antisymmetric with respect to principal axis } non-degenerate
- *E* Doubly degenerate
- *T* Triply degenerate
- *g* (as subscript) symmetric with respect to inversion through a centre
- *u* (as subscript) antisymmetric with respect to inversion through a centre

Further subscripts (1, 2, or 3) or superscripts (′ or ″) denote the behaviour with respect to subsidiary axes or planes of symmetry.

1. Cartesian coordinates

By employing cartesian displacement coordinates (Section II.A) and operating upon them with all the symmetry operations of the group, we may generate a *reducible representation* consisting of the characters χ_j^R which expresses the symmetry properties of the displacements from the equilibrium configuration of the molecule. In practice, the reducible representation is obtained by counting the number of atoms unshifted by each symmetry operation and multiplying by the contribution to character per unshifted atom[17]. As its name suggests, the reducible representation may be broken down into several irreducible representations; this process is conveniently accomplished by means of the formula

$$p = \frac{1}{g} \sum_j g_j \chi_j \chi_j^R \qquad (11)$$

in which p is the number of times that a given symmetry species will appear in the reducible representation; the sum runs over all symmetry operations of the group. No amount of description can clarify this process so well as an example, and the reader is recommended to study Table 3, where the symmetry classes of the normal modes for a tetrahedral molecule are worked out in full.

Because a cartesian coordinate system has been used, the representation so obtained, $A_1 + E + T_1 + 3T_2$, still contains contributions due to translation (x, y, and z) and rotation (R_x, R_y, R_z) of the molecule as a whole. These are readily subtracted by consultation of the table, and the vibrational representation results: $A_1 + E + 2T_2$, i.e. one totally symmetric, one doubly degenerate, and two triply degenerate normal modes.

2. Selection rules

The spectroscopic activity of these four normal modes, as fundamentals, is readily determined. The infrared selection rule requires that a normal mode has the same symmetry as one of the components of the dipole moment vector. The latter always transform in the same way as translations of the molecule x, y, or z. For our example, it is obvious that only the normal modes of symmetry class T_2 will be active as fundamentals in the infrared. The Raman active normal modes will be those transforming as components of the polarizability tensor $\alpha_{gg'}$. It may be shown[17] that the $\alpha_{gg'}$ have the same symmetry properties as quadratic functions gg' of the cartesian vectors. These are summarized on the right-hand side of Table 3. It follows that all four normal modes of the tetrahedral molecule are permitted as Raman fundamentals, the one of symmetry class A_1 being totally polarized.

When a molecule possesses a centre of symmetry, an important generalization can be made about its selection rules. The translations x, y, and z are obviously *ungerade* with respect to inversion; all infrared active transitions are therefore found only in u symmetry classes. The quadratic functions of x, y, and z, on the other hand, are intrinsically *gerade*, so Raman active transitions can only belong to g classes. A rule of *mutual exclusion* therefore applies, which may be stated: in centrosymmetric molecules, all transitions permitted in the infrared are forbidden in the Raman, and vice versa. The complementarity of Raman and infrared data in such cases is obvious. The selection rules for combination and overtones may be found by forming the direct product of the contributing fundamentals[17]. On a point of notation, it is widely accepted that the symmetry species of fundamentals should be written in lower-case (a_1, e, t_2) and of combination tones, etc., in upper-case (A_1, E, T_2), and this convention will be adopted henceforth.

3. *Internal coordinates*

A further insight into the nature of the normal modes is gained by using internal displacement coordinates which define the relative positions of the nuclei, instead of the space-fixed cartesian system. This has the advantage that translations and rotations of the molecule are automatically excluded at the outset but the danger exists that, if the internal coordinates are not all independent, redundancies (i.e. spurious modes) may be obtained. The chief restoring forces in a molecule usually coincide with the directions of chemical bonding. Internal coordinates are therefore most usefully chosen as incremental changes in bond lengths and interbond angles, taking all members of each symmetrically equivalent set. The irreducible representation formed by the internal coordinates is readily found by counting the number of coordinates in each set which are unshifted by each symmetry operation (multiplying by -1 if a coordinate changes sign). The reduction formula (11) is then applied to determine the symmetry classes of the

Table 4. Normal modes and selection rules for MX_4 (T_d).

Internal coordinates	Symmetry classes of fundamentals		
	a_1 (R,pol.)	e (R,depol.)	t_2 (i.r.,R,depol.)
σ(M—X)	ν_1		ν_3
δ(XMX)		ν_2	ν_4

R = Raman active, i.r. = infrared active, pol. = polarized, depol. = depolarized.
σ = bond-stretching vibration, δ = bond-angle deformation.

Table 5. Vibrational selecti

Type	Geometry	Point group	Numbers of normal modes in classes containing	
			stretching coordinates	deformation coordinates
MX_2	Linear	$D_{\infty h}$	$\sigma_g^+ + \sigma_u^+$	π_u
	Bent	C_{2v}	$a_1 + b_1$	a_1
MX_3	Planar trigonal	D_{3h}	$a_1' + e'$	$a_2'' + e'$
	Pyramidal	C_{3v}	$a_1 + e$	$a_1 + e$
	T-shape	C_{2v}	$2a_1 + b_1$	$a_1 + b_1 + b_2$
MX_4	Tetrahedral	T_d	$a_1 + t_2$	$e + t_2$
	Square planar	D_{4h}	$a_{1g} + b_{2g} + e_u$	$a_{2u} + b_{1g} + b_{1u} + e_u$
	Distorted tetrahedral	C_{2v}	$2a_1 + b_1 + b_2$	$2a_1 + a_2 + b_1 + b_2$
MX_5	Trigonal bipyramid	D_{3h}	$2a_1' + e' + a_2''$	$2e' + a_2'' + e''$
	Square pyramid	C_{4v}	$2a_1 + b_1 + e$	$a_1 + b_1 + b_2 + 2e$
MX_6	Octahedral	O_h	$a_{1g} + e_g + t_{1u}$	$e_g + t_{2g} + t_{1u} + t_{2u}$
MX_7	Pentagonal bipyramid	D_{5h}	$2a_1' + e_1' + e_2' + a_2''$	$2e_1' + e_2' + a_2'' + e_1'' + e_2''$
MX_8	Dodecahedron	D_{2d}	$2a_1 + 2b_2 + 2e$	$2a_1 + a_2 + 2b_1 + 2b_2 + 3e$
	Archimedian antiprism	D_{4d}	$a_1 + b_2 + e_1 + e_2 + e_3$	$a_1 + b_1 + b_2 + 2e_1 + 2e_2 +$

normal modes in which each set takes part. If only one set of bond-stretching (or deformation) coordinates contributes to a given symmetry class, the normal modes of that class may be described as bond-stretching (or deformation) vibrations. It is often the case, however, that more than one set contributes; in this event the normal modes will mix the internal coordinates to an extent depending on the atomic masses (mechanical coupling) and the potential field (electronic coupling); it is then no longer meaningful to speak of pure bond-stretching (or deformation) vibrations.

The internal coordinate treatment for a tetrahedral molecule is summarized in Table 3, using a set of four bond-length changes Δr and a set of six bond-angle changes $\Delta\theta$. It is unnecessary to examine the A_2 and T_1 classes because there are no normal modes of these species. The appearance of both Δr and $\Delta\theta$ sets in the A_1 class, which contains only one normal mode, is an example of a redundancy; it arises in this case because the $\Delta\theta$ coordinates are not independent. The spectroscopic predictions are summarized in Table 4, from which it may be seen that two of the Raman active s for simple molecules.

nan active lasses	Infrared active classes	*N*	*R*	*P*	*I*	*C*	Examples
	σ_u^+, π_u	3	1	1	2	0	$HgCl_2$[35], XeF_2[36]
b_1	a_1, b_1	3	3	2	3	3	SCl_2[37], NO_2^-[38]
a_2'', e'	a_2'', e'	4	4	1	3	3	BCl_3[39], SO_3[40], $C(CN)_3^-$[41a]
e	a_1, e	4	4	2	4	4	AsH_3[42], $SnCl_3^-$[43]
b_1, b_2	a_1, b_1, b_2	6	6	3	6	6	ClF_3[44]
e, t_2	t_2	4	4	1	2	2	OsO_4[45], BCl_4^-[46]
b_{1g}, b_{2g}	a_{2u}, e_u	7	3	1	3	0	$AuCl_4^-$, $PtCl_4^{2-}$, ICl_4^-[47,48]
a_2, b_1, b_2	a_1, b_1, b_2	9	9	4	8	8	SF_4[49]
e', e''	e', a_2''	8	6	2	5	3	$Sb(CH_3)_5$[50a], PF_5[51]
b_1, b_2, e	a_1, e	9	9	3	6	6	ClF_5, BrF_5[52]
e_g, t_{2g}	t_{1u}	6	3	1	2	0	UF_6[53], PtF_6^{2-}[54]
e_2', e_1''	e_1', a_2''	11	5	2	5	0	IF_7[55], ReF_7[56]
b_1, b_2, e	b_2, e	16	15	4	9	9	$Mo(CN)_8^{4-}$ solid[57,58a]
e_2, e_3	b_2, e_1	13	7	2	5	0	$Mo(CN)_8^{4-}$ solution?[57,58a]

Skeletal vibrations only.
= total number of normal modes, *R* = number of Raman active fundamentals, *P* = number olarized Raman lines, *I* = number of infrared active fundamentals, *C* = number of coinciden- between Raman and infrared spectra.

fundamentals (a_1 and e) are pure bond stretching σ(M—X) and deformation δ(XMX), respectively; the two remaining, ($2t_2$), which are coincident with the two infrared bands, have a 'mixed' character. The fundamentals are numbered ν_1 to ν_4 in order of decreasing symmetry. The selection rules for a number of simple molecular geometries of type MX_n are summarized in Table 5 and will be referred to later in the text.

III. EXPERIMENTAL TECHNIQUES

A. Infrared Instrumentation

Infrared spectroscopy above 300 cm^{-1} is a technique familiar to most practising chemists. It is not proposed to describe here the details of infrared spectrometers, for which the reader is referred to excellent accounts[18,59–61] elsewhere. The vibrational infrared region covers a wide range of energy (30–5,000 cm^{-1}); effective recording of the whole spectrum with a single optical system is not possible. The ranges of operation which are most widely used and the optical methods associated with them are summarized in Table 6. A number of alternative window and prism materials have been employed for special purposes, e.g. the examination of aqueous solutions.

Table 6. Ranges of the vibrational infrared.

Region (cm^{-1})	Window material	Means of dispersion	Detector	Energy source
5000–650	NaCl (CaF_2 or AgCl for aqueous samples)	NaCl prism (LiF prism or grating better above 1700 cm^{-1})	Thermo-couple or	Nernst filament or
700–400	KBr	KBr prism	Bolometer	globar
700–280	CsBr or polythene	CsBr prism		
300–30	Polythene	Gratings or interferometry	Golay cell	Mercury lamp

The recent use of scanning Michelson interferometers[62] for recording spectra in the far-infrared region (<300 cm^{-1}) deserves special comment. The optical simplicity of the interferometric method, which avoids using the several gratings found in conventional instruments, and the improved signal-to-noise ratio which may be achieved are commendable features. The spectrum is not obtained directly, however, but as a Fourier transform of the interferogram; to process it, computing facilities are needed. Interferometric spectrophotometers are gaining rapid acceptance by chemical spectroscopists and commercial models are now available.

B. Raman Instrumentation

Figure 6 shows a schematic diagram of a conventional Raman apparatus. When it is remembered that only a fraction in the order of 10^{-8} of the incident light is scattered as Raman information, it will be appreciated that there is no room for further inefficiency in the method of observation. The light source (1) must be very powerful, and place high intensity in a few monochromatic lines, which can be isolated by the prefilter (2). The atomic emission spectrum of mercury has these qualities and provides useful lines in the ultraviolet (2537, 3650 Å), violet (4057 Å), blue (4358 Å), and green (5461 Å) regions of the spectrum. The best source, which has a very low background continuum, is the 'Toronto' arc[63]: a high current d.c. discharge, water cooled to maintain low pressure and so avoid line broadening. After scattering by the sample (3), which must be as concentrated as possible and carefully prepared to avoid inhomogeneities, the intense Rayleigh line may be removed, at least partly, by a steep-cut postfilter (4). The light is then transferred as efficiently as possible by a suitable optical system[64] (5) to a spectrometer of high luminosity, where it is dispersed and the spectrum recorded either photographically or photoelectrically.

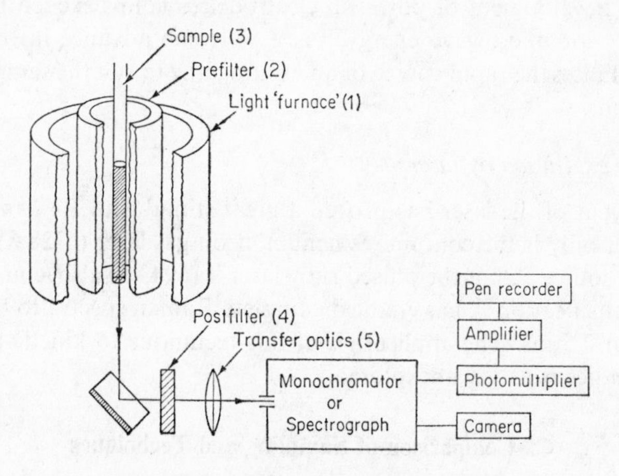

Figure 6. Raman apparatus (schematic).

1. Raman excitation at long wavelengths

Consider a Raman sample solution which has a non-zero absorbance at the wavelengths of excitation (λ_0) and Raman scattering (λ'). If the sample is made more concentrated, the intrinsic Raman scattering is greater but there is a concomitant increase in the absorbance of the solution at λ_0 and λ'. As a result of these mutually opposed effects, the flux of Raman light emitted by the sample reaches a maximum at a finite concentration. This is the optimum concentration for observing the Raman spectrum; its value may be calculated from the absorption spectrum of the sample and the geometry of the Raman excitation unit[65]. In practical terms, the optical density (1 cm path) of the sample should not exceed 0.2 at λ_0 and λ'. Since Raman spectra are difficult to obtain from samples more dilute than about 0.1 M, this places an upper limit of about 2 on the tolerable extinction coefficients at λ_0 and λ'. It is essential, therefore, to choose an exciting line which falls in a region of low sample absorption. The mercury arc is quite unsuited to the excitation of yellow and red substances; for this purpose, a helium discharge (5876 Å yellow and 6678 Å red) has proved most valuable[66], but use has also been made of cadmium[67] (6438 Å), argon[68] (8224 Å), sodium[69] (5890/5896 Å), potassium[70] (7665/7699 Å), and rubidium[71,72] (7800/7947 Å) lamps. Despite the difficulties which beset Raman spectroscopy at long wavelengths—for the Raman scattering, being proportional to $1/\lambda_0^4$, is intrinsically weaker and the photons, having lower energy, are inherently more difficult to disperse and record—these methods have recently yielded many Raman spectra which were hitherto inaccessible. The development of powerful electrodeless lamps excited by radio-frequency[72] or microwave energy[70] is a welcome advance in versatility, which facilitates the rapid choice of an exciting line to suit the sample under observation.

2. *Raman excitation by lasers*

The advent of the laser has proved a great stimulus to Raman spectroscopy. Not only is the continuous helium–neon gas laser (6328 Å) a useful excitation source[73], but the pulsed ruby laser[74] (6943 Å) also, coupled with image intensification[75], has enabled complete Raman spectra to be photographed in 0.5 ms. The application of this technique to kinetic and free-radical studies is as yet unexplored.

C. Comparison of Sampling and Techniques

There is a considerable disparity between the techniques demanded by infrared and Raman spectroscopy; the information most readily yielded by each method also differs. A comparison may be found in Table 7.

IV. AIDS TO ASSIGNMENT

Assignment consists in relating the observed Raman and infrared bands to the vibrational transitions which give rise to them; the aim is to corroborate the structure deduced from the selection rules and to provide a basis for force constant calculations, etc. Rigorously, we may only assign the bands to their various symmetry classes, but insofar as the normal modes may be approximately represented as vibrations of a particular group of atoms, the assignment can go further in labelling the bands as such. Fundamentals are usually assigned first and then the whole combination-overtone spectrum is explained in terms of them; occasionally the overtones may be used to determine inactive fundamentals. We shall begin by examining the criteria for 'group vibrations' and then review briefly the methods which assist the spectroscopist to reach an assignment.

Table 7. A comparison of infrared and Raman methods.

	Raman	Infrared
Sample size	1–10 g. Smaller samples need special equipment[76]	10 mg or less
Sample phase	Liquid or conc. solution. Solids[77] and gases[78] need special techniques	Spectra easily obtained from all phases
Sample colour	Must transmit at λ_0 and λ'	No problem
Sample photochemistry	Fluorescent or photosensitive samples need long wavelength excitation	No problem
Sample cells	Glass. Manipulation of reactive substances easy	Alkali halide windows may be attacked
Sample temperature	Readily controlled from −150 to +250°C[79]. High temperatures possible[80]	Less simple to control; special cells necessary
Aqueous solutions	Ideal over whole range. Raman bands of H_2O weak	Very limited[81]. Infrared bands of H_2O very strong
Range	0–4000 cm^{-1} in one scan	At least three separate ranges to cover 30–4000 cm^{-1}
Resolution	5–10 cm^{-1} usual. 1 cm^{-1} demands very high performance	1 cm^{-1} easily achieved
Presentation	Chart or photographic plate. Speed of best instruments comparable with infrared	Chart only

A. Group Vibrations and Frequencies

If most of the energy of a normal mode is localized in the stretching (or deformation) of a given bond (or set of symmetrically equivalent bonds), then a frequency characteristic of that bond will result. This will be so for *terminal groups*—(X—Y) if the mass of Y is much less than the mass of X, or if the X—Y bond has a force constant much higher than those of adjacent bonds. Thus (X—H), (X═O), (X═S), and (X≡N) groups usually perform localized vibrations. Similar considerations apply to the *molecular skeleton*: if the ligand vibrations are essentially localized, the ligands will behave approximately as rigid dynamic units during the vibrations of the skeleton.

For example, many of the normal modes of $HRe(CO)_5$ can be regarded as either σ(Re—H), σ(Re—C), or σ(C═O) vibrations[82], whereas the normal modes of species such as $AlH_3 \cdot 2NMe_3$[83] or $Mo_6Cl_8Cl_6^{2-}$[84] have been shown to involve extensive mixing of the various bond vibrations. In short, a normal mode may usually be considered as a group vibration when its frequency is well separated from others of the same symmetry class. Consider two such groups in different molecules: if their electronic environments are similar, the force constants will be roughly equal and consequently the group vibrations will have approximately the same frequency in each. The concept of characteristic group frequencies forms the basis of qualitative analysis and may greatly assist assignment because it can suggest the regions of the spectrum in which the fundamentals will probably be found. Inferences about chemical bonding which appeal to group frequency arguments must, however, be treated with caution, for they are at best approximate; in doubtful cases, there is no substitute for normal coordinate analysis.

B. Selection Rules

Table 5 shows how the selection rules may limit the possible assignments (e.g. by coincidences between infrared and Raman spectra, or the absence of them), and use is made of such arguments in Section V. It is more pertinent here to indicate some of the pitfalls which may be encountered due to the number of observed bands not being as expected.

Since symmetry theory alone cannot predict spectroscopic intensities, it is possible that a formally permitted band may be too weak to observe. On the other hand, the occurrence of Fermi resonance may confer an unusually high intensity on overtones or combination bands, which might then be mistaken for fundamentals. For an example, see the Raman spectrum of CCl_4 reproduced in Figure 7. Spectra obtained from samples in the solid state must be interpreted with care. If the symmetry of the lattice site is lower than that of the free molecule, modes previously forbidden can become active and degenerate bands may be split. Such effects are anticipated by using site group symmetry to determine the selection rules. But there may also appear in the spectra bands which are not due to intramolecular but intermolecular vibrations—the lattice modes[85]. Strictly, the normal modes of a solid should be obtained by the methods of factor group analysis (Section IV.D) which treats the unit cell rather than the isolated molecule.

Even in fluids difficulties may arise: solute–solvent interactions can cause frequency shifts and broadening of the bands, with an undesirable loss of resolution, and violations of the selection rules are not unknown. For instance, a number of linear species ($Hg(CH_3)_2$[86], CS_2[87], and UO_2^{2+}[88])

show a forbidden band in their Raman spectra, coincident with the infrared active deformation, $\nu_3(\pi_u)$. Either intermolecular effects or electrical anharmonicity of the vibrations can cause the selection rules to break down in this way.

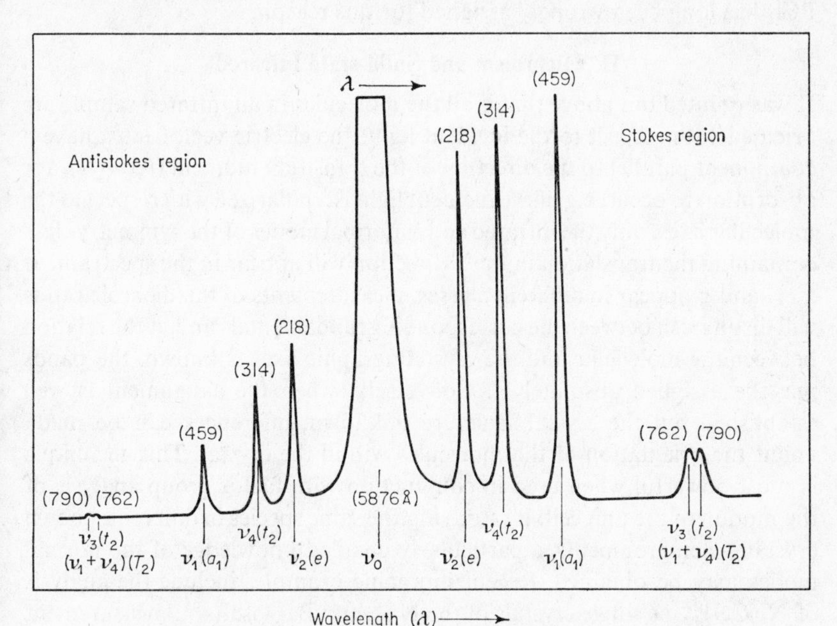

Figure 7. Raman spectrum of CCl_4, excited by He 5876 Å. Numbers in brackets are wave-number displacements from ν_0. The doublet at 762/790 cm^{-1} is due to Fermi resonance between ν_3 (775 cm^{-1}) and $\nu_1 + \nu_4$ (773 cm^{-1}).

C. Raman Polarization Data

The identification of totally symmetric normal modes (which are always Raman active) by the polarization of their fundamentals in the Raman spectrum is a most valuable aid. Because it can be applied to fluids, there is no counterpart to this method in infrared spectroscopy. As an illustration, we take the molecule $Ni(PF_3)_4$ [89]. The point group is T_d and of the nineteen fundamentals ($3a_1 + a_2 + 4e + 4t_1 + 7t_2$), fourteen are permitted in the Raman spectrum ($3a_1 + 4e + 7t_2$). Despite this complexity, the three a_1 fundamentals are readily picked out by their polarization ($\rho = 0$):

$$\nu_1(a_1),\ \sigma(PF) = 954\ cm^{-1};$$
$$\nu_2(a_1),\ \delta(PF_3) = 534\ cm^{-1};$$
$$\nu_3(a_1),\ \sigma(NiP) = 195\ cm^{-1}.$$

These values alone suffice for a simple force-constant calculation. This method is not without its difficulties, however. In non-cubic point groups, a ρ value close to but less than 6/7 may not be distinguished experimentally. It has recently been shown[90] that the spectrum of a molecule as simple as PCl_5 has long been wrongly assigned for this reason.

D. Dichroism and Solid-state Infrared

It was pointed out above that if all the molecules in an infrared sample are oriented with respect to the incident light, the electric vector must have a component parallel to the direction of the transition moment $(\partial\mu/\partial Q_k)_0$ for absorption to occur, e.g. if the incident light is x polarized with respect to the molecular axes, only the infrared active normal modes of the symmetry class containing the translation in the x direction will appear in the spectrum. If x, y, and z appear in different classes, measurements of the dichroic ratios will distinguish between the corresponding fundamentals and, if the relation between the molecular and the crystallographic axes is known, the bands may be assigned absolutely[91]. Conversely, when the assignment is well established but the crystal structure unknown, inferences can be made about the orientation of the molecules within the crystal. This technique is most powerful when used in conjunction with factor group analysis of the motion of the unit cell. By studying the same species in different, known crystalline environments, a particularly detailed knowledge of the normal modes may be obtained. Recent inorganic examples include the analysis of $Ni(CN)_4^{2-}$ as single crystals of the Na^+ and Ba^{2+} salts[92]; measurement of the infrared dichroism of *trans*-$[CoCl_2(en)_2](ClO_4)$ to establish the assignment of σ(Co—Cl) and δ(NCoN)[93]; and the examination of diborane[94] in three solid phases. A definitive study[95] of the ammine and nitro complexes of the Co^{III} by factor group analysis has also appeared.

E. Infrared Band Contours

For molecules in the vapour phase, vibrational transitions may be accompanied by relatively small changes in the rotational energy, which impart a fine structure to the vibrational bands. This is usually unresolved, except for very light molecules; but the bands still display characteristic contour shapes which depend on the rotational selection rules. It is convenient to classify molecules according to their symmetry as rotors; i.e. according to the moments of inertia I_A, I_B, and I_C, given in ascending order of magnitude. The axes A, B, and C usually coincide with symmetry axes in the molecule. For examples, see Table 8.

1. Linear molecules $I_A = 0$; $I_B = I_C$

The infrared active fundamentals may be classified as parallel ($\|$) or perpendicular ($\perp$) depending on the orientation of $(\partial\mu/\partial Q_k)_0$ with respect

Table 8. Molecular rotors and their vibrational band contours.

Rotor	Example	Infrared active classes	Type of vibration	Rotational selection rules	Typical contour[a]
Linear	CO_2[96]	σ_u^+	$\parallel$	$\Delta J = \pm 1$	*PR*
$I_A = 0, I_B = I_C$	$(D_{\infty h})$	π_u	$\perp$	$\Delta J = 0, \pm 1$	*PQR*
Spherical top	SiH_4[97]	t_2		$\Delta J = 0, \pm 1$	*PQR*
$I_A = I_B = I_C$	(T_d)				
Symmetric top	BF_3[98]	a_2''	$\parallel$	$\Delta J = 0, \pm 1; \Delta K = 0$	*PQR*
Oblate $I_A = I_B < I_C$	(D_{3h})	e'	$\perp$	$\Delta J = 0, +1; \Delta K = \pm 1$	*PQR*
Symmetric top	CH_3Cl[99]	a_1	$\parallel$	$\Delta J = 0, \pm 1; \Delta K = 0$	*PQR*
Prolate $I_A < I_B = I_C$	(C_{3v})	e	$\perp$	$\Delta J = 0, \pm 1; \Delta K = \pm 1$	*PQR*
Asymmetric top	CSF_2[100]	a_1	*A*	$\Delta J = 0, \pm 1; \Delta K = 0$	*PQR*
$I_A < I_B < I_C$	(C_{2v})	b_1	*B*	$\Delta J = 0, \pm 1; \Delta K = \pm 1$	*PR* or *PQQR*
		b_2	*C*	$\Delta J = 0, \pm 1; \Delta K = \pm 1$	*PQR*

[a] *PQR* = very prominent *Q* branch, *PQR* = *Q* branch comparable with *P* and *R* in intensity, *PQQR* = doublet *Q* branch, *PR* = no *Q* branch.

to the principal symmetry axis A. For parallel vibrations, the quantum number J governing the angular momentum about the A axis is subject to the selection rule $\Delta J = \pm 1$. The rotational fine structure appears as P $(J' = J'' - 1)$ and R $(J' = J'' + 1)$ branches on either side of the band origin. For perpendicular vibrations the selection rule is $\Delta J = 0, \pm 1$ so that, in addition to the P and R branches, a Q spike $(\Delta J = 0)$ appears at the band origin.

2. *Spherical tops* $I_A = I_B = I_C$

There is no distinction between $\parallel$ and $\perp$ bands; in all cases, $\Delta J = 0, \pm 1$ and the fundamentals all have PQR contours.

3. *Symmetric tops* $I_A = I_B < I_C$ (oblate), or $I_A < I_B = I_C$ (prolate)

The principal symmetry axis is of order 3 or more, and with respect to it we can define $\parallel$ and $\perp$ vibrations. Two rotational quantum numbers are required: J to define the total angular momentum, and K to define the momentum about the C axis. The selection rules and resulting band contours are indicated in Table 8.

4. *Asymmetric tops* $I_A < I_B < I_C$

Depending on which axis is parallel to the transition moment, the bands are defined as type A, B, or C. Their contour shapes are determined by the relative moments of inertia, but if these may be estimated the method is still useful, as indicated in Table 8. Not only are rotational contours of value as an aid to assignment, but they may also provide a direct guide to molecular geometry in deciding which type of rotor a molecule may be.

F. Isotopic Substitution

When a molecule is isotopically substituted, its force field remains unaffected to a high approximation. The relative vibrational frequencies of the isotopic species then depend only on the atomic masses. A diatomic hydride M—H provides a simple example: equation (2) gives, for the ratio of frequencies of the deuteride and hydride,

$$\frac{\nu_{\mathrm{H}}}{\nu_{\mathrm{D}}} = \left(\frac{\mu_{\mathrm{D}}}{\mu_{\mathrm{H}}}\right)^{\frac{1}{2}}$$

Since $m_{\mathrm{H}} < m_{\mathrm{M}}$, this ratio becomes approximately

$$\frac{\nu_{\mathrm{H}}}{\nu_{\mathrm{D}}} \simeq \left(\frac{m_{\mathrm{D}}}{m_{\mathrm{H}}}\right)^{\frac{1}{2}} = \sqrt{2}$$

If the isotopic atom executes group vibrations, as is usually the case with hydrogen, then frequency shifts on substitution provide a useful means of picking out the normal modes which involve motion of the isotopic atoms. In metal carbonyl hydrides, σ(M—H) falls in the same region (ca. 1900 cm^{-1}) as σ(CO). The bands may be readily distinguished by deuteration[101], when the σ(M—D) stretch appears at ca. 1400 cm^{-1}. Metal–nitrogen vibrations in nitrido complexes have similarly been identified in the infrared by using ^{15}N substitution[102], and σ(M—O) in binuclear complexes of type M—O—M has been located[103] by ^{18}O substitution. Interpretation of the Raman spectrum of boron trimethyl is complicated by the fact that σ(B—C) falls in the same region as methyl rocking vibrations $\rho(CH_3)$. The natural presence of ^{10}B and ^{11}B in the molecule, however, causes all the normal modes in which the boron atom moves to appear as split fundamentals in the spectra[104]; only the totally symmetric σ(B—C) fundamental is unshifted, providing further evidence for the planarity of the molecule.

The uses of isotopic substitution described above are essentially qualitative and rest on the assumption of group vibrations. The ratios of isotopic frequencies find quantitative expression in the Teller–Redlich product rule:

$$\prod_{k=1}^{3N-6} \frac{\nu'_k}{\nu_k} = \left(\frac{M'}{M}\right)^{\frac{3}{2}} \left(\frac{I'_x I'_y I'_z}{I_x I_y I_z}\right)^{\frac{1}{2}} \prod_{i=1}^{3N} \left(\frac{m_i}{m'_i}\right)^{\frac{1}{2}}$$

where a prime denotes the isotopic species, M and I are the masses and moments of inertia of the molecules as a whole, and m_i are the masses of the isotopic atoms. The product goes over all the normal modes, but if the two isotopic molecules have the same symmetry it may be factored into separate products for each symmetry class. Contributions of M or I to a given class are found from the group character table. For example, we may take two isotopic tetrahedral species, MX_4 and MX'_4: the product rules and the numerical results for CH_4 and CD_4 are shown in Table 9. Discrepancies

Table 9. Teller–Redlich product rules for MX_4 (T_d).

Class	Product ratio	Results for CH_4 and CD_4	
		Calculated	Observed
a_1	$\nu'_1/\nu_1 = (m_X/m'_X)^{\frac{1}{2}}$	0.707	0.715
e	$\nu'_2/\nu_2 = (m_X/m'_X)^{\frac{1}{2}}$	0.707	0.691
t_2	$\nu'_3\nu'_4/\nu_3\nu_4 = (m_X/m'_X)(M'/M)^{\frac{1}{2}}$	0.560	0.570

between the observed and the calculated frequencies, which are only about 1–2%, are due to anharmonicity.

Related to isotopic substitution is the use of chemical analogues, e.g. the substitution of Br for Cl[105]; since the force field changes as well as the masses, there is no simple quantitative rule, but the method is of great qualitative value in identifying group vibrations.

G. Solvent Shifts

This new technique has arisen chiefly from the work of Bellamy and associates[106]. If the group frequency of a polar linkage (e.g. C=O, C=S M—H) is ν_0 in the free molecule and ν_s in solution, it is found that the relative shift, $(\nu_0 - \nu_s)/\nu_0$, for a particular group varies in a linear manner over a wide range of solvents due to specific solute–solvent interaction. In this way, fundamentals associated with a given group can be distinguished, by their characteristic solvent variation, from other neighbouring bands. A recent application of this method to metal carbonyls and nitrosyls[107] offers a means of distinguishing between terminal and bridging carbonyl vibrations when the group frequency criterion is in doubt.

H. Use of Thermodynamic Data

The vibrational partition function Q_{vib} may be obtained from the frequencies (ν_i) and degeneracies (g_i) of all the normal modes:

$$Q_{\text{vib}} = \sum_i g_i \exp\left(-h\nu_i/kT\right)$$

When combined with translational and rotational contributions, the partition function enables the thermodynamic functions of the molecule to be calculated. Conversely, if all but one of the normal frequencies are known and measured thermodynamic data are available, the missing frequency may be calculated by a method of successive approximations. The form of Q_{vib} above dictates that its value is most sensitive to the lowest frequencies. An interesting example is afforded by carbon suboxide, C_3O_2. Its linear structure had been quite well established by earlier vibrational studies, but one bending fundamental (ν_7, π_u) appeared to be missing from the infrared spectrum. Calculation from the thermodynamic properties of the molecule[108] predicted an extraordinarily low value of $61.6 \pm 2.6\ \text{cm}^{-1}$ for this fundamental; this was confirmed by the subsequent observation[109] of a weak band at $63\ \text{cm}^{-1}$ in the far infrared.

I. Force Constants and Intensities

When the group frequency approximation breaks down, rough calculations using force-constant values transferred from related molecules can be used as a guide to assignment. In some systems (especially metal carbonyls), the relative magnitudes of the force constants and approximate relations between them may be inferred from molecular-orbital arguments (see Section VI.A). If an approximate picture of the normal modes can be formed, it is possible to assess the relative magnitudes of the changes in dipole moment, or polarizability, during vibration; a decision can then be made as to the likely intensities of certain fundamentals in the spectra.

V. STRUCTURAL STUDIES

The examples which follow have been subdivided in a rather formal way according to the type of molecular skeleton and the complexity of the attached ligands. It is hoped that the molecules which have been chosen, few though they are, will illustrate the application of vibrational spectroscopy to some of the fields of current inorganic interest.

A. Mononuclear Species

1. Krypton difluoride, KrF_2

The infrared and Raman spectra[110] or KrF_2 in the vapour phase are shown in Table 10. The absence of coincidences between the two spectra

Table 10. Infrared and Raman spectra of KrF_2 (in cm^{-1}).

Infrared	Raman	Assignment
233 s ($\perp$)		ν_2 π_u, δ(FKrF)
	449	ν_1 σ_g^+, σ(KrF)
588 vs ($\parallel$)		ν_3 σ_u^+, σ(KrF)
1032 m		$\nu_1 + \nu_3 = 1037$, Σ_u^+

immediately suggests a centrosymmetric structure. This is fully substantiated by an assignment based on a linear model ($D_{\infty h}$): the single Raman line at 449 cm^{-1} must be $\nu_1(\sigma_g^+)$, σ(Kr—F); of the two strong infrared bands, the higher one with a *PR* contour is clearly the parallel vibration $\nu_3(\sigma_u^+)$, σ(Kr—F) and the lower band with a *PQR* contour must be $\nu_2(\pi_u)$, δ(F—Kr—F). The remaining weak infrared band is then readily explained as the binary combination $\nu_1 + \nu_3$ (Σ_u^+), having a calculated value of 1037 cm^{-1}. From the separation of the *P* and *R* branches of ν_3, an approximate moment of inertia may be calculated, which yields a value of 1.9 ± 0.1 Å for the Kr—F bond length.

2. Disiloxane, $O(SiH_3)_2$

The interest here focuses on the geometry of the Si—O—Si skeleton. Examination of the fully deuterated compound[111] discriminates between the skeletal vibrations and those of the SiH_3 groups. The apparent lack of coincidences between the infrared and Raman spectra was naturally taken to imply a linear skeleton. But this finding is at variance with the results of electron diffraction, which indicates an Si—O—Si angle of about 140°. A force-constant calculation[112] also suggests an angle less than 180°. This example reveals a weakness of spectroscopic structure determination in that the wrong selection rules appear to be obeyed; but at the same time it provides an opportunity for a simple and elegant Raman experiment which confirms that the molecule is, in fact, bent: the Raman spectrum of disiloxane enriched in ^{18}O was recorded[113]. If the molecule were strictly linear, the totally symmetric Si—O stretching fundamental at 606 cm^{-1} should be unaffected by ^{18}O substitution because the oxygen atom does not move during this normal mode (see Figure 2). However, the band was seen to be displaced by 3 to 4 cm^{-1}, a shift which is consistent with an Si—O—Si angle of about 140°. A number of related molecules, silyl and germyl sulphides and selenides[114], have also been investigated but all give spectra characteristic of the bent skeleton. The implications of this quasilinear behaviour of disiloxane have been discussed[115] with reference to *p–d* π bonding and the potential function of the molecule.

3. Trisilylamine, $N(SiH_3)_3$

This compound bears an obvious relation to the foregoing example but the interpretation of its spectra is apparently without the peculiar complications present in disiloxane. Essentially, the choice has to be made between a planar or a pyramidal NSi_3 skeleton. Again, deuteration has been of great assistance. The observed skeletal vibrations[116] are shown in Table 11. The appearance of only three skeletal fundamentals in the

Table 11. Skeletal vibrations of $N(SiH_3)_3$ (in cm^{-1}).

	σ(NSi)		δ(SiNSi)
Infrared	987	—	not investigated
Raman	987	496 pol.	204

Raman spectrum and the absence in the infrared of a band corresponding to the polarized Raman line at 496 cm^{-1}, provide sufficient evidence for the planar structure; if pyramidal, the molecule would be expected to show all four fundamentals in both Raman and infrared (see Table 5). This result is in harmony with the planar skeleton found by electron diffraction. Interestingly, the related species, trisilylphosphine, appears also to have a planar PSi_3 configuration[117] but trigermylphosphine[118] and the compounds[119] $[(CH_3)_3Sn]_3X$ (where X is N, P, As, or Sb) revert to a pyramidal shape.

4. Xenon tetrafluoride, XeF_4

The infrared spectrum of the vapour and Raman spectrum[120] of solid XeF_4 are summarized in Table 12. The presence of only one (Xe—F) stretching fundamental (at 586 cm^{-1}) in the infrared immediately suggests a high symmetry for the molecule, either T_d (tetrahedral) or D_{4h} (square planar).

Table 12. Infrared and Raman spectra of XeF_4 (in cm^{-1}).

Infrared	Raman	Assignment
123 s		ν_7 e_u, δ(FXeF)
	235	ν_3 b_{1g}, δ(FXeF)
291 s (‖)		ν_2 a_{2u}, δ(FXeF)
	442 w	$2\nu_4$? A_{1g}
	502	ν_5 b_{2g}, σ(XeF)
	543	ν_1 a_{1g}, σ(XeF)
586 s		ν_6 e_u, σ(XeF)
1105		$\nu_5 + \nu_6 = 1088$, E_u
1136		$\nu_1 + \nu_6 = 1129$, E_u

The T_d model, however, predicts only one infrared active FXeF deformation, whereas two are observed in conformity with the D_{4h} selection rules (Table 5). Moreover, two coincidences would be expected for a tetrahedral model but none are observed, providing further evidence for the centrosymmetric square planar structure. A detailed assignment on the basis of point group D_{4h} substantiates this conclusion: the *PQR* contour of the band at 291 cm^{-1} marks it as the parallel (i.e. out of plane) deformation $\nu_2(a_{2u})$; similar arguments lead to the assignment of all the other fundamentals as indicated, except the inactive one $\nu_4(b_{1u})$. The weak Raman line at 442 cm^{-1} is tentatively assigned as the first overtone of this fundamental and the infrared bands at 1105 and 1136 cm^{-1} are readily attributed to binary combination tones. From the *Q–R* separation of the 291 cm^{-1} band, a Xe—F distance of 1.85 ± 0.2 Å may be calculated. X-ray diffraction has confirmed the planar structure in the solid state and gives a value of 1.92 Å for the Xe—F bond length.

5. Iron pentacarbonyl, $Fe(CO)_5$

The attempts to determine the structure of this important molecule make a long and somewhat contentious history. The electron-diffraction pattern[121] favours a trigonal bipyramidal structure (D_{3h}); the measured dipole moment of 0.64 D was at first thought to challenge this[122], being more indicative of a square-based pyramid (C_{4v}), but subsequent consideration[123] showed this value to be not inconsistent with the D_{3h} structure. Early incomplete studies of the infrared spectrum[124-5] were held to support both models. Very full Raman[126] and infrared[127] data for $Fe(CO)_5$ have now been obtained and there seems no question that the D_{3h} model is the correct one. The selection rules for the two disputed geometries are shown in Table 13 and the observed spectra are summarized in Table 14 (a large

Table 13. Selection rules for possible structures of $Fe(CO)_5$.

Activity	Internal coordinates: σ(CO)	σ(FeC)	δ(FeCO)	δ(CFeC)	Totals
D_{3h}					
Raman only	$2a_1'$	$2a_1'$	$2e''$	e''	13 R (4 pol.)
Raman and i.r.	e'	e'	$2e'$	$2e'$	6 coincidences
i.r. only	a_2''	a_2''	a_2''	a_2''	10 i.r.
Inactive			a_2'		1 inactive
Totals	4	4	6	4	18 normal modes
C_{4v}					
Raman only	b_1	b_1	$b_1 + b_2$	$b_1 + b_2$	19 R (6 pol.)
Raman and i.r.	$2a_1 + e$	$2a_1 + e$	$a_1 + 3e$	$a_1 + 2e$	13 i.r., all coincident
Inactive			a_2		1 inactive
Totals	4	4	7	5	20 normal modes

Table 14. Infrared and Raman spectra of $Fe(CO)_5$ (in cm^{-1}).

Infrared (vapour)	Raman (liquid)	Assignment	
	2114 m pol.	ν_1 a_1'	σ(CO)
2034 vs	2031 vs	ν_{10} e'	σ(CO)
2014 vs		ν_6 a_2''	σ(CO)
	1984 s	ν_2 a_1'	σ(CO)
	753 w	ν_{16} e''	δ(FeCO)
646 vs	640 w	ν_{11} e'	δ(FeCO)
620 vs		ν_7 a_2''	δ(FeCO)
544 w	558 w	ν_{12} e'	δ(FeCO)
	492 w	ν_{17} e''	δ(FeCO)
474 s		ν_8 a_2''	σ(FeC)
431 s	441 w	ν_{13} e'	σ(FeC)
	414 vs pol.	ν_3 a_1'	σ(FeC)
	377 w	ν_4 a_1'	σ(FeC)
111	112 w	ν_{14} e'	δ(CFeC)
	104 vs	ν_{18} e''	δ(CFeC)
74?	68 m	ν_{15} e'	δ(CFeC)

Totals: 13 Raman (2 polarized), 9 infrared (6 coincidences).

number of weak infrared features have been omitted from this table). Comparison of the two tables will show that the data fit the D_{3h} model very well, entirely ruling out the C_{4v} structure. A discussion of the detailed assignment, which illustrates several of the points made in Section IV, may be found in the original references.

B. Species with Complicated Ligands

The vibrational analysis of complexes with chelating ligands (e.g. acetylacetonates) is described in detail by Nakamoto[22]. Of more relevance to current inorganic developments is the large class of organometallic compounds which bear an unsaturated organic species coordinated to a metal —the π-alkenyl and π-aryl complexes. Reviews[128] indicate that few of these have been thoroughly studied spectroscopically, owing to their vibrational complexity and the difficulties which are so frequently encountered in recording their Raman spectra. However the vibrational analyses of the important archetypes, ferrocene[129] and bisbenzenechromium[130], have reached a high state of refinement and represent something of a vibrational *tour de force*.

1. π-Cyclopentadienyl compounds

The Raman and infrared spectra of ferrocene, ferrocene-*d*-10, and ruthenocene have been recorded[129] for the vapour and solution phases. The relative simplicity of the spectra of these 21-atomic molecules betokens a high symmetry. With the aid of band contours and the product rule, assignments have been made on the basis of the D_{5d} staggered 'sandwich' structure found by x-ray diffraction for solid $Fe(C_5H_5)_2$. In particular, the metal–ring stretching frequencies are interesting: these are found at 303, 330 cm^{-1} (symmetric), and 478, 476 cm^{-1} (antisymmetric) for the Fe and Ru compounds, respectively. A recent comparison with other biscyclopentadienyl derivatives[131] suggests that the metal–ring stretching force constant is significantly higher in the Os, Fe, and Ru compounds than in those of Cr, Co, V, and Ni. The vibrations of the coordinated $C_5H_5^-$ rings also offer strong evidence for their planar aromatic nature; a normal coordinate analysis shows the $C_5H_5^-$ rings to have force constants similar to those of the other 6 π-electron systems, benzene and the tropylium cation, $C_7H_7^+$.

The discovery of an eclipsed (D_{5h}) sandwich structure for solid ruthenocene[132] raises the interesting question of the conformation of this molecule in the vapour or solution states. It is doubtful whether the existing vibrational data would suffice for an unequivocal decision on this point but the barrier opposing rotation about the C_5 axis in ferrocene has recently been discussed[133] in relation to the infrared and Raman selection rules. In contrast to the highly symmetrical π-bonded cyclopentadienyl complexes, those with σ-bonded rings (e.g. $Hg(C_5H_5)_2$) are distinguished by a markedly different, more complex spectrum.

2. Cyclooctatetraeneiron tricarbonyl, $C_8H_8Fe(CO)_3$

The appearance of a single, sharp proton resonance in the n.m.r. spectrum of this compound at room temperature, together with certain chemical and infrared evidence, was generally thought to establish the planarity of the C_8H_8 ring. A subsequent x-ray diffraction study, however, revealed a highly unsymmetrical structure in the solid: the $Fe(CO)_3$ moiety is bonded to a butadiene-like portion of the ring which lies in a different plane to the rest, producing four non-equivalent proton sites. This finding naturally prompts the question: is the inconsistency between the n.m.r. and x-ray data due to a different geometry for this molecule in solution and solid? Vibrational studies[134] have provided the answer. The complicated infrared and Raman spectra prove that the molecule has the same low symmetry (C_s) in both phases. The interesting implication is that the protons of the C_8H_8 ring can achieve a tautomeric equivalence relative to the $Fe(CO)_3$ group. This is an example of the time scale of a physical measurement influencing the apparent structure of a molecule (Section II.C).

C. Polynuclear Species

A number of complete vibrational analyses have been performed for molecules of more complicated skeletal structures which often involve bonding between like atoms. This general class of inorganic compound has a renewed significance in the light of recent developments in the chemistry of metal atom cluster compounds and the cage-like derivatives of the boron hydrides.

1. Diphosphorus tetraiodide, P_2I_4

X-ray diffraction studies have established a *trans* (C_{2h}) structure in the solid state; but the dipole moment (0.45 D) and infrared spectra suggested a *gauche* (C_2) arrangement of the two PI_2 groups in solution which was therefore supposed to be the stable conformation of the free molecule, unsubject to the constraints of crystal packing. However, a thorough study[135] of the vibrational spectra of solid and solution convincingly supports the *trans* (C_{2h}) structure in both phases. The selection rules for the three conformations, *trans*, *gauche*, and *cis* (depending on the azimuthal angle between the two PI_2 groups), are summarized, together with the numbers of observed bands, in Table 15. The lack of coincidences between the spectra, the number of polarized Raman lines in the stretching region, and the close similarity of spectra recorded from solid and solution,

Table 15. Selection rules for P_2I_4.

Structure	Azimuthal angle	Point group	Raman	Infrared	Coincidences	Raman pol.	σ's dep.
trans	$\theta = 180°$	C_{2h}	6	6	0	2	1
gauche	$0 < \theta < 180°$	C_2	12	12	12	3	2
cis	$\theta = 0°$	C_{2v}	12	9	9	2	3
	Observed spectra:		6	5 or 6?	0	2	1

all weigh in favour of the C_{2h} structure. It is relevant to recall the conformations determined for other binuclear species: P_2Cl_4 behaves similarly to P_2I_4 in showing the *trans* conformation in vapour, liquid, and solid phases; N_2F_4 is *gauche* in the vapour and solid, but the liquid contains a mixture of the *gauche* and *trans* forms; however, the subhalides of boron, B_2F_4 and B_2Cl_4, are planar (D_{2h}) in the crystal but staggered (D_{2d}) in the vapour phase. The interesting stereochemical behaviour of P_2I_4 may be ascribed to steric overcrowding, which causes a high potential barrier to rotation about the P—P axis.

2. Compounds with metal–metal bonds

Apart from the classic proof[136] by Raman spectroscopy of the diatomic nature of the mercurous ion in aqueous solution, developments in this field are very recent and most vibrational studies are as yet incomplete. The region of metal–metal stretching frequencies (where these have meaning) has been established[137] as 100–220 cm^{-1}, with interesting implications regarding the force constants of metal–metal bonds. Two examples of structure determination may be noted. A Raman study of cadmium ions in a tetrachloroaluminate melt[138] has demonstrated the existence of a rather feebly bonded Cd_2^{2+} ion for the first time; and the Raman and infrared spectra of $Cd[Co(CO)_4]_2$ and $Hg[Co(CO)_4]_2$ strongly favour[139] a linear Co—M—Co skeleton. The vibrational spectra of the well-known molybdenum cluster compounds, $(Mo_6X_8)Y_6^{2-}$ (X, Y = halogen), are consistent with O_h symmetry for these species in solution. Normal coordinate analysis[140] suggests a distinct metal–metal interaction in the cluster, with a force constant of about 1.8 mdyn/Å; but the normal modes have such a 'mixed' character that none can be referred to as 'pure' metal–metal vibrations.

3. Boranes

An especially pleasing example of the application of group theory to molecular vibrations is afforded[141] by the dodecaborane anion, $B_{12}H_{12}^{2-}$. The most likely structural model is a regular icosahedron having a boron atom at each apex and no bridging hydrogen atoms. The point group of such a molecule is I_h and the selection rules predict only three infrared fundamentals ($3t_{1u}$) and six Raman lines ($2a_g + 4h_g$), of which two should be totally polarized and four are five-fold degenerate. The observed spectra conform exactly to these expectations (Raman: 2518(pol.), 2475, 949, 770, 743(pol.), and 584 cm^{-1}; infrared: 2480, 1070, and 720 cm^{-1}).

D. Diagnostic Uses

In this category there fall many studies which, while not comprising a full vibrational analysis, have enabled structural inferences to be made. A very few examples must serve as illustration of these widespread techniques. *Cis* and *trans* stereoisomers in square planar (MX_2L_2) or octahedral (MX_4L_2) complexes may often be distinguished by their spectra in the *skeletal stretching region.* For instance, the *cis* and *trans* isomers of $(Et_3P)_2PtX_2$ have been identified[142] by infrared spectroscopy (*trans* (D_{2h}): 1σ(PtX), 1σ(PtP); *cis* (C_{2v}): 2σ(PtX), 2σ(PtP)); stereoisomerism in complexes $SnCl_4L_2$ has similarly been investigated[143]. With regard to the use of *ligand vibrations* alone as an indication of molecular geometry, the dangers inherent in relying on the unsupported data have already been mentioned. Deductions about the overall geometry of metal carbonyls, in particular, based solely on the (CO) stretching frequencies in the infrared, have been commonly made in the past and have often proved erroneous. The ease of making these measurements gives them a value to the preparative chemist which cannot be denied but they should only be regarded as pointers for more exact methods of structure determination. Quite reliable criteria, on the other hand, have been established to detect *functional group isomerism* and *coordination isomerism* in ligands. For example, the various modes of coordination of the NO_2^- ion—O-bonded, N-bonded, bidentate, or bridging—may be distinguished by the number and positions of the (NO) stretching frequencies[144]. Pseudohalide ligands such as NCO^- and NCS^- may also show a coordination isomerism which is readily detected by the infrared spectrum[145].

VI. ELECTRONIC EFFECTS

A. Force Constants and their Significance

To obtain the relationship between the force constants and frequencies of a polyatomic molecule, we must revert to a consideration of the cartesian displacement coordinates q_i. It follows from equation (5) that, for the kth normal mode, any q_i coordinate is given by

$$q_i = A_{ik} \cos (\lambda_k^{\frac{1}{2}} t + \epsilon_k) \tag{12}$$

where A_{ik} is an amplitude constant. The equations of motion in terms of the

q_i are obtained by substituting the expressions for the kinetic and potential energies (equations 3 and 4) in Lagrange's equation, with the result

$$\ddot{q}_i + \sum_j f_{ij} q_j = 0 \tag{13}$$

We have seen how the existence of solutions to these $3N$ equations may be demonstrated by transforming to normal coordinates. Equation (12) is derived from such a solution, and by substituting it in (13), we obtain

$$-\lambda_k A_{ik} + \sum_{j=1}^{3N} f_{ij} A_{jk} = 0 \qquad i = 1, 2, 3, \ldots, 3N$$

Thus for each normal mode k there exists a set of $3N$ linear simultaneous equations in the $3N$ unknown amplitudes A_i. If each such set is to produce non-zero solutions for the A_i, a condition of compatibility must be satisfied, namely, the determinant of the coefficients must vanish:

$$\begin{vmatrix} f_{11} - \lambda & f_{12} & f_{13} & \cdots \\ f_{21} & f_{22} - \lambda & f_{23} & \cdots \\ f_{31} & f_{32} & f_{33} - \lambda & \cdots \\ \vdots & \vdots & \vdots & \end{vmatrix} = 0$$

This is the secular equation for vibrational motion. On expansion, it gives a polynominal in λ of degree $3N$, whose $3N$ roots λ_k are related to the frequencies of the normal modes by $\lambda_k = 4\pi^2 c^2 \nu_k^2$. For a given set of force constants f_{ij}, the secular equation may be solved for the frequencies by extraction of its roots. In spectroscopic practice, it is the reverse operation which is needed, i.e. calculation of the coefficients f_{ij} from the roots λ_k. This is not possible analytically in the general case, so iterative methods[146] must be adopted which converge on a 'best fit' set of frequencies.

Cartesian coordinates are not convenient if we wish to adopt a chemically realistic force field. Internal coordinates (Section II.D) are the natural choice, in terms of which potential energy expressions become quite simple. For example, a linear triatomic molecule of type XY_2 ($D_{\infty h}$) has two sets of internal coordinates: Δr_1 and Δr_2 (the two bond-length changes) and $\Delta\theta$ (the YXY bond-angle change). If a general quadratic potential function is assumed, the potential energy may be written

$$2V = f_r(\Delta r_1^2 + \Delta r_2^2) + 2f_{rr}\,\Delta r_1\,\Delta r_2 + r_0^2 f_\theta\,\Delta\theta^2 + r_0 f_{r\theta}\,(\Delta r_1 + \Delta r_2)\,\Delta\theta$$

f_r is the bond-stretching and f_θ is the bond-angle deformation force constant; f_{rr} and $f_{r\theta}$ are interaction constants between the respective internal coordinates*. This example typifies a situation most commonly

* The equilibrium bond length r_0 is included to preserve the units in dyn/cm or mdyn/Å.

encountered in force-constant calculations: there are four unknown f values, but only three frequencies (see Table 5) to determine them. Unless extra data are available (from isotopic species, vibrational–rotational interaction, or vibrational amplitudes), simplifying assumptions must be made about the force field in order to reduce the number of unknown parameters. In order to set up and solve the secular equation in terms of internal coordinates, the powerful G–F matrix method developed by Wilson[17] is usually employed. The symmetry properties of the normal modes often make factorizing of the secular equation possible, thus further simplifying the calculation.

A bond-stretching force constant is a measure of bond strength at internuclear distances close to the equilibrium value (e.g. for a diatomic molecule, $f = (\partial^2 V/\partial r^2)_0$) and may be correlated with other characteristic bond parameters. Depending on the choice of the potential function, various semiempirical relationships have been obtained. Most potential functions (e.g. the Morse curve) suggest a direct proportionality between f and the bond-dissociation energy D_0. This has been confirmed experimentally[147] in a few simple systems and is often implicitly assumed in drawing inferences from force-constant data, but there are no grounds *a priori* for believing it to be generally true. Many expressions relating f to the equilibrium bond length r_0 have been put forward, notably by Badger[148] who proposed a relationship of the form $f = a/(r_0 - b)^3$, where a and b depend on the position of the elements in the periodic table. Morse's original rule, $f \propto r_0^{-6}$, has received a recent confirmation[149] for N—N and C—O bonds. Gordy[150] has shown evidence for a correlation of f with the electronegativities χ of the constituent atoms and the bond order N: $f = aN(\chi_A\chi_B/r_0^2)^{\frac{3}{4}} + b$. The relationship of f to orbital overlap integrals[151] has also been discussed.

Interpretations of the significance of force constants to chemical bonding are numerous. Mention can only be made of the work of Linnett[25] and Mills[24] on small molecules, of Jones[152,153] on cyanide and carbonyl complexes, and the successful approximate treatment of metal carbonyls by Cotton and Kraihanzel[154], in which the high frequency of the CO stretching vibrations permits these groups to be considered as dynamically independent of the rest of the molecule. Force constants of CO bonds provide a useful index of electronic character. For example, in proceeding along the isoelectronic series[155] $Ni(CO)_4$, $Co(CO)_4^-$, and $Fe(CO)_4^{2-}$, f(MC) rises as f(CO) falls, an observation which is in keeping with the changes of bond order that should result from increasing $d \rightarrow \pi^*$ electron delocalization as the oxidation state of the metal becomes more negative.

Interaction constants may also give interesting information. Calculations for the inert gas difluorides[110] have brought to light a surprising difference in the interaction constants f_{rr} (see above): for XeF_2, $f_{rr} = +0.13$

mdyn/Å, which suggests that the stretching of one XeF bond tends to induce a contraction in the other; for KrF_2, on the other hand, $f_{rr} = -0.20$ mdyn/Å, the negative sign implying that as one KrF bond is stretched, the other tends likewise to lengthen. This puzzling result has been rationalized by Coulson[156] in terms of a resonance between the forms **1** to **3**.

$$\underset{\mathbf{1}}{F\text{—}M^+F^-} \qquad \underset{\mathbf{2}}{F^-{}^+M\text{—}F} \qquad \underset{\mathbf{3}}{\overset{\frown}{F\ M\ F}}$$

Forms **1** and **2**, which favour a positive f_{rr}, have greater weight in XeF_2. Form **3**, which favours a negative f_{rr}, tends to predominate in the less stable KrF_2 due to the relatively higher ionization potential of krypton.

B. Intensities and Related Phenomena

The intensity of a Raman line can be related[157], at least in principle, to the derived polarizability tensor $(\partial \boldsymbol{\alpha}/\partial Q_k)_0$ for the normal mode. Wolkenstein[158] has assumed that $(\partial \boldsymbol{\alpha}/\partial Q_k)_0$ may be regarded as the sum of contributions $(\partial \alpha_{nm}/\partial Q_k)_0$ from the individual bonds. Furthermore, if the transformation from normal to internal coordinates is known (which entails solution of the molecular force field), the Raman intensities may be expressed in terms of *derived bond polarizabilities* $(\partial \alpha/\partial r)_{nm}$. The calculation of values for this interesting bond parameter is hampered by the implicit assumptions and the prerequisite need for a normal coordinate analysis. But in favourable cases significant results have been obtained and correlations with electronegativity[159] and bond order[160] in inorganic compounds have been demonstrated.

Infrared intensities may be similarly treated, and related to derived bond dipole moments $(\partial \mu/\partial r)_{nm}$. When appropriate assumptions can be made, the relative intensities of infrared bands may yield structural information, e.g. about the bond angles in cyanide, carbonyl, and nitrosyl compounds[161].

It has long been recognized that the classical theory of Raman intensities becomes invalid (i) if the exciting line lies close to an electronic absorption band of the molecule, or (ii) if the molecule possesses a degenerate electronic ground state. In case (i), certain Raman lines may be unusually intensified by the *resonance Raman effect*. This is dramatically apparent in some organic dyestuffs but the occurrence in inorganic molecules[162] has not yet been fully investigated. The extent of resonance enhancement depends on the symmetry of the normal mode and the type of electronic absorption band[163]; for this reason the effect is potentially useful as a source of information about excited electronic states.

Intensity anomalies stemming from a degenerate electronic ground term, case (ii), are a manifestation of vibronic Jahn–Teller coupling[164].

According to the Jahn–Teller theorem, the presence of electronic degeneracy in a molecule destabilizes its most symmetrical configuration with respect to certain distorted forms. When the barrier to interconversion of these distorted forms is comparable with the zero-point vibrational energy, no permanent distortion of the structure is detected by diffraction methods or by a departure from the vibrational selection rules for the most symmetrical configuration. Nevertheless the 'inherent' Jahn–Teller instability may reveal itself by the unusual appearance of the infrared and Raman spectra. In this respect, a study[165] of metal hexafluorides (e.g. ReF_6) has been particularly fruitful and anomalies have also been detected in the vibrational spectra of VCl_4[166], $V(CO)_6$[167], WCl_5[168], and $ReCl_6^{2-}$[169].

In conclusion, mention must be made of the recently observed Raman spectra of crystalline Ce^{3+} and Pr^{3+} compounds[170]. In addition to the expected vibrational transitions, Raman shifts were obtained which could be identified with pure electronic transitions between the multiplet terms of the $4f^1$ and $4f^2$ configurations. Further discussion of these *electronic Raman shifts* would be inappropriate to a chapter devoted to vibrational spectroscopy, but they represent a most important advance which may be of future value in the study of electronic states.

VII. REFERENCES

1. E. O. Bodger and J. W. White, *Chem. Commun.*, **1967**, 74.
2. R. G. Denning, *Chem. Commun.*, **1967**, 120.
3. R. A. Satten, D. J. Young, and D. M. Gruen, *J. Chem. Phys.*, **33**, 1140 (1960); M. J. Reisfeld and G. A. Crosby, *Inorg. Chem.*, **4**, 65 (1965).
4. M. I. Al-Joboury and D. W. Turner, *J. Chem. Soc.*, **1963**, 5141; D. W. Turner, Chap. 3 of this book.
5. B. P. Stoicheff, *Phys. Letters*, **7**, 186 (1963).
6. S. Dumartin, B. Oksengorn, and B. Vodar, *Compt. Rend.*, **261**, 3767 (1965).
7. S. J. Cyvin, J. E. Rauch, and J. C. Decius, *J. Chem. Phys.*, **43**, 4083 (1965); R. W. Terhune, P. D. Maker, and C. M. Savage, *Phys. Rev. Letters*, **14**, 681 (1965).
8. A. D. Cross, *An Introduction to Practical Infrared Spectroscopy*, Butterworths, London, 1960.
9. E. A. V. Ebsworth, 'Inorganic applications of infrared spectroscopy' in *Infrared Spectroscopy and Molecular Structure* (Ed. M. Davies), Elsevier, Amsterdam, 1963.
10. K. E. Lawson, *Infrared Absorption of Inorganic Substances*, Reinhold, New York, 1961.
11. J. J. Turner, *Chem. Ind.* (*London*), **1966**, 109.
12. F. A. Miller and C. H. Wilkins, *Anal. Chem.*, **24**, 1253 (1952).
13. F. A. Miller, G. L. Carlson, F. F. Bentley, and W. H. Jones, *Spectrochim. Acta*, **16**, 135 (1960).
14. L. A. Woodward, *Quart. Rev.* (*London*), **10**, 185 (1956).

15. R. E. Dodd, *Chemical Spectroscopy*, Elsevier, Amsterdam, 1962.
16. G. Herzberg, *Infrared and Raman Spectra of Polyatomic Molecules*, Van Nostrand, New York, 1945.
17. E. B. Wilson, J. C. Decius, and P. C. Cross, *Molecular Vibrations*, McGraw-Hill, New York, 1955.
18. A. E. Martin, 'Instrumentation and general experimental methods' in *Infrared Spectroscopy and Molecular Structure* (Ed. M. Davies), Elsevier, Amsterdam, 1963.
19. J. C. Evans, 'Raman spectroscopy' in *Infrared Spectroscopy and Molecular Structure* (Ed. M. Davies), Elsevier, Amsterdam, 1963.
20. L. H. Daly, N. B. Colthup, and S. E. Wiberley, *Introduction to Infrared and Raman Spectroscopy*, Academic Press, New York, 1964.
21. F. A. Cotton, 'The infrared spectra of transition metal complexes' in *Modern Coordination Chemistry* (Ed. J. Lewis and R. Wilkins), Interscience, New York, 1960.
22. K. Nakamoto, *Infrared Spectra of Inorganic and Coordination Compounds*, Wiley, New York, 1963.
23. J. A. Ibers, private communication.
24. I. M. Mills, 'Force-constant calculations for small molecules' in *Infrared Spectroscopy and Molecular Structure* (Ed. M. Davies), Elsevier, Amsterdam, 1963.
25. J. W. Linnett, *Quart. Rev.* (*London*), **1,** 73 (1947).
26. T. V. Long and R. A. Plane, *J. Chem. Phys.*, **43**, 457 (1965).
27. D. Steele, *Quart. Rev.* (*London*), **18**, 21 (1964).
28. L. Pauling and E. B. Wilson, *Introduction to Quantum Mechanics*, McGraw-Hill, New York, 1935.
29. E. Fermi, *Z. Physik*, **71**, 250 (1931).
30. H. Eyring, J. Walter, and G. E. Kimball, *Quantum Chemistry*, Wiley, New York, 1944.
31. C. V. Raman, *Indian J. Phys.*, **2**, 387 (1928).
32. F. A. Cotton, *Chemical Applications of Group Theory*, Interscience, New York, 1963.
33. R. McWeeney, *Symmetry—An Introduction to Group Theory*, Pergamon, Oxford, 1963.
34. D. Schonland, *Molecular Symmetry*, Van Nostrand, New York, 1965.
35. W. Klemperer and L. Lindemann, *J. Chem. Phys.*, **25**, 397 (1956); H. Braune and G. Engelbrecht, *Z. Physik. Chem.*, **B19**, 303 (1932).
36. D. F. Smith, *J. Chem. Phys.*, **38**, 270 (1963).
37. H. Stammreich, R. Forneris, and K. Sone, *J. Chem. Phys.*, **25**, 972 (1955).
38. R. E. Weston and T. F. Brodasky, *J. Chem. Phys.*, **27**, 683 (1957).
39. D. A. Dows and G. Bottger, *J. Chem. Phys.*, **34**, 689 (1961).
40. R. W. Lovejoy, J. H. Colwell, D. F. Eggers, and G. D. Halsey, *J. Chem. Phys.*, **36**, 612 (1962).
41. D. A. Long, R. A. G. Carrington, and R. B. Gravenor, *Nature*, **196**, 371 (1962); F. A. Miller and W. K. Baer, *Spectrochim. Acta*, **19**, 73 (1963).
42. E. Lee and C. K. Wu, *Trans. Faraday Soc.*, **35**, 1366 (1939).
43. L. A. Woodward and M. J. Taylor, *J. Chem. Soc.*, **1962**, 407.
44. H. H. Claassen, B. Weinstock, and J. G. Malm, *J. Chem. Phys.*, **28**, 285 (1958).
45. R. E. Dodd, *Trans. Faraday Soc.*, **55**, 1480 (1959).

46. J. A. Creighton, *J. Chem. Soc.*, **1965**, 6589.
47. A. Sabatini, L. Sacconi, and V. Schettino, *Inorg. Chem.*, **3**, 1775 (1964).
48. H. Stammreich and R. Forneris, *Spectrochim. Acta*, **16**, 363 (1960).
49. I. W. Levin and C. V. Berney, *J. Chem. Phys.*, **44**, 2557 (1966); R. E. Dodd, L. A. Woodward, and H. L. Roberts, *Trans. Faraday Soc.*, **52**, 1052 (1956).
50. A. J. Downs, R. Schmutzler, and I. A. Steer, *Chem. Commun.*, **1966**, 221.
51. J. E. Griffiths, R. P. Carter, and R. R. Holmes, *J. Chem. Phys.*, **41**, 863 (1964).
52. G. M. Begun, W. H. Fletcher, and D. F. Smith, *J. Chem. Phys.*, **42**, 2236 (1965).
53. H. H. Claassen, B. Weinstock, and J. G. Malm, *J. Chem. Phys.*, **25**, 426 (1955).
54. L. A. Woodward and M. J. Ware, *Spectrochim Acta*, **19**, 775 (1963).
55. R. C. Lord, M. A. Lynch, W. C. Schumb, and E. J. Slowinskii, *J. Am. Chem. Soc.*, **72**, 522 (1950).
56. H. H. Claassen, and H. Selig, *J. Chem. Phys.*, **43**, 103 (1965).
57. H. Stammreich and O. Sala, *Z. Elektrochem.*, **64**, 741 (1960); **65**, 149 (1961).
58. S. F. A. Kettle and R. V. Parish, *Spectrochim Acta*, **21**, 1087 (1965).
59. G. R. Wilkinson, 'Low-frequency infrared spectroscopy', in *Infrared Spectroscopy and Molecular Structure* (Ed. M. Davies), Elsevier, Amsterdam, 1963.
60. J. L. Wood, *Quart. Rev. (London)*, **17**, 362 (1963).
61. F. F. Bentley, E. F. Wolfarth, N. E. Srp, and W. R. Powell, *Spectrochim. Acta*, **13**, 1 (1958).
62. H. A. Gebbie, *National Physical Laboratory Report, Basic Physics Group*, 1962.
63. H. L. Welsh, M. F. Crawford, T. R. Thomas, and G. R. Love, *Can. J. Phys.*, **30**, 577 (1952).
64. J. R. Nielsen, *J. Opt. Soc. Am.*, **40**, 89 (1950).
65. E. R. Lippincott and R. D. Fisher, *Anal. Chem.*, **26**, 435 (1954); E. R. Lippincott, J. P. Sibilia, and R. D. Fisher, *J. Opt. Soc. Am.*, **49**, 83 (1959).
66. H. Stammreich, *Spectrochim. Acta*, **8**, 41 (1956).
67. F. X. Powell, E. R. Lippincott, and D. Steele, *Spectrochim. Acta*, **17**, 880 (1961).
68. R. Forneris, *Thesis*, University of Sao Paulo, 1961.
69. F. T. King and E. R. Lippincott, *J. Opt. Soc. Am.*, **46**, 661 (1956).
70. N. S. Ham and A. Walsh, *J. Chem. Phys.*, **36**, 1096 (1962).
71. H. Stammreich, R. Forneris, and Y. Tavares, *Spectrochim. Acta*, **17**, 1173 (1961).
72. F. X. Powell, O. Fletcher, and E. R. Lippincott, *Rev. Sci. Instr.*, **34**, 36 (1963); J. A. Creighton, E. R. Lippincott, F. X. Powell, and D. G. Jones, *Develop. Appl. Spectry.*, **3**, 106 (1964).
73. R. C. C. Leite and S. P. S. Porto, *J. Opt. Soc. Am.*, **54**, 981 (1964); J. A. Koningstein and R. G. Smith, *J. Opt. Soc. Am.*, **54**, 1061 (1964).
74. M. Delhaye and M. Migeon, *Compt. Rend.*, **261**, 263 (1965).
75. M. Crunelle-Cras, and M. Delhaye, *Compt. Rend.*, **257**, 2823 (1963).
76. B. Schrader, *Z. Anal. Chem.*, **197**, 295 (1963); B. Schrader and W. Meier, *Z. Naturforsch.*, **21a**, 437 (1965).

77. J. R. Ferraro, *Spectrochim. Acta*, **20**, 901 (1964); H. Busey and O. L. Keller, *J. Chem. Phys.*, **41**, 215 (1964).
78. H. L. Welsh, E. J. Stansbury, J. Romanko, and T. Feldman, *J. Opt. Soc. Am.*, **45**, 338 (1955).
79. A. R. Gee and D. C. O'Shea, *Rev. Sci. Instr.*, **37**, 670 (1966); J. R. Ferraro, J. S. Ziomek, and K. Puckett, *Rev. Sci. Instr.*, **35**, 754 (1964).
80. G. J. Janz and S. C. Wait, *Quart. Rev.* (*London*), **17**, 225 (1963).
81. J. D. S. Goulden, *Spectrochim. Acta*, **15**, 657 (1959).
82. D. K. Huggins and H. D. Kaesz, *J. Am. Chem. Soc.*, **86**, 2734 (1964).
83. I. R. Beattie and T. Gilson, *J. Chem. Soc.*, **1964**, 3528.
84. D. Hartley and M. J. Ware, to be published.
85. D. M. Adams and H. A. Gebbie, *Spectrochim. Acta*, **19**, 925 (1963).
86. L. A. Woodward, *Spectrochim. Acta*, **19**, 1963 (1963).
87. J. C. Evans and H. J. Bernstein, *Can. J. Chem.*, **34**, 1127 (1956).
88. G. K. T. Conn and C. K. Wu, *Trans. Faraday Soc.*, **34**, 1483 (1938).
89. L. A. Woodward and J. R. Hall, *Spectrochim. Acta*, **16**, 654 (1960).
90. L. A. Woodward and M. J. Taylor, *J. Chem. Soc.*, **1963**, 4670.
91. J. P. Mathieu, *J. Phys. Rad.*, **16**, 219 (1955).
92. R. L. McCullogh, L. H. Jones, and G. A. Crosby, *Spectrochim. Acta*, **16**, 929 (1960).
93. I. Nakagawa and T. Shimanouchi, *Spectrochim. Acta*, **22**, 759 (1966).
94. I. Freund and R. S. Halford, *J. Chem. Phys.*, **43**, 3795 (1965).
95. I. Nakagawa and T. Shimanouchi, *Spectrochim, Acta*, **22**, 1707 (1966).
96. E. F. Barker and A. Adel, *Phys. Rev.*, **44**, 185 (1933).
97. W. B. Stewart and H. H. Nielsen, *Phys. Rev.*, **47**, 828 (1935).
98. D. M. Gage and E. F. Barker, *J. Chem. Phys.*, **7**, 455 (1939).
99. S. L. Gerhard and D. M. Dennison, *Phys. Rev.*, **43**, 197 (1932).
100. A. J. Downs, *Spectrochim. Acta*, **19**, 1165 (1963).
101. A. P. Ginsberg, 'Hydride complexes of the transition metals', in *Transition Metal Chemistry* (Ed. R. Carlin), Arnold, London, 1965.
102. J. Lewis and G. Wilkinson, *J. Inorg. Nucl. Chem.*, **6**, 12 (1958).
103. D. J. Hewkin and W. P. Griffith, *J. Chem. Soc.*, *A*, 472 (1966).
104. L. A. Woodward, J. R. Hall, R. N. Dixon, and N. Sheppard, *Spectrochim. Acta*, **14**, 249 (1959).
105. D. M. Adams and P. J. Chandler, *Chem. Ind.* (*London*), **1965**, 269.
106. L. J. Bellamy, *Spectrochim. Acta*, **14**, 192 (1959); L. J. Bellamy and P. E. Rogasch, *J. Chem. Soc.*, **1960**, 2218.
107. W. D. Horrocks and R. H. Mann, *Spectrochim. Acta*, **21**, 399 (1965).
108. L. A. McDougall and J. E. Kilpatrick, *J. Chem. Phys.*, **42**, 2311 (1965).
109. F. A. Miller, D. H. Lemmon, and R. E. Witkowski, *Spectrochim. Acta*, **21**, 1709 (1965).
110. H. H. Claassen, G. L. Goodman, J. G. Malm, and F. Schreiner, *J. Chem. Phys.*, **42**, 1229 (1965).
111. R. C. Lord, D. W. Robinson, and W. C. Schumb, *J. Am. Chem. Soc.*, **78**, 1327 (1956).
112. D. C. McKean, *Spectrochim. Acta*, **13**, 38 (1958).
113. D. C. McKean, R. Taylor, and L. A. Woodward, *Proc. Chem. Soc.*, **1959**, 321.

114. T. D. Goldfarb and S. Sujishi, *J. Am. Chem. Soc.*, **86**, 1679 (1964); E. A. V. Ebsworth, R. Taylor, and L. A. Woodward, *Trans. Faraday Soc.*, **55**, 211 (1959).
115. E. A. V. Ebsworth, *Chem. Commun.*, **1966**, 530.
116. E. A. V. Ebsworth, J. R. Hall, M. J. Mackillop, D. C. McKean, N. Sheppard, and L. A. Woodward, *Spectrochim. Acta*, **13**, 202 (1958); D. W. Robinson, *J. Am. Chem. Soc.*, **80**, 5924 (1958).
117. G. Davidson, E. A. V. Ebsworth, G. M. Sheldrick, and L. A. Woodward, *Chem. Commun.*, **1965**, 122.
118. S. Cradock, G. Davidson, E. A. V. Ebsworth, and L. A. Woodward, *Chem. Commun.*, **1965**, 515.
119. R. E. Hester and K. Jones, *Chem. Commun.*, **1966**, 317.
120. H. H. Claassen, C. L. Chernick, and J. G. Malm, *J. Am. Chem. Soc.*, **85**, 1927 (1963).
121. R. V. G. Evans and M. W. Lister, *Trans. Faraday Soc.*, **35**, 681 (1939).
122. E. Bergmann and L. Engel, *Z. Physik. Chem.*, **B15**, 377 (1932).
123. E. Weiss, *Z. Anorg. Allgem. Chem.*, **287**, 223 (1956).
124. W. G. Fateley and E. R. Lippincott, *Spectrochim. Acta*, **10**, 8 (1957).
125. M. F. O'Dwyer, *J. Mol. Spectry.*, **2**, 144 (1958).
126. H. Stammreich, O. Sala, and Y. Tavares, *J. Chem. Phys.*, **30**, 856 (1959).
127. W. F. Edgell, W. E. Wilson, and R. Summitt, *Spectrochim. Acta*, **19**, 863 (1963).
128. H. P. Fritz, *Advan. Organomet. Chem.*, **1**, 240 (1964); D. K. Huggins and H. D. Kaesz, *Progr. Solid-state Chem.*, **1**, 417 (1964).
129. E. R. Lippincott and R. D. Nelson, *Spectrochim. Acta*, **10**, 307 (1958).
130. H. P. Fritz, W. Luttke, H. Stammreich, and R. Forneris, *Spectrochim. Acta*, **17**, 1068 (1961).
131. H. P. Fritz and R. Schneider, *Chem. Ber.*, **93**, 1171 (1960).
132. G. L. Hardgrove and D. H. Templeton, *Acta Cryst.*, **12**, 28 (1959).
133. P. R. Bunker, *Mol. Phys.*, **9**, 247 (1965).
134. R. T. Bailey, E. R. Lippincott, and D. Steele, *J. Am. Chem. Soc.*, **87**, 5346 (1965).
135. S. G. Frankiss, F. A. Miller, H. Stammreich, and Th. Teixeira Sans, *Spectrochim. Acta*, **23A**, 543 (1967).
136. L. A. Woodward, *Phil. Mag.*, **18**, 823 (1934).
137. H. M. Gager, J. Lewis, and M. J. Ware, *Chem. Commun.*, **1966**, 616.
138. J. D. Corbett, *Inorg. Chem.*, **1**, 700 (1962).
139. H. Stammreich, K. Kawai, O. Sala, and P. Krumholz, *J. Chem. Phys.*, **35**, 2175 (1961).
140. D. Hartley and M. J. Ware, to be published.
141. E. L. Muetterties, R. E. Merrifield, H. C. Miller, W. H. Knoth, and J. R. Downing, *J. Am. Chem. Soc.*, **84**, 2506 (1962).
142. P. L. Goggin and R. J. Goodfellow, *J. Chem. Soc.*, *A*, **1966**, 1462.
143. I. R. Beattie and L. Rule, *J. Chem. Soc.*, **1964**, 3267.
144. D. M. L. Goodgame, private communication.
145. D. Forster and D. M. L. Goodgame, *J. Chem. Soc.*, **1965**, 1286; M. F. A. Dove, *Chem. Commun.*, **1965**, 23.
146. H. Hunziker, *J. Mol. Spectry*, **17**, 131 (1965); D. A. Long, R. B. Gravenor, and M. Woodger, *Spectrochim. Acta*, **19**, 937 (1963).

147. G. W. Chantry, A. Finch, P. N. Gates, and D. Steele, *J. Chem. Soc., A*, **1966**, 896.
148. R. M. Badger, *J. Chem. Phys.*, **2**, 128 (1934).
149. J. C. Decius, *J. Chem. Phys.*, **45**, 1069 (1966).
150. W. Gordy, *J. Chem. Phys.*, **14**, 305 (1946).
151. M. Scrocco, *Spectrochim. Acta*, **22**, 201 (1966).
152. L. H. Jones, *J. Chem. Phys.*, **41**, 856 (1964).
153. L. H. Jones, *J. Chem. Phys.*, **43**, 594 (1965).
154. F. A. Cotton and C. S. Kraihanzel, *J. Am. Chem. Soc.*, **84**, 4432 (1962).
155. H. Stammreich, K. Kawai, O. Sala, and P. Krumholz, *J. Chem. Phys.*, **35**, 2168 (1961).
156. C. A. Coulson, *J. Chem. Phys.*, **44**, 468 (1966).
157. G. Placzek, *Handbuch der Radiologie*, Akademische Verlag, Leipzig, 1934.
158. M. Wolkenstein, *Dokl. Akad. Nauk SSSR*, **32**, 185 (1941).
159. L. A. Woodward and D. A. Long, *Trans. Faraday Soc.*, **45**, 1131 (1949).
160. R. E. Hester and R. A. Plane, *Inorg. Chem.*, **3**, 513 (1964); G. W. Chantry and R. A. Plane, *J. Chem. Phys.*, **35**, 1027 (1961).
161. W. Beck, A. Melnikoff, and R. Stahl, *Angew. Chem.*, **4**, 692 (1965).
162. W. Hofman and H. Moser, *Ber. Bunsen Gesell. Phys. Chem.*, **68**, 129 (1964).
163. A. Albrecht, *J. Chem. Phys.*, **34**, 1476 (1961).
164. M. S. Child and H. C. Longuet-Higgins, *Phil. Trans. Roy. Soc.*, **254**, 259 (1961).
165. B. Weinstock and G. L. Goodman, *Advan. Chem. Phys.*, **9**, 169 (1965), and references cited therein.
166. C. J. Ballhausen and A. D. Liehr, *Acta Chem. Scand.*, **15**, 775 (1961); C. J. Ballhausen and J. de Meer, *J. Chem. Phys.*, **43**, 4304 (1965).
167. H. Hass and R. K. Sheline, *J. Am. Chem. Soc.*, **88**, 3219 (1966).
168. R. F. W. Bader and K. P. Huang, *J. Chem. Phys.*, **43**, 3760 (1965).
169. L. A. Woodward and M. J. Ware, *Spectrochim. Acta*, **20**, 711 (1964).
170. J. T. Hougen and S. Singh, *Proc. Roy. Soc.* (*London*), *Ser. A*, **227**, 193 (1964); J. Y. H. Chau, *J. Chem. Phys.*, **44**, 1708 (1966).

7

Electron Paramagnetic Resonance

E. KÖNIG

I. INTRODUCTION

A. Nature and Scope of Method

Electron paramagnetic resonance (e.p.r.) is a method of considerable utility in studies of all systems having uncompensated electronic spin. In essence, paramagnetic resonance deals with transitions between Zeeman levels which may occur whenever a paramagnetic substance is placed in a magnetic field and, at the same time, exposed to electromagnetic radiation. If the magnetic field strength is 3000 G, the resonant absorption occurs at about 0.3 cm^{-1} or 9000 Mc/s, and thus microwave techniques are usually applied in order to observe the transitions. However, the operation of e.p.r. spectrometers in the microwave range is more a matter of convenience than one of necessity. In fact, the first successful resonance study was made in the radiofrequency region by Zavoisky[1], who employed rather weak magnetic fields. Transition metal ions have been given particular attention from the beginning. Zavoisky's first measurements, for example, involved the study of Zeeman splittings in manganese salts. Soon afterwards, similar experiments were reported by Cummerow and Halliday[2], Bagguley and Griffiths[3], and Bleaney and Penrose[4], and the investigations were extended through the whole transition metal series. For a review of the early e.p.r. studies, see Bagguley and colleagues[5].

The most complete results are obtained, in general, from the resonance study of dilute single crystals. In this case, we observe the dependence of the spectrum on the orientation of the crystal with respect to the external magnetic field. Dilution is necessary in order to reduce magnetic interactions between different paramagnetic ions. If the crystal structure is known, the orbitals of the unpaired electrons may be determined as the final result of such study. It should be mentioned that methods have also been developed to obtain extensive information on the distribution of the unpaired electrons from a study of dilute polycrystalline samples and from the investigation of glasses prepared by the freezing of suitable solutions. If the crystal structure is not known, e.p.r. may be used to determine the number of magnetically inequivalent ions within the unit cell, the symmetry at the crystal sites of these ions, and similar things. Electron resonance is thus a useful supplement to crystallographic studies.

Paramagnetic resonance has provided the first reliable evidence for the dynamic Jahn–Teller effect in transition metal compounds[6,7]. At room temperature, the Jahn–Teller effect partially lifts the orbital degeneracy of the ^2E ground state in trigonally distorted hexahydrated copper(II) compounds, while at lower temperatures the action of the Jahn–Teller effect is replaced by static distortions of tetragonal symmetry.

The resonance signal is highly affected by the presence of odd isotopes of the paramagnetic ion. If the isotope has a nuclear spin I, the signal shows a hyperfine structure consisting of $2I + 1$ equally spaced lines. This fact has been used for determining the nuclear spin and the nuclear magnetic moment of various isotopes, e.g. ^{50}V, ^{57}Fe, ^{60}Co, ^{99}Ru, and ^{101}Ru. The size of the hyperfine splitting is different for the free ion and for the ion in a complex compound. From the relative change of the splitting, valuable information about the orbitals of the unpaired electrons may be obtained.

If the ligands comprise nuclei having spin and magnetic moments, an additional super-hyperfine structure is found. Such structure, especially if observed on single crystals, may provide the most complete knowledge about the distribution of the unpaired electrons throughout the molecule. Once the MO coefficients in the various molecular orbitals are determined, it becomes possible to assess the fraction of time each unpaired electron spends on the central ion and on the ligands, respectively. The first spectacular results of this sort were reported by Griffiths and Owen[8]. These authors have shown that in the $[IrCl_6]^{2-}$ ion the unpaired electron spends about 3% of the time on each Cl atom or a total of about 20% of the time on the ligands. In addition, these results provided the first direct proof of the presence of π bonding in a complex compound. A considerable number of compounds containing various transition metal ions have been studied in this way; selected examples are discussed in Section V.

B. Outline and Coverage of the Field

An essential prerequisite for the observation of paramagnetic resonance is that the compounds studied should be paramagnetic, since pure diamagnetic substances exhibit no such resonance. The following types of systems may thus be studied by this method:

1. atoms and molecules having an odd number of electrons as, for example, atomic hydrogen, molecules like NO, and free radicals like CH_3;

2. ions having partly filled inner electron shells (*d* or *f* shells), such as ions in the transition metal groups, the rare earths, and the actinides;

3. a small number of molecules with an even number of electrons but a resultant angular momentum (e.g. a triplet ground state), such as O_2;

4. molecules with a diamagnetic ground state but a low-lying triplet state which may be populated under certain conditions;

5. colour centres; and

6. metals and semiconductors, where resonance is caused by conduction electrons.

The resonance properties of any particular of the groups listed are interesting in one way or another. However, in this chapter, we shall limit our discussion essentially to compounds within group 2. This restriction is for the simple reason that, in the realm of inorganic chemistry, theoretical concepts such as ligand-field theory have been developed and successfully applied to this type of compounds. Extensive information about various structural and bonding properties may be deduced, as briefly outlined in Section I.A, if these theories are applied to experimental e.p.r. results. The paramagnetic resonance of free radicals in general has been treated by Ingram[9] and that of inorganic radicals by Atkins and Symons[9a]. The available experimental data were recently compiled by Fischer[10]. The e.p.r. of molecules in their lowest excited triplet states has been studied mainly on organic systems (cf. Brinen and coworkers[11] and the references listed therein). The resonance properties of colour centres and metals are rather different and thus beyond the scope of this contribution. Finally, a review of e.p.r. studies in semiconductors has been given by Ludwig and Woodbury[12].

C. Resonance Condition

Electronic states of free atoms and ions may usually be well characterized by the resultant angular momentum *J*. If such an atom or ion is paramagnetic, i.e. if it belongs to group 1 or 2 as listed above, and if it is placed in a magnetic field of strength *H*, its energy levels will be split; the energy of the resulting sublevels is given by

$$E = g\beta HM_J \tag{1}$$

In (1), *g* is the Landé factor, given by

$$g = 1 + \frac{J(J+1) + S(S+1) - L(L+1)}{2J(J+1)} \tag{2}$$

assuming for a free electron the value $g_s = 2.00229$, and β is the Bohr magneton*, given by

$$\beta = \frac{eh}{4\pi mc} = 0.92731 \times 10^{-20}\ \mathrm{erg/G}$$

e and m are the charge and mass of the electron, h is the Planck constant, and c is the velocity of light. Also, S and L are the spin and orbital angular momenta and M_J is the component of J along the field acting on the ion. The situation is illustrated in the lower part of Figure 1 for the most

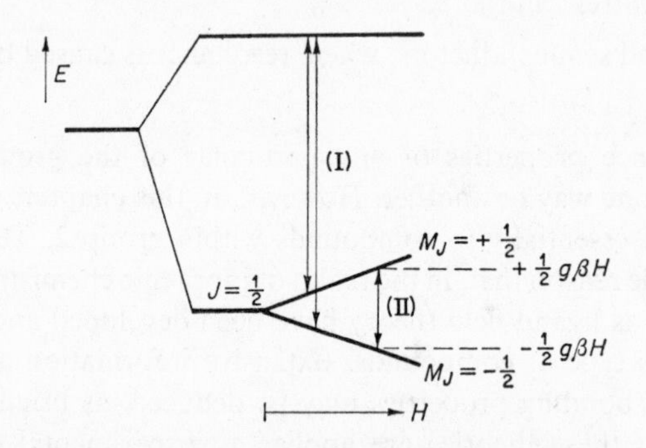

Figure 1. Splitting of an electronic ground state $J = \frac{1}{2}$ by a magnetic field. Zeeman transitions of electronic spectroscopy denoted by (I) correspond to energies of the order of 10,000 cm^{-1}. Paramagnetic resonance transitions (II) have energies of about 1 cm^{-1} or less.

simple case, $J = \frac{1}{2}$. An example is provided by the hydrogen atom in its ground state $^2S_{\frac{1}{2}}$ ($S = \frac{1}{2}$, $L = 0$). Clearly, this splitting is the reason for the Zeeman effect of electronic spectroscopy, where transitions between such Zeeman components of different electronic levels are observed (cf. Figure 1). In paramagnetic resonance, an alternating field of frequency ν is applied at right angles to H. According to the selection rule

$$\Delta M_J = \pm 1 \tag{3}$$

allowed magnetic dipole transitions are produced between the Zeeman components of the same, usually the lowest, electronic level. The energy separation of the components and thus the energy of the resonance transition is

$$\Delta E = h\nu = g\beta H \tag{4}$$

(the resonance condition). Transitions corresponding to $\Delta M_J = \pm 2$,

* In e.p.r. work, it sometimes is more convenient to list the Bohr magneton β as $e/4\pi mc^2 = 4.6688 \times 10^{-5}\ \mathrm{cm}^{-1}\ \mathrm{G}^{-1}$.

$\pm 3, \ldots$, which are generally not allowed, may be observed, however, in special cases with much lower intensity than $\Delta M_J = \pm 1$ transitions.

This simple picture may be considerably changed if the paramagnetic ion is present as a substituent in a diamagnetic solid or if it is at the centre of the ligand polyhedron in a coordination compound. The most important interactions are of three types: (*a*) exchange interaction between neighbouring magnetic dipoles, (*b*) ligand-field and spin–orbit coupling effects, and (*c*) hyperfine interactions. The three topics will be discussed separately in the following sections.

D. Exchange Interaction

Magnetic dipolar coupling and exchange effects between ions in a solid may radically change the lowest electronic levels and thus the resonance absorption. It seems that the assumption of no coupling between spins of neighbouring paramagnetic ions is never strictly correct for, at sufficiently low temperatures, many paramagnetic substances become antiferromagnetic. The transition to an antiferromagnetic state usually occurs at a well-defined temperature Θ_N, the Néel temperature. Θ_N is characterized by a maximum in the susceptibility curve and a pronounced change of the resonance spectrum. Thus paramagnetic resonance may contribute significantly to the study of exchange interactions.

Theoretically, the exchange interaction between the spins of electrons localized in orbitals, two of which are $\psi_i(r)$ and $\psi_j(r)$, may be described by the hamiltonian

$$\mathscr{H}_{ex} = \sum_{i,j} J_{ij}\mathbf{s}_i \cdot \mathbf{s}_j \tag{5}$$

where the exchange integral J_{ij} is negative for a ferromagnetic interaction and positive for an antiferromagnetic interaction, and where $\mathbf{s}_i$ and $\mathbf{s}_j$ are the spin quantum number operators for the respective electrons. The simplest antiferromagnetic system which occurs in compounds like copper(II) acetate monohydrate (cf. Section V) is that of a pair of ions each having a spin of $\frac{1}{2}$. In this case, $\mathscr{H}_{ex} = J_{12}\mathbf{s}_1 \cdot \mathbf{s}_2$, the pair has a singlet and a triplet state separated in energy by J_{12}, and the singlet lies lower if $J_{12} > 0$, which is the case for antiferromagnetic ordering.

However, since our main interest is the study of molecular properties rather than that of cooperative phenomena, it is imperative that exchange interactions be reduced efficiently. It is for this reason that paramagnetic resonance measurements are most frequently carried out on diluted single crystals, in very dilute solutions, or in polycrystalline samples diluted by a suitable diamagnetic substance. An isomorphous diamagnetic salt is usually used as diluent in growing single crystals. A dilution of the paramagnetic substance of one part in a thousand is in most cases sufficient to reduce any interaction between magnetic ions to a negligible amount.

E. Ligand-field Effects and Fine Structure

If paramagnetic coordination compounds are considered, the interaction of the central ion with the diamagnetic ligands has to be taken into account. These ligands set up strong internal electric fields which usually change the energy of the electronic levels of the ion and split them into a number of components. The number of such components and the magnitude of the splitting depend critically on the initial level, the symmetry of the electric

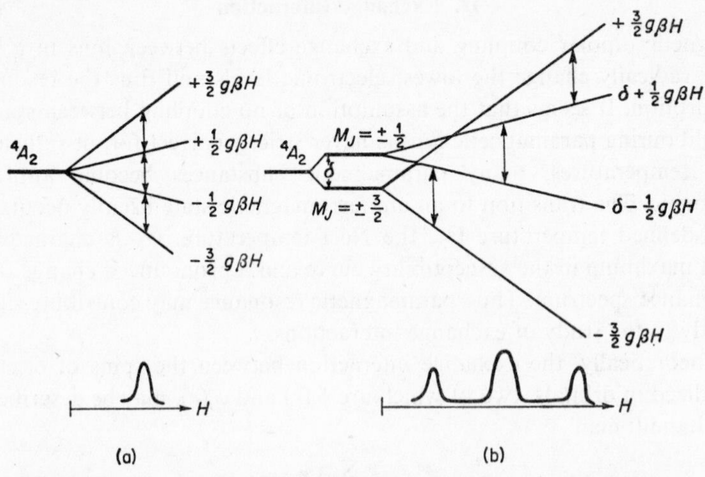

Figure 2. Lowest level $^4A_{2g}$ of a Cr^{3+} ion in an external magnetic field (a) assuming perfect octahedral symmetry, and (b) assuming an additional small axial field. The zero-field splitting is denoted by δ. Energies of the individual components are listed on the right.

field, and on its strength. The situation is similar to the Stark effect in electronic spectroscopy. There is, however, a difference in that the symmetry of the field, which depends on the local disposition of the ligands around the paramagnetic ion, may be quite complicated.

If the ligand field to which the central ion in a complex is subject is of cubic symmetry, the g value is isotropic and equation (4) is still valid. Since the g value usually deviates from the free-electron value g_s, it is determined from experiment according to $g = h\nu/\beta H$. Such a field of high symmetry may leave the ground state degenerate, even if spin–orbit coupling is included. An example is provided by the ground state $^4A_{2g}$ of a Cr^{3+} ion in an octahedral field (Figure 2a). The four-fold spin degeneracy is lifted only by an external magnetic field. Since the components of the original level diverge linearly with increasing field strength H, the three possible transitions coincide and one single line is observed. However, the degeneracy of the spin levels may be lifted by the combined action of an axial field component and spin–orbit coupling in absence of any magnetic field. The mechanism responsible for this splitting is a second-order spin–orbit coupling interaction. In the case of an A_{2g} ground state, the M_J levels are connected via spin–orbit coupling to split orbital components of energetically higher T_{2g} states. The separation of the M_J levels in absence of a magnetic field varies from zero to a few cm^{-1} and is called 'zero-field splitting'. Figure 2b illustrates the situation for the $^4A_{2g}$ ground state of a Cr^{3+} ion. The two doublets are split again in a magnetic field and there are three $\Delta M_J = \pm 1$ transitions. Thus three lines are observed in the resonance spectrum instead of one line as in perfect cubic symmetry. The apparent splitting in the spectrum is usually referred to as 'fine structure'.

Another consequence of low-symmetry ligand-field components is that the g value becomes anisotropic. Since the splitting of the M_J levels varies with direction, the g value is now orientation dependent. Therefore, g is in general a tensor quantity which reflects the anisotropy of the ligand field.

F. Hyperfine Interactions

If the metal ion which contains unpaired electrons also possesses a nuclear spin I, each level specified by M_J is $(2I + 1)$-fold degenerate with respect to the nuclear spin. This degeneracy is lifted if an external magnetic field is applied and thus each M_J level is split into $2I + 1$ components. Since the applied magnetic field is much stronger than the magnetic field due to nuclear spin, the magnitude of the splitting is, in general, *independent* of the magnetic field. However, the amount of the splitting does depend upon the magnetic moment of the nucleus and upon the density of the unpaired electrons at the position of the nucleus. When the applied magnetic field becomes extremely weak, the nuclear spin angular momentum **I** and the total angular momentum **J** couple to give

$$\mathbf{F} = \mathbf{I} + \mathbf{J} \tag{6}$$

In this case, the splitting does depend on the strength of the external magnetic field.

Transitions are allowed according to $\Delta M_J = \pm 1$ and $\Delta M_I = 0$. Thus, for each fine structure line, there are $2I + 1$ equally spaced and equally intense lines in the resonance spectrum, an observation usually referred to as 'hyperfine structure'. The resonance condition (4) has to be extended to

$$h\nu = g\beta H M_J + \sum A M_I M_J \tag{7}$$

the position of the individual hyperfine lines is thus given by

$$H = \frac{h\nu}{g\beta} + \frac{AM_I}{g\beta} \tag{8}$$

and the distance between each two lines is

$$\Delta H = \frac{A}{g\beta} \tag{9}$$

Here, A is the hyperfine splitting constant. Figure 3 illustrates the hyperfine splitting for the case $J = \frac{1}{2}, I = \frac{3}{2}$.

Hyperfine splitting may arise not only from the nuclear spin of the central metal ion in a complex compound, but also from ligand nuclei. This occurs only if the unpaired electron has a finite probability density

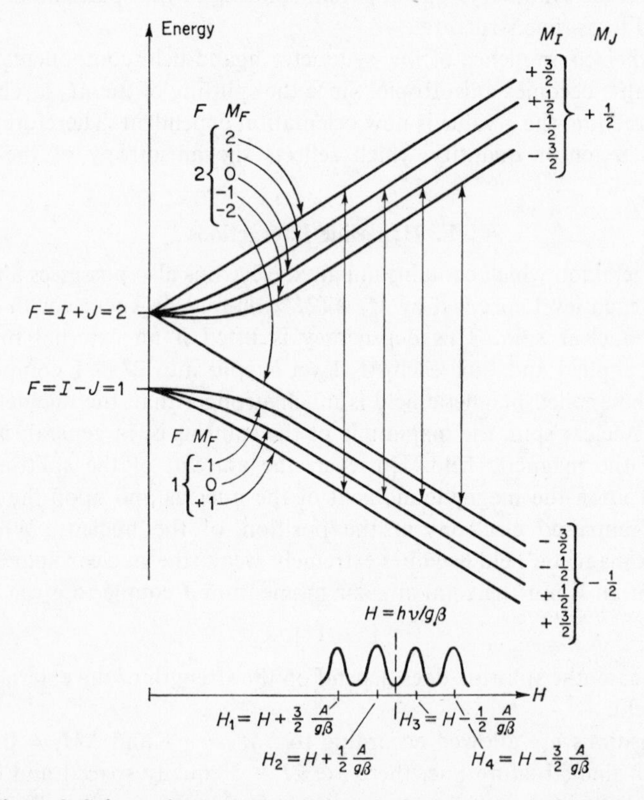

Figure 3. Schematic energy level diagram for $J = \frac{1}{2}, I = \frac{3}{2}$. Allowed transitions correspond to $\Delta M_J = \pm 1, \Delta M_I = 0$. Centre of spectrum at $H = h\nu/g\beta$. Positions of individual hyperfine lines at fields H_1, H_2, H_3, and H_4.

at the position of the corresponding ligand nucleus. The resulting additional splitting of lines in the paramagnetic resonance spectrum is called 'super hyperfine splitting'.

II. BASIC THEORY

A. The Free Ion

The effect of ligand fields is usually studied as a perturbation of the energy levels of the free ion. It is therefore convenient to start with a discussion of the appropriate free-ion hamiltonian. The most important part of the hamiltonian, including all interactions within the free ion insofar as they are independent of the spin variables, is given by

$$\mathscr{H}_0 = \sum_{k=1}^{n} \left(\frac{\mathbf{p}_k^2}{2m} - \frac{Ze^2}{r_k} \right) + \sum_{k>j=1}^{n} \frac{e^2}{r_{kj}} \tag{10}$$

The three terms in (10) describe the kinetic energy, the electron-core potential, and the electron–electron repulsion, respectively. Here, $\mathbf{p}_k$ is the linear momentum of the kth electron, r_k its distance from the nucleus, and r_{kj} the distance between the kth and jth electrons. Also, m is the mass and $-e$ the charge of the electron, $+Ze$ the charge on the nucleus, and the summation extends over all n electrons.

Assuming that the basic orbitals of the free atom or ion are well approximated by hydrogen-like orbitals, the aufbau principle is used to assign the n electrons to the orbitals in order of increasing energy (i.e. in the sequence 1*s*, 2*s*, 2*p*, 3*s*, 3*p*, 3*d*, etc.). In this procedure, the characteristic degeneracy numbers (viz. 1, 3, 5 for *s*, *p*, *d* orbitals, respectively, etc.) and the Pauli principle have to be taken into account. The total energy of the resulting electron configuration is given by the sum of the energies of the occupied orbitals.

The orbital energies obtained are modified by the electron–electron repulsion term in (10). In Russell–Saunders coupling, different electronic terms result, each one being characterized by definite values of L and S. The term energies may be expressed as linear functions of certain radial integrals F_k ($k = 0, 2, 4$ for d electrons) or Racah parameters A, B, C, defined as

$$\begin{aligned} A &= F_0 - 49F_4 \\ B &= F_2 - 5F_4 \\ C &= 35F_4 \end{aligned} \tag{11}$$

Numerical values for the repulsion parameters are usually assessed empirically by fitting the resulting energy expressions to the optical spectra. An alternative method consists of directly calculating the integrals F_k, using non-empirical (self-consistent field) radial functions.

Obviously (10) is not the complete hamiltonian of the free ion. Thus, each electron has a magnetic moment associated with its spin. On the other hand, a magnetic moment is also produced by the orbital motion of the electrons. The spin–orbit interaction, acting between these moments, introduces an extra term into (10). In addition, the nucleus also may have a magnetic moment and so additional terms representing the magnetic interaction between electrons and the nucleus may have to be taken into account.

However, the effect of electric fields in the environment of the metal ion is usually larger, in terms of energy, than the various magnetic interactions, but comparable to or little smaller than the Coulomb interaction in (10). The discussion of the magnetic and additional effects will therefore be deferred until the ligand-field interaction has been incorporated into the theory.

More details about the basic theory of free ions may be found in the classical book of Condon and Shortley[13], as well as in the treatments by Griffith[14] and by Ballhausen[15].

B. Theory of the Ligand Field

In paramagnetic ions of the transition metals, the rare earths, and the actinides, the unpaired electrons occupy partly filled *d* or *f* shells. If the positively charged ion is at the centre of a polyhedron of negatively charged diamagnetic ions or of electric dipoles with their negative ends pointing inwards, the behaviour of these electrons, originally represented by (10), will be modified. The change is effected by the strong electric fields set up by the surrounding ligands. Such a situation is encountered in complex ions, e.g. $[CuCl_4]^{2-}$ or $[Cr(H_2O)_6]^{3+}$, as well as in metal ions present as substituents in the lattice of a salt-like compound, viz. Fe^{2+} in MgO. The problem is therefore to determine the energy levels of such an ion in a Stark field of certain strength and symmetry. The additional term describing the effect of the ligand field may be written as

$$\mathscr{H}_{\mathrm{LF}} = \sum_k -eV(x_k,y_k,z_k) \tag{12}$$

where V is the potential of the electric field set up by the ligands and x_k, y_k, z_k are the coordinates of the kth electron of the partly filled shell.

In calculating the effect of $\mathscr{H}_{\mathrm{LF}}$ on the levels by perturbation theory, it is important to know how the effect of the electrostatic potential compares in magnitude with the interelectronic repulsions in (10), namely with $\mathscr{H}_{\mathrm{e}} = \sum_{k>j} e^2/r_{kj}$ as well as with the effect of spin–orbit coupling represented by $\mathscr{H}_{LS}$ (see Section II.D). Three cases have been distinguished in this respect:

1. $\mathscr{H}_e > \mathscr{H}_{LS} > \mathscr{H}_{LF}$ rare earths and actinide compounds
2. $\mathscr{H}_e > \mathscr{H}_{LF} > \mathscr{H}_{LS}$ weak-field case } transition metal
3. $\mathscr{H}_{LF} > \mathscr{H}_e > \mathscr{H}_{LS}$ strong-field case } compounds

Thus, in rare earth and actinide compounds, ligand-field splittings are small compared to the separations between individual multiplet levels. Both the interelectronic repulsion $\mathscr{H}_e$ and the spin–orbit interaction $\mathscr{H}_{LS}$ have to be diagonalized before the influence of the ligand field is taken into account in a perturbation calculation. In weak ligand fields, the field splittings are comparable to intervals between the multiplets. $\mathscr{H}_{LF}$ is considered as a perturbation on the terms of the free ion which are obtained from (10), and spin–orbit coupling is introduced only after that. This case is encountered in most compounds of the first transition metal series. Since the bulk of available paramagnetic resonance results belongs to this class, the weak-field case will be subsequently discussed in more detail. Finally, in strong ligand fields, $\mathscr{H}_{LF}$ is comparable with the interelectronic repulsion energy, i.e. of the order of 10^4 cm^{-1}. Some compounds of the first transition metal series, sometimes called 'covalent' compounds, and most compounds of the second and third series, belong to this class.

In calculating the ligand-field potential V contained in (12), it is assumed that the orbitals of the paramagnetic ion do not overlap the surrounding ligands which are regarded as point charges. Consequently, the potential V is supposed to satisfy the Laplace equation $\Delta V = 0$ and can therefore be expanded in a series of spherical harmonics

$$V = \sum_n \sum_{m=n}^{-n} A_n^m r^n Y_n^m(\theta_k, \varphi_k) = \sum_n \sum_m V_n^m \tag{13}$$

Here, the harmonics $Y_n^m(\theta, \varphi)$ are normalized to unity and defined as

$$Y_n^m(\theta, \varphi) = (-1)^n \left[\frac{2n+1}{2} \frac{(n-|m|)!}{(n+|m|)!}\right]^{\frac{1}{2}} \frac{1}{\sqrt{2\pi}} P_n^{|m|}(\cos\theta) e^{im\varphi} \tag{14}$$

where

$$P_n^{|m|}(x) = \frac{(1-x^2)^{m/2}}{2^n n!} \frac{d^{n+m}}{dx^{n+m}} (x^2 - 1)^n \tag{15}$$

The number of terms in the expansion (13) can be reduced considerably by using orthogonality relations and symmetry arguments. Firstly, for d electrons, only terms up to $n = 4$ and, similarly, for f electrons, only terms up to $n = 6$ need to be considered. The reason is that in calculating matrix elements such as $\int \psi_i^{l*} V_n^m \psi_i^l$, the wave functions ψ_i^l can also be expanded in spherical harmonics, $\psi_i^l \sim R(r_i) Y_l^l(\theta_i, \varphi_i)$, and thus, by orthogonality, the matrix element is zero for all potentials having $n > 2l$. Secondly, all terms for which n is odd may be dropped. The reason is that again the relevant matrix elements are zero, since the product $\psi_i^{\dagger}\psi_i^{\dagger}$ is unchanged by inversion of the coordinates, whereas the potential V reverses sign. Thirdly, the term for $n = 0$ is a constant and therefore of no significance with respect to the optical and magnetic properties. Finally, since V is real, it follows that $A_n^m = (A_n^{-m})^*$.

A few symmetry properties of the surface spherical harmonics may be helpful in deriving the potential expression for a special disposition of ligands. Any surface harmonic $Y_n^m(\theta, \varphi)$ shows no dependence on φ and therefore has axial symmetry when $m = 0$. On the other hand, if $m = \pm 4$, $Y_n^m(\theta, \varphi)$ has tetragonal symmetry, if $m = \pm 3$, trigonal symmetry, if $m = \pm 6$, hexagonal symmetry, and finally if $m = \pm 2$, rhombic symmetry. Any expression in $Y_n^m(\theta, \varphi)$ containing terms of different symmetry is said to have an overall symmetry corresponding to the highest symmetry common to all terms. For a field of cubic symmetry, the form which V takes is dependent on the direction which is chosen for the polar axis. If the polar axis (z axis) coincides with a four-fold axis of symmetry, the potential takes the form

$$V_{\text{cub}} = A_4^0 r^4 \{Y_4^0(\theta, \varphi) + \sqrt{\tfrac{5}{14}}\,[Y_4^4(\theta, \varphi) + Y_4^{-4}(\theta, \varphi)]\} \tag{16}$$

However, if the polar axis is taken along the (111) direction, i.e. if it is a three-fold axis of symmetry, the potential has to be written as

$$V_{\text{cub}} = D_4 r^4 \{Y_4^0(\theta, \varphi) + \sqrt{\tfrac{10}{7}}\,[Y_4^3(\theta, \varphi) + Y_4^{-3}(\theta, \varphi)]\} \tag{17}$$

where $D_4 = \frac{2}{3}A_4^0$. A tetragonal potential may be represented by the sum of an axial term and a cubic term corresponding to the expression (16). Likewise, a trigonal potential may be written as the sum of an axial term and a cubic term of the form (17). Sometimes preference is given to a potential expressed in cartesian coordinates. In this case, (16) becomes

$$V_{\text{cub}} = C_4(x^4 + y^4 + z^4 - \tfrac{3}{5}r^4) \tag{18}$$

where $C_4 = (15/4\sqrt{\pi})A_4^0$. Potential functions for various crystal symmetries are listed by Low[16].

The energy due to the ligand field is usually calculated by a first-order perturbation procedure using (12) as perturbation operator. The effect of a cubic field, represented by the potential (16), (17), or (18), may be studied in a more qualitative way, considering a single d electron. In the free transition metal ion, the d electron occupies a set of five degenerate orbitals. Dropping in $d_{m_l} = R(r)Y_2^{m_l}$ the common factor $\sqrt{5/4\pi}$ as well as the radial part $R(r)$, and writing the spherical harmonics in cartesian coordinates, the angular functions of d electrons are

$$
\begin{aligned}
d_0 &= Y_2^0 = \tfrac{1}{2}3z^2 - r^2 \\
d_1 &= Y_2^1 = -\sqrt{\tfrac{3}{2}}(x + iy)z \\
d_{-1} &= Y_2^{-1} = \sqrt{\tfrac{3}{2}}(x - iy)z \\
d_2 &= Y_2^2 = \sqrt{\tfrac{3}{8}}(x + iy)^2 \\
d_{-2} &= Y_2^{-2} = \sqrt{\tfrac{3}{8}}(x - iy)^2
\end{aligned}
\tag{19}
$$

Since the cubic potential (16) contains only terms having $m = 0$ or ± 4, the state d_0 should have non-zero matrix elements to no other states than those which have $m = 0$ or ± 4. For d electrons, $l = 2$, the highest $m = \pm 2$, and thus only the diagonal element of d_0 will exist, making d_0 an eigenfunction in cubic symmetry. The same result is obtained for d_1 and d_{-1}, whereas d_2 and d_{-2} have interconnecting matrix elements, thus making $(d_2 + d_{-2})/\sqrt{2}$ and $(d_2 - d_{-2})/\sqrt{2}$ suitable eigenfunctions. For many purposes it is more convenient to use real functions which are obtained as linear combinations of (19):

$$
\begin{aligned}
d_{z^2} &= d_0 = \tfrac{1}{2}(3z^2 - r^2) \\
d_{x^2-y^2} &= (d_2 + d_{-2})/\sqrt{2} = (\sqrt{3}/2)\,(x^2 - y^2) \\
d_{xy} &= (d_2 - d_{-2})/i\sqrt{2} = \sqrt{3}\,(xy) \\
d_{xz} &= -(d_1 - d_{-1})/\sqrt{2} = \sqrt{3}\,(xz) \\
d_{yz} &= -(d_1 + d_{-1})/i\sqrt{2} = \sqrt{3}(yz)
\end{aligned}
\tag{20}
$$

In an octahedral field, the d_γ orbitals (d_{z^2} and $d_{x^2-y^2}$) which have their lobes of electron density pointing towards the negatively charged ligands will have a larger interaction than the d_ϵ orbitals (d_{xy}, d_{xz}, and d_{yz}) concentrated between the x, y, and z axes. The degeneracy of d orbitals will be thus partially lifted into a lower-lying orbital triplet d_ϵ (or t_{2g}) and a higher-lying doublet d_γ (or e_g) with an energy separation usually called Δ or $10Dq$. From optical spectra, $10Dq$ is of the order of 10,000 cm^{-1} for divalent and 20,000 cm^{-1} for trivalent cations.

More details may be found in this book in the chapter on electronic absorption spectroscopy by Schmidtke. For supplementary reading, the texts of Griffith[14] and Ballhausen[15] as well as several other recent treatments[17,18,18a] may be consulted.

A qualitative account about the number of Stark levels into which a given energy term is split in a ligand field may be obtained by the group theoretical methods introduced by Bethe[19]. In the free ion, each term, which is characterized by a value L of the angular momentum, induces a $(2L + 1)$-dimensional representation D^L of the rotation group. If the ion is placed in a crystal, this representation becomes reducible due to the lowering of symmetry. The reduction of D^L yields in general several irreducible representations Γ_n of the symmetry group of the crystal, each of which corresponds to one sublevel in the crystal. The resulting levels may thus be characterized by S (provided S is a good quantum number) and the ligand-field quantum number Γ_n. Table 1 lists the resolution of D^L into irreducible representations of cubic symmetry. More details about application of group theory may be found in various textbooks [14-16].

Table 1. Resolution of D^L into irreducible representations of cubic symmetry.

Representation D^L of the rotation group		Corresponding irreducible representations of cubic symmetry	
L	Spectroscopic notation	Notation according to Bethe	Notation according to Mulliken
0	S	Γ_1	A_1
1	P	Γ_4	T_1
2	D	$\Gamma_3 + \Gamma_5$	$E + T_2$
3	F	$\Gamma_2 + \Gamma_4 + \Gamma_5$	$A_2 + T_1 + T_2$
4	G	$\Gamma_1 + \Gamma_3 + \Gamma_4 + \Gamma_5$	$A_1 + E + T_1 + T_2$
5	H	$\Gamma_3 + 2\Gamma_4 + \Gamma_5$	$E + 2T_1 + T_2$
6	I	$\Gamma_1 + \Gamma_2 + \Gamma_3 + \Gamma_4 + 2\Gamma_5$	$A_1 + A_2 + E + T_1 + 2T_2$

C. Calculation of Matrix Elements and Energies

This section will present a short account of calculation methods for matrix elements of the ligand field. A more extensive treatment may be found in the outlines by Watanabe [20] and by Polo [21].

The calculation of matrix elements of the ligand field is particularly simple if we use the method of operator equivalents introduced by Stevens [22], Elliott [23,24], and Judd [25]. The method is based on the fact that the matrix elements of any operator **T** are proportional, within a given level, to the matrix elements of the angular momentum operator **J**. Consider the set of spherical harmonics $Y_n^m(\theta, \varphi)$ which forms the basis of an irreducible representation of the $(2n + 1)$-dimensional rotation group. Since each potential V may be expressed either as a function of the $Y_n^m(\theta, \varphi)$ or as a function of the cartesian coordinates of the electrons, it may be associated with an equivalent operator, i.e. an analogous function of the operators J_x, J_y, J_z which has the same transformation properties. The functions $\sum_i (x_i^2 - y_i^2)$ and $\sum_i (3z_i^2 - r_i^2)$, for example, correspond to the operators $J_x^2 - J_z^2$ and $3\ J_z^2 - J(J + 1)$. In constructing such equivalent operators, proper regard has to be given to the commutation rules of these operators, since J_x, J_y, and J_z do not commute, whereas x, y, and z do. The function $\sum_i x_i y_i$ therefore corresponds to the operator $\frac{1}{2}\{J_x J_y + J_y J_x\}$ or, more generally, the operator equivalent to the expression $\sum_i x_i^k y_i^l z_i^m$ is obtained as the arithmetic means of the $(k + l + m)!/k!\,l!\,m!$ possible permutations of k operators J_x, l operators J_y and m operators J_z. Let us, as an example, construct the operator equivalent to the cubic potential (18) within the manifold spanned by the $(2L + 1)$ orbital states of given L, S, and M_s. It is

$$\begin{aligned}\sum_i (x_i^4 + y_i^4 + z_i^4 - \tfrac{3}{5}r_i^4) \\ &= \beta\langle r^4\rangle\{L_x^4 + L_y^4 + L_z^4 - \tfrac{1}{5}L(L+1)[3L(L+1) - 1]\} \\ &= \frac{\beta\langle r^4\rangle}{20}\{35L_z^4 - 30L(L+1)L_z^2 + 25L_z^2 - 6L(L+1) \\ &\qquad + 3L^2(L+1)^2\} + \frac{\beta\langle r^4\rangle}{8}\{L_+^4 + L_-^4\}\end{aligned} \tag{21}$$

where $L_\pm = L_x \pm iL_y$. Since the radial part $R(r)$ of the functions is usually unknown, the following abbreviation has been used:

$$\langle r^n\rangle = \int [R(r)]^2 r^n r^2 \, dr \tag{22}$$

Thus, instead of the direct calculation of matrix elements of the ligand-field potential, the more simple calculation of matrix elements of polynomials in J_x, J_y, and J_z may be performed. The proportionality constant β in (21) and similarly the constants α and γ which occur in operator equivalent expressions are determined by evaluating one single matrix element in both perturbation matrices. For d^n electron ions

$$\alpha = \mp \frac{2(2l + 1 - 4S)}{(2l - 1)(2l + 3)(2L - 1)}$$

$$\beta = \mp \frac{3(2l + 1 - 4S)[-7(l - 2S)(l - 2S)(l - 2S + 1) + 3(l - 1)(l + 2)]}{(2l - 3)(2l - 1)(2l + 3)(2l + 5)(L - 1)(2L - 3)} \tag{23}$$

where the minus sign applies for $n < 5$ and the plus sign for $n > 5$. It should be noted that the operator equivalent method cannot be used between states differing in J because then all products of components of **J** have zero matrix elements.

Table 2 gives the linearly independent operator equivalents which correspond to polynomials of degrees $n = 2$, 4, and 6. The transformation

Table 2. Operator equivalents within a manifold for which J is constant.

(Y_2^0)	$\sum (3z^2 - r^2) = \alpha\langle r^2\rangle[3J_z^2 - J(J+1)]$
(Y_2^1)	$\sum xz = (\alpha\langle r^2\rangle/2)[J_xJ_z + J_zJ_x] = (\alpha\langle r^2\rangle/4)[J_z(J_+ + J_-) + (J_+ + J_-)J_z]$
(Y_2^2)	$\sum (x^2 - y^2) = (\alpha\langle r^2\rangle/2)[J_+^2 + J_-^2]$
(Y_4^0)	$\sum (35z^4 - 30r^2z^2 + 3r^4) = \beta\langle r^4\rangle[35J_z^4 - 30J(J+1)J_z^2 + 25J_z^2 - 6J(J+1) + 3J^2(J+1)^2]$
(Y_4^1)	$\sum (7z^2 - 3r^2)xz = (\beta\langle r^4\rangle/4)\{[7J_z^2 - 3J(J+1)J_z - J_z](J_+ + J_-) + (J_+ + J_-)[7J_z^2 - 3J(J+1)J_z - J_z]\}$
(Y_4^2)	$\sum (7z^2 - r^2)(x^2 - y^2) = (\beta\langle r^4\rangle/4)\{[7J_z^2 - J(J+1) - 5](J_+^2 + J_-^2) + (J_+^2 + J_-^2)[7J_z^2 - J(J+1) - 5]\}$
(Y_4^3)	$\sum (x^2 - 3y^2)xz = (\beta\langle r^4\rangle/4)[J_z(J_+^3 + J_-^3) + (J_+^3 + J_-^3)J_z]$
(Y_4^4)	$\sum (x^4 - 6x^2y^2 + y^4) = (\beta\langle r^4\rangle/2)(J_+^4 + J_-^4)$
(Y_6^0)	$\sum (231z^6 - 315r^2z^4 + 105r^4z^2 - 5r^6) = \gamma\langle r^6\rangle[231J_z^6 - 315J(J+1)J_z^4 + 735J_z^4 + 105J^2(J+1)^2J_z^2 - 525J(J+1)J_z^2 + 294J_z^2 - 5J^3(J+1)^3 + 40J^2(J+1)^2 - 60J(J+1)]$
(Y_6^1)	$\sum (33xz^5 - 30xz^3r^2 + 5r^4xz) = (\gamma\langle r^6\rangle/4)\{[33J_z^5 - 30J(J+1)J_z^3 + 15J_z^3 + 5J^2(J+1)^2J_z - 10J(J+1)J_z + 12J_z] \times (J_+ + J_-) + (J_+ + J_-)[33J_z^5 - 30J(J+1)J_z^3 + 15J_z^3 + 5J^2(J+1)^2J_z - 10J(J+1)J_z + 12J_z]\}$

(Y_6^2) $\quad \sum (16z^4(x^2 - y^2) - 16(x^4 - y^4)z^2 + x^6 + x^4y^2 - y^4x^2 - y^6) = (\gamma\langle r^6\rangle/4)\{[33J_z^4 - 18J(J+1)J_z^2 - 123J_z^2 + J^2(J+1)^2 + 10J(J+1) + 102] \times (J_+^2 + J_-^2) + (J_+^2 + J_-^2) \times [33J_z^4 - 18J(J+1)J_z^2 - 123J_z^2 + J^2(J+1)^2 + 10J(J+1) + 102]\}$

(Y_6^3) $\quad \sum (11z^2 - 3r^2)(x^2 - 3y^2)xz = (\gamma\langle r^6\rangle/4)\{[11J_z^3 - 3J(J+1)J_z - 59J_z](J_+^3 + J_-^3) + (J_+^3 + J_-^3)[11J_z^3 - 3J(J+1)J_z - 59J_z]\}$

(Y_6^4) $\quad \sum (11z^2 - r^2)(x^4 - 6x^2y^2 + y^4) = (\gamma\langle r^6\rangle/4)\{[11J_z^2 - J(J+1) - 38](J_+^4 + J_-^4) + (J_+^4 + J_-^4)[11J_z^2 - J(J+1) - 38]\}$

(Y_6^5) $\quad \sum (x^5z - 10x^3y^2z + 5xy^4z) = (\gamma\langle r^6\rangle/4)[J_z(J_+^5 + J_-^5) + (J_+^5 + J_-^5)J_z]$

(Y_6^6) $\quad \sum (x^6 - 15x^4y^2 + 15x^2y^4 - y^6) = (\gamma\langle r^6\rangle/2)[J_+^6 + J_-^6]$

$$\sum (x^4 + y^4 + z^4 - \tfrac{3}{5}r^4) = \beta\langle r^4\rangle[J_x^4 + J_y^4 + J_z^4 - \tfrac{1}{5}J(J+1)(3J^2 + 3J - 1)]$$
$$= (\beta\langle r^4\rangle/20)[35J_z^4 - 30J(J+1)J_z^2 + 25J_z^2 - 6J(J+1) + 3J^2(J+1)^2] + (\beta\langle r^4\rangle/8)[J_+^4 + J_-^4$$

$$\sum (x^3y - xy^3) = -(i\beta\langle r^4\rangle/8)[J_+^4 - J_-^4]$$

$$\sum z(x^3 - 3xy^2) = (\beta\langle r^4\rangle/4)[J_z(J_+^3 + J_-^3) + (J_+^3 + J_-^3)J_z]$$

$$\sum z(y^3 - 3x^2y) = (i\beta\langle r^4\rangle/4)[J_z(J_+^3 - J_-^3) + (J_+^3 - J_-^3)J_z]$$

Here $J_+ = J_x + iJ_y$ and $J_- = J_x - iJ_y$. For manifolds where L is constant, J has to be replaced by the corresponding L.

properties of the polynomials have been indicated by the symbol of the corresponding spherical harmonic Y_n^m. A few other useful operator equivalents have been appended. The summation indices in the polynomials have been omitted throughout. Values of the proportionality constants α, β, and γ for the various d^n and f^n ground states are listed in Table 3.

An alternative method of calculating matrix elements may be based on vector coupling relations. Since the ligand field potential V_n^m transforms

Table 3. Numerical values of coefficients α, β, and γ for the various d^n and f^n ground states.

Ground state		α	β	γ
d^1	^2D	$-\frac{2}{21}$	$\frac{2}{63}$	0
d^2	^3F	$-\frac{2}{105}$	$-\frac{2}{315}$	0
d^3	^4F	$\frac{2}{105}$	$\frac{2}{315}$	0
d^4	^5D	$\frac{2}{21}$	$-\frac{2}{63}$	0
d^5	^6S	0	0	0
d^6	^5D	$-\frac{2}{21}$	$\frac{2}{63}$	0
d^7	^4F	$-\frac{2}{105}$	$-\frac{2}{315}$	0
d^8	^3F	$\frac{2}{105}$	$\frac{2}{315}$	0
d^9	^2D	$\frac{2}{21}$	$-\frac{2}{63}$	0
f^1	$^2F_{\frac{5}{2}}$	$-\frac{2}{35}$	$\frac{2}{7}\cdot 45$	0
f^2	$^3H_4$	$-\frac{52}{11}\cdot 15^2$	$-\frac{4}{55}\cdot 33\cdot 3$	$17\cdot\frac{16}{7}\cdot 11^2\cdot 13\cdot 5\cdot 3^4$
f^3	$^4I_{\frac{9}{2}}$	$-\frac{7}{33}\cdot 33$	$-8\cdot\frac{17}{11}\cdot 11\cdot 13\cdot 297$	$-17\cdot 19\cdot\frac{5}{13^2}\cdot 11^3\cdot 3^3\cdot 7$
f^4	$^5I_4$	$\frac{14}{11}\cdot 11\cdot 15$	$\frac{952}{13}\cdot 3^3\cdot 11^3\cdot 5$	$\frac{7584}{11^2}\cdot 13^2\cdot 3\cdot 63$
f^5	$^6H_{\frac{5}{2}}$	$\frac{13}{7}\cdot 45$	$\frac{26}{33}\cdot 7\cdot 45$	0

Table 3 (continued)

Ground state		α	β	γ
f^6	$^7F_0$	0	0	0
f^7	^8S	0	0	0
f^8	$^7F_6$	$-\frac{1}{99}$	$\frac{2}{11}\cdot 1485$	$-\frac{1}{13}\cdot 33\cdot 2079$
f^9	$^6H_{\frac{15}{2}}$	$-\frac{2}{9}\cdot 35$	$-\frac{8}{11}\cdot 45\cdot 273$	$\frac{4}{11^2}\cdot 13^2\cdot 3^3\cdot 7$
f^{10}	$^5I_8$	$-\frac{1}{30}\cdot 15$	$-\frac{1}{11}\cdot 2730$	$-\frac{5}{13}\cdot 33\cdot 9009$
f^{11}	$^4I_{\frac{15}{2}}$	$\frac{4}{45}\cdot 35$	$\frac{2}{11}\cdot 15\cdot 273$	$\frac{8}{13^2}\cdot 11^2\cdot 3^3\cdot 7$
f^{12}	$^3H_6$	$\frac{1}{99}$	$\frac{8}{3}\cdot 11\cdot 1485$	$-\frac{5}{13}\cdot 33\cdot 2079$
f^{13}	$^2F_{\frac{7}{2}}$	$\frac{2}{63}$	$-\frac{2}{77}\cdot 15$	$\frac{4}{13}\cdot 33\cdot 63$

according to an irreducible representation of the rotation group, matrix elements may be written as

$$\langle JM|\, V_n^m \,|J'M'\rangle = c_{mM'}^J \langle J\| V_n \|J'\rangle \tag{24}$$

(Wigner–Eckart theorem). Here $\langle J\| V_n \|J'\rangle$ is the reduced matrix element, which is independent of M and M', and $c_{mM'}^J$ is a Wigner coefficient defined by

$$c_{m_1 m_2}^J = \sum_{\nu} (-1)^\nu \delta_{M, m_1 + m_2} \times \frac{[(j_1 + m_1)!\,(j_1 - m_1)!\,(j_2 + m_2)!\,(j_2 - m_2)!\,(J + M)!\,(J - M)!]^{\frac{1}{2}}}{(j_1 - m_1 - \nu)!\,(j_1 + m_1 - \lambda + \nu)!\,(j_2 + m_2 - \nu)! \times (j_2 - m_2 - \lambda + \nu)!\,\nu!\,(\lambda - \nu)!} \tag{25}$$

where $J = j_1 + j_2 - \lambda$. Both λ and ν assume all integral values that make the arguments of the factorials non-negative. The reduced matrix element is proportional to α, β, and γ according to

$$\langle J\| V_n \|J'\rangle \sim \begin{cases} \langle r^2\rangle/\alpha & \text{if} \quad n = 2 \\ \langle r^4\rangle/\beta & \text{if} \quad n = 4 \\ \langle r^6\rangle/\gamma & \text{if} \quad n = 6 \end{cases} \tag{26}$$

Using one of these methods, the secular equation can be set up and solved in a straightforward way. However, considerable simplifications are often possible. Firstly, many of the matrix elements $\langle JM|\ V_n^m\ |J'M'\rangle$ vanish unless $M = m + M'$. Secondly, symmetry considerations may be used to partly reduce the matrix. In addition, the ligand field may be decomposed into a main component of cubic and an additional component of lower symmetry. The low-symmetry component may then be treated as a perturbation on the cubic ligand field.

D. Spin–Orbit Coupling

The electron has a magnetic moment due to its spin. On the other hand, a magnetic moment is associated with the orbital motion of the electron. The magnetic interaction between the electron spins $\mathbf{s}_k$ and the orbital moments $\mathbf{l}_k$ gives rise to an additional term in the hamiltonian which has the form

$$\mathcal{H}_{LS} = \sum_{j,k} a_{jk}\mathbf{l}_j\mathbf{s}_k + b_{jk}\mathbf{l}_j\cdot\mathbf{l}_k + c_{jk}\mathbf{s}_j\cdot\mathbf{s}_k \tag{27}$$

where a_{jk}, b_{jk}, and c_{jk} are constants. The last term in (27) describes the spin–spin magnetic interaction which will be discussed in Section II.E. The remaining part of the expression for $\mathcal{H}_{LS}$ is usually simplified to

$$\mathcal{H}_{LS} = \sum_i \xi(r_i)\mathbf{l}_i\cdot\mathbf{s}_i \tag{28}$$

In writing (28), we have neglected the magnetic interaction between the two orbital moments (orbit–orbit coupling) of an electron pair as well as that between the spin moment of one electron and the orbital moment of the other (spin–other-orbit coupling). These terms are usually much smaller than the spin–orbit interaction. In addition, they give rise to no new effects, but rather add small contributions to other terms already present in the hamiltonian.

In the case of Russell–Saunders coupling, i.e. within states of definite L and S belonging to the same electron configuration, the spin–orbit interaction assumes the simple form

$$\mathcal{H}_{LS} = \lambda\mathbf{L}\cdot\mathbf{S} \tag{29}$$

where λ is the spin–orbit coupling constant for the lowest term,

$$\lambda = \pm\frac{\zeta_{nd}}{2S} \tag{30}$$

whereas ζ_{nd} is the spin–orbit coupling constant for a single d electron. The negative sign in (30) applies to a d shell more than half full. In principle, ζ could be taken over from atomic spectroscopy, where numerical values were compiled by Dunn[26]. However, it has been observed that ζ in complexes is reduced as compared to free ions. Thus ζ and similarly λ are usually treated as semiempirical parameters.

In calculating matrix elements of $\mathbf{L}\cdot\mathbf{S}$, it has to be observed that $\langle M_L M_S|\,\mathbf{L}\cdot\mathbf{S}\,|M_L' M_S'\rangle$ is different from zero only if

$$M_L + M_S = M_L' + M_S' \tag{31}$$

Thus, there may be, for example, off-diagonal elements between states differing in S by $\Delta S = \pm 1$. Writing

$$\mathbf{L}\cdot\mathbf{S} = L_z S_z + \tfrac{1}{2}L_+S_- + \tfrac{1}{2}L_-S_+ \tag{32}$$

the matrix elements may be calculated according to

$$\begin{aligned}
&\lambda\langle M_L M_S|\,\mathbf{L}\cdot\mathbf{S}\,|M_L M_S\rangle = \lambda M_L M_S \\
&\lambda\langle M_L M_S|\,\mathbf{L}\cdot\mathbf{S}\,|M_L \pm 1, M_S \mp 1\rangle \\
&\quad = \tfrac{1}{4}\lambda\{L(L+1) - M_L(M_L \pm 1)\}^{\frac{1}{2}}\{S(S+1) - M_S(M_S \mp 1)\}^{\frac{1}{2}}
\end{aligned} \tag{33}$$

If only number and type of the resulting spin–orbit levels are needed, group theoretical methods again may be used. Assuming a weak ligand field, i.e. $\mathcal{H}_{\mathrm{LF}} > \mathcal{H}_{LS}$, we first proceed as in Section II.C and determine the ligand-field levels Γ_k into which the original term of given L is split. The additional splitting due to the effect of spin–orbit interaction is determined by the direct product

$$\Gamma_k \times D^S = \sum_i p_{k_i}\Gamma_i \tag{34}$$

where D^S is the appropriate irreducible representation of the rotation group for the spin S. It should be observed that in cases where D^S is characterized by a half-integer value of S, the Γ_i will contain irreducible representations of the double groups[19,27–29]. All these double-valued representations are of even dimension. Consequently, all levels having half-integer values of J are at least two-fold degenerate. This is the statement of the Kramers[29] theorem which holds for any system containing an odd number of electrons, provided that no magnetic field is present.

E. Spin–Spin, Nuclear Spin, and Quadrupole Interaction

The mutual interaction between the magnetic moments due to electron spin may be described according to Pryce[30] by

$$\mathcal{H}_{SS} = \beta^2 \sum_{j,k}\left[\frac{\mathbf{s}_j\cdot\mathbf{s}_k}{r_{jk}^3} - \frac{3(\mathbf{r}_{jk}\cdot\mathbf{s}_j)(\mathbf{r}_{jk}\cdot\mathbf{s}_k)}{r_{jk}^5}\right] \tag{35}$$

Within terms which are eigenstates of $\mathbf{L}$ and $\mathbf{S}$, this perturbation assumes the form

$$\mathcal{H}_{SS} = -\rho[(\mathbf{L}\cdot\mathbf{S})^2 + \tfrac{1}{2}(\mathbf{L}\cdot\mathbf{S}) - \tfrac{1}{3}L(L+1)S(S+1)] \tag{36}$$

where ρ is a proportionality constant. The value of ρ is difficult to ascertain but, for the ground terms of divalent and trivalent ions within the first transition metal series, the value is probably not greater than 0.1 cm^{-1}.

If the nucleus has a spin I, there is an interaction between the magnetic moment of the nucleus and the magnetic field resulting from the orbital and spin moments of the electrons. The form of the interaction has been derived from the Dirac theory of the electron[31] and may be written

$$\mathscr{H}_N = 2\gamma\beta\beta_N \sum_k \left[\left\{\frac{(\mathbf{l}_k - \mathbf{s}_k)\cdot I}{r_k^3} + \frac{3(\mathbf{r}_k\cdot\mathbf{s}_k)\cdot(\mathbf{r}_k\cdot I)}{r_k^5}\right\} + \frac{8\pi}{3}\delta(r_k)(\mathbf{s}_k\cdot\mathbf{I})\right] \quad (37)$$

where β_N and γ are the nuclear magneton and the nuclear gyromagnetic ratio, respectively. The last term, involving a delta function, denotes the anomalous interaction of s electrons with the nuclear spin (Fermi interaction[31]). There is no such interaction within a pure d^n configuration. However, the analysis of the hyperfine structure of supposedly d-electron terms, and especially of S-state ions like Mn^{2+}, requires the assumption of an extra isotropic interaction proportional to $\mathbf{S}\cdot\mathbf{I}$. It has been assumed[32] that the Fermi term is due to excited terms involving unpaired s electrons which are mixed into the ground state via configuration interaction. According to another suggestion made by Sternheimer[33], the polarization of core (e.g. 1s, 2s, and 3s) electrons by the spin of the outer electrons (3d electrons in the first transition group) is made responsible for the observed hyperfine splittings. The physical reason is that if there is an unpaired electron (with $\uparrow$ spin say) present, the core s electrons of $\uparrow$ spin will experience different exchange interactions than electrons of $\downarrow$ spin. Consequently, the s electrons with $\uparrow$ spin and $\downarrow$ spin have different radial functions, leading to a net s-electron spin density

$$\sum_{s \text{ shells}} \{|\psi\uparrow(0)|^2 - |\psi\downarrow(0)|^2\} \quad (38)$$

Thus, the electron spin density for closed s-electron shells is now non-vanishing, and is the origin of the non-zero Fermi interaction. Wood and Pratt[34] and Heine[35], in independent investigations, estimated the magnitude of the effect. Watson and Freeman[36] made extensive free-ion spin-polarized Hartree–Fock calculations and showed that reasonable agreement with experiment is achieved by this mechanism.

If we consider only states which form part of the ground term of the ion, and if the terms of (37) are transformed to the appropriate angular momentum operators **L**, **S**, **I**, the nuclear magnetic interaction becomes

$$\mathscr{H}_N = (2\gamma\beta\beta_N\langle r^{-3}\rangle)[(\mathbf{L}\cdot\mathbf{I}) + \xi\{L(L+1)(\mathbf{S}\cdot\mathbf{I}) - \tfrac{3}{2}(\mathbf{L}\cdot\mathbf{S})(\mathbf{L}\cdot\mathbf{I}) - \tfrac{3}{2}(\mathbf{L}\cdot\mathbf{I})(\mathbf{L}\cdot\mathbf{S})\} - \kappa(\mathbf{S}\cdot\mathbf{I})] \quad (39)$$

Here,

$$\xi = \frac{(2l + 1) - 4S}{S(2l - 1)(2l + 3)(2L - 1)} \tag{40}$$

for $l = 2$, and κ is a dimensionless constant. In the iron group, κ is positive. It is usual to write $P = 2\gamma\beta\beta_N\langle r^{-3}\rangle$, thus putting the anomalous s-electron interaction equal to $P\kappa(\mathbf{S}\cdot\mathbf{I})$.

Provided the nucleus has also a quadrupole moment Q, there is an electrostatic interaction between the electrons and the nuclear quadrupole moment. The additional term has the form

$$\mathscr{H}_Q = \frac{e^2Q}{2I(2I - 1)} \sum_k \left[\frac{I(I + 1)}{r_k^3} - \frac{3(\mathbf{r}_k\cdot\mathbf{I})^2}{r_k^5}\right] \tag{41}$$

Considering an eigenstate of **L** and **S**, this expression may be transformed to the form

$$\mathscr{H}_Q = \eta q[3(\mathbf{L}\cdot\mathbf{I})^2 + \tfrac{3}{2}(\mathbf{L}\cdot\mathbf{I}) - L(L + 1)I(I + 1)] \tag{42}$$

where

$$q = e^2Q\langle r^{-3}\rangle/[2I(2I - 1)]$$

and

$$\eta = \pm 2S\xi \tag{43}$$

and where the negative sign in η applies when the shell is more than half full.

F. Effect of an External Magnetic Field

The remaining degeneracy is completely removed by the interaction of the electrons with an external magnetic field H described by

$$\mathscr{H}_H = \sum_k \beta(\mathbf{l}_k + 2\mathbf{s}_k)\cdot\mathbf{H} \tag{44}$$

The interaction produces equidistant sublevels of spacing $g\beta H$. In (44), the additional term $(e^2H^2/8mc^2)\sum_k (x_k^2 + y_k^2)$ has been omitted, since it only gives rise to diamagnetism. For terms of definite **L** and **S**, (44) may be simplified to

$$\mathscr{H}_H = \beta(\mathbf{L} + 2\mathbf{S})\cdot\mathbf{H} \tag{45}$$

Finally, there is a direct interaction of the nuclear moment with the external field H

$$\mathscr{H}_h = -\gamma\beta_N\mathbf{H}\cdot\mathbf{I} \tag{46}$$

G. The General 'Theoretical' Hamiltonian

Collecting all the terms which contribute to the energy (cf. equations 10, 12, 28, 35, 37, 41, 44, and 46) we obtain for the general 'theoretical' hamiltonian of the ion

$$\mathscr{H} = \mathscr{H}_0 + \mathscr{H}_{\mathrm{LF}} + \mathscr{H}_{LS} + \mathscr{H}_{SS} + \mathscr{H}_{\mathrm{N}} + \mathscr{H}_{\mathrm{Q}} + \mathscr{H}_{H} + \mathscr{H}_{h} \qquad (47)$$

An estimate of the order of magnitude of the single contributions may be obtained from optical spectra. In terms of energy, the values are: $\mathscr{H}_0 \sim 10^5\ \mathrm{cm}^{-1}$; $\mathscr{H}_{\mathrm{LF}} \sim 10^4\ \mathrm{cm}^{-1}$; $\mathscr{H}_{LS} \sim 10^2\ \mathrm{cm}^{-1}$ for the first transition metal group, $\sim 10^2$ to $10^3\ \mathrm{cm}^{-1}$ for the second, and $\sim 10^3\ \mathrm{cm}^{-1}$ for the third transition metal group, the rare earths, and the actinides; $\mathscr{H}_{SS} \sim 1\ \mathrm{cm}^{-1}$; $\mathscr{H}_{\mathrm{N}} \sim 10^{-1}$ to $10^{-3}\ \mathrm{cm}^{-1}$; and $\mathscr{H}_{\mathrm{Q}} \sim 10^{-3}\ \mathrm{cm}^{-1}$. Paramagnetic resonance experiments are usually carried out at such temperatures that only energy levels which are less than $100\ \mathrm{cm}^{-1}$ above the ground state are occupied. Thus, in many cases, only the lowest ligand-field level needs to be considered. It is then convenient to write (47) as

$$\mathscr{H} = \mathscr{H}_0 + \mathscr{H}_{\mathrm{LF}} + V \qquad (48)$$

Considering the lowest level of definite **L** and **S**, the perturbation V may be written in the simple form

$$\begin{aligned} V = (\lambda - \tfrac{1}{2}\rho)(\mathbf{L}\cdot\mathbf{S}) - \rho(\mathbf{L}\cdot\mathbf{S})^2 + \beta\mathbf{H}\cdot(\mathbf{L} + 2\mathbf{S}) \\ + P[(\mathbf{L}\cdot\mathbf{I}) + \{\xi L(L+1) - \kappa\}(\mathbf{S}\cdot\mathbf{I}) \\ - \tfrac{3}{2}\xi(\mathbf{L}\cdot\mathbf{S})(\mathbf{L}\cdot\mathbf{I}) - \tfrac{3}{2}\xi(\mathbf{L}\cdot\mathbf{I})(\mathbf{L}\cdot\mathbf{S})] \\ + q'[(\mathbf{L}\cdot\mathbf{I})^2 + \tfrac{1}{2}(\mathbf{L}\cdot\mathbf{I})] - \gamma\beta_{\mathrm{N}}(\mathbf{H}\cdot\mathbf{I}) \end{aligned} \qquad (49)$$

in which constant terms like $-\tfrac{1}{3}\rho L(L+1)S(S+1)$ have been omitted, since they produce only an equal shift of all levels. In (49), $q' = 3\eta q$ has been used.

III. THE SPIN-HAMILTONIAN

A. Concept of the Spin-Hamiltonian

The hamiltonian given in equation (47) is, in general, very complicated. If we take into account the terms $\mathscr{H}_0 + \mathscr{H}_{\mathrm{LF}} + \mathscr{H}_{LS}$ only, the original energy levels of the free ion will be split by the electrostatic field of the ligands and by spin–orbit interaction. The resulting wave functions will therefore be a very complicated mixture of various orbital and spin functions of the free ion. The calculation of the effect produced by the remaining interactions will thus become extremely cumbersome.

Abragam and Pryce[37,38] have developed a perturbation procedure for the calculation of splittings within the ground state of a paramagnetic ion. This method, which employs the so-called spin-hamiltonian, has found extensive application in experimental studies of paramagnetic resonance.

The concept of the spin-hamiltonian is based on the fact that the lowest energy levels between which transitions occur under conditions of paramagnetic resonance are usually well separated from all higher lying levels. If, in an e.p.r. experiment, transitions between $2S' + 1$ levels are observed, these levels may be considered as originating from a fictitious state characterized by the effective spin S'. This treatment parallels that of a free ion, where a state of quantum number J is split into $2J + 1$ components by an external magnetic field. The effective spin S' may, in some cases, equal the free-ion spin S, e.g. for the $[Ni(H_2O)_6]^{2+}$ ion where the lowest orbital level is $^3A_2$ (see Section V.H), so that $S = S' = 1$. In general, however, $S' \neq S$; for the $[Co(H_2O)_6]^{2+}$ ion, e.g., the lowest cubic field level is $^4T_1$ (see Section V.G). Lower symmetry fields and spin–orbit coupling lift the degeneracy, so that only transitions between the levels of the lowest Kramers doublet are observed. Thus $S = \frac{3}{2}$ but $S' = \frac{1}{2}$. Using an effective spin S', the paramagnetic ion is treated like a magnetic dipole which has $2S' + 1$ possible orientations in an external magnetic field, each level being associated with one orientation. The effective magnetic moment of the dipole is not given by the Landé factor ('spin-only' value g_s); rather the spectroscopic splitting factor g is determined from experiment and may differ considerably from g_s.

Actually, the higher lying levels may influence the ground state considerably. The lowest spin levels then display initial splittings in zero magnetic field. If the initial splitting is small compared to the applied microwave frequency, transitions between these spin states may still be observed. The interaction of the lowest level with the applied magnetic field is not sufficient to account for this and other details of e.p.r. spectra, but this deficiency is remedied if additional energy terms are taken into consideration. The most important of these terms are those representing the electrostatic interaction between d electrons and the ligand field, as well as those pertaining to the magnetic interaction between electrons and the nucleus. The interaction between the ground state and states of higher energy is then absorbed into empirical parameters occurring in these extra energy terms.

The sum of all the terms, written in form of energy operators, defines the spin-hamiltonian $\mathcal{H}(S)$ of the system. The energy eigenvalues E_0 satisfy the equation $\mathcal{H}(S)\psi = E_0\psi$, where ψ is representing the wave functions of the effective spin states.

The advantage of using a spin-hamiltonian is that it becomes possible to characterize the paramagnetic resonance spectrum by specifying the effective spin (which, for convenience, will be subsequently written as S rather than S') and a small number of parameters as g_x, g_y, g_z, D, E, etc. These parameters measure the magnitudes of the various terms in the spin-hamiltonian which will be discussed in more detail in Section III.C and subsequent sections. The main object of paramagnetic resonance experiments is to determine these parameters from the observed spectrum. The theory, on the other hand, aims at deducing the same parameters as well as the spin-hamiltonian employed from a reasonable model of the ligand field.

B. Derivation

The formal derivation of the spin-hamiltonian from the theoretical hamiltonian has been worked out by Abragam and Pryce[38]. Since the method is essentially based on a perturbation calculation, these authors distinguish two possible cases. Case (a), which will be the only one considered here, deals with orbital singlet ground states. In case (b), the ground state is orbitally degenerate.

We have already derived in equation (49) a perturbation operator V particularly suitable for a ground state of definite L and S. The formal perturbation treatment starts by assuming that the ground state is an orbital singlet. This assumption implies that $\langle 0|\, L\, |0\rangle = 0$ and also

$$\langle 0|\, L_i L_j + L_j L_i\, |0\rangle = \tfrac{2}{3}L(L+1)\delta_{ij} + l_{ij} \tag{50}$$

where, especially, $l_{ii} = 0$. The only non-zero diagonal elements are those which do not contain L. The result to first order in perturbation theory, defined by the diagonal elements, thus contains terms like $2\beta\mathbf{H}\cdot\mathbf{S}$, $-\gamma\beta_N\mathbf{H}\cdot\mathbf{I}$, a spin–spin contribution $-\rho l_{ij}\cdot S_i S_j$, and a nuclear magnetic and nuclear quadrupole contribution

$$-P(\kappa\delta_{ij} + 3\xi l_{ij})S_i I_j + q' l_{ij} I_i I_j$$

Proceeding to second order in perturbation theory, we have to calculate

$$\sum_{n\neq 0} \frac{\langle 0|\, V\, |n\rangle\langle n|\, V\, |0\rangle}{E_n - E_0} \tag{51}$$

where the ground state is taken as $|0\rangle$ and the excited states as $|n\rangle$, with energies E_0 and E_n, respectively ($n = 1, 2, 3$, etc.). Terms which do not involve orbital variables vanish. The non-vanishing elements $\langle 0|\, L_i\, |n\rangle$ have to be evaluated separately in every specific case. It is therefore convenient to define the tensors

$$\Lambda_{ij} = \sum_{n\neq 0} \frac{\langle 0|\, L_i\, |n\rangle\langle n|\, L_i\, |0\rangle}{E_n - E_0} \tag{52}$$

$$u_{ij} = -\tfrac{1}{2}i\epsilon_{ikl} \sum_{n\neq 0} \frac{\langle 0|\, L_l\, |n\rangle\langle n|\, L_j L_k + L_k L_j\, |0\rangle}{E_n - E_0} \tag{53}$$

where $u_{ii} = 0$. The result of the second-order perturbation treatment takes the form

$$-\Lambda_{ij}(2\beta\lambda H_i S_j + \lambda^2 S_i S_j + 2\lambda P S_i I_j + 2P\beta H_i I_j + \beta^2 H_i H_j) - u_{ij} \cdot 3\xi P\lambda S_i I_j$$

where small terms in ρ^2, q'^2 and P^2 are neglected. It is usually not necessary to go to any higher order in perturbation theory and thus we obtain by collecting the terms in first and second order

$$\begin{aligned}\mathscr{H}(S) = 2\beta(\delta_{ij} - \lambda\Lambda_{ij})\, H_i S_j + \{-\lambda^2\Lambda_{ij} - \rho l_{ij}\} S_i S_j \\ - P\{\kappa\delta_{ij} + 3\xi l_{ij} + 2\lambda\Lambda_{ij} - 3\xi\lambda u_{ij}\} S_i I_j \\ + q' l_{ij} I_i I_j - (\gamma\beta_N + 2P\beta\Lambda_{ij}) H_i I_j - \beta^2\Lambda_{ij} H_i H_j \quad (54)\end{aligned}$$

This spin-hamiltonian may be written in the abbreviated form

$$\begin{aligned}\mathscr{H}(S) = \beta H_i g_{ij} S_j + S_i D_{ij} S_j + S_i A_{ij} I_j + I_i P_{ij} I_j \\ - \gamma\beta_N H_i I_j - H_i R_{ij} I_j - \beta^2 H_i \Lambda_{ij} H_j \quad (55)\end{aligned}$$

where the following quantities have been introduced:

$$\begin{aligned} g_{ij} &= 2(\delta_{ij} - \lambda\Lambda_{ij}) \\ D_{ij} &= -\lambda^2\Lambda_{ij} - \rho l_{ij} \\ A_{ij} &= -P\{\kappa\delta_{ij} + 3\xi l_{ij} + 2\lambda\Lambda_{ij} - 3\xi\lambda u_{ij}\} \\ P_{ij} &= q' l_{ij} \\ R_{ij} &= 2P\beta\Lambda_{ij} \end{aligned} \quad (56)$$

Dropping the small anisotropic sixth term as well as the last term, which gives rise only to a uniform displacement of the levels, and using vectorial notation, it follows that*

$$\mathscr{H} = \beta\mathbf{H}\cdot(\mathbf{g})\cdot\mathbf{S} + \mathbf{S}\cdot(\mathbf{D})\cdot\mathbf{S} + \mathbf{S}\cdot(\mathbf{A})\cdot\mathbf{I} + \mathbf{I}\cdot(\mathbf{P})\cdot\mathbf{I} - \gamma\beta_N\mathbf{H}\cdot\mathbf{I} \quad (57)$$

The quantities **(g)**, **(D)**, **(A)**, and **(P)** which occur in (57) are, in general, tensors. Their components are used as characteristic parameters of the relevant spin-hamiltonian. The interpretation of these quantities is as follows. g_{ij} is the spectroscopic splitting factor. It contains, besides the 'spin-only' value $g_s = 2.0023$, contributions from higher lying states, $\lambda\Lambda_{ij} \sim \lambda/\Delta_{ij}$, Δ_{ij} being the energy separation between the ground singlet and higher lying orbitals. D_{ij} describes the zero-field splitting of the ground state in absence of nuclear interaction. This term arises from lower symmetry ligand fields, spin–orbit coupling, and spin–spin contributions. The zero-field splitting causes the fine structure of the spectrum. The term in A_{ij} characterizes the magnetic interactions between the electrons and the nucleus which give rise to the hyperfine structure of the spectrum. The term containing P_{ij} describes the quadrupole interaction, and the

* Here and in the following the symbol $\mathscr{H}(S)$ has been replaced by $\mathscr{H}$, which is the symbol usually written for the spin-hamiltonian.

last term is the common form of the direct interaction between the nuclear moment and the magnetic field. The term $-\beta^2\mathbf{H}\cdot(\Lambda)\cdot\mathbf{H}$, which has been neglected (cf. equation 55), needs to be considered only in calculating magnetic susceptibilities where it is responsible for the temperature-independent paramagnetism.

The spin-hamiltonian has to conform to the local symmetry at the crystal site of the paramagnetic ion. We list here the form of the spin-hamiltonian in a field of axial symmetry:

$$\begin{aligned}\mathscr{H} &= g_{\parallel}\beta H_z S_z + g_{\perp}\beta(H_x S_x + H_y S_y) + D[S_z^2 - \tfrac{1}{3}S(S+1)] \\ &\quad + AS_z I_z + B(S_x I_x + S_y I_y) + P[I_z^2 - \tfrac{1}{3}I(I+1)] - \gamma\beta_{\mathrm{N}}\mathbf{H}\cdot\mathbf{I} \qquad (58)\end{aligned}$$

and that of the spin-hamiltonian in presence of a rhombic field component:

$$\begin{aligned}\mathscr{H} &= \beta(g_x H_x S_x + g_y H_y S_y + g_z H_z S_z) + D[S_z^2 - \tfrac{1}{3}S(S+1)] \\ &\quad + E(S_x^2 - S_y^2) + A_x S_x I_x + A_y S_y I_y + A_z S_z I_z \\ &\quad + P[I_z^2 - \tfrac{1}{3}I(I+1)] + P'(I_x^2 - I_y^2) - \gamma\beta_{\mathrm{N}}\mathbf{H}\cdot\mathbf{I} \qquad (59)\end{aligned}$$

In the general case, the principal axes of the tensors (**g**) and (**A**) will not coincide and the spin-hamiltonian may contain cross-terms of the form

$$\sum_{i,j=x,y,z} F_{ij}(S_i I_j + S_j I_i) + \sum_{i,j=x,y,z} G_{ij}(S_i I_j - S_j I_i) \qquad (60)$$

A specific example of such behaviour occurs with $(NH_4)_2VO(SO_4)_2\cdot 6H_2O$ (see Section V.A).

In the case where the ligands of a complex compound contain nuclei with non-zero magnetic moment and spin, an additional hyperfine structure ('super-HFS') may be observed. The spin-hamiltonian (57) then contains the extra term

$$\sum_n \mathbf{S}\cdot(\mathbf{A}^n)\cdot\mathbf{I}^n \qquad (61)$$

where the summation is extended over all nuclei n of the ligand which contribute to the hyperfine splitting. Sometimes (61) is written as the sum of isotropic and anisotropic contributions to the hyperfine splitting,

$$\sum_n \{\mathbf{S}\cdot(\mathbf{A}_s^n)\cdot\mathbf{I}^n + \mathbf{S}\cdot(\mathbf{A}_p^n)\cdot\mathbf{I}^n(3\cos^2\theta - 1)\} \qquad (62)$$

Here, $\mathbf{A}_s^n$ describes the s-orbital interaction, and $\mathbf{A}_p^n$ describes the p_σ orbital and dipole–dipole interactions.

C. Properties of Spin-Hamiltonian Parameters

1. The g factor

The spectroscopic splitting factor g, although analogous to the Landé factor g_s, is defined in terms of the experimentally observed splittings

$g\beta H$ between the $2S + 1$ equally spaced effective spin levels produced in an applied magnetic field H. The $2S$ allowed transitions ($\Delta M_S = \pm 1$) then coincide at the same field $H = h\nu/g\beta$, giving

$$g = 21.4178/H\lambda \tag{63}$$

if g is isotropic. Here, H is measured in kilogauss and the wavelength λ in centimeters. This simple relationship holds true only if no other interactions are present. In general, allowance must be made for the additional energy terms and consequently the transitions do not all occur at the same field.

Since (**g**) is a tensor, its magnitude is usually different for different directions of H, and it shows an angular variation which follows the symmetry of the ligand field. It may always be represented by three principal values g_x, g_y, g_z which are measured in three mutually perpendicular directions. For a general direction, specified by the direction cosines l, m, n with respect to the axes x, y, z, the g value is given by

$$g = \{l^2g_x^2 + m^2g_y^2 + n^2g_z^2\}^{\frac{1}{2}} \tag{64}$$

2. *Fine structure*

The fine structure of the spectrum is due to initial splittings of the effective spin levels when $H = 0$. It may be formally represented by the term $D\{S_z^2 - \frac{1}{3}S(S + 1)\}$ in the spin-hamiltonian if the field has axial symmetry. A field of symmetry lower than axial requires the additional term $E(S_x^2 - S_y^2)$. The zero-field separations of levels may be obtained directly from these operator expressions. Thus, in axial symmetry and assuming $S = 1$, the levels specified by $M_S = \pm 1$ are separated from the level $M_S = 0$ by $D(1^2 - 0^2) = D$. If $H \parallel z$, the spin-hamiltonian may be written

$$\mathscr{H} = g_z\beta HS_z + D\{S_z^2 - \tfrac{1}{3}S(S + 1)\}$$

with energies $\pm g_z\beta H + \frac{1}{3}D$ and $-\frac{2}{3}D$ for the $M_S = \pm 1$ and $M_S = 0$ levels, respectively. The allowed transitions are given by $\Delta M_S = \pm 1$ and thus have transition energies $h\nu = g_z\beta H \pm D$. The spectrum shows a fine structure consisting of two absorption lines having a distance of $2D/g_z\beta$ in a magnetic field.

3. *Hyperfine structure*

When the nucleus possesses a magnetic moment $\mu_N = I\gamma\beta_N$, there will be an interaction between μ_N and the magnetic moment of the electrons giving rise to a hyperfine structure of the spectrum. In the spin-hamiltonian, this interaction is represented by the term $\mathbf{S}\cdot(\mathbf{A})\cdot\mathbf{I}$, and in accord with the $2I + 1$ possible orientations of the nuclear spin I, each resonance line is split up into $2I + 1$ hyperfine components. (**A**) is a tensor and thus there will be an anisotropy in A, if the g value has been found to be anisotropic. The angular variation of the hyperfine structure is then described by the three principal values A_x, A_y, A_z. Let us consider as an example the isotropic hamiltonian

$$\mathscr{H} = g\beta \mathbf{H}\cdot\mathbf{S} + A\mathbf{S}\cdot\mathbf{I}$$

with $S = \frac{1}{2}$, $I = \frac{3}{2}$, $H \parallel z$, and $g\beta H \gg A$. The energies of the $M_S = +\frac{1}{2}$ and $M_S = -\frac{1}{2}$ electronic levels are $\frac{1}{2}g\beta H + \frac{1}{2}AM_I$ and $-\frac{1}{2}g\beta H - \frac{1}{2}AM_I$, respectively, where M_I assumes the values $M_I = \frac{3}{2}, \frac{1}{2}, -\frac{1}{2}, -\frac{3}{2}$. The allowed transitions are now those for which $\Delta M_S = \pm 1$, $\Delta M_I = 0$, the transition energies are $h\nu = g\beta H + AM_I$. The spectrum shows a hyperfine structure of four absorption lines with equal spacings $A/g\beta$ (Figure 3). It should be realized that, when $H = 0$, the hyperfine levels collapse into two levels characterized by the values of $F = I + S = 2$ and $I - S = 1$ with respective degeneracies $2F + 1 = 5$ and 3.

An additional hyperfine structure due to ligand nuclei, which is represented by the term $\sum_n \mathbf{S}\cdot(\mathbf{A}^n)\cdot\mathbf{I}$ in the spin-hamiltonian, shows up as an extra splitting of each hyperfine line. The additional splitting will be completely resolved if $A^n \ll A$ and if the linewidth is sufficiently small. In general, however, the lines will overlap and the resolution will thus be only partial.

4. Quadrupole interaction

The nucleus of a paramagnetic ion is not necessarily spherically symmetric and may have an electric quadrupole moment Q. If such an ion is placed in an electric field with gradient $\partial E/\partial z$, certain nuclear orientations will be preferred to others and there will be an electrostatic interaction between electrons and nucleus. The effect is represented by the operator $P\{I_z^2 - \frac{1}{3}I(I + 1)\}$ in the spin-hamiltonian, if the field gradient has axial symmetry. An electric field gradient of less than axial symmetry requires the extra term $P'(I_x^2 - I_y^2)$. Thus, in axial symmetry, the quadrupole interaction may be described by the spin-hamiltonian

$$\mathscr{H} = g\beta \mathbf{H}\cdot\mathbf{S} + P\{I_z^2 - \tfrac{1}{3}I(I + 1)\}$$

Assuming $S = \frac{1}{2}$, $I = \frac{3}{2}$, and $H \parallel z$, the energies of the $M_S = \pm\frac{1}{2}$ electronic levels are $\pm\frac{1}{2}g\beta H + P(M_I^2 - \frac{5}{4})$, where $M_I^2 = (\pm\frac{3}{2})^2, (\pm\frac{1}{2})^2$. Each electronic level splits up into two components with spacing $2P$. However, this does not affect the energies of the allowed transitions, since for these $\Delta M_I = 0$. Thus the quadrupole interaction influences the resonance spectrum in a rather complicated way. In general, the intensities and spacings of the usual $2I + 1$ hyperfine lines corresponding to $\Delta M_I = 0$ transitions are modified, and since the selection rule $\Delta M_I = 0$ is no longer valid, there are additional lines corresponding to transitions $\Delta M_I = \pm 1$, ± 2. The effects are small and have been analysed only in a few cases (see Section V, especially V.I).

D. *S*-State Ions

For paramagnetic ions having a half-filled shell, the ground state is an orbital singlet. Such a situation occurs in the first transition metal group for $3d^5$ (Mn^{2+}, Fe^{3+}), ground state $^6S_{\frac{5}{2}}$; in the rare earths and actinides for $4f^7$ (Gd^{3+}, Eu^{2+}) and $5f^7$ (Cm^{3+}), respectively, ground state $^8S_{\frac{7}{2}}$. Since the resultant orbital angular momentum is zero, ligand fields cannot produce any splitting. Also, spin–orbit coupling by itself cannot remove the six- or eight-fold spin degeneracy. Actually, small *S*-state splittings have been observed for all the metal ions mentioned above. Van Vleck and Penney[39] found that it is necessary to go to fifth order in perturbations, involving simultaneously a ligand field of cubic symmetry and spin–orbit coupling, to obtain a splitting of the ground level. They estimated the constant a of the spin-hamiltonian (see below) as

$$a \sim \frac{K\lambda^4}{E(^4P) - E(^6S)} \tag{65}$$

where $K = \langle 3d|\, V_{\text{cub}}\, |3d\rangle$ is a matrix element of the ligand-field potential and ^4P and ^6S are terms of the free ion. Taking $K = 10^4$ cm^{-1}, $\lambda = 300$ cm^{-1}, and $E(^4P) - E(^6S) = 2.5 \times 10^4$ cm^{-1}, we obtain $a \sim 10^{-4}$ cm^{-1}.

The origin of the spin-hamiltonian term in D has been investigated by Abragam and Pryce[38]. They suggested that it should be due to spin–spin interaction and argued as follows: although the ion is in an S ground state, a small ligand-field component of tetragonal or trigonal symmetry may cause a slight distortion of the orbits. Thus, instead of being a perfect sphere, the charge cloud acquires a slightly ellipsoidal shape. The spin–spin interaction energy then depends on the spin orientation. The resulting splitting of the ground state is obtained in second approximation and is proportional to $D(S_z^2 - \frac{35}{12})$, where

$$D \sim \frac{U(\beta^2/r^3)}{E(^6D) - E(^6S)} \tag{66}$$

Here,

$$U = \langle 3d \mid U_2^0 \mid 4s\rangle, \quad U_2^0 = \frac{1}{4}\sqrt{\frac{5}{\pi}}\, A_2^0(3z^2 - r^2)$$

and the denominator in (66) gives the energy difference between the $3d^44s\,^6D$ and $3d^5\,^6S$ terms. Another explanation for the term in D has been advanced by Watanabe[40].

Next, we try to obtain an idea about the spin-hamiltonian expression in S_x, S_y, S_z for $S = \frac{5}{2}$. In the absence of a magnetic field, we should expect that only even powers of S_x, S_y, and S_z will occur. Since the spin-hamiltonian must reflect the symmetry of the ligand field, we assume, for convenience, that the symmetry is cubic. This implies that the spin-hamiltonian will contain both S_y^2 and S_z^2 if it is supposed that it should contain the term S_x^2. Since $S_x^2 + S_y^2 + S_z^2 = S(S+1)$ is a constant, such a term would be omitted. Likewise, S_x^4 implies $S_x^4 + S_y^4 + S_z^4$; however, this expression will be retained. On the other hand, $S_x^2 S_y^2$ implies all the terms obtained by permutation of the subscripts x, y, z; however, their sum can be reduced to the two earlier polynomials. Terms in sixth and higher powers of S_x, S_y, S_z can also be reduced to one of those terms already considered. Thus, the spin angular momentum part of the spin-hamiltonian will be

$$\tfrac{1}{6}a[S_x^4 + S_y^4 + S_z^4 - \tfrac{1}{5}S(S+1)(3S^2 + 3S - 1)]$$

where the constant terms are added in order to conform with the corresponding operator equivalent in Table 2.

The complete spin-hamiltonian of a $^6S_{\frac{5}{2}}$ ground-state ion in a cubic field is thus

$$\mathscr{H} = g\beta\mathbf{H}\cdot\mathbf{S} + \tfrac{1}{6}a[S_x^4 + S_y^4 + S_z^4 - \tfrac{1}{5}S(S+1)(3S^2 + 3S - 1)] + A\mathbf{S}\cdot\mathbf{I} - \gamma\beta_{\mathrm{N}}\mathbf{H}\cdot\mathbf{I} \quad (67)$$

If the symmetry is only axial, the general spin-hamiltonian becomes more complicated and is given by

$$\begin{aligned}\mathscr{H} = {} & g_{\parallel}\beta H_1 S_1 + g_{\perp}\beta(H_2 S_2 + H_3 S_3) \\ & + \tfrac{1}{6}a[S_x^4 + S_y^4 + S_z^4 - \tfrac{1}{5}S(S+1)(3S^2 + 3S - 1)] \\ & + b[35S_1^4 - 30S(S+1)S_1^2 + 25S_1^2 - 6S(S+1) + 3S^2(S+1)^2] \\ & + D[S_1^2 - \tfrac{1}{3}S(S+1)] + AS_1I_1 + B(S_2I_2 + S_3I_3) \\ & + P[I_1^2 - \tfrac{1}{3}I(I+1)] - \gamma\beta_{\mathrm{N}}\mathbf{H}\cdot\mathbf{I} \qquad (68)\end{aligned}$$

where subscripts 1, 2, 3 denote orthogonal axes with axis 1 in the direction of the distortion.

E. Molecular-orbital Theory

A generalization of the ligand-field approach which takes into account the orbital structure of the ligands is obtained by formation of molecular orbitals. These orbitals may be constructed by adding a linear combination of ligand orbitals of the appropriate symmetry to the central ion d function (MO–LCAO method) according to

$$\phi = \alpha d_{\mathrm{M}} + (1-\alpha)^{\frac{1}{2}} \sum_{\mathrm{L}} \gamma_{\mathrm{L}} \psi_{\mathrm{L}} \quad (69)$$

Thus, in octahedral symmetry, the σ-bonding orbitals may be written as

$$\phi_{z^2} = N_\sigma\left\{d_{z^2} + \frac{1}{\sqrt{12}}\Lambda_\sigma(2\sigma_6 + 2\sigma_3 + \sigma_1 - \sigma_4 + \sigma_2 - \sigma_5)\right\}$$

$$\phi_{x^2-y^2} = N_\sigma\{d_{x^2-y^2} + \tfrac{1}{2}\Lambda_\sigma(-\sigma_1 + \sigma_4 + \sigma_2 - \sigma_3)\} \qquad (70)$$

where the subscripts 1, 2, 3 and 4, 5, 6 refer to ligand atoms located on the positive and negative *x*, *y*, *z* axes, respectively. Similarly, the π-bonding orbitals are

$$\begin{aligned}\phi_{xy} &= N_\pi\{d_{xy} + \tfrac{1}{2}\Lambda_\pi(p_{y1} - p_{y4} + p_{x2} - p_{x5})\}\\ \phi_{yz} &= N_\pi\{d_{yz} + \tfrac{1}{2}\Lambda_\pi(p_{z2} - p_{x5} + p_{y3} - p_{y6})\}\\ \phi_{xz} &= N_\pi\{d_{xz} + \tfrac{1}{2}\Lambda_\pi(p_{z1} - p_{z4} + p_{x3} - p_{x6})\}\end{aligned} \qquad (71)$$

The corresponding antibonding orbitals ϕ^* are obtained from (70) and (71) by changing the positive sign of the admixture coefficients Λ_σ and Λ_π into a negative sign. If $\Lambda_\sigma = 0$, the metal–ligand bond in (70) is purely ionic in character. If $\Lambda_\sigma = 1$, the electrons are equally distributed between the central ion and the ligands. Corresponding statements hold for Λ_π. N_σ and N_π are normalization constants defined by

$$\frac{1}{N^2} = 1 + 4\Lambda S + \Lambda^2 \qquad (72)$$

where

$$S_\sigma = \langle d_{z^2}|\sigma_1\rangle \qquad (73)$$

and

$$S_\pi = \langle d_{xy}|p_{y1}\rangle \qquad (74)$$

are overlap integrals. More details on molecular-orbital theory may be found in the chapter on electronic absorption spectroscopy and in the book by Ballhausen and Gray[41].

The MO functions (70) and (71) have to be used in the derivation of a modified spin-hamiltonian. It can be shown that the matrix elements involving the orbital functions ϕ are related to the corresponding matrix elements of *d* functions by multiplication with a factor. This is due to the fact that the matrices of angular momentum, spin–orbit coupling, and hyperfine interaction have the same transformation properties within the two manifolds. In particular, we obtain for the angular momentum **L** and for λ**L** of the spin–orbit coupling operator

$$\langle\phi_n|\,\mathbf{L}\,|\phi_m\rangle = k_{nm}\langle d_n|\,\mathbf{L}\,|d_m\rangle \qquad (75)$$

and

$$\langle\phi_n|\,\lambda\mathbf{L}\,|\phi_m\rangle = N_nN_m\langle d_n|\,\lambda\mathbf{L}\,|d_m\rangle \qquad (76)$$

The factor k_{nm} is a tensor quantity and may be written in the case of pure π bonding as

$$k_{nm} = 1 - \tfrac{1}{2}\Lambda_\pi^2 N_\pi^2 \tag{77}$$

For the case where both σ and π bondings are present, the reader should consult Tinkham[42].

These changes of the matrix elements modify the magnitude of, and the relation between, the spin-hamiltonian parameters, which influence the spectroscopic splitting factor g as well as the fine structure and hyperfine structure constants. Specific examples are given in Section V.

IV. EXPERIMENTAL ASPECTS

A. Spectrometers

In principle, paramagnetic resonance spectrometers are set up in a similar way to spectrometers designed for operation in other regions of the electromagnetic spectrum. The equipment consists of a radiation source, an absorption cell containing the substance under investigation, and a suitable detector. However, unlike spectral investigations in other frequency regions, a source of a steady magnetic field is required in addition.

Most paramagnetic resonance spectrometers operate in the microwave region, although radiofrequency spectrometers have occasionally been used[43–45]. In this chapter, we shall emphasize only the most basic facts about the design and operation of microwave e.p.r. spectrometers. More details may be found in the books by Ingram[46], Low[16], and Altshuler and Kozyrev[47]. Sensitivity problems have been discussed in detail by Feher[48] and construction details of microwave equipment are dealt with in several texts[49–52].

In the frequency range between 3 and 140 kMc/s, a klystron is usually employed as source of microwave radiation. The resultant microwave frequency has to be stabilized by comparing the klystron frequency with the frequency of a standard, and by subsequently adjusting the reflector voltage of the klystron. In order to obtain maximum sensitivity, the instrument should be operated at the highest frequency possible[48]. From equation (4), it follows that the microwave frequency, measured in megacycles per second, should be about 2.8 times larger than the magnetic field strength in gauss (G)

$$\nu\,(\text{Mc/s}) = 2.8 \times H\,(\text{G}) \tag{78}$$

Thus the employment of high frequencies is limited, in practice, by the very high magnetic field strengths required.

To avoid an unnecessary loss of energy, silver-plated waveguides are used for the transfer of microwaves. A small part of the energy is branched off for control and adjustment to maximum power; another small part is used for measuring the frequency.

The absorption cell is usually in the form of a resonance cavity. The size and form of the cavity are determined by the space between the pole pieces of the magnet, the 'quality factor' Q of the cavity, and the size of the sample. A discussion of various forms of rectangular and cylindrical cavities is given by Schoffa[53]. For paramagnetic resonance measurements at hydrogen and helium temperatures, resonance cavities of special design are required[46,54,55]. The sample is adjusted within the cavity in such a way that the microwave magnetic field is at its maximum and the electric field is at its minimum. The cavity with sample is then placed between the poles of the magnet such that the condition of external magnetic field and microwave magnetic field being mutually perpendicular is satisfied.

The detector most widely used at microwave frequencies is the silicon crystal diode. The sensitivity of a spectrometer using such a crystal detector is limited mainly by the low-frequency noise of the diode. Beringer and Castle[56] succeeded in avoiding this source of noise and thus increased the sensitivity by using a bolometer as detector. Another possibility to minimize the crystal noise is by means of superheterodyne detection (see below).

There are two basic types of paramagnetic resonance spectrometers in use. In the transmission-type spectrometer, described for the first time by Cummerow, Holliday, and Moore[57], the resonance absorption is detected by measuring the change in power transmitted through the cavity. In the reflection-type spectrometer, the change in the reflection coefficient is measured by the unbalance of the reflected and input powers of a 'magic-tee' bridge[58]. The resonance absorption is thus obtained essentially from the change in power which is reflected from the cavity. Both methods have been frequently applied in practical spectrometers.

The various experimental arrangements of paramagnetic resonance spectrometers differ mainly in the method of detection of the transmitted or reflected power. We shall describe briefly some of the more common systems, namely (*a*) d.c. detection, (*b*) magnetic field modulation, (*c*) double modulation, and (*d*) superheterodyne detection.

If d.c. detection methods are employed, the microwave signal is rectified and the absorption measured directly point by point using a galvanometer. Because of low sensitivity, spectrometers of this kind were used only for measurements on undiluted paramagnetic substances, which show rather broad and intense absorption lines.

A greater sensitivity is achieved if the external magnetic field is subject to a modulation by an audiofrequency magnetic field[43]. If the magnetic field sweeps across an absorption line, there will be a corresponding modulation of the power which is transmitted or reflected by the resonant cavity. The modulated power can be amplified by a low-frequency receiver and displayed on an oscilloscope. The horizontal sweep of the oscilloscope has to be synchronized with the field modulation. This method has been used, for example, by Bleaney and Ingram[59], employing a modulation frequency of 50 c/s. The main limitation of the method arises from the noise output of the crystal rectifier.

An additional increase in sensitivity may be accomplished by a double modulation of the magnetic field[60]. In a spectrometer designed by Buckmaster and Scovil[61], the amplitude modulation is carried out at the audiofrequency of 60 c/s. The second modulation of the magnetic field at 462.5 c/s is sufficiently high for the low-frequency noise to become negligibly small. A difficulty with the application of this method is that the high-frequency modulation cannot penetrate through ordinary waveguides. In order to introduce the radiofrequency magnetic field inside the resonance cavity, the cavity is partially slotted in a plane which passes through its axis. The modulation current flowing along the interior of the cavity then produces the modulation r.f. field.

The superheterodyne detection system was used in paramagnetic resonance for the first time by Schneider and England[62,63] and subsequently described by several authors[64,65,48]. The method uses a 'magic-tee' or similar bridge which receives power both from the measuring klystron (frequency ν_1) and from an auxiliary klystron (frequency ν_2). The signal, which arises as a result of unbalancing the bridge by absorption, is amplified at the intermediate frequency $\nu_1 - \nu_2$. The low-frequency noise of the crystal diode becomes negligibly small. A block diagram of a superheterodyne spectrometer is shown in Figure 4.

To characterize the paramagnetic resonance absorption, e.g. according to equation (4), it is necessary to determine the magnetic field intensity. For this purpose, magnetometers based on the principle of proton magnetic resonance are employed[66]. The determination of the magnetic field strength is thus reduced to the measurement of the proton resonance frequency in the applied magnetic field.

B. General Conditions for Observation of Paramagnetic Resonance

In Section II.D, we have shown that according to the Kramers theorem[29] each system containing an odd number of electrons will show an even (at least two-fold) degeneracy in absence of a magnetic field. On the other hand, the theorem of Jahn and Teller[67,68] states that, in a molecule having an electronic energy level which is degenerate, the geometrical configuration of its nuclei cannot be stable except (*a*) in linear molecules and (*b*) in systems having Kramers degeneracy. Excluding these two cases, nuclear

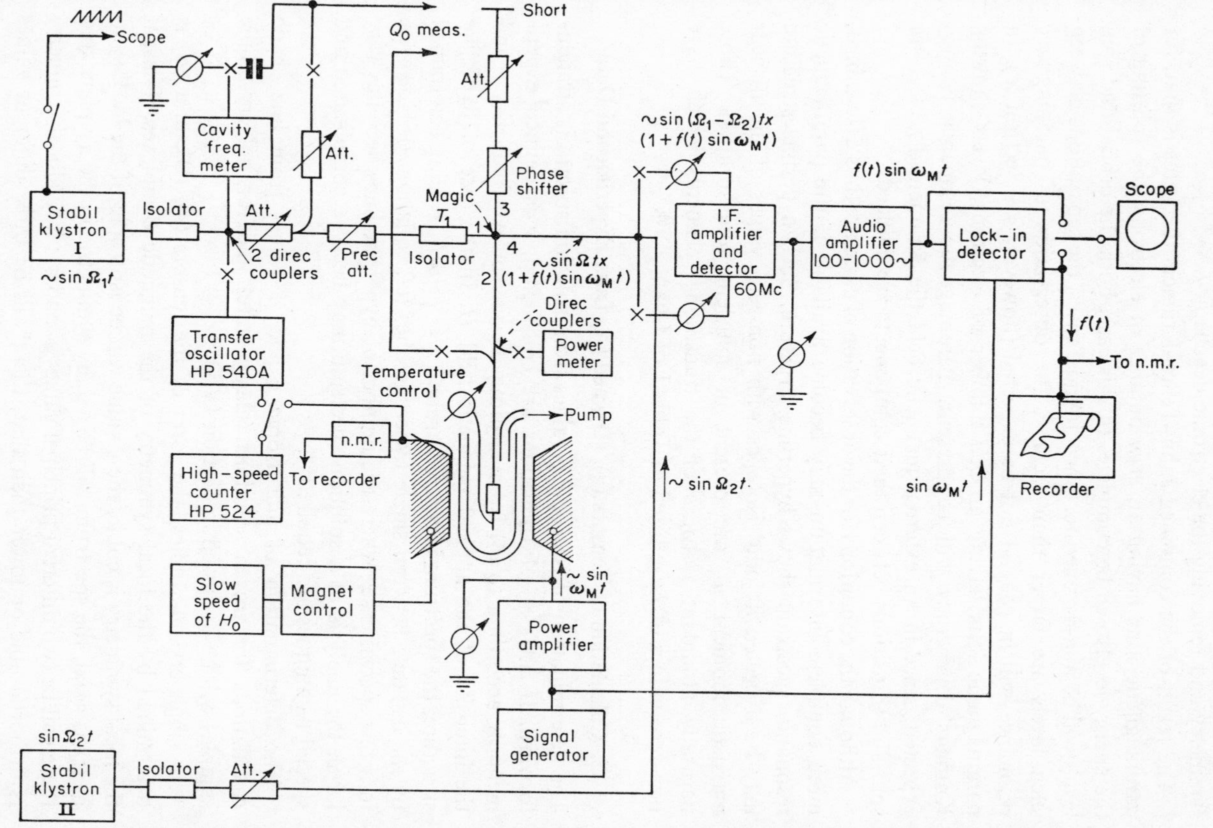

Figure 4. Block diagram of a superheterodyne e.p.r. spectrometer [according to Feher[48]].

displacements will spontaneously take place, thus lowering the molecular symmetry and removing the electronic degeneracy.

The result of the combined Jahn–Teller–Kramers theorems is that, for paramagnetic ions having an *even* number of electrons, the splitting of the energy levels will be complete. In this case, both the orbital and spin levels will be non-degenerate. Since, in most cases, the separations between such levels are larger than $1\ \text{cm}^{-1}$, the observation of paramagnetic resonance will in general not be expected (however, see Section V). In paramagnetic ions having an *odd* number of electrons, levels showing Kramers degeneracy will be present. Paramagnetic resonance is then expected, since in an external magnetic field the Kramers doublets are split and transitions between their components are induced.

Although the conditions for the observation of e.p.r. are thus determined, more complete information may become available if the paramagnetic resonance spectrum shows hyperfine structure. Such hyperfine structure may be expected if 'odd' isotopes with non-zero values of the nuclear magnetic moment μ_N are present in sufficient concentration. Those naturally abundant isotopes of the transitional elements which have non-zero values of I are listed in Table 1 of Chapter 10.

C. Calculation of Tensors (g), (D), and (A) from Experimental Data

The determination of isotropic g values from solution spectra is straightforward. If $\Delta M_S = \pm 1$ transitions are considered, g is defined experimentally according to (4) by measuring the microwave frequency ν and the intensity of the applied magnetic field H. In the case of hyperfine structure, the spectrum is centred around $g = h\nu/\beta H$, A being determined by the distance between single hyperfine lines (cf. equation 9). However, in a few special cases where the number of hyperfine lines becomes very large, the analysis of a solution spectrum may become complicated and special techniques are required[69].

The determination of anisotropic g and A values from line shape calculations for frozen solutions (glassy state) and for polycrystalline samples will be discussed in Section IV.D.

In single-crystal studies, the form of the tensors (**g**), (**D**), and (**A**) is determined by the local symmetry of the magnetic complexes. Thus, if the local symmetry is cubic, the g value will be isotropic; if the local symmetry is axial, the spectrum will be axially symmetric in the xy plane and there will be two different principal values $g_\parallel$ and $g_\perp$; if the local symmetry is orthorhombic or lower than that, there will be three different values g_x, g_y, g_z. However, without additional information, the paramagnetic resonance spectrum cannot distinguish between orthorhombic and lower symmetry.

In axial symmetry, $g_{\parallel}$ and $g_{\perp}$ may be obtained from a study of the angular dependence of g. Denoting by θ the angle between the external magnetic field H and the symmetry axis of the crystal, then

$$g^2 = g_{\parallel}^2 \cos^2 \theta + g_{\perp}^2 \sin^2 \theta \tag{79}$$

When the axes of symmetry are known from x-ray analysis, crystal faces, or the paramagnetic resonance spectrum, only two measurements are required ($\theta = 0°$ determines $g_{\parallel}$, $\theta = 90°$ determines $g_{\perp}$). If the axes are unknown, several measurements at different θ values are necessary. Always one principal value will be the maximal, the other the minimal g value observed.

Considerably more difficult is the evaluation of the principal axes of a completely asymmetric g tensor. Three rotations of the crystal (or the magnetic field) about three mutually perpendicular axes, not necessarily about the symmetry axes, need to be made. If the orientation of the magnetic field is determined by $\cos\varphi \sin\theta$, $\sin\varphi \sin\theta$, $\cos\theta$, we have

$$\begin{aligned} g^2 = {} & \sin^2\theta[(g^2)_{11}\cos^2\varphi + 2(g^2)_{12}\sin\varphi\cos\varphi + (g^2)_{22}\sin^2\varphi] \\ & + 2\sin\theta\cos\theta[(g^2)_{13}\cos\varphi + (g^2)_{23}\sin\varphi] + (g^2)_{33}\cos^2\theta \end{aligned} \tag{80}$$

where $(g^2)_{ij}$, the elements of the tensor $(\mathbf{g}^2)$, are readily determined by suitable choice of the polar angle θ and the azimuthal angle φ. By diagonalizing $(\mathbf{g}^2)$, we obtain the principal values g_{ii}^2 and, since $g \geqslant 0$, the tensor $(\mathbf{g})$ will be diagonalized by the same transformation. In the actual application, several equations containing the measured angles and g values have to be solved for g_x, g_y, g_z[70].

Another method has been suggested by Park[71]. By rotating the magnetic field through 180° around the crystal, a maximal g value g_+ and a minimal g value g_- can be determined. The crystal is then rotated about an axis perpendicular to the magnetic field. For each position, the outlined procedure is carried through. The resultant values of g_+ and g_- again will be orientation dependent, yielding a maximal $g_+^{\max}$ value as well as a minimal $g_-^{\min}$ value. Choosing $g_x > g_y > g_z$, $g_+^{\max}$ becomes equal to g_x and $g_-^{\min}$ equal to g_z. A rotation of the crystal around an axis perpendicular to both z and x axes finally determines g_y. The same problem has also been considered by Geusic and Brown[72] and by Schonland[73].

To determine the fine structure constants D and E, formulae describing the angular variation of the spectrum have been obtained. Also, if one of the principal g values is very small, it becomes very difficult to determine this quantity with accuracy by rotation about the relevant axis. In this case, the g components are also determined from the angular variation formulae. Bleaney[74] has derived the expression for the angular dependence of the $\Delta M_S = \pm 1$ transitions in axial symmetry, neglecting hyperfine structure.

His result is

$$h\nu = g\beta H + D(M_S - \tfrac{1}{2})\left[\frac{3g_{\parallel}^2}{g^2}\cos^2\theta - 1\right]$$

$$- \left[\frac{Dg_{\parallel}\, g_{\perp} \cos\theta \sin\theta}{g^2}\right]^2 \frac{1}{2g\beta H_0} [4S(S+1) - 24M_S(M_S - 1) - 9]$$

$$+ \left[\frac{Dg_{\perp}^2 \sin^2\theta}{g^2}\right]^2 \frac{1}{8g\beta H_0} [2(S+1) - 6M_S(M_S - 1) - 3] \qquad (81)$$

where g^2 is determined by equation (79) and θ is defined as previously. The fine structure thus produces $2S$ equidistant lines to first order, corresponding to the different values of M_S. These lines are equally separated in the magnetic field if $\theta = 0$, since all the second-order terms in D^2 are then zero. In this case

$$H = H_0 - \frac{2D}{g\beta}(M_S - \tfrac{1}{2}) \qquad (82)$$

and thus D can be determined directly from the separation of such lines.

At low fields, it may be possible to observe transitions corresponding to $\Delta M_S = \pm 2$, $\Delta M_S = \pm 3$, etc. These transitions arise from off-diagonal elements in D, and thus they are very weak if D is small, and vanish if $H \parallel z$. There are $2S - 1$ transitions corresponding to $\Delta M_S = \pm 2$, $2S - 2$ transitions corresponding to $\Delta M_S = \pm 3$, etc. Complete formulae for the angular dependence of these transitions have been given by Buckmaster[75].

Finally, formulae for the angular variation of the energy levels and of the $\Delta M_S = \pm 1$ transition in rhombic symmetry have been derived by Weger and Low (see ref. 16, p. 57). These equations may especially be used to evaluate the fine structure constant E from the angular dependence of the spectrum.

In the presence of hyperfine interaction, each energy level and thus each transition is split again. According to Bleaney[74], this splitting can be taken into account by adding the following term to equation (81)

$$KM_I + \frac{B^2}{4g\beta H_0}\left[\frac{A^2 + K^2}{K^2}\right][I(I+1) - M_I^2] + \frac{B^2}{2g\beta H_0}\left(\frac{A}{K}\right)M_I(2M_S - 1)$$

$$+ \frac{1}{2g\beta H_0}\left[\frac{A^2 - B^2}{K}\right]^2\left(\frac{g_{\parallel}g_{\perp}}{g^2}\right)^2 \sin^2\theta\cos^2\theta M_I^2$$

$$+ \frac{Q'^2 \cos^2\theta \sin^2\theta}{2KM_S(M_S - 1)}\left(\frac{ABg_{\parallel}g_{\perp}}{K^2 g^2}\right)^2 M_I[4I(I+1) - 8M_I^2 - 1]$$

$$- \frac{Q'^2 \sin^4\theta}{8KM_S(M_S - 1)}\left(\frac{Bg_{\perp}}{Kg}\right)^4 M_I[2I(I+1) - 2M_I^2 - 1] \qquad (83)$$

where

$$K^2g^2 = A^2g_{\parallel}^2 \cos^2\theta + B^2g_{\perp} \sin^2\theta$$

Neglecting all second-order terms, the transitions $\Delta M_S = \pm 1, \Delta M_I = 0$ are given by KM_I. The nuclear interaction thus produces $2I + 1$ lines of equal separation K to first order, and the nuclear spin I can be determined by simply counting these lines. This equal splitting is changed in second order, the limiting cases of (83) being for $\theta = 0°$

$$AM_I + \frac{B^2}{2g\beta H_0}[I(I+1) - M_I^2] + \frac{B^2}{2g\beta H_0} M_I(2M_S - 1) \qquad (84)$$

and for $\theta = 90°$

$$BM_I + \frac{A^2 + B^2}{4g\beta H_0}[I(I+1) - M_I^2] + \frac{AB}{2g\beta H_0} M_I(2M_S - 1)$$
$$- \frac{Q'^2}{8BM_S(M_S - 1)} M_I[2I(I+1) - 2M_I^2 - 1] \qquad (85)$$

The separation between hyperfine components thus determines the hyperfine splitting constants for a given direction, e.g. A, B in axial symmetry. The changes introduced in second order can be used to determine the relative signs of D, A, and B. Finally, very careful measurements provide information regarding the term in P^2 and thus the quadrupole coupling constant.

An important method of determining the quadrupole interaction is to study transitions corresponding to $\Delta M_I = \pm 1$, $\Delta M_I = \pm 2$, etc., which may be observed especially when H makes an angle with the axis of the quadrupole interaction. The angular variation of these lines has been studied by Bleaney[74] who lists explicit formulae. Since these transitions are linear in P, they permit a more accurate determination of this quantity, and thus actual values of the nuclear quadrupole moment Q may be derived.

All the interactions discussed so far do not permit us to determine the absolute signs of D, A, B, or P, only relative signs being obtained. This deficiency may, however, be remedied by a careful investigation of the direct interaction between the external magnetic field and the nucleus (cf. last term in the hamiltonian 57). This interaction again produces shifts of the various hyperfine lines which, according to Bleaney[74], may be used to determine the absolute sign of the quadrupole moment and subsequently of the other spin-hamiltonian parameters.

More recently, computer methods have been used for determining the various spin-hamiltonian parameters from the angular variation of the spectrum. For more details, the reader should consult the literature listed by Swalen and Gladney[69].

D. Line Shape Calculations for Polycrystalline Substances

In a polycrystalline sample, the g value of each microcrystal will depend on the angle θ between the symmetry axis and the external magnetic field and, assuming axial symmetry, may be given by (79) or similarly by

$$g^2 = g_\perp^2 + (g_\parallel^2 - g_\perp^2)\cos^2\theta \tag{86}$$

Since $g_\perp$ occurs for both the x and y axes and $g_\parallel$ for the z axis only, there will be a statistical accumulation of signals around the position of $g_\perp$. With random orientation, the fraction of crystals whose axes lie between θ and $\theta + d\theta$ with the magnetic field will be $\frac{1}{2}\sin\theta\, d\theta$. The absorption line shape for a single microcrystal may be assumed to be either lorentzian

$$I = \frac{AK}{1 + \pi^2K^2(x - c)^2} \tag{87}$$

or gaussian

$$I = \pi^{\frac{1}{2}}AK\exp\left[-(x - c)^2\pi^2K^2\right] \tag{88}$$

and I may then be integrated over all θ, giving the macroscopic line shape

$$J = \int_0^{\pi/2} \tfrac{1}{2}I\sin\theta\, d\theta \tag{89}$$

In (87) and (88), A is proportional to the total number of spins, $K = 2/\pi\Delta x$ for a lorentzian line shape and $K = 2(\ln 2)^{\frac{1}{2}}/\pi\Delta x$ for a gaussian line shape. Also, Δx is the linewidth $x = \beta H/h\nu$, and $c = 1/g$ if the absorption is obtained experimentally by sweeping the field H.

If $g_\parallel - g_\perp$ is very small, (86) may be replaced by

$$g = g_\perp + (g_\parallel - g_\perp)\cos^2\theta \tag{90}$$

In this case, J may be integrated exactly to give a closed expression for the derivative line shape dJ/dx[76]. Figure 5 gives an example of line shapes calculated in this way.

For larger $g_\parallel - g_\perp$, (86) has to be used and the integration has to be performed numerically. Also, the variation of A with g, which according to Bleaney[77] is proportional to

$$\tfrac{1}{2}g_\perp^2[(g_\parallel/g)^2 + 1]$$

has to be taken into account. Such calculations have been performed by, e.g., Searl, Smith, and Wyard[78].

Bleaney[79] and Sands[80] assumed a delta function for the line shape of each microcrystal; the absorption as function of the magnetic field may be written as

$$I(H) = \int_0^{\pi/2} f(H - H')\sin\theta\, d\theta \tag{91}$$

where $f(H - H')$ is an arbitrary shape function with a centre at the position H'. Taking $f(H - H')$ as the delta function, it follows that

$$I(H) \propto \left.\frac{d(\cos\theta)}{dH'}\right|_{H=H'} \tag{92}$$

where H lies between $H_{\parallel}$ and $H_{\perp}$. Using (79), we obtain

$$I(H) = H_{\perp}^2 H_{\parallel}(H_{\perp}^2 - H_{\parallel}^2)^{-\frac{1}{2}} \cdot H^{-2}(H_{\perp}^2 - H^2)^{-\frac{1}{2}} \tag{93}$$

where $H_{\perp} = h\nu/g_{\perp}\beta$. This gives only a crude approximation to the expected line shape. Improvements may be achieved by computer calculation using a lorentzian line shape for $f(H - H')$. Compare, e.g., Ibers and Swalen[81].

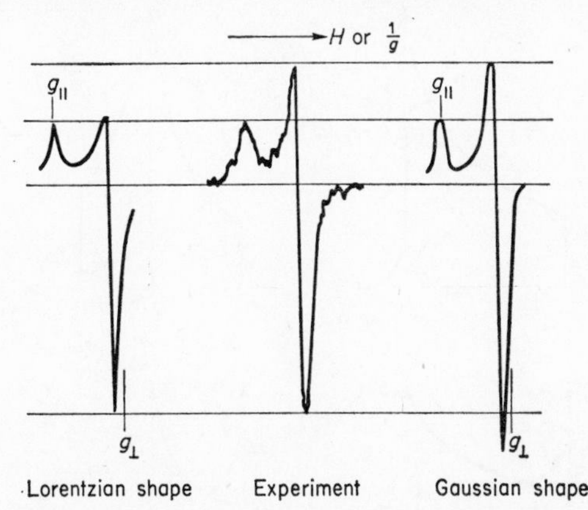

Figure 5. Calculated lorentzian and gaussian line shapes for small $g_{\parallel} - g_{\perp}$ and experimental line shape for 80% D_2O irradiated by u.v. radiation at 90°K [according to Wyard and Smith[76]].

The extension to orthorhombic and lower symmetry ($g_x \neq g_y \neq g_z$) has been treated by Kneubühl[82]. In this case, the total absorption is obtained by a double integral as

$$I(H) \propto \int_0^{2\pi} d\phi \int_0^{\pi} K(\theta, \phi, \psi) f(H - H'(\theta, \phi)) \sin\theta \, d\theta \tag{94}$$

where $K(\theta, \phi, \psi)$ is the transition probability. Proceeding as above, we obtain the normalized absorption for $g_3 > g_2 > g_1$ and for $H \leqslant H_2$ as

$$I(H) = (2/\pi) \frac{H_1 H_2 H_3 H^{-2}}{(H_1^2 - H^2)^{\frac{1}{2}}(H_2^2 - H_3^2)^{\frac{1}{2}}} K(k) \tag{95}$$

where

$$k^2 = \frac{(H_1^2 - H_2^2)(H^2 - H_3^2)}{(H_1^2 - H^2)(H_2^2 - H_3^2)} \tag{96}$$

and for $H \geqslant H_2$ as

$$I(H) = (2/\pi)\frac{H_1 H_2 H_3 H^{-2}}{(H_1^2 - H_2^2)^{\frac{1}{2}}(H^2 - H_3^2)^{\frac{1}{2}}} K(1/k) \tag{97}$$

$K(k)$ is the complete elliptic integral of the first kind,

$$K(k) = \int_0^{\pi/2} \mathrm{d}b/(1 - k^2 \sin^2 b)^{\frac{1}{2}}$$

$$K(0) = \pi/2, \qquad K(1) = \infty \tag{98}$$

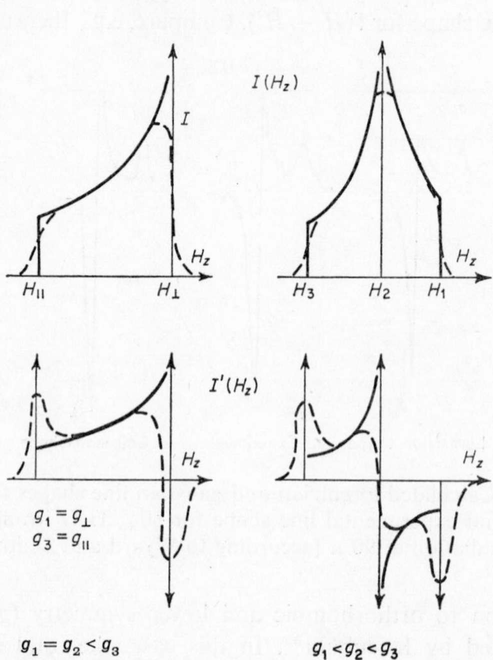

Figure 6. Line shapes $I(H)$ and derivatives $I'(H)$ for axial and orthorhombic symmetry [according to Kneubühl[82]].

The line shapes and their derivatives calculated according to (93) and (97) are shown in Figure 6. Full lines represent the ideal curves; broken lines are an approach to the real shapes. It is clear that we can distinguish between shapes which are due to two and those due to three different principal g values. The derivative shape function corresponding to axial symmetry has five, and that corresponding to orthorhombic symmetry has seven points of inflexion. In both cases, all g values may be obtained from the curves; $g_{\|}$, $g_{\perp}$, g_1, and g_3 correspond to maxima, and g_2 corresponds to a zero point of the derivatives. Figure 7 shows two examples of actual g-value determination from line shapes of polycrystalline samples[82].

Hyperfine structure has been included by Neiman and Kivelson[83], Gersmann and Swalen[84], and others. The result is that $g_{\parallel}$ and $|A|$ are easily found from the spectrum, axial symmetry provided, since the maxima of the hyperfine lines around $g_{\parallel}$ are close to the fields $H = h\nu/g_{\parallel}\beta - |A|M_I/g_{\parallel}\beta$.

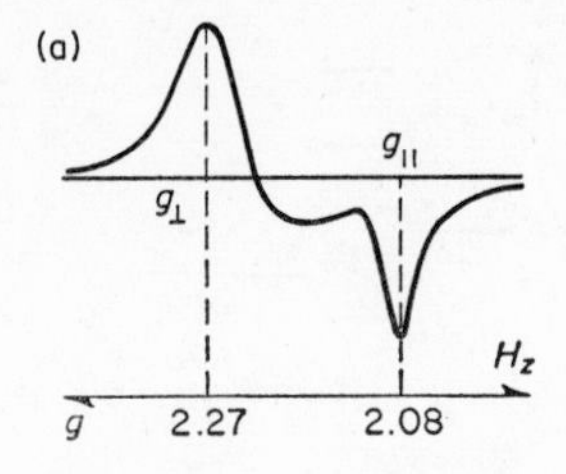

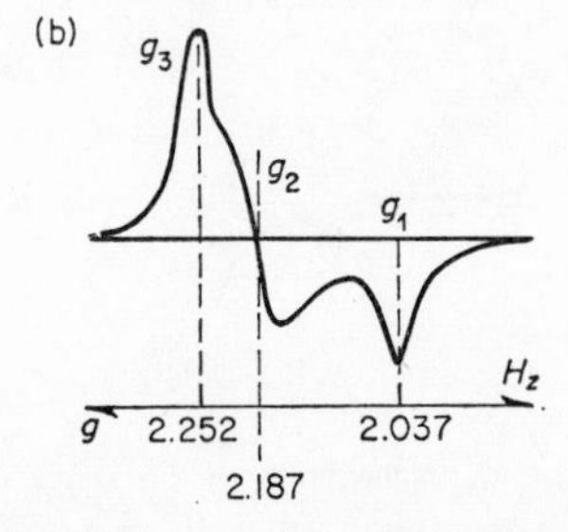

Figure 7. Determination of g values from experimental line shapes of polycrystalline samples (a) $CuSO_4 \cdot 5H_2O$, (b) $CuCl_2 \cdot 2H_2O$ [according to Kneubühl[82]].

It is considerably more difficult to obtain a reasonable estimate for $g_{\perp}$ and $|B|$ for which we must refer to the discussion by Vänngard and Aasa[85]. More complete listings of references concerning line shape calculations may be found in the paper by Swalen and Gladney[69] and in a recent compilation by the author[86].

V. EXAMPLES OF FIRST TRANSITION SERIES COMPOUNDS

This section contains selected examples of paramagnetic resonance results from the first transition metal series. The results are employed to demonstrate the application of the theoretical methods outlined above.

The reason that preference has been given to transition metal compounds is that the partly filled d-electron shells are less shielded than the partly filled f shells of the rare earths and the actinides. The d electrons are profoundly influenced by the surrounding ligands and the effects of metal–ligand bonding, electron transfer, etc., which are of some general interest, may be studied with ease.

Results from single-crystal studies have been used almost exclusively. In addition, tables containing the experimental data of single-crystal investigations in iron series compounds, which are discussed in the text, have been

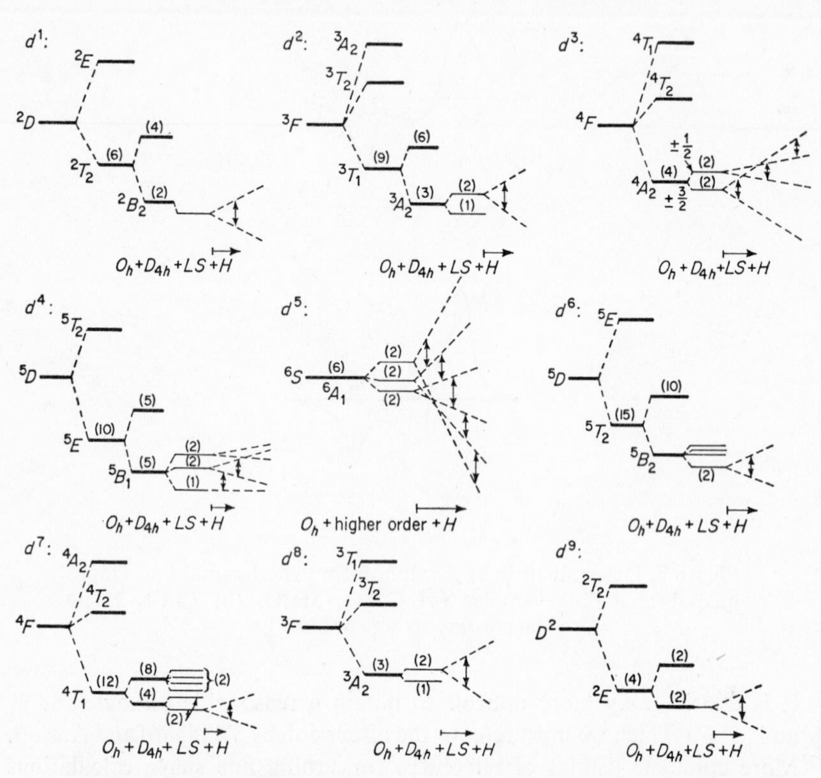

Figure 8. Lowest energy levels of paramagnetic d^1 to d^9 transition metal ions in a magnetic field (not to scale). O_h, octahedral ligand field; D_{4h}, tetragonal field component; LS, spin–orbit interaction; H, magnetic field[86].

appended. Results of e.p.r. from dilute solutions and polycrystalline samples are considered only in few cases. For more complete information, a recent compilation of data[86] should be consulted.

The situation in which paramagnetic resonance is observed in the examples treated here is quite different for each specific case and depends on electronic configuration, symmetry, diluent, concentration of paramagnetic ions, etc. Figure 8 may be used to obtain a preliminary view.

It gives a schematic disposition of the lowest energy levels for the configurations d^1 to d^9 in an octahedral environment with a small axial field present. More details for each specific configuration will be found below.

A. Six- and Four-coordinated d^1 Compounds: Ti^{III}, V^{IV}, Cr^{V}, Mn^{VI}

The ^2D ground state of the ionic d^1 configuration is split by a field of octahedral symmetry into the states $^2T_2$ and ^2E. The ^2E state is usually higher in energy by more than 10^4 cm^{-1} and thus does not influence the properties of the $^2T_2$ state to a first approximation. Spin–orbit interaction splits the $^2T_2$ state into the lower quartet Γ_8 with energy $-\frac{1}{2}\lambda$ and the higher doublet Γ_7 with energy λ. On application of a magnetic field, the energies of the resulting levels are given by[87]:

$$\begin{aligned} \Gamma_7 &\begin{cases} \lambda + \beta H + \frac{4}{3}\beta^2 H^2/\lambda \\ \lambda - \beta H + \frac{4}{3}\beta^2 H^2/\lambda \end{cases} \\ \Gamma_8 &\begin{cases} -\frac{1}{2}\lambda & \text{(doubly degenerate)} \\ -\frac{1}{2}\lambda \quad -\frac{4}{3}\beta^2 H^2/\lambda & \text{(doubly degenerate)} \end{cases} \end{aligned} \tag{99}$$

There is no first-order Zeeman effect for the two levels of Γ_8, and thus the g factor is zero.

A field of tetragonal (D_{4h}) symmetry splits the cubic $^2T_2$ state into $^2B_2$ and ^2E with a separation δ. Spin–orbit coupling removes the remaining orbital degeneracy, producing three Kramers doublets according to $^2B_2 \rightarrow \Gamma_{t7}$, $^2E \rightarrow \Gamma_{t6} + \Gamma_{t7}$. Similarly, a trigonal distortion splits the $^2T_2$ into $^2A_1$ and ^2E. These states are subsequently transformed by spin–orbit interaction according to $^2A_1 \rightarrow \Gamma_6^T$, $^2E \rightarrow \Gamma_4^T + \Gamma_5^T + \Gamma_6^T$. Since in the absence of a magnetic field Γ_4^T and Γ_5^T are degenerate, three Kramers doublets again are produced. The splitting δ is taken as positive whenever the orbital singlet is lowest. We can distinguish two cases: if $\delta < 0$, the two possible ground doublets are part of Γ_8 and thus $g = 0$. If $\delta > 0$, the g values for both tetragonal and trigonal distortion are[79]

$$\begin{aligned} g_{\parallel} &= -1 + 3S^{-1}(\delta + \tfrac{1}{2}\lambda) \\ g_{\perp} &= 1 + S^{-1}(\delta - \tfrac{3}{2}\lambda) \end{aligned} \tag{100}$$

where $S = (\delta^2 + \delta\lambda + \frac{9}{4}\lambda^2)^{\frac{1}{2}}$. Taking the energy of the lowest level to be $E_0 = 0$, the other two levels have energies $E_1 = \frac{1}{2}(\delta - \frac{3}{2}\lambda + S)$ and $E_2 = S$, respectively.

In deriving the expressions (100), the contribution of the upper ^2E state has been neglected[79]. However, in the case of a trigonal field, the ground state and the upper manifold have interconnecting matrix elements of the trigonal potential and of spin–orbit coupling. This interaction has been considered in detail by Gladney and Swalen[88] who performed extensive numerical calculations of $g_{\parallel}$ and $g_{\perp}$. Covalency of the metal–ligand bond influences the g values and was also taken into account[89,90].

Experimentally, it is found that, for slightly distorted octahedral complexes, the separation between the lowest doublets is about 100 cm^{-1} and, due to line broadening by rapid spin-lattice relaxation, e.p.r. is observed only at very low temperatures. This is especially the case for the caesium[89] and rubidium titanium alums[91] where the distortion of the field is of trigonal symmetry. (For the x-ray structure of alums isomorphous to caesium titanium alum, see Lipson[92].) For rubidium titanium alum, the direction cosines between the axes x, y, z of the (g) tensor and the crystallographic axes were determined (see Table 4)[91]. For caesium titanium alum, the best fit to the experimental g values (Table 4) which was obtained by Bleaney and coworkers[89] using equation (100) is $g_{\parallel} = 1.40$ and $g_{\perp} = 0.95$, $\delta = 215$ cm^{-1}. If interaction with the upper ^2E is taken into account, a better agreement with theory may be achieved, especially if orbital reduction factors $k_{\parallel}$ and $k_{\perp}$ are introduced. However, no unique set of parameters has been obtained[88], values of δ between 170 and 500 cm^{-1} producing an equally good fit. In these calculations, the off-diagonal elements specifying the interaction of the ground manifold with the upper manifold (^2E) have been assigned values between 2400 and -2800 cm^{-1}.

Similarly, single crystals of $AlCl_3 \cdot 6H_2O$ containing Ti^{3+} ions show resonance only below 77°K. To account for the observed isotropic g value, a large axial field splitting[93] of the order of 10^4 cm^{-1} or considerable covalency[88] have been assumed.

In a complex having a molecular symmetry distinctly lower than cubic, e.p.r. may be observed even at room temperature. The only single crystal containing the Ti^{3+} ion in a site of low symmetry which has been studied extensively is titanium(III) acetylacetonate[94]. The point symmetry of the molecule is D_3 and the axial splitting δ has been estimated to be between 2000 and 4000 cm^{-1}. No reliable estimate of the amount of π bonding has been made, not even after the interaction with ^2E has properly been included[88]. As in solutions containing some highly unsymmetrical Ti^{III} complexes[95], the hyperfine interaction due to the isotopes ^{47}Ti and ^{49}Ti has been observed (Table 4). For the titanium(III) oxalato complex, see Table 4.

In solutions containing Ti^{3+} ions in presence of complexing agents, resonance is often observed and is due to the formation of complex species of low symmetry[96,97].

In vanadyl and chromyl compounds, the strong metal–oxygen bond is the reason for an approximately C_{4v} symmetry and thus e.p.r. may easily be observed at room temperature. Since, in all these compounds, only transitions between components of the lowest doublet are observed, the results are interpreted using an effective spin $S = \frac{1}{2}$ and the spin-hamiltonian

$$\mathscr{H} = g_{\|}\beta H_z S_z + g_{\perp}\beta(H_x S_x + H_y S_y) + AS_z I_z + B(S_x I_x + S_y I_y) \quad (101)$$

In $(NH_4)_2VO(SO_4)_2 \cdot 6H_2O$, it has been possible to determine separately[98] g_x and g_y as well as B_x and B_y (Table 5). Here, off-diagonal terms of the form $\sum_{i,j} F_{ij}(S_i I_j + S_j I_i)$ had to be included in (101) and were characterized by $G_{xy} = F_{xy} + F_{yx} = -4.6 \times 10^{-4}\ \text{cm}^{-1}$. These terms result from non-coincident principal axes of the (**g**) and (**A**) tensors. The principal values of the tensor (**A**) have been obtained by diagonalization, using the experimental values of A, B_x, B_y, and G_{xy}, and are given by

$$A_{zz} = A$$
$$A_{xx} = 71.200 \times 10^{-4}\ \text{cm}^{-1}$$
$$A_{yy} = 72.439 \times 10^{-4}\ \text{cm}^{-1}$$

In addition, a quadrupole interaction term $P\{I_z^2 - \frac{1}{3}I(I + 1)\}$ had to be included in (101).

Ballhausen and Gray[99] have used a semiempirical LCAO–MO method[100] to propose a general bonding scheme for transition metal oxycations of C_{4v} symmetry. According to this study, as applied to the $[VO(H_2O)_5]^{2+}$ ion, a strong σ bond of symmetry A_1 is formed between the $(2p_z + 2s)$ oxygen hybrid orbital and the $(3d_{z^2} + 4s)$ hybrid orbital of the vanadium ion. In addition, there are two π bonds of symmetry E between oxygen $2p_x$ and $2p_y$ orbitals and vanadium $3d_{xz}$ and $3d_{yz}$ orbitals, completing a total of three vanadium to oxygen bonds in the VO^{2+} ion. The remaining five bonds involve sp_σ hybrid orbitals of the four equivalent water oxygens as well as vanadium $A_1(3d_{z^2} - 4s)$, $E(4p_x, 4p_y)$, and $B_1(3d_{x^2-y^2})$ orbitals and, finally, a bond between the sp_σ orbital of the axial water oxygen and the vanadium $A_1(4p_z)$ orbital. The vanadium $3d_{xy}$ orbital of symmetry B_2 is non-bonding. The situation is illustrated by Figure 9.

Since the unpaired electron has to be assigned to the non-bonding molecular orbital of symmetry B_2, the description of the magnetic properties requires evaluation of matrix elements between the ground state B_2 and antibonding orbitals of symmetry B_1 and E which may be written (using the ket notation) as

$$|B_2\rangle = \beta\ |d_{xy}\rangle - \beta'\ |\phi_{b_2}\rangle$$
$$|E\rangle = \begin{cases} \gamma\ |d_{yz}\rangle - \gamma'\ |\phi_{e_y}\rangle \\ \gamma\ |d_{xz}\rangle - \gamma'\ |\phi_{e_x}\rangle \end{cases} \quad (102)$$
$$|B_1\rangle = \beta_1\ |d_{x^2-y^2}\rangle - \beta_1'\ |\phi_{b_1}\rangle$$

Table 4[a]. Titanium Ti^{3+}.

Crystal	Temp. (°K)	g	HFS: $(A, B)^{47,49}$ (10^{-4} cm^{-1})	Remarks	Ref.
$[Ti(H_2O)_6]Cl_3$ $d > 10^4$ in $[Al(H_2O)_6]Cl_3$	77, 4	1.930 ± 0.002		Point group of host crystal D_{3d}^6, point symmetry of Al^{3+} sites C_{3i}, crystal field approx. O_h; g isotropic	93, C.S. 220
$RbTi(SO_4)_2 \cdot 12H_2O$ $d > 100$ in $RbAl(SO_4)_2 \cdot 12H_2O$	4.2	$g_x = 1.715 \pm 0.002$ $g_y = 1.767 \pm 0.002$ $g_z = 1.895 \pm 0.002$		Point group of host crystal T_h, α-alum, $Z = 4$; point symmetry of Ti^{3+} sites in diluted crystal orthorhombic or lower, $M_m = 12$, direction cosines determined[b]	91, C.S. 92, 124
$CsTi(SO_4)_2 \cdot 12H_2O$ und.	4.2–2.5	$g_{\parallel} = 1.25 \pm 0.02$ $g_{\perp} = 1.14 \pm 0.02$		α-Alum, $Z = 4$; g values may be fitted in several ways assuming a trigonal field splitting of the $^2T_2$ level, orbital mixing with ^2E via spin–orbit coupling, and anisotropic orbital reduction	89, 88

[Ti aca_3] $d > 100$ in [Al aca_3]	77	$g_{\parallel} = 2.000 \pm 0.002$ $g_{\perp} = 1.921 \pm 0.001$	$A = 6.3 \pm 3.0$ $B = 17.5 \pm 0.5$	Host crystal monoclinic, space group C_{2h}^5, $Z = 4$; point symmetry of the molecule D_3; angle between b axis and trigonal axis of the molecule, $\theta = 31 \pm 1°$; results suggest A_1 ground state with an E state being 2000 to 4000 cm^{-1} higher; π bonding present	94, C.S. 221, 222
K[Ti ox_2]$\cdot 2H_2O$	90, 20	$g_{\parallel} = 1.86$ $g_{\perp} = 1.96$		Not octahedral; $M_m = 2$	223

[a] In this and the following tables, the abbreviations listed below are used:

d = dilution, expressed as the ratio of number of ions of diluent to number of paramagnetic ions;
d. = diluted
und. = undiluted;
Z = number of molecules per unit cell as determined by x-ray methods;
M_m = number of magnetically distinct complexes per unit cell as determined by e.p.r.;
C.S. = reference to crystal structure;
d.c. = direction cosine.

[b] For the magnetic complex 1, the d.c. were determined as:

	x	y	z
$\cos\alpha$	−0.4482	−0.3242	0.8330
$\cos\beta$	0.8937	−0.1421	0.4255
$\cos\gamma$	−0.0196	0.9352	0.3535

The d.c. of the remaining 11 complexes follow from those of the complex 1 by application of the symmetry operations of the T_h point group.

Table 5. Vanadium V^{4+}.

Crystal	Temp. (°K)	g	HFS: $(A, B)^{51}$ (10^{-4} cm^{-1})	Remarks	Ref.
$(NH_4)_2VO(SO_4)_2 \cdot 6H_2O$ $d = 3 \times 10^3$ in $(NH_4)_2Zn(SO_4)_2 \cdot 6H_2O$	293	$g_z = 1.9331$ $g_x = 1.9813$ $g_y = 1.9801$	$A = 182.81$ $B_x = 71.37$ $B_y = 72.56$	The V–O axis has three possible orientations with populations in the ratio 20:5:1; parameters listed are for most populous position; off-diagonal terms in **S·I** characterized by $G_{xy} = -4.6 \times 10^{-4}$ cm^{-1}; x, y principal axes of g and A tensors displaced from each other by 23°20′; $P = 0.7 \times 10^{-4}$ cm^{-1}	98, C.S. 224
$(NH_4)_3[VOCl_5]$ d. in $(NH_4)_2[InCl_5(H_2O)]$	293	$g_\parallel = 1.9450$ $g_\perp = 1.9847$	$A = 173$ $B = 63.8$	Host crystal orthorhombic, $Z = 4$; $M_m = 2$	105, C.S. 225
$K_2[VO\ ox_2] \cdot 2H_2O$ $d = 2$ to 200 in $K_2[TiO\ ox_2] \cdot 2H_2O$	293	$g_\parallel = 1.940 \pm 0.002$ $g_\perp = 1.972 \pm 0.002$	$A = 163 \pm 2$ $B = 60 \pm 2$	No crystal structure available, uniaxial; $\|P\| = 0.65 \pm 0.09$ 10^{-4} cm^{-1}	107

In (102), the ligand functions ϕ_Γ are group orbitals of the appropriate symmetry[99]. Due to normalization and overlap relations, only three of the six coefficients in (102), namely β, β_1, and γ are independent.

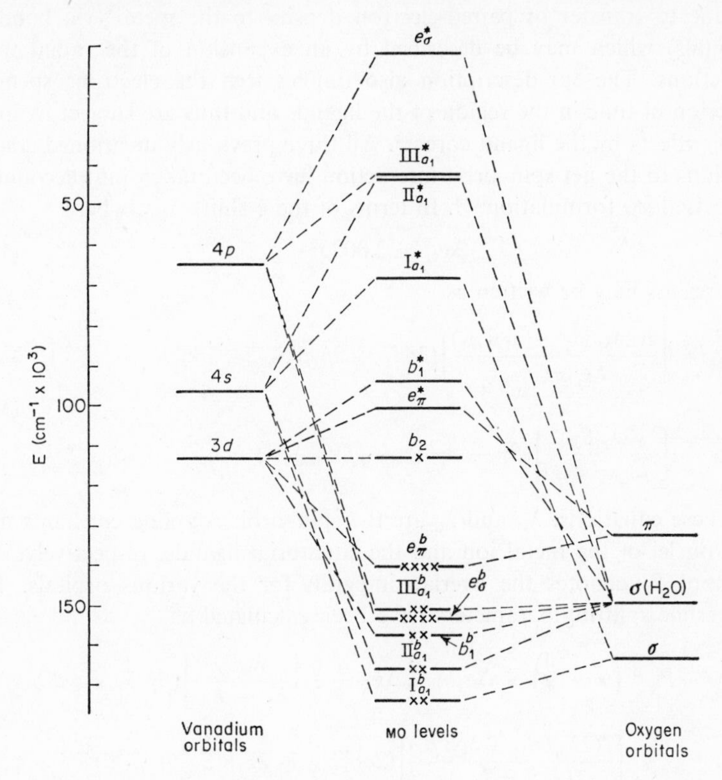

Figure 9. Bonding scheme for the VO^{2+} ion [according to Ballhausen and Gray[99]].

On the basis of their MO description of the $[VO(H_2O)_5]^{2+}$ ion, Ballhausen and Gray[99] obtained for the g values the formulae

$$g_{\parallel} = 2.0023\left(1 - \frac{4\beta_1^2\lambda}{E_{b_1} - E_{b_2}}\right)$$
$$g_{\perp} = 2.0023\left(1 - \frac{\gamma^2\lambda}{E_e - E_{b_2}}\right) \tag{103}$$

where the denominator contains the energies of known crystal-field transitions. The inclusion of metal MO coefficients (orbital reduction[101,102]) in (103) attempts to explain the fact that the reduction of g values is smaller than predicted from the free-ion spin–orbit parameter. However, the amount of covalency estimated from the reduction in λ is excessive. Marshall and Stuart[103] proposed that the decrease in spin–orbit mixing is due to transfer of paired-electron density to the metal (via bonding orbitals) which may be described by an expansion of the radial wave functions. The MO description also implies that the electrons spend a fraction of time in the region of the ligands and thus are subject to spin–orbit effects by the ligand core[104]. All three previously mentioned contributions to the net spin–orbit interaction have been taken into account in a revised MO formulation[105]. In terms of the g shifts Δg_{ij}, where

$$\Delta g_{ij} = 2.0023 - g_{ij} \tag{104}$$

the results may be written as

$$\begin{aligned} \Delta g_{\|} &= \left[\frac{2(2\lambda_{\mathrm{M}}\beta\beta_1 - \lambda_{\mathrm{L}}\beta'\beta_1')}{E_{b_1} - E_{b_2}}\right](2\beta\beta_1 - 2\beta_1\beta' S_{b_2} - 2\beta\beta_1' S_{b_1} - \beta_1'\beta') \\ \Delta g_{\perp} &= \left[\frac{2\lambda_{\mathrm{M}}\beta\gamma}{(E_e - E_{b_2})}\right](\beta\gamma - \beta\gamma' S_e - \gamma\beta' S_{b_2}) \end{aligned} \tag{105}$$

In these equations, λ_{M} and λ_{L} are the spin–orbit coupling constants near the nuclei of the metal ion and the equatorial ligands, respectively. The symbol S_{Γ} denotes the overlap integrals for the various orbitals. The hyperfine splitting parameters have been calculated as

$$\begin{aligned} A = -P\Big\{\beta^2\left(\kappa + \frac{4}{7}\right) + \Delta g_{\|} + \frac{3}{7}\Delta g_{\perp} + \frac{6}{7}\left[\frac{\lambda_{\mathrm{M}}\beta\gamma}{E_e - E_{b_2}}\right](\beta\gamma' S_e + \gamma\beta' S_{b_2}) \\ + \left[\frac{2(2\lambda_{\mathrm{M}}\beta\beta_1 - \lambda_{\mathrm{L}}\beta'\beta_1')}{(E_{b_1} - E_{b_2})}\right](2\beta\beta_1' S_{b_1} + 2\beta_1\beta' S_{b_2} + \beta_1'\beta')\Big\} \\ B = -P\Big\{\beta^2\left(\kappa - \frac{2}{7}\right) + \frac{11}{14}\Delta g_{\perp} + \frac{11}{7}\left[\frac{\lambda_{\mathrm{M}}\beta\gamma}{(E_e - E_{b_2})}\right](\beta\gamma' S_e + \gamma\beta' S_{b_2})\Big\} \end{aligned} \tag{106}$$

where $P = 2\beta_0\gamma_{\mathrm{N}}\beta_{\mathrm{N}}\langle r^{-3}\rangle$ and where κ is the isotropic hyperfine interaction (Fermi contact term).

Application of the equations (105) and (106) to the e.p.r. data of the ions $[VO(H_2O)_5]^{2+}$ and $[VOCl_5]^{3-}$ (see Table 5) results in values of the MO coefficients β, γ, and β_1 as listed in Table 6. The estimate $\lambda_{\mathrm{L}} = 587\ \mathrm{cm}^{-1}$ for chlorine has been taken from McClure[106], λ_{L} for equatorial oxygens in $[VO(H_2O)_5]^{2+}$ has been assumed to be negligible ($\lambda_{\mathrm{L}} = 85\ \mathrm{cm}^{-1}$[104]), and κ has been chosen as 0.85. A comparison of the bonding coefficients in the two compounds leads to the conclusion that the V—O π bonding is similar in strength, whereas the equatorial σ bonding seems to be stronger in the $[VOCl_5]^{3-}$ ion.

The only other single-crystal study on a vanadyl compound has been effected on $K_2[VO\,ox_2]\cdot 2H_2O$ diluted in the corresponding titanium(IV) compound[107]. To account for the experimental results (see Table 5), a quadrupole interaction term had to be included in (101). The results are consistent with the order of energy levels as discussed above.

Table 6.

Coefficient	$[VO(H_2O)_5]^{2+}$	$[VOCl_5]^{3-}$
β	1.0	$\leqslant 1.0$
γ	0.962 ± 0.014	0.992 ± 0.014
β_1	0.981 ± 0.009	< 1.0

There are a number of investigations of various vanadyl compounds performed on dilute solutions which, in general, support the bonding scheme of Figure 9. Within these, we would like to refer to measurements on vanadyl acetylacetonate[108], vanadyl tetraphenylporphyrin[108,109], and vanadyl phthalocyanine[110]. The results for $[VO\,aca_2]$ yield $\beta = 1.0$, $\gamma = 0.950$ to 0.975, and $\beta_1 = 0.915$. The extrahyperfine splitting in the spectrum of the porphyrin, which arises from interactions with the nitrogens, indicates that the in-plane π bonding is practically ionic, i.e. $\beta = 1.02$. Also, $\gamma \sim 0.78$ and $\beta_1 \sim 0.85$. Thus, the in-plane σ bonding (β_1) and out-of-plane π bonding (γ) vary from compound to compound and $\beta = 1.0$ for all the complexes.

There are no single-crystal studies available on chromyl compounds. However, measurements on frozen solutions, particularly on $K_2[CrOCl_5]$ enriched with ^{53}Cr have demonstrated[111] that the relative magnitudes of $g_{\parallel}$ and $g_{\perp}$ are opposite to the prediction[112] by equation (103), the experimental values being $g_{\parallel} = 2.008$, $g_{\perp} = 1.977$. An explanation is provided if a contribution from the excited configuration $b_1 \rightarrow b_2^*$ is included in the MO description of the bonding. The unpaired electron again occupies the b_2^* orbital. Since the bonding orbital b_1 may be written as

$$|B_1^b\rangle = \beta_1' \, |d_{x^2-y^2}\rangle + \beta_1 \, |\phi_{b_1}\rangle \qquad (107$$

the shift in $g_{\parallel}$ is calculated as

$$\Delta g_{\parallel} = \left[\frac{(2\lambda_M \beta\beta_1 - \lambda_L \beta'\beta_1)}{E_{b_2^* \rightarrow b_1^*}}\right](2\beta\beta_1 - \beta_1'\beta') + \left[\frac{(2\lambda_M \beta\beta_1' + \lambda_M \beta_1 \beta')}{E_{b_1 \rightarrow b_2*}}\right](2\beta\beta_1' + \beta_1\beta') \qquad (108)$$

The in-plane π bonding involving the metal $3d_{xy}$ orbital appears to be important, since $\beta = 0.930$, $\beta' = 0.115$. If, arbitrarily, we set $\beta = 1.0$, a very large in-plane σ-bond covalency is required, viz. $(\beta_1')^2 \sim 0.5$, which is unlikely. The presence of π bonding is also indicated by the observation of an anisotropic chlorine extrahyperfine splitting[113] in frozen solutions of $K_2[CrOCl_5]$ enriched with ^{52}Cr, which results in $A^{Cl} = 7.8 \times 10^{-4}\ cm^{-1}$. Two contributions have been discussed: a dipole interaction between a $3p$ in-plane π electron and Cl nuclei and an isotropic p polarization. Near cancellation of these interactions is assumed for the parallel position.

Two compounds, where the metal ion is in a tetrahedral field of ligands, have been studied: tetrakis-*t*-butoxyvanadium(IV) and the manganate ion.

In frozen $[V\{(CH_3)_3CO\}_4]$ magnetically diluted by the corresponding titanium(IV) compound, the spin-hamiltonian (101) may be applied and the parameters found are[114]

$$g_{\parallel} = 1.940 \pm 0.005 \qquad A_{\parallel} = (125 \pm 50) \times 10^{-4}\ cm^{-1}$$
$$g_{\perp} = 1.984 \pm 0.005 \qquad A_{\perp} = (36 \pm 40) \times 10^{-4}\ cm^{-1}$$

Using a molecular-orbital description, where the antibonding orbitals are defined as

$$\begin{aligned} |xz\rangle &= \gamma\,|d_{xz}\rangle - \gamma'\,|\phi_{xz}\rangle \\ |yz\rangle &= \gamma\,|d_{yz}\rangle - \gamma'\,|\phi_{yz}\rangle \\ |xy\rangle &= \beta\,|d_{xy}\rangle - \beta'\,|\phi_{xy}\rangle \\ |z^2\rangle &= \delta\,|d_{z^2}\rangle - \delta'\,|\phi_{z^2}\rangle \\ |x^2 - y^2\rangle &= \alpha\,|d_{x^2-y^2}\rangle - \alpha'\,|\phi_{x^2-y^2}\rangle \end{aligned} \tag{109}$$

and assuming $|x^2 - y^2\rangle$ as the ground state, the low-temperature g and A values may be related to the bonding parameters. The resulting values $\alpha = 0.95$, $\beta = 0.91$, and $\gamma = 0.97$ suggest a moderately covalent bonding. The spin–orbit coupling parameter λ is reduced by about 62% as compared to the free-ion value.

The manganate ion has been investigated[115] in a single crystal of K_2CrO_4. X-ray analysis[116] has shown that each chromium atom of the host crystal is at the centre of a distorted tetrahedron of oxygen atoms. Because of the low symmetry of the equally distorted $[MnO_4]^{2-}$ ion, the principal axes of the **(g)**, **(A)**, and **(P)** tensors will not coincide. If the x axis is the principal axis of **(g)**, **(A)**, and **(P)** with origin at the manganese nucleus, and if we choose the y and z axes so as to diagonalize **(g)**, then **(A)** and **(P)** will contain off-diagonal elements (see Table 7). Since an MO analysis[117] has shown that the ground state is a doublet, $S = \frac{1}{2}$ and the spin-hamiltonian takes the form

$$\begin{aligned} \mathscr{H} = \beta(g_x S_x H_x + g_y S_y H_y + g_z S_z H_z) + A_x S_x I_x + A_y S_y I_y + A_z S_z I_z \\ + A_{yz} S_y I_z + A_{zy} S_z I_y + P_x I_x^2 + P_y I_y^2 \\ + P_z I_z^2 + P_{yz}(I_y I_z + I_z I_y) \end{aligned} \tag{110}$$

Table 7. Manganese Mn^{6+}.

Compound	Temp. (°K)	g	HFS: $(A_x, A_y, A_z)^{55}$	Remarks	Ref.
K_2MnO_4 $d = 100$ in K_2CrO_4	20	$g_x = 1.970 \pm 0.005$ $g_y = 1.966 \pm 0.001$ $g_z = 1.938 \pm 0.001$	$A_x = 25 \pm 5 \times 10^{-4}$ cm^{-1} $A_z^2 + A_{zy}^2 = 186.9 \times 10^{-6}$ cm^{-2} $A_y^2 + A_{yz}^2 = 15.9 \times 10^{-6}$ cm^{-2} $A_zA_{yz} + A_yA_{zy} = -36.7 \times 10^{-6}$ cm^2	Host crystal orthorhombic, $Z = 4$; A_y, A_z not separately determined; additional lines due to $\Delta M_I = \pm 1$ observed; quadrupole moment of ^{55}Mn determined as $Q = 0.2 \times 10^{-24}$ cm^2	115, C.S. 116

A discussion of the hyperfine splitting has shown that the orbital of the unpaired electron (approximately of symmetry type E) is largely concentrated on the Mn ion with a comparatively small overlap on to the oxygen ligands. The main effect of the quadrupole interaction in (110) is to allow transitions in which the z component of I changes according to $\Delta M_I = \pm 1$ or ± 2. Only $\Delta M_I = \pm 1$ lines were observed. The resulting parameters were used to calculate the electric quadrupole moment of ^{55}Mn, resulting in $Q = 0.2 \times 10^{-24}$ cm^2.

B. Six- and Four-coordinated d^2 Compounds: V^{III}, Mn^{V}, Fe^{VI}

The ground state ^3F of the $3d^2$ configuration of the free ion is split by a field of octahedral symmetry into $^3T_1$, $^3T_2$, and $^3A_2$. The lowest level is the nine-fold degenerate $^3T_1$ which is split by spin–orbit interaction into $\Gamma_1 + \Gamma_3 + \Gamma_4 + \Gamma_5$. Since Γ_3 and Γ_5 are accidentally degenerate, actually a five-fold degenerate level ($\Gamma_3 + \Gamma_5$) is obtained at $-\alpha\lambda$, with a three-fold degenerate level (Γ_4) at $+\alpha\lambda$ and a non-degenerate level (Γ_1) at $+2\alpha\lambda$, where $\alpha \sim \frac{3}{2}$. The degeneracy of Γ_3 and Γ_5 is resolved only in second approximation via spin–orbit interaction with the next higher Γ_5 level. However, it is expected that there will always be a small axial field present, either due to a static distortion or to the Jahn–Teller effect, which will be acting on the $^3T_1$ level. If the field is trigonal, the $^3T_1$ splits into $^3A_2$ and ^3E, the $^3A_2$ being the lowest. Spin–orbit coupling on $^3A_2$ finally produces a non-degenerate lower level ($M_S = 0$) and a fairly close lying (~ 10 cm^{-1}) doubly degenerate higher level ($M_S = \pm 1$), all the remaining levels being more than 100 cm^{-1} higher. The energies of the lowest three levels have been given[38] for large zero-field splitting δ (i.e. small λ/δ)

$$\begin{aligned} M_S = 0: &\quad E_0 = \rho - 2\alpha'^2\lambda^2/\delta \\ M_S = \pm 1: &\quad E_{\pm 1} = -\tfrac{1}{2}\rho - \alpha'^2\lambda^2/\delta \end{aligned} \tag{111}$$

Only these three levels contribute to e.p.r. at the low temperatures necessary to overcome spin-lattice relaxation. Thus $S = 1$ and

$$\begin{aligned} \mathscr{H} = g_\parallel \beta H_z S_z + g_\perp \beta (H_x S_x + H_y S_y) + D S_z{}^2 + A S_z I_z \\ + B(S_x I_x + S_y I_y) + P[I_z{}^2 - \tfrac{1}{3} I(I+1)] \end{aligned} \tag{112}$$

Electronic spectra of Al_2O_3 containing V^{3+} ions[16] indicate a splitting between the lowest singlet and doublet of about 8 ± 1 cm^{-1}. The large zero-field splitting makes the observation of $\Delta M_S = \pm 1$ transitions impossible. Only transitions between the levels $M_S = +1$ and $M_S = -1$ ($\Delta M_S = 2$) are expected. From measurements on sapphire (V^{3+} in Al_2O_3), $g_\parallel = 1.92 \pm 0.01$, $A = 193 \pm 2 \times 10^{-4}$ cm^{-1} and $D \sim 10$ cm^{-1} were obtained[118]. The resonance was observed in parallel orientation ($H \parallel z$) at 4.2°K only.

Since the number of unpaired electrons is even, an additional small splitting of the $M_S = +1$ and $M_S = -1$ levels at zero magnetic field is expected. This splitting is accounted for by adding a term ΔS_x to the hamiltonian (112). It is assumed that the strong trigonal field present in Al_2O_3 makes the observation of e.p.r. possible.

In a tetrahedral field, the ground state is the $^3A_2$ state, which goes over into Γ_5 under influence of spin–orbit coupling. If an orthorhombic distortion is present, the Γ_5 level is split into three components characterized by $M_S = +1, 0, -1$. Only transitions having $\Delta M_S = \pm 1$ are allowed.

Single-crystal results are available[115] for the hypomanganate and ferrate ions (see Table 8). The $[MnO_4]^{3-}$ ion gives an almost completely isotropic spectrum showing two sets of hyperfine lines. This indicates that two electronic transitions were observed. Since both $g = 2$ and $A = 62.5 \times 10^{-4}$ cm^{-1} are isotropic, the spin-hamiltonian is simply

$$\mathcal{H} = g\beta\mathbf{H}\cdot\mathbf{S} + (D_xS_x^2 + D_yS_y^2 + D_zS_z^2) + A\mathbf{S}\cdot\mathbf{I} \qquad (113)$$

with $S = 1$, where the zero-field splittings D_i have not been determined. The ferrate ion, $[FeO_4]^{2-}$, similarly shows two lines which again correspond to the $|0\rangle \leftrightarrow |1\rangle$ and $|0\rangle \leftrightarrow |-1\rangle$ transitions within the lowest level. Again $g = 2$ and isotropic. Thus the hamiltonian (113) without the hyperfine term is applicable.

C. Six-coordinated d^3 Compounds: V^{II}, Cr^{III}, Mn^{IV}

The ground state ^4F of a free d^3 ion is split in an octahedral field into $^4A_2$, $^4T_2$, and $^4T_1$ in the order of increasing energy. The four-fold spin degeneracy of the lowest $^4A_2$ term is not removed by the action of spin–orbit coupling ($D_{\frac{3}{2}} \times A_2 = \Gamma_8$). However, the spin quadruplet is split into two Kramers doublets by fields of lower symmetry. The spin quantum numbers for the two levels are $\pm\frac{3}{2}$ and $\pm\frac{1}{2}$. The results are interpreted in terms of the spin-hamiltonian

$$\mathcal{H} = g\beta(H_xS_x + H_yS_y + H_zS_z) + D(S_z^2 - \tfrac{5}{4}) + E(S_x^2 - S_y^2) + A(S_xI_x + S_yI_y + S_zI_z) \qquad (114)$$

with $S = \frac{3}{2}$, since the hamiltonian parameters are almost isotropic.

Experimentally, single-crystal investigations on the V^{2+} ion are limited to the vanadium Tutton salts and the hexacyano complex. Vanadium has isotopes ^{50}V and ^{51}V with nuclear spins $I = 6$ and $I = \frac{7}{2}$ and abundances 0.25 and 99.75% respectively. The wave function of the $^4A_2$ ground state corresponds to a half-filled t_{2g} orbital and, since this distribution of spin moments produces no magnetic field at the nucleus, the hyperfine structure should be zero. However, the spectrum consists of three electronic transitions according to $\Delta M_S = \pm 1$, of relative intensities 3:4:3, each of

Table 8. Manganese Mn^{5+}, iron Fe^{6+}.

Crystal	Temp. (°K)	g	HFS: A^{55} (10^{-4} cm^{-1})	FS: D (10^{-4} cm^{-1})	Remarks	Ref.
$Na_3MnO_4 \cdot 12H_2O$ d = 50 to 100 in $Na_3VO_4 \cdot 12H_2O$	90, 20	2	62.5		$Z = 12$; g and A isotropic; two transitions observed	115, C.S. 226
K_2FeO_4 d = 100 in K_2CrO_4	90	2.000 ± 0.004		$D_x = 504$ $D_y = 180$ $D_z = 684$	Host crystal orthorhombic, $Z = 4$; g isotropic; two transitions observed	115, C.S. 116

them showing a hyperfine structure of $(2I + 1) = 8$ components due to the isotope ^{51}V as indicated in Figure 10. The hyperfine splitting is completely isotropic, $A = 88 \times 10^{-4}$ cm^{-1} for $(NH_4)_2V(SO_4)_2 \cdot 6H_2O$[119] and $A = -55.5 \times 10^{-4}$ cm^{-1} for $K_4[V(CN)_6] \cdot 3H_2O$[120]. In the same way as for *S*-state ions, the hyperfine structure may be accounted for by the spin polarization of inner core *s* electrons[33,36,121]. A small anisotropy in $(NH_4)_2V(SO_4)_2 \cdot 6H_2O$ has been observed[98] both using an ordinary single crystal as well as one where the V^{2+} ions were obtained by x-ray irradiation of the corresponding vanadyl Tutton salt (see Table 9). The hyperfine

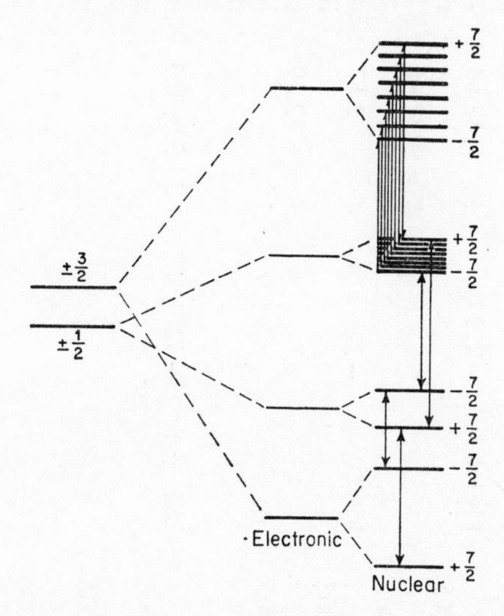

Figure 10. Schematic diagram for the lowest energy levels of a V^{2+} ion in a field of symmetry lower than cubic. The hyperfine structure of the three allowed transitions in a strong magnetic field is indicated on the right [according to Bleaney and colleagues[119]].

splitting due to the isotope ^{50}V has been resolved[122] using a Tutton salt crystal enriched in ^{50}V. Thus the nuclear spin $I(^{50}V) = 6$ and the ratio $g(^{50}V)/g(^{51}V) = 0.3792 \pm 0.00008$ were determined. Practically the same results were obtained[120, 123] in an analogous investigation carried out on $[K_4V(CN)_6] \cdot 3H_2O$.

There is a large number of single-crystal data available for compounds of chromium(III). Chromium has an isotope ^{53}Cr with $I = \frac{3}{2}$ and an abundance of 9.6%. The hyperfine structure has been resolved only in some of the compounds studied and, as in vanadium(II), it mainly arises from the *s*-electron Fermi interaction.

Table 9. Vanadium V^{2+}.

Crystal	Temp. (°K)	g	D (10^{-4} cm^{-1})	E (10^{-4} cm^{-1})	HFS: $(A,B)^{51}$ (10^{-4} cm^{-1})	Remarks	Ref.
$(NH_4)_2V(SO_4)_2 \cdot 6H_2O$ $d = 1000$ in $(NH_4)_2Zn(SO_4)_2 \cdot 6H_2O$	90, 20	1.951 ± 0.002	1580 ± 100	490 ± 50	88 ± 2	Monoclinic, $Z = 2$, $\psi = +4°$, $\alpha = 23.5°$; $A^{50}/A^{51} = 0.3792 \pm 0.0008$; $\theta = 68°$, $\phi = 2°$ [a]	119, 122, C.S. 201, 202
$(NH_4)_2V(SO_4)_2 \cdot 6H_2O$ $d = 200$ in $(NH_4)_2Zn(SO_4)_2 \cdot 6H_2O$	293	$g_z = 1.9717 \pm 0.0005$ $g_y = 1.9733 \pm 0.0005$	-1561.3	-229.0	$A = -82.63$ $B = -82.46$	$\theta = 69.5° \pm 0.5°$, $\phi = 2° \pm 5°$	98
$(NH_4)_2V(SO_4)_2 \cdot 6H_2O$ d. in $(NH_4)_2Zn(SO_4)_2 \cdot 6H_2O$ [b]	293	$g_z = 1.9718 \pm 0.0005$ $g_y = 1.9733 \pm 0.0005$	-1560.9	-229.7	$A = -82.67$ $B = -82.49$	Sample obtained by x-ray irradiation of $(NH_4)_2VO(SO_4)_2 \cdot 6H_2O$ in the same matrix; $\theta = 69.5° \pm 0.5°$, $\phi = 3° \pm 1°$	98
$K_4[V(CN)_6] \cdot 3H_2O$ $d = 20$ to 1000 in $K_4[Fe(CN)_6] \cdot 3H_2O$	90	$g_x = 1.9919 \pm 0.0006$ $g_y = 1.9920 \pm 0.0006$ $g_z = 1.9920 \pm 0.0006$	-264 ± 4	-72 ± 4	-55.5 ± 0.3	Monoclinic, $Z = 2$; space group C_{2h}^6; $A^{50}/A^{51} = 0.380 \pm 0.001$; direction cosines determined[c]	120, 123

[a] The angles θ and ϕ locate the z axis of the spin hamiltonian in the crystal. θ is the polar angle from the b axis, ϕ is the azimuthal angle from the c axis in the ac plane.

[b] For spin-hamiltonian parameters of x-ray irradiated VO^{2+} in different Tutton salts, see ref. 98.

[c] The d.c. were determined as (upper signs for complex 1, lower signs for complex 2):

	x	y	z
a	0.707	0.523	0.470
b	0	± 0.676	∓ 0.737

The alums, in particular, do not show hyperfine structure, which is apparently due to large linewidth. However, a number of interesting observations were made in relation to their molecular symmetry. The alums all have the same space group *Pa*3, with four molecules per unit cell[92,124]. Each trivalent ion is surrounded by an octahedron of water molecules. In the so-called α-alums, the octahedron shows a small trigonal distortion along the crystal [111] axis, and the cubic axes of the octahedron are displaced by an angle ϕ relative to the cubic axes of the crystal. For the β-alums, x-ray results show that the octahedra should be perfectly regular and that the cubic axes of the octahedra should coincide with those of the crystal.

Paramagnetic resonance shows that both α- and β-alums have trigonal distortions of the same order of magnitude[125] (for example: $KCr(SO_4)_2 \cdot 12H_2O$, an α-alum, $D = 600 \times 10^{-4}$ cm^{-1}; $CsCr(SO_4)_2 \cdot 12H_2O$, a β-alum, $D = 725 \times 10^{-4}$ cm^{-1}, both at 290°K). The ground-state splittings are strongly dependent on temperature and, in general, decrease linearly with temperature (see Table 10). In several alums, the resonance results indicate crystal transitions to lower symmetries at a certain temperature. In potassium alum, a gradual transition occurs below 160°K[126]. Below the transition point, two magnetically different complexes I and II are observed, both having larger initial splittings than the complex above 160°K, as listed in Table 10. In ammonium alum, a sudden transition is observed at about 80°K, two magnetically different complexes being present below this temperature[126]. Again, much larger values for the parameter D are obtained below the transition temperature. Methylamine alum undergoes a crystal transition at 157°K (170°K if diluted 1:100 in aluminium alum). Above 157°K, the crystalline electric field is of trigonal symmetry; below 157°K, the symmetry is rhombic[127]. It is assumed that the hexaquo ion in the methylamine alum undergoes a random tumbling motion to produce a time-averaged spherical symmetry in order to be accommodated in a lattice site which, in other alums, is occupied by spherically symmetrical ions. At low temperatures, the energy of tumbling would not be sufficient to overcome the potential barrier, and thus the molecule is frozen in some fixed orientation which may be followed by a lowering of symmetry. The pressure dependence of the spin-hamiltonian parameters has been studied on ammonium chromium alum[128]. Whereas g remains constant, D increases rapidly with rising pressure and temperature.

Potassium chromicyanide, $K_3[Cr(CN)_6]$, has been studied in two single crystals, the diluents being $K_3[Co(CN)_6]$ and $K_3[Mn(CN)_6]$, respectively, as well as in undiluted crystals[120]. In diluted crystals, the relative signs of the parameters D, E, and A were determined from

Table 10. Chromium Cr^{3+}.

Crystal	Temp. (°K)	g	D (10^{-4} cm^{-1})	E (10^{-4} cm^{-1})	HFS: $(A, B)^{53}$ (10^{-4} cm^{-1})	Remarks	Ref.
$KCr(SO_4)_2 \cdot 12H_2O$ und.	290	1.98 ± 0.02	600 ± 30			α-Alum, $Z = 4$; below 160°K two magnetically different complexes involving change in point symmetry at Cr^{3+} site; crystal transition is gradual	126, 125, 227, C.S. 92, 124
	160	1.98 ± 0.02	175				
	90	1.98 ± 0.02	I: 1300				
			II: 750 ± 50				
$NH_4Cr(SO_4)_2 \cdot 12H_2O$ und.	290	1.98 ± 0.02	675 ± 30			α-Alum, $Z = 4$; below 80°K two magnetically different complexes; crystal transition is sudden	126, 125, 227 C.S. 92, 124
	90	1.98 ± 0.02	175				
	80	1.98 ± 0.02	I: 1570 ± 20				
			II: 1210 ± 20				
$NH_4Cr(SO_4)_2 \cdot 12H_2O$ $d = 17$ in $NH_4Al(SO_4)_2 \cdot 12H_2O$	298	1.9771 ± 0.0010	492 ± 5			α-Alum, $Z = 4$; same results obtained for $d = 47$ and $d = 100$; D increases with rising pressure and temperature	228, 128, C.S. 92, 124
$RbCr(SO_4)_2 \cdot 12H_2O$ und.	290	1.98 ± 0.02	825 ± 30			α-Alum, $Z = 4$	126, 125, 227, C.S. 92, 124
	193	1.98 ± 0.02	630				
	90, 20, 4	1.98 ± 0.02	540 ± 10				
$CsCr(SO_4)_2 \cdot 12H_2O$ und.	290	1.98 ± 0.02	725 ± 30			β-Alum, $Z = 4$	126, 125, 227, C.S. 92, 124
	193	1.98 ± 0.02	-670^{a}				
	90, 20, 4	1.98 ± 0.02	−665 ± 10				

$(CH_3NH_3)Cr(SO_4)_2\cdot$ $12H_2O$ und.	290 90 20	1.98 ± 0.01 1.975 ± 0.01 1.976 ± 0.007	825 ± 30 870 ± 20 −871 ± 7	 90 ± 10 −92 ± 8		β-Alum, $Z = 4$; crystal transition at 157° ± 3°K; spectrum has rhombic symmetry below transition temperature; direction cosines determined[b]	126, 125, 227, 127, C.S. 92, 124
$(CH_3NH_3)Cr(SO_4)_2\cdot$ $12H_2O$ $d = 100$ in $(CH_3NH_3)Al(SO_4)_2\cdot$ $12H_2O$	90 20	1.975 ± 0.01 1.977 ± 0.003	959 ± 20 −958 ± 4	90 ± 10 −92 ± 8		β-Alum, $Z = 4$; crystal transition at 170° ± 3°K; spectrum has rhombic symmetry below transition temperature; at low temperature, space group is D_{4h}	127, 229, C.S. 92, 124
$KCr(SeO_4)_2\cdot 12H_2O$ $d = 100$ in $KAl(SeO_4)_2\cdot 12H_2O$	90 20	1.976 ± 0.002 1.976 ± 0.002	900 ± 3 983 ± 3		18.5 ± 1 18.5 ± 1	$d = 10$ and und. crystals also measured; HFS only in crystals enriched in [53]Crand containing D_2O observed	229, 230
$K_3[Cr(CN)_6]$ und.	90, 20, 4	1.998 ± 0.008	~450	~130		Monoclinic, $Z = 2$; width and shape of resonance line anisotropic; exchange interaction	120, 130, 231, C.S. 232, 233

(*continued*)

Table 10 (*continued*)

Crystal	Temp. (°K)	g	D (10^{-4} cm^{-1})	E (10^{-4} cm^{-1})	HFS: $(A, B)^{53}$ (10^{-4} cm^{-1})	Remarks	Ref.
$K_3[Cr(CN)_6]$ d = 100 to 1000 in $K_3[Co(CN)_6]$	90, 20	$g_x = 1.993 \pm 0.001$ $g_y = 1.991 \pm 0.001$ $g_z = 1.991 \pm 0.001$	$+831 \pm 10$	$+108 \pm 10$	$+14.7 \pm 0.5$	Monoclinic, $M_m = 2$; direction cosines determined[c]; in ^{13}C-enriched sample, ligand (^{13}C) HFS observed: $A^{13} = -7.98 \times 10^{-4}$ cm^{-1}, $B^{13} = -9.80 \times 10^{-4}$ cm^{-1}	120, 231, C.S. 232, 233
$K_3[Cr(CN)_6]$ d = 100 to 1000 in $K_3[Mn(CN)_6]$	90, 20	$g_x = 1.992 \pm 0.002$ $g_y = 1.995 \pm 0.002$ $g_z = 1.993 \pm 0.002$	$+538 \pm 10$	$+120 \pm 10$	$+14.7 \pm 0.5$	Monoclinic, $M_m = 2$; direction cosines determined[d]	120, 231, C.S. 232, 233
[Cr aca$_3$] $d = 2000$; $d = 250$ in [Co aca$_3$]	298	1.9802 ± 0.0005	6000 ± 10	85 ± 5	$A = 16.2 \pm 0.1$ $B = 16.9 \pm 0.1$	Monoclinic, $Z = 4$;	133, C.S. 221, 222
[Cr aca$_3$] $d = 5000$; $d = 250$ in [Al aca$_3$]	298	1.9820 ± 0.0005	5920 ± 20	96 ± 5		Monoclinic, $Z = 4$; D obtained from $d = 50$ crystal; value of E originally reported by Singer is incorrect	133, 132, C.S. 221, 222

$[Cr\{(C_2H_5O)_2P(S)S\}_3]$ $d = 100$ in $[Co\{(C_2H_5O)_2P(S)S\}_3]$	298	$g_{xx} = 1.9903 \pm 0.0010$ $g_{yy} = 1.9914 \pm 0.0010$ $g_{zz} = 1.9901 \pm 0.0010$	$\pm 138 \pm 1$	$\mp 814 \pm 1$		Monoclinic	234
$[Cr(H_2O)_6]Cl_3$ d. in $AlCl_3 \cdot 6H_2O$	298	$g_{\parallel} = 1.9769 \pm 0.0004$ $g_{\perp} = 1.9764 \pm 0.0004$	326 ± 1		17.0 ± 0.1	Rhombohedral, local symmetry C_{3i}; deuterated crystal also investigated	235, 93, C.S. 220
$[Cr(NH_3)_6]Cl_3$ $d = 10$ in $[Co(NH_3)_6]Cl_3$	298	I: 1.986 ± 0.003 II: 1.984 ± 0.005 III: 1.984 ± 0.007	313 ± 2 537 ± 3 688 ± 7	19 ± 1 119 ± 1 203 ± 2		Monoclinic, $M_m = 3$; direction cosines were also determined	139
$[Cr(NH_3)_6]I_3$ $d = 400$ in $[Co(NH_3)_6]I_3$	298	1.9842 ± 0.0005				Cubic; parameter $u = -19 \pm 4 \times 10^{-4}$ cm^{-1} determined	140, C.S. 236–239
$[Cr\ en_3]Cl_3$ $d = 1000$ in $[Co\ en_3]Cl_3$	298	1.9871 ± 0.0003	413 ± 1	0			135
$[Cr\ en_3]Cl_3 \cdot 3H_2O$ $d = 10$; $d = 1000$ in $[Co\ en_3]Cl_3 \cdot 3H_2O$	298	1.9900 ± 0.0004	360 ± 30	0		Hexagonal, D_{3d}^4, $Z = 4$; H_2O molecules in channels parallel c axis at random causing large linewidth	135, C.S. 240

(continued)

Table 10 (*continued*)

Crystal	Temp. (°K)	g	D (10^{-4} cm^{-1})	E (10^{-4} cm^{-1})	HFS: $(A, B)^{53}$ (10^{-4} cm^{-1})	Remarks	Ref.
$[Cr\ en_3]Cl_3 \cdot NaCl \cdot 6H_2O$ $d = 300$ in $[Co\ en_3]Cl_3 \cdot NaCl \cdot 6H_2O$	298	1.9874 ± 0.0002	49.5 ± 7	0	16.2 ± 0.7	Hexagonal	135
trans-$[CrCl_2\ en_2]Cl \cdot HCl \cdot 2H_2O$ $d = 500$ in $[CoCl_2\ en_2]Cl \cdot HCl \cdot 2H_2O$	298	1.9765 ± 0.0005	5040 ± 10	360 ± 10	16.5 ± 1.5	Monoclinic, C^5_{2h}, $Z = 2$	135, C.S. 241, 242

[a] No signs listed for D, E, A, B means that they have not been determined.

[b] d.c. of rhombic axes with respect to the tetragonal axes a, b, c:

	a	b	c
x	-0.35	-0.35	$+0.87$
y	$+0.71$	-0.71	0
z	$+0.61$	$+0.61$	$+0.50$

[c] d.c. of the principal magnetic axes x, y, z at 20°K:

	a	b	c
x	0.104	± 0.994	0
y	0	0	1
z	0.994	± 0.104	0

[d] d.c. of the principal magnetic axes x, y, z at 20°K:

	a	b	c
x	0	± 0.996	0.087
y	0	∓ 0.087	0.996
z	1	0	0

second-order effects in spacings of the hyperfine components. The absolute sign of D is found from an observation of the relative intensities of the two outer transitions $-\frac{3}{2} \leftrightarrow -\frac{1}{2}$ and $+\frac{1}{2} \leftrightarrow +\frac{3}{2}$ as function of temperature. As can be seen from Table 10, g is practically isotropic. The splitting constant A is significantly reduced in magnitude, $A = +14.7 \times 10^{-4}$ cm^{-1} (cf. $A = 18.5$ in $KCr(SeO_4)_2 \cdot 12D_2O$, $A = 16.2$ in [Cr aca_3], $A = 16.2$ in [Cr $en_3]Cl_3 \cdot NaCl \cdot 6H_2O$, in units of 10^{-4} cm^{-1}), which is taken as an indication of strong covalent bonding. Ligand hyperfine structure due to ^{13}C has been observed in a crystal enriched in the isotope and, using an MO description of the bonding, the σ- and π-bond characters of the metal–ligand bond have been calculated[129]. In undiluted crystals, the resonance line is very anisotropic[120] and additional lines are observed[130,131]. Exchange interaction and polytypism were suggested to account for it.

In chromium(III) acetylacetonate, a very large value for D has been reported ($D = 6000$ for [Cr aca_3] diluted in [Co aca_3], $D = 5920$ for [Cr aca_3] diluted in [Al aca_3], both in units of 10^{-4} cm^{-1})[132,133]. Since the molecular symmetry is D_3, a configuration interaction between the ground state and a quartet excited state is allowed, producing a distortion in the ground state. The appropriate D_3 symmetry orbitals have the form

$$\begin{aligned} a_1 &= d_0 \\ e_{a1} &= ad_{-2} + bd_1 \\ e_{a2} &= ad_2 - bd_{-1} \\ e_{b1} &= bd_{-2} - ad_1 \\ e_{b2} &= bd_2 + ad_{-1} \end{aligned} \tag{115}$$

where the orbitals d_m are determined by equation (19) and where $a^2 = \frac{2}{3}$, $b^2 = \frac{1}{3}$, when the symmetry is regularly octahedral. To calculate the spin-hamiltonian parameter D, we use the following equation which has been obtained from second-order perturbation theory

$$D = -\tfrac{1}{2}\sum_n{}' \langle 0\tfrac{3}{2}|\, \mathscr{H}_{SO}\, |n\rangle\langle n|\, \mathscr{H}_{SO}\, |0\tfrac{3}{2}\rangle/(E_n - E_0) + \tfrac{1}{2}\sum_n{}' \langle 0\tfrac{1}{2}|\, \mathscr{H}_{SO}\, |n\rangle\langle n|\, \mathscr{H}_{SO}\, |0\tfrac{1}{2}\rangle/(E_n - E_0) \tag{116}$$

where

$$\mathscr{H}_{SO} = \sum_j \left[\sum_{i=1}^{3} \xi_j(r_{ij})\mathbf{l}_{ij}\cdot\mathbf{s}_i\right] \tag{117}$$

$|0M_S\rangle$ is the ground state function with $S_z = M_S$; j in (117) refers to different nuclear centres and $\mathbf{l}_{ij}$ is the single-electron angular momentum operator centred on nucleus j[134]. Since the ground state is a quartet from the t_2^3 configuration, the spin–orbit coupling has a non-zero interaction only with the quartet t_2^2e states and the doublet t_2^3 and t_2^2e states. Consequently, D is given by the equation[135]

$$\begin{aligned} D = &-18\lambda^2a^2b^2(\{\Delta E[^4A_1(^4T_{2g})]\}^{-1} - \{\Delta E[^2A_1'(^2T_{2g})]\}^{-1}) \\ &+ \lambda^2[1 + (\tfrac{3}{2})^{\frac{1}{2}}a]^2(\{\Delta E[^4E(^4T_{2g})]\}^{-1} - \{\Delta E[^2E'(^2T_{2g})]\}^{-1}) \\ &+ \lambda^2[1 - (\tfrac{3}{2})^{\frac{1}{2}}a]^2(\{\Delta E[^4E(^4T_{1g})]\}^{-1} - \{\Delta E[^2E'(^2T_{1g})]\}^{-1}) \\ &- 9\lambda^2b^2\{\Delta E[^2E(^2T_{2g})]\}^{-1} + 3\lambda^2(b^2 - 2a^2)\{\Delta E[^2A_1(^2T_{2g})]\}^{-1} \end{aligned} \quad (118)$$

The various energies have been estimated from the optical spectrum of the complex[135,136] as

$$\begin{aligned} \Delta E[^4A_1(^4T_{2g})] &= 17{,}600 \text{ cm}^{-1} \\ \Delta E[^4E(^4T_{2g})] &= 18{,}400 \text{ cm}^{-1} \\ \Delta E[^4E(^4T_{1g})] &= 26{,}000 \text{ cm}^{-1} \\ \Delta E[^2E(^2T_{2g})] &= \Delta E[^2A_1(^2T_{2g})] = 22{,}500 \text{ cm}^{-1} \\ \Delta E[^2A_1'(^2T_{2g})] &= 30{,}600 \text{ cm}^{-1} \\ \Delta E[^2E'(^2T_{2g})] &= 31{,}400 \text{ cm}^{-1} \\ \Delta E[^2E'(^2T_{1g})] &= 35{,}600 \text{ cm}^{-1} \end{aligned}$$

Using the free ion value of $\lambda = 91$ cm^{-1}, we obtain $a^2 = 0.620$. If $\lambda = 70$ cm^{-1}, it follows $a^2 = 0.585$. On the other hand, first-order perturbation theory may be used to calculate the hyperfine splittings as[133]

$$\begin{aligned} A &= P\{\tfrac{4}{21}[1 - 2b^2 + a^2] - \kappa\} \\ B &= P\{-\tfrac{2}{21}[1 - 2b^2 + a^2] - \kappa\} \\ P &= 2g\beta\beta_N\langle 1/r^3\rangle_{av} \end{aligned} \quad (119)$$

where κ is the isotropic *s*-electron interaction. Estimating $\langle 1/r^3\rangle_{av}$ = 4.0 a.u., and thus $P = -40 \times 10^{-4}$ cm^{-1} and using the experimental results for A and B, we arrive at $a^2 = 0.65$. A value of a^2 smaller than $\frac{2}{3}$ indicates that in [Cr aca_3] the d electrons are concentrated along the molecular D_3 axis to a larger degree than in the case of perfect O_h symmetry. Such a change in *d*-electron distribution would be expected if the distortion from O_h symmetry would occur by a compression of the octahedron along the trigonal axis. As a consequence, the a_1 orbital should be lower in energy than the e orbital which is consistent with the results from [Ti aca_3][94]. Also, a compression along the D_3 axis has been assumed to explain the sign of the trigonal field parameter in the electronic spectrum of [Cr aca_3][137]. Similar conclusions probably hold in the case of the chromium trifluoro- and hexafluoroacetylacetonates[138] ($D = 5900 \times 10^{-4}$ cm^{-1} and 7000×10^{-4} cm^{-1}, respectively), since the initial splittings are of the same order of magnitude as in [Cr aca_3]. For results on chromium tris(diethyldithiophosphate) and the chromium hexaquo ion, consult Table 10.

Single crystals of $[Cr(NH_3)_6]Cl_3$ diluted in $[Co(NH_3)_6]Cl_3$ were shown to contain three magnetically inequivalent complexes[139] which have different orientations of their y and z axes as well as different initial splittings (different values of the D and E parameters). $[Cr(NH_3)_6]I_3$ (diluted 1:400 in $[Co(NH_3)_6]I_3$), on the other hand, gives a single e.p.r. absorption[140] which is assumed to be the $-\frac{1}{2} \rightarrow +\frac{1}{2}$ transition. The angular variation of the spectrum can be described by the spin-hamiltonian

$$\mathscr{H} = g\beta\mathbf{H}\cdot\mathbf{S} + u\beta\{S_x^3H_x + S_y^3H_y + S_z^3H_z - \tfrac{1}{5}(\mathbf{S}\cdot\mathbf{H})[3S(S+1) - 1]\} \quad (120)$$

where $S = \frac{3}{2}$, $g = 1.9842$, and $u = -0.0019\ \text{cm}^{-1}$. Equation (120) is the most general form of the spin-hamiltonian for a $\Gamma_8(3d^3)$ ground state[141,142]. The form (120) is also required to explain the resonance of Re^{4+} in cubic fields[143]. An explanation for the relatively large value of the parameter u may be given by assuming that each Cr^{3+} site has a small axial field of the same magnitude, and also that the orientation of the axes is randomly distributed among the three cubic axes of the crystal. The axial field gives rise to a term $D[S_z^2 - \frac{1}{3}S(S+1)]$ in the spin-hamiltonian and, if the observed resonance is the average of the three frequencies expected,

$$u = -\tfrac{1}{2}(D/g\beta H)^2 \quad (121)$$

With these assumptions, $g = 1.9857 \pm 0.0005$ and $|D| = (190 \pm 20) \times 10^{-4}\ \text{cm}^{-1}$ is obtained.

The ion $[Cr\ en_3]^{3+}$ has been studied in several different single crystals[135], viz. $[Cr\ en_3]Cl_3$, $[Cr\ en_3]Cl_3 \cdot 3H_2O$, and $[Cr\ en_3]Cl_3 \cdot NaCl \cdot 6H_2O$, all of which were diluted in the corresponding cobalt(III) compounds. The g values were found to be isotropic, the fine structure parameter D assumes the values 413, 360, and $49.5 \times 10^{-4}\ \text{cm}^{-1}$, respectively, and hyperfine structure due to ^{53}Cr is observed only in $[Cr\ en_3]Cl_3 \cdot NaCl \cdot 6H_2O$. To account for the observed values of D, the same approach as in the case of $[Cr\ aca_3]$ may be used. From the electronic spectrum of the $[Cr\ en_3]^{3+}$ ion, the energy differences of the various levels are estimated as[144]

$$\begin{aligned}
\Delta E[^4A_1(^4T_{2g})] &= \Delta E[^4E(^4T_{2g})] = 21{,}880\ \text{cm}^{-1}\\
\Delta E[^4E(^4T_{1g})] &= 28{,}490\ \text{cm}^{-1}\\
\Delta E[^2A_1(^2T_{2g})] &= \Delta E[^2E(^2T_{2g})] = 25{,}200\ \text{cm}^{-1}\\
\Delta E[^2A_1'(^2T_{2g})] &= \Delta E[^2E'(^2T_{2g})] = 37{,}000\ \text{cm}^{-1}\\
\Delta E[^2E'(^2T_{1g})] &= 40{,}300\ \text{cm}^{-1}
\end{aligned}$$

Using $\lambda = 91\ \text{cm}^{-1}$, (118) yields $a^2 = 0.6627$ for $[Cr\ en_3]Cl_3$ and $a^2 = 0.6662$ for $[Cr\ en_3]Cl_3 \cdot NaCl \cdot 6H_2O$.

A single crystal of *trans*-$[CrCl_2\ en_2]Cl \cdot HCl \cdot 2H_2O$ diluted in the corresponding cobalt(III) compound gave the values[135] of $D = 5040$ and $E = 360$ in $10^{-4}\ \text{cm}^{-1}$ for the axial splitting. From the e.p.r. results it follows also that the symmetry of the electric field around the Cr^{3+} ion is at most D_{2h}, the main distortion being in the plane of the two ethylenediamine ligands. On the other hand, the electronic spectra of both the Cr^{III} and Co^{III} compounds in solution[145–147] as well as the polarized single-crystal spectra of *trans*-$[CoCl_2\,en_2]Cl \cdot HCl \cdot 2H_2O$[148–150] have been interpreted assuming D_{4h} symmetry with the four-fold axis along the Cl–Co–Cl direction. To reconcile these results, it has been suggested that a minor distortion from D_{4h} to D_{2h} symmetry might produce a much larger distortion of the spin–spin interaction than the large distortion from O_h symmetry along the Cl–M–Cl axis. This assumption has been supported by a calculation of D and E from both the crystal field model and from an extended Hückel MO treatment[135]. According to this calculation, the reason for the fact that a minor distortion $D_{4h} \rightarrow D_{2h}$ overcomes a major distortion $O_h \rightarrow D_{4h}$ can be found in that only in the first case a configuration interaction occurs between the ground state and an excited quartet state.

Only one compound containing the Mn^{4+} ion has been studied, namely $Cs_2[MnF_6]$ diluted in $Cs_2[GeF_6]$[151]. Since both the g value and the hyperfine structure due to the Mn nucleus are isotropic, whereas the fluorine hyperfine splitting is anisotropic, the following spin-hamiltonian has been used:

$$\mathscr{H} = g\beta\mathbf{H}\cdot\mathbf{S} + A^{\mathrm{Mn}}\mathbf{S}\cdot\mathbf{I} + A_s^{\mathrm{F}}\sum_1^6 \mathbf{S}\cdot\mathbf{I}$$

$$+ A_p^{\mathrm{F}}\left\{\sum_1^6 f_i(\alpha)S_zI_z + f_i'(\alpha)S_zI_x + f_i''(\alpha)S_zI_y\right\} + B\sum_1^6 I_z \qquad (122)$$

Here, $f_i(\alpha)$, $f_i'(\alpha)$, and $f_i''(\alpha)$ are factors describing the angular dependence of the $\mathbf{S}\cdot\mathbf{I}$ interaction. The results are $A^{\mathrm{Mn}} = 72.0 \pm 1.0 \times 10^{-4}\ \mathrm{cm}^{-1}$, $A_s^{\mathrm{F}} = \pm 5.1 \pm 1.0 \times 10^{-4}\ \mathrm{cm}^{-1}$, and $A_p^{\mathrm{F}} = \mp 8.9 \pm 1.0 \times 10^{-4}\ \mathrm{cm}^{-1}$.

D. Six-coordinated d^4 in Cr^{II}

The ionic ^5D ground state is split by a weak ligand field of octahedral symmetry into ^5E and $^5T_2$. In the high-spin $3d^4$ ion, the ground state is the non-magnetic ^5E (in the low-spin case, a $^3T_1$ becomes the ground state) and this term breaks up into $^5A_1$ and $^5B_1$ under the action of a tetragonal field. There is no non-zero spin–orbit coupling energy within ^5E, and the admixture of various excited terms into ^5E under the influence of spin–orbit coupling has to be considered. If rhombic symmetry is present, the five-fold spin degeneracy of the lowest $^5B_1$ level is completely lifted.

Only in one compound, $CrSO_4 \cdot 5H_2O$, has resonance been observed[152,153]. Hyperfine splitting due to ^{53}Cr has not been resolved and the results have been interpreted in terms of the following spin-hamiltonian with $S = 2$:

$$\mathscr{H} = g_{\parallel}\beta H_z S_z + g_{\perp}\beta(H_x S_x + H_y S_y) + D(S_z^2 - 2) + E(S_x^2 - S_y^2) \quad (123)$$

Here $g_{\parallel} = 1.95$, $g_{\perp} = 1.99$, $|D| = 2.24 \text{ cm}^{-1}$, and $|E| = 0.10 \text{ cm}^{-1}$. Due to the large initial splitting, 5 mm wavelength had to be used to detect the transition. In the Tutton salt, $(NH_4)_2Cr(SO_4)_2 \cdot 6H_2O$, no resonance has been observed at 3 cm wavelength[38].

E. High-spin and Low-spin d^5 Compounds: Ti^{-I}, V^0, Cr^I, Mn^{II}, Fe^{III}

In the free $3d^5$ ions, the ground state is ^6S. For ions which are subject to the electric field of ligands, the multiplicity of the ground state is dependent on the strength of the field. In weak fields, $S = \frac{5}{2}$, whereas in strong fields, $S = \frac{1}{2}$. The two cases will be discussed separately.

1. Weak ligand fields, $S = \frac{5}{2}$

In a weak octahedral field, the lowest state $^6A_1$ is split into one four-fold (Γ_8) and one two-fold (Γ_7) level, as indicated by specific-heat measurements. Since crystal fields cannot split the A_1 state and since spin–orbit coupling cannot by itself remove the six-fold spin degeneracy, it has been suggested[39] that the splitting is due to higher order interactions. As shown in Section III.D, such interactions may be accounted for by the addition of the term

$$\tfrac{1}{6}a[S_x^4 + S_y^4 + S_z^4 - \tfrac{1}{5}S(S+1)(3S^2 + 3S - 1)] \quad (124)$$

to the spin-hamiltonian. If an axial, say a trigonal, field is also present (like in the alums), an additional splitting of the four-fold Γ_8 level occurs. Thus the $^6A_1$ level is finally resolved into three Kramers doublets, usually separated by less than 1 cm^{-1}. It is convenient to work in cartesian coordinates, even though this involves the use of different coordinate systems for the cubic and the trigonal terms. The spin-hamiltonian may then be written with $S = \frac{5}{2}$ as[154]

$$\begin{aligned}\mathscr{H} = {} & g\beta(H_x S_x + H_y S_y + H_z S_z) + \tfrac{1}{6}a(S_\xi^4 + S_\eta^4 + S_\zeta^4 - \tfrac{707}{16}) \\ & + D(S_z^2 - \tfrac{35}{12}) + \tfrac{7}{36}F(S_z^4 - \tfrac{95}{14}S_z^2 + \tfrac{81}{16}) \\ & + A(S_x I_x + S_y I_y + S_z I_z) \quad (125)\end{aligned}$$

Here (ξ, η, ζ) is a set of three mutually perpendicular axes which are the cubic axes of the crystalline field. The z axis is along the trigonal axis of the crystal, i.e. along the (111) axis of the $\xi\eta\zeta$ system. The terms in D and F correspond to trigonal fields of the second and fourth degree, respectively. The g value usually is found to be isotropic and very close to 2. Also, the anisotropy in the A values reported for some compounds is not greater than the maximum error and has thus been neglected in equation (125).

If the symmetry of the complex is not higher than orthorhombic, as in the Tutton salts, the following spin-hamiltonian with $S = \frac{5}{2}$ is used:

$$\begin{aligned}\mathscr{H} = g\beta(H_xS_x + H_yS_y + H_zS_z) + \tfrac{1}{6}a(S_x^4 + S_y^4 + S_z^4 - \tfrac{707}{16}) \\ + D(S_z^2 - \tfrac{35}{12}) + E(S_x^2 - S_y^2) + A(S_xI_x + S_yI_y + S_zI_z) \quad (126)\end{aligned}$$

The hyperfine term in (125) and (126) is required for compounds of manganese, the only natural occurring isotope ^{55}Mn having $I = \frac{5}{2}$, whereas it may be dropped for iron compounds, since the hyperfine structure of ^{57}Fe is not resolved in most iron compounds[154]. For the origin of the extensive hyperfine structure observed for Mn^{2+}, the reader should refer to the discussion in Section II.E.

A consequence of the zero-field splitting is that the transitions $M_S \rightarrow M_S + 1$ do not all have the same energy and thus five lines are obtained. Each of these lines is split into six components by the nuclear hyperfine interaction and the resulting spectrum is in general very complicated. Figure 11 shows an energy level diagram for manganese ammonium sulphate with the magnetic field H parallel to the zero-field splitting axis[155]. Details of the spin-hamiltonian parameters for this and some other representative salts are listed in Table 11. For additional data, consult ref. 86.

Ligand ^{19}F hyperfine structure has been observed for the complex $[MnF_6]^{4-}$ which is present in crystals of MnF_2, $KMnF_3$, and K_2MnF_4[156,157]. Crystals of $MnBr_2$ and MnI_2 contain the complexes $[MnBr_6]^{4-}$ and $[MnI_6]^{4-}$ and show hyperfine splitting due to $^{79,81}Br$ and ^{127}I nuclei, respectively[158]. The interaction with the ligands has been described using (61), i.e. assuming for each ligand n a hamiltonian of the form[156]

$$\mathscr{H}' = A^n S_z I_z^n + B^n(S_x I_x^n + S_y I_y^n) \quad (127)$$

where $S = \frac{5}{2}$ and $I^n = \frac{1}{2}, \frac{3}{2}$, and $\frac{5}{2}$ for the fluoride, bromide, and iodide ions, and where the z axis lies along the bond axis in each case. The splitting parameters A^n and B^n are further decomposed according to

$$\begin{aligned}A^n &= A_s^n + 2(A_D^n + A_{p_\sigma}^n - A_{p_\pi}^n) \\ B^n &= A_s^n - (A_D^n + A_{p_\sigma}^n - A_{p_\pi}^n)\end{aligned} \quad (128)$$

Here A_s^n is representing the interaction with ligand s orbitals, A_D^n the dipolar interaction with d electrons and $A_{p_\sigma}^n - A_{p_\pi}^n$ the interaction with the admixed ligand p orbitals. Within iron(III) compounds, a ligand ^{19}F hyperfine structure has been observed in $K_2Na[FeF_6]$[159]. In this case, off-diagonal terms in $\mathbf{S}\cdot\mathbf{I}$ were found to be important[160].

In iron(III) compounds, hyperfine splitting due to naturally abundant ^{57}Fe (2.2%) has been observed in ZnO only[161]. From paramagnetic resonance[162] and Endor results[163] on iron (enriched with ^{57}Fe) in silicon, the nuclear magnetic moment of ^{57}Fe has been determined as $+0.0903 \pm 0.0007\ \beta_N$ and the nuclear spin as $I = \frac{1}{2}$. The only other instance of a resolved hyperfine structure has been reported for iron enriched to 92.8% ^{57}Fe in cerium zinc double nitrate crystals[164].

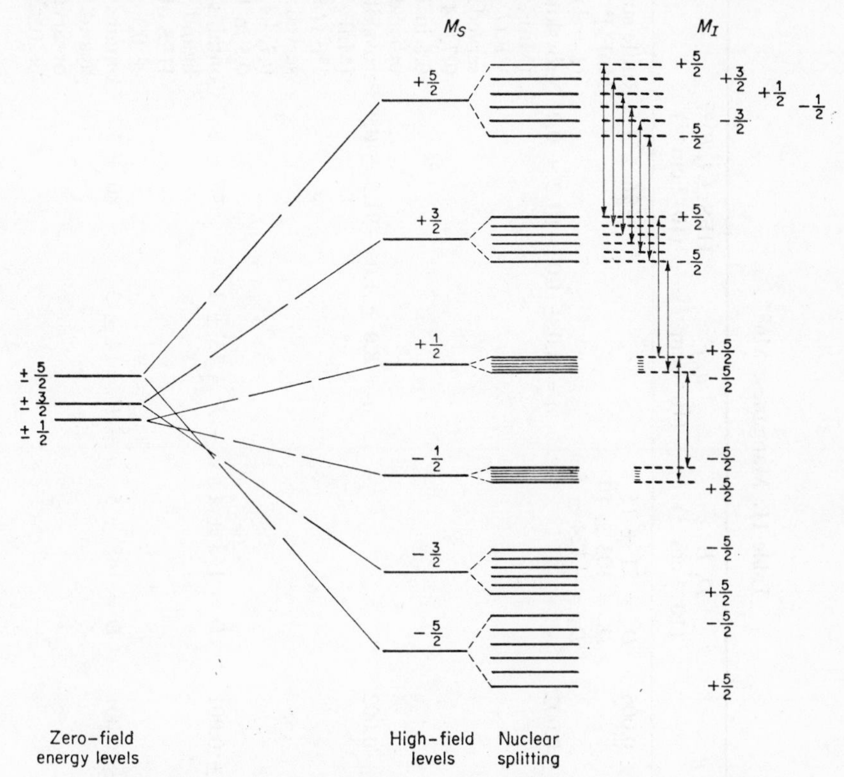

Figure 11. Schematic diagram for the lowest energy levels of manganese ammonium sulphate in zero and in strong magnetic field. Some of the hyperfine transitions indicated at right [according to Bleaney and Ingram[155]].

Various hydrated iron(III) salts, especially the alums, were extensively studied (see Table 12). An inspection of the results obtained for potassium, rubidium, and ammonium sulphate alum and on potassium selenate alum shows that D varies from 22 to 160×10^{-4} cm^{-1} for the four materials and is temperature dependent, whereas $|a|$ varies only from 127 to 134×10^{-4} and is temperature independent. The term F is much

Table 11. Manganese Mn^{2+}.

Crystal	Temp. (°K)	g	D, E (10^{-4} cm^{-1})	a, F (10^{-4} cm^{-1})	HFS: $(A, B)^{55}$ (10^{-4} cm^{-1})	Remarks	Ref.
MnF_2 $d = 2 \times 10^5$ in ZnF_2	293	2.002 ± 0.005	$D_x = 21 \pm 15$ $D_y = 103 \pm 10$ $D_z = -124 \pm 5$		-96 ± 3	Rutile structure; contains $[MnF_6]^{4-}$ ions; ligand ^{19}F ($I = \frac{1}{2}$) HFS observed	156, C.S. 243
$KMnF_3$ $d = 1000$ in $KMgF_3$	293, 77	2.000 ± 0.002		$a = 8.0 \pm 1.0$	-91.2 ± 0.9	Perovskite; contains $[MnF_6]^{4-}$ ions; ligand ^{19}F ($I = \frac{1}{2}$) HFS observed: $\|A_s\| = 18.1 \pm 0.7$, $\|A_p + A_D\| = 4.2 \pm 0.5$ in 10^{-4} cm^{-1}; for other diluents, see ref.	157
K_2MnF_4 $d = 1000$ in K_2MgF_4	293, 77	2.000 ± 0.002		$a = 8.0 \pm 1.0$	-91.5 ± 0.9	Perovskite; contains $[MnF_6]^{4-}$ ions; ligand ^{19}F ($I = \frac{1}{2}$) HFS observed: $\|A_s\| = 18.2 \pm 0.7$, $\|A_p + A_D\| = 4.2 \pm 0.5$ in 10^{-4} cm^{-1}	157
$MnBr_2$ $d = 1000$ in $CdBr_2$	80	2.001 ± 0.001	$D = 168 \pm 1$	$a - F = 13 \pm 1$	78 ± 1	Contains $[MnBr_6]^{4-}$ ions; ligand $^{79,81}Br$ ($I = \frac{3}{2}$) HFS observed: $A_s = 6.7 \pm 0.1 \times 10^{-4}$ cm^{-1}	158
MnI_2 $d = 1000$ in CdI_2	80	2.002 ± 0.001	$D = 148 \pm 5$	$a - F = 14 \pm 3$	80 ± 3	Contains $[MnI_6]^{4-}$ ions; ligand ^{127}I ($I = \frac{5}{2}$) HFS observed: $A_s = 6.9 \pm 0.1 \times 10^{-4}$ cm^{-1}	158

$MnSO_4 \cdot 7H_2O$ d. in $MgSO_4 \cdot 7H_2O$	290	2.000 ± 0.005	$D = 400$ $E \sim 0$		88	Rhombic, $Z = 4$	244, C.S. 245
$(NH_4)_2Mn(SO_4)_2 \cdot 6H_2O$ $d = 1000$ in $(NH_4)_2Zn(SO_4)_2 \cdot 6H_2O$	230	2.000 ± 0.001	$D = +243 \pm 5$ $E = 100 \pm 20$	$a = +5 \pm 1$	-91.1 ± 1	Monoclinic, $Z = 2$; $\psi = +58°$, $\theta = 32°$[a]; for different dilutions and temperatures see ref.	155
$Mn_3Bi_2(NO_3)_{12} \cdot 24H_2O$ $d = 200$ in $Mg_3Bi_2(NO_3)_{12} \cdot 24H_2O$	90	$g_{\parallel} = 1.98 \pm 0.02$ $g_{\perp} = 2.00 \pm 0.02$	I: $D = -211 \pm 1$, $E = 0$ II: $D = -64 \pm 1$, $E = 0$	$a = +8 \pm 1$ $a = +10 \pm 1$	$A = -90$, $B = -89$ $A = -88$, $B = -90$	Spectrum corresponds to one ion of type I and two ions of type II	186
$MnCO_3$ $d = 2000$ in $CaCO_3$	290	$g_{\parallel} = 2.0022 \pm 0.0006$ $g_{\perp} = 2.0014 \pm 0.0006$	$D = 75$ $E = 0$	$F = 58$	87.8 ± 0.1		
$MnSiF_6 \cdot 6H_2O$ $d = 1000$ in $ZnSiF_6 \cdot 6H_2O$	290	2.000 ± 0.001	$D = -179 \pm 3$ $E = 0$	$a = +7 \pm 1$	$A = -96.1$ $B = -93.3$	Trigonal, $Z = 1$; for different dilutions and temperatures, see ref. 86	155

(continued)

Table 11 (*continued*)

Crystal	Temp. (°K)	g	D, E (10^{-4} cm^{-1})	a, F (10^{-4} cm^{-1})	HFS: $(A, B)^{55}$ (10^{-4} cm^{-1})	Remarks	Ref.
$K_4[Mn(CN)_6]\cdot 3H_2O$ d = 1000 to 20 in $K_4[Fe(CN)_6]\cdot 3H_2O$	12	$g_x = 2.624 \pm 0.008$ $g_y = 2.182 \pm 0.008$ $g_z = 0.63 \pm 0.10$			$A_x = 84.5 \pm 0.5$ $A_y = 46.5 \pm 0.5$ $A_z = 104 \pm 20$	Low-spin, $S = \frac{1}{2}$; monoclinic, $Z = 2$	120, C.S. 247

[a] The angles ψ and θ are defined in the following way: for monoclinic crystals, the principal magnetic axis corresponding to the principal susceptibility χ_3 is uniquely fixed along the crystallographic b axis by symmetry. The remaining two axes with $\chi_1 > \chi_2$, which lie in the (010) plane, are inclined at angles ψ and θ to the c and a axes, respectively, such that the monoclinic angle $\beta = \psi + \theta + \pi/2$.

smaller, usually of the order of 1×10^{-4} cm^{-1}. It has been shown that the cubic component of the crystalline potential determines $|a|$ through the term (124) in the spin-hamiltonian. The behaviour of the hamiltonian parameters is thus consistent with the expectation that the cubic field is mainly due to the octahedron of water molecules surrounding the Fe^{3+} ion and thus is relatively insensitive to changes of constituents outside the octahedron. On the other hand, the value of D is determined by the axial field component and therefore is much more susceptible to such changes.

The results obtained on methylamine alum using a fixed frequency[154] have shown a large number of lines and were not completely understood. Bogle and Symmons[165] have measured the zero-field splitting in a direct way using a polycrystalline sample diluted 1:100 in the corresponding aluminium compound. The separation of the $S_z = \frac{5}{2}$ from the $\pm\frac{3}{2}$ states is 22,043 $\pm$ 10 Mc/s, and that of the $\pm\frac{3}{2}$ from the $\pm\frac{1}{2}$ states is 12,227 $\pm$ 5 Mc/s, as shown in Figure 12. The values of the parameters ($D = +1893$, $a = +130$, $F = -40 \times 10^{-4}$ cm^{-1}) show that D is from ten to hundred times larger, and F about twenty times larger than for the other alums.

Iron(III) acetylacetonate, [Fe aca$_3$], has been measured as single crystals diluted in [Co aca$_3$] at 300°K by Jarrett[138] who obtained $D = 0.085$ cm^{-1} (the value of $|D'|$ is 0.170 cm^{-1}, not 0.140 cm^{-1} as quoted[166], and also $D' = 2D$). In a reinvestigation performed at 80°K, Symmons and Bogle[166] have shown that the results can be well accounted for in terms of the spin-hamiltonian

$$\mathscr{H} = g\beta\mathbf{H}\cdot\mathbf{S} + b_2^0O_2^0 + b_2^2O_2^2 + b_4^0O_4^0 + b_4^2O_4^2 + b_4^4O_4^4 \qquad (129)$$

where the b_m^n are empirical parameters, and the O_m^n are spin operators derived from the corresponding surface spherical harmonics. The values obtained at 80°K in units of 10^{-4} cm^{-1} are

$$b_2^0 = -1389 \pm 20 \qquad b_4^0 = +14 \pm 10$$
$$b_2^2 = -222 \pm 50 \qquad b_4^2 = -202 \pm 50$$
$$b_4^4 = +802 \pm 100$$

also, $g = 2.003 \pm 0.001$. The coefficients b_m^n are related to the conventional spin-hamiltonian parameters by

$$D = b_2^0, \qquad E = b_2^2/3, \qquad F = 3b_4^0 \qquad (130)$$

the values of which are listed in Table 12. The zero-field resonance frequencies have been measured directly[166] and are listed in Table 13 in megacycles per second.

Table 12. Iron Fe^{3+}.

Crystal	Temp. (°K)	g	D, E (10^{-4} cm^{-1})	a (10^{-4} cm^{-1})	F (10^{-4} cm^{-1})	Remarks	Ref.
$FeCl_3 \cdot 6H_2O$ d. in $AlCl_3 \cdot 6H_2O$	77	2.002 ± 0.002	$D = 1500 \pm 20$	160 ± 50	310 ± 20	At 293°K no resonance observed	93, C.S. 220
$KFe(SO_4)_2 \cdot 12H_2O$ $d = 385$ in $KAl(SO_4)_2 \cdot 12H_2O$	293	2.00	$D = 160$				248
$RbFe(SO_4)_2 \cdot 12H_2O$ $d = 300$ in $RbAl(SO_4)_2 \cdot 12H_2O$	90	2.003 ± 0.001	$D = +22 \pm 2$	-134 ± 2	-3 ± 2	α-Alum; $Z = 4$	154, 249 C.S. 92, 124
$NH_4Fe(SO_4)_2 \cdot 12H_2O$ $d = 100$ in $NH_4Al(SO_4)_2 \cdot 12H_2O$	293	2.00	$D \sim 50$	120		$\Delta M_S = \pm 2$ transitions observed at low temperatures	250, 248
$(CH_3NH_3)Fe(SO_4)_2 \cdot 12H_2O$ $d = 200$ in $(CH_3NH_3)Al(SO_4)_2 \cdot 12H_2O$	90		$D = +1880 \pm 140$	$+100 \pm 40$		Zero-field splitting measured directly[165]; for more accurate values of D, a, F, see text	154
$KFe(SeO_4)_2 \cdot 12H_2O$ $d = 300$ in $KAl(SeO_4)_2 \cdot 12H_2O$	90	2.003 ± 0.001	$D = -103 \pm 1$	-127 ± 2	-2 ± 2	α-Alum	154, C.S 92, 124
$K_2Na[FeF_6]$ $d > 100$ in $K_2Na[AlF_6]$	293					Ligand ^{19}F ($I = \frac{1}{2}$) HFS observed: $A_s = 23.4 \pm 1$, $A_p = 6.5 \pm 1$ in 10^{-4} cm^{-1}	160, 159

$K_3[Fe(CN)_6]$ d = 1000 to 100 in $K_3[Co(CN)_6]$	20	$g_x = 2.35 \pm 0.02$ $g_y = 2.10 \pm 0.02$ $g_z = 0.915 \pm 0.01$			Low-spin, $S = \frac{1}{2}$; monoclinic, $Z = 2$	120, 177
[Fe aca_3] d = 100 in [Co aca_3]	80	2.003 ± 0.001	$D = -1389 \pm 20$ $E = -74 \pm 16$	+42 ± 30	Orthorhombic; additional spin-hamiltonian coefficients obtained; zero-field splittings measured directly (Table 13)	166, 138, C.S. 221, 222, 251
	4		I: $D = 1410$ II: $D = 1370$ III: $D = 1280$		Three different cystallographic sites observed	
[Fe aca_3] d = 100 in [Al aca_3]	300		$D = 890$		At 80°K, three different crystallographic sites	166, 138, C.S. 221, 222
	80		I: $D = 1430$ II: $D = 1300$ III: $D = 1050$			
Na[Fe enta]·$4H_2O$ d = 30 in Na[Co enta]·$4H_2O$	77	2.00	$D = 825 \pm 270$ $E = 5500 \pm 1000$		Orthorhombic, $Z = 4$; three different transitions observed corresponding to g' values listed in Table 14	167

Table 13.

Frequency (Mc/s)	Temperature (°K)			
	4.2	80	275	300
$\nu(\frac{3}{2} \leftrightarrow \frac{1}{2})$	unresolved	8562 ± 5	6550 ± 50	6325 ± 50
$\nu(\frac{5}{2} \leftrightarrow \frac{3}{2})$	16630; 16230; 15130, all ±20	16417 ± 15	12565 ± 50	11770 ± 50

The associated level system at 80°K is shown in Figure 12. The discrepancy in the values of D obtained by the different authors probably arises from the simple hamiltonian

$$\mathscr{H} = g\beta\mathbf{H}\cdot\mathbf{S} + D[S_z^2 - \tfrac{1}{3}S(S+1)] \tag{131}$$

assumed by Jarrett. [Fe aca$_3$] diluted in [Al aca$_3$] shows at 80°K a separation into three different level systems[166], whereas at 300°K the spectrum has been reported as being similar to that in the cobalt salt[138].

The iron(III) complex of ethylenediaminetetraacetic acid, Na[Fe edta]·4H_2O, diluted in a single crystal of the corresponding cobalt(III) compound has been studied at 77°K[167]. The spectrum shows one almost isotropic transition (I) at an apparent g value $g' = 4.27$ and two very anisotropic transitions (II, III) with g' values between 1.33 and 9.80 as shown in Table 14.

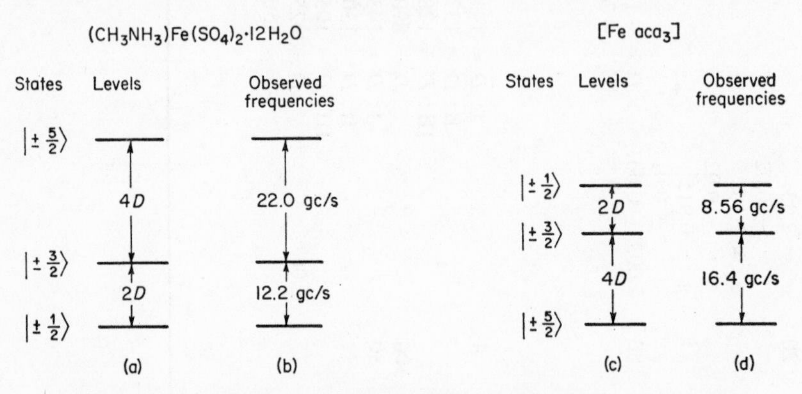

Figure 12. (a) Lowest energy levels of Fe^{3+} in iron(III) methylamine alum for the simplified case of a spin-hamiltonian

$$D[S_z^2 - \tfrac{1}{3}S(S+1)] \text{ with } S = \tfrac{5}{2};$$

(b) measured zero-field frequencies;

(c) lowest energy levels of Fe^{3+} in iron(III) trisacetylacetonate for the same hamiltonian as in (a);

(d) measured zero-field frequencies.

[According to Symmons and Bogle[165,166].]

Table 14. Principal values of g' for Na[Fe edta]·$4H_2O$.

Transition	Species	Single crystal			Powder		
		x	y	z	x	y	z
I	1	3.98	4.64	4.08	3.94	4.64	4.07
	2	4.00	4.62	4.07			
	3	3.82	4.78	3.94			
II		9.48		1.31	9.48		1.33
III			9.78			9.80	

The energy levels and the expected g values for the situations when different terms in the spin-hamiltonian

$$\mathscr{H} = g\beta\mathbf{H}\cdot\mathbf{S} + D[S_z^2 - \tfrac{1}{3}S(S+1)] + E(S_x^2 - S_y^2) \qquad (132)$$

are dominant have been widely studied[168–170]. The three limiting cases are illustrated in Figure 13. Whenever D or E are considerably larger than the magnetic field energy, it is possible to use a new spin-hamiltonian with $S' = \frac{1}{2}$ for the lowest doublet, as will be shown below. The g values introduced in this hamiltonian are the apparent g values denoted by g'.

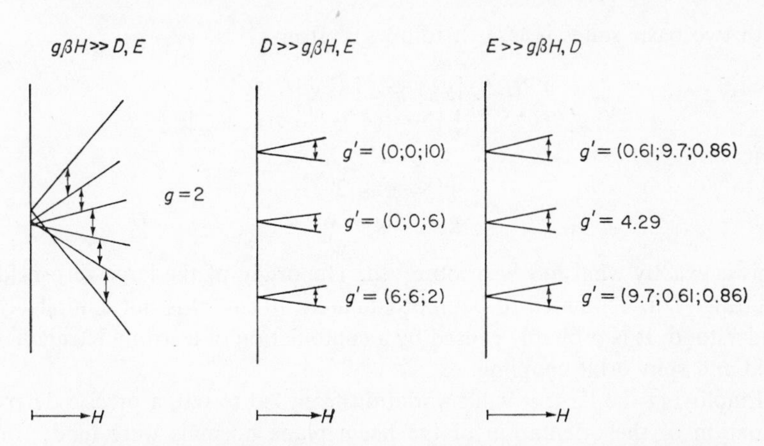

Figure 13. Energy levels and apparent g values assuming that the Zeeman term (left), the axial fine structure term (centre), and the rhombic fine structure term (right) are much larger than the other terms in the Fe^{3+} spin-hamiltonian

$$\mathscr{H} = g\beta\mathbf{H}\cdot\mathbf{S} + DS_z^2 + E(S_x^2 - S_y^2)$$

$S = \frac{5}{2}$, $g = 2.0$ [according to Aasa and Vänngard[167]].

The spectrum observed for $Na[Fe\ edta]\cdot 4H_2O$ is consistent with a large rhombic fine structure splitting, $E \gg g\beta H, D$. The experimental data were therefore fitted to (132) with $S = \frac{5}{2}$ and $g = 2.00$, which results in $|E| = 0.55 \pm 0.10\ \text{cm}^{-1}$ and $D/E = 0.15 \pm 0.05$ for the most common of the three observed species. However, exact agreement within the experimental errors for the g' values has not been achieved, not even after the cubic fourth-order term $(a/6)(S_x^4 + S_y^4 + S_z^4)$ has been included. The complex is probably one of a very low symmetry, the inner coordination being very likely of pentagonal bipyramidal symmetry (two N and four carboxylate O atoms from the enta ligand plus the O atom from a H_2O molecule)[171]. Thus fourth-order axial terms in the spin-hamiltonian might be required.

The limiting case where D is much larger than the magnetic field energy, $D \gg g\beta H$, has been observed in several derivatives of ferrihaemoglobin[172,173]. The ground doublet is again well separated from the excited doublets (cf. Figure 13) and therefore, considering just the parts which depend on the magnetic field, the hamiltonian

$$\mathscr{H}(\tfrac{5}{2}) = g\beta \mathbf{H}\cdot\mathbf{S}$$

with six basic states $|\frac{5}{2}M_S\rangle$ may be replaced by a new spin-hamiltonian with $S' = \frac{1}{2}$

$$\mathscr{H}(\tfrac{1}{2}) = g'_{\|}\beta H_z S_z + g'_{\perp}\beta(H_x S_x + H_y S_y)$$

with two basic states $|\frac{1}{2}M_S\rangle$. It follows[174] from

$$\langle\tfrac{1}{2}\tfrac{1}{2}|\ g'_{\|}\beta H_z S_z\ |\tfrac{1}{2}\tfrac{1}{2}\rangle = \langle\tfrac{5}{2}\tfrac{1}{2}|\ g\beta H_z S_z\ |\tfrac{5}{2}\tfrac{1}{2}\rangle$$
$$\langle\tfrac{1}{2}\ -\tfrac{1}{2}|\ g'_{\perp}\beta H^+ S^-\ |\tfrac{1}{2}\tfrac{1}{2}\rangle = \langle\tfrac{5}{2}\ -\tfrac{1}{2}|\ g\beta H^+ S^-\ |\tfrac{5}{2}\tfrac{1}{2}\rangle$$

that

$$g'_{\|} = g \sim 2$$
$$g'_{\perp} = 3g \sim 6$$

This is exactly what has been observed. The origin of the large zero-field splitting (D is estimated to be approximately $10\ \text{cm}^{-1}$) is not completely understood. It is probably caused by a combination of a strong tetragonal field and spin–orbit coupling.

Employing the large g value variation from 2.0 to 6.0, a precise determination of the orientation of the haem plane normals were made for ferrihaemoglobin[175] as well as for ferrimyoglobin crystals of types A, B, C, D, and F[176].

2. Strong ligand fields, $S = \frac{1}{2}$

This situation is encountered in the low-spin ions Ti^-, V^0, Cr^+, Mn^{2+}, and Fe^{3+}. In a strong ligand field of octahedral symmetry, the ground term is $^2T_2$, and the results are interpreted using the spin-hamiltonian

$$\mathcal{H} = \beta(g_x H_x S_x + g_y H_y S_y + g_z H_z S_z) + A_x S_x I_x + A_y S_y I_y + A_z S_z I_z \quad (133)$$

with $S = \frac{1}{2}$.

Single-crystal results are available only for the complex hexacyanides of Mn^{2+} and Fe^{3+}. In crystals of $K_4[Mn(CN)_6]\cdot 3H_2O$ diluted in the corresponding iron(II) hexacyanide, the ^{55}Mn hyperfine structure has been resolved only at 12°K. Within experimental error, the principal axes of the (g) and (A) tensors were found to be identical[120]. $K_3[Fe(CN)_6]$ has been investigated in a diluted single crystal of $K_3[Co(CN)_6]$ at 20°K[120], as well as in undiluted form[120,177]. The spectrum of undiluted $K_3[Fe(CN)_6]$ shows a complicated behaviour and anisotropic exchange interaction between pairs of iron(III) ions resulting in a lower lying singlet level and an excited triplet level has been involved to account for the anomalous resonance observed.

The tris-(2,2′-dipyridyl) complexes $[Cr\ dip_3]^+$, $[V\ dip_3]$, and $[Ti\ dip_3]^-$ were studied in polycrystalline form and in solution[178,179]. Due to the D_3 symmetry of the molecules, the $^2T_2$ ground state is split into $^2A_1$ and ^2E. The results given in Table 15 are consistent with an $^2A_1$ ground state with 5.47% and 1.09% 4*s* admixture for the V^0 and Cr^+ complexes, respectively. The magnitude of the additional hyperfine structure due to the ligand ^{14}N nuclei suggest that strong σ bonding exists between metal 4*s* and ligand orbitals of symmetry A_1. It may be of some interest to compare these results with those obtained for dibenzene compounds of vanadium(O)[180] and chromium(I)[181,182]. From the hyperfine splitting constants (cf. Table 15), we obtain a 4*s* contribution to the A_1 ground state of 7.6% and 2.2% for the V^0 and Cr^+ compounds, respectively.

Table 15.

Compound	g	A^{Me} (10^{-4} cm^{-1})	A^N (10^{-4} cm^{-1})	A^H (10^{-4} cm^{-1})
$[Cr\ dip_3]ClO_4$	1.9972 ± 0.0002	20.4 ± 0.5	2.85 ± 0.05	—
$[V\ dip_3]$	1.9831 ± 0.0002	77.5 ± 0.9	2.13 ± 0.10	—
$[Cr(C_6H_6)_2]I$	1.9863 ± 0.0002	15.8 ± 0.3	—	3.02 ± 0.05
$[V(C_6H_6)_2]$	1.9866 ± 0.0002	58.9 ± 0.9	—	3.71 ± 0.10

F. High-spin d^6 Compounds: Fe^{II}

The ^5D ground term of the free ferrous ion is split in a weak ligand field of octahedral symmetry into a higher ^5E and a lower $^5T_2$ state. Spin–orbit coupling partly removes the fifteen-fold degeneracy of the $^5T_2$ state, splitting it to first approximation into a lowest triplet with energy $+3\lambda$, a quintet at $+\lambda$, and a septet at -2λ.

An axial field separates the orbital degeneracy of $^5T_2$ into an orbital singlet and a doublet ($^5B_2$ and ^5E in tetragonal, $^5A_1$ and ^5E in trigonal symmetry), and spin-orbit coupling leaves a singlet ($M_S = 0$) and a doublet ($M_S = \pm 1$) lowest. Writing the wave functions of these levels as

$$\begin{aligned} \psi_1 &= a_1|1, 0\rangle + b_1|0, 1\rangle + c_1|-1, 2\rangle \\ \psi_{-1} &= a_1|-1, 0\rangle + b_1|0, -1\rangle + c_1|1, -2\rangle \\ \psi_0 &= a_0|1, -1\rangle + b_0|0, 0\rangle + a_0|-1, 1\rangle \end{aligned} \tag{134}$$

we obtain for the g values

$$\begin{aligned} g_{\parallel} &= 2\{(4 + k_{\parallel})c_1^2 + 2b_1^2 - k_{\parallel}a_1^2\} \\ g_{\perp} &= 0 \end{aligned} \tag{135}$$

The magnetic dipole transition between the components of $|\pm 1\rangle$ is rigorously forbidden. A non-zero intensity may arise from matrix elements mixing the $|\pm 1\rangle$ state with the $|0\rangle$ state under the influence of rhombic symmetry components of the field. For $g_{\perp}$ this gives

$$g_{\perp} = \{2\sqrt{3}\, a_1 a_0 - k_{\perp}(a_1 b_0 + b_1 a_0) + 2\sqrt{3}\, b_1 b_0 + 2\sqrt{2}\, c_1 a_0\} \tag{136}$$

In (135) and (136), $k_{\parallel}$ and $k_{\perp}$ are orbital-reduction factors.

The only single-crystal study which has been reported has been for FeF_2, diluted 1:10^4 in ZnF_2[156]. The host crystal shows macroscopic tetragonal symmetry and has the rutile structure, space group D_{4h}^{14}, $Z = 2$[183]. The resonance results were fitted to an $S = \frac{1}{2}$ spin-hamiltonian, taking $g'_{\perp} = 0$:

$$\mathscr{H} = g'_{\parallel}\beta H_z S_z + \Delta S_x \tag{137}$$

The parameter values obtained are $g'_{\parallel} = 8.97 \pm 0.02$, $\Delta = 0.203\ \text{cm}^{-1}$ at 20°K, and $\Delta = 0.224\ \text{cm}^{-1}$ at 90°K. The resonance occurs from the ground doublet which shows the separation Δ. Each Fe^{2+} ion is surrounded by a distorted octahedron of six fluoride ligands. Below 12°K, ^{19}F hyperfine structure from two types of F^- ions (I and II) has been observed, yielding the splitting parameters $A_z^{I} = 99.8 \pm 5.2$, $A_z^{II} = 63.9 \pm 4.3$ in units of $10^{-4}\ \text{cm}^{-1}$. The large values of the constants A_z and $g'_{\parallel}$ result from the use of the fictitious $S = \frac{1}{2}$ formalism. Also, since $g'_{\perp} = 0$, $A_x = A_y = 0$.

G. High-spin and Low-spin d^7 Compounds: Co^{II}, Ni^{III}

In the free $3d^7$ ions, the ground state is ^4F. For ions in coordination compounds, the multiplicity of the ground state is determined by the strength of the ligand field. In weak ligand fields, $S = \frac{3}{2}$, whereas in strong fields, $S = \frac{1}{2}$.

1. Weak ligand fields, $S = \frac{3}{2}$

An octahedral field leaves the orbital triplet $^4T_1$ lowest. $^4T_2$ and $^4A_2$ which also originate from the free ion ^4F state are considerably higher in energy. Spin–orbit coupling removes part of the degeneracy and splits

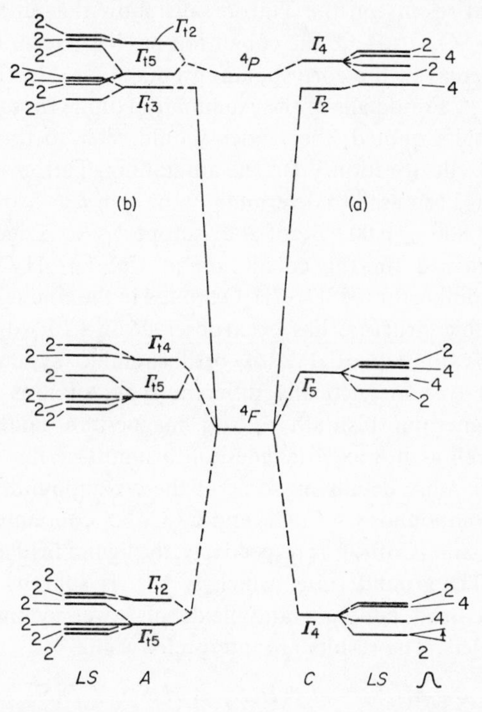

Figure 14. Energy levels of a Co^{2+} ion in an octahedral field (a) and in an axial field (b). *LS* denotes the spin–orbit interaction [according to Low[16]].

the $^4T_1$ into a lowest doublet, a higher quartet, and even higher a doublet and quartet which are accidentally degenerate ($T_1 \times D_{\frac{3}{2}} = \Gamma_6 + \Gamma_7 + 2\Gamma_8$) (see Figure 14). Usually, only transitions between the two components of the lowest Kramers doublet can be observed. The spectrum is isotropic with $S = \frac{1}{2}$. In a field of lower symmetry, the spectrum consists of six Kramers doublets and becomes highly anisotropic, although again $S = \frac{1}{2}$.

Low temperatures are generally required, since at high temperatures rapid spin-lattice relaxation broadens the lines. Since the cobalt nucleus (^{59}Co) has a spin $I = \frac{7}{2}$, there is usually an extensive hyperfine structure. The spectra are interpreted in terms of the spin-hamiltonian (133) using $S = \frac{1}{2}$. In regular octahedral symmetry in the weak-field limit, an isotropic $g = 4.33$ is expected. Experimental g values, however, often show a considerable anisotropy and the average $\bar{g}$ deviates from the theoretical value. These deviations are caused either by crystalline fields of lower than cubic symmetry or by internal fields arising from hyperfine interaction between electrons and nuclei.

Experimental results on the Tutton salts show that in the ammonium salt $(NH_4)_2Co(SO_4)_2 \cdot 6H_2O$, the cobalt ion has very accurately tetragonal symmetry, whereas in the corresponding potassium salt, the symmetry is only rhombic[184]. For details of the examination of the rhombic component and of the results quoted, the reader should refer to the discussion on copper Tutton salts (Section V.I). The ammonium Tutton salt, enriched to 2% in ^{60}Co, has been used to determine[185] the spin $I = 5$ and the magnetic moment $\mu = 3.800 \pm 0.007\ \beta_N$ of the isotope ^{60}Co. Trigonal symmetry has been confirmed for the cobalt ion in $CoSiF_6 \cdot 6H_2O$, whereas the symmetry is rhombic for $CoSO_4 \cdot 7H_2O$ diluted in the zinc salt[184]. Extensive fluorine hyperfine structure has been observed in CoF_2 diluted in ZnF_2, which contains ions $[CoF_6]^{4-}$ of orthorhombic symmetry[156]. Axial symmetry and two magnetically different types of ions are found for cobalt in magnesium bismuth[186] and magnesium lanthanum double nitrate[187] as well as in mixed magnesium bismuth–cerium double nitrate crystals[187]. For more details on some of these compounds, see Table 16.

In the two compounds Cs_3CoCl_5 and Cs_3CoBr_5 containing the complex ions $[CoCl_4]^{2-}$ and $[CoBr_4]^{2-}$, respectively, the ligand field is of tetrahedral symmetry[188]. The ground state, which is $^4A_2$, is split by the combined action of spin–orbit coupling and fields of lower symmetry into two Kramers doublets. The results are interpreted using

$$\mathcal{H} = g_{\parallel}\beta H_z S_z + g_{\perp}\beta(H_x S_x + H_y S_y) + D(S_z^2 - \tfrac{5}{4}) \qquad (138)$$

with $S = \frac{3}{2}$.

2. Strong ligand fields, $S = \frac{1}{2}$

Low-spin ground states are found in planar compounds of cobalt(II) like cobalt(II) phthalocyanine and related compounds[109,189–190]. Measurements on α-[Co pc], diluted in α-[Zn pc] powder[189], result in $g_{\parallel} = 2.007$, $g_{\perp} = 2.422$, and $A = 116$, $B = 66$ in units of $10^{-4}\ \text{cm}^{-1}$. The different crystallographic site of β-[Co pc] is demonstrated by the values $g_{\parallel} = 1.89$, $g_{\perp} = 2.94$, and $A = 150$, $B = 280$ in units of $10^{-4}\ \text{cm}^{-1}$ obtained from

Table 16. Cobalt Co^{2+}.

Crystal	Temp. (°K)	g	HFS: $(A, B)^{59}$ (10^{-4} cm^{-1})	Remarks	Ref.
CoF_2 d. in ZnF_2	20	$g_x = 2.6$ $g_y = 6.05 + 0.01$ $g_z = 4.1 \pm 0.1$	$A_x = -43$ $A_y = 217 \pm 2$ $A_z = 67$	Rutile structure; contains ions $[CoF_6]^{4-}$; ligand ^{19}F ($I = \frac{1}{2}$) HFS observed: $A_y^F = 32 \pm 1$, $A_z^F = 21 \pm 5$ in 10^{-4} cm^{-1}	156
Cs_3CoCl_5 und.	77, 4	$g_\parallel = 2.41 \pm 0.02$ $g_\perp = 2.33 \pm 0.04$		Tetrahedral coordination, $S = \frac{3}{2}$; $M_m = 1$; $D = -4.31 \pm 0.04$ cm^{-1}, $E = 0$	188
Cs_3CoBr_5 und.	77, 4	$g_\parallel = 2.6$ $g_\perp = 2.4$		Tetrahedral coordination, $S = \frac{3}{2}$; $M_m = 1$; $D = -5.4$ cm^{-1}	188
$CoSO_4 \cdot 7H_2O$ $d = 100$ in $ZnSO_4 \cdot 7H_2O$	20	$g_x = 2.30 \pm 0.05$ $g_y = 3.30 \pm 0.07$ $g_z = 6.90 \pm 0.14$	$A_x = 28 \pm 1$ $A_z = 254 \pm 5$	Rhombic, $Z = 4$	184, C.S. 245
$K_2Co(SO_4)_2 \cdot 6H_2O$ $d = 500$ to 10^5 in $K_2Zn(SO_4)_2 \cdot 6H_2O$	20	$g_z = 6.56 \pm 0.13$ $g_{min} = 2.50 \pm 0.05$ $g_2 = 3.35 \pm 0.07$	$A_z = 286 \pm 6$ $A_{min} = 65 \pm 3$ $A_2 = 80 \pm 4$	Monoclinic, $\psi = +163°$, $\theta = 35°$; g_{min}, A_{min} determined in K_1K_3 plane	184, C.S. 201, 202

Table 16 (*continued*)

Crystal	Temp. (°K)	g	HFS: $(A, B)^{59}$ (10^{-4} cm^{-1})	Remarks	Ref.
$Rb_2Co(SO_4)_2 \cdot 6H_2O$ d. in $Rb_2Zn(SO_4)_2 \cdot 6H_2O$	20	$g_z = 6.65$ $g_{min} = 2.7$ $g_2 = 3.3$	$A_z = 293 \pm 3$ $A_{min} = 49 \pm 5$	Monoclinic, $\psi = +157°$, $\theta = 37°$; g_{min}, A_{min} determined in K_1K_3 plane	252, C.S. 201, 202
$(NH_4)_2Co(SO_4)_2 \cdot 6H_2O$ d = 500 to 10^5 in $(NH_4)_2Zn(SO_4)_2 \cdot 6H_2O$	20	$g_z = 6.45 \pm 0.13$ $g_{min} = 3.06 \pm 0.06$ $g_2 = 3.06 \pm 0.06$	$A_z = 245 \pm 5$ $A_{min} = 20 \pm 1$ $A_2 = 20 \pm 1$	Monoclinic, $\psi = +130°$, $\theta = 34°$; from sample enriched to 2% ^{60}Co, $I(^{60}Co) = 5$, $\mu(^{60}Co) = 3.800 \pm 0.007\ \beta_N$ determined; g_{min}, A_{min} determined in K_1K_3 plane	184, C.S. 201, 202
$Co_3Bi_2(NO_3)_{12} \cdot 24H_2O$ d = 100 in $Mg_3Bi_2(NO_3)_{12} \cdot 24H_2O$	20	I: $g_{\parallel} = 7.29 \pm 0.01$ $g_{\perp} = 2.338 \pm 0.004$ II: $g_{\parallel} = 4.108 \pm 0.003$ $g_{\perp} = 4.385 \pm 0.003$	$A = 283 \pm 1$ $B \leqslant 1$ $A = 85 \pm 1$ $B = 103 \pm 1$	Unit cell contains two different types of magnetic ions (I, II)	187
$CoSiF_6 \cdot 6H_2O$ d = 500 to 10^5 in $ZnSiF_6 \cdot 6H_2O$	20	$g_{\parallel} = 5.82 \pm 0.12$ $g_{\perp} = 3.44 \pm 0.07$	$A = 184 \pm 4$ $B = 47 \pm 2$	Trigonal, $Z = 1$	184, C.S. 202, 253

polycrystalline samples[189] diluted in β-[Zn pc], β-[Ni pc], and β-H_2pc as well as from single-crystal results. For dithiolate complexes of cobalt(II) and nickel(III), the reader should refer to the literature[86, 191].

H. Six-coordinated d^8 Compounds: Ni^{II}

The ground state ^3F of the free Ni^{2+} ion is split by an octahedral ligand field into $^3A_2$, $^3T_2$, and $^3T_1$, the $^3A_2$ term being lowest. Spin–orbit coupling on $^3A_2$ produces Γ_5, and thus does not remove the three-fold spin degeneracy. If a trigonal or tetragonal field is present, the spin triplet is split into a singlet and a doublet. A rhombic field produces three close-lying singlets having a separation in the order of 1 cm^{-1}. No hyperfine structure due to the isotope ^{61}Ni has been detected, since its natural abundance is only 1.25%. The results are therefore interpreted in terms of the following spin-hamiltonian with $S = 1$:

$$\mathscr{H} = g\beta(H_xS_x + H_yS_y + H_zS_z) + D(S_z^2 - \tfrac{2}{3}) + E(S_x^2 - S_y^2) \quad (139)$$

According to the available resonance results, g is taken to be isotropic.

Resonance has been observed in rhombic $NiSO_4 \cdot 7H_2O$ at 0.54 and 0.65 cm^{-1} and the direction cosines of the four inequivalent ions in the unit cell have been determined[192]. In the nickel Tutton salts with two inequivalent ions in the unit cell, the resonance results[193] have been interpreted by assuming that the non-cubic part of the ligand field again has rhombic symmetry. The splittings between the adjacent levels of the spin triplet Γ_5 in zero-field are $\delta_1 = |D| - |E|$ and $\delta_2 = |2E|$. Representative values for the spin-hamiltonian parameters as obtained for $(NH_4)_2Ni(SO_4)_2 \cdot 6H_2O$ at 290°K are $g = 2.25$, $D = -2.24$ cm^{-1}, and $E = -0.387$ cm^{-1}. D and E are sensitive to temperature and dilution. In $NiSiF_6 \cdot 6H_2O$ with a single ion per unit cell, the ligand field is of trigonal symmetry[194] and thus $E = 0$. The pressure dependence of D has been studied[195,196] and its temperature dependence has been investigated down to 0.4°K[197]. The g value appears to be independent of temperature. In the tris(acetylacetonates) Na[Ni aca_3]$\cdot C_4H_4O_2$ and Na[Ni aca_3]$\cdot C_6H_6$, an orthorhombic component of the ligand field has been found in the former ($E = -0.0833$ cm^{-1}), whereas in the latter the field is of trigonal symmetry ($E = 0$)[198]. For details of the experimental results, see Table 17.

I. Tetragonal and Square Planar d^9 Compounds: Cu^{II}

The ^2D state of the free ion is split in an octahedral field into $^2T_2$ and the non-magnetic ^2E which becomes the ground state. Spin–orbit interaction on ^2E produces Γ_8 and the four-fold degeneracy is not lifted. No resonance has been reported for Cu^{2+} ions in cubic single crystals.

Table 17. Nickel Ni^{2+}.

Crystal	Temp. (°K)	g	D (10^{-4} cm^{-1})	E (10^{-4} cm^{-1})	Remarks	Ref.
$NiSO_4 \cdot 7H_2O$ und.	290	2.20	−35600	−15000	Rhombic, $Z = 4$; direction cosines determined[a]	192, C.S. 245
$K_2Ni(SO_4)_2 \cdot 6H_2O$ und.	290	2.25 ± 0.05	−33000	−5170	Monoclinic, $Z = 2$; $\psi = -12.5°$, $\theta = 11°$	193, C.S. 201, 202
$(NH_4)_2Ni(SO_4)_2 \cdot 6H_2O$ und. and $d = 50$ in $(NH_4)_2Zn(SO_4)_2 \cdot 6H_2O$	290	2.25 ± 0.05	−22400	−3870	Monoclinic, $Z = 2$; $\psi = -14°$, $\theta = 3.5°$	193, 254, C.S. 201, 202
$Tl_2Ni(SO_4)_2 \cdot 6H_2O$ und.	290	2.25 ± 0.05	−26000	−1000	Monoclinic, $Z = 2$; $\psi = -11°$, $\theta = 11°$	193, C.S. 201, 202
$K_2Ni(SeO_4)_2 \cdot 6H_2O$ und.	290	2.25 ± 0.05	−30000	−10000	Monoclinic, $Z = 2$; $\psi = -13°$, $\theta = 0°$	193, C.S. 201, 202
$(NH_4)_2Ni(SeO_4)_2 \cdot 6H_2O$ und.	290	2.25 ± 0.05	−18900	−7900	Monoclinic, $Z = 2$; $\psi = -28°$, $\theta = 0°$	193, C.S. 201, 202
$NiSiF_6 \cdot 6H_2O$ und.	290	2.3	−5000	0	Rhombohedral, $Z = 1$; D increases by ~20% in crystals of $d = 4$ to 16 in $ZnSiF_6 \cdot 6H_2O$	197, 194, 193, C.S. 202, 253
$Na[Ni\ aca_3] \cdot C_4H_4O_2$ und.	4.2	2.20	−18700	−833		198
$Na[Ni\ aca_3] \cdot C_6H_6$ und.	4.2	2.20	−21900	0		198

[a] For one complex, the d.c. are:

	a	b	c
z	0.95	0.31	0
y	−0.31	0.95	0.09

The d.c. of the remaining 3 complexes are determined by reflection symmetry.

In a tetragonal field, the ^2E breaks up into two terms, $^2A_1 + {}^2B_1$ which, under the influence of spin–orbit interaction, are transformed into two Kramers doublets $\Gamma_{t6} + \Gamma_{t7}$. Paramagnetic resonance is observed for the lower one of the doublets. Two Kramers doublets are formed likewise in the slightly more complicated case of rhombic symmetry[199,200]. The results are usually interpreted in terms of the following spin-hamiltonian with $S = \frac{1}{2}$:

$$\mathscr{H} = \beta(g_x H_x S_x + g_y H_y S_y + g_z H_z S_z) + A_x S_x I_x + A_y S_y I_y + A_z S_z I_z + P(I_z^2 - \tfrac{5}{4}) + P'(I_x^2 - I_y^2) - \gamma\beta_N \mathbf{H}\cdot\mathbf{I} \quad (140)$$

There are actually two naturally occurring isotopes of copper, ^{63}Cu and ^{65}Cu, and $I = \frac{3}{2}$ for each. Since, in most cases, their hyperfine structure is not resolved separately, only average values of the splitting constants (viz. A_x, A_y, and A_z) have been included in (140). The behaviour of the energy levels according to the spin-hamiltonian (140) is illustrated schematically in Figure 15. The allowed transitions are those in which $\Delta M_I = 0$.

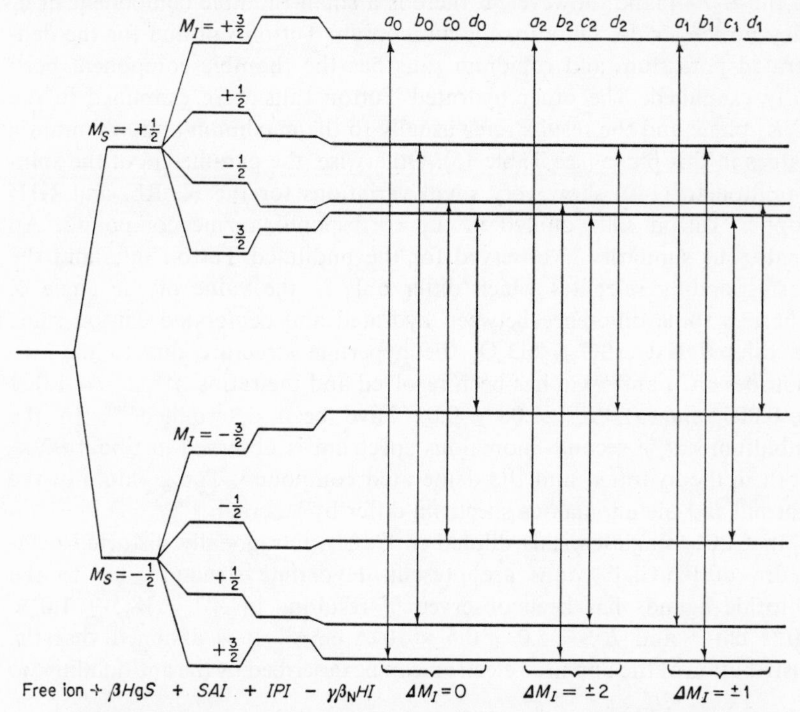

Figure 15. Energy levels of a Cu^{2+} ion ($S = \frac{1}{2}, I = \frac{3}{2}$). The applied spin-hamiltonian is shown in the lowest line. $\Delta M_I = 0$ and $\Delta M_I = \pm 2$ transitions are allowed if H is along a rhombic axis, and $\Delta M_I = \pm 1$ transition are allowed in intermediate directions [according to Bleaney and colleagues[199]].

If the quadrupole interaction is comparable with the magnetic interaction, $\Delta M_I = \pm 2$ transitions are also possible. In addition, $\Delta M_I = \pm 1$ transitions are allowed if the external magnetic field is not along a rhombic axis.

This type of situation may be exemplified by the carefully investigated copper Tutton salts. The space group has been determined to be $P2_1a$, with two molecules in the unit cell[201,202]. Each divalent ion is surrounded by a distorted octahedron of six water molecules. The crystal structure suggests that the ligand field, and thus the g tensor, has approximately tetragonal symmetry. The principal axes of the g tensor for either of the copper ions in a unit cell are not directly related to the monoclinic axes of the crystal. However, the three axes of one ion are derived from those of the other by reflection in the *ac* plane. If the magnetic field is in the *ac* plane, the spectra of the two ions coincide and the maximum and minimum g values correspond to the two directions of the K_1 and K_2 magnetic axes. If the g value for each ion has axial symmetry, the two spectra coincide in the K_2K_3 plane; however, if there is a small rhombic component in $g_\perp$ they no longer do. Only for the ammonium Tutton salt and for the deuterated potassium and rubidium salts has the rhombic component been fully examined. The other hydrated Tutton salts were examined in the K_1K_3 plane and the results refer usually to the maximum and minimum g values in this plane (see Table 15). Otherwise, the parameters of the spin-hamiltonian (140) show very small variations for the K, Rb, and NH_4 copper Tutton salts diluted in the corresponding zinc compound. An analogous similarity is observed for the undiluted Tutton salts and the corresponding selenates which differ only in the value of the angle ψ. There is some difference between hydrated and deuterated Tutton salts. In diluted $K_2Cu(SO_4)_2 \cdot 6D_2O$, the hyperfine structure due to the two isotopes ^{63}Cu and ^{65}Cu has been resolved and the ratios $A^{65}/A^{63} = 1.069 \pm 0.003$ and P^{63}/P^{65} 1.08 ± 0.02 have been determined[199]. In the rubidium salt, a second anomalous spectrum is observed at about 90°K, both in the hydrated and the deuterated compound. The g values of the normal and the anomalous spectrum differ by less than 1%.

In $CuCl_2$ single crystals diluted in $CdCl_2$, tetragonally distorted octahedra of $[CuCl_6]^{4-}$ ions are present. Hyperfine structure due to the chloride ligands has been observed[203] resulting in $A^{Cl} = 18.5 \pm 1.0 \times 10^{-4}$ cm^{-1} and $B^{Cl} = 5.0 \pm 0.5 \times 10^{-4}$ cm^{-1}. It is assumed that the distribution of the unpaired electron can be described by the antibonding MO

$$\psi_{x^2-y^2} = N_\sigma[d_{x^2-y^2} - \alpha_s(s_1 - s_2 + s_3 - s_4) - \alpha_{p\sigma}(-p_1 + p_2 + p_3 - p_4)] \quad (141)$$

where s and p refer to chlorine $3s$ and $3p\sigma$ orbitals respectively, N_σ is the normalization constant, and the overlap with chlorine $2s$ and $2p\sigma$ orbitals is neglected. The chlorine hyperfine splitting parameters due to the $3s$ and $3p\sigma$ contributions

$$A_s = \tfrac{16}{3}\pi\gamma\beta\beta_N N_\sigma^2 \alpha_s^2 |s(0)|_{3s}^2 \tag{142}$$

and

$$A_{p_\sigma} = \tfrac{4}{5}\gamma\beta\beta_N \langle r_{3p}^{-3} \rangle N_\sigma^2 \alpha_{p\sigma}^2 \tag{143}$$

are related to the experimentally determined parameters by

$$A^{\text{Cl}} = A_s + 2A_{p\sigma}, \qquad B^{\text{Cl}} = A_s - A_{p\sigma} \tag{144}$$

If it is assumed that[203] $|s(0)|_{3s}^2 = 69 \times 10^{24}$ and $\langle r_{3p}{}^{-3} \rangle = 55.6 \times 10^{24}$ cm^{-3}, the effective fraction of unpaired spin in a chloride $3s$ or $3p\sigma$ orbital is estimated as

$$f_s = N_\sigma^2 \alpha_s^2 = 0.67\%$$

$$f_{p\sigma} = N_\sigma^2 \alpha_{p\sigma}^2 = 8.3\%$$

These values indicate the presence of relatively strong σ bonding in the $[CuCl_6]^{4-}$ ion.

There is a large number of copper(II) compounds showing exchange interactions of different intensity. $CuSO_4 \cdot 5H_2O$ assumes an ordered antiferromagnetic structure at very low temperatures which is characterized by the Néel temperature $\Theta_N = 0.03°\text{K}$. The exchange energy between neighbouring ions is of the order of a microwave quantum, and thus the spectrum depends on the microwave frequency used. $CuCl_2 \cdot 2H_2O$ becomes antiferromagnetic below 4.3°K, and its resonance spectrum changes at the transition temperature. Strong exchange interactions between isolated pairs of Cu^{2+} ions, which result in the formation of a lower singlet and a higher triplet state, only the latter being paramagnetic, are often encountered. A well-known example for this sort of behaviour is copper acetate monohydrate, $Cu(CH_3COO)_2 \cdot H_2O$. The crystal is monoclinic, with four molecules per unit cell[204]. Two molecules form a dimeric bridged structure with strong Cu—Cu interaction. The exchange energy is $J/hc = 116\ \text{cm}^{-1}$. The resonance results are interpreted in terms of the spin-hamiltonian

$$\mathscr{H} = \beta(g_x H_x S_x + g_y H_y S_y + g_z H_z S_z) + D(S_z^2 - \tfrac{2}{3}) + E(S_x^2 - S_y^2) + A_x S_x I_x + A_y S_y I_y + A_z S_z I_z \tag{145}$$

with $S = 1$, $D = 3450 \pm 50$, and $E = 70 \pm 50$ in units of 10^{-4} cm^{-1} [205,206]. Similar behaviour is found in $Cu(CH_3CH_2COO)_2 \cdot H_2O$ and many other copper salts. For a complete list of similar compounds, see ref. 86.

Another interesting effect is associated with fluorosilicates and other crystals of trigonal symmetry. The orbital doublet ^2E, which is the ground state in octahedral symmetry, is not split by a superimposed trigonal field. The orbital degeneracy is lifted only by the Jahn–Teller effect[67,68] which distorts the magnetic complex. There are several distortions having the same energy, e.g. three distortions of tetragonal symmetry with mutually perpendicular axes. The system resonates through these potential minima, and possibly also through other potential minima formed by distortions of rhombic symmetry. Thus a dynamical Jahn–Teller effect is created leading to the expectation of an isotropic resonance spectrum[207]. Experimentally, in the case of $CuSiF_6 \cdot 6H_2O$ diluted in the corresponding zinc compound, there is a nearly isotropic g value and an almost isotropic hyperfine structure at high temperatures. At sufficiently low temperatures, the ion becomes frozen into one of the potential minima and the dynamic Jahn–Teller effect is replaced by a static distortion. The spectrum corresponds to three magnetic complexes, each with tetragonal symmetry; the three tetragonal axes are the edges of a cube whose trigonal axis is the crystal axis. For $CuSiF_6 \cdot 6H_2O$[6,7] at 90°K, $g_{\parallel} = 2.221$ and $g_{\perp} = 2.230$, whereas at 20°K, $g_{\parallel} = 2.46$ and $g_{\perp} = 2.10$. The transition between the two forms is very gradual; in the intermediate range of temperature both spectra are found simultaneously. The transition takes place over the temperature range 12 to 50°K[7].

Similar observations were made on several other cupric salts with trigonal symmetry. These include the double nitrates $Cu_3La_2(NO_3)_{12} \cdot 24D_2O$ and $Cu_3Bi_2(NO_3)_{12} \cdot 24H_2O$, both diluted in the corresponding manganese compounds[7]. The transition takes place in the former salt in the range 33 to 45°K. In $Cu(BrO_3)_2 \cdot 6H_2O$, the spectrum is isotropic at 90°K with a g value of about 2.22. At 20°K, the isotropic spectrum is still present, but with decreasing temperature several anisotropic spectra are observed which have not been analyzed in detail. The transition is not completed at 7°K.

In a large number of copper(II) complexes, paramagnetic resonance results were used to investigate the bonding between the Cu^{2+} ion and the ligands. As an example, we shall discuss the case of square planar complexes with D_{4h} symmetry. Each ligand has available, e.g., $2s$, $2p_x$, $2p_y$, and $2p_z$ orbitals for the formation of molecular orbitals with d orbitals of the central ion. The following antibonding molecular orbitals are then constructed and are listed in the order of increasing energy:

$$
\begin{aligned}
|B_{1g}\rangle &= \alpha d_{x^2-y^2} - \tfrac{1}{2}\alpha'(-\sigma_x^1 + \sigma_y^2 + \sigma_x^3 - \sigma_y^4) \\
|B_{2g}\rangle &= \beta_1 d_{xy} - \tfrac{1}{2}(1 - \beta_1^2)^{\frac{1}{2}}(p_y^1 + p_x^2 - p_y^3 - p_x^4) \\
|A_{1g}\rangle &= \alpha_1 d_{z^2} - \tfrac{1}{2}(1 - \alpha_1^2)^{\frac{1}{2}}(\sigma_x^1 + \sigma_y^2 - \sigma_x^3 - \sigma_y^4) \\
|E_g\rangle &= \begin{cases} \beta d_{xz} - [(1 - \beta^2)^{\frac{1}{2}}/\sqrt{2}](p_z^1 - p_z^3) \\ \beta d_{yz} - [(1 - \beta^2)^{\frac{1}{2}}/\sqrt{2}](p_z^2 - p_z^4) \end{cases}
\end{aligned}
\qquad (146)
$$

Here, $\sigma^i = (1/\sqrt{3})(\sqrt{2}\,p^i \mp s^i)$ are sp^2-hybridized directed orbitals centred on the ith nucleus of the ligands (the $\pm$ signs refer to ligands situated on the positive and negative axes); s^i and p^i refer to $2s$ and $2p$ orbitals on the ith ligand nucleus (1 and 3 refer to ligands on the positive and negative x axis, 2 and 4 refer to ligands on the positive and negative y axis). Applying the normalization condition on the ground-state orbital $|B_{1g}\rangle$, we find

$$\alpha^2 + \alpha'^2 + 2\alpha\alpha' S = 1 \tag{147}$$

where S is the overlap integral between $d_{x^2-y^2}$ and the normalized ligand orbital of B_{1g} symmetry,

$$S = \tfrac{1}{2}\langle d_{x^2-y^2}|(-\sigma_x^1 + \sigma_y^2 + \sigma_x^3 - \sigma_y^4)\rangle \tag{148}$$

Only in the case of the B_{1g} orbital is the overlap taken into account, since, for the other orbitals, it is likely to be small. The unpaired electron is placed into the lowest antibonding MO of B_{1g} symmetry. Using the molecular orbitals (146) as a basis, the crystal-field hamiltonian may be solved to second order by application of perturbation theory. If the observed spectrum is fitted to a spin-hamiltonian characteristic of tetragonal symmetry,

$$\begin{aligned}\mathscr{H} = g_\parallel \beta H_z S_z + g_\perp \beta(S_x H_x + H_y S_y) + A S_z I_z + B(S_x I_x + S_y I_y)\\ + P[I_z^2 - \tfrac{1}{3}I(I+1)] - \gamma\beta_{\mathrm{N}}\mathbf{H}\cdot\mathbf{I}\end{aligned} \tag{149}$$

with $S = \frac{1}{2}$ and $I = \frac{3}{2}$, the following expressions for the spin-hamiltonian parameters are obtained:

$$\begin{aligned} g_\parallel &= 2.0023 - 8\rho\{\alpha\beta_1 - \alpha'\beta_1 S - \alpha'(1-\beta_1^2)^{\frac{1}{2}}\tfrac{1}{2}[T(n)]\}\\ g_\perp &= 2.0023 - 2\mu[\alpha\beta - \alpha'\beta S - \alpha'(1-\beta^2)^{\frac{1}{2}}T(n)/\sqrt{2}]\end{aligned} \tag{150}$$

$$\begin{aligned} A = P(&-\tfrac{4}{7}\alpha^2 - \kappa + (g_\parallel - 2) + \tfrac{3}{7}(g_\perp - 2) -\\ &- 8\rho\{\alpha'\beta_1 S + \alpha'(1-\beta_1^2)^{\frac{1}{2}}\tfrac{1}{2}[T(n)]\}\\ &- \tfrac{6}{7}\mu\{\alpha'\beta S + \alpha'(1-\beta^2)^{\frac{1}{2}}T(n)/\sqrt{2}\})\end{aligned} \tag{151}$$

$$B = P(\tfrac{2}{7}\alpha^2 - \kappa + \tfrac{11}{14}(g_\perp - 2) - \tfrac{22}{14}\mu\{\alpha'\beta S + \alpha'(1-\beta^2)^{\frac{1}{2}}T(n)/\sqrt{2}\})$$

where

$$\rho = \lambda_0\alpha\beta_1/(E_{xy} - E_0) \tag{152}$$

$$\mu = \lambda_0\alpha\beta/(E_{xz,yz} - E_0) \tag{153}$$

and

$$T(n) = n - \tfrac{1}{2}(1-n^2)^{\frac{1}{2}}R\Omega \tag{154}$$

$$P = 2\gamma\beta\beta_{\mathrm{N}}\langle d_{x^2-y^2}|\, r^{-3}\, |d_{x^2-y^2}\rangle \tag{155}$$

In these expressions, λ_0 is the spin–orbit coupling constant for the free Cu^{2+} ion, E_n is the energy of the appropriate molecular orbital, κ is the isotropic hyperfine coupling parameter, and

$$\Omega = \frac{2}{\sqrt{3}} \int_0^\infty r^2 R_{21}(r) \frac{\mathrm{d}}{\mathrm{d}r} [R_{20}(r)] \,\mathrm{d}r \tag{156}$$

$R_{21}(r)$ and $R_{20}(r)$ in (156) are normalized radial $2p$ and $2s$ ligand functions and R in (154) is the metal–ligand distance. Ω has been evaluated for hydrogen-like radial functions and, if Z_s and Z_p denote the effective nuclear charges on the s and p orbitals and a_0 the Bohr radius, it is

$$\Omega = 16(Z_p Z_s)^{\frac{5}{2}}(Z_s - Z_p)/(Z_s + Z_p)^5 a_0 \tag{157}$$

Assuming reasonable values for Z_s, Z_p and putting $n = (\frac{2}{3})^{\frac{1}{2}}$ for sp^2 hybridization, $T(n)$ is readily calculated.

The theory has been applied to single-crystal results obtained for [Cu aca_2] diluted in the monoclinic [Pd aca_2]. In this crystal, the acetyaceto-nate ligands are bonded to the central ion through their oxygen atoms. The Cu^{2+} ion has four oxygens as nearest neighbours. The analysis yields the following values of the bonding parameters and of κ[208]:

$$\alpha^2 = 0.81; \qquad \beta^2 = 0.99; \qquad \beta_1^2 = 0.85; \qquad \kappa = 0.33$$

The overlap integral S (equation 148) has been estimated using hydrogen-like orbitals as $S = 0.094$. $\langle r^{-3}\rangle_0$ has been taken as 7.25 a.u. according to Abragam and Pryce[209]. The orbital excitation energies $E_{xy} - E_0 = 15{,}000$ cm^{-1} and $E_{xz,yz} - E_0 = 25{,}000\ \mathrm{cm}^{-1}$ were taken from the electronic spectra, $\lambda_0 = -828\ \mathrm{cm}^{-1}$, $R = 3.7a_0$, and $Z_s + Z_p$ and $Z_s - Z_p$ were assumed to be 4.5 and 0.5, respectively. Terms in S occurring in the expressions (150) and (151) were neglected. The parameters obtained indicate that the in-plane σ and π bonding in the B_{1g} and B_{2g} orbitals has covalent character, whereas the out-of-plane π bonding in the E_g orbitals is ionic. Little interaction with the π-electron system of the ligands is therefore present.

In copper diethyldithiocarbamate, $[Cu\{(C_2H_5)_2NC(S)S\}_2]$, diluted in the corresponding zinc compound, the bonding parameters were evaluated as[210]

$$\alpha^2 = 0.504; \qquad \beta^2 = 0.703; \qquad \beta_1^2 \approx 1; \qquad \kappa = 0.23$$

In this complex, the copper ion has four sulphurs as nearest neighbours with a Cu—S distance of 2.3 Å. The values of $T(n)$ and the overlap integral S were calculated by Gersman and Swalen[84] and are given as $T(n) = 0.44$, $S = 0.005$. From the electronic spectra, $E_{xz,yz} - E_0 = 22{,}000\ \mathrm{cm}^{-1}$ and $E_{xy} - E_0$, however, could not be determined to any high reliability. The bonding parameters indicate that in this compound, the σ bonding in the B_{1g} orbital is purely covalent. The out-of-plane π bonding ($\beta = 0.70$) has some covalent character, whereas the in-plane π bonding ($\beta_1 \approx 1$) is purely ionic.

A similar analysis has been carried out for copper(II) bis(salicylaldimine), $[Cu\{2\text{-}OC_6H_4CH{=}NH\}_2]$, diluted in the corresponding nickel(II) compound which is diamagnetic[211]. Although the molecular field at the position of Cu^{2+} ion is expected to be of rhombic symmetry, since there are two oxygen and two nitrogen atoms as nearest neighbours, the g factor is about the same in the direction of the O as in the direction of the N atoms. Thus, the bonding with the oxygen and nitrogen atoms should be very similar and the system may be regarded as one of tetragonal symmetry. From the previously mentioned theoretical treatment the following parameters were calculated:

$$\alpha^2 = 0.83; \qquad \beta^2 \geqslant 0.91; \qquad \beta_1^2 = 0.72; \qquad \kappa = 0.34$$

These values are very similar to those obtained for [Cu aca$_2$] except that the in-plane π bonding is more covalent in the present case ($\beta_1 = 0.85$).

Another example which achieved much attention is copper phthalocyanine. Recently, single crystals of [Cu pc] diluted 1:1000 in [Zn pc][212] and those of [Cu pc] diluted 1:1000 in metal-free H_2 pc[213,214] were studied by independent groups of workers. Both crystals are monoclinic, space group $P2_1/a$, with two molecules per unit cell[215,216]. The Cu^{2+} ion is at the centre of a square of four nitrogen atoms with a Cu—N distance of 1.83 Å. The octahedral array is completed by two nitrogen atoms from neighbouring [Cu pc] molecules at a distance of 3.38 Å from the central Cu^{2+} ion. The symmetry may be again considered as tetragonal. The spectra show a superhyperfine structure from the four nearest surrounding nitrogen atoms of the ligand. The hamiltonian (149) has therefore to be extended to include the hyperfine interaction

$$A^N S_z I_z^N + B^N(S_x I_x^N + S_y I_y^N) \tag{158}$$

The nitrogen-splitting parameters are then calculated to be

$$\begin{aligned} A^N &= \left(\frac{\alpha'}{2}\right)^2 2\gamma\beta\beta_N\{\tfrac{8}{9}\pi|s(0)|^2 - \tfrac{8}{15}\langle r_p^{-3}\rangle\} \\ B^N &= \left(\frac{\alpha'}{2}\right)^2 2\gamma\beta\beta_N\{\tfrac{8}{9}\pi|s(0)|^2 + \tfrac{4}{15}\langle r_p^{-3}\rangle\} \end{aligned} \tag{159}$$

where $s(0)$ is the value of the ligand nitrogen $2s$ function at the nitrogen nucleus and $\langle r_p^{-3}\rangle$ is an average over the ligand nitrogen $2p$ wave function.

The spin-hamiltonian parameters obtained for the two crystals are very similar. However, the two groups of investigators arrive at different conclusions as far as the amount of in-plane π bonding is concerned. The weak point of the analysis is that β^2 and β_1^2 can be ascertained only if the excitation energies $E_{xy} - E_0$ and $E_{xz,yz} - E_0$ are known from the electronic spectra. Unfortunately, it is very difficult to identify the crystal-field transitions, since they are overlapped and obscured by the stronger π–π transitions of the heterocyclic ring system. The extinction coefficients of the *d*–*d* transitions are about 5–50 while those of the π–π transitions are of the order of 10^4. Thus only α' can be readily evaluated from equations (150) and (151). Using $|s(0)|^2 = 33.4 \times 10^{24}\ \text{cm}^{-3}$[13], it follows $\alpha' = 0.6$, and if[208] $S = 0.047$, $\alpha^2 = 0.72$ is obtained. According to Harrison and Assour[213,214], the equations (150) and (151) together with $P = 0.036\ \text{cm}^{-1}$ and $\kappa = 0.33$ result in $\beta_1 \sim 1$. This would mean that although there is appreciable covalent character of in-plane σ bonding in the B_{1g} orbital, $\alpha = 0.85$, no in-plane π bonding of B_{2g} symmetry is indicated. Deal and colleagues[212] have carefully examined the difference between optical spectra of single crystals of [Cu pc] and metal-free phthalocyanines. These authors report two weak absorptions in [Cu pc] at 17,000 and 14,500 cm^{-1} which they assign to excitations from the B_{1g} ground state to the B_{2g} and E_g antibonding MO, respectively. Taking $E_{xy} - E_0 = 17{,}000$ and $E_{xz,yz} - E_0 = 14{,}500\ \text{cm}^{-1}$, they arrive at

$$\alpha^2 = 0.79; \qquad \beta^2 = 0.63; \qquad \beta_1^2 = 0.65$$

These values indicate that the in-plane as well as out-of-plane π bonds are highly covalent as well.

A comparison of the bonding parameters for different copper(II) complexes is interesting particularly with respect to σ bonding. Inspection of the listed values of α^2 shows that the Cu—S σ bond in copper diethyldithiocarbamate is much more covalent than the Cu—O or Cu—N bonds in copper acetylacetonate, copper salicylaldimine or copper phthalocyanine. The coordinated ligand atoms thus give rise to an increasing covalent character of the metal–ligand bond in the order of O, N, S. The covalency of the B_{1g} σ bond influences also the order of the excited antibonding orbitals. In a purely ionic tetragonal complex, the degenerate d_{xz}, d_{yz} orbital of symmetry E_g is expected to be lower in energy than the $B_{2g}(d_{xy})$ orbital. This is also observed in both copper acetylacetonate[208] and copper salicylaldimine[211]. In copper phthalocyanine[212–214] and copper diethyldithiocarbamate[210], however, this order is reversed. This change probably results in copper phthalocyanine from the large covalency of the π bonds. On the other hand, in copper diethyldithiocarbamate, the same change seems to be effected by the increased covalency of the σ bond, since the B_{2g} orbital is purely metal d_{xy}, and the E_g orbital

Table 18. Copper Cu^{2+}.

Crystal	Temp. (°K)	g	HFS: $(A, B)^{63,65}$ (10^{-4} cm^{-1})	P (10^{-4} cm^{-1})	Remarks	Ref.
$CuCl_2$ $d = 200$ in $CdCl_2$	20	$g_{\parallel} = 2.339 \pm 0.002$ $g_{\perp} = 2.070 \pm 0.002$	$A = 113 \pm 3$ $B = 0.0 \pm 4$		Tetragonally distorted $[CuCl_6]^{4-}$ ions present; ligand $^{35,37}Cl$ ($I = \frac{3}{2}$) HFS observed: $A^{Cl} = 18.5 \pm 1.0$, $B^{Cl} = 5.0 \pm 0.5$ in 10^{-4} cm^{-1}	203
$Cu(BrO_3)_2 \cdot 6H_2O$ d. in $Zn(BrO_3)_2 \cdot 6H_2O$	90	2.217 ± 0.01	28 ± 5		Octahedral, $Z = 4$; dynamic Jahn–Teller effect replaced by static distortion below the transition temperature (7 to 35°K)	7
$K_2Cu(SO_4)_2 \cdot 6H_2O$ $d = 2000$ to 100 in $K_2Zn(SO_4)_2 \cdot 6H_2O$	290	$g_z = 2.05 \pm 0.03$ $g_1 = 2.26 \pm 0.03$ $g_{max} = 2.25 \pm 0.03$	$A = 85$ $B = 78$		Monoclinic, $Z = 2$; g_{max} in K_2K_3 plane; $\psi = 15°$, $\alpha = 32°$; $g_{\parallel} = 2.05$, $g_{\perp} = 2.25$	255, C.S. 201, 202
	20	$g_{\parallel} = 2.44 \pm 0.02$ $g_{\perp} = 2.13 \pm 0.02$	$A = 103 \pm 5$ $B = 34 \pm 5$	11 ± 1	$g_{\parallel} = g_z$, $g_{\perp} = g_{min}$ in K_1K_3 plane; $\psi = +105°$, $\alpha = 42°$	199

(continued)

Table 18 (*continued*)

Crystal	Temp. (°K)	g	HFS: $(A, B)^{63,65}$ (10^{-4} cm^{-1})	P (10^{-4} cm^{-1})	Remarks	Ref.
$K_2Cu(SO_4)_2 \cdot 6D_2O$ $d = 1000$ to 50 in $K_2Zn(SO_4)_2 \cdot 6D_2O$	20	$g_x = 2.16 \pm 0.02$ $g_y = 2.04 \pm 0.02$ $g_z = 2.42 \pm 0.02$	$A_x < 17$ $A_y = +61 \pm 3$ $A_z = -99 \pm 1$	$P = +11.0 \pm 0.5$ $P' = +1.3 \pm 0.6$	$A^{65}/A^{63} = 1.069 \pm 0.003$, $P^{63}/P^{65} = 1.08 \pm 0.02$; $\psi = +105°$, $\alpha = 43°$	199
$Rb_2Cu(SO_4)_2 \cdot 6H_2O$ $d = 1000$ to 50 in $Rb_2Zn(SO_4)_2 \cdot 6H_2O$	290	$g_z = 2.08 \pm 0.03$ $g_1 = 2.25 \pm 0.03$ $g_{max} = 2.27 \pm 0.03$	$A = 85$ $B = 78$		Monoclinic, $Z = 2$; g_{max} in K_2K_3 plane; $\psi = 15°$, $\alpha = 33°$; $g_{\parallel} = 2.08$, $g_{\perp} = 2.27$	255, C.S. 201, 202
	20	$g_{\parallel} = 2.44 \pm 0.02$ $g_{\perp} = 2.12 \pm 0.02$	$A = 116 \pm 5$ $B = 30 \pm 5$	11 ± 1	$g_{\parallel} = g_z$, $g_{\perp} = g_{min}$ in K_1K_3 plane; $\psi = +105°$, $\alpha = 42 \pm 2°$; anomalous second spectrum at temp. above 90°K	199
$Rb_2Cu(SO_4)_2 \cdot 6D_2O$ $d = 1000$ to 50 in $Rb_2Zn(SO_4)_2 \cdot 6D_2O$	20	$g_x = 2.15 \pm 0.02$ $g_y = 2.04 \pm 0.02$ $g_z = 2.43 \pm 0.02$	$A_x < 20$ $A_y = +59 \pm 4$ $A_z = -110 \pm 2$	$+12 \pm 1$	$\psi = +105°$, $\alpha = 42 \pm 2°$; anomalous second spectrum at temp. above 90°K	199

$(NH_4)_2Cu(SO_4)_2 \cdot 6H_2O$ d = 1000 to 50 in $(NH_4)_2Zn(SO_4)_2 \cdot 6H_2O$	20	$g_x = 2.12 \pm 0.02$ $g_y = 2.05 \pm 0.02$ $g_z = 2.46 \pm 0.02$	$A_x = 25 \pm 5$ $A_y = 35 \pm 5$ $A_z = 130 \pm 5$	11 ± 1	Monoclinic, $Z = 2$; $\psi = +65°$, $\alpha = 38°$	199, C.S. 201, 202
$Cu_3Bi_2(NO_3)_{12} \cdot 24H_2O$ d = 100 in $Mg_3Bi_2(NO_3)_{12} \cdot 24H_2O$	90	$g_\parallel = 2.219 \pm 0.003$ $g_\perp = 2.217 \pm 0.003$	$A = 27 \pm 1$ $B = 26 \pm 1$	10	Dynamic Jahn–Teller effect above 90°K (spectrum isotropic) replaced by static distortion at 20°K ($M_m = 3$)	186, 7
	20	$g_\parallel = 2.454 \pm 0.003$ $g_\perp = 2.096 \pm 0.003$	$A = 110 \pm 1$ $B = 17 \pm 2$	10		
$Cu_3La_2(NO_3)_{12} \cdot 24D_2O$ d = 500 in $Mg_3La_2(NO_3)_{12} \cdot 24D_2O$	90	$g_\parallel = 2.219 \pm 0.003$ $g_\perp = 2.218 \pm 0.003$	$A = 29.0 \pm 0.5$ $B = 27.5 \pm 0.5$	$P = +11.1 \pm 0.5$ $P' = -0.4 \pm 0.1$	Dynamic Jahn–Teller effect above 90°K (spectrum almost isotropic) replaced by static distortion at 20°K ($M_m = 3$)	7, 200
	20	$g_x = 2.097 \pm 0.002$ $g_y = 2.097 \pm 0.002$ $g_z = 2.470 \pm 0.002$	$A_x = +19.0 \pm 0.5$ $A_y = +12.3 \pm 0.5$ $A_z = -113 \pm 0.5$			

(continued)

Table 18 (*continued*)

Crystal	Temp. (°K)	g	HFS: $(A, B)^{63,65}$ (10^{-4} cm^{-1})	P (10^{-4} cm^{-1})	Remarks	Ref.
$CuSiF_6 \cdot 6H_2O$ d. in $ZnSiF_6 \cdot 6H_2O$	90	$g_{\parallel} = 2.221 \pm 0.005$ $g_{\perp} = 2.230 \pm 0.005$	$A = 21 \pm 5$ $B = 28 \pm 5$		Trigonal, $Z = 1$; dynamic Jahn–Teller effect replaced by static distortion below ~ 50°K ($M_m = 3$)	7, 6
	20	$g_{\parallel} = 2.46 \pm 0.01$ $g_{\perp} = 2.10 \pm 0.01$	$A = 110 \pm 3$ $B < 30$			
$Cu(CH_3COO)_2 \cdot H_2O$ und.	300	$g_x = 2.053 \pm 0.005$ $g_y = 2.093 \pm 0.005$ $g_z = 2.344 \pm 0.01$			Monoclinic, $Z = 4$; strong exchange interaction between pairs of Cu^{2+} ions: $S = 1$, $D = 3450 \pm 50$, $E = 70 \pm 50$ in 10^{-4} cm^{-1}; exchange energy $J/hc = 116$ cm^{-1}	205, 206, C.S. 204
[Cu aca_2] $d = 200$ in [Pd aca_2]	300, 77	$g_{\parallel} = 2.2661$ $g_{\perp} = 2.0535$	$A = -160$ $B = -19.5$	7	Monoclinic, $Z = 2$; additional lines due to $\Delta M_I = \pm 1$ transitions	208, C.S. 256

$[Cu\{(C_2H_5)_2NC(S)S\}_2]$ $d = 500$ in $[Zn\{(C_2H_5)_2NC(S)S\}_2]$	298, 90	$g_{\parallel} = 2.1085 \pm 0.0005$; $g_{\perp} = 2.023 \pm 0.002$	$A^{63} = 142.4 \pm 1$; $A^{65} = 152.0 \pm 1$; $B^{63,65} = 22.4 \pm 1$	3	Monoclinic, $Z = 4$	210, C.S. 257
$[Cu\{NH_2CH_2COO\}_2]$ d. in $[Cd\{NH_2CH_2COO\}_2]$	300	$g_{\parallel} = 2.2674 \pm 0.002$; $g_{\perp} = 2.055 \pm 0.003$	$A = 141 \pm 1.5$		Monoclinic, $Z = 2$; ligand ^{14}N HFS observed: $A^N = 10.0 \pm 0.5$, $B^N = 7.2$ in 10^{-4} cm^{-1}	258, C.S. 259
$[Cu\{2\text{-}OC_6H_4CH{=}NH\}_2]$ $d = 200$ in $[Ni\{2\text{-}OC_6H_4CH{=}NH\}_2]$ containing isotopically pure ^{63}Cu	300	$g_x = 2.0402$; $g_y = 2.0500$; $g_z = 2.2004$	$A^{63} = -185$; $B^{63} = -21$		Monoclinic, $Z = 2$; ligand ^{14}N HFS observed: $A^N = 10.4$; ^1H HFS from CH group protons: $A^H = 5.15$ in 10^{-4} cm^{-1}	211, C.S. 260, 261
[Cu pc] $d = 1000$ in [Zn pc]	77	$g_{\parallel} = 2.162 \pm 0.001$; $g_{\perp} = 2.047 \pm 0.001$	$A^{63} = 215 \pm 1$; $A^{65} = 231 \pm 1$; $B^{63} = 28 \pm 1$; $B^{65} = 30 \pm 1$	5	Monoclinic, $Z = 2$; ligand ^{14}N HFS observed: $A^N = 19.4 \pm 0.03$, $B^N = 15.9 \pm 0.03$ in 10^{-4} cm^{-1}	212
[Cu pc] $d = 1000$ in H_2pc	300	$g_{\parallel} = 2.179$; $g_{\perp} = 2.050$	$A = 202$; $B = 19$	6	Monoclinic, $Z = 2$; ligand ^{14}N HFS observed: $A^N = 14.5$, $B^N = 17.8$ in 10^{-4} cm^{-1}	213, 214, C.S. 215, 216

has predominantly metal-ion character. The resulting MO scheme for covalent tetragonal copper(II) complexes is illustrated in Figure 16.

Since the parameters of the spin-hamiltonian (149) may be obtained from a study of frozen solutions, such results have also been used to calculate bonding cofficients for a number of copper(II) complexes. In the same way, measurements on diluted polycrystalline samples have been utilized to this end. Bonding parameters were determined, e.g. for copper(II)

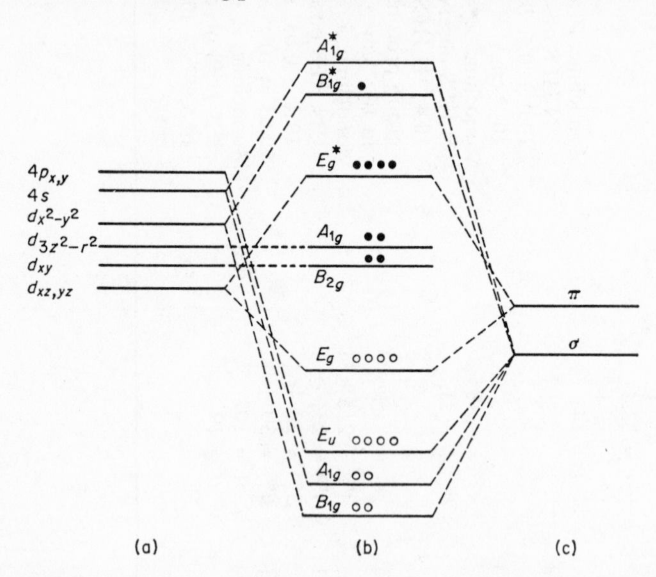

Figure 16. Molecular-orbital scheme for copper diethyldithiocarbamate. (a) Cu^{2+} levels; (b) molecular levels; (c) ligand levels. ○ denotes electrons with predominantly ligand character, ● denotes electrons with predominantly *d*-electron character [according to Reddy and Srinivasan [210]].

bis(salicylamide) [217], copper(II) bis(salicylaldoxime) [84], copper(II) bis-(8-hydroxyquinolate) [218], copper(II) etioporphyrin II [219] and others. For complete listing of data and references, the reader should consult ref. 86.

VI. REFERENCES

1. E. J. Zavoisky, *J. Phys. USSR*, **9**, 211 (1945).
2. R. L. Cummerow and D. Halliday, *Phys. Rev.*, **70**, 433 (1946).
3. D. M. S. Bagguley and J. H. E. Griffiths, *Nature*, **160**, 532 (1947).
4. B. Bleaney and R. P. Penrose, *Proc. Phys. Soc.* (*London*), **60**, 395 (1948).
5. D. M. S. Bagguley, B. Bleaney, J. H. E. Griffiths, R. P. Penrose, and B. I. Plumpton, *Proc. Phys. Soc.* (*London*), **61**, 542, 551 (1948).
6. B. Bleaney and K. D. Bowers, *Proc. Phys. Soc.* (*London*), *A*, **65**, 667 (1952).

7. B. Bleaney, K. D. Bowers, and R. S. Trenam, *Proc. Roy. Soc. (London), Ser. A*, **228**, 157 (1955).
8. J. H. E. Griffiths and J. Owen, *Proc. Roy. Soc. (London), Ser. A*, **226**, 96 (1954).
9. D. J. E. Ingram, *Free Radicals as Studied by Electron Spin Resonance*, Butterworths, London, 1958.
9a. P. W. Atkins and M. C. R. Symons, *The Structure of Inorganic Radicals*, Elsevier, Amsterdam, 1967.
10. H. Fischer, *Magnetic Properties of Free Radicals*, Vol. II/1 of *Landolt–Börnstein, New Series* (Ed. K. H. Hellwege and A. M. Hellwege), Springer, Berlin, 1965.
11. J. D. Brinen, J. G. Koren, and W. G. Hodgson, *J. Chem. Phys.*, **44**, 3095 (1966).
12. G. W. Ludwig and H. H. Woodbury, *Solid State Physics*, Vol. 13, Academic Press, New York, 1962, p. 223.
13. E. U. Condon and G. H. Shortley, *Theory of Atomic Spectra*, Cambridge University Press, Cambridge, 1959.
14. J. S. Griffith, *The Theory of Transition-Metal Ions*, Cambridge University Press, Cambridge, 1961.
15. C. J. Ballhausen, *Introduction to Ligand Field Theory*, McGraw-Hill, New York, 1962.
16. W. Low, *Paramagnetic Resonance in Solids*, Suppl. 2 to *Solid State Physics*, Academic Press, New York, 1960.
17. C. K. Jørgensen, *Absorption Spectra and Chemical Bonding in Complexes*, Pergamon Press, Oxford, 1962.
18. B. N. Figgis, *Introduction to Ligand Fields*, Interscience, New York, 1966.
18a. H. L. Schläfer and G. Gliemann, *Einführung in die Ligandenfeldtheorie*, Akademische Verlagsgesellschaft, Frankfurt, 1967.
19. H. A. Bethe, *Ann. Phys.*, [5] **3**, 133 (1929).
20. H. Watanabe, *Operator Methods in Ligand Field Theory*, Prentice-Hall, New Jersey, 1966.
21. R. S. Polo, *Studies on Crystal Field Theory*, RCA Preprint, Princeton, 1959.
22. K. W. H. Stevens, *Proc. Phys. Soc. (London), A*, **65**, 209 (1952).
23. R. J. Elliott and K. W. H. Stevens, *Proc. Roy. Soc. (London), Ser. A*, **218**, 553 (1953).
24. R. J. Elliott and K. H. W. Stevens, *Proc. Roy. Soc. (London), Ser. A*, **219**, 387 (1953).
25. B. R. Judd, *Proc. Roy. Soc. (London), Ser. A*, **227**, 552 (1955).
26. T. N. Dunn, *Trans. Faraday Soc.*, **57**, 1441 (1961).
27. W. Opechowski, *Physica*, **7**, 552 (1940).
28. R. J. Elliott, *Phys. Rev.*, **96**, 280 (1954).
29. H. A. Kramers, *Proc. Acad. Sci. Amsterdam*, **33**, 959 (1930).
30. M. H. L. Pryce, *Phys. Rev.*, **80**, 1107 (1950).
31. E. Fermi, *Z. Phys.*, **60**, 320 (1930).
32. A. Abragam, J. Horowitz, and M. H. L. Pryce, *Proc. Roy. Soc. (London), Ser. A*, **230**, 169 (1955).
33. R. M. Sternheimer, *Phys. Rev.*, **86**, 316 (1952).
34. J. H. Wood and G. W. Pratt, *Phys. Rev.*, **107**, 995 (1957).
35. V. Heine, *Phys. Rev.*, **107**, 1002 (1957).

36. R. E. Watson and A. J. Freeman, *Phys. Rev.*, **123**, 2027 (1961).
37. M. H. L. Pryce, *Proc. Phys. Soc.* (*London*), *A*, **63**, 25 (1950).
38. A. Abragam and M. H. L. Pryce, *Proc. Roy. Soc.* (*London*), *Ser. A*, **205**, 135 (1951).
39. J. H. Van Vleck and W. G. Penney, *Phil. Mag.*, [7] **17**, 961 (1934).
40. H. Watanabe, *Progr. Theoret. Phys.* (*Kyoto*), **18**, 405 (1957).
41. C. J. Ballhausen and H. B. Gray, *Molecular Orbital Theory*, Benjamin, New York, 1964.
42. M. Tinkham, *Proc. Roy. Soc.* (*London*), *Ser. A*, **236**, 549 (1956).
43. Ye. K. Zavoisky, *Zh. Eksperim. i Teor. Fiz.*, **16**, 603 (1946).
44. Ye. K. Zavoisky, *J. Phys. USSR*, **9**, 245 (1945).
45. G. Feher and A. F. Kip, *Phys. Rev.*, **98**, 337 (1955).
46. D. J. E. Ingram, *Spectroscopy at Radio and Microwave Frequencies*, Butterworths, London, 1955.
47. S. A. Altshuler and B. M. Kozyrev, *Electron Paramagnetic Resonance*, Academic Press, New York, 1964.
48. G. Feher, *Bell System Tech. J.*, **36**, 449 (1957).
49. W. Gordy, W. Smith, and P. Trambarulo, *Microwave Spectroscopy*, Wiley, New York, 1953.
50. V. Strandberg, *Microwave Spectroscopy*, 1956.
51. C. Montgomery, *Technique of Microwave Measurements*, McGraw-Hill, New York, 1948.
52. D. D. King, *Measurements at Centimeter Wavelength*, Van Nostrand, New York, 1952.
53. G. Schoffa, *Elektronenspinresonanz in der Biologie*, G. Braun, Karlsruhe, 1964.
54. B. Bleaney and K. W. H. Stevens, *Rept. Progr. Phys.*, **16**, 108 (1953).
55. G. S. Bogle, A. H. Cooke, and S. Whitley, *Proc. Phys. Soc.* (*London*), *A*, **64**, 931 (1951).
56. R. Beringer and J. G. Castle, *Phys. Rev.*, **78**, 581 (1950).
57. R. L. Cummerow, D. Holliday, and G. E. Moore, *Phys. Rev.*, **72**, 1233 (1947).
58. C. A. Whitmer, R. T. Weidner, J. S. Hsiang, and P. R. Weiss, *Phys. Rev.*, **74**, 1478 (1948).
59. B. Bleaney and D. J. E. Ingram, *Nature*, **164**, 116 (1949).
60. B. Smaller and E. L. Yasaitis, *Rev. Sci. Instr.*, **24**, 337 (1953).
61. H. A. Buckmaster and H. E. D. Scovil, *Can. J. Phys.*, **34**, 711 (1956).
62. T. S. England and E. E. Schneider, *Nature*, **166**, 437 (1950).
63. E. E. Schneider and T. S. England, *Physica*, **17**, 221 (1951).
64. J. M. Hirshon and G. E. Fraenkel, *Rev. Sci. Instr.*, **26**, 34 (1955).
65. M. Misra, *Rev. Sci. Instr.*, **29**, 590 (1958).
66. R. V. Pound and V. D. Knight, *Rev. Sci. Instr.*, **21**, 219 (1950).
67. H. A. Jahn and E. Teller, *Proc. Roy. Soc.* (*London*), *Ser. A*, **161**, 220 (1937).
68. H. A. Jahn, *Proc. Roy. Soc.* (*London*), *Ser. A*, **164**, 117 (1938).
69. J. D. Swalen and H. M. Gladney, *IBM J. Res. Develop.*, **8**, 515 (1964).
70. J. A. Weil and J. H. Anderson, *J. Chem. Phys.*, **28**, 864 (1958).
71. J. G. Park, *Proc. Phys. Soc.* (*London*), **74**, 513 (1959).
72. J. E. Geusic and L. C. Brown, *Phys. Rev.*, **112**, 64 (1958).
73. D. S. Schonland, *Proc. Phys. Soc.* (*London*), **73**, 788 (1959).

74. B. Bleaney, *Phil. Mag.*, [7] **42**, 441 (1951).
75. H. A. Buckmaster, *Can. J. Phys.*, **39**, 1073 (1961).
76. J. W. Searl, R. C. Smith, and S. J. Wyard, *Proc. Phys. Soc.* (*London*), *A*, **74**, 491 (1959); *Arch. Sci.* (*Geneva*), **13**, Fasc. Spéc., 236 (1960).
77. B. Bleaney, *Proc. Phys. Soc.* (*London*), **75**, 621 (1960).
78. J. W. Searl, R. C. Smith, and S. J. Wyard, *Proc. Phys. Soc.* (*London*), *A*, **78**, 1174 (1961).
79. B. Bleaney, *Proc. Phys. Soc.* (*London*), *A*, **63**, 407 (1950).
80. R. H. Sands, *Phys. Rev.*, **99**, 1222 (1955).
81. J. A. Ibers and J. B. Swalen, *Phys. Rev.*, **127**, 1914 (1962).
82. F. K. Kneubühl, *J. Chem. Phys.*, **33**, 1074 (1960).
83. R. Neiman and D. Kivelson, *J. Chem. Phys.*, **35**, 156 (1961).
84. H. R. Gersmann and J. D. Swalen, *J. Chem. Phys.*, **36**, 3221 (1962).
85. T. Vänngard and R. Aasa in *Proc. Intern. Conf. Paramagnetic Resonance, 1st, Jerusalem, 1962*, Vol. 2 (Ed. W. Low), Academic Press, New York, 1963, p. 509.
86. E. König, *Magnetic Properties of Coordination and Organo-Metallic Transition Metal Compounds*, Vol. II/2 of *Landolt–Börnstein, New Series* (Ed. K. H. Hellwege and A. M. Hellwege), Springer, Berlin, 1966.
87. M. Kotani, *J. Phys. Soc. Japan*, **4**, 293 (1949).
88. H. M. Gladney and J. D. Swalen, *J. Chem. Phys.*, **42**, 1999 (1965).
89. B. Bleaney, G. S. Bogle, A. H. Cooke, R. J. Duffus, M. C. M. O'Brien, and K. W. H. Stevens, *Proc. Phys. Soc.* (*London*), *A*, **68**, 57 (1955).
90. D. K. Réi, *Fiz. Tverd. Tela*, **3**, 2525 (1962).
91. G. F. Dionne, *Can. J. Phys.*, **42**, 2419 (1964).
92. W. Lipson, *Proc. Roy. Soc.* (*London*), *Ser. A*, **151**, 347 (1935).
93. E. Y. Wong, *J. Chem. Phys.*, **32**, 598 (1960).
94. B. R. McGarvey, *J. Chem. Phys.*, **38**, 388 (1963).
95. E. L. Waters and A. H. Maki, *Phys. Rev.*, **125**, 233 (1962).
96. S. Fujiwara and M. Codell, *Bull. Chem. Soc. Japan*, **37**, 49 (1964).
97. S. Fujiwara, K. Nagashima, and M. Codell, *Bull. Chem. Soc. Japan*, **37**, 773 (1964).
98. R. H. Borcherts and C. Kikuchi, *J. Chem. Phys.*, **40**, 2270 (1964); R. H. Borcherts, *Dissertation Abstr.*, **26**, 2287 (1965).
99. C. J. Ballhausen and H. B. Gray, *Inorg. Chem.*, **1**, 111 (1962).
100. M. Wolfsberg and L. Helmholz, *J. Chem. Phys.*, **20**, 837 (1952).
101. K. W. H. Stevens, *Proc. Roy. Soc.* (*London*), *Ser. A*, **219**, 542 (1953).
102. J. Owen, *Proc. Roy. Soc.* (*London*), *Ser. A*, **227**, 183 (1955).
103. W. Marshall and R. Stuart, *Phys. Rev.*, **123**, 2048 (1961).
104. R. Lacroix and G. Emch, *Helv. Phys. Acta*, **35**, 592 (1962).
105. K. De Armond, B. B. Garrett, and H. S. Gutowsky, *J. Chem. Phys.*, **42**, 1019 (1965).
106. D. S. McClure, *J. Chem. Phys.*, **17**, 905 (1949).
107. R. M. Golding, *Mol. Phys.*, **5**, 369 (1962).
108. D. Kivelson and S. K. Lee, *J. Chem. Phys.*, **41**, 1896 (1964).
109. J. M. Assour, *J. Chem. Phys.*, **43**, 2477 (1965).
110. J. M. Assour, J. Goldmacher, and S. E. Harrison, *J. Chem. Phys.*, **43**, 159 (1965).
111. H. Kon and N. E. Sharpless, *J. Chem. Phys.*, **42**, 906 (1965).
112. C. R. Hare, I. Bernal, and H. B. Gray, *Inorg. Chem.*, **1**, 831 (1962).

113. H. Kon and N. E. Sharpless, *J. Chem. Phys.*, **43**, 1081 (1965).
114. G. F. Kokoszka, H. C. Allen, and G. Gordon, *Inorg. Chem.*, **5**, 91 (1966).
115. A. Carrington, D. J. E. Ingram, K. A. K. Lott, D. S. Schonland, and M. C. R. Symons, *Proc. Roy. Soc. (London), Ser. A*, **254**, 101 (1960).
116. W. H. Zachariesen and G. E. Ziegler, *Z. Krist.*, **80**, 164 (1931).
117. D. Schonland, *Proc. Roy. Soc. (London), Ser. A*, **254**, 111 (1960).
118. G. M. Zverev and A. M. Prokhorov, *Zh. Eksperim. i. Teor. Fiz.*, **34**, 1023 (1958).
119. B. Bleaney, D. J. E. Ingram, and H. E. D. Scovil, *Proc. Phys. Soc. (London), A*, **64**, 601 (1951).
120. J. M. Baker, B. Bleaney, and K. D. Bowers, *Proc. Phys. Soc. (London), B*, **69**, 1205 (1956).
121. A. Abragam, *Phys. Rev.*, **79**, 534 (1950).
122. O. Kikuchi, M. H. Sirvetz, and V. W. Cohen, *Phys. Rev.*, **92**, 109 (1953).
123. J. M. Baker and B. Bleaney, *Proc. Phys. Soc. (London), A*, **65**, 952 (1952).
124. H. Lipson and C. A. Beevers, *Proc. Roy. Soc. (London), Ser. A*, **148**, 664 (1935).
125. D. M. S. Bagguley and J. H. E. Griffiths, *Proc. Roy. Soc. (London), Ser. A*, **204**, 188 (1950).
126. B. Bleaney, *Proc. Roy. Soc. (London), Ser. A*, **204**, 203 (1950).
127. J. M. Baker, *Proc. Phys. Soc. (London), B*, **69**, 633 (1956).
128. W. M. Walsh, *Phys. Rev.*, **114**, 1485 (1959).
129. H. A. Kuska and M. T. Rogers, *J. Chem. Phys.*, **41**, 3802 (1964).
130. P. Swarup, *Can. J. Phys.*, **37**, 848 (1959).
131. T. Mitsuma, *J. Phys. Soc. Japan*, **16**, 1796 (1961).
132. L. S. Singer, *J. Chem. Phys.*, **23**, 379 (1955).
133. B. R. McGarvey, *J. Chem. Phys.*, **40**, 809 (1964).
134. D. S. McClure, *J. Chem. Phys.*, **20**, 682 (1952).
135. B. R. McGarvey, *J. Chem. Phys.*, **41**, 3743 (1964).
136. A. Chakravorty and S. Basu, *J. Chem. Phys.*, **33**, 1266 (1961).
137. T. S. Piper and R. L. Carlin, *J. Chem. Phys.*, **36**, 3330 (1962).
138. H. S. Jarrett, *J. Chem. Phys.*, **27**, 1298 (1957).
139. K. Okumura, *J. Phys. Soc. Japan*, **17**, 1341 (1962).
140. B. R. McGarvey, *J. Chem. Phys.*, **37**, 3020 (1962).
141. B. Bleaney, *Proc. Phys. Soc. (London), A*, **73**, 939 (1959).
142. G. F. Koster and H. Statz, *Phys. Rev.*, **113**, 445 (1959).
143. R. O. Rahn and P. B. Dorain, *J. Chem. Phys.*, **41**, 3249 (1964).
144. H. L. Schäffer, *Z. Physik. Chem. (Frankfurt)*, **11**, 65 (1957).
145. F. Basolo, C. J. Ballhausen, and J. Bjerrum, *Acta Chem. Scand.*, **9**, 810 (1955).
146. M. Linhard and M. Weigel, *Z. Physik. Chem. (Frankfurt)*, **5**, 20 (1955).
147. L. E. Slaten and C. S. Garner, *J. Phys. Chem.*, **63**, 1214 (1959).
148. S. Yamada and R. Tsuchida, *Bull. Chem. Soc. Japan*, **25**, 127 (1952).
149. S. Yamada, A. Nakahara, Y. Shimura, and R. Tsuchida, *Bull. Chem. Soc. Japan*, **28**, 222 (1955).
150. C. J. Ballhausen and W. Moffitt, *J. Inorg. Nucl. Chem.*, **3**, 178 (1956).
151. L. Helmholz, A. V. Guzzo, and R. N. Sanders, *J. Chem. Phys.*, **35**, 1349 (1961).
152. K. Ono, S. Koide, H. Sekiyama, and H. Abe, *Phys. Rev.*, **96**, 38 (1954).

153. K. Ono, *J. Phys. Soc. Japan*, **12**, 1231 (1957).
154. B. Bleaney and R. S. Trenam, *Proc. Roy. Soc.* (*London*), *Ser. A*, **223**, 1 (1954).
155. B. Bleaney and D. J. E. Ingram, *Proc. Roy. Soc.* (*London*), *Ser. A*, **205**, 336 (1951).
156. M. Tinkham, *Proc. Roy. Soc.* (*London*), *Ser. A*, **236**, 535 (1956).
157. S. Ogawa, *J. Phys. Soc. Japan*, **15**, 1475 (1960).
158. C. G. Windsor, J. H. E. Griffiths, and J. Owen, *Proc. Phys. Soc.* (*London*), **81**, 373 (1963).
159. L. Helmholz, *J. Chem. Phys.*, **31**, 172 (1959).
160. L. Helmholz and A. V. Guzzo, *J. Chem. Phys.*, **32**, 302 (1960).
161. W. M. Walsh and L. W. Rupp, *Phys. Rev.*, **126**, 952 (1962).
162. G. W. Ludwig, H. H. Woodbury, and R. O. Carlson, *Phys. Rev. Letters*, **1**, 295 (1958).
163. G. W. Ludwig and H. H. Woodbury, *Phys. Rev.*, **117**, 1286 (1960).
164. J. W. Culvahouse and L. C. Olsen, *J. Chem. Phys.*, **43**, 1145 (1965).
165. G. S. Bogle and H. F. Symmons, *Proc. Phys. Soc.* (*London*), **78**, 812 (1961).
166. H. F. Symmons and G. S. Bogle, *Proc. Phys. Soc.* (*London*), **82**, 412 (1963).
167. R. Aasa and T. Vänngard, *Arkiv Kemi*, **24**, 331 (1965).
168. T. Castner, G. S. Newell, W. C. Holton, and C. P. Slichter, *J. Chem. Phys.*, **32**, 668 (1960).
169. J. S. Griffith, *Mol. Phys.*, **8**, 217 (1964).
170. J. S. Griffith, *Mol. Phys.*, **8**, 213 (1964).
171. M. D. Lind, J. L. Hoard, M. J. Hamor, and T. A. Hamor, *Inorg. Chem.*, **3**, 34 (1964).
172. D. J. E. Ingram and J. E. Bennett, *Discussions Faraday Soc.*, **19**, 140 (1955).
173. J. E. Bennett and D. J. E. Ingram, *Nature*, **177**, 275 (1956).
174. J. S. Griffith, *Proc. Roy. Soc.* (*London*), *Ser. A*, **235**, 23 (1956).
175. J. E. Bennett, J. F. Gibson, and D. J. E. Ingram, *Proc. Roy. Soc.* (*London*) *Ser. A*, **240**, 67 (1957).
176. J. E. Bennett, J. F. Gibson, D. J. E. Ingram, T. M. Haughton, G. A. Kerkut, and K. A. Munday, *Proc. Roy. Soc.* (*London*), *Ser. A*, **262**, 395 (1961).
177. B. Bleaney and D. J. E. Ingram, *Proc. Phys. Soc.* (*London*), *A*, **65**, 953 (1952).
178. E. König, *Z. Naturforsch.*, **19a**, 1139 (1964).
179. B. Elschner and S. Herzog, *Arch. Sci.* (*Geneva*), **11**, Fasc. Spéc., 160 (1958).
180. K. H. Hausser, *Z. Naturforsch.*, **16a**, 1190 (1961).
181. K. H. Hausser, *Naturwissenschaften*, **48**, 666 (1961).
182. K. H. Hausser, *Naturwissenschaften*, **48**, 426 (1961).
183. J. W. Stout and S. A. Reed, *J. Am. Chem. Soc.*, **76**, 5279 (1954).
184. B. Bleaney and D. J. E. Ingram, *Proc. Roy. Soc.* (*London*), *Ser. A*, **208**, 143 (1951).
185. W. Dobrowolski, R. V. Jones, and C. D. Jeffries, *Phys. Rev.*, **101**, 1001 (1956).
186. R. S. Trenam, *Proc. Phys. Soc.* (*London*), *A*, **66**, 118 (1953).
187. W. B. Gager, P. S. Jastram, and J. G. Daunt, *Phys. Rev.*, **111**, 803 (1958).
188. H. G. Beljers, P. F. Bongers, R. P. Van Stapele, and H. Zijlstra, *Phys. Letters*, **12**, 81 (1964).

189. J. M. Assour and K. Kahn, *J. Am. Chem. Soc.*, **87**, 207 (1965).
190. J. M. Assour, *J. Am. Chem. Soc.*, **87**, 4701 (1965).
191. H. B. Gray in *Transition Metal Chemistry*, Vol. 1 (Ed. R. L. Carlin), Dekker, New York, 1965, p. 240.
192. K. Ono, *J. Phys. Soc. Japan*, **8**, 802 (1953).
193. J. H. E. Griffiths and J. Owen, *Proc. Roy. Soc.* (*London*), *Ser. A*, **213**, 459 (1952).
194. R. P. Penrose and K. W. H. Stevens, *Proc. Phys. Soc.* (*London*), *A*, **63**, 29 (1950).
195. W. M. Walsh, *Phys. Rev.*, **114**, 1473 (1959).
196. W. M. Walsh and N. Bloembergen, *Phys. Rev.*, **107**, 904 (1957).
197. I. Svare and G. Seidel, *Phys. Rev.*, **134**, A172 (1964).
198. M. Peter, *Phys. Rev.*, **116**, 1432 (1959).
199. B. Bleaney, K. D. Bowers, and D. J. E. Ingram, *Proc. Roy. Soc.* (*London*), *Ser. A*, **228**, 147 (1955).
200. B. Bleaney, K. D. Bowers, and M. H. L. Pryce, *Proc. Roy. Soc.* (*London*), *Ser. A*, **228**, 166 (1955).
201. W. Hofmann, *Z. Krist.*, **78**, 279 (1931).
202. R. W. Wyckoff, *Crystal Structures*, Vol. 2, Interscience, New York, 1957, Chap. 10.
203. J. H. M. Thornley, B. N. Mangum, J. H. E. Griffiths, and J. Owen, *Proc. Phys. Soc.* (*London*), **78**, 1263 (1961).
204. J. N. Van Niekerk and F. R. L. Schoening, *Acta Cryst.*, **6**, 227 (1953).
205. H. Abe and J. Shimada, *Phys. Rev.*, **90**, 316 (1953).
206. B. Bleaney and K. D. Bowers, *Proc. Roy. Soc.* (*London*), *Ser. A*, **214**, 451 (1952).
207. A. Abragam and M. H. L. Pryce, *Proc. Phys. Soc.* (*London*), *A*, **63**, 409 (1950).
208. A. H. Maki and B. R. McGarvey, *J. Chem. Phys.*, **29**, 31 (1958).
209. A. Abragam and M. H. L. Pryce, *Proc. Roy. Soc.* (*London*), *Ser. A*, **206**, 164 (1951).
210. T. R. Reddy and R. Srinivasan, *J. Chem. Phys.*, **43**, 1404 (1965).
211. A. H. Maki and B. R. McGarvey, *J. Chem. Phys.*, **29**, 35 (1958).
212. R. M. Deal, D. J. E. Ingram, and R. Srinivasan, *Electronic Magnetic Resonance and Solid Dielectrics, Proc. 12th Colloq. Ampère, Bordeaux, 1963*, 1964, p. 239.
213. S. E. Harrison and J. M. Assour in *Proc. Intern. Conf. Paramagnetic Resonance, 1st, Jerusalem, 1962*, Vol. 2 (Ed. W. Low), Academic Press, New York, 1963, p. 855.
214. S. E. Harrison and J. M. Assour, *J. Chem. Phys.*, **40**, 365 (1964).
215. J. M. Robertson, *J. Chem. Soc.*, **1935**, 615; **1935**, 1195.
216. J. M. Robertson and I. Woodward, *J. Chem. Soc.*, **1937**, 219.
217. A. K. Wiersema and J. J. Windle, *J. Phys. Chem.*, **68**, 2316 (1964).
218. G. F. Kokoszka, H. C. Allen, and G. Gordon, *J. Chem. Phys.*, **42**, 3730 (1965).
219. E. M. Roberts and W. S. Koski, *J. Am. Chem. Soc.*, **82**, 3006 (1960).
220. K. R. Andress and C. Carpenter, *Z. Krist.*, *Ser. A*, **87**, 446 (1934).
221. W. T. Astbury, *Proc. Roy. Soc.* (*London*), *Ser. A*, **112**, 448 (1926).
222. G. T. Morgan and H. D. K. Drew, *J. Chem. Soc.*, **1921**, 1059.
223. G. S. Bogle and J. Owen in K. D. Bowers and J. Owen, *Rept. Progr. Phys.*, **18**, 304 (1955).

224. H. Montgomery, *Thesis*, University of Washington, 1961.
225. H. P. Klug, E. Kummer, and E. L. Alexander, *J. Am. Chem. Soc.*, **70**, 3064 (1948).
226. G. L. Clark and S. T. Gross, *Z. Krist.*, **98**, 107 (1937).
227. B. Bleaney, *Phys. Rev.*, **75**, 1962 (1949).
228. C. F. Davis and M. W. P. Strandberg, *Phys. Rev.*, **105**, 447 (1957).
229. J. M. Baker and B. Bleaney in K. D. Bowers and J. Owen, *Rept. Progr. Phys.*, **18**, 304 (1955).
230. B. Bleaney and K. D. Bowers, *Proc. Phys. Soc.* (*London*), *A*, **64**, 1135 (1951).
231. K. D. Bowers, *Proc. Phys. Soc.* (*London*), *A*, **65**, 860 (1952).
232. V. Barkhatov, *Acta Phys. Chim. URSS*, **16**, 123 (1942).
233. V. Barkhatov and H. Zhdanov, *Acta Phys. Chim. URSS*, **16**, 43 (1942).
234. S. Gregorio, J. Weber, and R. Lacroix, *Helv. Phys. Acta*, **38**, 172 (1965).
235. G. Emch and R. Lacroix, *Helv. Phys. Acta*, **33**, 1021 (1960).
236. H. Hentschel and F. Rinne, *Ber. Verhandl. Saechs. Akad. Wiss. Leipzig, Math. Phys. Kl.*, **79**, 3, 5 (1927).
237. K. Meisel and W. Tiedje, *Z. Anorg. Allg. Chem.*, **164**, 223 (1927).
238. R. W. G. Wyckoff and R. P. McCutcheon, *Am. J. Sci.*, **13**, 223 (1927).
239. J. Hentschel, *Z. Krist.*, **66**, 466 (1928).
240. K. Nakatsu, Y. Saito, and H. Kuroya, *Bull. Chem. Soc. Japan*, **29**, 428 (1956).
241. S. Ooi, Y. Komiyama, and H. Kuroya, *Bull. Chem. Soc. Japan*, **33**, 354 (1960).
242. A. Nakahara, Y. Saito, and H. Kuroya, *Bull. Chem. Soc. Japan*, **25**, 331 (1952).
243. A. Ferrari, *Rend. Accad. Nazl. Lincei*, **3**, 224 (1926).
244. I. Hayashi and K. Ono, *J. Phys. Soc. Japan*, **8**, 270 (1953).
245. C. A. Beevers and C. M. Schwartz, *Z. Krist.*, **91**, 157 (1935).
246. F. K. Hurd, M. Sachs, and W. D. Herschberger, *Phys. Rev.*, **93**, 373 (1954).
247. V. A. Pospelov and G. S. Zhdanov, *Zh. Fiz. Khim.*, **21**, 405 (1947).
248. M. Date, *Sci. Rept. Res. Inst. Tohoku Univ., Ser. A*, **6**, 497 (1954).
249. B. Bleaney and R. S. Trenam, *Proc. Phys. Soc.* (*London*), *A*, **65**, 560 (1952).
250. S. Maekawa, *J. Phys. Soc. Japan*, **16**, 2337 (1941).
251. R. B. Roof, *Acta Cryst.*, **9**, 791 (1956).
252. K. D. Bowers, in K. D. Bowers and J. Owen, *Rept. Progr. Phys.*, **18**, 304 (1955).
253. L. Pauling, *Z. Krist.*, **72**, 482 (1930).
254. J. H. E. Griffiths and J. Owen, *Proc. Phys. Soc.* (*London*), *A*, **64**, 583 (1951).
255. D. M. S. Bagguley and J. H. E. Griffiths, *Proc. Phys. Soc.* (*London*), *A*, **65**, 594 (1952).
256. H. Koyama, Y. Saito, and H. Kuroya, *J. Inst. Polytech. Osaka City Univ.*, **4**, 43 (1953).
257. R. Bally, *Compt. Rend.*, **257**, 425 (1963).
258. H. A. Kuska and M. T. Rogers, *J. Chem. Phys.*, **43**, 1744 (1965).
259. K. Tomita and I. Nitta, *Bull. Chem. Soc. Japan*, **34**, 286 (1961).
260. M. A. Porai-Koshits, and P. M. Zorkij, *Zh. Strukt. Khim.*, **2**, 20 (1961).
261. J. D. Breazedale, *Thesis*, University of Washington, 1955.

8

Mössbauer Spectroscopy

J. DANON

I. INTRODUCTION

Mössbauer spectroscopy is based on the phenomenon of recoilless emission and resonance absorption of low-energy γ-radiation discovered in 1958 by Mössbauer[1,2]. Its application to chemistry is a consequence of the fact that this nuclear phenomenon is markedly influenced by interactions between the nucleus and the orbital electrons.

The resonance absorption of photons emitted in electronic transitions has been known since the beginning of the century when Wood made the first demonstration in 1904 using the yellow monochromatic line of sodium[3]. The possible observation of the phenomenon in nuclear transitions involving γ-radiation was suggested in 1929 by Kuhn[4] but for many years unsuccessful attempts were reported[5]. In order to understand the reason for these difficulties and the significance of the Mössbauer discovery, it is useful to compare the main features of the resonance absorption of photons emitted in electronic and nuclear transitions, as was recently done by Goldanskii[6].

II. RESONANCE SCATTERING OF PHOTONS AND RECOIL-FREE PHENOMENA

Consider a source S of sodium atoms, emitting the monochromatic yellow line of frequency ν, and an absorber A, containing sodium atoms in the vapour state. By electronic deexcitation, S emits light of energy $h\nu$. The photons are absorbed by A and reradiated in all directions with the same energy $h\nu$ as the incoming ones (Figure 1).

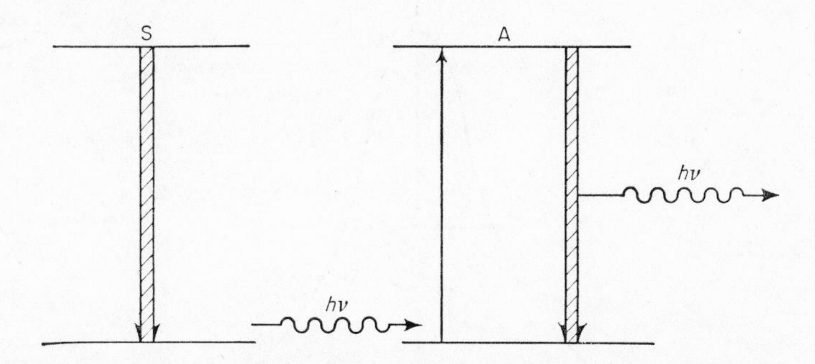

Figure 1. Schematic illustration of resonance scattering of photons: decay of the excited level of the source S with emission of a photon $h\nu$ which is absorbed and reradiated by the absorber A.

Let us examine more closely the dependence of the resonance scattering on photon energy. The fact that the lifetime τ of the excited state is finite means that the transition has a given linewidth Γ; moreover, the two quantities τ and Γ are connected by the Heisenberg uncertainty relation

$$\tau \times \Gamma = \hbar \qquad (1)$$

A plot of excitation probability $W(E)$ versus energy of resonance radiation (Figure 2) has the Lorentzian shape and is described by the Breit–Wigner relationship[7]

$$W(E) = \frac{\Gamma^2/4}{(E - E_0)^2 + \Gamma^2/4} \tag{2}$$

which has been normalized to give $W(E_0) = 1$. The condition that the probability for resonance be high is then

$$E_0 - \frac{\Gamma}{2} \leqslant h\nu \leqslant E_0 + \frac{\Gamma}{2} \tag{3}$$

The scattering cross-section σ is connected with (1) by the relation

$$\sigma = \sigma_0 W(E) \tag{4}$$

where the factor σ_0 is given in its simplest form by

$$\sigma_0 = \frac{2I_{ex} + 1}{2(2I_{gr} + 1)} 4\pi\lambda_0^2 \tag{5}$$

I_{ex} and I_{gr} being the spins of the excited and ground states of the transition and λ_0 the wavelength at the resonance. However, the photon energy

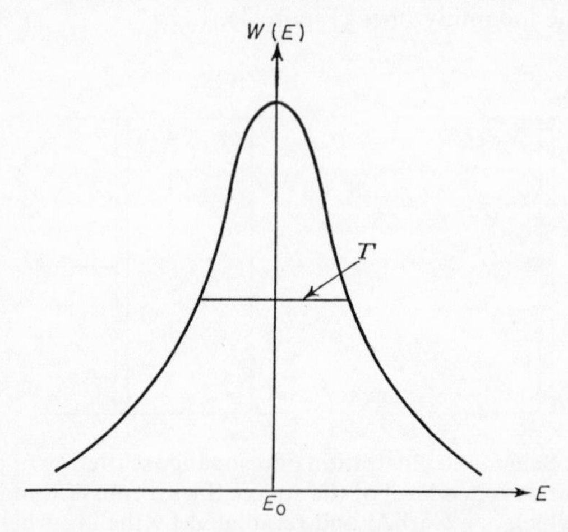

Figure 2. Excitation probability $W(E)$ versus energy of resonance radiation.

$h\nu$ is not exactly equal to the total energy E_0 of the transition, since momentum conservation requires a non-vanishing recoil energy of the emitting system. For a system at rest, the recoil energy at the instant of the transition E_R is given by

$$E_R = \frac{(h\nu)^2}{2mc^2} \simeq \frac{E_0^2}{2mc^2} \tag{6}$$

where m is the mass of the recoiling system and c is the velocity of light. Since the same recoil energy is involved in the absorption of the photon, the total energy loss in the emission–absorption is $2E_R$ and the energy available for the photon is $E_0 - 2E_R$. Let us now investigate how large $2E_R$ is compared with the linewidth Γ in typical electronic and nuclear transitions. Table 1 compares the recoil energies and linewidths of the

Table 1. Comparison between atomic and nuclear transitions.

Transition (ev)	Electronic	Nuclear
E_0	2.1	23,800
Γ	4.4×10^{-8}	2.2×10^{-8}
E_R	10^{-10}	2.5×10^{-3}

yellow spectral line of sodium and of the 23.8 kev γ-ray of ^{119}Sn. The difference in energy for the electronic transition $2E_R$ is so small compared with Γ that the photon energy certainly fulfills condition (3). On the other hand, in the nuclear transition, E_R is very large compared with Γ and as a consequence the γ-ray will not have the correct energy for resonance absorption.

The preceding conclusion is strictly valid only for a free nucleus. For nuclei bound in a solid, it is necessary to take into account the coupling between the atoms of the crystal lattice. The effect of lattice coupling can be represented by means of a dynamic effective mass m_{eff} which can be expressed by

$$E_R = \frac{(h\nu)^2}{2m_{eff}c^2} \tag{7}$$

Due to the fact that the emitting nucleus is bound to the crystal lattice, the recoil momentum is necessarily transferred to the centre of mass of the whole crystal and consequently the value of M_{eff} may be very large and the energy loss by recoil may become negligible. However, the available energy can be dissipated in processes other than recoil. The binding energy of an atom in a lattice is of the order of several electronvolts, but the recoil energy due to a 100 kev γ-transition is only about 10^{-3} ev. This energy can be dissipated by altering the vibrational state of the lattice or, in other words, by phonon propagation through the lattice.

If, as a consequence of the emission of the γ-ray, phonons are created in the lattice with frequencies up to ω_i, the energy available for the emitted γ-ray will be

$$h\nu = E_0 - \sum_i n\hbar\omega_i \tag{8}$$

Again, if $n\hbar\omega_i$ is large compared with Γ, condition (3) cannot be fulfilled and the γ-ray will not be resonantly absorbed. What must be considered is the possibility of a zero-phonon process. In such a process, all the energy of the nuclear transition E_0 is given to the γ-ray and the resonance condition (3) is completely fulfilled.

The fact that transitions can occur in which the lattice remains in its initial state and the γ-ray receives the full energy arises because the lattice vibrations are quantized. Energy cannot be transferred to the lattice in an arbitrary fashion and for those transitions which cannot excite phonons in the lattice a recoilless γ-emission occurs. The Mössbauer effect is thus the recoilless emission and resonance absorption of low-energy γ-rays.

The quantitative description of the Mössbauer effect requires a quantum-mechanical treatment[8,9]. However, it is interesting to observe that the possibility of phonon scattering without lattice excitation was known long before the discovery of the Mössbauer effect. Thus, in a beam of scattered x-rays, a fraction of the total intensity will have the same frequency as the incoming radiation. This fraction f is known as the Debye–Waller factor. The close analogy between x-ray scattering and the recoilless process with γ-radiation suggested an interpretation of the Mössbauer effect in terms of classical electromagnetic theory[10]. An electromagnetic wave train emitted or scattered by a nucleus suffers a frequency modulation as a result of the vibrational motion of the nucleus around its equilibrium position. This frequency modulation leads to a central unshifted line and to side components. The central line represents the recoilless processes and we can define the Debye–Waller factor as the fractional intensity of the central line. The general expression for the recoil-free function may be written

$$f = \exp\left(-4\pi^2\langle x^2\rangle/\lambda^2\right) \tag{9}$$

where $\langle x^2\rangle$ is the mean square amplitude of the vibration in the direction of emission of the γ-ray averaged over an interval equal to the lifetime of the nuclear level involved in the γ-ray emission process and λ is the wavelength of the γ-quantum.

Equation (9) shows why the Mössbauer effect cannot take place in a liquid, where the molecular motion is characterized by unbounded $\langle x^2\rangle$, making the recoil-free fraction vanish. However, as explained in Wertheim's book[27], in viscous liquids $\langle x^2\rangle$ may remain sufficiently small to allow the detection of the Mössbauer effect.

In order to express f in terms of the usual experimental variables, the mean square deviation $\langle x^2\rangle$ is calculated using the Debye model of a solid. In this model, the crystal is represented by $3N$ oscillators of frequency ω_j; the distribution of frequencies is limited by a maximum value

ω_{max} having a characteristic value for a given solid. Usually there is associated with this frequency a Debye temperature θ_D expressed by

$$k\theta_D = \hbar\omega_{max} \tag{10}$$

With the usual approximations of the Debye specific heat theory of solids, we find for the Mössbauer factor

$$f = \exp\left(-\frac{E_R}{k\theta_D}\left\{\frac{3}{2} + \frac{\pi^2 T^2}{\theta_D^2}\right\}\right) \tag{11}$$

The following conclusions can be derived from (11):

1. The Mössbauer effect is limited to relatively low-energy γ-radiation emission, since (11) decreases with the exponential square of γ-ray energy. The effect has been observed up to 155 kev γ-radiation, which is approximately the upper limit of detectability.
2. The higher the Debye temperature θ_D of the solid, the larger is the Mössbauer recoilless fraction. The physical meaning of this conclusion is that larger f factors are observed with rigid solids since the Debye temperature is inversely proportional to the compressibility of the solid.
3. The recoilless fraction increases with decreasing temperature as first observed by Mössbauer with the 129 kev transition in ^{191}Ir.

Expression (11) is only valid for a monatomic lattice of identical atoms. Two situations, however, occur frequently: (*a*) the emitting nucleus is an impurity in a host lattice, and (*b*) there are solids in which, besides the acoustic modes of vibration, optical modes can also be excited. Neither case can be described by (11); instead, a detailed knowledge of the form of the frequency distribution must be available in order to derive the appropriate relations for the recoilless fraction f.

III. EXPERIMENTAL ASPECTS

A. Detection

The Mössbauer spectrometer consists essentially of a source which emits γ-radiation in recoil-free conditions, an absorber (or scatterer) which contains the substance to be investigated, and a detector of low-energy γ-radiation. The source (or absorber) is mounted on a velocity drive to change the energy of the γ-radiation by utilizing the Doppler effect. A γ-quantum which is emitted from a source moving at the velocity v receives a Doppler energy E_D such that

$$\frac{E_D}{E_0} = \frac{v}{c} \tag{12}$$

where E_0 is the energy of the γ-quantum emitted by the same source at rest and c is the velocity of light.

Due to the very narrow linewidth Γ of the resonance, it is generally sufficient to produce a small energy change $E_D = \Gamma$ by Doppler shifting the source in order to sweep over the resonance. Some typical values of Doppler velocities corresponding to the width of the resonance are given below.

Nucleus	$v = \frac{c\Gamma}{E_0}$ (cm/s)
^{57}Fe	9.6×10^{-3}
^{67}Zn	1.5×10^{-5}
^{197}Au	9.4×10^{-2}

Various types of Mössbauer spectra which occur are illustrated in Figures 10, 13, 15, 19, and 20. We are thus faced with the typical problems of low-energy γ-radiation measurements and that of producing uniform velocities, whilst carefully avoiding vibrations in the system.

Most of the work so far on the Mössbauer effect has used simple transmission geometry. This is the most versatile arrangement, but other possibilities exist which may offer certain advantages in special cases. One of the most important is the scattering method, in which the reradiated γ-rays are counted. By this method, it is possible to reduce considerably the background count rate, and thus for high-energy transitions (above 100 kev) scattering may be the only way by which the Mössbauer effect can be observed. The detailed application of the scattering method has been demonstrated in the Mössbauer effect with the 155 kev transition of ^{188}Os, the measurements being achieved with recoilless fractions f as low as 0.0063[11]. Other methods utilize the internal conversion electrons, which are always intense in these low-energy Mössbauer transitions. This may be done either by counting directly the conversion electrons or by counting the x-rays which follow the internal conversion. These methods were successfully applied[12,13] to ^{182}W and ^{57}Fe.

B. Source, Absorber, and Detectors

The source for the Mössbauer effect is a radioactive isotope of reasonable half-life. By radioactive disintegration, the isotope populates an excited level which decays to the ground state by emitting low-energy γ-radiation. Table 2 lists the isotopes in which the Mössbauer effect has been observed.

In order to emit γ-radiation under recoil-free conditions, the radioactive isotope must be incorporated in a crystal lattice. As a typical example,

Table 2. Isotopes in which the Mössbauer effect has been observed.

Mössbauer isotope	Abundance of stable element (%)	γ-Ray energy (kev)	Half-life of Mössbauer transition (10^{-9} s)	Parent isotope	Half-life of parent isotope
^{40}K	0.0118	29.4	3.9	^{39}K	nuclear reaction
^{57}Fe	2.19	14.4	98	^{57}Co	270 d
^{61}Ni	1.19	67.4	5.3	^{61}Cu	33 h
^{67}Zn	4.11	93	9400	^{67}Ga	78 h
^{83}Kr	11.55	9.3	147	^{83}Rb	83 d
^{99}Ru	12.72	90	20	^{99}Rh ^{99}Tc	16 d 2.1×10^5 y
^{119}Sn	8.58	23.8	18.5	^{119m}Sn	245 d
^{121}Sb	57.25	37.2	3.5	^{121m}Te	154 d
^{125}Te	6.99	35.6	1.4	^{125}I	57 d
^{127}I	100	59	1.8	^{127}Te ^{127}Xe	105 d 36 d
^{129}I	0	26.8	16.3	^{129}Te	33 d
^{129}Xe	26.44	40	0.96	^{129}I	1.6×10^7 y
^{131}Xe	12.18	80.2	0.50	^{131}I	8.1 d
^{133}Cs	100	81	6.23	^{133}Ba	7.2 y
^{141}Pr	100	145.0	2.0	^{141}Ce	33 d
^{149}Sm	13.83	22	7.6	^{149}Eu	106 d
^{151}Eu	47.82	21.6	8.8	^{151}Gd ^{151}Sm	120 d 93 y
^{152}Sm	26.63	122	1.4	^{152}Eu	9 h
^{153}Eu	52.18	97.5 103.2	0.14 3.8	^{153}Gd	242 d
^{155}Gd	14.73	86.5	5.86	^{155}Tb ^{155}Eu	5 d 1.81 y
^{156}Gd	20.47	89	1.9	^{155}Gd	nuclear reaction
^{158}Gd	15.68	79.5	—	^{157}Gd	nuclear reaction
^{159}Tb	100	58	0.13	^{159}Dy	144 d
^{160}Dy	2.29	86.8	2.05	^{160}Tb	72.4 d
^{161}Dy	18.88	25.6 74.5	28 3.0	^{161}Tb	6.9 d
^{166}Er	33.41	80.6	1.83	^{166m}Ho	> 30 y
^{169}Tm	100	8.41	3.9	^{169}Er ^{169}Y	9.4 d 32 d
^{170}Yb	3.03	84.2	1.61	^{170}Tm	127 d
^{171}Yb	14.31	66.7	0.5	^{171}Tm	1.9 y
^{177}Hf	18.50	113	0.52	^{177}Lu	68 d
^{181}Ta	39.99	6.25	6800	^{181}Hf ^{181}W	45 d 140 d

Table 2 (*continued*)

Mössbauer isotope	Abundance of stable element (%)	γ-Ray energy (kev)	Half-life of Mössbauer transition (10^{-9} s)	Parent isotope	Half-life of parent isotope
^{182}W	26.41	100.1	1.4	^{182}Ta	115 d
^{183}W	14.40	{46.5 99.1	{0.15 0.52	^{183}Ta	5 d
^{186}Os	1.64	137.2	0.84	^{186}Re	90 h
^{187}Re	62.93	134.2	0.01	^{187}W	24 h
^{188}Os	13.3	155.0	7.2	^{188}Ir	41 h
^{191}Ir	37.3	129.4	0.13	{^{191}Os ^{191}Pt	15 d 3 d
^{193}Ir	62.7	73	6.0	{^{193}Pt ^{193}Os	<500 y 32 h
^{195}Pt	33.8	{98.9 129.7	{0.17 0.55	{^{195}Au ^{195}Ir	192 d 2.3 h
^{197}Au	100	77.3	1.8	^{197}Hg	65 h
^{237}Np	0	59.6	63	{^{237}U ^{241}Am	6.75 d 458 y

Figure 3 shows the disintegration scheme of ^{57}Co which populates the 14.4 kev Mössbauer level of ^{57}Fe. Sources of ^{57}Co are usually prepared[14] by electrochemically depositing the carrier-free isotope on metallic supports and then diffusing it into the solid at high temperatures.

Simple spectra result from using sources which give an unsplit line with a width as nearly equal to the natural value as possible; this requires that the solid matrix must be such that no hyperfine splittings are induced in the Mössbauer transitions. Even using diamagnetic supports having high point-charge symmetry at the Mössbauer nuclear sites, it is only in a few cases that emission with natural linewidth has been achieved. In most cases, inhomogeneous line broadening occurs. Thus, for ^{57}Fe in a Pt matrix, the linewidth is [15] about 10% higher than the natural values; with ^{119}Sn the natural linewidth is observed with a Mg_2Sn alloy support[16].

More recently, other methods have been developed for obtaining Mössbauer transitions. These methods make use of nuclear reactions such as Coulomb excitation and (n, γ) reactions. In Coulomb excitation, the incident charged particle on passing through the electric field of the nucleus excites its low-lying levels. Thus, with ^{57}Fe it is possible by this method to excite the 137 kev level which will populate the 14.4 kev level, as illustrated in Figure 4. Successful Mössbauer effects following Coulomb excitation have been realized with ^{57}Fe using a 1.5 μA beam from a 3 Mev

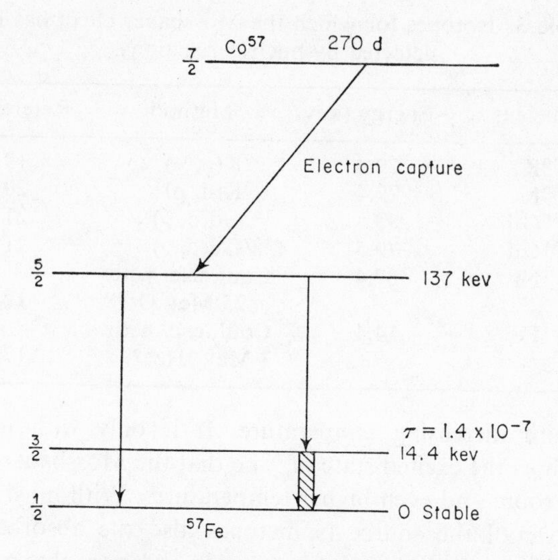

Figure 3. Decay scheme of ^{57}Co.

α-particle Van der Graaff accelerator[17] and in ^{61}Ni with 25 Mev O^{4+} ions[18]. The use of (n, γ) reactions to populate the excited nuclear level for the Mössbauer effect was first demonstrated with the 29.4 kev level which occurs in the ^{39}K(n, γ)^{40}K reaction[19]. Table 3 lists the nuclei for which the Mössbauer effect has been detected by nuclear reactions.

An important difference between the isotope-source method and those using nuclear reactions is the fact that in the latter the Mössbauer effect is observed under conditions characterized by the intense radiation due to the incoming beam. This can be of interest in radiation-damage investigations using the Mössbauer effect.

The yield of a source in terms of intensity of recoil-free radiation can be measured by the *f* factor, which, for a given isotope in a given matrix,

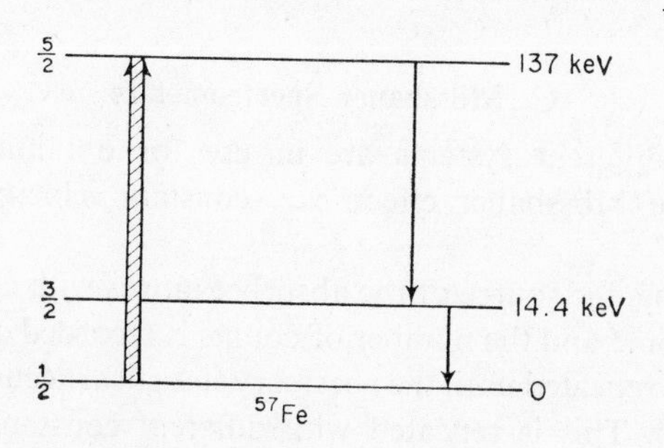

Figure 4. Coulomb excitation of the 137 kev level of ^{57}Fe.

Table 3. Isotopes for which the Mössbauer effect has been detected by nuclear reactions.

Nucleus	γ-Energy (kev)	Method	Reference
^{40}K	29.4	$^{39}K(n, \gamma)$	19
^{40}K	29.4	$^{39}K(d, p)$	20
^{156}Gd	89	$^{155}Gd(n, \gamma)$	21
^{158}Gd	79.5	$^{157}Gd(n, \gamma)$	21
^{61}Ni	67.4	Coul. exc. with 25 Mev O^{4+}	18
^{57}Fe	14.4	Coul. exc. with 3 Mev $^{4}He^{2+}$	17

decreases with increasing temperature. It is only with low-energy γ-emitters such as the excited state of ^{57}Fe that the Mössbauer effect can be observed at room and even higher temperatures. With most systems, it is necessary to cool the source (sometimes also the absorber) at liquid-nitrogen and liquid-helium temperatures in order to obtain a significant effect. Several types of Dewar flasks for these purposes have been described[22].

The absorber must contain a minimum number of Mössbauer nuclei in the ground state in order to give an appreciable resonance absorption of the γ-rays from the source. For ^{57}Fe with natural isotopic abundance of 2.17%, this amounts to 4–30 mg/cm^2.

The effect of source and absorber thickness on observed Mössbauer spectra is of great practical importance. Corrections for a finite thickness have been derived, and analytical and graphical results for the various cases of interest are given in the original publications[23,24].

The usual detectors for low-energy γ-radiation are the thin NaI(Tl) scintillation crystals coupled to a photomultiplier, and gas-filled proportional counters. Marked improvements in energy resolution have recently been obtained with solid-state detectors such as the lithium-drift germanium counter[25].

C. Mössbauer Spectrometers

Two basically different systems are in use for obtaining the velocity spectrum of the Mössbauer effect, viz. constant-velocity and velocity-sweep devices.

In the first type, the source or the absorber moves with constant velocity during a fixed time and the number of counts is recorded during this time. The operation is repeated until the necessary statistical accuracy of counting rate is obtained. This is repeated with different constant velocities and hence the Mössbauer spectrum is obtained point-by-point in terms of count rate as a function of velocity. Drivers for constant velocity such as rotating disks, cams, hydraulic devices, etc., are described in the literature[26]. In general, such spectrometers are of simple design and quite reliable, but since they are used in connexion with single-channel analysers, problems of drift of the electronic systems limit the obtainable accuracy.

In velocity-sweep systems, instead of a single velocity in each cycle of the driver, the system passes through an entire range of velocities during one cycle, and the source (or absorber) repeatedly sweeps this range of velocities. Usually the instantaneous velocity is a triangle function, sweeping linearly from $-v_{min}$ to $+v_{max}$, which can be obtained by parabolic driver motion. An electromechanical feedback system for generating motion by parabolic displacement in two mechanically coupled loudspeakers is described in Wertheim's book[27].

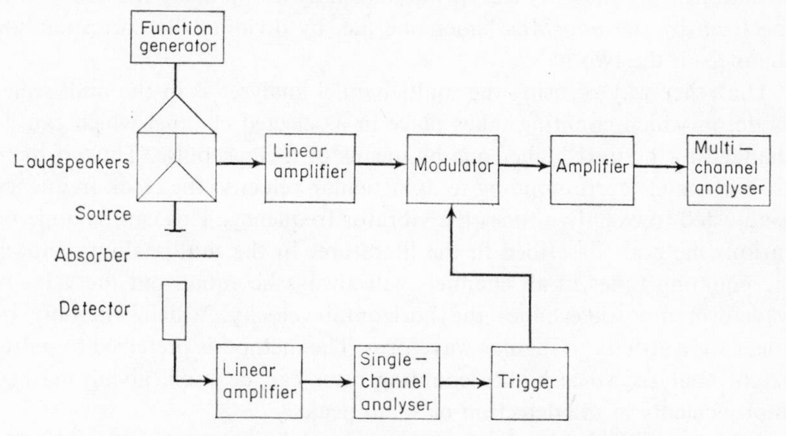

Figure 5. Block diagram of a Mössbauer spectrometer.

Other types of motion of the driver can be used, such as the sinusoidal one, which can be obtained with greater ease than any other. They have, however, the drawback of giving a non-linear velocity scale to the spectrum. For a miniaturized low-temperature Mössbauer drive giving sinusoidal motion, see ref. 22.

The problem of storing the counting information is solved by multichannel analysers. These instruments can be used as pulse-height analysers by making the signal from a velocity pick-up, which is connected to the moving source, modulate the output pulses from the detector. Figure 5 is a diagram of the basic set-up for this type of Mössbauer spectrometer. Two permanent-magnet loudspeakers are mounted in a light metal cylinder passing through their common axis. The two voice coils, attached rigidly to the tube, are free to move together axially in the magnet gaps. One coil connected to a power amplifier is fed by a low-frequency triangle function generator and is used as the velocity driver. The other coil is used as velocity pick-up, since it develops an e.m.f. proportional to the instantaneous velocity. This signal is amplified and used to modulate linearly the amplitudes of the standardized pulses from a single-channel pulse-height analyser set on the peak of interest. The modulated pulses, whose amplitudes are proportional to velocity, are then sorted by the multichannel analyser. In this method, the channel number will be proportional to velocity, provided both modulator and analyser operate linearly.

Usually the modulator is fed by a non-Mössbauer signal which is stored in one of the halves of the memory of the multichannel analyser, the other half being used for the modulated Mössbauer signal. Thus, deviations from linearity can be minimized by normalizing the Mössbauer spectrum by the non-Mössbauer one, i.e. by dividing the corresponding channels in the two halves.

The other way of using the multichannel analyser is in the multiscaler mode, in which counting takes place in a selected channel which can be changed to the next higher one by means of a clock pulse. Thus, if there are n channels corresponding to a particular velocity, the clock frequency is adjusted to exactly n times the vibrator frequency. This can be done by various methods described in the literature. In the multiscaling method, the counting times in all channels will always be equal, but the velocity waveform now determines the horizontal velocity, which will only be linear for a strictly triangular waveform. The method is preferred to pulse-height analysis, since higher counting rates can be used, giving marked improvements in the detection of small peaks.

A high-precision Mössbauer spectrometer has been developed at the United States National Bureau of Standards, details being available in a recent publication[28].

IV. HYPERFINE INTERACTIONS

Due to the narrow linewidth of the Mössbauer nuclear transitions, the resonance spectrum is extremely sensitive to energy variations of the γ-radiation. For this reason, small interactions between the nucleus and the orbital electrons manifest themselves very markedly in the Mössbauer effect. It is thus the influence of the electronic environment on the emission and absorption nuclear γ-transition which determines the hyperfine structure of the Mössbauer spectra. Essentially, all applications of the Mössbauer effect to chemistry and solid-state physics are based on the hyperfine structure of the spectra.

As we shall see in what follows, these interactions can to a first approximation be expressed by the product of a term containing only nuclear parameters and another term with parameters of electronic origin. These electronic parameters refer to electric and magnetic effects caused by the orbital electrons in the region of the nucleus and they are susceptible of an interpretation in terms of the electronic structure of the atom.

The main interactions which manifest themselves in the Mössbauer effect are:

A. Nuclear isomer shift.
B. Nuclear quadrupole coupling.
C. Magnetic hyperfine interaction.

In the Coulomb approximation, the nucleus is described as a point charge Ze. The next approximation assumes that:

1. The nuclear dimensions are finite and the electrostatic potential viewed by electrons is different inside and outside the nucleus.

2. The nuclear charge distribution is not spherical. The deviation from sphericity is described by electric multipole moments of order n. Parity conservation requires even values for n, and for the present purpose only $n = 2$, the quadrupole moment of the nucleus, is important.

3. The magnetic properties of the nucleus can be described by magnetic multipole moments of order n (n odd), the dipole moment $n = 1$ being the important one for describing the present properties.

A. Nuclear Isomer Shift

The nuclear isomer shift is the nuclear analogue of two hyperfine interactions observed in the optical spectra of free ions, viz. the isotope shift and the isomer shift. These effects are changes in wavelength of a given spectral term for different isotopes of the same element and in nuclear isomeric states of one and the same isotope of an element. These effects, as well as the nuclear isomer shift, which is defined as the displacement from zero velocity of the centroid of the Mössbauer resonance spectrum, are due to the electrostatic interaction between the charge distribution of the nucleus and that of electrons which have a finite probability at the region of the nucleus. Thus, the isomer shift arises from the fact that the nucleus of an atom occupies a finite volume[29].

An expression for the interaction between the electronic charge density $-e\rho$ and the nuclear charge characterized by a radius R will be derived with the assumptions that (*a*) both charge distributions have spherical symmetry, and (*b*) the electronic charge density remains unchanged inside the nuclear dimensions.

The energy difference δE arising from the finite nuclear radius can be evaluated in a simple way by calculating the difference between the electrostatic interaction of a point nucleus and one of actual radius R, both having the same charge Ze.

$$\delta E = \int_0^\infty \rho(V - V_{pt}) 4\pi r^2 \, dr \tag{13}$$

where

$$V_{pt} = \frac{Ze}{R} \tag{14}$$

for the point nucleus, and

$$V = \frac{Ze}{R}\left[\frac{3}{2} - \left(\frac{r^2}{2R^2}\right)\right] \quad \text{for} \quad r \leqslant R$$

$$V = \frac{Ze}{R} \quad \text{for} \quad r \geqslant R \tag{15}$$

for the finite nucleus of radius R.

Substituting (14) and (15) into (13), we get

$$\delta E = -\frac{2\pi}{5} Ze\rho R^2 \tag{16}$$

Calling $-e\psi^2(0)$ the electronic charge density at the nucleus,

$$\delta E = \frac{2\pi}{5} Ze^2\psi^2(0)R^2 \tag{17}$$

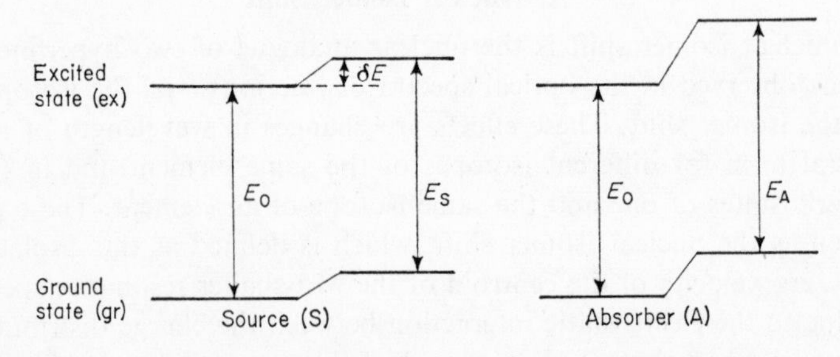

Figure 6. Nuclear isomer shift. The interaction shifts the nuclear levels in the source and the absorber.

Since nuclear excited and ground states do not have the same radius, the electronic density $\psi^2(0)$ will interact to a different extent with the two nuclear levels (Figure 6). This difference is

$$\delta E_{ex} - \delta E_{gr} = \frac{2\pi}{5} Ze^2\psi^2(0)(R_{ex}^2 - R_{gr}^2) \tag{18}$$

Expression (18) measures the energy shift of the γ-ray emitted in the transition from the excited to the ground nuclear state due to the nuclear electrostatic interaction. The shift actually observed in a Mössbauer spectrum is the difference of the energy shift (18) for an absorber and a source which are in different chemical states and characterized by two different values of the electronic density at the nucleus $\psi^2(0)$. From (18) this difference is

$$\text{Isomer shift} = \frac{2\pi}{5} Ze^2[\psi^2_{\text{abs}}(0) - \psi^2_{\text{source}}(0)](R^2_{\text{ex}} - R^2_{\text{gr}}) \qquad (19)$$

Thus, the isomer shift is expressed by a nuclear term and an electronic one, which measures the electronic density at the nucleus of an absorber *relative to a given source.*

In a non-relativistic approximation, the electron density at the nucleus is large only for electrons with zero angular momentum or *s* electrons and can be approximated by $\psi_s^2(0)$. Thus, the isomer shift of the Mössbauer spectra is a relative measure of the total *s*-electron density at the nucleus in a compound. It is important here to observe that the isomer shift is a

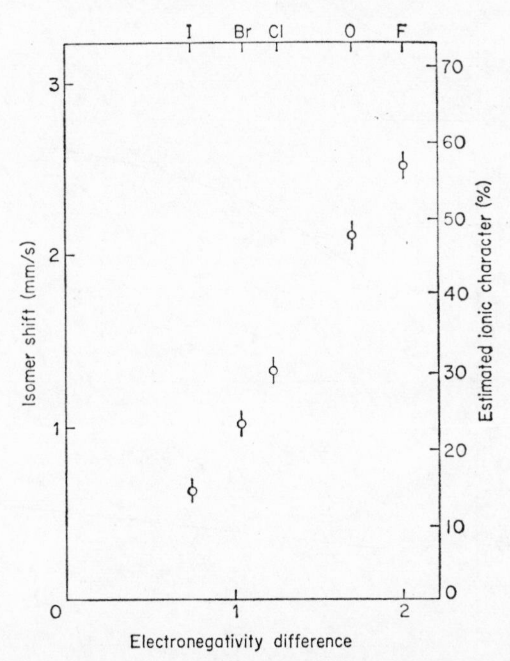

Figure 7. The isomer shifts (relative to grey tin) and the corresponding ionic character of the Sn—X bonds for quadrivalent inorganic compounds. [From Ref. A.7 of Bibliography.]

hyperfine interaction of electrical origin and $\psi_s^2(0)$ measures the total *charge* density at the nucleus. It must not be confused with the $\psi_s^2(0)$ derived from *magnetic* hyperfine interactions which originate from the Fermi contact term as observed in the isotropic hyperfine interaction of the electron spin resonance spectra, and the internal fields in metals which measure unpaired *spin* density at the nucleus.

We can, as an approximation, consider that the total *s*-electron density of an atom in a chemical compound is given by

$$\psi_s^2(0) = \psi_s^2(0)|_{\text{inner shells}} + \psi_s^2(0)|_{\text{outer shells}}$$

The contribution from inner electron shells comes from the filled *s* orbitals of the atom. The contribution from the outer electron shells arises from those external *s* orbitals of the element which are partially

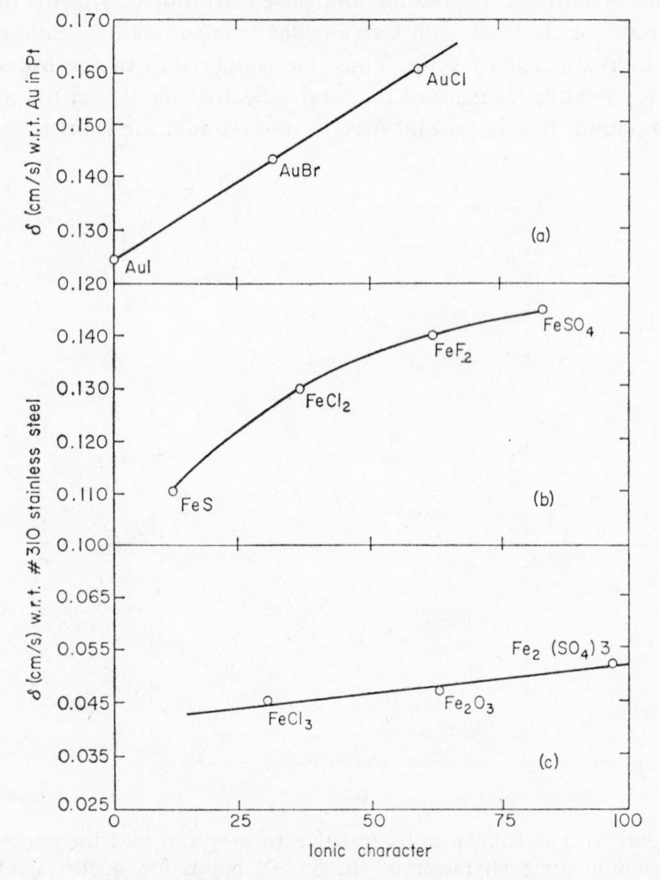

Figure 8. Isomer shift versus ionic character[31].

occupied by electrons from the ligands. This last contribution is sensitive to changes in the chemical environment of the element.

Since the isomer shift is a measure of the tendency of bond formation through the *s*-orbital occupation, we expect to observe correlations with other chemical bonding parameters such as the electronegativity, ionicity, and the nephelauxetic effect. For example, the observed isomer shifts for quadrivalent tin compounds clearly show increasing ionic character with increasing electronegativity difference[30] (Figure 7). Correlation between the isomer shift and ionic character for other elements is shown in Figure 8[31]. We shall analyse in Section VI the interpretation for these correlations in terms of the electronic structure of bonded atoms.

B. Nuclear Quadrupole Coupling

The quadrupole splitting of the Mössbauer spectra is the result of the interaction of the nuclear quadrupole moment Q with the gradient of the electric field at the nucleus. The value of Q expresses the deviation of the nucleus from spherical symmetry; an oblate nucleus has a negative moment, whilst a prolate one has a positive moment. Nuclei with $I = 0$ or $\frac{1}{2}$ are spherically symmetrical, so that $Q = 0$.

The field gradient can be completely specified by the three components

$$\varphi_{xx} = \frac{\partial^2 \varphi}{\partial x^2}, \quad \varphi_{yy} = \frac{\partial^2 \varphi}{\partial y^2}, \quad \varphi_{zz} = \frac{\partial^2 \varphi}{\partial z^2} \tag{20}$$

Since these components must obey the Laplace equation $\nabla \varphi = 0$, only two of them are independent. They are usually given in terms of φ_{zz} or q, where $eq \equiv \varphi_{zz}$, and the asymmetry parameter which is defined by

$$\eta = \frac{\varphi_{xx} - \varphi_{yy}}{\varphi_{zz}} \tag{21}$$

The theoretical description of nuclear quadrupole coupling is discussed in detail in Chapter 9. The quadrupole interaction in the Mössbauer effect differs from the pure quadrupole coupling only by the fact that in the Mössbauer effect the interaction involves a short-lived nuclear excited state. Thus, as is shown by Sillescu in the next chapter (equation 19, p. 441), the nuclear quadrupole coupling is expressed by the hamiltonian

$$\mathscr{H}_{Q} = \frac{e^2 Qq}{4I(2I - 1)} [3\mathbf{I}_z^2 - I(I + 1) + \tfrac{1}{2}\eta(\mathbf{I}_+^2 + \mathbf{I}_-^2)] \tag{22}$$

For a field gradient with axial symmetry, $\eta = 0$, the eigenstates of (22) are the states $\mathbf{I}_z = M$, where M is the nuclear magnetic quantum number; the $\pm M$ states are degenerate since the square of $\mathbf{I}_z$ appears in (22). For $\eta \neq 0$, it is necessary to solve the secular equation corresponding to (22).

For levels with spins $\frac{3}{2}$ (Figure 9), as in ^{57}Fe and ^{119}Sn first excited states, we have

$$E_Q = \pm \frac{e^2 Qq}{4}\left(1 + \frac{\eta^2}{3}\right)^{\frac{1}{2}} \tag{23}$$

The interaction is positive if the $\pm\frac{3}{2}$ state of the excited level is higher than, and negative when it is lower than, the $\pm\frac{1}{2}$ state of this level. It is important to observe that the quadrupole splitting yields only the absolute value of the product of the nuclear moment and the field gradient, not its sign.

The contributions to the field gradient are of three types:

1. Charges external to the atom; for example the ionic charges of the lattice create a field gradient at a given atomic site.

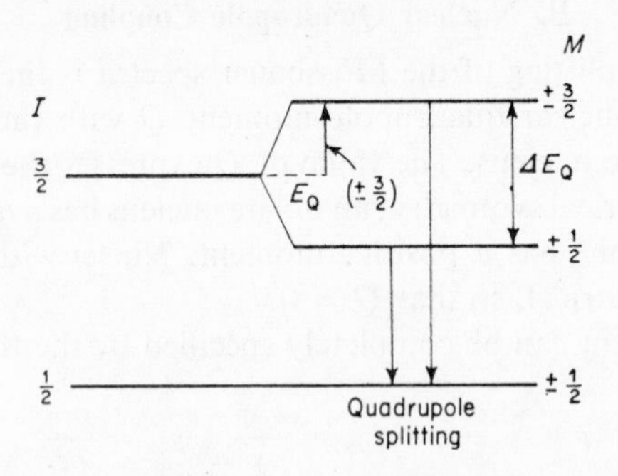

Figure 9. Nuclear quadrupole coupling. The interaction lifts the spin degeneracy of the $I = \frac{3}{2}$ excited level.

2. Electrons of the atom; for example in transition ions the inner *d* or *f* subshells are not completely filled, causing a field gradient at the nucleus.

3. Polarization of inner shells of the atom. The filled inner shells have spherical symmetry and make no contribution to the field gradient; however, both the charges external to the atom and its unpaired electrons can polarize these inner shells which will then contribute to the field gradient at the nucleus. This magnification of the electric field gradient due to external charges is called 'antishielding', and is expressed by multiplying the field gradient by the so-called Sternheimer factors [32,33].

It is thus possible to write that, in general, the field gradient is given by

$$q = (1 - \mathscr{R})q_{\text{val}} + (1 - \gamma_\infty)q_{\text{lat}} \tag{24}$$

where $(1 - \mathscr{R})$ and $(1 - \gamma_\infty)$ are the Sternheimer antishielding factors which correct for the polarization due to the valence electrons and the lattice charge distribution.

The relative importance of these factors depends on the type of solid considered. For instance, the lattice contribution will predominate in ionic crystals of non-transition elements, whereas with transition elements the electrons in the unfilled shells will make the largest contribution to the electric field gradient. The influence of antishielding factors, which is probably large for heavy ions, is not adequately known, as is discussed by Sillescu (p. 449).

Other interesting information can be derived from the *intensity* of the quadrupole peaks. If the quadrupole measurements are made with single crystals, for which it is possible to define the field gradient axis, we observe a change in the intensities of the hyperfine components of the spectrum with the angle of incidence of the γ-radiation. This angular correlation between the incident γ-radiation and the field gradient axis is due to the difference in polarization of the hyperfine transitions. Such is the case for ^{57}Fe, in which the $(\pm\frac{3}{2} \rightarrow \pm\frac{1}{2})$ transition is fully circularly polarized and the $(\pm\frac{1}{2} \rightarrow \pm\frac{1}{2})$ transition is partly circularly and partly linearly polarized. In consequence, it can be shown that when the axis of an axially symmetrical electric field gradient of a single crystal is oriented at an angle θ to the direction of the γ-radiation, the intensity ratio of the transitions is[6]

$$\frac{I_1(\pm\frac{3}{2} \rightarrow \pm\frac{1}{2})}{I_2(\pm\frac{1}{2} \rightarrow \pm\frac{1}{2})} = \frac{1 + \cos^2\theta}{\frac{5}{3} - \cos^2\theta} \tag{25}$$

More general expressions have been derived by Zory[34] for the frequent case of the single crystal having more than one direction for the field gradient, due to the presence of more than one molecule in the unit cell with different orientations.

When measuring the quadrupole interaction in polycrystalline materials with isotropic distribution of the field gradients, it is usually found that the two peaks have the same intensity. This can be interpreted by integrating equation (25) over all directions, which gives $I_1 = I_2$. However, it has been shown by Goldanskii[35] that if the Mössbauer recoil-free factor f is anisotropic, then

$$\frac{I_1}{I_2} = \frac{\int_0^\pi (1 + \cos^2\theta) f(\theta) \sin\theta \, \mathrm{d}\theta}{\int_0^\pi (\frac{5}{3} - \cos^2\theta) f(\theta) \sin\theta \, \mathrm{d}\theta} \neq 1 \tag{26}$$

and in consequence we should see two peaks with different intensities even for polycrystalline materials.

In the Goldanskii effect, which has been demonstrated for tin and iron compounds, the two peaks have different areas but the same half-width. It has been shown by Karyagin[36] that if, instead of a single-valued electric field gradient, a distribution of electric field gradient values occur, we may have peaks with different shapes; this could be the case for a solid with a disordered structure, e.g. a glass. The influence of fluctuating internal fields on the shape of the Mössbauer spectrum has also been investigated[37,38] and this is of importance for systems where slow spin-lattice relaxation occurs, as we shall discuss in connexion with the magnetic hyperfine interactions.

The determination of the sign of the electric field gradient is of interest, since it can be correlated with the electronic structure of the molecule. This is relatively straightforward if single crystals are available, i.e. by applying relation (25) to the angular variation of the Mössbauer spectrum. When the γ-ray direction is along the principal axis ($\theta = 0$) of the axially symmetric electric field gradient, then the intensity of the $\frac{3}{2}$ line is 3 times that of the $\frac{1}{2}$ line; in a perpendicular direction this ratio changes to 0.6. If the $\frac{3}{2}$ line is at the higher energy position in the Mössbauer spectrum (at higher Doppler velocities), then the interaction is positive; if the $\frac{1}{2}$ line is at the higher energy position, then it is negative. This is exemplified by the Mössbauer spectrum[39] in Figure 10, obtained with a single crystal of sodium nitroprusside $Na_2[Fe(CN)_5NO]\cdot 2H_2O$ at $\theta = 90°$. The ratio of the intensities is close to 0.6 with the $\frac{3}{2}$ line located at positive velocities. The quadrupole interaction is by consequence positive, and since $Q_{^{57}Fe} > 0$, we have $q > 0$ for the iron nucleus in this complex. The sign of the field gradient using single-crystal Mössbauer spectra and its significance for molecular structure has been demonstrated with $Fe(CO)_5$[40] and $XeCl_4$[41].

For nuclear spins greater than $\frac{3}{2}$, the quadrupole-split levels are not uniformly spaced and it is therefore possible, from inspection of the Mössbauer spectrum, to decide the sign of the quadrupole interaction for both the excited and the ground-state levels. This has been described[42] in detail in the Mössbauer transition from a level with $I = \frac{5}{2}$ to a ground state with $I = \frac{7}{2}$ in ^{129}I.

For polycrystalline samples, it is possible to perturb the quadrupole-split levels with strong external magnetic fields, as first suggested by Ruby and Flinn, and to obtain information on the sign[43] of the electric field gradient from the resulting spectra. A successful application of this method to ferrocene[44] has been described by Collins. The theory for this perturbation is the same as that described by Sillescu (p. 442) for the Zeeman splitting of n.q.r. frequencies. Using the same first-order perturbation treatment exemplified with the $I = \pm\frac{3}{2}$ level by Sillescu, Collins investigated the case of a ($\pm\frac{3}{2} \rightarrow \pm\frac{1}{2}$) transition. The energy levels and wave functions obtained are expressed as a function of the angle θ between the axis of molecular symmetry and the axis along the magnetic field. At any given value of θ, the spectrum is expressible by multiplying the square of the coefficients of these eigenstates by the appropriate transition probability between the magnetically perturbed levels. For the powder pattern, it is necessary to weight each transition by the solid angle between θ and $\theta + d\theta$. The result obtained, taking into account the effect of the

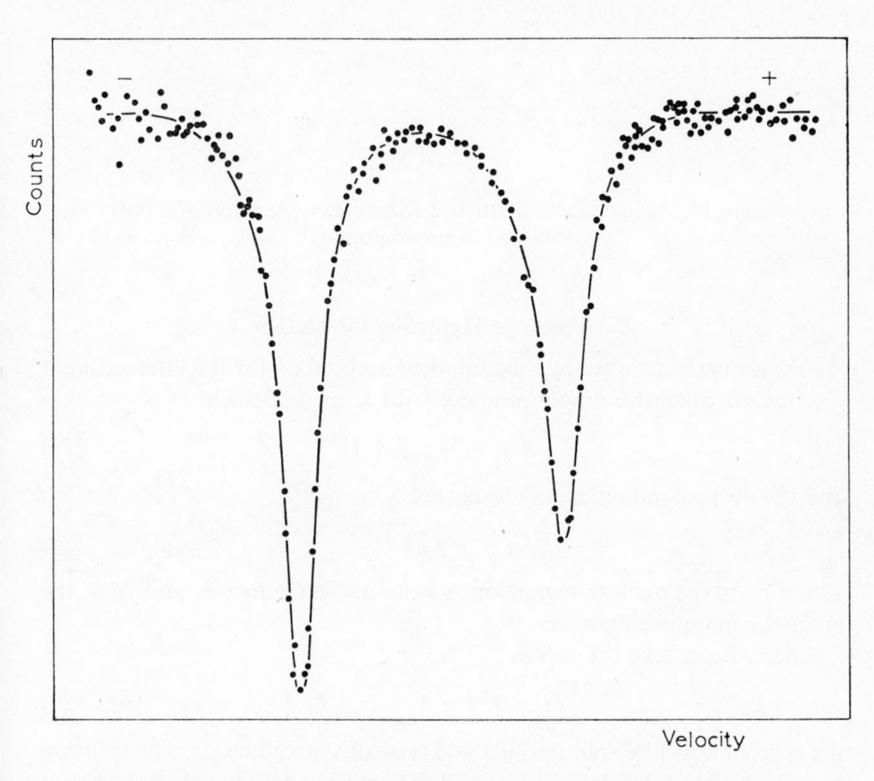

Figure 10. Quadrupole splitting of a sodium nitroprusside single crystal oriented at $\theta = 90°$ with the incident radiation.

finite linewidth of the transitions, is illustrated in Figure 11. This shows that, upon application of an adequate magnetic field, one line splits into a triplet and the other into a doublet. If the doublet is of higher energy than the triplet, the quadrupole interaction is positive, and if the doublet is of lower energy, the interaction is negative. Magnetic fields with intensity not less than 20 kG are required for this type of investigation. The results obtained by Collins with polycrystalline ferrocene are discussed in Section VI.

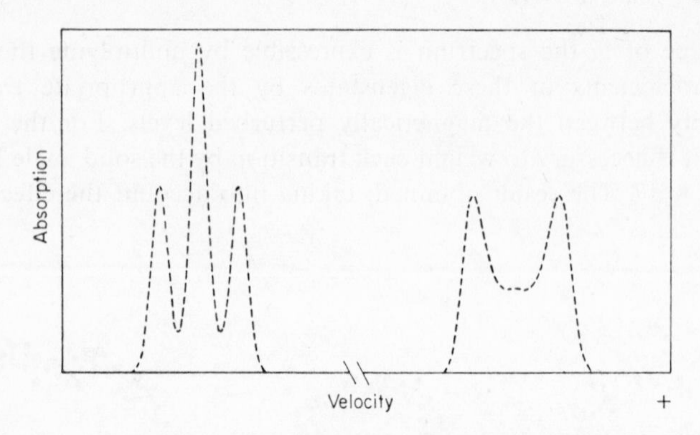

Figure 11. Magnetically perturbed Mössbauer spectrum of a polycrystalline iron compound[44].

C. Magnetic Hyperfine Interaction

As shown in the chapter by Sillescu, the hamiltonian for the interaction of the nuclear magnetic dipole moment with a magnetic field **H** is

$$\mathscr{H}_{\mathrm{M}} = -g_{\mathrm{N}}\beta_{\mathrm{N}}\mathbf{I}\cdot\mathbf{H} \tag{27}$$

and the corresponding energy levels are

$$E_{\mathrm{N}} = -g_{\mathrm{N}}\beta_{\mathrm{N}}HM \tag{28}$$

where β_{N} is the nuclear magneton, g_{N} the nuclear g factor, and M is the magnetic quantum number.

Since M can take the values

$$M = I, I - 1, \ldots, -I \tag{29}$$

the spectrum will be split into $(2I + 1)$ equally spaced levels. The splitting between adjacent levels is $g_{\mathrm{N}}\beta_{\mathrm{N}}H$ and the splitting between the lowest and the highest level is $2g_{\mathrm{N}}\beta_{\mathrm{N}}HI$ for a given I. Figure 12 illustrates the magnetic transitions in the Mössbauer spectrum of ^{57}Fe. In this Mössbauer spectrum, we observe γ-transitions between two nuclear levels, both of which exhibit magnetic hyperfine splitting*. Thus, as with the two other Mössbauer hyperfine interactions, the magnetic splitting is expressed by the product of a nuclear term, $g_{\mathrm{N}}\beta_{\mathrm{N}}$, and an electronic one, H.

The effective magnetic field acting on the nucleus arises from the electrons of the atom itself and is usually called the internal field. The

* The selection rules for these transitions between magnetic levels of nuclear states as well as the transition probabilities and angular dependence are described in Wertheim's book[27].

main contribution to this field is given by the Fermi contact interaction, i.e direct coupling between the nucleus and the unpaired *s*-electron density[45],

$$H_s = -\frac{16\pi}{3}\beta\langle\sum(\uparrow\psi_s^2(0) - \downarrow\psi_s^2(0))\rangle \quad (30)$$

where β is the Bohr magneton and $\uparrow\psi_s(0)$ is the spin density at the nucleus of *s* electrons of spin parallel to the *d* shell and $\downarrow\psi_s(0)$ those with spin antiparallel to the *d* shell. Other contributions arising from the interactions between the electron and nuclear spins and the electron angular momentum are usually smaller[46].

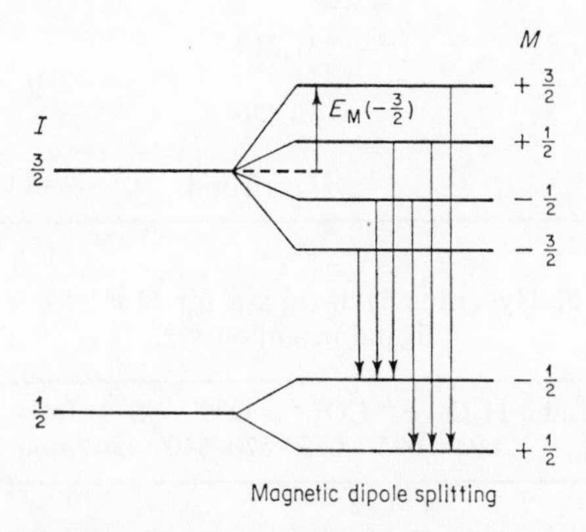

Figure 12. Magnetic hyperfine splitting of the ground and excited state in a $I = \frac{3}{2}$ to $\frac{1}{2}$ nuclear transition.

It is important to observe that unpaired spin densities appear even in filled *s* shells of transition ions. This is due to the fact that the exchange interaction between the spin-up polarized *d* shell and the spin-up *s* electron is attractive, while that between the *d* shell and a spin-down *s* electron is repulsive. As a consequence, the radial parts of the two *s* electrons are distorted, one being closer to the nucleus and the other being more distant.

Watson and Freeman[47] calculated the total unpaired *s* density arising from this exchange-interaction mechanism for several transition ions. The result of these calculations for the individual *s* electron in kilogauss to the internal field are illustrated for the Mn^{2+} ion in Table 4. The value obtained is in remarkably good agreement with the results obtained by e.s.r. with the more ionic Mn^{2+} salts but, as is illustrated in Table 5, covalent bonding strongly affects this result[48].

Table 4. Individual *s*-electron contribution (in kG) to the internal field; ↑ denotes electrons with spin parallel (and ↓ antiparallel) to the 3*d* subshell spin[47].

Electron	Contribution	$ns\uparrow + ns\downarrow$
1*s*↑	2,502,840	
		~30
1*s*↓	−2,502,870	
2*s*↑	226,670	
		−1400
2*s*↓	−228,080	
3*s*↑	31,210	
		+740
3*s*↓	−30,470	
	Total	~690 kG

Table 5. Hyperfine fields (in kG) for Mn^{2+} for various ligand neighbours[48].

Ligand	H_2O	F^-	CO_3^{2-}	O^{2-}	S^{2-}	Se^{2-}	Te^{2-}
H_i	695	695	665	570–640	490	460	420

The hyperfine structure of Mössbauer spectra depends essentially on the magnetic state of the sample: paramagnetic, ferromagnetic, or antiferromagnetic. With ferro-, ferri-, and antiferromagnetic materials, the electronic spin–spin coupling is much greater than the nuclear Zeeman coupling. As a consequence, the nuclear spin interacts with the average value of the internal field and the Mössbauer spectrum shows magnetic hyperfine structure. For paramagnetic samples, it is important to consider the time t which elapses between two successive flips of the electronic spin. These flips can be due to electronic relaxation or, above the Curie point, to exchange coupling between neighbouring electron spins. If the frequency $1/t$ of the flips is much larger than the Larmor frequency of the nuclear spin in the internal field, the hyperfine structure is absent, since the mean value of the internal field seen by the nucleus averages to zero. In contrast, if the frequency of the spin flips is comparable to or smaller than the frequency of the nuclear spin, the spectrum will show a magnetic hyperfine pattern. Examples of the latter situation have been reported for the Fe^{3+} ions diluted in corundum[49] and in various glasses[50], in some trivalent iron compounds[51], and also in rare earth Mössbauer spectra[52]. For S state ions such as Fe^{3+}, the orbital angular momentum is rigorously zero and for this reason they give relatively long spin-lattice relaxation times. Moreover, these times become longer at lower temperature. The first systematic investigation of the hyperfine splitting of ^{57}Fe in a paramagnetic material was made with highly diluted Fe^{3+} ions in α-Al_2O_3

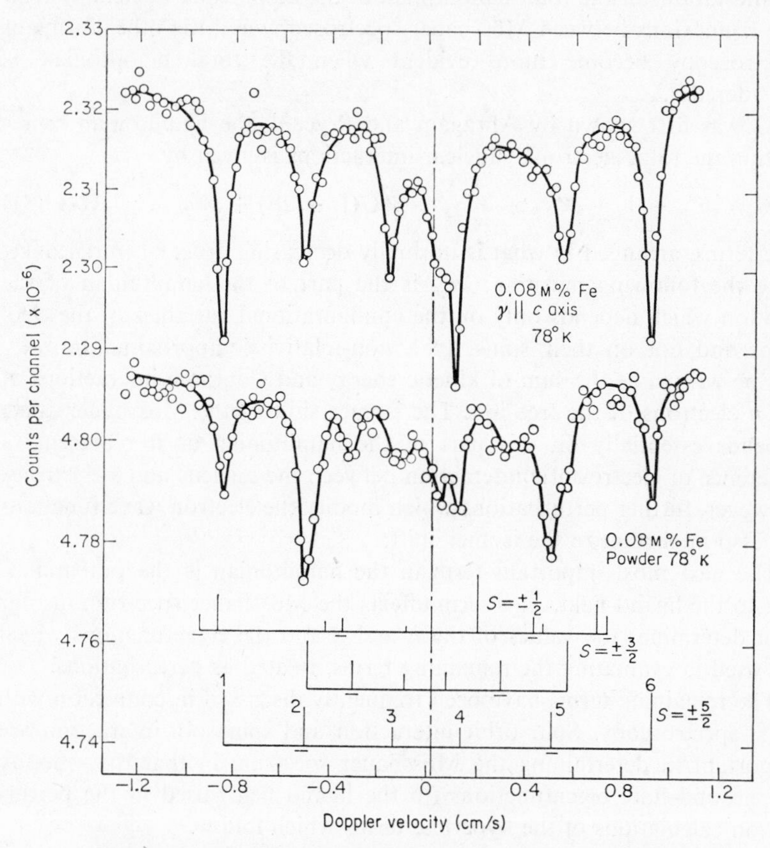

Figure 13. The hyperfine splitting of ^{57}Fe in corundum[49].

(corundum). The six ^6S states of Fe^{3+} will form three Kramers doublets $S_z = \pm\frac{1}{2}$, $\pm\frac{3}{2}$, and $\pm\frac{5}{2}$. The hyperfine structures corresponding to slow electron spin relaxation in these three crystal-field states are apparent in Figure 13, obtained with a single crystal of α-Al_2O_3, 0.08% M in ^{57}Fe at 78°K and with the same material in the polycrystalline state. These hyperfine splittings are observable only if the iron is sufficiently dilute so that spin–spin interaction remains unimportant. There has been an increasing interest in this type of phenomenon, which provides an opportunity for the study of the spin-relaxation times in solids and molecules.

D. Combined Hyperfine Interactions

We have discussed the Mössbauer parameters as isolated energy terms but this is not the most general view, since the splittings of the levels are manifestations of the total hamiltonian of the atomic and nuclear system. The connexions between Mössbauer spectroscopy and the other forms of spectroscopy become more evident when the total hamiltonian is considered[53].

As was first shown by Abragam and Pryce[54], the hamiltonian representing the total electronic–nuclear interactions is given by

$$\mathscr{H} = \mathscr{H}_{\mathrm{F}} + V + \mathscr{H}_{LS} + \mathscr{H}_{SS} + \beta\mathbf{H}\cdot(\mathbf{L} + 2\mathbf{S}) + \mathscr{H}_{\mathrm{N}} - \gamma\beta_{\mathrm{N}}\mathbf{H}\cdot\mathbf{I} \quad (31)$$

The terms, arranged in what is normally decreasing order of importance, have the following meaning: $\mathscr{H}_{\mathrm{F}}$ is the part of the hamiltonian of the free ion which depends only on the configurational variables of the electrons and not on their spins. In a non-relativistic approximation, $\mathscr{H}_{\mathrm{F}}$ can be written as the sum of kinetic energy and Coulomb interactions of the n electrons of the free ion. The isomer shift of the Mössbauer effect depends essentially on this part of the hamiltonian as it represents a difference of electrostatic interaction between the nucleus and s electrons. However, further perturbations which modify the electron wave functions are also important to the isomer shift.

The next most important term in the hamiltonian is the potential V due to the ligand field. This term affects the Mössbauer spectrum insofar as it determines the values of the L and S and the eigenfunctions which are used in evaluating the remaining terms, treated as perturbations.

The remaining terms have been frequently discussed in connexion with e.s.r. spectroscopy. Spin–orbit interaction and spin–spin interaction are important in determining the Mössbauer spectrum in that they modify the ground-state eigenfunctions (in the ligand field) used in the perturbation calculations of the hyperfine terms which follow.

$\mathscr{H}_{\mathrm{N}}$ represents the interaction of the nuclear spin with the orbital and spin electronic moments and the Fermi contact term which give rise to effective magnetic field measured by the Mössbauer effect. $\mathscr{H}_{\mathrm{N}}$ includes also the electrostatic interaction with the electric quadrupole moment of the nucleus which, although being a small perturbation of the ground-state eigenfunctions, is important for Mössbauer spectra because of the electric field gradient it represents.

Finally, the terms in **H** represent the interaction with an external magnetic field. As shown in König's chapter, since Abragam and Pryce it has been customary to sum up all these contributions in a first-order perturbation hamiltonian called the 'spin-hamiltonian'. This is the current way of handling the results from electron spin resonance spectroscopy. An illustrative example of the use of the spin-hamiltonian for interpreting Mössbauer spectra has recently been given by Lang and Marshall for iron–porphyrin systems[55].

V. APPLICATIONS TO CHEMISTRY

The possible Mössbauer elements are shown in Figure 14; the basic limitation for the extension to other elements arises from the uncommon occurrence of low-energy γ-transitions. The effect has been measured by

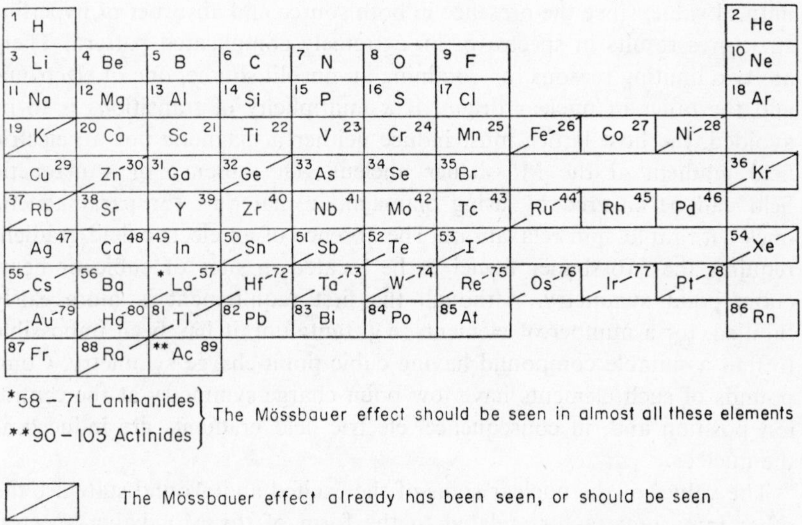

1 H										2 He
3 Li	4 Be	5 B	6 C	7 N	8 O	9 F				10 Ne
11 Na	12 Mg	13 Al	14 Si	15 P	16 S	17 Cl				18 Ar
19 K	20 Ca	Sc 21	Ti 22	V 23	Cr 24	Mn 25	Fe 26	Co 27	Ni 28	
Cu 29	Zn 30	31 Ga	32 Ge	33 As	34 Se	35 Br				36 Kr
37 Rb	38 Sr	Y 39	Zr 40	Nb 41	Mo 42	Tc 43	Ru 44	Rh 45	Pd 46	
Ag 47	Cd 48	49 In	50 Sn	51 Sb	52 Te	53 I				54 Xe
55 Cs	56 Ba	* La 57	Hf 72	Ta 73	W 74	Re 75	Os 76	Ir 77	Pt 78	
Au 79	Hg 80	81 Tl	82 Pb	83 Bi	84 Po	85 At				86 Rn
87 Fr	88 Ra	** Ac 89								

*58 – 71 Lanthanides
**90 – 103 Actinides
The Mössbauer effect should be seen in almost all these elements

(boxed diagonal) The Mössbauer effect already has been seen, or should be seen

Figure 14. Possible Mössbauer elements. [After Goldanskii[6], with more recent data.]

scattering techniques up to 155 kev with ^{188}Os, and this appears to be the limit of detectability, corresponding to an f factor as small as 0.0063. The low-energy γ-transitions occur with nuclei at the middle of the periodic table and, more frequently, with those of the heavy elements. However, it is important to realize that, for these low-energy transitions, internal conversion considerably decreases the intensity of the γ-emission. With ^{73}Ge, which has a favourable γ-transition (13.5 kev), the Mössbauer effect is difficult to detect, since the internal conversion factor is very high ($\alpha < 3600$).

The lifetime of the excited state is of great importance in possible applications of the Mössbauer effect. For short lifetimes, the ratio of level width to the energy of the transition, Γ/E_0, is large and the resonance selectivity is lost. This occurs, for instance, with the 99 kev transition ^{195}Pt, the half-life of the excited state being 0.16 nanosecond, but the linewidth is larger than the characteristic values for the hyperfine interactions[56]. For long lifetimes of the nuclear transition, the ratio Γ/E_0 is very small and hence the probability of the Mössbauer effect is low. Moreover, the very narrow linewidths make observations of the effect difficult. The optimum range of Γ/E_0 is 10^{-10} to 10^{-14}, although the effect has already been observed outside these limits.

For practical reasons, it is very important to use a source emitting an unsplit Mössbauer transition with a linewidth as close as possible to the natural value, since the presence in both source and absorber of hyperfine structures results in spectra of an extremely complicated pattern. There are two limiting reasons for obtaining an unsplit source, one of electronic and the other of nuclear origin. If a multiplicity of transitions is to be avoided, the host lattice must induce neither a magnetic nor an electric field gradient at the Mössbauer nucleus. The absence of a magnetic field can be ensured by using diamagnetic materials (or paramagnetic ones with rapid spin relaxation). The absence of an electric field gradient requires the Mössbauer nuclei to be located in sites of cubic or near-cubic point symmetry. Although the first requirement is more easily fulfilled, for a number of elements, e.g. tantalum, it has been impossible to find a suitable compound having cubic point-charge symmetry. Compounds of such elements have low point-charge symmetry at the central ion position and, in consequence, electric field gradients are induced at the nucleus.

The values of the nuclear spins of the excited and ground state are the other important factors related to the form of the Mössbauer spectra. The multiplicity of lines of the hyperfine structure increases with the value of the nuclear spins. Thus, even in the absence of large hyperfine fields, Mössbauer transitions involving nuclei with large spin values will give rise to a broadening of the linewidth. This, for example, is the case for ^{237}Np which gives an intense Mössbauer effect[57] due to the 59.6 kev transition between a $\frac{5}{2}^-$ excited level to $\frac{5}{2}^+$ ground state. The usefulness of the spectrum, however, is reduced by the fact that the observed linewidth is about 30 times the natural value.

Possible complications can be introduced in the spectrum by the radioactive decay process which necessarily precedes the Mössbauer γ-transition. The Mössbauer nucleus may be formed by β-decay (as in $^{61}Cu \xrightarrow{\beta} {}^{61}Ni$), K capture (as in $^{57}Co \xrightarrow{K} {}^{57}Fe$), γ-transition (as in $^{119}Sn^* \rightarrow {}^{119}Sn$), or even α-decay (as in $^{241}Am \xrightarrow{\alpha} {}^{237}Np$). As a consequence of these various modes of radioactive decay, the Mössbauer nucleus may be formed in a variety of oxidation states or chemical forms of varying stability. The nature and mechanism of formation of these states inside the solids is still largely unknown. The important question, however, is the time for which they survive in the solid. It has been shown that in metals and some semiconductors, the lifetimes for the higher oxidation states are less than the lifetime of the Mössbauer emission. Thus, for ^{57}Co the mixture of chemical forms which are almost certainly produced following the *K* capture survive for a time less than about 10^{-7}s. However, evidence for relatively long-lasting effects was obtained with CoO[58] and SnO_2[59] and more recently for hydrated salts such as ferrous ammonium sulphate[60,61]. The existence of these effects clearly limits the choice of the emitter, for which the greatest possible uniformity of the Mössbauer atoms is required.

It is, however, important to note that the Mössbauer effect has been observed following such strong perturbations as α-decay in $^{241}Am \xrightarrow{\alpha} {}^{237}Np$ and the intense bombardment of the source in methods involving nuclear reactions. Thus, although the features of the Mössbauer spectra can be altered by preceding nuclear transitions, their effect is certainly less dramatic than has sometimes been suggested.

Another important question regarding the possible applications of the Mössbauer effect to chemical systems is the probability of resonant capture of the γ-radiation in complex lattices. The probability of the Mössbauer effect for a simple monoatomic lattice can be evaluated by means of equation (11). However, for a more complex lattice, e.g. having the Mössbauer atom bound in a lattice with much lighter atoms, such as in organometallic compounds, we should expect a negligible f factor, since these compounds have a very low Debye temperature. Nevertheless, it is important to recall that the Debye model is no longer valid for these complex lattices and it is possible to reach high f factors even with these light-atom lattices. This is particularly true in molecular crystals, where the vibrational properties of the molecule are largely preserved in the solid. The validity of these considerations is supported by the fact that the Mössbauer effect has been successfully applied to a large number of organic iron and tin compounds, most of which have an average atomic weight much lower than that of the Mössbauer atom[62].

We can summarize this section by observing that the possibility of chemical applications of the Mössbauer effect has been well demonstrated with ^{57}Fe, ^{119}Sn, ^{129}I, ^{129}Xe, ^{197}Au, ^{83}Kr, ^{151}Eu and with some more recent work on ^{133}Cs, ^{40}K, and ^{125}Te.

VI. EXAMPLES OF APPLICATIONS IN INORGANIC CHEMISTRY

A. Symmetry and Structure of Molecules

The Mössbauer parameters relevant to chemical investigations are the linewidth, the magnitude of the resonance effect, the isomer shift, the quadrupole splitting, the magnetic hyperfine interaction, the line asymmetry, and temperature coefficients of these parameters. Of these parameters, the greatest amount of chemical information has been derived from quadrupole splitting and isomer shift measurements.

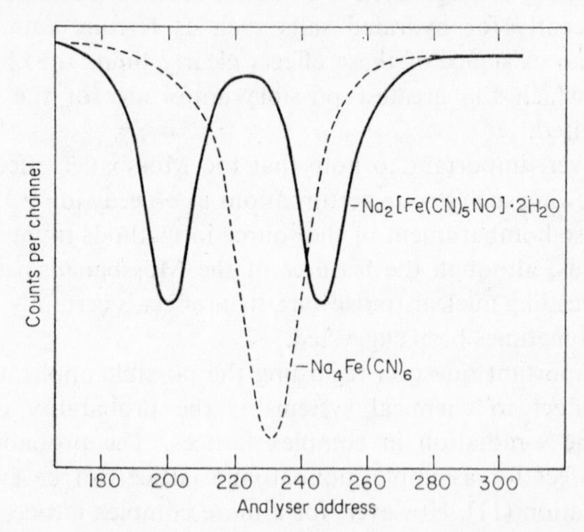

Figure 15. Mössbauer spectra of an octahedrally symmetric complex $Na_4[Fe(CN)_6]$ and of a distorted octahedral complex $Na_2[Fe(CN)_5NO]\cdot 2H_2O$, showing the effect of a non-vanishing electric field gradient at the Mössbauer lattice point.

The quadrupole splitting is directly related to the point symmetry of the local environment of the atom under observation. The absence of quadrupole splitting indicates a cubic or near-cubic symmetry; its presence indicates a significant distortion from cubic symmetry[24]. This is clearly exemplified by the spectra[63] presented in Figure 15. Sodium ferrocyanide, $Na_4[Fe^{II}(CN)_6]$, with nearly perfect octahedral symmetry around the iron atom, gives a characteristic single-line spectrum. The substitution of an axial CN by an NO ligand introduces a tetragonal distortion in the sodium nitroprusside molecule, $Na_2[Fe(CN)_5NO]\cdot 2H_2O$, which gives a well-resolved doublet in the Mössbauer spectrum. This splitting arises from the strong electric field gradient created by the molecular distortion at the site of the Fe atom. We shall analyse this important spectrum in more detail later.

The use of the quadrupole splitting for elucidating molecular structure is exemplified by the tin(IV) halide spectra[64,65]. In the $Sn^{IV}Hal_4$ series, where Hal = F, Cl, Br, and I, only $Sn^{IV}F_4$ shows a quadrupole splitting of the Mössbauer spectrum. The relatively large value of the quadrupole interaction, $\Delta E = 0.17–0.2$ cm/s, indicates that the point symmetry for $Sn^{IV}F_4$ is lower than in the other Sn^{IV} halides.

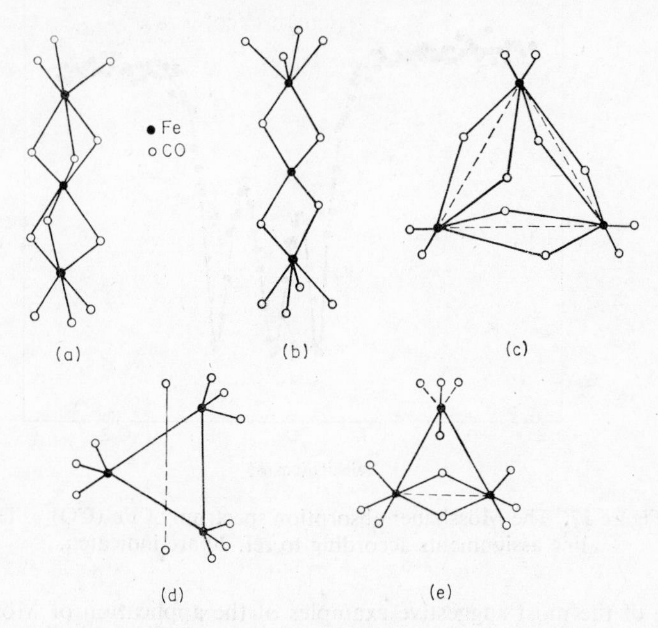

Figure 16. Some proposed structures for $Fe_3(CO)_{12}$[68].

The quadrupole splitting in the inorganic tin compounds can be interpreted by using the relationship between electric field gradient and atomic wave function. Undisturbed closed shells and subshells are spherically symmetrical and cannot contribute to the quadrupole interaction; similarly, the contribution of *s* electrons in outer closed shells is zero. It is only the asymmetric charge distribution of valence *p* electrons which produces large gradients at the nucleus[66]. Furthermore, when all three *p* orbitals are equally populated, their superposition forms a spherically symmetric electronic charge distribution. Thus, the absence of quadrupole splitting in the Sn^{IV} halides other than $Sn^{IV}F_4$ supports the sp^3 hybridized structure for these compounds with four equivalent bonds.

$Sn^{IV}F_4$ cannot be a tetrahedral molecular solid like the other Sn^{IV} halides. In fact, recent results[67] have shown that stannic(IV) fluoride has an octahedral structure containing four bridging fluorine atoms, whilst two further fluorine atoms are bound directly to each tin atom to complete the distorted octahedral environment (**D_4** symmetry). The quadrupole splitting in $Sn^{IV}F_4$ is thus a consequence of the inequivalence of the fluorine bonds, and the Mössbauer spectrum is supporting evidence for the polymeric structure of the compound[6].

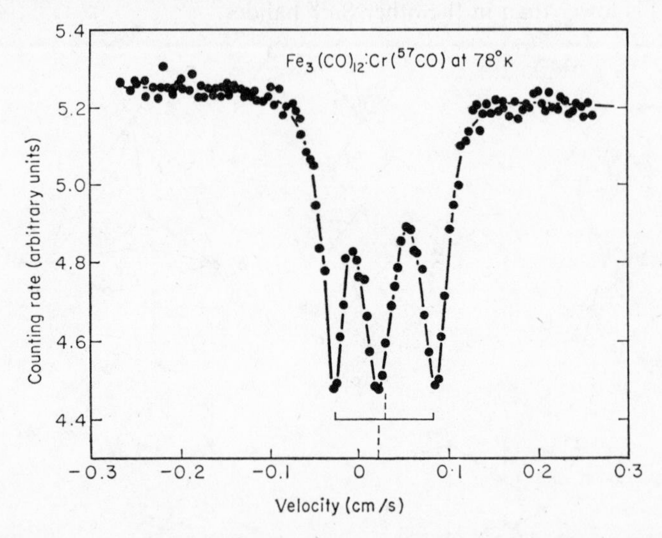

Figure 17. The Mössbauer absorption spectrum of $Fe_3(CO)_{12}$. The line assignments according to ref. 70 are indicated.

One of the most suggestive examples of the application of Mössbauer spectroscopy to molecular structure investigations is presented by the structure of triiron dodecacarbonyl, $Fe_3(CO)_{12}$. Six widely different models have been proposed for the structure of this molecule[68] (Figure 16), the Mössbauer spectrum of which[69,70] (Figure 17) proves that all three iron atoms are not equivalent because, if so, either a single line or a doublet, but not three lines, would be obtained. Any of the symmetrical triangular structures, such as shown in (c) or (d) in Figure 16, are therefore ruled out. Analysis of the Mössbauer spectrum shows that two iron atoms have equivalent positions with rather large field gradients, whilst the third iron atom is on a different site with no, or only a small, field gradient[69,70]. Recent x-ray work[71] has shown that the probable configuration for $Fe_3(CO)_{12}$ possesses ideally $\mathbf{C}_{2v}$–2 mm point symmetry and can be regarded as being formed by the insertion of a *cis*-$Fe(CO)_4$ group at one of the three bridging carbonyl positions of $Fe_2(CO)_9$ as in structure (e) of Figure 16. The Mössbauer spectrum of $Fe_3(CO)_{12}$ is consistent with this configuration in that the isomeric shift for the quadrupole-split peaks of

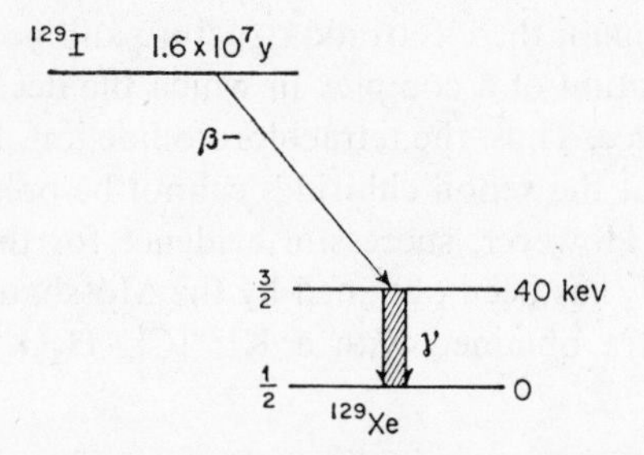

Figure 18. Decay scheme of $^{129}Xe_{54}$.

the two iron atoms of lower symmetry is nearly identical with the isomeric shift found for the iron atoms in $Fe_2(CO)_9$. Also, the other iron atom is in an essentially octahedral environment, and consequently any quadrupole splitting would be expected to be negligible.

The decay scheme for ^{129}Xe is shown in Figure 18. The Mössbauer effect of the 40 kev γ-transition may readily be observed by means of a variety of sources and absorbers in experiments at 4.2°K[72,73].

It is known that most β-decays cause no change in the electron number of the atom containing the decaying nucleus. In the β-decay $^{129}I \xrightarrow{\beta} {}^{129}Xe$

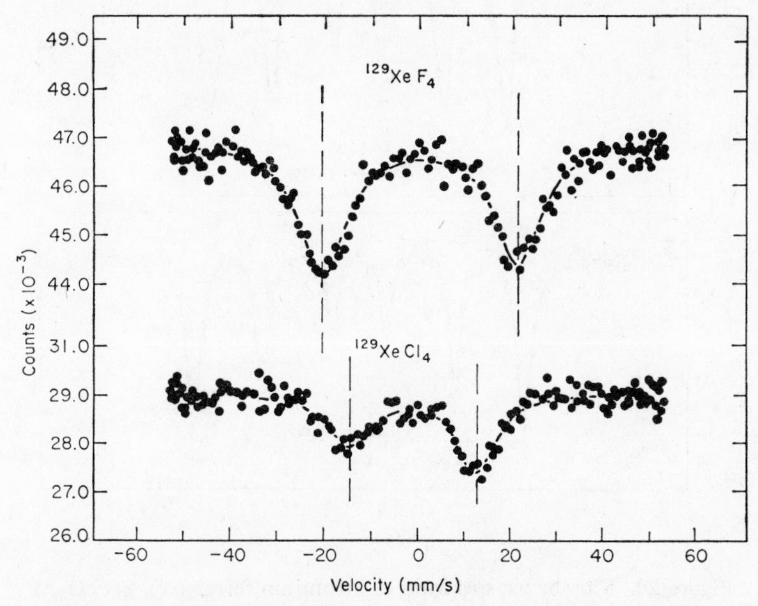

Figure 19. $XeCl_4$ produced in the decay of $^{129}ICl_4^-$. XeF_4 as absorber is shown for comparison[74].

of an iodine compound, the electronic structure and geometry are appropriate for the formation of a complex in which the decay product acts as the Mössbauer source. Thus, the tetrachloroiodide ion, ICl_4^-, has a square planar structure, but the xenon chlorides cannot be prepared by ordinary chemical methods. However, successful evidence for the existence of the square planar $XeCl_4$ has been obtained by the Mössbauer effect[74]. Figure 19 shows the spectra obtained with a $K^{129}ICl_4 \cdot H_2O$ source and ^{129}Xe

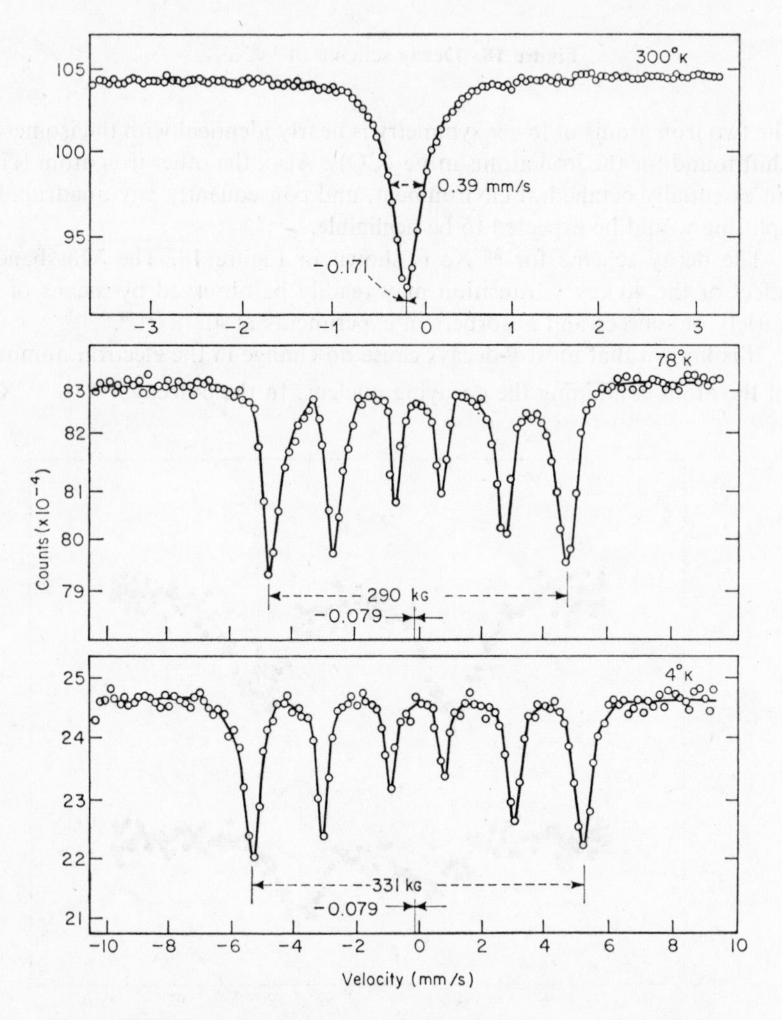

Figure 20. Mössbauer spectrum of strontium ferrate(IV), $SrFeO_3$, at different temperatures. Values of the hyperfine interactions are indicated[75].

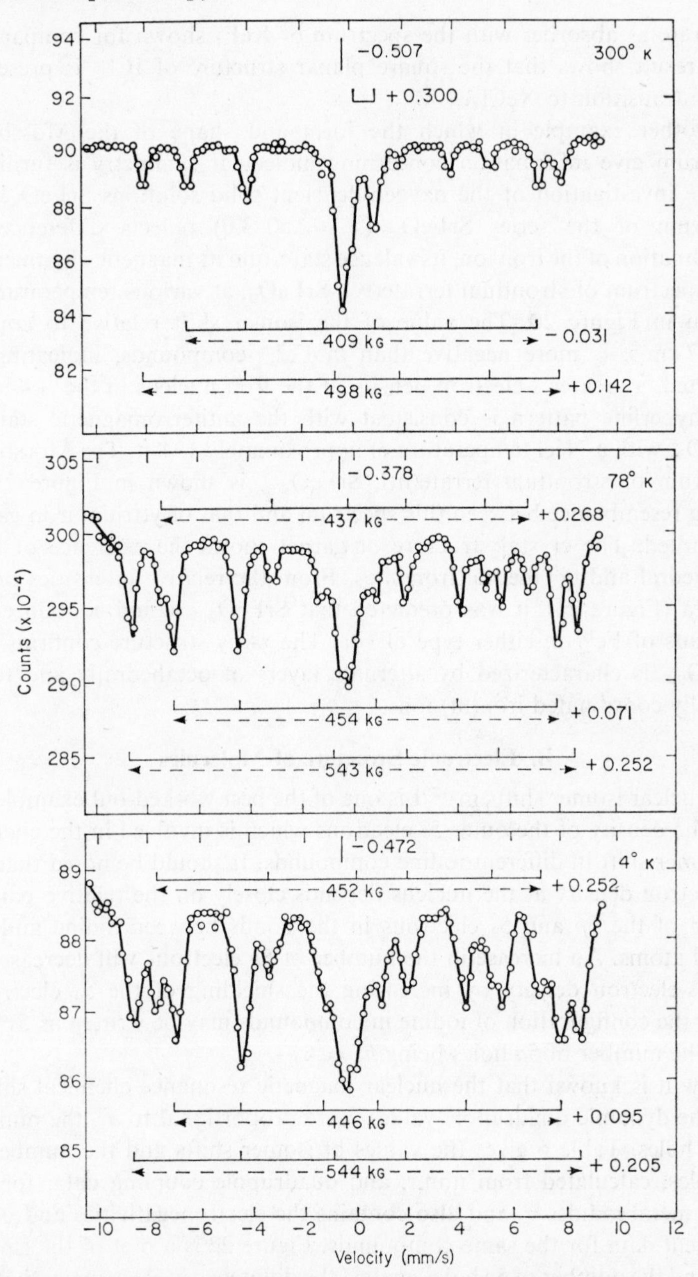

Figure 21. Mössbauer spectrum of strontium ferrate(III), $SrFeO_{2.5}$, at different temperatures. The hyperfine interactions for trivalent iron in two different sites are indicated[76].

clathrate as absorber with the spectrum of XeF_4 shown for comparison. This result shows that the square planar structure of ICl_4^- is preserved in the transition to $XeCl_4$.

Another example in which the form and shape of the Mössbauer spectrum give information concerning molecular symmetry is furnished by the investigation of the oxygen-deficient solid solutions $SrFeO_x$. The spectrum of the series $SrFeO_x$ ($x = 2.50$–3.0) reflects differences in coordination of the iron ion, its valence state, and its magnetic alignment[75]. The spectrum of strontium ferrate(IV), $SrFeO_3$, at various temperatures is shown in Figure 20. The value of the isomer shift relative to copper, -0.17 cm/s, is more negative than in Fe^{3+} compounds, indicating, as expected, a greater *s*-electron density at the iron nucleus in the $+4$ state. The hyperfine pattern is consistent with the antiferromagnetic state of $SrFeO_3$, with a Néel temperature of approximately 134°K. The Mössbauer spectrum of strontium ferrate(III), $SrFeO_{2.5}$, is shown in Figure 21. A strong resemblance between this spectrum and that of yttrium iron garnet was noted. The crystal structure of garnet shows the existence of both tetrahedral and octahedral iron sites. From the relative intensities of the spectra (Figure 21), it was predicted that $SrFeO_{2.5}$ would contain equal amounts of Fe^{III} in either type of site. The x-ray structure confirms that $SrFeO_{2.5}$ is characterized by alternate layers of octahedrally and tetrahedrally coordinated iron(III) ions.

B. Electronic Structure of Molecules

The nuclear isomer shifts in ^{129}I is one of the best worked-out examples[42]. It is the density of the outer 5*s* electrons which is involved in the changes of isomer shift in different iodine compounds. It should be noted that the 5*s*-electron density at the nucleus depends closely on the relative participation of the 5*p* and 5*s* electrons in the bonds between iodine and the ligand atoms. An increase in the number of 5*p* electrons will decrease the total *s*-electron density by increasing the shielding of the 5*s* electrons. Thus, the configuration of iodine in compounds may be written as $5s^25p^y$, with the number of 5*p* holes being $h_p = 6 - y$.

Now it is known that the nuclear magnetic resonance chemical shift[76] and the dynamic quadrupole values[77] are proportional to h_p, the number of 5*p* holes. Table 6 gives the values of isomer shifts and the number of 5*p* holes, calculated from n.m.r. and quadrupole coupling data, for the alkali metal iodides[42], and also contains the electronegativities and other pertinent data for the same compounds. Figure 22 is a plot of the isomer shift and the number of 5*p* holes against the difference in electronegativities. It can be seen that the behaviour of h_p is similar to that of the isomer shift which shows that the 5*s* electron density as measured by the isomer shift

Table 6. The alkali metal iodide isomeric shifts relative to ZnTe; number of iodine *p* holes calculated from chemical shift and dynamical quadrupole resonance; the alkali metal iodine electronegativity difference and the fractional deviation of the iodide ion 5*s* density from the $5s^25p^6$ closed-shell iodide ion density.

Compound	Isomer shift	Number of *p* holes		Electro-negativity difference	$\frac{\Delta\|\psi_{5s}(0)\|^2}{\|\psi_0(0)\|^2}$
		From chem. shift	Dynamic quadrupole		
LiI	−0.038 ± 0.0025	0.112	—	1.5–1.55	+0.014
NaI	−0.046 ± 0.0025	0.040	0.085	1.6	+0.0075
KI	−0.051 ± 0.0025	0.033	0.035	1.7	+0.0034
RbI	−0.043 ± 0.0025	0.055	0.070	1.7	+0.010
CsI	−0.037 ± 0.0025	0.165	0.165	1.75–1.8	+0.015
ZnTe	0	—	—	0.9–1.0	+0.046

is linearly dependent on the number of 5*p* holes. This important result has been used for calibrating the isomer shift in terms of relative 5*s* density. Table 6 gives the fractional deviation of the iodide ion 5*s* density from the $5s^25p^6$ closed-shell iodide density computed for the alkali metal iodides from the isomer shift data.

It is important to observe that a simple theory of electronegativity cannot interpret the results from Figure 22. Such a theory, based on electron-transfer energy, necessarily predicts a monotonic behaviour of the isomer shift along the series and cannot explain the serious deviation observed with RbI and CsI. Recently, a method has been used to compute the isomer shift[78] in alkali metal iodides starting with the free-ion wave functions and deforming these functions by overlap and electrostatic effects. The conclusions are that the isomer shift is primarily caused by the overlap deformation of the free-ion wave functions and calculated relative shifts agree well with the experimental values.

The analysis of the isomer shift in iron complexes has provided one of the best pieces of evidence for the mechanism of back donation in low-spin iron complexes[79,80]. The observed values of the isomer shift with respect to stainless steel in $K_4Fe(CN)_6 \cdot 3H_2O$ and $K_3Fe(CN)_6$ are $83 \pm 10 \times 10^{-4}$ cm/s and $0 \pm 10 \times 10^{-4}$ cm/s, respectively[81]. The values for typical ionic compounds like $FeSO_4 \cdot 7H_2O$ and $Fe_2(SO_4)_3 \cdot 6H_2O$ are $140 \pm 5 \times 10^{-4}$ cm/s and $52 \pm 5 \times 10^{-4}$ cm/s, with respect to the same source[82]. As was observed by Shulman and Sugano[83], the similarity between ferro- and ferricyanides is most impressive compared to the Fe^{2+} and Fe^{3+} shifts in the more ionic compounds.

It was also observed that the usual tendency of the isomer shift, indicating more *s*-electron covalency with decreasing ligand electronegativity, is not followed in low-spin iron complexes with ligands containing empty π-bonding orbitals[84]. For these complexes, the *s*-electron density at the Fe nucleus increases with increasing ligand electronegativity, as for example along the isoelectronic series $CN^- < CO < NO^+$. This tendency is so

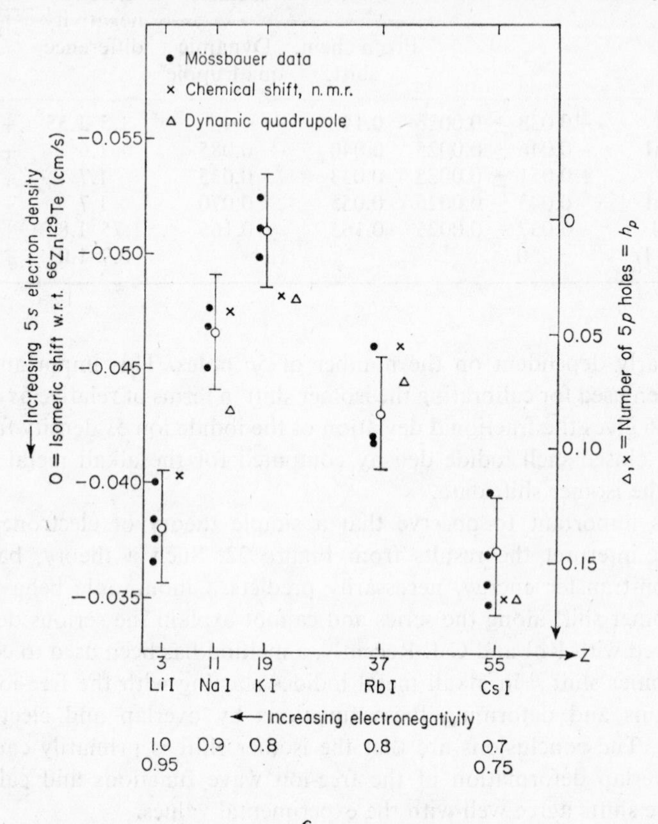

Figure 22. The isomer shift (○) and the number of iodine *p* holes h_p (× and △) versus alkali atomic number for the alkali metal iodides[42].

strong that the isomer shift of iron in sodium nitroprusside, $Na_2[Fe(CN)_5NO]\cdot 2H_2O$, corresponds to the largest *s*-electron density observed among the iron compounds of common oxidation state. A correlation has been proposed[84] between the isomer shift and the nephelauxetic effect, which is the decrease of the parameters of interelectronic repulsion in transition ion complexes compared with the corresponding gaseous ion. This decrease corresponds to an expanded radial function of the partly filled shell[85].

The nephelauxetic series of ligands usually corresponds to their increasing polarizability, decreasing electronegativity, and increasing tendency to covalent-bond formation: $F^- < H_2O < NH_3 < Cl^- < CN^- < Br^- < S^{2-}$. In high-spin ferrous and ferric compounds, it has been frequently found that the isomer shift has the same ordering as the nephelauxetic series[84]. However, this is not true when spin-paired iron complexes are considered, e.g. with the CN ligand, since the order for isomer shifts has been found to be $CN^- > Br^- > S^{2-}$.

In order to analyse the contributions to the total *s*-electron density of iron in its compounds, let us begin with the information available for the free iron ion in different configurations. The free iron ion has the $1s + 2s + 3s$ shells filled with paired electrons. The sum $\sum_1^3 \psi_{is}^2(0)$ can be evaluated for different d^n configurations of iron using the restricted Hartree–Fock wave functions calculated by Watson[86] or the more recent calculations of Clementi[87]. These calculations show that the total *s*-electron density is different for different d^n configurations of the free ion. This notable finding has been interpreted by Walker, Wertheim, and Jaccarino[82] as a consequence of the shielding of the 3*s* electrons by the 3*d* electrons, since the removal of a 3*d* electron *increases* the total *s*-electron density at the nucleus.

The isomer shift of the free ion will be modified by bonding with ligands. The possible mechanisms for this modification are: (*a*) the increase of 4*s* density resulting from the partial occupation of outer orbitals of iron by electrons from the ligands; or (*b*) an indirect contribution from 3*d* bonding, which may be due to two causes, viz.

1. Covalency effects between *d* electrons and filled ligand orbitals, which would increase the number of *d* electrons on the metal. The increase in *d*-electron density can decrease the *s*-electron density, if we assume that the same shielding mechanism observed with the free ion operates with these complexes.

2. The bonding of the electrons with empty ligand orbitals would decrease the *d*-electron density at the metal ion because of back donation. The delocalization of *d* electrons through π bonding tends to decrease the shielding of 3*s* electrons giving rise to an increased electron density at the nucleus.

As mentioned above, the results suggest a marked difference in the bonding mechanism which determines the isomer shift in the two types of iron complexes. In high-spin complexes, the contribution from bonding to the total *s*-electron density is mainly due to the partial occupation of the 4*s* orbitals of the central ion by electrons from the ligands. This tendency is significantly overcome in low-spin complexes which have empty π orbitals on the ligands such as CN^- and withdraw 3*d* electrons from the iron by back donation. In consequence, the 3*s* electrons are less shielded and the total *s*-electron density at the iron nucleus increases with the tendency of the ligand to *withdraw* electrons from the central ion. The back-donation mechanism is thus the main cause for the variations of isomer shift in low-spin iron complexes with empty π orbitals.

This result explains why the *s*-electron density increases in the order of increasing ligand electronegativity in iron complexes with CN^-, CO, and NO^+, since this is the order for d_π-electron delocalization in the metal–ligand bonds.

The break in the parallelism between the nephelauxetic effect and the isomer shift when the low-spin cyanide complexes are considered can be explained by taking into account that the delocalization of the d_π electrons toward the ligands affects essentially the inner 3*s* electrons. The *s*-electron density at the iron nucleus is thus increased by a mechanism connected with the inner filled electron shells. The latter do not manifest themselves in the nephelauxetic effect, which is dependent on the orbitals involved in chemical bonding.

Shulman and Sugano presented a semiempirical molecular-orbital analysis of the ferro- and ferricyanides which correlates Mössbauer, electron spin resonance, and optical absorption experiments and explains the similar isomer shifts for iron in these complexes[83]. Their basic idea is to calculate the effective number of *d* electrons which remain in the iron atom after bonding. This number can be expressed as

$$n_{\text{eff}} = n_1 + n_2 - n_3 \tag{32}$$

where n_1 is the number of *d* electrons in the limiting ionic case, n_2 the electrons which came from ligand–metal bonding, and n_3 the electrons removed from the metal by back donation.

The three molecular orbitals which are formed from the d_{xy}, d_{xz}, and d_{yz} metal orbitals and the π bonding and π^* antibonding orbitals of CN can be written with the usual LCAO approximations

$$\begin{aligned} \Psi_1 &= N^{-\frac{1}{2}}(\chi_1 + \gamma_1\varphi) \\ \Psi_2 &= N^{-\frac{1}{2}}(\varphi - \lambda_1\chi_1 + \gamma_2\chi_2) \\ \Psi_3 &= N^{-\frac{1}{2}}(\chi_2 - \lambda_2\varphi) \end{aligned} \tag{33}$$

where χ_1 and χ_2 are linear combinations of π and π^* cyanide orbitals and

φ is the metal d orbital. Mutual orthogonality between these molecular orbitals requires the following relations:

$$\lambda_1 = \gamma_1 + S_1, \quad \lambda_2 = \gamma_2 + S_2$$

where

$$S_1 = (\varphi \mid \chi_1), \quad S_2 = (\varphi \mid \chi_2) \tag{34}$$

The effective value of the orbital angular momentum in a complex is reduced by covalency, and Stevens introduced the reduction factor k, defined by

$$k = (\psi_t|l|\psi_t)/(\varphi_t|l|\varphi_t) \tag{35}$$

This covalency parameter can be expressed as a function of the coefficients of χ_1 and χ_2 in (33) by the approximate relation

$$k \simeq 1 - \tfrac{1}{2}(\lambda_1^2 + \gamma_2^2) \tag{36}$$

If n_e and n_t represent respectively the number of holes in the e_g^* and t_{2g}^* antibonding orbitals we have

$$n_2 = n_e\gamma_e^2 + n_t\gamma_t^2 \tag{37}$$

where γ_e and γ_t are the coefficients of $3d$ orbitals in bonding e_g and t_{2g}, respectively.

The number of electrons which move to the ligands, n_3, is given by

$$n_3 = (6 - n_t)\gamma_2^2 \tag{38}$$

Values of k are available for $Fe(CN)_6^{3-}$ and $Mn(CN)_6^{4-}$ from the analysis of electron spin resonance spectra of these complexes. These values are

$$\begin{aligned} k &= 0.87 \quad \text{for} \quad Fe^{III}(CN)_6^{3-} \\ k &= 0.74 \quad \text{for} \quad Mn^{II}(CN)_6^{4-} \end{aligned} \tag{39}$$

Introducing these values into (36) gives

$$(\lambda_1^2 + \gamma_2^2)_{2+} = (\lambda_1^2 + \gamma_2^2)_{3+} + 0.26 \tag{40}$$

Since it is possible to assume in general that

$$|\lambda_1^2|_{2+} \simeq |\lambda_1^2|_{3+} \tag{41}$$

we find after substituting (41) into (40) and into (38)

$$n_3(Mn^{2+}) - n_3(Fe^{3+}) \sim 1.3 \tag{42}$$

The same result is obtained by assuming $\lambda_1 \simeq \gamma_2$ for both the $+2$ and $+3$ complexes and that $|\gamma_2^2|_{2+}$ is the same for Mn^{II} and Fe^{II}. Making these approximations in (40) and taking the values of k from (39), we find

$$n_3(Fe^{2+}) - n_3(Fe^{3+}) \sim 0.9 \tag{43}$$

This shows that, quite generally, we have a difference of approximately one electron between the back donation of Fe^{2+} and Fe^{3+}. Since the values of n_2 in Fe^{2+} and Fe^{3+} must be similar[83], we find from (32) that

$$n_{eff}(Fe^{2+}) \simeq n_{eff}(Fe^{3+}) \tag{44}$$

The iron ion is approximately isoelectronic in these complexes and this explains why no difference in isomer shift is observed between ferro- and ferricyanide complexes.

A remarkable difference was observed in quadrupole splitting between ionic compounds of the two common oxidation states of iron[88]. For high-spin Fe^{III}, the splitting values range from 0.03 to 0.07 cm/s, whereas higher values are observed for Fe^{II}, ranging from 0.11 to 0.36 cm/s. The splittings in Fe^{II} are, in general, temperature dependent and vary from compound to compound[89].

These results have been interpreted in terms of crystal-field theory[88,90]. In an electric field produced by an octahedron of negative charges around the transition ion, the five-fold spatial degeneracy of the orbital $3d$ wave functions is lifted. The value of the electric field gradient produced by the different $3d$ wave functions at the nucleus[89] is listed in Table 7.

Table 7. Values of field gradient q and asymmetry parameter η for $3d$ electrons.

Orbital		q	η
d_γ	$d_{x^2-y^2}$	$+\frac{4}{7}\langle r^{-3}\rangle$	0
	d_{z^2}	$-\frac{4}{7}\langle r^{-3}\rangle$	0
d_ϵ	d_{xy}	$+\frac{4}{7}\langle r^{-3}\rangle$	0
	d_{xz}	$-\frac{2}{7}\langle r^{-3}\rangle$	+3
	d_{yz}	$-\frac{2}{7}\langle r^{-3}\rangle$	−3

The Fe^{III} ion has an outer electron configuration of $3d^5$ and is in a $^6S_{\frac{5}{2}}$ ionic state. The half-filled $3d$ subshell forms a spherically symmetric charge distribution and does not therefore contribute to the electric field gradient. The small splittings observed with some ionic ferric salts are probably due to field gradients produced by the ionic charges in the lattice.

The Fe^{II} ion has an outer electron configuration of $3d^6$ and is in a ^5D state. The main contribution to the field gradient at the nucleus is made by an extra electron with spin antiparallel to the other five. The orbital which this electron will occupy depends on the deviation from cubic symmetry of the crystal field. Axial and rhombic fields lift the degeneracy within the d_γ and d_ε shells and further splitting of the energy levels occurs by spin–orbit coupling as shown in Figure 23. The relative population

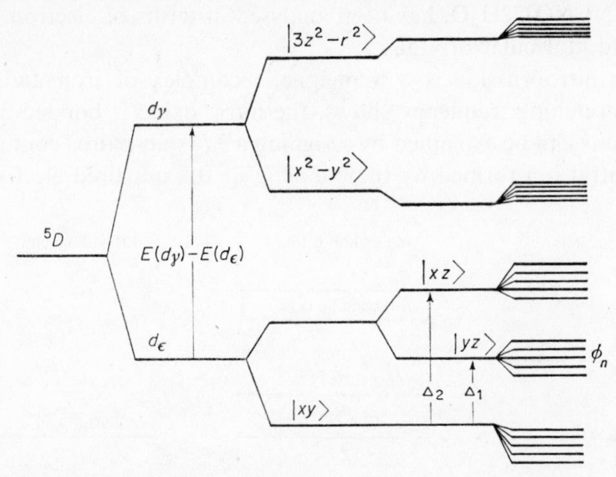

Figure 23. Energy-level scheme for the ferrous ion under the action of the crystalline field plus spin–orbit coupling[89].

of these levels accounts for the temperature dependence of the quadrupole splitting. Covalent bonding contributes to the decrease of the quadrupole splitting by expanding the radial part of the 3*d* wave function. Reasonable estimates of the energy splitting of the d_ϵ orbitals by axial and rhombic fields and of the ground-state orbital wave function in several iron(II) compounds have been obtained by Ingalls[89] from the temperature dependence of the quadrupole splitting, taking into account spin–orbit coupling and covalency factors (Table 8).

Table 8. Parameters used in describing the ^{57}Fe quadrupole splitting observed in several ferrous compounds.

Compound	α^2	Δ_1 (cm^{-1})	Δ_2 (cm^{-1})	Ground-state orbital wave function
$FeSiF_6 \cdot 6H_2O$	0.80	760	760	$\|3z^2 - r^2\rangle$
$FeSO_4 \cdot 7H_2O$	0.80	480	1300	$\|xy\rangle$
$Fe(NH_4SO_4)_2 \cdot 6H_2O$	0.80	240	320	$\|xy\rangle$
$FeC_2O_4 \cdot 2H_2O$	0.80	100	960	$\|xy\rangle$
$FeSO_4$	0.80	360	1680	$\|x^2 - y^2\rangle + 0.09\,\|3z^2 - r^2\rangle$
$FeCl_2 \cdot 4H_2O$	0.80	750	2900	$\|x^2 - y^2\rangle + 0.10\,\|3z^2 - r^2\rangle$
FeF_2	0.67	1000	2200	$\|x^2 - y^2\rangle + 0.14\,\|3z^2 - r^2\rangle$

The interpretation of the quadrupole splittings in low-spin iron complexes is possible when a detailed knowledge of the molecular-orbital bonding scheme is available. This is the case for transition metal–nitrosyl complexes of the form MX_5NO, and the quadrupole splitting of sodium nitroprusside, $Na_2[Fe(CN)_5NO]\cdot 2H_2O$, has been analysed in terms of electron delocalization and molecular orbitals.

Sodium nitroprusside is a diamagnetic complex of iron and, as the infrared stretching frequency shows, the nitric oxide is bonded as NO^+. These results can be explained by assigning a $3d^6$ spin-paired configuration of the central ion formed by the coupling of the unpaired electron from

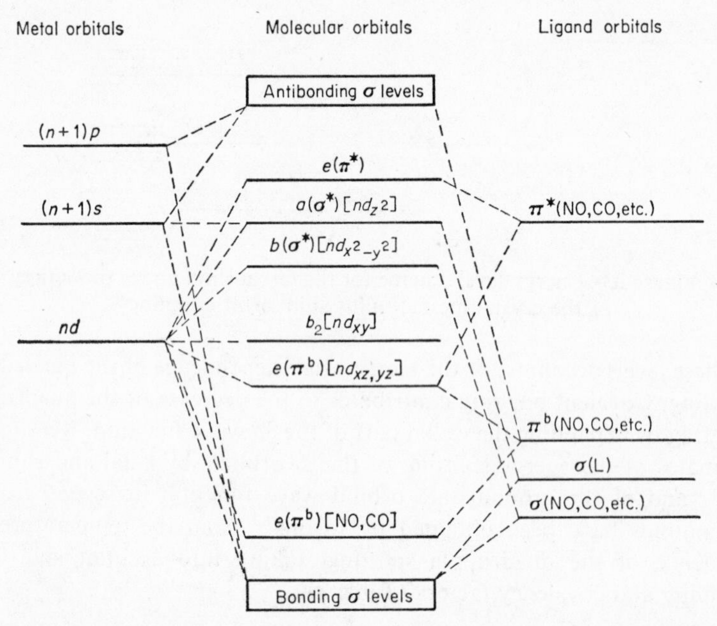

Figure 24. Molecular-orbital energy level diagram for MX_5NO complexes[91].

spin-paired Fe^{III} with that of NO. Thus, the molecular orbitals to be considered are those formed by Fe^{II} and NO^+. The molecular-orbital bonding scheme for transition metal–nitrosyl complexes, MX_5NO, established on the basis that the NO^+ dominates the ligand field at the central ion, is reproduced in Figure 24.

The value of the quadrupole splitting in $Na_2[Fe(CN)_5NO]\cdot 2H_2O$ is practically ($\Delta E = 0.185$ cm/s or 43.1 Mc at room temperature) the same in single crystals, polycrystalline samples, and frozen solutions of the complex at liquid-nitrogen temperatures, showing that the electric field gradient is essentially of molecular origin and that the contributions of the crystal lattice are negligible. Using a single crystal of the complex as absorber, with the symmetry axis oriented at different angles to the direction of the incident γ-ray, it was found[39] that the ratio of intensities of the doublet follows the relation (25), valid for an axially symmetric electric field gradient with the $\pm\frac{3}{2}$ state assumed to be the higher energy hyperfine component, as is illustrated in Figure 10. This shows that the quadrupole coupling is positive, since the axially symmetric field gradient at the iron nucleus in this complex is positive.

In order to interpret this result, we assume that the field gradient is due to the asymmetric expansion of the filled $3d$ subshell toward the ligands. According to the molecular-orbital bonding scheme reproduced in Figure 24, two different covalency parameters characterize the d subshell in the nitrosyl complexes: α_1 for the two pairs of electrons placed in the d_{xz}, d_{yz} lower doublet which is involved in the strong π bonds to NO, and α_2 for d_{xy} electrons which are assumed to be non-bonding[91]. The electric field gradient due to the $3d$ electrons is given by

$$q = \alpha_1^2(d_{xz}) + \alpha_1^2(d_{yz}) + \alpha_2^2(d_{xy}) \tag{45}$$

Using Table 7, we express the relation (45) in terms of $q' = \frac{4}{7}\langle r^{-3}\rangle$

$$q = 2(\alpha_2^2 - \alpha_1^2)q' \tag{46}$$

Since the d_{xy} orbital is assumed non-bonding and the d_{xz}, d_{yz} involved in strong π bonds we have, necessarily, $\alpha_2 > \alpha_1$ and in consequence $q > 0$, as observed experimentally. It is possible to use expression (46) to estimate the covalency parameters α^2.

The quadrupole interaction in $Fe(CN)_5NO^{2-}$ corresponds to 43.1 Mc. Taking $Q = 0.15$ barn (1 barn $= 10^{-24}$ cm^2), we find for the electric field gradient the value $q = +1.212$ a.u. The average of the inverse cube of the distance between a $3d$ electron and the nucleus, $\langle r^{-3}\rangle_{3d}$ ^{57}Fe, has been estimated to be 4.4 a.u. Substituting in (46) gives

$$(\alpha_2^2 - \alpha_1^2) = 0.24 \tag{47}$$

If the d_{xy} orbital is assumed non-bonding (neglecting delocalization to the cyanide ligands), $\alpha_2 = 1$ and we find $\alpha_1^2 = 0.76$.

It has recently been shown by Fortman and Hayes[92] that this is in agreement with the results obtained by quadrupole coupling splitting of the electron spin resonance spectrum of $Mn^{II}(CN)_5NO^{2-}$. This spectrum has been investigated in solution and in single-crystal environment[92,93]. Satellite lines of low intensity were observed between the $\Delta M = 0$ absorptions. These $\Delta M = \pm 1$ transitions were interpreted as arising from quadrupole coupling with the ^{55}Mn nucleus[92]. From the quadrupole coupling constant obtained, −62.4 Mc, and the quadrupole moment of ^{55}Mn = 0.35 barn, we obtain a field gradient of −0.754 a.u. The anisotropic coupling constant of the hyperfine interaction with ^{55}Mn gives $\langle r^{-3}\rangle^{55}$Mn = 3.10 a.u.[92].

According to the molecular-orbital scheme from Figure 24, the ground-state configuration for the manganese ion in this complex is $(d_{xz})^2(d_{yz})^2(d_{xy})^1$ and we have the following contributions to the field gradient:

$$q = -\alpha_1^2 q' - \alpha_1^2 q' + \alpha_2^2 q' = (\alpha_2^2 - 2\alpha_1^2)q' \qquad (48)$$

Substituting the values of q and $\langle r^{-3}\rangle$ given before, we find

$$(\alpha_2^2 - 2\alpha_1^2) = -1.43 \qquad (49)$$

Putting $\alpha_2 = 1$, we find $\alpha_1^2 = 0.72$, which compares well with that obtained for the nitroprusside complex from Mössbauer spectroscopy and indicates a strong delocalization of the d_{xz}, d_{yz} orbitals in the pentacyanonitrosyls, in agreement with the theoretical predictions.

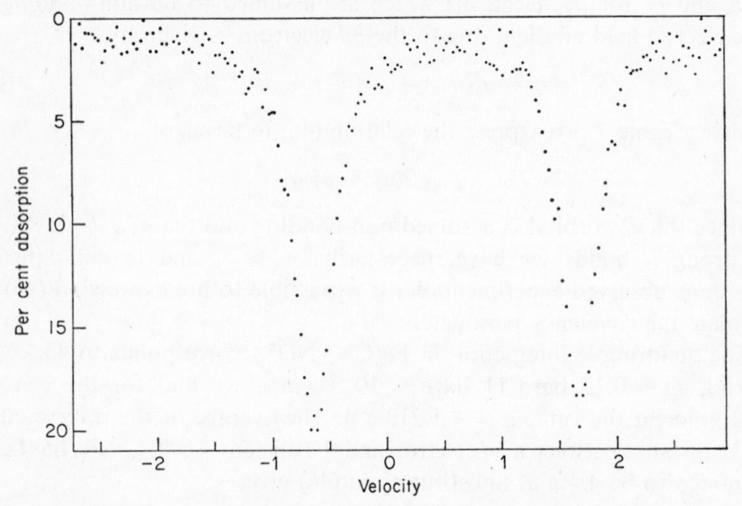

Figure 25. Mössbauer absorption spectrum of ferrocene at 4.2°K in absence of magnetic field[44].

As mentioned in Section IV.B, ferrocene was the first complex in which the sign of the electric field gradient was determined in polycrystalline samples by using the perturbation of the quadrupole lines in a strong external magnetic field. The spectra of polycrystalline ferrocene at 4.2°K and 40 kG[44] using a superconducting solenoid is shown in Figures 25 and 26. Comparison with the theoretical magnetically perturbed spectrum reproduced in Figure 11 shows unequivocally that the electric field gradient is positive.

Let us now analyse the predictions about the sign of the field gradient on the basis of the different treatments for the ferrocene molecule. Crystal-field calculations using an ellipsoidal potential with the iron as Fe^{2+} and unit negative charge on the rings predict a ground-state configuration $(d_{z^2})^2(d_{xy}, d_{yz})^4$. The value of the electric field gradient for the six d electrons is

$$q = 2(d_{z^2}) + 2(d_{xy}) + 2(d_{yz}) \tag{50}$$

which, by taking the values of the electric field gradients of each type of d orbital listed in Table 7, leads to $q < 0$, contrary to the observed sign of the field gradient.

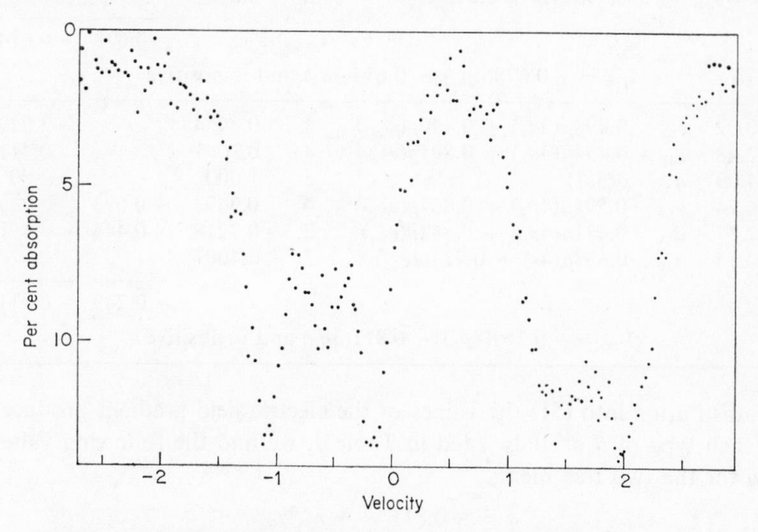

Figure 26. Mössbauer absorption spectrum of ferrocene at 4.2°K perturbed by a magnetic field of 40 kG[44].

Two SCF–LCAO molecular-orbital treatments have been given for ferrocene. The first one by Shustorovich and Dyatkina[94] uses Slater orbitals; the second one by Dahl and Ballhausen[95] uses Watson's Hartree–Fock orbitals for iron rather than Slater functions. Table 9 summarizes the results obtained. Another calculation[96], which uses the Wolfsberg–Helmholz method, gives the same order of the energy levels of the bonding orbitals as Shustorovich and Dyatkina.

The electric field gradient due to d electrons is in all cases given by

$$q = \alpha_1^2(d_{z^2}) + \alpha_2^2(d_{xy}) + \alpha_3^2(d_{xy}) \tag{51}$$

where α_1^2, α_2^2, and α_3^2 are the squares of the coefficients of the different d orbitals in the a_{1g}, e_{1g}, and e_{2g} molecular orbitals.

Table 9. Molecular orbitals by Shustorovich and Dyatkina and by Dahl and Ballhausen for ferrocene.

Energy (ev)	Sym.	Molecular orbitals	No. of electrons	(Metal coeff.)^2	Net $4p_0$	Net $3d_0$
−6.39	e_{2g}	$0.85\mu(3d_2) + 0.52p(e_{2g^*})$	4	0.722		−2.888
−8.44	$a_{1g'}$	$\mu(3d_0)$	2	1.000		2.000
−11.02	e_{1g}	$0.37\mu(3d_1) + 0.93p(e_{1g^*})$	4	0.137		0.274
−12.62	e_{1u}	$0.59\mu(4p_1) + 0.81p(e_{1u^*})$	4	0.349	−0.698	
−13.74	a_{2u}	$0.10\mu(4p_0) + 0.99p(e_{2u})$	2	0.010	0.020	
−16.05	a_{1g}	$0.49\mu(4s) + 0.87p(e_{1g})$	2	0.240		
					−0.678	−0.614
		$V_{zz} = -0.678(4p_0) - 0.614(3d_0)$ and is positive				
−10.92	e_{2g}	$0.898\mu(3d_2) + 0.440p(e_{2g^*})$	4	0.8064		−3.222
−12.48	e_{1g}	$0.454\mu(3d_1) + 0.891p(e_{1g^*})$	4	0.2061		0.411
−14.03	$a_{1g'}$	$\mu(3d_0)$	2	1.000		2.00
−14.74	e_{1u}	$0.591\mu(4p_1) + 0.807p(e_{1u^*})$	4	0.3493	−0.693	
−17.77	a_{2u}	$0.471\mu(4p_0) + 0.882p(e_{2u})$	2	0.2218	0.444	
−20.15	a_{1g}	$0.633\mu(4s) + 0.774p(e_{1g})$	2	0.4007		
					−0.249	−0.811
		$V_{zz} = -0.249(4p_0) - 0.811(3d_0)$ and is positive				

Substituting into (51) the values of the electric field gradient produced by each type of d orbitals listed in Table 7, we find the following values of q for the two treatments:

$$q = +0.811 \times \tfrac{4}{7}\langle r^{-3}\rangle_{3d}$$
$$q = +0.614 \times \tfrac{4}{7}\langle r^{-3}\rangle_{3d} \quad (52)$$

All three treatments predict a positive sign for the electric field gradient at the iron nucleus in ferrocene, in agreement with experimental results. It is important to observe that this conclusion remains unchanged if we disregard differences of $\langle r^{-3}\rangle$ between the $3d$ and the $4p$ contribution to the electric field gradient, since the $4p$ contribution has also positive sign in the MO treatments. However, the assumption that the main contribution to the electric field gradient is given by the $3d$ electrons which have larger $\langle r^{-3}\rangle$ is consistent with the interesting fact that the ferricinium ion exhibits a negligible quadrupole splitting. The removal of one electron from the highest bound state in the molecular-orbital bonding scheme leads to a very small net d-electron density in the molecule, whereas the $4p$ density remains unchanged. The collapse of the quadrupole splitting on going from ferrocene to the ferricinium ion is thus consistently explained by the MO treatments on the basis that the $3d$ electrons make the largest contribution to the quadrupole splittings at the iron nucleus.

C. Concluding Remarks

In the present chapter, we have confined ourselves to those aspects of Mössbauer spectroscopy useful in chemical-bond investigations. The Mössbauer effect appears in this respect to be a powerful tool, since it combines the results of two hyperfine interactions of electric origin which can give information about electron delocalization in molecules. In this respect, it is important to observe that these interactions are not subject to restrictions of the necessary presence of unpaired spin density for detection as in the usual magnetic resonance methods.

The Mössbauer effect hyperfine interactions can also be used in a variety of other applications in inorganic chemistry. The fact that characteristic values of the isomer shift are observed for different oxidation states of an element allows the use of the Mössbauer spectra for identifying these states in solids. This has found a number of interesting geochemical and mineralogical applications, such as the studies of natural iron silicates[97], the composition of several types of iron-containing meteorites[98], etc. The same reasoning has been used to point out potential applications to radiolytic decomposition of solids, since the different oxidation states induced by the radiation can be identified *in situ*[99].

A promising field for applications appears to be recoil chemistry. Here the Mössbauer effect provides a unique tool for the study of non-equilibrium charge states formed in nuclear decay processes, provided that the lifetime of these states are comparable with that of the Mössbauer transition[60,61].

The magnetic hyperfine interaction has been determined for a large number of compounds. Most interesting results have been recently obtained with iron compounds of biological importance such as the haemoproteins[55]. These recent measurements of internal fields and spin-relaxation times in molecules containing paramagnetic ions and their interpretation in connexion with electron spin resonance data constitute an example of how the Mössbauer effect can be a powerful method for deducing the electronic state of molecules.

VII. BIBLIOGRAPHY

A. Books

1. H. Frauenfelder, *The Mössbauer Effect*, Benjamin, New York, 1962.
2. A. Abragam, *L'Effet Mössbauer et ses Applications à l'Etude des Champs Internes*, Gordon and Breach, New York, 1964.

3. G. K. Wertheim, *Mössbauer Effect, Principles and Applications*, Academic Press, New York, 1964.
4. A. H. Muir, Jr., K. J. Ando, and H. M. Coogan, *Mössbauer Effect Data Index*, North-American Aviation Science Center, Thousand Oaks, California, 1965.
5. D. M. J. Compton and A. H. Shoen (Eds.), *Proc. Intern. Conf. Mössbauer Effect, 2nd, Saclay, France, 1961*, Wiley, New York, 1962.
6. *Proc. Conf. Mössbauer Effect, Dubna, July, 1962* (English Transl.), Consultants Bureau, New York, 1963.
7. *Applications of the Mössbauer Effect in Chemistry and Solid State Physics*, Technical Report Series No. 50, International Atomic Energy Agency, Vienna, 1966.

B. Reviews

1. H. Lustig, *Am. J. Phys.*, **29**, 1 (1961).
2. A. J. F. Boyle and H. E. Hall, 'Mössbauer effect', *Rept. Progr. Phys.*, **25**, 441 (1962).
3. E. F. Hammel, W. E. Keller, and P. P. Craig, 'The Mössbauer effect', *Ann. Rev. Phys. Chem.*, **13**, 295 (1962).
4. V. I. Goldanskii, 'The Mössbauer effect and its applications to chemistry', *Atomic Energy Rev.*, **1**, 3 (1963).
5. E. Fluck, W. Kerker, and W. Neuwirth, *Angew. Chem.* (*Intern. Ed. Engl.*), **2**, 277 (1963).
6. G. K. Wertheim, *Science*, **144**, 253 (1964).
7. E. Fluck, 'The Mössbauer effect and its application in chemistry', *Advan. Inorg. Chem. Radiochem.*, **6**, 433 (1964).
8. J. F. Duncan and R. M. Golding, *Quart. Rev.* (*London*), **19**, 36 (1965).

VIII. REFERENCES

1. R. L. Mössbauer, *Z. Phys.*, **151**, 124 (1958); *Naturwissenschaften*, **45**, 538 (1958).
2. R. L. Mössbauer, *Z. Naturforsch.*, **14**, 211 (1959).
3. R. W. Wood, *Physical Optics*, Macmillan, New York, 1954.
4. W. Kuhn, *Phil. Mag.*, **8**, 625 (1929).
5. K. G. Malmfors, 'Resonant scattering of gamma rays', in *Beta–Gamma Ray Spectroscopy* (Ed. K. Siegbahn), North Holland, Amsterdam, 1964.
6. V. I. Goldanskii, 'The Mössbauer effect and its applications to Chemistry' *Atomic Energy Rev.*, **1**, 3 (1963).
7. G. Breit and E. Wigner, *Phys. Rev.*, **49**, 519 (1936).
8. R. L. Mössbauer and D. H. Sharp, *Rev. Mod. Phys.*, **36**, 410 (1964).
9. A. A. Maradudin, *Rev. Mod. Phys.*, **36**, 417 (1964).
10. F. L. Shapiro, *Usp. Fiz. Nauk*, **72**, 68 (1960); *Soviet Phys.–Usp.* (*English Transl.*), **3**, 881 (1961).
11. R. J. Morrison, M. Atac, P. Debrunner, and H. Frauenfelder, *Phys. Letters*, **12**, 35 (1964).
12. E. Kankeleit, *Z. Phys.*, **164**, 442 (1961).
13. H. Frauenfelder, D. R. F. Cochran, D. E. Nagle, and R. D. Taylor, *Nuovo Cimento*, **19**, 183 (1961).

14. J. Stephen, *Nucl. Instr. Methods*, **26**, 269 (1964).
15. W. Kerler and W. Neuwirth, *Z. Phys.*, **167**, 176 (1962).
16. V. A. Bryukhanov, N. N. Delyagin and R. N. Kulzmin, *Zh. Eksperim. i Teor. Fiz.*, **46**, 137 (1964); *Soviet Phys. JETP* (*English Transl.*), **19**, 98 (1964).
17. Y. K. Lee, P. W. Keaton, Jr., E. T. Ritter, and J. C. Walker, *Phys. Rev. Letters*, **14**, 957 (1965).
18. D. Seyboth, F. E. Obenshain and G. Czjzek, *Phys. Rev. Letters*, **14**, 954 (1965).
19. D. W. Hafemeister and E. Brooks-Shera, *Phys. Rev. Letters*, **14**, 593 (1965).
20. S. L. Ruby and K. E. Holland, *Phys. Rev. Letters*, **14**, 591 (1965).
21. S. Fink and P. Kienle in *Panel on the Applications of the Mössbauer Effect in Chemistry and Solid State Physics*, International Atomic Energy Agency, Vienna, 1965.
22. *Mössbauer Effect Methodology* (Ed. I. Gruberman), Plenum Press, New York, 1965.
23. S. Margulies and J. R. Ehrman, *Nucl. Instr. Methods*, **12**, 131 (1961).
24. R. M. Housley, N. E. Erikson, and J. G. Dash, *Nucl. Instr. Methods*, **27**, 29 (1964).
25. D. A. Shirley, *Nucleonics*, **23**, 62 (1965).
26. H. Frauenfelder, *The Mössbauer Effect*, Benjamin, New York, 1962.
27. G. K. Wertheim, *Mössbauer Effect, Principles and Applications*, Academic Press, New York, 1964.
28. J. J. Spijkerman, F. C. Ruegg, and J. R. DeVoe in *Applications of the Mössbauer Effect in Chemistry and Solid State Physics*, Technical Report Series No. 50, International Atomic Energy Agency, Vienna, 1966.
29. D. A. Shirley, *Rev. Mod. Phys.*, **36**, 339 (1964).
30. V. I. Goldanskii, *Dokl. Akad. Nauk SSSR*, **147**, 127 (1962).
31. V. G. Bhide, G. K. Shenoy and M. S. Multani, *Solid State Commun.*, **2**, 221 (1964).
32. R. M. Sternheimer and H. M. Foley, *Phys. Rev.*, **92**, 1460 (1953); R. M. Sternheimer, *Phys. Rev.*, **96**, 951 (1954); **105**, 158 (1957).
33. A. J. Freeman and R. E. Watson, *Phys. Rev.*, **131**, 2566 (1963).
34. P. Zory, *Phys. Rev.*, **140**, A 1401 (1965).
35. V. I. Goldanskii, E. F. Makarov and V. V. Khrapov, *Phys. Letters*, **3**, 344 (1963).
36. S. V. Karyagin, *Soviet Phys.–Solid State* (*English Transl.*), **5**, 1552 (1964).
37. A. M. Afanasiev and Yu. Kagan, *Soviet Phys. JETP* (*English Transl.*), **18**, 1139 (1964).
38. M. Blume, *Phys. Rev. Letters*, **14**, 36 (1965).
39. J. Danon, *J. Chem. Phys.*, **41**, 3378 (1964).
40. P. Kienle, *Phys. Verhandl.*, **213**, 33 (1963).
41. G. J. Perlow and M. R. Perlow, *J. Chem. Phys.*, **41**, 1157 (1964).
42. D. W. Hafemeister, G. De Pasquali, and H. de Waard, *Phys. Rev.*, **135**, B 1089 (1964).
43. S. L. Ruby and P. A. Flinn, *Rev. Mod. Phys.*, **36**, 351 (1964).
44. R. L. Collins, *J. Chem. Phys.*, **42**, 1072 (1965).
45. W. Marshall, *Phys. Rev.*, **110**, 1280 (1958).
46. W. Marshall and C. E. Johnson, *J. Phys. Radium*, **23**, 733 (1962).

47. R. E. Watson and A. J. Freeman, *Phys. Rev.*, **123**, 2027 (1961).
48. J. S. van Wieringen, *Discussions Faraday Soc.*, **19**, 118 (1955).
49. G. K. Wertheim and J. P. Remeika, *Phys. Letters*, **10**, 14 (1964).
50. C. R. Kurkjian and D. N. E. Buchanan, *Phys. Chem. Glasses*, **5**, 63 (1964).
51. H. H. Wickman and A. M. Trozzolo, *Phys. Rev. Letters*, **15**, 156 (1965).
52. S. Ofer, B. Khurgin, M. Rakavy and I. Nowik, *Phys. Letters*, **11**, 205 (1964).
53. T. H. Moss, *Thesis*, Cornell University, 1965.
54. A. Abragam and M. H. L. Pryce, *Proc. Roy. Soc.* (*London*), *Ser. A*, **205**, 135 (1951).
55. G. Lang and W. Marshall, *Proc. Phys. Soc.* (*London*), **87**, 3 (1966).
56. G. M. Rothberg, N. Benczer-Koller, and J. R. Harris, *Rev. Mod. Phys.*, **36**, 357 (1964).
57. J. A. Stone and W. L. Pillinger, *Phys. Rev. Letters*, **13**, 200 (1964); J. A. Stone and W. L. Pillinger in *Applications of the Mössbauer Effect in Chemistry and Solid State Physics*, Technical Report No. 50, International Atomic Energy Agency, Vienna, 1966.
58. G. K. Wertheim, *Phys. Rev.*, **124**, 764 (1961).
59. R. H. Herber and H. A. Stockler in *Proc. Symp. Chem. Effects Associated with Nuclear Reactions and Radioactive Transformations*, Vol. 2, Vienna, 7th Dec., 1964, International Atomic Energy Agency, Vienna, 1965, p. 403.
60. G. K. Wertheim and H. J. Guggenheim, *J. Chem. Phys.*, **42**, 3873 (1965).
61. R. Ingalls and G. De Pasquali, *Phys. Letters*, **15**, 262 (1965).
62. V. A. Bryukhanov, V. I. Goldanskii, N. N. Delyagin, L. A. Korytko, E. F. Makarov, I. P. Suzdalev, and V. S. Shpinel, *Zh. Eksperim. i Teor. Fiz.*, **43**, 448 (1962); *Soviet Phys. JETP* (*English Transl.*), **16**, 321 (1963).
63. R. H. Herber, *J. Chem. Educ.*, **42**, 181 (1965).
64. A. J. F. Boyle, D. St. P. Bunbury, and C. Edwards, *Proc. Phys. Soc.* (*London*), **79**, 416 (1962).
65. V. I. Goldanskii, E. F. Makarov, R. A. Stukan, T. N. Sumarokava, V. A. Trukhtanov, and V. V. Khrapov, *Dokl. Akad. Nauk SSSR*, **156**, 400 (1964); *Proc. Acad. Sci. USSR* (*English Transl.*), **156**, 474 (1964).
66. C. H. Townes and B. P. Dailey, *J. Chem. Phys.*, **17**, 782 (1949).
67. W. Dahne in *Intern. Conf. Coordination Chem., 7th, Stockholm, 1962.* Almquist and Wiksell, A. B., Uppsala, 1962.
68. N. E. Erickson and A. W. Fairhall, *Inorg. Chem.*, **4**, 1320 (1965).
69. M. Kalvius, U. Zahn, P. Kienle, and H. Eicher, *Z. Naturforsch.*, **17a**, 494 (1962).
70. R. H. Herber, W. R. Kingston, and G. K. Wertheim, *Inorg. Chem.*, **2**, 153 (1963).
71. L. F. Dahl and J. F. Blount, *Inorg. Chem.*, **4**, 1373 (1965).
72. C. L. Chermick, C. E. Johnson, J. G. Malm, G. J. Perlow, and M. R. Perlow, *Phys. Letters*, **5**, 103 (1963).
73. G. J. Perlow and M. R. Perlow, *Proc. Symp. Chem. Effects Associated with Nuclear Reactions and Radioactive Transformations, Vienna, Dec., 1964*, International Atomic Energy Agency, Vienna, 1965, Vol. 2, p. 443.
74. G. J. Perlow and M. R. Perlow, *J. Chem. Phys.*, **41**, 1157 (1964).
75. P. K. Gallagher, J. B. MacChesney, and D. N. E. Buchanan, *J. Chem. Phys.*, **41**, 2429 (1964).

76. N. Bloembergen and P. Sorokin, *Phys. Rev.*, **110**, 865 (1958).
77. M. Menes and D. Bolef, *Phys. Chem. Solids*, **19**, 79 (1961).
78. W. H. Flygare and D. W. Hafemeister, *J. Chem. Phys.*, **43**, 789 (1965).
79. J. Danon, *J. Chem. Phys.*, **39**, 236 (1963); J. Danon, *Rev. Mod. Phys.*, **36**, 459 (1964).
80. J. Danon in *Proc. Intern. Conf. Coordination Chem.*, *8th*, Springer, Vienna, 1964.
81. W. Kerler, W. Neuwirth, E. Fluck, P. Kuhn, and B. Zimmermann, *Z. Phys.*, **173**, 321 (1963).
82. L. R. Walker, G. K. Wertheim, and V. Jaccarino, *Phys. Rev. Letters*, **6**, 98 (1961).
83. R. G. Shulman and S. Sugano, *J. Chem. Phys.*, **42**, 39 (1965).
84. J. Danon, *Rev. Mod. Phys.*, **36**, 460 (1964).
85. C. K. Jørgensen, 'The nephelauxetic series' in *Progr. Inorg. Chem*, **4**, 73 (1962).
86. R. E. Watson, *Solid State and Molecular Theory Group*, Technical Report No. 12, Massachusetts Institute of Technology, 1959; R. E. Watson, *Phys. Rev.*, **119**, 1934 (1960).
87. E. Clementi, C. C. J. Roothan, and M. Yoshimine, *Phys. Rev.*, **127**, 1618 (1962).
88. S. DeBenedetti, G. Lang, and R. I. Ingalls, *Phys. Rev. Letters*, **6**, 60 (1961).
89. R. I. Ingalls, *Phys. Rev.*, **133**, A 787 (1964).
90. N. L. Costa, J. Danon, and R. Moreira-Xavier, *Phys. Chem. Solids*, **23**, 1783 (1962).
91. H. B. Gray, I. Bernal, and E. Billig, *J. Am. Chem. Soc.*, **84**, 3404 (1962); P. T. Manahoran and H. B. Gray, *J. Am. Chem. Soc.*, **89**, 3340 (1965).
92. J. J. Fortman and R. G. Hayes, *J. Chem. Phys.*, **43**, 15 (1965).
93. J. Danon, H. Panepucci, and A. A. Misetich, *J. Chem. Phys.*, **44**, 4154 (1966).
94. E. M. Shustorovich and M. E. Dyatkina, *Dokl. Akad. Nauk SSSR*, **128**, 1324 (1959); *Zh. Strukt. Khim.*, **1**, 109 (1960).
95. J. P. Dahl and C. J. Ballhausen, *Kgl. Danske Vidensk. Selskab, Mat.-Fys. Medd.*, **33**, 5 (1961).
96. R. D. Fischer, *Theoret. Chim. Acta*, **1**, 418 (1963).
97. M. De Coster, H. Pollak, and S. Amelinckx, *Phys. Status Solidi*, **3**, 283 (1963).
98. E. L. Sprenkel-Segel and S. S. Hanna, *Geochim. Cosmochim. Acta*, **28**, 1913 (1964).
99. N. Saito, T. Tominaga, F. Ambe, and H. Sano in *Applications of the Mössbauer Effect in Chemistry and Solid State Physics*, Technical Report Series No. 50, International Atomic Energy Agency, Vienna, 1966.

9

Nuclear Quadrupole Resonance

H. SILLESCU

I. INTRODUCTION

Nuclear quadrupole resonance (n.q.r.) is a means of measuring the inhomogeneity of internal electric fields in molecules at the sites of nuclei which have electric quadrupole moments. The frequencies corresponding to the energy of coupling between nuclear quadrupole moments and the gradients of intramolecular electric fields lie in the range of radio waves. Transitions between the energy levels can be detected by slightly modified nuclear magnetic resonance (n.m.r.) techniques.

A simple example will illustrate the application of nuclear quadrupole resonance to problems of chemical bonding. In the Cl^- ion, the closed electron shell of spherical symmetry will produce a vanishing field gradient at the chlorine nucleus. Therefore no n.q.r. is to be expected in completely ionic compounds if we neglect the influence of neighbouring ions. In the free chlorine atom, on the other hand, the *p*-electron hole will produce a large field gradient at the nucleus. From atomic beam experiments[1], we know the magnitude of this field gradient, and we can calculate from it an n.q.r. frequency of 54.87 Mc/s for ^{35}Cl in the chlorine atom. In typical organic compounds, n.q.r. frequencies of about 30–40 Mc/s are encountered, e.g. 34.6 Mc/s for C_6H_5Cl. In most inorganic compounds, lower n.q.r. frequencies are found, e.g. 12.8 Mc/s for $CrCl_3$. The scale of n.q.r. frequencies in chlorine compounds gives some indication of how 'ionic' a compound is, but difficulties arise if we try to fix the correlation of n.q.r. frequencies and ionicity in a more quantitative manner. We shall discuss these problems in Sections II.C and IV.A. Any environmental effects that alter the electric field gradient at the nucleus will change its n.q.r. frequency, so that n.q.r. is a very sensitive means of studying phase transitions or any changes in the crystalline environment of molecules.

We start with a thorough introduction to the physical principles of nuclear quadrupole coupling in order to give a clear basis from which applications to chemical problems may be understood.

II. THEORY

A. Hamiltonian of Nuclear Quadrupole Coupling

Let us consider the coupling of a nuclear charge cloud of density $\rho^{(n)}(\mathbf{R})$ with an electron charge cloud $\rho^{(e)}(\mathbf{r})$, and assume $\mathbf{R} < \mathbf{r}$. Then according to Figure 1, the nuclear charge is separated from the electron charge. It is not necessary to consider the density of *s* electrons* at the nucleus, since they do not produce a field gradient because of their spherical symmetry.

The external charge of density $\rho^{(e)}$ produces a potential

$$\varphi(\mathbf{R}) = \int \frac{\rho^{(e)}(\mathbf{r})\,\mathrm{d}\tau^{(e)}}{|\mathbf{r} - \mathbf{R}|} \tag{1}$$

inside the sphere of radius R (Figure 1). The interaction potential of the two charge clouds $\rho^{(e)}$ and $\rho^{(n)}$ is thus†

$$V = \int \varphi(\mathbf{R})\rho^{(n)}(\mathbf{R})\,\mathrm{d}\tau^{(n)} = \int\!\!\int \frac{\rho^{(e)}(\mathbf{r})\rho^{(n)}(\mathbf{R})}{|\mathbf{r} - \mathbf{R}|}\,\mathrm{d}\tau^{(e)}\,\mathrm{d}\tau^{(n)} \tag{2}$$

* The very small density of *p* electrons at the nucleus is neglected.

† This is an extension of the well-known potential energy, e^2/r, of two point charges at a distance r.

Let X_1, X_2, X_3 be the cartesian coordinates of **R**, and x_1, x_2, x_3 those of **r**. Expansion of $\varphi(\mathbf{R})$ in a Taylor series around the origin gives

$$\varphi(\mathbf{R}) = \varphi_0 + \sum_{j=1}^{3} \varphi_j X_j + \tfrac{1}{2} \sum_{j=1}^{3} \sum_{k=1}^{3} \varphi_{jk} X_j X_k + \cdots \tag{3}$$

where we have used the abbreviations

$$\varphi(0) \equiv \varphi_0, \quad \left(\frac{\partial \varphi}{\partial X_j}\right)_{R=0} \equiv \varphi_j, \quad \text{and} \quad \left(\frac{\partial^2 \varphi}{\partial X_j\, \partial X_k}\right)_{R=0} \equiv \varphi_{jk}$$

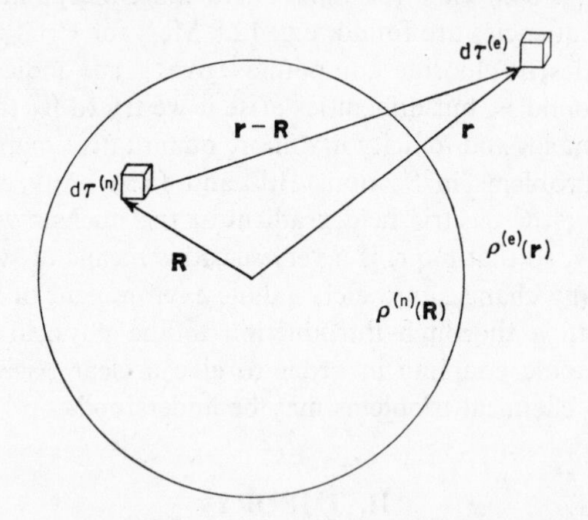

Figure 1. Interaction of two charge clouds. The vector from the origin to any point of the charge $\rho^{(n)}$ (**R**) within the sphere (indicated by a circle) is called **R**. The vector to any point outside the sphere is called **r**.

From (2) and (3), we obtain

$$\begin{aligned} V &= V_0 + V_D + V_Q + \cdots \\ V_0 &= \varphi_0 \int \rho^{(n)}(\mathbf{R})\, d\tau^{(n)} \\ V_D &= \sum_j \varphi_j \int X_j \rho^{(n)}(\mathbf{R})\, d\tau^{(n)} \\ V_Q &= \tfrac{1}{2} \sum_j \sum_k \varphi_{jk} \int X_j X_k \rho^{(n)}(\mathbf{R})\, d\tau^{(n)} \end{aligned} \tag{4}$$

V_0 is the coupling of a point charge at the origin with the potential φ_0 of the external charge cloud. V_D is the coupling of a dipole moment with the electric field at the origin

$$\mathbf{E}(0) = -(\text{grad } \varphi)_{R=0} = -\{\varphi_1, \varphi_2, \varphi_3\} \tag{5}$$

As may be seen from parity considerations[3], atomic nuclei have a vanishing dipole moment; accordingly $V_D = 0$. V_Q is the coupling of the quadrupole moment of the nuclear charge $\rho^{(n)}$ with the gradient of the electric field at the nucleus, which is given by

$$\nabla\mathbf{E} = \begin{pmatrix} \text{grad}\ \varphi_1 \\ \text{grad}\ \varphi_2 \\ \text{grad}\ \varphi_3 \end{pmatrix} = \begin{pmatrix} \varphi_{11} & \varphi_{12} & \varphi_{13} \\ \varphi_{21} & \varphi_{22} & \varphi_{23} \\ \varphi_{31} & \varphi_{32} & \varphi_{33} \end{pmatrix} \tag{6}$$

The field gradient tensor $\nabla\mathbf{E}$ is symmetric, i.e. $\varphi_{jk} = \varphi_{kj}$, and its trace vanishes because of the validity of the Laplace equation

$$\Delta\varphi = \sum_j \varphi_{jj} = 0 \tag{7}$$

In order to get a quadrupole moment tensor of vanishing trace, we add

$$-\tfrac{1}{2}\sum_j\sum_k \varphi_{jk}\int \delta_{jk}\tfrac{1}{3}R^2\rho^{(n)}\,\mathrm{d}\tau^{(n)} = -\tfrac{1}{6}\sum_j \varphi_{jj}\int R^2\rho^{(n)}\,\mathrm{d}\tau^{(n)} = 0$$

to V_Q, and thus we have from equation (4)

$$V_Q = \tfrac{1}{6}\sum_{j=1}^{3}\sum_{k=1}^{3} \varphi_{jk}Q_{jk} \equiv \tfrac{1}{6}\nabla\mathbf{E}:\mathbf{Q} \tag{8}$$

with

$$Q_{jk} = \int (3X_jX_k - \delta_{jk}R^2)\rho^{(n)}(\mathbf{R})\,\mathrm{d}\tau^{(n)} \tag{9}$$

being the components of the nuclear quadrupole moment tensor **Q**. **Q** is a symmetric tensor and its trace vanishes according to (9), that is

$$\sum_{j=1}^{3} Q_{jj} = \int (3X_1^2 - R^2 + 3X_2^2 - R^2 + 3X_3^2 - R^2)\rho^{(n)}(\mathbf{R})\,\mathrm{d}\tau^{(n)} = 0 \tag{9a}$$

We have from (1)

$$\varphi_{jk} = \left(\frac{\partial^2\varphi}{\partial X_j\,\partial X_k}\right)_{R=0} = \int\left(\frac{\partial^2}{\partial X_j\,\partial X_k}\frac{1}{|\mathbf{r}-\mathbf{R}|}\right)_{R=0}\rho^{(e)}(\mathbf{r})\,\mathrm{d}\tau^{(e)} = \int \frac{3x_jx_k - \delta_{jk}r^2}{r^5}\rho^{(e)}(\mathbf{r})\,\mathrm{d}\tau^{(e)} \tag{10}$$

where we have used the identity

$$|\mathbf{r} - \mathbf{R}| = (r^2 + R^2 - 2 \sum_{i=1}^{3} x_i X_i)^{\frac{1}{2}}$$

Let us consider a simple example to illustrate the sense of equations (8) to (10). In Figure 2, we have two point charges $-e$ at a distance $2a$

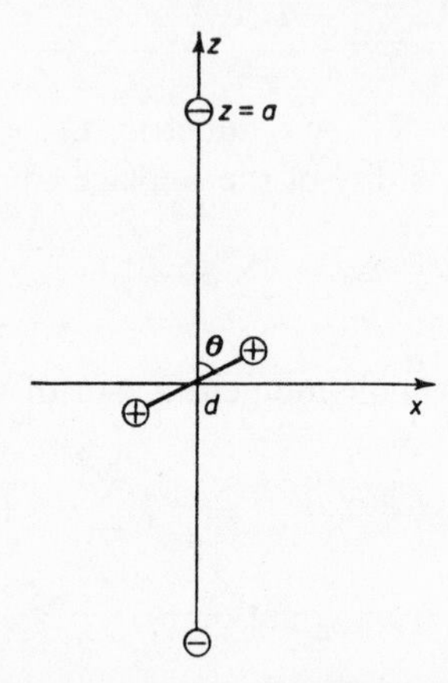

Figure 2. Linear quadrupole in the field of two point charges.

aligned on the z axis. Therefore we have

$$\rho^{(e)}(\mathbf{r}) = -e[\delta(\mathbf{r} - \mathbf{a}) + \delta(\mathbf{r} + \mathbf{a})]$$

for the charge density, $\delta(\mathbf{r} - \mathbf{r}')$ being the Dirac δ function, and $\mathbf{a}$ being a vector with components $\{0, 0, a\}$. It is easily seen from (10) that the only non-vanishing components of the field gradient tensor are

$$\begin{aligned}\varphi_{zz} &= -\tfrac{1}{2}\varphi_{xx} = -\tfrac{1}{2}\varphi_{yy} \\ &= -e \int \frac{3z^2 - r^2}{r^5} [\delta(\mathbf{r} - \mathbf{a}) + \delta(\mathbf{r} + \mathbf{a})] \, d\tau^{(e)} \\ &= -4e/a^3\end{aligned}$$

In Figure 2, the two positive charges are held at a fixed distance by a rod of length $d \ll a$ which is rotatable around the origin in the xz plane. The coordinates of the charges are $\mathbf{d}/2 = \{\sin\theta, 0, \cos\theta\}$ and $-\mathbf{d}/2$. The charge density is

$$\rho^{(n)}(\mathbf{R}) = e[\delta(\mathbf{R} - \mathbf{d}/2) + \delta(\mathbf{R} + \mathbf{d}/2)]$$

We have to calculate only the component

$$Q_{zz} = e\int(3Z^2 - R^2)\rho^{(n)}(\mathbf{R})\,\mathrm{d}\tau^{(n)}$$

$$= \frac{ed^2}{2}(3\cos^2\theta - 1)$$

of the quadrupole moment since, from (9a),

$$Q_{xx} + Q_{yy} + Q_{zz} = 0$$

and thus

$$V_Q = \tfrac{1}{4}\varphi_{zz}Q_{zz}$$

$$= -\frac{e^2d^2}{2a^3}(3\cos^2\theta - 1) \tag{11}$$

is the coupling energy for the field gradient of the two negative charges with the quadrupole moment of the two positive charges in Figure 2. Note that a neutral quadrupole is obtained if the charge $-2e$ is attached to the centre. The potential energy $V_Q(\theta)$ has two minima at $\theta = 0$ and π, and two maxima at $\theta = \pi/2$ and $3\pi/2$.

The calculation of V_Q for the example (Figure 2) was especially simple because we have chosen as the coordinate system the principal axes of the field gradient tensor. In the further treatment of (8), we also use as coordinates the principal axes of $\nabla\mathbf{E}$, i.e. we rotate the axes x_1, x_2, x_3 into the principal axes x, y, z of $\nabla\mathbf{E}$. Then we have, from equation (8),

$$V_Q = \tfrac{1}{6}(\varphi_{xx}Q_{xx} + \varphi_{yy}Q_{yy} + \varphi_{zz}Q_{zz}) \tag{12}$$

the non-diagonal elements of $\nabla\mathbf{E}$ being zero. It is convenient to define*

$$eq \equiv \varphi_{zz}$$

$$\eta \equiv (\varphi_{xx} - \varphi_{yy})/\varphi_{zz} \tag{13}$$

$\nabla\mathbf{E}$ is completely determined by q, η and the directions of the principal axes. η is called the asymmetry parameter of the field gradient. The directions of x, y, z are chosen to have $|\varphi_{xx}| \leqslant |\varphi_{yy}| \leqslant |\varphi_{zz}|$, and thus $0 \leqslant \eta \leqslant 1$. From (12) and (13), we get

$$V_Q = \frac{eq}{6}\left[\tfrac{1}{2}(\eta - 1)Q_{xx} - \tfrac{1}{2}(\eta + 1)Q_{yy} + Q_{zz}\right] \tag{14}$$

$\eta = 0$ yields $V_Q = \frac{1}{4}eqQ_{zz}$ from which (8) is a special case.

In the quantum-mechanical treatment, we replace q, η, Q_{xx}, Q_{yy}, and Q_{zz} by the quantum-mechanical operators to get the hamiltonian of

* $e = 4.8 \times 10^{-10}$ e.s.u. is the elementary charge.

nuclear quadrupole coupling. As we are interested in the n.q.r. of solids, we consider only the averages $\langle \Psi_0 | q | \Psi_0 \rangle$ and $\langle \Psi_0 | \eta | \Psi_0 \rangle$ over the electronic ground state Ψ_0. In what follows, we write these averages as q and η. The hamiltonian is then a pure nuclear operator

$$\mathscr{H}_Q = \frac{eq}{6}\left[\tfrac{1}{2}(\eta - 1)\mathbf{Q}_{xx} - \tfrac{1}{2}(\eta + 1)\mathbf{Q}_{yy} + \mathbf{Q}_{zz}\right] \tag{15}$$

$\mathbf{Q}_{xx}$, $\mathbf{Q}_{yy}$, and $\mathbf{Q}_{zz}$ operating on the nuclear states ψ_{IM}. The nuclear states ψ_{IM} are chosen to be eigenstates of the nuclear spin operators $\mathbf{I}^2$ and $\mathbf{I}_z$. Then we have[2]

$$\begin{aligned} \mathbf{I}^2\psi_{IM} &= I(I+1)\psi_{IM} \\ \mathbf{I}_z\psi_{IM} &= M\psi_{IM} \\ \mathbf{I}_+\psi_{IM} &= [(I-M)(I+M+1)]^{\frac{1}{2}}\psi_{IM+1} \\ \mathbf{I}_-\psi_{IM} &= [(I+M)(I-M+1)]^{\frac{1}{2}}\psi_{IM-1} \end{aligned} \tag{16}$$

$\mathbf{I}_\pm = \mathbf{I}_x \pm i\mathbf{I}_y$ are the so-called 'shift operators' as they shift the states from M to $M \pm 1$.

To find out how $\mathbf{Q}_{xx}$, $\mathbf{Q}_{yy}$, and $\mathbf{Q}_{zz}$ operate on the states ψ_{IM}, we use the Wigner–Eckhart theorem*, but to do so we have to construct from $\mathbf{Q}_{xx}, \mathbf{Q}_{xy}, \mathbf{Q}_{xz}, \mathbf{Q}_{yy}, \ldots$ linear combinations, say $\mathbf{Q}_{\pm 2}$, $\mathbf{Q}_{\pm 1}$, and $\mathbf{Q}_0$, that behave as spherical harmonics of order 2 when the coordinate system is rotated. It is easy to construct operators $\mathbf{T}_{\pm m}$ with the same transformation behaviour using products of the nuclear spin operators $\mathbf{I}_x, \mathbf{I}_y, \mathbf{I}_z$. Then the proportionality of the matrix elements $\langle \psi_{IM} | \mathbf{Q}_{\pm m} | \psi_{IM'} \rangle$ and $\langle \psi_{IM} | \mathbf{T}_{\pm m} | \psi_{IM'} \rangle$ follows from the Wigner–Eckart theorem, and the proportionality constant depends only on the nuclear spin I. From this proportionality, it is not difficult to get the equations

$$\begin{aligned} \mathbf{Q}_{xx} &= \alpha(I)[3\mathbf{I}_x^2 - I(I+1)] \\ \mathbf{Q}_{yy} &= \alpha(I)[3\mathbf{I}_y^2 - I(I+1)] \\ \mathbf{Q}_{zz} &= \alpha(I)[3\mathbf{I}_z^2 - I(I+1)] \end{aligned} \tag{17}$$

which define the operation of $\mathbf{Q}_{xx}$, $\mathbf{Q}_{yy}$, and $\mathbf{Q}_{zz}$ on the nuclear states ψ_{IM}. The constant $\alpha(I)$ is related to the quadrupole moment eQ of the nucleus which is defined by the average of Q_{zz} over ψ_{II}, the nuclear state with $M = I$:

$$\begin{aligned} eQ &\equiv \langle \psi_{II} | \mathbf{Q}_{zz} | \psi_{II} \rangle \\ &= \alpha(I)[3I^2 - I(I+1)] \\ &= \alpha(I)I(2I-1) \end{aligned} \tag{18}$$

No accurate calculation of the average $\langle \psi_{II} | \mathbf{Q}_{zz} | \psi_{II} \rangle$ is possible at the present time. However, experimental values of the quadrupole coupling

* For details of the calculation leading to equation (17) see, e.g., ref. 3.

constant e^2qQ are known for many atoms from atomic-beam measurements, and calculations of the field gradient eq in these atoms yield approximate values for the nuclear quadrupole moment eQ.

From (15) to (18), we get the hamiltonian

$$\mathscr{H}_Q = \frac{e^2Qq}{4I(2I-1)}[3\mathbf{I}_z^2 - I(I+1) + \tfrac{1}{2}\eta(\mathbf{I}_+^2 + \mathbf{I}_-^2)] \tag{19}$$

Noticing that the nuclear states are orthonormal, i.e. $\langle\psi_{IM} \mid \psi_{IM'}\rangle = \delta_{MM'}$, we get the matrix elements of $\mathscr{H}_Q$ from (19) and (16)

$$\begin{aligned} \langle\psi_{IM}|\,\mathscr{H}_Q\,|\psi_{IM}\rangle &= \frac{e^2Qq}{4I(2I-1)}[3M^2 - I(I+1)] \\ \langle\psi_{IM\pm 2}|\,\mathscr{H}_Q\,|\psi_{IM}\rangle &= \frac{e^2Qq}{8I(2I-1)}\eta[(I\mp M)(I\pm M+1) \\ &\qquad \times (I\mp M-1)(I\pm M+2)]^{\frac{1}{2}} \end{aligned} \tag{20}$$

all other matrix elements being zero. The eigenvalues of $\mathscr{H}_Q$ are calculated from the secular equation

$$\|\langle\psi_{IM'}|\,\mathscr{H}_Q\,|\psi_{IM}\rangle - E\delta_{MM'}\| = 0 \tag{21}$$

the left-hand side being an abbreviation for the secular determinant of degree $2I + 1$. For nuclei with half-integral spin, the secular determinant is the product of two identical determinants of degree $I + \frac{1}{2}$. We have thus a quadratic equation for $I = \frac{3}{2}$ with solutions

$$\begin{aligned} E_{\pm\frac{3}{2}} &= \tfrac{1}{4}e^2Qq(1+\eta^2/3)^{\frac{1}{2}} \\ E_{\pm\frac{1}{2}} &= -\tfrac{1}{4}e^2Qq(1+\eta^2/3)^{\frac{1}{2}} \end{aligned} \tag{22}$$

The subscripts on E refer to the quantum numbers M for the limiting case $\eta = 0$, in which $\mathscr{H}_Q$ is diagonal with respect to the basis ψ_{IM}. Magnetic transitions are possible between $E_{\pm\frac{1}{2}}$ and $E_{\pm\frac{3}{2}}$, and these may be detected in an n.q.r. spectrometer (Section III.A). The n.q.r. frequency is

$$\begin{aligned} \nu_Q &= h^{-1}(E_{\pm\frac{3}{2}} - E_{\pm\frac{1}{2}}) \\ &= \frac{e^2Qq}{2h}(1+\eta^2/3)^{\frac{1}{2}} \end{aligned} \tag{23}$$

The solution of the secular equation for $I = 1$ gives three frequencies $\nu_1 = e^2Qq(3+\eta)/4h$, $\nu_2 = e^2Qq(3-\eta)/4h$, and $\nu_3 = e^2Qq\eta/2h$, which allow a separate determination of e^2Qq and η. N.Q.R. frequencies for spins $I > \frac{3}{2}$ have been tabulated by Cohen[4].

B. Influence of a Static Magnetic Field

1. Zeeman splitting of n.q.r. frequencies

In a static magnetic field **H**, there is a magnetic coupling with the nuclear magnetic dipole moment $g_N\beta_N\mathbf{I}$, which has to be added to the nuclear quadrupole coupling $\mathscr{H}_Q$. The hamiltonian of the magnetic coupling is

$$\mathscr{H}_M = -g_N\beta_N\mathbf{H}\cdot\mathbf{I} \tag{24}$$

Hence, in the coordinate system that coincides with the principal axes x,y,z of the field gradient tensor, we have the hamiltonian

$$\mathscr{H} = \mathscr{H}_Q + \mathscr{H}_M$$

$$\mathscr{H}_Q = \frac{e^2Qq}{4I(2I-1)}[3\mathbf{I}_z^2 - I(I+1) + \tfrac{1}{2}\eta(\mathbf{I}_+^2 + \mathbf{I}_-^2)] \tag{25}$$

$$\mathscr{H}_M = -g_N\beta_N(\mathbf{H}_z\mathbf{I}_z + \tfrac{1}{2}\mathbf{H}_-\mathbf{I}_+ + \tfrac{1}{2}\mathbf{H}_+\mathbf{I}_-)$$

$\mathbf{H}_+$ and $\mathbf{H}_-$ are defined in analogy to $\mathbf{I}_+$ and $\mathbf{I}_-$:

$$\mathbf{H}_\pm = \mathbf{H}_x \pm i\mathbf{H}_y$$

In weak magnetic fields ($g_N\beta_N H \ll e^2Qq$), we may treat $\mathscr{H}_M$ as a small perturbation of $\mathscr{H}_Q$. We shall consider in some detail the special example $I = \frac{3}{2}$ and $\eta = 0$, and give only some results for the general case. In this example, $\mathscr{H}_Q$ is diagonal with respect to ψ_{IM} and the $\pm M$ degeneracy of the energy levels of $\mathscr{H}_Q$ is lifted by the perturbation $\mathscr{H}_M$. Hence, we have to solve quadratic secular equations to get the energy levels of $\mathscr{H}$ to first order. The result for the $\pm\frac{1}{2}$ states follows from (22) and (24):

$$\begin{aligned} E_+ &= -\tfrac{1}{4}e^2Qq - \tfrac{1}{2}g_N\beta_N H\cos\theta(4\tan^2\theta + 1)^{\frac{1}{2}} \\ E_- &= -\tfrac{1}{4}e^2Qq + \tfrac{1}{2}g_N\beta_N H\cos\theta(4\tan^2\theta + 1)^{\frac{1}{2}} \end{aligned} \tag{26}$$

where θ is the angle between the magnetic field and the z axis. The calculation for the $\pm\frac{3}{2}$ states (and any $\pm M$ states with $M > \frac{1}{2}$ in the general case of half-integral spin I) is very simple, as $\mathscr{H}_M$ only mixes states with $\Delta M \leqslant 1$. Hence the non-diagonal elements of the secular determinant for $M = \pm\frac{3}{2}$ vanish, and we have

$$\begin{aligned} E_{\frac{3}{2}} &= \tfrac{1}{4}e^2Qq - \tfrac{3}{2}g_N\beta_N H\cos\theta \\ E_{-\frac{3}{2}} &= \tfrac{1}{4}e^2Qq + \tfrac{3}{2}g_N\beta_N H\cos\theta \end{aligned} \tag{27}$$

In Figure 3, the energy levels and transition frequencies ν_α, $\nu_{\alpha'}$, ν_β, $\nu_{\beta'}$ are shown.

The Zeeman splitting of the n.q.r. frequency in a small magnetic field **H** may be used to determine the z direction of the field gradient tensor in a single crystal. From an experimental point of view, the method of 'zero splitting' of the α,α' lines is most convenient. In this method, we determine, with respect to any fixed crystalline frame, special directions of **H** which are given by the equation

$$\nu_\alpha = \nu_{\alpha'} = \nu_Q$$

These directions of **H** form the surface of a right circular cone having a semivertical angle of θ_0. The symmetry axis of the cone coincides with the

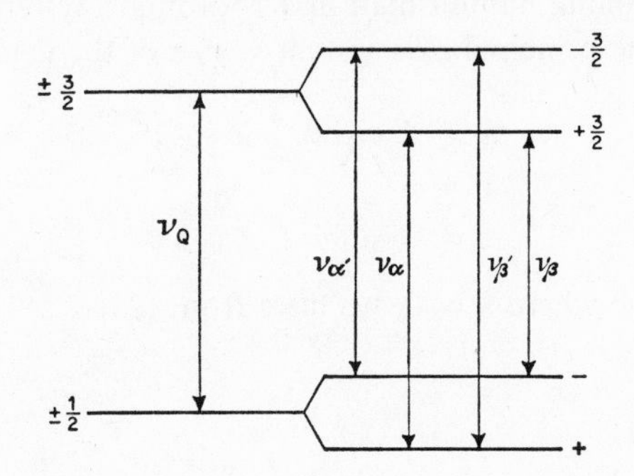

Figure 3. Zeeman splitting of the nuclear quadrupole levels for $I = \frac{3}{2}$.

z axis of the field gradient tensor. θ_0 is given from (26) and (27) by the equation

$$E_{-\frac{3}{2}} - E_{-} = E_{+\frac{3}{2}} - E_{+}$$

resulting in

$$\theta_0 = \tan^{-1}\sqrt{2} = 54° \, 44'$$

In the more general case ($\eta \neq 0, I = \frac{3}{2}$), θ_0 is a function of the azimuthal angle ϕ. A somewhat lengthy calculation gives

$$\sin^2 \theta_0 = \frac{2}{3 - \eta \cos 2\phi} \tag{28}$$

for the **H** directions of zero splitting. The asymmetry parameter η as well as the directions of the x and y axes of the field gradient tensor may be determined from (28). The x axis is given by the maximum of θ_0, that is at $\phi = 0$, and the y axis by the minimum at $\phi = \pi/2$. For the asymmetry parameter, we have from (28)

$$\eta = 3\,\frac{\sin^2[\theta_0(0)] - \sin^2[\theta_0(\pi/2)]}{\sin^2[\theta_0(0)] + \sin^2[\theta_0(\pi/2)]} \tag{29}$$

Further details on Zeeman splitting may be found in the book by Das and Hahn[5].

2. Quadrupole splitting in strong and intermediate magnetic fields

In many ionic crystals, $e^2Qq \ll g_N\beta_N H$ in strong magnetic fields. It is then more convenient to measure the quadrupole splitting of the n.m.r. lines instead of the pure n.q.r., which may be unobservable at very low frequencies. We now wish to treat $\mathscr{H}_Q$ as a small perturbation of $\mathscr{H}_M$, by taking the magnetic field direction to be the quantization axis, and calculate the coupling hamiltonian in a coordinate system X, Y, Z which is related with the principal axis system x, y, z of the field gradient tensor by

$$\begin{aligned} x &= X\cos\theta - Z\sin\theta \\ y &= Y \\ z &= X\sin\theta + Z\cos\theta \end{aligned}$$

In the simple case where $\eta = 0$, we have from (25)

$$\begin{aligned} \mathscr{H} &= \mathscr{H}_M + \mathscr{H}_Q \\ \mathscr{H}_M &= -g_N\beta_N H\mathbf{I}_Z \\ \mathscr{H}_Q &= \frac{e^2Qq}{4I(2I-1)}\{\tfrac{1}{2}(3\cos^2\theta - 1)[3\mathbf{I}_Z^2 - I(I+1)] + \tfrac{3}{2}\sin\theta\cos\theta \\ &\quad \times [\mathbf{I}_Z(\mathbf{I}_+ + \mathbf{I}_-) + (\mathbf{I}_+ + \mathbf{I}_-)\mathbf{I}_Z] \\ &\quad + \tfrac{3}{4}\sin^2\theta(\mathbf{I}_+^2 + \mathbf{I}_-^2)\} \end{aligned} \tag{30}$$

$\mathbf{I}_\pm = \mathbf{I}_X \pm \mathbf{I}_Y$ being defined in the coordinate system X, Y, Z. To a first order, the perturbation treatment yields

$$\begin{aligned} \nu_M &= h^{-1}(E_{M-1} - E_M) \\ &= \nu_L + \tfrac{1}{2}\nu_Q(M - \tfrac{1}{2})(3\cos^2\theta - 1) \end{aligned} \tag{31}$$

E_M is the eigenstate of $\mathscr{H}$ that corresponds to the quantum state ψ_{IM}. ν_L and ν_Q are the Larmor and n.q.r. frequencies

$$\begin{aligned} \nu_L &= g_N\beta_N H/h \\ \nu_Q &= 3e^2Qq/2I(2I-1)h \end{aligned} \tag{32}$$

We see from (31) that nuclear quadrupole coupling splits the n.m.r. frequency into $2I + 1$ components, the central line $\nu_{\frac{1}{2}}$ being equal to ν_L in first order. The splittings $\nu_M - \nu_{-M}$ $(M \neq \frac{1}{2})$ are not influenced by a second-order perturbation treatment, but the central line $(M = \frac{1}{2})$ is shifted. The result is

$$\nu_{\frac{1}{2}} = \nu_L - \frac{\nu_Q^2[I(I+1) - \frac{3}{4}]}{16\nu_L}(1 - 3\cos^2\theta)(9\cos^2\theta - 1) \tag{33}$$

The general case of a non-vanishing asymmetry parameter and unknown principal axes of the field gradient tensor has been treated by Volkoff[6] up to the second order of perturbation theory.

No perturbation treatment is possible if e^2Qq and $g_N\beta_N H$ are comparable in magnitude. In principle, we could make the magnetic field as small as necessary for a perturbation calculation, but sometimes the n.q.r. lines are not observable because the signal-to-noise ratio is too low, whereas a resonance line is observed in a higher magnetic field. It is then possible to get the quadrupole coupling tensor by a method[7] which has some similarity to the method of zero splitting described above. Instead of the curves of zero splitting, we determine the curves of constant frequency which are given by the directions of the magnetic field vector with respect to any fixed crystalline frame. The principal axes of the field gradient tensor may be determined from the symmetry of the curves. To get e^2Qq and η, we have to diagonalize the matrix of $\mathscr{H} = \mathscr{H}_Q + \mathscr{H}_M$ for some special direction. In paramagnetic compounds, the quadrupole coupling tensor as well as the paramagnetic chemical shift tensor may be determined separately from curves of constant frequency.

C. Field Gradient Tensor

The previous sections have demonstrated that, provided the nuclear quadrupole moment is known, the observables extracted from n.q.r. experiments are the components of the field gradient tensor. We shall now discuss the way in which these are related to the electronic structures of molecules. In (15) to (33), q, η, and consequently $\mathscr{H}_Q$ were considered to be averages over the electronic ground state Ψ_0. In this section, we write $\mathbf{q}$ and $\boldsymbol{\eta}$ for the operators to distinguish them from the averages over Ψ_0.

From any approximate wave function of the ground state, we can calculate approximate values of q and η which may be compared with experiment. Hence, n.q.r. data provide an experimental test of approximate wave functions, which has the advantage that only the electronic ground state is involved, in contrast to optical spectroscopy, where measured differences between energies are also averages over excited states.

The molecular-orbital approximation (see chapter by H. H. Schmidtke, p. 131ff.) is often used to correlate experimental data with chemical concept. First let us consider a simple diatomic molecule AB. By linear combination of two non-degenerate atomic orbitals ψ_A and ψ_B, we obtain two orthogonal molecular orbitals

$$\begin{aligned} \psi_1 &= N_1(\psi_A + \kappa\psi_B) \\ \psi_2 &= N_2(\psi_B - \lambda\psi_A) \end{aligned} \tag{34}$$

with normalization constants

$$N_1 = (1 + \kappa^2 + 2\kappa S)^{-\frac{1}{2}}$$
$$N_2 = (1 + \lambda^2 - 2\lambda S)^{-\frac{1}{2}} \tag{35}$$

and the overlap integral

$$S = \langle \psi_A \mid \psi_B \rangle \tag{36}$$

We see that

$$\lambda = \frac{\kappa + S}{1 + \kappa S} \tag{37}$$

from the orthogonality of ψ_1 and ψ_2. If the one-electron energies are as shown in Figure 4, we call ψ_1 a bonding and ψ_2 an antibonding molecular orbital.

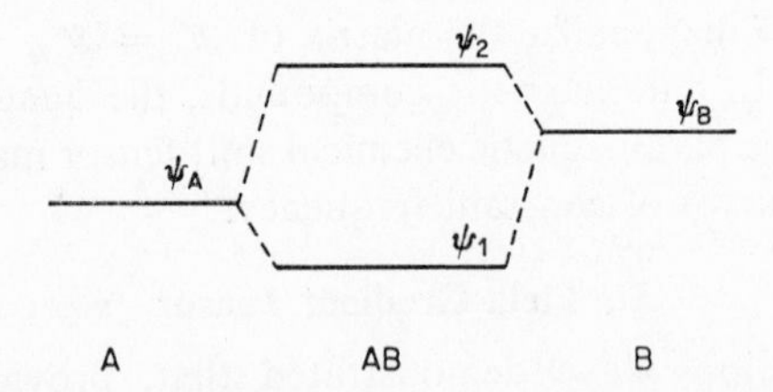

Figure 4. Energy levels for a diatomic molecule.

In a two-electron bond of a diatomic molecule, we have two electrons of opposite spin in ψ_1. Let us calculate the field gradient produced by these two electrons at the nucleus A, which may have a nuclear quadrupole moment. The field gradient tensor is axially symmetric around the A—B bond ($\eta = 0$), because ψ_1 is an orbitally non-degenerate Σ state of the molecule. Hence we have only to calculate the average

$$q_{AB} = \langle \Psi_0 | \mathbf{q} | \Psi_0 \rangle \tag{38}$$

over the electronic ground state

$$\Psi_0 = \frac{1}{\sqrt{2}}\{\alpha(1)\beta(2) - \beta(1)\alpha(2)\}\psi_1(1)\psi_1(2) \tag{39}$$

From (13) and (10), we have

$$\mathbf{q} = \int \frac{3\cos^2\theta - 1}{r^3}\left[\sum_j \delta(\mathbf{r} - \mathbf{r}_j)\right] d\tau = \sum_j \frac{3\cos^2\theta_j - 1}{r_j^3} \tag{40}$$

the sum being taken over all the electrons (two in our example). θ_j is the angle of the position vector of electron j to the AB direction, and $\mathbf{r}_j$ is the distance of electron j from the nucleus A. Insertion of (39) and (40) into (38) yields

$$q_{AB} = 2\langle\psi_1| \frac{3\cos^2\theta - 1}{r^3} |\psi_1\rangle$$

$$= 2N_1^2\left[\langle\psi_A| \frac{3\cos^2\theta - 1}{r^3} |\psi_A\rangle + 2\kappa\langle\psi_A| \frac{3\cos^2\theta - 1}{r^3} |\psi_B\rangle + \kappa^2\langle\psi_B| \frac{3\cos^2\theta - 1}{r^3} |\psi_B\rangle\right] \quad (41)$$

In the approximation of Townes and Dailey[8,9], we assume the validity of the equation

$$q_{AB} = \frac{2}{1 + \kappa^2}\langle\psi_A| \frac{3\cos^2\theta - 1}{r^3} |\psi_A\rangle \quad (42)$$

The overlap integral S, and the second and third term of (41), are considered to be negligible*. This approximation has proved very useful for molecules with a covalent bond between atoms which are not too small, e.g. ICl. There is also a large number of applications[5] to polyatomic molecules AR, where R stands for the rest of the molecule. In small molecules, the two-centre integrals $\langle\psi_A|\,(3\cos^2\theta - 1)/r^3\,|\psi_B\rangle$ contribute an appreciable amount to the field gradient. Improved (MO–LCAO–SCF) wave functions have been used to compute the field gradient in, e.g, HCl[10], Li_2, N_2, CO, and BF[11,12].

The total field gradient eq at the nucleus A consists of $-eq_{AB}$ ($e > 0$), and contributions $-eq_A$ and eq_B of the cores of the atoms A and B. $-eq_A$ may contribute a large amount to the field gradient if the core of A is not spherically symmetrical. In halogen compounds, for instance, we have contributions from electrons in p_x and p_y orbitals, if ψ_A is taken to be a p_z orbital. Usually, ψ_A is assumed to be an sp hybrid

$$\psi_A = s\psi_s + (1 - s^2)^{\frac{1}{2}}\psi_{p_z}$$

We then have in addition a core contribution to $-eq_A$ from two electrons in the orbital

$$\psi'_A = (1 - s^2)^{\frac{1}{2}}\psi_s - s\psi_{p_z}$$

that is orthogonal to ψ_A. The contribution of the core B to the field gradient may be approximated by

$$eq_{\text{ion}} = 2Zer_{AB}^{-3} \quad (43)$$

* It should be mentioned that S and $\langle\psi_A|\,(3\cos^2\theta - 1)/r^3\,|\psi_B\rangle$ contribute amounts of opposite sign to the field gradient that partly cancel.

which is the field gradient produced by a point charge Ze at the nucleus B if the distance between A and B is r_{AB}.

To summarize the result for a halogen compound AR in the Townes–Dailey approximation, we assume that there are two electrons in each of the orbitals

$$\begin{aligned}\psi_1 &= (1 + \kappa^2)^{-\frac{1}{2}}(\psi_A + \kappa\psi_R) \\ &= (1 + \kappa^2)^{-\frac{1}{2}}[s\psi_s + (1 - s^2)^{\frac{1}{2}}\psi_{p_z} + \kappa\psi_R]\end{aligned} \tag{44}$$

ψ'_A, ψ_{p_x}, and ψ_{p_y}. ψ_R stands for the orbitals of the rest of the molecule. We can see from symmetry considerations that the integrals

$$\langle\psi_s|\frac{3\cos^2\theta - 1}{r^3}|\psi_s\rangle, \qquad \langle\psi_s|\frac{3\cos^2\theta - 1}{r^3}|\psi_{p_z}\rangle$$

vanish. Hence the field gradient at the halogen nucleus A is

$$\begin{aligned}eq &= -\left(2\frac{1 - s^2}{1 + \kappa^2} + 2s^2 - 2\right)eq_{p_z}^{at} + eq_{ion} \\ &= (1 - s^2)\frac{2\kappa^2}{1 + \kappa^2}eq_{p_z}^{at} + eq_{ion}\end{aligned} \tag{45}$$

We have used the abbreviation

$$-eq_{p_z}^{at} = -e\langle\psi_{p_z}|\frac{3\cos^2\theta - 1}{r^3}|\psi_{p_z}\rangle \tag{46}$$

for the field gradient of one electron in the orbital ψ_{p_z}, and the obvious relation

$$q_{p_x}^{at} + q_{p_y}^{at} + q_{p_z}^{at} = 0$$

For halogens, $eq_{p_z}^{at}$ has been determined by Jaccarino and King[1] from atomic-beam resonances.

In polyatomic molecules, the field gradient at the halogen nucleus need not be of axial symmetry. Sometimes p_x and p_y orbitals form covalent π bonds

$$\begin{aligned}\psi_x &= N_x(\psi_{p_x} + \kappa_x\psi_{R,x}) \\ \psi_y &= N_y(\psi_{p_y} + \kappa_y\psi_{R,y})\end{aligned} \tag{47}$$

and it may be that $\kappa_x \neq \kappa_y$. $\psi_{R,x}$ and $\psi_{R,y}$ stand for orbitals of the rest of the molecule, which are assumed to have the appropriate symmetry to form π bonds. From a simple calculation, we obtain the field gradient for a molecule with two electrons in each of the orbitals ψ_1, ψ'_A, ψ_x, and ψ_y:

$$\begin{aligned}eq &= [(1 - s^2)\sigma - \tfrac{1}{2}(\pi_x + \pi_y)]eq_{p_z}^{at} + eq_{ion} \\ \eta &= 3(\pi_x - \pi_y)q_{p_z}^{at}/2q + \eta_{ion} \\ \eta_{ion} &= (\varphi_{xx}^{ion} - \varphi_{yy}^{ion})/eq\end{aligned} \tag{48}$$

The abbreviations

$$\sigma = \frac{2\kappa^2}{1 + \kappa^2}, \qquad \pi_x = \frac{2\kappa_x^2}{1 + \kappa_x^2}, \qquad \pi_y = \frac{2\kappa_y^2}{1 + \kappa_y^2} \tag{49}$$

which we have used in (48) are called the covalent characters of the σ, π_x, and π_y bonds. Sometimes other abbreviations are used, such as the ionic character (i) of the σ bond

$$i = 1 - \sigma \tag{50}$$

and the total ionic character (I).

$$I = 1 - \sigma - \pi_x - \pi_y \tag{51}$$

Following the Townes–Dailey approximation, all overlap integrals have been neglected in (48).

In (48), $\varphi_{xx}^{\text{ion}}$, $\varphi_{yy}^{\text{ion}}$, and $\varphi_{zz}^{\text{ion}} = eq_{\text{ion}}$ are the components of the ionic field gradient given by (43) in a diatomic molecule. In an ionic crystal, all ionic charges outside A contribute to the ionic field gradient, and (48) is valid only if the principal axes of the ionic and covalent field gradients coincide. In a point-charge model of the ionic crystal, the components of the ionic field gradient are the sums

$$\varphi_{xx}^{\text{ion}} = e \sum_j Z_j \sum_i \frac{3x_{ij}^2 - r_{ij}^2}{r_{ij}^5}$$
$$\varphi_{xy}^{\text{ion}} = e \sum_j Z_j \sum_i \frac{3x_{ij}y_{ij}}{r_{ij}^5} \tag{52}$$

and corresponding expressions are found for $\varphi_{yy}^{\text{ion}}$, $\varphi_{zz}^{\text{ion}}$, $\varphi_{xz}^{\text{ion}}$, and $\varphi_{yz}^{\text{ion}}$. eZ_j is the charge of j-type ions. $\mathbf{r}_{ij}$ is the position vector of the ith ion of type j, the nucleus A being the origin of any cartesian coordinate system fixed in the crystal. Generally, the sums (52) are determined by numerical summation over about 10,000 ions within a sphere around A[13]. Sometimes planewise summation is more convenient[14].

In ionic compounds, the field gradient calculated from a point-charge model is in poor agreement with experiment unless the polarizability of the ions is taken into consideration. This amounts to multiplying q_{ion} by the so-called 'antishielding factor' $1 - \gamma_\infty$ which generally reinforces the field gradient. The values of $1 - \gamma_\infty$ which have been calculated by Sternheimer and others (see, e.g., ref. 15) are very large for halogen ions, e.g. 57.6 for Cl^-. This large polarization effect leads to difficulties in the chemical interpretation of quadrupole coupling constants of ionic halogen compounds. Let us consider the example of $CrCl_3$ (see Section IV.A.2) where the point-charge model yields a field gradient of $q_{ion} = -0.47 \times 10^{24}\ cm^{-3}$ whereas the experimental value is $|q| = 9.4 \times 10^{24}\ cm^{-3}$. Multiplication of q_{ion} by $1 - \gamma_{\infty} = 57.6$ leads to a marked disagreement with experiment. A consistent MO treatment, on the other hand, does not allow for any polarization of ions and predicts a very small field gradient of $\sim q_{ion}$ in the ionic limit ($\kappa = 0$). A somewhat inconsistent compromise would be to account for antishielding in ionic compounds by an empirical factor (e.g.[16] $1 - \gamma'_{\infty} = 10$ for Cl^-), and to multiply in (48) the ionic field gradient q_{ion} by this factor. Bershohn and Shulman[17] propose to describe any polarization effect by some covalent bonding character within the MO approximation. De Wijn[18] has pointed out that the polarization of the halogen ion leads to an expansion of the electron density into a region where repulsion with the electrons of the neighbouring positive ion should be considered. This electron repulsion effectively quenches the quadrupole polarization.

III. EXPERIMENTAL METHODS

A. N.Q.R. Spectrometers

In principle, any n.m.r. spectrometer* may be used to detect nuclear quadrupole resonances. In such an experiment, magnetic transitions are induced between energy levels which are eigenvalues of the quadrupole coupling hamiltonian (19). The response to these transitions is the observable n.q.r. There is no marked difference in an n.m.r. experiment where magnetic transitions are induced between the energy levels of the magnetic coupling hamiltonian (24).

There are, however, practical aspects that make some types of spectrometers more suitable for n.q.r. work. When searching for an unknown quadrupole resonance, an oscillator is required which is capable of an easy frequency variation over a large frequency range. Sensitivity should be high because of the low intensity of the usually broad n.q.r. lines. To increase line intensity, we should be able to use high r.f. power. Generally, no saturation is to be expected unless the relaxation is slow at low temperatures.

In most n.q.r. spectrometers, superregenerative oscillators are used, which fit well the requirements of high sensitivity, high r.f. power, and easy frequency variation. In a superregenerative circuit, oscillations are periodically quenched at a rate of about 0.1% of the oscillator frequency. Roughly speaking, the oscillator acts as a transmitter during the 'on' periods of the oscillations and induces a transverse nuclear magnetization in the sample which is 'received' during the subsequent 'off' periods.

* See chapter by D. R. Eaton.

The reader should refer to the literature for a detailed description of super-regenerative oscillators[19,20].

Other oscillator types[5,21] have been used for n.q.r. line shape studies and measurements at low frequencies. For instance, most ^{14}N quadrupole resonances (frequency below 4 Mc/s) have been measured with the marginal oscillator of Watkins[22].

It is sometimes argued in the literature (e.g. ref. 23) that the crossed-coil induction method is not suitable for n.q.r. detection. The argument is as follows. While in an n.m.r. experiment the transverse nuclear magnetization vector rotates clockwise around the external magnetic field direction, in an n.q.r. experiment ($\eta = 0$ for simplicity), the nuclear magnetization vector may be split into two rotating components, one rotating clockwise and the other counterclockwise around the z axis of the field gradient. In zero magnetic field, both components rotate with the same frequency ν_Q because of the $\pm M$ degeneracy of the quadrupole energy levels (22), and they add up to zero in the direction of the receiver coil of a conventional crossed-coil probe[24]. But in a small static magnetic field the $\pm M$ degeneracy is lifted; therefore, the two components rotate with different frequencies, resulting in a non-vanishing signal in the direction of the receiver coil. As magnetic field modulation is used in commercial crossed-coil n.m.r. spectrometers, they are well suited to the detection of pure n.q.r., especially at low frequencies[25,26].

B. Other Techniques for Studying Nuclear Quadrupole Coupling

Apart from pure n.q.r., there are a number of other experimental methods yielding information on nuclear quadrupole coupling. A short description of n.m.r. methods of measuring the nuclear quadrupole coupling constant e^2Qq in solids is given in Section II.B.2 of this chapter. In favourable cases, it is possible to obtain e^2Qq from the n.m.r. relaxation times in liquids[27,28]. Hartmann and Hahn[29] have used a nuclear double-resonance technique to determine e^2Qq at nuclei of low natural abundance when the substance contains simultaneously nuclei of high abundance giving a strong signal (e.g. e^2Qq of ^{41}K in $KClO_3$). In some cases, it is possible to determine the magnitude and even the sign of e^2Qq from the hyperfine structure of e.p.r. spectra[30]. The electron nuclear double resonance (Endor) technique[30] permits a better resolution and an accurate measurement of the hyperfine separations and thus of e^2Qq. In free molecules, e^2Qq may be determined from the rotational spectra[31] in gases. The molecular- or atomic-beam techniques[32] yield e^2Qq in small molecules and atoms. It should be mentioned that nuclear quadrupole coupling was discovered by Schüler and Schmidt[33], when they investigated small shifts in the hyperfine structure of optical spectra. There are some other experimental methods that fall into the region of nuclear physics. A well-known example is the Mössbauer effect which is described by J. Danon in Chapter 8. Others are the angular correlation of successive nuclear radiations[34] and the γ-emission by Coulomb excitation[35].

IV. APPLICATIONS

A. Information on Chemical Bonding

Nuclear quadrupole resonance frequencies have been measured in a large number of halogen compounds. In the following section, we shall consider whether quantitative conclusions can be drawn from experimental field gradients about the ionicity of halogen bonds in inorganic compounds.

1. Diatomic halides

We shall discuss in some detail the field gradient in ICl, where the quadrupole coupling is known for both the ^{35}Cl and ^{127}I nuclei from the microwave spectrum of the molecule[36], and only give some results for other diatomic halides. Chlorine and iodine have the electron configurations $3s^2 3p^5$ and $5s^2 5p^5$, respectively. According to MO theory, we put the $3p_z$ and $5p_z$ electrons of Cl and I into the σ-bonding molecular orbital (34)

$$\psi_1 = (1 + \kappa^2)^{-\frac{1}{2}}(\psi_{\text{Cl}} + \kappa\psi_{\text{I}})$$

The overlap integral $S = \langle\psi_{\text{Cl}} \mid \psi_{\text{I}}\rangle$ is assumed to vanish in the Townes–Dailey approximation. Neglecting possible π bonding, the field gradient at the chlorine nucleus is given by (45), (49), and (50) as

$$eq(\text{Cl}) = (1 - s_{\text{Cl}}^2)(1 - i)eq_{p_z}^{\text{at}}(\text{Cl}) \tag{53}$$

The small term eq_{ion} of (45) has been dropped in (53).

We obtain the field gradient at the iodine nucleus by a calculation similar to that in Section II.C. The only difference is that we have to keep the last term in (41) instead of the first one, because r is now the distance of the electron to the I nucleus rather than to the Cl nucleus. We therefore have

$$q_{\text{ICl}}(\text{I}) = \frac{2\kappa^2}{1 + \kappa^2}\langle\psi_{\text{I}}|\,\frac{3\cos^2\theta - 1}{r^3}\,|\psi_{\text{I}}\rangle$$

instead of (42) and

$$eq(\text{I}) = (1 - s_{\text{I}}^2)(1 + i)eq_{p_z}^{\text{at}}(\text{I}) \tag{54}$$

instead of (53).

The experimental values of the nuclear quadrupole coupling constant at ^{35}Cl and ^{127}I in ICl as well as in the free I and Cl atoms are given in Table 1. To determine the ionic character i of the I—Cl bond, we must

Table 1. Nuclear quadrupole coupling and ionic character in ICl.

	^{35}Cl	^{127}I
$\lvert e^2Qq/h\rvert$ (Mc/s)	82.5	2944
$\lvert e^2Qq_{p_z}^{at}/h\rvert$ (Mc/s)	109.74	2292.8
i	0.25 ($s_{Cl}^2 = 0$)	0.28 ($s_I^2 = 0$)
	0.11 ($s_{Cl}^2 = 0.15$)	

know the amount of s hybridization for Cl or I. To a rough approximation, we may consider the hybridization to be small ($s_{Cl}^2 = s_I^2 = 0$)[17,37], and obtain an ionic character of 0.25 or 0.28 from (53) or (54), respectively. Dailey and Townes[9] argued that it would be more realistic to take some hybridization into account at least for the more electronegative Cl atom. Hybridization may be negligible at the I atom, because the promotional energy required for sp hybridization is higher in the positive ionic state. The following rule has been proposed[9] to hold for all diatomic halides: in Cl, Br, and I bonds, the s hybridization is 15% for all cases where these atoms are more electronegative than the atoms to which they are bonded by 0.25 units, otherwise hybridization is zero. As the electronegativity difference between I and Cl is 0.50[38,39], this rule requires $s_{Cl}^2 = 0.15$ and $s_I^2 = 0$, yielding $i = 0.11$ from (53). The ionic character of 0.28 calculated from (54) may be too high because $eq^{at}(I^+)$ of the I^+ ion is larger than $2eq^{at}(I)$ of the neutral atom by a fraction of 0.12[9]. Taking this effect into account, Dailey and Townes obtain $i = 0.23$ from the experimental quadrupole coupling constant at the ^{127}I nucleus.

The ionic character has been calculated[9] in a large number of diatomic halides by the method exemplified with ICl. The scale of increasing ionic character has been compared with those scales calculated from electronegativity differences and from dipole moments. The good qualitative agreement indicates that the concept of ionicity rests on a satisfactory experimental basis at least in diatomic molecules. Whitehead and Jaffé[40] have also discussed the relation between quadrupole coupling and ionic character. They calculated the s character of halogen bonds from the experimental quadrupole coupling constants and from an ionic character defined through its relation to valence-state energies.

2. Transition metal halides

Nuclear quadrupole coupling has been studied in a relatively small number of transition metal halides. This is due in part to experimental difficulties, as n.q.r. signals are often broad and not easy to find in these substances. We shall discuss nuclear quadrupole coupling for three examples typical of metal–halogen bonding in transition metal complexes.

In $CrCl_3$, the n.q.r. of Cl has been investigated in polycrystalline material[41–43]. Morosin and Narath have also reinvestigated the crystal structure of this compound (see Section IV.B), and they have calculated the field gradient at the chlorine nucleus from a point-charge model. The Cr^{3+} ions are surrounded in the crystal by a distorted octahedron of Cl^- ions, the Cr—Cl distances being 2.340, 2.342, and 2.347 Å for two Cl^- ions each. Each Cl^- ion has two nearest Cr^{3+} neighbours. The $Cr\widehat{Cl}Cr$ angle is somewhat larger than 90°. From their field gradient calculation using the ionic model, Morosin and Narath find that the dominant contribution to the chlorine field gradient comes from the two nearest-neighbour Cr^{3+} point charges.

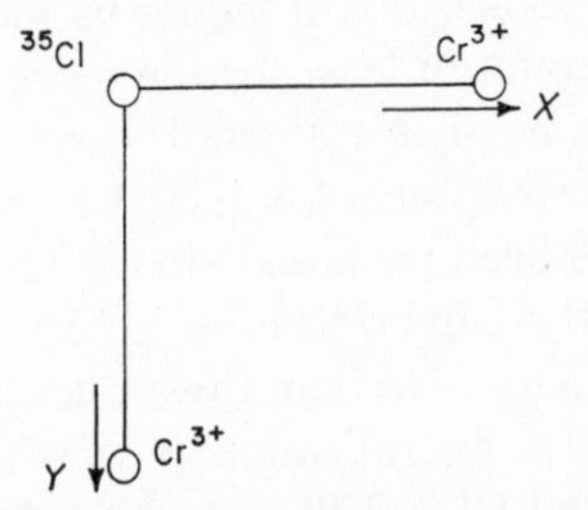

Figure 5. ^{35}Cl nucleus in the field of two Cr^{3+} charges.

We consider the simplified model of two Cr^{3+} ions at a distance of 2.34 Å from the Cl nucleus, and we assume $\pi/2$ for the $Cr\widehat{Cl}Cr$ angle. Then, according to (10), (13), and Figure 5, we have

$$\frac{1}{e}\varphi_{zz} \equiv q' = (+3)\frac{(-2)}{(2.34)^3} \times 10^{24}\ \text{cm}^{-3} = -0.47 \times 10^{24}\ \text{cm}^{-3} \qquad (55)$$

and $\eta = 0$ for the field gradient produced by the two Cr^{3+} point charges at the Cl nucleus. We now assume that there is one σ bond from Cl^- to each of the Cr^{3+} ions. According to the Townes–Dailey approximation (equations 48, 50), the field gradient (in units of e) of each Cr—Cl bond is

$$q = (1 - s^2)(1 - i)q_{p_z}^{\text{at}} + q_{\text{ion}} \qquad (56)$$

the X axis being the principal (z) axis of the field gradient for the Cr(X)—Cl bond (the Y axis for the Cr(Y)—Cl bond). The orthogonality of the two

σ bonds requires $s^2 = 0$*. In the point-charge model ($i = 1$), the field gradient of each Cr—Cl bond would be $q_{ion}(i = 1) = -\frac{1}{2}q'(1 - \gamma_\infty)$. In the case of partially covalent bonding ($i \neq 1$), the charge transfer from the six Cl^- ions to the Cr^{3+} ion leads to

$$q_{ion} = \frac{3 - 6(1 - i)}{3}[-\tfrac{1}{2}q'(1 - \gamma_\infty)]$$

$$= (1 - \gamma_\infty)(2i - 1)0.24 \times 10^{24}\ \text{cm}^{-3} \qquad (57)$$

We have $q_{p_z}^{at} = 39.99 \times 10^{24}$ cm^{-3} from Cl atomic-beam measurements[1] and $q = 4.7 \times 10^{24}$ cm^{-3} calculated for the Cr—Cl bond from the experimental quadrupole frequency of 12.9 Mc/s in $CrCl_3$. If we assume† an 'empirical' Sternheimer factor of 10^{17}, we obtain from (55), (56), and (57) an ionic character of $i = 0.93$ for each Cr—Cl bond in $CrCl_3$. In this approximation, there is a covalent contribution of 2.6×10^{24} cm^{-3} and an ionic contribution of 2.1×10^{24} cm^{-3} to the experimental field gradient of 4.7×10^{24} cm^{-3} of each Cr—Cl bond.

The ionic character of the metal–halogen bond has been determined[7] in a similar manner from the experimental n.q.r. data on *trans*-(Cr en_2Cl_2)Cl·HCl·$x$$H_2O$ and *trans*-(Co en_2Cl_2)Cl·HCl·$x$$H_2O$. For the ^{35}Cl nuclei of the complex ions, the n.q.r. frequencies are 10.3 and 16.1 Mc/s. The resultant ionic characters are 0.90 and 0.77 for the Cr—Cl and the Co—Cl bonds, respectively.

3. *Hexahalides of heavier atoms*

Nakamura and coworkers[44–50] have investigated the n.q.r. in the hexahalides of heavier atoms in which covalency is known to be larger than in transition metal halides. Some of their results are reproduced in Table 2. In calculating the ionic character i from (45), the assumption has been made that the ionic contribution eq_{ion} to the field gradient at the nucleus is negligible. That is, the quadrupole coupling constants of Table 2, which are taken from n.q.r. frequencies measured at 77°K, are considered to result only from the covalent contribution of the M—X σ bonds.

In some cases, Nakamura and coworkers[48,50] have discussed the possibility of π bonding. As it is possible to obtain the π character of the Ir—Cl bond from the ligand hyperfine structure of e.p.r. in hexachloroiridates[51], the total ionic character $I = 1 - \sigma - \Pi$ ($\Pi = \pi_x + \pi_y$, see equation 51) may be calculated from (48). The result is $I = 0.47$ when $\Pi = 0.054$ is the π character obtained from e.p.r. measurements. To estimate the π bonding in other $5d^n$ halides, Nakamura and coworkers

* See ref. 5, p. 141.

† Morosin and Narath[43] assume $i = 1$ and thus have $1 - \gamma_\infty = 22$ from the n.q.r. frequency in $CrCl_3$.

Table 2. Nuclear quadrupole coupling and ionic character in hexahalides[a].

Hexahalide	X = ^{35}Cl		X = ^{79}Br		X = ^{127}I	
	$\lvert e^2Qq/h \rvert$ (Mc/s)	i	$\lvert e^2Qq/h \rvert$ (Mc/s)	i	$\lvert e^2Qq/h \rvert$ (Mc/s)	i
K_2WX_6	20.4[b]	0.78 (0.43)				
K_2ReX_6	27.9	0.72 (0.45)	231	0.66 (0.39)	822	0.57 (0.32)
K_2OsX_6	33.8	0.62 (0.47)				
K_2IrX_6	41.7	0.55 (0.47)				
K_2PtX_6	52.0	0.44	406	0.38	1360	0.30
K_2PdX_6	53.5	0.43	411	0.37		
K_2SnX_6	31.5	0.66	261	0.60	884[c]	0.55
$(NH_4)_2PbX_6$	34.5	0.63				
K_2SeX_6	41.2	0.56	346	0.47		
K_2TeX_6	30.3[d]	0.68	271	0.58	1020	0.48

[a] The data are collected from refs. 41–47. The numbers in brackets stand for the total ionic character I where π bonding has been considered (see text).
[b] At 22.5°C.
[c] From n.q.r. of Rb_2SnI_6.
[d] From n.q.r. of $(NH_4)_2TeCl_6$.

assume* that 'the number of electrons migrating from a halogen ion to the central metal ion through π bonding is proportional to the number of electronic vacancies in the t_{2g} orbitals of the central metal ion.' In MO theory, this means that Π is proportional to $6 - n$, where n is the number of electrons in the antibonding t_{2g} orbital. As $n = 2$, 3, and 4 for W^{IV}, Re^{IV}, and Os^{IV} complexes, we get $\Pi = 0.22$, 0.16, and 0.11 from the π character of the Ir—Cl bond ($n = 5$ in Ir^{IV}). If we use these Π values to calculate I from (48), we obtain the numbers shown in brackets in Table 2. They are appreciably smaller than the values of i calculated for pure σ bonding. We mention that n is equal to 3 and 6 in Cr^{III} and Co^{III} complexes, respectively, and therefore π bonding should be possible in the former but not in the latter within the simplified MO picture (see Section IV.A.2).

We have mentioned above (Section IV.A.1) that, in the case of diatomic molecules, the ionic characters calculated from nuclear quadrupole coupling constants within the frame of the Townes–Dailey approximation fit well with the overall concept of ionicity which has developed from other physical quantities such as dipole moments, bond lengths, force constants, etc. In polyatomic molecules, it is far more difficult to develop an unambiguous general concept of ionicity. The ionic character i (and also I) calculated from n.q.r. is restricted to a special approximation within the molecular-orbital theory. But the MO theory itself gives only an approx-

* Within the frame of MO theory, this assumption is valid when the overlap integrals between the central ion and the ligands vanish. But even in the rather ionic NiF_6^{4-} ion, there is considerable overlap between Ni^{2+} and F^- [52].

imation to the wave function of the molecule. Observables calculated from this approximate wave function are often in poor agreement with experiment[52]. The interpretation of n.q.r. data, in addition, is hampered by the uncertainty of the Sternheimer polarization (see Section II.C). It is rather astonishing, therefore, that there exists a linear relation between the ionic characters given in Table 2 and the electronegativity differences of the central atoms and the ligand halogen atoms, if the central atoms belong to the same family of the periodic table[49]. This indicates, perhaps, that the empirical concept of ionicity, as represented by the electronegativities of the elements, is useful irrespective of the difficulties inherent in a theoretical treatment of polyatomic molecules.

4. Quadrupole coupling of nuclei other than halogens

There are many nuclei having an electric quadrupole moment. Most of them are of low natural abundance so that n.q.r. measurements are difficult if not impossible, but there remains an appreciable number of isotopes which are well suited to n.q.r. experiments. We give a list (Table 3) of

Table 3. Nuclear quadrupole moments[a].

Isotope	Natural abundance (%)	Spin I	Quadrupole moment Q (10^{-24} cm^2)
^{2}H	1.56×10^{-2}	1	2.783×10^{-3}
^{7}Li	92.57	$\frac{3}{2}$	0.035
^{9}Be	100	$\frac{3}{2}$	0.03
^{11}B	81.17	$\frac{3}{2}$	3.55×10^{-2}
^{14}N	99.635	1	1.06×10^{-2}
^{23}Na	100	$\frac{3}{2}$	0.097
^{27}Al	100	$\frac{5}{2}$	0.149
^{35}Cl	75.4	$\frac{3}{2}$	-7.894×10^{-2}
^{39}K	93.08	$\frac{3}{2}$	0.07
^{55}Mn	100	$\frac{5}{2}$	0.3
^{57}Fe	2.245	$\frac{3}{2}$[b]	0.15[b]
^{59}Co	100	$\frac{7}{2}$	0.5
^{63}Cu	69.09	$\frac{3}{2}$	-0.157
^{75}As	100	$\frac{3}{2}$	0.27
^{79}Br	50.57	$\frac{3}{2}$	0.33
^{93}Nb	100	$\frac{9}{2}$	-0.2
^{121}Sb	57.25	$\frac{5}{2}$	-0.52
^{127}I	100	$\frac{5}{2}$	-0.59
^{201}Hg	13.24	$\frac{3}{2}$	0.42

[a] Quadrupole moments have been selected from a table in ref. 53.

[b] Values for the excited state are given (see chapter by J. Danon).

some nuclei which have been most frequently the subject of experimental quadrupole coupling studies*, and refer the reader to Section 3.9 of ref. 5 for a review of n.q.r. investigations. The more recent literature is accessible through the excellent Documentation of Molecular Spectroscopy (D.M.S.) title list (combined with a punched card index) on n.m.r., n.q.r., and e.p.r.[54].

B. Structural Information

There are influences of the crystalline environment on the n.q.r. lines that have been neglected in the previous section. In $CrCl_3$, for instance, we have seen that the main contribution to the field gradient at the chlorine nucleus comes from the two nearest-neighbour Cr^{3+} ions. Nevertheless, we obtain valuable structural information from the contribution of the further environment to the field gradient. Let us, again, take the n.q.r. in polycrystalline $CrCl_3$ [43] as an example. At 273.4°K, there are two resonances of ^{35}Cl with frequencies of 12.9080 Mc/s and 12.9472 Mc/s, and relative intensities of 1 : 2, respectively. On cooling, a third resonance appears below about 220°K, midway between the high-temperature lines. On further cooling, the intensity of the new resonance increases while the other two resonances simultaneously decrease in intensity. Below about 80°K, only the low-temperature resonance is observable. On warming, the low-temperature resonance is observable up to about 270°K. The following explanation can be given for this behaviour of n.q.r. At low temperatures, $CrCl_3$ crystallizes in a structure in which all chlorine ions are equivalent; that is, the crystalline environment is identical for all chlorine ions, and so only one n.q.r. line is observable. At room temperature, the crystal structure of $CrCl_3$ is different from that in the low-temperature phase. There are now two inequivalent lattice sites for the chlorine ions with an occupation ratio of 1 : 2. No sharp transition point is observed for the phase transition, because the microcrystals may exist in a metastable state of either phase, the probability for the phase transition of an individual crystal being determined by crystal imperfections. When the temperature dependence of n.q.r. was observed in a dense sample[43] obtained by compression of the previously loosely packed crystalline powder, the phase transition occurred over temperature intervals of only 1°; on cooling and warming the transformation took place at 234°K and 242°K, respectively.

Morosin and Narath also performed an X-ray diffraction study[43] of the crystal structure of $CrCl_3$. While a former crystal structure determination of Wooster predicts three inequivalent Cl lattice sites, the X-ray data of Morosin and Narath confirm the results of n.q.r. spectroscopy.

* Only the most abundant isotope having a non-vanishing quadrupole moment is listed for each element.

Furthermore, the observation of only two lines of equal width ruled out the space groups **C**m and **C**2, which would result in three inequivalent Cl sites, and confirmed the space group **C**2/m which was considered to be the correct symmetry of $CrCl_3$.

The foregoing example has given evidence of the value of n.q.r. spectroscopy in the study of structural problems. Zeeman measurements in single crystals will give further information from the directions of the principal axes of the field gradient. These directions may be different for nuclei at equivalent lattice sites*. A systematic study of the relation between crystal symmetry and the number of n.q.r. lines as well as the number of different directions of the principal axes of the field gradients has been published by Shimomura[55]. We mention that the temperature dependence of the amplitudes of lattice vibrations lead to a temperature dependence of n.q.r. frequencies[5], which may conversely give some information on lattice vibrations.

The linewidth of n.q.r. is determined in an ideal crystal by magnetic couplings between the nuclear dipole moments. Any imperfections (impurities, dislocations, etc.) within the crystal will lead to an additional line broadening due to changes of the field gradient at the nuclei in the neighbourhood of the imperfections. Some systematic studies of the influence of impurities (or imperfections created by irradiation) on the n.q.r. linewidth have been published[5]. In many cases, however, no definite explanation can be given for the broadening of n.q.r. lines due to imperfections. For instance, the linewidths[56] in the isomorphous substances $CrCl_3$ and $FeCl_3$ are 5 kc/s and $\sim$50 kc/s, respectively.

V. REFERENCES

1. V. Jaccarino and J. G. King, *Phys. Rev.*, **83**, 471 (1951).
2. L. I. Schiff, *Quantum Mechanics*, 2nd ed., McGraw-Hill, New York, 1955.
3. C. P. Slichter, *Principles of Magnetic Resonance*, Harper and Row, New York, 1963, Section 6.3.
4. M. H. Cohen, *Phys. Rev.*, **96**, 1278 (1954).
5. T. P. Das and E. L. Hahn, *Nuclear Quadrupole Resonance Spectroscopy*, Suppl. 1 to *Solid State Physics*, Academic Press, New York, 1958.
6. G. M. Volkoff, *Can. J. Phys.*, **31**, 820 (1953).
7. H. Hartmann, M. Fleissner, G. Gann, and H. Sillescu, *Theoret. Chim. Acta*, **3**, 347 (1965).
8. C. H. Townes and B. P. Dailey, *J. Chem. Phys.*, **17**, 782 (1949).
9. B. P. Dailey and C. H. Townes, *J. Chem. Phys.*, **23**, 118 (1955).
10. E. Scrocco and J. Tomasi, *Theoret. Chim. Acta*, **2**, 386 (1964).

* Sometimes nuclei at inequivalent lattice sites are called 'chemically inequivalent' while nuclei at equivalent lattice sites are called 'physically inequivalent' when the principal axes of the field gradient do not coincide.

11. J. W. Richardson, *Rev. Mod. Phys.*, **32**, 461 (1960).
12. W. M. Huo, *J. Chem. Phys.*, **43**, 624 (1965).
13. R. Bersohn, *J. Chem. Phys.*, **29**, 326 (1958).
14. F. W. DeWette and G. E. Schacher, *Phys. Rev.*, **137A**, 78 (1965).
15. R. M. Sternheimer, *Phys. Rev.*, **130**, 1423 (1963).
16. G. Burns and E. G. Wikner, *Phys. Rev.*, **121**, 155 (1961).
17. R. Bersohn and R. G. Shulman, *J. Chem. Phys.*, **45**, 2298 (1966).
18. H. W. De Wijn, *J. Chem. Phys.*, **44**, 810 (1966).
19. C. Dean and M. Pollak, *Rev. Sci. Instr.*, **29**, 630 (1958).
20. A. Narath, W. J. O'Sullivan, W. A. Robinson, and W. W. Simmons, *Rev. Sci. Instr.*, **35**, 476 (1964).
21. T. C. Wang, *Phys. Rev.*, **99**, 566 (1955).
22. G. D. Watkins and R. V. Pound, *Phys. Rev.*, **85**, 1062 (1952).
23. R. Livingston, *Methods of Experimental Physics*, Vol. 3, Academic Press, New York, 1962.
24. C. Dean, *Phys. Rev.*, **96**, 1053 (1954).
25. L. B. Robinson, *Can. J. Phys.*, **35**, 1344 (1957).
26. H. Hartmann, M. Fleissner, and H. Sillescu, *Theoret. Chim. Acta*, **2**, 63 (1964).
27. H. Hartmann and H. Sillescu, *Theoret. Chim. Acta*, **2**, 371 (1964).
28. M. D. Zeidler, *Ber. Bunsenges. Phys. Chem.*, **69**, 659 (1965).
29. S. R. Hartmann and E. L. Hahn, *Phys. Rev.*, **128**, 2042 (1962).
30. W. Low, *Paramagnetic Resonance in Solids*, Suppl. 2 to *Solid State Physics*, Academic Press, New York, 1960.
31. C. H. Townes and A. L. Schawlow, *Microwave Spectroscopy*, McGraw-Hill, New York, 1955.
32. N. F. Ramsey, *Molecular Beams*, Clarendon Press, Oxford, 1956.
33. H. Schüler and Th. Schmidt, *Z. Phys.*, **98**, 239 (1935).
34. A. J. Freeman and R. E. Watson in *Magnetism*, Vol. II (Ed. G. T. Rado and H. Suhl), Academic Press, New York, 1965.
35. C. H. Townes, *Handbuch der Physik*, **38**/1, Springer, Berlin, 1958.
36. C. H. Townes, F. R. Merritt, and B. D. Wright, *Phys. Rev.*, **73**, 1334 (1948).
37. W. Gordy, *Discussions Faraday Soc.*, **19**, 14 (1955).
38. M. J. Huggins, *J. Am. Chem. Soc.*, **75**, 4123 (1953).
39. W. Gordy and W. J. O. Thomas, *J. Chem. Phys.*, **24**, 439 (1956).
40. M. A. Whitehead and H. H. Jaffé, *Theoret. Chim. Acta*, **1**, 209 (1963).
41. R. G. Barnes and S. L. Segel, *Phys. Rev. Letters*, **3**, 462 (1959).
42. R. G. Barnes, S. L. Segel, and W. H. Jones, Jr., *J. Appl. Phys., Suppl.*, **33**, 296 (1962).
43. B. Morosin and A. Narath, *J. Chem. Phys.*, **40**, 1958 (1964).
44. D. Nakamura, I. Kurita, K. Ito, and M. Kubo, *J. Am. Chem. Soc.*, **82**, 5783 (1960).
45. K. Ito, D. Nakamura, Y. Kurita, K. Ito, and M. Kubo, *J. Am. Chem. Soc.*, **83**, 4526 (1961).
46. D. Nakamura, K. Ito, and M. Kubo, *J. Am. Chem. Soc.*, **84**, 163 (1962).
47. D. Nakamura, K. Ito, and M. Kubo, *Inorg. Chem.*, **2**, 61 (1963).
48. K. Ito, D. Nakamura, and M. Kubo, *Inorg. Chem.*, **2**, 690 (1963).
49. D. Nakamura, *Bull. Chem. Soc. Japan*, **36**, 1662 (1963).
50. R. Ikeda, D. Nakamura, and M. Kubo, *J. Phys. Chem.*, **69**, 2101 (1965).
51. J. Owen, *Discussions Faraday Soc.*, **19**, 127 (1955).

52. S. Sugano and Y. Tanabe, *J. Phys. Soc. Japan*, **20**, 1155 (1965).
53. J. Michielsen-Effinger, *J. Chim. Phys.*, **58**, 533 (1961).
54. Documentation of Molecular Spectroscopy, Butterworths, London, and Verlag Chemie, Weinheim, 1963.
55. K. Shimomura, *J. Phys. Soc. Japan*, **12**, 652 (1957).
56. A. Narath, *J. Chem. Phys.*, **40**, 1169 (1964).

10

Nuclear Magnetic Resonance

D. R. EATON

I. INTRODUCTION

The physical phenomenon of nuclear magnetic resonance (n.m.r.) was discovered in 1946. As soon as the instrumentation had been developed sufficiently to give high-resolution spectra, it became apparent that this phenomenon could form the basis of a very useful technique for studying molecular structure problems. Since then n.m.r. has become invaluable in many areas of chemistry, and the inorganic field is no exception to this generalization.

There have been a number of excellent monographs on the theory of n.m.r. and its application to problems of molecular structure. The most comprehensive text on the subject is *High-Resolution Nuclear Magnetic Resonance* by J. A. Pople, W. G. Schneider, and H. J. Bernstein[1]*. The basic theory has been developed in greater depth in C. P. Slichter's *Principles of Magnetic Resonance*[2], and in *The Principles of Nuclear Magnetism* by A. Abragam[3]. The reader is referred to these sources for a detailed discussion of the field as a whole. Smaller but useful monographs on the subject have been written by Roberts[4] and by Jackman[5]. These are directed more toward the organic chemist. Most recently, a monograph by Fluck[6] (in German) has appeared which covers inorganic aspects of the subject. In the present chapter, no attempt will be made to do more than lightly sketch in the theoretical background of those parts of the subject which seem most likely to prove useful to inorganic chemists. A greater emphasis will be placed on discussion of the types of problems which can be most profitably attacked by the use of n.m.r., and these topics will be illustrated by means of selected examples taken from the literature. The 1962 review article by Muetterties and Phillips[7] and also

* A more recent text along the same lines is by Emsley, Feeney, and Sutcliffe[1a].

the chapter by P. C. Lauterbur[8] in *Determination of Organic Structures by Physical Methods*, Vol. 2, may be consulted for additional examples.

A. Theoretical Basis

A spin quantum number I can be assigned to any atomic nucleus. In n.m.r., we are concerned only with nuclei in their ground states so that I is a constant quantity for each isotope. In Mössbauer spectroscopy, though, the spin quantum numbers of nuclei in excited states become important, and these are in general different from those of the same nuclei in their ground states. I can have any integral or half-integral value including zero. The spin quantum number determines the angular momentum and the resulting magnetic moment of the nucleus. For a nucleus of spin I, the angular momentum in units of $h/2\pi$ can have only the values M where

$$M = I, I - 1, \ldots, -I + 1, -I$$

The magnetic moment is similarly quantized in units of $M\mu/I$, where μ is the maximum observable component of the magnetic moment. There is associated, therefore, with each nucleus a spin degeneracy of $(2I + 1)$. This degeneracy is split by a magnetic field, and it is the transitions between these split levels which are observed in n.m.r. Nuclei with $I = 0$ are non-degenerate and therefore do not give rise to n.m.r. spectra. All nuclei with $I \neq 0$ are potentially available for the experiment.

It is next necessary to ask what are the energy levels of a nucleus with $I \neq 0$ and hence what will be the absorption frequencies corresponding to transitions between them. The energy of a magnetic moment in a uniform magnetic field H_0 applied in the z direction is

$$W = -\mu_z H_0 \tag{1}$$

where μ_z is the component of the moment in the z direction. This component is specified by the quantum number M. It is convenient to define the magnetogyric ratio γ such that

$$\mu = \gamma(I\hbar) \tag{2}$$

which gives the energy levels

$$W = -\gamma\hbar M H_0 \tag{3}$$

and the transition frequencies for $\Delta M = 1$

$$h\nu = \gamma\hbar H_0 \qquad \text{or} \qquad \nu = \gamma H_0/2\pi \tag{4}$$

It may be noted that γ, the magnetogyric ratio, is a constant characteristic of a given nucleus.

Equation (4) refers to an isolated nucleus and must be modified to take account of the surroundings. Most importantly, the nucleus is part of an atom or molecule which also contains electrons. The induced orbital motions of these electrons in a magnetic field produce a secondary magnetic field at the nucleus. The effective field at the nucleus H_{eff} differs from the applied magnetic field H_0, i.e.

$$H_{eff} = H_0(1 - \sigma)$$

where σ is known as the screening constant. Nuclei in chemically different environments in general have different screening constants. Resonance of the ith nucleus will now occur at a frequency given by

$$\nu_i = \frac{\gamma_i H_0}{2\pi}(1 - \sigma_i) \tag{5}$$

Experimentally, the quantity usually determined is the chemical shift δ_i where

$$\delta_i = \sigma_i - \sigma_{ref} = \frac{H_i - H_{ref}}{H_{ref}} \times 10^6 \tag{6}$$

σ_{ref} and H_{ref} are the shielding constant and the magnetic field at the nucleus of an arbitrarily chosen reference substance. δ_i is a dimensionless quantity and is usually given in units of parts per million (ppm). Thus our initial expectation is that there should be discrete resonances for each chemically distinct type of nucleus in a molecule and that each such type can be characterized by a chemical shift δ_i.

There are, however, several other effects which have to be considered to obtain a complete description of an n.m.r. spectrum. Firstly, since the nuclei of interest in n.m.r. are magnetic dipoles, magnetic fields are produced by direct dipolar interactions between nearby nuclei. Such fields can vary between $\pm 2\mu/R^3$, where R is the distance between the two nuclei, depending on the orientation of the line joining the two nuclei with respect to the applied magnetic field. If the molecules are rotating or tumbling rapidly, as they are in the gas or liquid phases, the dipolar field at a given nucleus due to the neighboring nuclei averages to zero. This effect is not therefore of importance for gases or liquids but is a major factor in the n.m.r. of solids. A second effect arises from quadrupole coupling. All nuclei with $I \geqslant 1$ possess electric quadrupole moments which can interact with electric field gradients. Such gradients will be present at any nucleus which does not have a spherically symmetric arrangement of surrounding charges. The quadrupole coupling can have a very large effect on the nuclear magnetic energy levels, but, as with the dipolar coupling, this effect is angle dependent and is averaged to zero in solution spectra. The presence of such quadrupolar interactions is, however, readily apparent in solid-state spectra. In solution, even though the transition frequencies are unaffected, quadrupole relaxation mechanisms have an important bearing on the practical aspects of the subject.

Finally, one further type of interaction must be mentioned in this brief introductory survey. These are electron coupled spin–spin interactions. They differ from dipole–dipole and quadrupole effects in that the interaction is isotropic and is not averaged to zero by the rapid tumbling of molecules in solution. The effect is usually much smaller than those mentioned above but nevertheless can yield a great deal of structurally important information. The form of the interaction is such that the coupling is proportional to the scalar product $\mathbf{I}_i \cdot \mathbf{I}_j$, where $\mathbf{I}_i$ and $\mathbf{I}_j$ are the nuclear spin vectors of the *i*th and *j*th nuclei. For molecules in solution where dipole–dipole and quadrupole effects can be neglected, the calculation of the n.m.r. spectrum of a given molecule is based on the hamiltonian

$$\mathscr{H} = \mathscr{H}^{(0)} + \mathscr{H}^{(1)} \tag{7}$$

where

$$\mathscr{H}^{(0)} = (2\pi^{-1}) \sum_i \gamma_i H_0(1 - \sigma_i) I_{zi}$$

and

$$\mathscr{H}^{(1)} = \sum_{i<j} J_{ij} \mathbf{I}_i \cdot \mathbf{I}_j$$

$\mathscr{H}^{(0)}$ represents the interaction of the nuclear moments with the applied magnetic field and depends on the shielding parameters σ_i introduced above. $\mathscr{H}^{(1)}$ takes account of the electron coupled spin–spin interactions and introduces the spin coupling constants J_{ij}. The appropriate wave functions can be specified by the spin quantum numbers of each of the nuclei, and the operation of the above hamiltonian on these wave functions leads to the energy levels of the system. Selection rules for magnetic dipole transitions may also be derived and the required spectrum calculated. In general, the problem is one of some complexity when more than two or three nuclei are involved, and the reader is referred to the book of Pople, Schneider, and Bernstein[1] for detailed discussion. However, no matter how complex the spectrum is, analysis will eventually lead to a set of shielding parameters σ_i (or in practice chemical shifts δ_i) for the different nuclei and a set of coupling constants J_{ij} describing the interactions between them. Fortunately, in many cases there are simplifying circumstances which lead to rather simple ('first-order') spectra. In such cases, spectral analysis is trivial. The n.m.r. spectra of very many inorganic compounds fall into this category, for reasons which will become apparent, and we will restrict our discussion of spectral analysis to a description of these first-order spectra. It should, however, be borne in mind that not all spectra are amenable to this simplified treatment and that attempts to interpret spectra using first-order rules where these are not applicable can lead to misleading results.

B. First-order Spectra

Briefly, a 'first-order' spectrum will occur if the relative chemical shifts of the different nuclei in the molecule are large compared with the spin–spin coupling constants between the nuclei. It may be noted at this point that chemical shifts are field dependent, whereas spin–spin coupling constants are field independent so that the condition may be satisfied if a high-frequency spectrometer is used but not satisfied at a lower frequency. Thus a proton chemical shift of 1 ppm at 200 Mc/s corresponds to 200 cps, and if the two nuclei have a coupling of $J = 20$ cps the spectrum will be first order to a rather good approximation. However, at 60 Mc/s the chemical shift will only be 60 cps but J will remain 20 cps, and additional complications will begin to appear in the spectrum. At 30 Mc/s with $\delta = 30$ cps the spectrum will probably bear little resemblance to the simple first-order pattern. The condition is always satisfied if the nuclei in question represent different elements. This situation is often encountered in inorganic compounds. Thus HF, for example, will give a first-order spectrum because the difference in resonance frequency depends not on a chemical shift effect but on the difference in the magnetogyric ratios of ^{1}H and ^{19}F. First-order spectra are also much more common in the n.m.r. of elements other than hydrogen because, as will be discussed in rather more detail below, chemical shifts in these elements are usually much larger than those for protons.

The distinguishing features of a first-order spectrum are as follows:

1. For each chemically distinct type of magnetic nucleus, there is a separate line or group of lines in the spectrum.
2. The groups of lines (multiplets) are well separated, and their integrated intensities are proportional to the numbers of each type of magnetic nucleus in the molecule.
3. The spacing within each multiplet give the spin–spin coupling constants.
4. The number of lines and their relative intensities within each multiplet are simply related to the number and type of the equivalent nuclei giving rise to the splitting. For nuclei with $I = \frac{1}{2}$, spin coupling with a group of n equivalent nuclei will give rise to a multiplet with $n + 1$ components, the intensities of which will have a binomial distribution. Thus the fluorine resonance of CH_3F, for example, will be a quartet of intensity 1:3:3:1 due to the splitting of the three equivalent

protons. The proton resonance will be a doublet (intensities 1:1) due to the splitting of the single fluorine. The total intensity of the proton resonance will be three times that of the fluorine resonance (obtained at the same resonance frequency) reflecting the ratio of three protons to one fluorine in the molecule. The appearance of this spectrum is shown diagrammatically in Figure 1.

5. The resonances are not split by spin interactions between chemically equivalent nuclei. Thus in the above examples no splittings due to spin–spin couplings between the protons of the methyl group are observed. The validity of this rule is not restricted to first-order spectra.

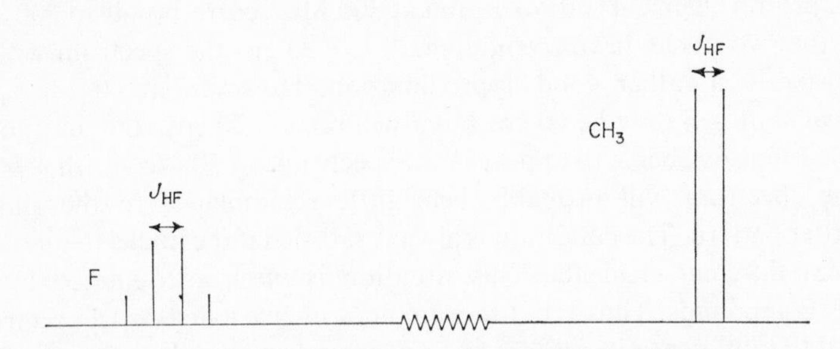

Figure 1. N.M.R. spectrum of methyl fluoride (diagrammatic).

The rationale behind the simple multiplet structure given by rule 4 is easy to see and may be illustrated by the case of CH_3F given above. The three protons each with $I = \frac{1}{2}$ must be considered together and a total spin quantum number F_z for the group used to label the wave functions. F_z can have the values $\frac{3}{2}, \frac{1}{2}, -\frac{1}{2}$, and $-\frac{3}{2}$. The corresponding wave functions have degeneracies of 1, 3, 3, and 1 since there is only one combination of the individual spins $(\frac{1}{2},\frac{1}{2},\frac{1}{2})$ which will give a total spin of $\frac{3}{2}$, but there are three combinations $(-\frac{1}{2},\frac{1}{2},\frac{1}{2},\ \frac{1}{2},-\frac{1}{2},\frac{1}{2}$, and $\frac{1}{2},\frac{1}{2},-\frac{1}{2})$ which will give a total spin of $\frac{1}{2}$. The combinations for $F_z = -\frac{1}{2}$ and $-\frac{3}{2}$ are similar. These eigenfunctions of F_z must now be combined with the ^{19}F spin wave functions of $\frac{1}{2}$ and $-\frac{1}{2}$ to give product wave functions. The energy levels of the system are then given by (8), where M_H and M_F are the total z spin

$$E(M_H, M_F) = \nu_H M_H + \nu_F M_F + J_{HF} M_H M_F \tag{8}$$

components for the hydrogen and fluorine, respectively. The proton spectrum results from the selection rule $\Delta M_H = \pm 1$, $\Delta M_F = 0$ and the ^{19}F spectrum from $\Delta M_H = 0$, $\Delta M_F = \pm 1$. The energy levels and transitions are illustrated in Figure 2, and the origin of the multiplet patterns and the relative intensities is readily apparent.

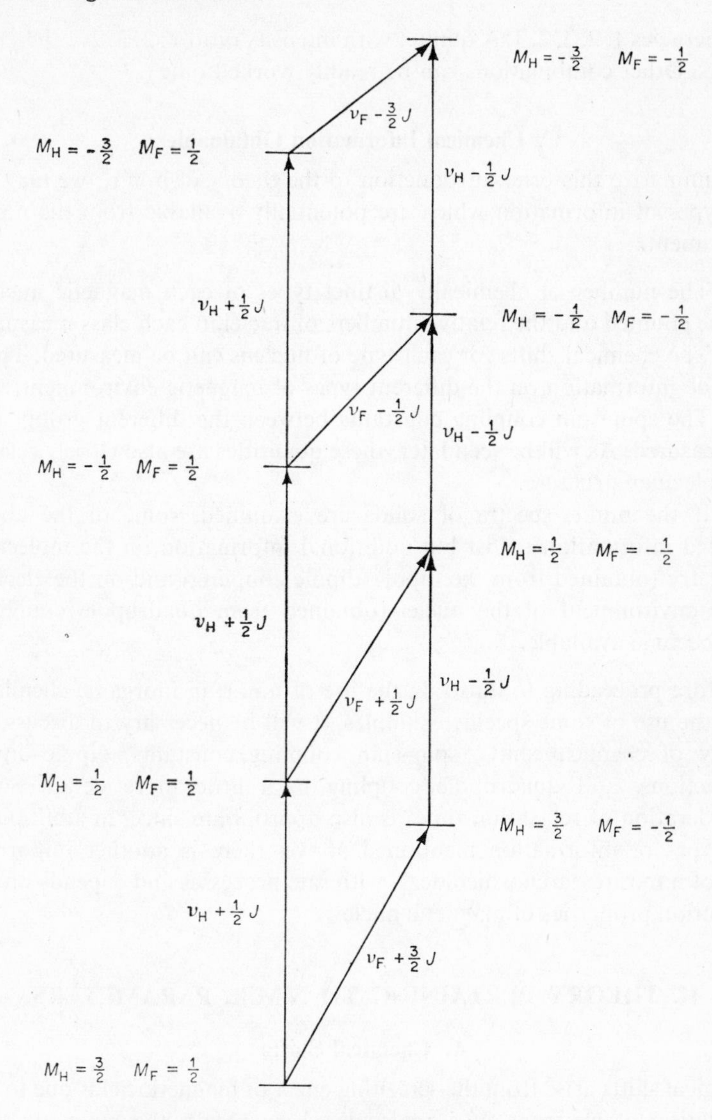

Figure 2. Energy levels and transitions for CH_3F.

These simple considerations can be readily extended to nuclei with $I > \frac{1}{2}$. Thus a single nucleus of spin I, if spin coupled to another nucleus or group of nuclei, will give rise to a splitting into $2I \pm 1$ components of equal intensity. Two equivalent nuclei with $I = 1$ (e.g. two ^{14}N) lead to states with total spin quantum numbers 2, 1, 0, −1, and −2 with degeneracies 1, 2, 3, 2, 1. A quintet with intensity ratio 1:2:3:2:1 therefore results. Other combinations can be readily worked out.

C. Chemical Information Obtainable

To summarize this brief introduction to the theory of n.m.r., we may list the types of information which are potentially available from the n.m.r. experiment.

1. The number of chemically distinct types of each magnetic nucleus can be counted and the relative numbers of nuclei in each class measured.
2. The chemical shifts for each type of nucleus can be measured. These provide information on the different types of magnetic environment.
3. The spin–spin coupling constants between the different groups can be measured. As will be seen later, these quantities are also closely related to molecular structure.
4. If the n.m.r. spectra of solids are examined, some of the above detailed information is lost but additional information on the molecular geometry (obtained from the dipole–dipole coupling) and on the electrostatic environment of the nuclei (obtained from quadrupole coupling) may become available.

Before proceeding to illustrate the use of n.m.r. in inorganic chemistry with the use of some specific examples, it will be necessary to discuss the theory of chemical shifts, spin–spin coupling constants, dipole–dipole interactions, and quadrupole coupling in a little more detail. Some consideration of relaxation times is also appropriate since, in addition to the types of information mentioned above, there is another important area of n.m.r. research which deals with rate processes and depends on the relaxation properties of magnetic nuclei.

II. THEORY PERTAINING TO N.M.R. PARAMETERS

A. Chemical Shifts

Chemical shifts arise from the screening effect of magnetic fields due to the circulation of electrons in a molecule. As a result, the magnetic field experienced by the nucleus is not the same as the magnetic field externally applied. It is not easy to apply a rigorous theory to these effects. However, useful qualitative theories have been developed for some cases, and even in cases where the present theory is clearly inadequate there are many empirical correlations between chemical shifts and other features relevant to molecular structure. Thus, even in these latter cases, chemical shift data can provide much useful information.

Initially, it will be useful to define some of the terms commonly employed in this area. A diamagnetic shift arises if the field induced at the nucleus is opposed to the applied field. In the magnetic resonance experiment, it will therefore be necessary to apply a higher field than would otherwise have been needed to bring about the resonance condition. Alternatively, if the applied field is held constant, resonance will be attained at a lower radiofrequency. Conversely, a paramagnetic shift is one to lower field or to higher frequency. In the usual n.m.r. experiment, the frequency is held constant and the field varied. Since, however, it is experimentally impractical to measure the n.m.r. frequency of a completely unshielded nucleus, these terms can only be used in a relative way. Thus to say that a shift is paramagnetic does not necessarily imply that the induced field is in the same direction as the applied field—more probably it means simply that there is a smaller diamagnetic effect than in the reference compound. It might be noted in passing that the terms diamagnetic and paramagnetic as used above have no direct relationship to the bulk magnetic properties of the substance in question. Another convention which is generally adopted is to call shifts to high field of the reference positive and those to low field, negative. For proton magnetic resonance, tetramethylsilane (TMS) is now generally accepted as a reference material. For other nuclei, there are no universally accepted standards, and chemical shifts relative to a variety of reference materials are likely to be encountered.

The theory for the magnetic shielding of an *S* state atom was developed by Lamb[9]. A diamagnetic shift is predicted, and the theory has a rather high degree of reliability. However, if the atom is incorporated into a molecule so that the spherical symmetry of the atom is lost, the theory becomes much more complicated. Following Saika and Slichter[10], it is usual to consider the total shielding as arising from several additive effects. Firstly, there is a diamagnetic shielding arising from the electrons of the atom in question. This term is analogous to the diamagnetic shielding calculated by Lamb for atoms. It will, however, be partly neutralized by a paramagnetic term arising from these same electrons. This term reflects the fact that in a molecule, since the spherically symmetric environment of the free atom has been lost, the electrons are no longer entirely free to rotate about the direction of the applied field. As a result, the field induced in opposition to the applied field is correspondingly diminished. An alternative, but equivalent, statement of the origin of the paramagnetic term is that in the presence of a magnetic field there will be some mixing of the ground state of the molecule with excited electronic states of appropriate symmetry. The latter approach defines the problem in a manner more amenable to calculation. Thus Griffith and Orgel[11] considered specifically the chemical shifts of ^{59}Co in octahedral Co^{III}

complexes and obtained a paramagnetic contribution of the form (9),

$$\Delta\sigma = -\frac{8e^2h^2}{m^2r^2\,\Delta E}\langle r^{-3}\rangle_{av} \tag{9}$$

where ΔE is an electronic excitation energy and $\langle r^{-3}\rangle_{av}$ is a mean value for the distance of the shielding electrons from the nucleus. The important qualitative result of such calculations is that the magnitude of the paramagnetic term is inversely proportional to the energy separation of the ground state and the relevant excited states. This suggests the possibility of correlations between chemical shift data and spectroscopic data. As will be seen below, such correlations have indeed been made in certain cases.

In addition to the above terms arising from the shielding effects of electrons of the atom under examination, there is also the possibility of shielding by the electrons of neighboring atoms. These latter effects might be expected to be intrinsically smaller than the former, but nevertheless it has proved necessary to invoke them in some cases. For aromatic compounds and similar molecules containing delocalized electrons, the shielding due to the circulation of these mobile electrons must also be considered. This is often referred to as the ring current effect. Finally, for paramagnetic molecules, there will be fields due to the spins of the unpaired electrons, and these will also markedly affect the chemical shifts.

Theoretical treatment of chemical shift data has proved most profitable when one of the above effects dominates the remainder. In the field of proton magnetic resonance, this is rarely true unless the considerations are restricted to the comparison of a relatively small number of compounds of closely related structure. It has been found, though, that for many of the heavier elements the local paramagnetic term appears to be dominant. For example, Saika and Slichter[10] found this to be the case for ^{19}F chemical shifts. Experimentally, it is found that a covalently bonded fluorine gives rise to a resonance at lower field than does an ionically bonded fluorine. Thus the ^{19}F resonance for HF is 625 ppm to high field of that for F_2. The simplest explanation of this effect would be to argue that in the ionic HF there is a greater electron density around the fluorine and hence the diamagnetic shielding due to the additional electrons will be greater. However, Saika and Slichter showed that the assumption that the change from F_2 to F^- arose from the presence of an additional $2p$ electron in the F^- leads to a calculated diamagnetic shift of only 23 ppm. This is clearly inadequate to account for the experimental result. They concluded that the correct explanation lay in the following line of argument. The F^- ion has a spherically symmetric electron distribution, and there is therefore no paramagnetic contribution to the shift. On passing to the covalent F_2, this spherical symmetry is lost and the paramagnetic contribution becomes large and of the right order of magnitude to account for the observed shift. Thus the observed difference between HF and F_2 reflects not an increased diamagnetic contribution in HF but an increased paramagnetic contribution in F_2. Using this line of reasoning, ^{19}F chemical shifts can be interpreted to give information on the covalency of the bonding. However, many anomalies remain, e.g. the resonance for F^- in solution is to low field of that for HF. This probably reflects solvation effects, but no unambiguous interpretation of this and many other anomalies has yet been proposed.

Probably the greatest value of chemical shift data still lies in empirical correlations with molecular structure. These have been exploited most widely for proton resonances, and it is often possible to identify a functional group with very little ambiguity from the frequency of the proton resonance. Correlation charts for other nuclei have, however, been drawn up and an example for fluorine is shown in Figure 3. Within a given type of compound, it is often possible to make correlations with other physical parameters. Thus Taft[12] has exploited correlations between ^{19}F chemical shifts in substituted fluorobenzenes and Hammett σ factors (a number of related parameters have also been used) to rationalize aromatic substituent effects. This type of approach has not yet been widely applied to inorganic systems, but there is little doubt that it will be found to be of value. An example along these lines will be mentioned below.

On turning to other nuclei, the data upon which to base correlations become increasingly sparse. This will no doubt be rectified with the passage of time. The compilation of ^{31}P chemical shifts by Katritzky and Jones[13] represents a good start for this nucleus. These data indicate that the resonance frequencies of trivalent and pentavalent phosphorus fall in reasonably well-separated regions. A similar statement can be made with respect to the tetrahedral and planar bonded environments of ^{11}B nuclei. However, it might be noted that there is very little separation of divalent and tetravalent tin in a list of ^{119}Sn chemical shifts. In this case, extreme caution would be necessary in making such structural deductions from chemical shift data.

There is one further class of compounds for which chemical shifts give unambiguous structural information. This class comprises certain paramagnetic transition metal complexes. For these compounds, the chemical shifts are again determined by a single dominant type of mechanism, in this case electron spin–nuclear spin interactions. It is usual to subdivide these interactions into two types. The first of these is the Fermi contact interaction. This is the effect which gives rise to the hyperfine splittings in the electron spin resonance spectra of free radicals in solution. In

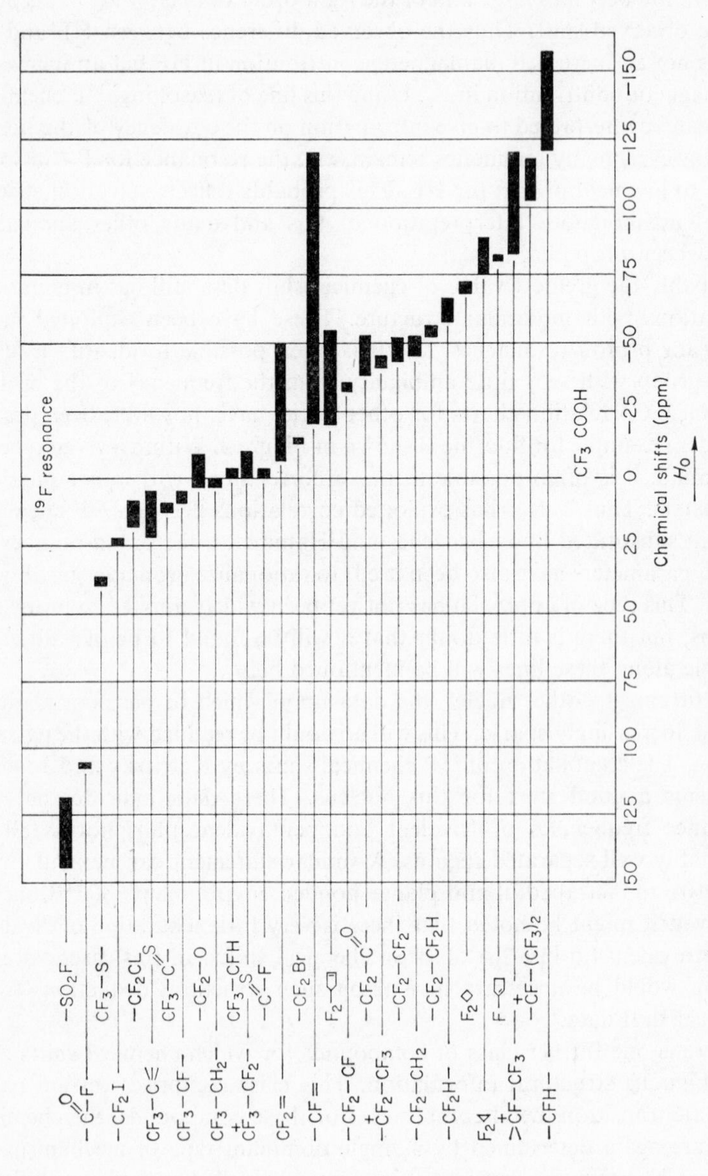

Figure 3. Correlation chart for fluorine chemical shifts. [By courtesy of Dr. E. G. Brame.]

order to be operative, it is necessary that there be a finite probability of finding the unpaired electron at the nucleus in question. This implies that there is spin density in an *s*-type orbital of the atom, since *p*, *d*, etc., orbitals all have nodes at the nucleus. The hyperfine coupling constant a is, in fact, given by the expression (10), where g and g_{N} are the electron

$$a = \frac{8\pi}{3h} g g_{\text{N}} \beta \beta_{\text{N}} |\Psi(0)|^2 \tag{10}$$

and nuclear 'g values', β and β_{N} are the Bohr magneton and nuclear magneton, respectively, and $\Psi(0)$ is the wave function describing the orbital containing the unpaired electron evaluated at the nucleus. This hyperfine coupling will give rise to a chemical shift, the magnitude of which is given by (11), where S is the electron spin quantum number. The second type

$$\left(\frac{\Delta H}{H}\right)_i = a_i \frac{\gamma}{\gamma_{\text{N}}} g \frac{\beta S(S+1)}{3kT} \tag{11}$$

is an electron dipole–nuclear dipole interaction and is known as the pseudo-contact effect. In contrast to nuclear dipole–dipole interactions, this effect is not always completely averaged to zero by the rapid tumbling of the molecules in solution. It has been shown that it will give rise to a chemical shift in solution if the electron g tensor is not isotropic. The presence of these chemical shift effects in paramagnetic molecules leads to useful molecular structure information at two levels of sophistication. Firstly, the fact that the chemical shifts produced are large compared with those usually encountered (especially for protons) can greatly facilitate the analysis of an n.m.r. spectrum of a complex molecule and the determination of the structure. This advantage is realized regardless of whether the exact mechanism producing the shifts is correctly identified or not. Secondly, a more detailed interpretation of the shifts opens the way to much information on the electronic structure and chemical bonding of the molecules. These topics will be discussed in more detail in a later section.

B. Spin–Spin Coupling Constants

The second type of parameters obtained from high-resolution n.m.r. spectra are the spin–spin coupling constants. These can provide structural information of several types. Firstly, the mere counting of the number of lines in the different multiplets suffices to determine the numbers of nuclei of the various chemically distinct types. Secondly, the rough magnitudes of the coupling constants can often indicate, for example, whether two nuclei are directly bonded or are separated by other atoms. Thirdly, the actual numerical values of the coupling constants may tell us something about the nature of the bonding. These points can perhaps be best illustrated by means of a specific example.

Muetterties and colleagues[14] have examined the ^{19}F spectra of a number of compounds of the type XPF_5, where X is a base such as trimethylamine, pyridine, ether, or dimethyl sulfoxide. They report spectra of the type shown diagrammatically in Figure 4. There are two sets of resonances, each comprising a doublet and a quintet with relative intensities 4:1. Thus it is deduced firstly that there is one set of four equivalent fluorines split into a doublet with $J = 740$ cps by a single nucleus of spin $\frac{1}{2}$ with a second

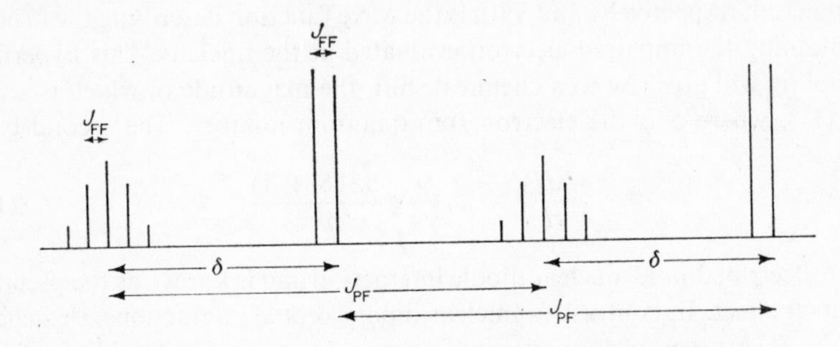

Figure 4. ^{19}F n.m.r. spectrum of XPF_5 (diagrammatic).

doubling by another nucleus of spin $\frac{1}{2}$ with $J \sim 55$ cps. There is another set of a single fluorine split by a group of four equivalent nuclei of spin $\frac{1}{2}$ (which must be the above set of four fluorines, since J for the resulting quintet is about 55 cps) and by one other single nucleus of spin $\frac{1}{2}$ with $J \sim 740$ cps. The 740 cps splitting arises most logically in both cases from coupling with the ^{31}P nucleus. This point could be proved by observation of the same coupling constant in the ^{31}P spectrum. Thus, from purely qualitative observations of the spin–spin coupling, we can say that the data are consistent with structure **1**. The set of four equivalent nuclei are

1

the equatorial fluorines, and the single nucleus is the axial fluorine. The large value of J_{PF} for both types of fluorine provides additional evidence in favor of this structure. Thus phosphorus–fluorine coupling constants in the range 600 cps to 1400 cps have been observed for fluorines directly bonded to phosphorus, whereas the longer range coupling interactions (through two or more bonds) are characteristically 20–40 cps. It is therefore a very reasonable deduction that all five fluorines are directly bonded to the phosphorus. Still more information can, in principle, be obtained from the magnitude of the ^{31}P–^{19}F coupling constants. Thus in the present case the authors conclude that, since X–F coupling constants appear to reflect the fractional p character of the bond, the similarity of J_{PF} to the value of 710 cps observed for the PF_6^- ion suggests that the phosphorus hybridization is also similar, i.e. d^2sp^3 octahedral. However, to develop arguments of this kind it is necessary to consider the theory of spin–spin coupling a little more closely.

Three types of interactions have been considered as possible contributors to the spin–spin coupling constants; namely, dipole–dipole interactions, interactions of the nuclear magnetic moments with orbital electronic currents, and Fermi or contact interactions. Fortunately, it appears that the contact interactions are usually much larger than the others so that it suffices to consider only their contributions. In this respect, the theory is more straightforward than that for chemical shifts and holds forth a better hope for useful practical application. Perturbation theory leads to the expression (12) for the Fermi contact contribution to the

$$J_{NN'} = -\frac{2}{3h}\left(\frac{16\pi\beta\hbar}{3}\right)^2 \gamma_N\gamma_{N'}\frac{1}{\Delta E}\left[0\left|\sum_k\sum_j \delta(\mathbf{r}_{kN})\delta(\mathbf{r}_{jN'})S_k\cdot S_j\right|0\right] \quad (12)$$

spin–spin coupling constant. In this expression, ΔE is a mean triplet excitation energy and $\delta(\mathbf{r}_{kN})$, $\delta(\mathbf{r}_{jN'})$ are Dirac delta functions. Thus only the values of the wave functions of the electrons (k and j) at the nuclei (N and N′) are important in determining the magnitude of J. It may be noted that only ground-state wave functions are required for the evaluation of the matrix element in equation (12). Both molecular-orbital and valence-bond approximations have been widely used to obtain the required ground-state wave functions. The use of molecular-orbital theory leads to a rather simple physical interpretation in which the coupling constants are proportional to $P^{\alpha\beta}(N,N') - P^{\alpha\alpha}(N,N')$. Here $P^{\alpha\beta}(N,N')$ is the probability of finding an electron of spin α at nucleus N and one of spin β at nucleus N′, while $P^{\alpha\alpha}(N,N')$ is the probability of finding an α electron at N and another α electron at N′. Thus, it is immediately clear that the problem of calculating spin–spin coupling constants demands the correct assessment of electron spin correlation effects. Since spin correlation is introduced directly into simple valence-bond calculations but indirectly into simple molecular-orbital calculations, the former approach has proved more popular for qualitative calculations. Most of the calculations so far made have been for proton–proton coupling constants, since wave functions for molecules involving heavy nuclei are more difficult to come by. The theory has been relatively successful in calculating coupling constants which agree fairly well with experiment (e.g. J_{HH} for methane: calculated 10.4 cps, observed 12.4 cps)[15] and in predicting qualitative trends, e.g. proton *trans* coupling constants are larger than *cis* coupling constants in ethylenic compounds[16]. Wider application to inorganic molecules can probably be expected in the future.

C. Dipole–Dipole Interactions

The local magnetic field produced at a neighboring nucleus by a nuclear dipole is given by

$$H' = \pm \mu r^{-3}(3 \cos^2 \theta - 1) \tag{13}$$

The important characteristics of this expression are firstly that it depends on internuclear separations *r* and hence is a potential source of information on the geometry of the molecule, and secondly that it is a function of the

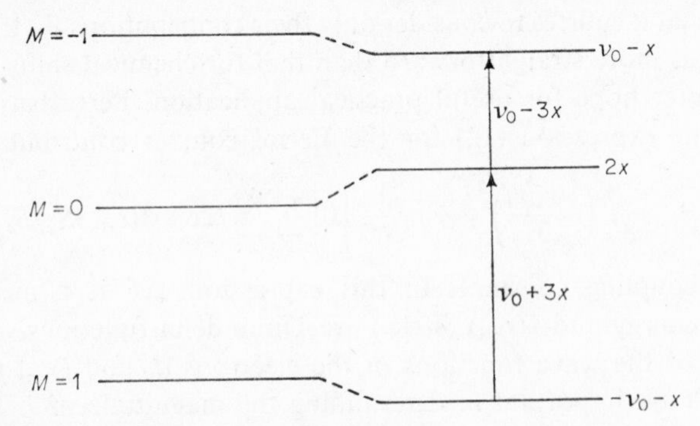

Figure 5. Dipolar interaction of two nuclei with spin $\frac{1}{2}$; $x = \mu r^{-3}(3 \cos^2 \theta - 1)$.

angle θ which represents the inclination of a line joining the two nuclei to the direction of the applied magnetic field. This latter factor causes the dipole–dipole interactions to average to zero if the molecules are tumbling rapidly in solution. These interactions have an important influence, though, in determining the spectra observed in solids. If a single crystal is examined, discrete lines will be observed, the resonance frequencies of which will change as the orientation of the crystal with respect to the applied magnetic field is changed. Thus, consider the simplest possible case of a molecule containing two identical nuclei of spin $\frac{1}{2}$ in a unit cell of the crystal. The energy-level diagram will be that of Figure 5. Two lines will be observed, the separation of which will vary from $6\mu r^{-3}$ to zero as the orientation of the crystal is changed. From the maximum separation, the distance between the magnetic nuclei is readily calculated. As the number of magnetic nuclei per unit cell is increased, the spectra rapidly become more complicated but the basis of the analysis remains the same. In practice, only a limited number of rather simple molecules have been examined in this way, partly because of the experimental difficulty that large single crystals are required. This requirement arises from the inherently rather low sensitivity of n.m.r. experiments, and sufficiently large crystals are often not available. Very many more solids have been examined as polycrystalline powders. In this case, molecules with all orientations relative to the magnetic field will be present so that there will be simultaneous absorption over a wide range of energies. The observed spectrum, therefore, represents the envelope of these individual absorptions. In some cases, especially when there are small groups of equivalent magnetic nuclei well separated from their nearest neighbors (e.g. CH_3 groups in some molecules), the envelope will still show sufficient characteristic structure to allow the calculation of interatomic distances (see Andrew[17]). More usually, though, only a broad and featureless resonance is observed, but some information is still available by measuring the second moment of the absorption defined as

$$S_2 = \int_{-\infty}^{\infty} h^2 g(h)\,\mathrm{d}h \tag{14}$$

where $h = H_0 - H_0^*$ (H_0^* is the applied field at the center of the resonance) and $g(h)$ is the normalized line shape function. Van Vleck showed that this second moment is related to the internuclear distances by

$$S_2 = \tfrac{6}{5}I(I+1)\gamma^2\hbar^2N^{-1}\sum_{j>k} r_{jk}^{-6} \tag{15}$$

providing only identical nuclei are involved. N is the number of magnetic nuclei in a unit crystal cell with different environments, and the r_{jk} are the corresponding internuclear distances. If more than one type of magnetic nucleus is present, an additional but similar term must be added to this expression, since the dipolar interaction between two unlike nuclei is less than that between two like nuclei by a numerical factor of $\frac{2}{3}$. Measurement of the second moment provides only one piece of information and so can be used to obtain only one internuclear distance. The method can be used most profitably, therefore, in conjunction with other experimental data. Thus, in the case of solid ammonium chloride[17,18], the positions of the nitrogen and chlorine atoms can be determined by x-ray diffraction and combination of this data with a broad-line n.m.r. study sufficed to fix the positions of the hydrogen atoms.

It may be noted that the above line shape calculations are based on the assumption that the atomic positions are rigidly fixed. If there is rotation of molecules or atomic groups in the crystal, the interactions between the magnetic dipoles will be modified leading in general to narrower lines. Study of the variation of n.m.r. linewidth with temperature can, therefore, yield much information on hindered rotation in solids, and the method has been profitably applied to a large number of problems[17].

A recent development which promises to extend the utility of dipole–dipole interactions as a tool for molecular structure determinations involves the study of molecules oriented in liquid-crystal matrices. Certain substances possess the property of existing over a discrete temperature range in a so-called nematic liquid crystalline phase in which the molecules possess a one-dimensional order. Such a state of matter is intermediate between the true crystalline state with three-dimensional order

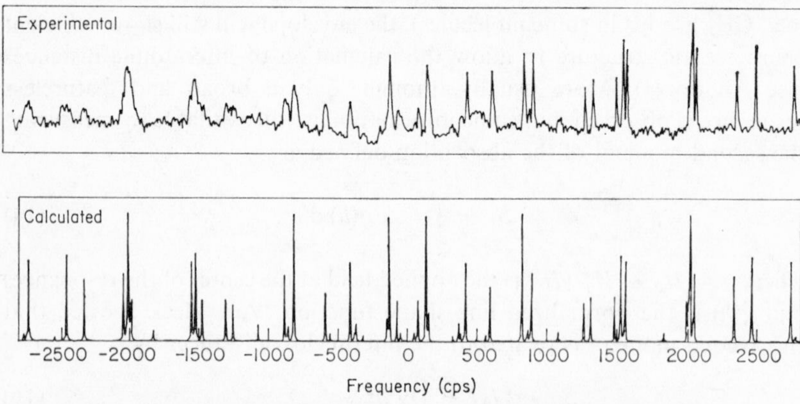

Figure 6. Experimental and calculated ^{19}F spectra of perfluorobenzene in the nematic liquid crystal *p*-hexyloxyazoxybenzene.

and the liquid state in which the orientation of the molecules is completely random. One of the unique properties of the nematic phase is that the ordered arrays of molecules tend to take up a definite orientation in a magnetic field. It has been found that this preferred orientation is imparted to a greater or a lesser extent to solute molecules dissolved in the liquid crystal. In these circumstances, the dipole–dipole interactions of the solute will not be averaged to zero as they are in a true isotropic solution, but the situation is quite a lot simpler than in a single crystal in that the average orientation of the molecules relative to the magnetic field is fixed. The term average orientation is used because in no case is the alignment of solute molecules in a liquid crystal perfect—the situation is better regarded as involving a statistical distribution in which one particular orientation with respect to the field is favored over others. It can be shown that the total spread of the n.m.r. spectrum is determined by the degree of ordering of the solute molecules but that the pattern of lines is unaffected by this parameter. As an example[19], the spectrum of perfluorobenzene dissolved in *p*-hexyloxyazoxybenzene together with a theoretical reconstruction is illustrated in Figure 6. A multiline spectrum such as this contains, in principle, sufficient information to completely determine the molecular geometry. Analysis of the spectrum can be carried out by generating the spectra resulting from model geometries with an electronic computer, and Snyder[19] has shown that such spectra are sensitive to rather small changes in bond distances and angles so that the precision of the method is high even for rather complicated molecules. This method has not yet been applied to inorganic molecules, but providing suitable liquid crystalline solvents can be found it should eventually prove useful.

D. Quadrupolar Coupling

Nuclei with spin greater than $\frac{1}{2}$ possess quadrupole moments. Interaction of these quadrupole moments with the electric field gradients produced

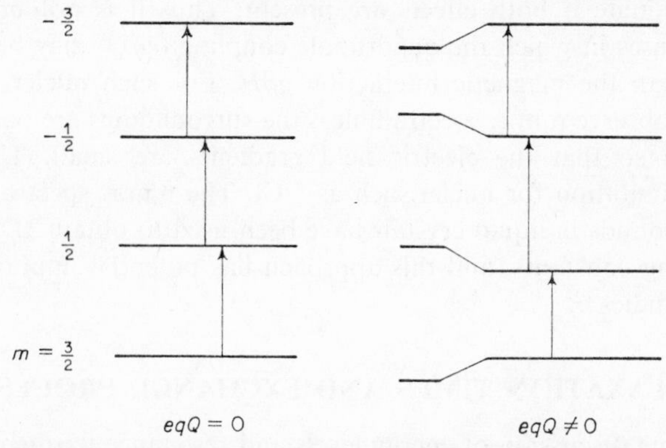

Figure 7. Energy-level diagram for nucleus with spin $\frac{3}{2}$.

by surrounding electrons and positive ions modifies the n.m.r. spectra obtained from these nuclei. The quadrupole coupling constant (eqQ) is proportional to the product of the nuclei quadrupole moment and the electric field gradient. The theory of such quadrupolar coupling has been developed in the excellent monograph of Das and Hahn[20] and the effects observed in n.m.r. spectra discussed in detail by Cohen and Reif[21] in a review article. It will suffice here to point out that both the nuclear energy levels and the relaxation properties are affected. The effect on the energy levels is illustrated qualitatively by the diagram in Figure 7 which applies to a nucleus of spin $\frac{3}{2}$. In the absence of quadrupolar coupling, a magnetic field splits the levels in such a way that the separations between the $\frac{3}{2}$

and $\frac{1}{2}$, the $\frac{1}{2}$ and $-\frac{1}{2}$, and the $-\frac{1}{2}$ and $-\frac{3}{2}$ levels are equal. The three n.m.r. transitions arising from the selection rule $\Delta M = 1$ are therefore degenerate and a single line results. The addition of quadrupolar coupling splits this degeneracy, and three lines are possible. However, the situation is rather more complex than this, since the electric field gradient and hence the quadrupole coupling has an angular dependence entirely analogous to the dipole–dipole coupling discussed above. Thus in a single-crystal experiment, the observed splittings will depend on the orientation of the crystal with respect to the applied magnetic field. In solution, the splittings will be averaged to zero by the rapid tumbling of the molecules, and only the relaxation effects will be important. For a polycrystalline sample, the spectrum can be calculated by summing the contributions from all possible orientations. This leads in general to a complex line shape from which the quadrupole coupling constant can be obtained in favorable cases. It may be noted that, although qualitatively the quadrupole coupling gives rise to the same type of effects as dipole–dipole interactions, quantitatively the effects are usually much larger and the quadrupole interaction will usually dominate if both effects are present. Thus it is not unusual to encounter cases in which the quadrupole coupling (eqQ) may be as large or larger than the magnetic interaction $g\beta H$. For such nuclei, it is not possible to observe n.m.r. spectra unless the surroundings are particularly symmetrical so that the electric field gradients are small. This is an important limitation for nuclei such as ^{35}Cl. The n.m.r. spectra of deuterium compounds in liquid crystals have been used to obtain ^{2}D quadrupole coupling constants, and this approach has potential importance for structural studies [22].

III. RELAXATION TIMES AND EXCHANGE PROCESSES

The preceding discussion of energy levels and resonance frequencies does not suffice to describe the spectra which may be encountered for some inorganic compounds. It is necessary to introduce some brief consideration of relaxation processes. Such consideration is of importance for two reasons. Firstly, there is the negative reason that unfavorable relaxation properties place a limitation on the spectra that can be observed. In extreme cases, the lines may be either too broad to detect or may saturate very easily, i.e. application of r.f. power quickly equalizes the population of the upper and lower states so that there is no longer any net absorption of energy. In either event, useful n.m.r. spectra can no longer be observed. Secondly, there is the more positive feature that processes, chemical or otherwise, which involve rapid changes in the magnetic environment of the nuclei affect their relaxation properties and produce significant changes in the spectra. This latter type of process is quite common and is a valuable source of new information. A number of examples from inorganic chemistry will be described in subsequent sections.

A. Spin Relaxation

Two parameters are commonly employed to describe nuclear relaxation. The first, T_1, is a measure of the rate at which the spin system comes into thermal equilibrium with the other degrees of freedom. Thus, if a system containing a single type of nucleus of $I = \frac{1}{2}$ is placed in a magnetic field at equilibrium, there will be slightly more nuclei with $M = +\frac{1}{2}$ than with $M = -\frac{1}{2}$. This difference in population follows from the usual Boltzman law. The net energy absorbed is proportional to the population difference and for this reason resonances obtained at high radiofrequency are inherently stronger than those obtained at low frequency. The approach to equilibrium is exponential with time, i.e. is proportional to $\exp(-t/T_1)$. T_1 is commonly called the spin-lattice relaxation time but is also sometimes known as the longitudinal relaxation time, since it describes the rate of change of the longitudinal magnetization M_z. A long T_1 is associated with an easily saturated resonance, since the rate of excitation to the upper state will become equal to the rate of relaxation from the upper state at rather small applied r.f. fields. The second relaxation time, T_2, has been defined in several ways. Experimentally it is often taken to be the reciprocal of the width of half-height (in cps) of the resonance line. Since some of the linewidth is due to instrumental factors, i.e. inhomogeneities in the applied field H_0, it is more correct to obtain T_2 from the half-width in excess of the instrumental half-width. In solids, the width of the line arises predominantly from the local magnetic fields resulting from dipolar interactions between the spins. For this reason, T_2 is sometimes referred to as the spin–spin relaxation time. An alternative approach uses T_2 to describe the decay of the magnetizations M_x and M_y, and for this reason it is also known as the transverse relaxation time. It has been shown that these definitions are equivalent. The distinction between the liquid and the solid in this context is that in the former case there are rapid fluctuations in the local environment due to tumbling of the molecules, while in the latter case there are not. These rapid fluctuations can be described by a correlation time τ_c. For the condition that

$$\omega_0 \tau_c \ll 1 \tag{16}$$

where $\omega_0 = \gamma H_0$, T_1 will equal T_2. This condition is usually met for non-viscous liquids. For most work in solutions, therefore, the theoretical considerations which deal with T_1 will lead directly to changes in the linewidth (T_2).

The relaxation of nuclei with $I = \frac{1}{2}$ can only occur through fluctuating magnetic fields. In the simplest case, the magnetic field is provided by the moments of neighboring spins and the fluctuations by molecular motion in the liquid. Thus, interaction with a nuclear spin located in the same molecule gives

$$\frac{1}{T_1} = \frac{\frac{3}{2}\hbar^2\gamma^4}{b^6}\,\tau_c \tag{17}$$

where b is the internuclear distance and τ_c, the correlation time, characterizes the random motion. τ_c is a function of viscosity and of temperature. Rather similar expressions can be derived for intermolecular interactions and the final T_1 obtained by summing the effects. Additional terms due to anisotropic electronic shielding have also been introduced. More importantly, if paramagnetic ions are present there is a term due to the fluctuating field produced by the moment of the electron. This is given by

$$\frac{1}{T_1} = \frac{4\pi^2\gamma^2\eta N_p\mu_{\text{eff}}^2}{kT} \tag{18}$$

where N_p is the concentration of paramagnetic ions of moment μ_{eff}, expressed in ions per cm^3. η is the viscosity of the solution. Since the magnetic moment of an unpaired electron is of the order 10^3 larger than a nuclear moment, for reasonably large N_p this mechanism is usually dominant. There is also a second mechanism for relaxation by paramagnetic molecules based on the Fermi contact term in the hamiltonian which contributes to T_2 but not to T_1. It may be noted that the dipolar term (equation 18) becomes inapplicable when T_{1e} for the electron becomes shorter than τ_c, since the rate of fluctuation is then determined by T_{1e}. Thus, ions with short T_{1e} values such as Co^{2+} are relatively ineffective in producing broadening of nuclear resonances. The importance of this consideration will become apparent during the discussion of paramagnetic molecules (Section V).

Nuclei with $I > \frac{1}{2}$ possess quadrupole moments and can also undergo relaxation by interaction with fluctuating electric fields. This is almost always the dominant interaction when present. It leads to a relaxation time given by

$$\frac{1}{T_1} = \tfrac{3}{8}e^2\hbar^2(eqQ)^2\tau_c \tag{19}$$

For nuclei with reasonably large quadrupole moments in asymmetric surroundings, this interaction is often large enough to broaden the resonance beyond detection. This places an important restriction on n.m.r. experiments with many of the heavier nuclei.

B. Exchange Phenomena

Nuclei in different chemical environments experience different magnetic fields. If the environments are changed by some rate process, by analogy with the above considerations of spin-relaxation effects, changes in the n.m.r. line shapes are to be expected. There is an excellent recent review of the topic by Johnson[23] to which the reader is referred for a complete treatment of the subject. Thus a typical case is illustrated by the equilibrium[23a]

$$\text{(Ph)N{=}C(CF_3)_2 \;\rightleftharpoons\; (Ph)N{=}C(CF_3)_2} \qquad (20)$$

This molecule contains two types of CF_3 groups, one *cis* to the phenyl group and one *trans*, and each will experience a different effective magnetic field which gives rise to a different chemical shift. The reaction (20) exchanges these environments. If this reaction is fast, each CF_3 group will experience the same average environment, and a single resonance will appear. If the reaction is slow, two separate resonances, one from the *cis* CF_3 and one from the *trans* CF_3, will be observed. The rate of the reaction can be changed by varying the temperature. It is important to establish the criterion for defining 'fast' and 'slow' in the preceding statements. To be fast the rate of field fluctuation should be greater than the *difference* in precessional frequencies of the two types of spins. In other words, the lifetime τ of one of the isomers must be less than the reciprocal of the difference in chemical shifts in cps between the two CF_3 groups. Slow exchange is defined similarly. Some experimental spectra for the system of equilibrium (20) are shown in Figure 8. These results are typical of the spectra obtained for exchanging systems: at low temperature (slow exchange) there are separate sharp resonances for each type of nuclei, at high temperature (fast exchange) there is a single sharp resonance, and at intermediate temperatures there is initial broadening followed by collapse of the lines to the average frequency. The rate of the reaction can be obtained from data of this kind, though it is not easy to do accurately. Measurement of the rates at different temperatures leads to an activation energy and entropy for the exchange reaction.

Analyses of the spectra of exchanging systems have been carried out at three levels of sophistication. The first and most common consists of qualitative observations leading to the conclusion that exchange is taking place together with a note of the temperature at which coalescence of the lines occurs. The condition for such coalescence in a simple two-spin system is

$$\tau = \frac{1}{\sqrt{2}\pi(\nu_A^0 - \nu_B^0)} \tag{21}$$

where ν_A^0 and ν_B^0 are the positions of the resonances in the nonexchanging system. Since the difference in chemical shifts is commonly in the range of 10 to 1000 cps, the measured lifetimes are usually in the region 10^{-1}

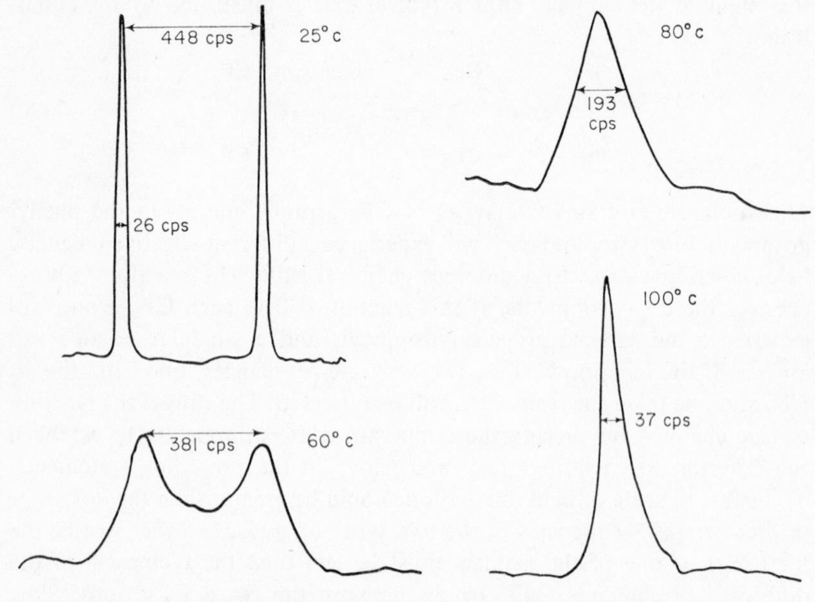

Figure 8. Temperature dependence of the ^{19}F spectrum of $PhN{=}C(CF_3)_2$. Solvent toluene, frequency 56.4 Mc/s.

to 10^{-3} cps. τ can then be converted into a rate coefficient if the order of the reaction is known or surmised. The preferable way to treat the data would be to carry out a complete line shape analysis of the spectra obtained over a range of temperatures. The theory which enables this to be done has been worked out, but the mathematics are rather cumbersome and computing facilities are required. The reader is referred to the article of Johnson[23] for a complete discussion and leading references. As yet, rather few systems have been analyzed in this way. A more common procedure has been to use approximations valid over a limited range of rates. Thus for two-spin systems, in the region where the exchange is fast, the linewidth of the single resonance can be measured. This is given by

$$\frac{1}{T_2'} = \frac{\chi_A}{T_A} + \frac{\chi_B}{T_B} + 4\pi^2\chi_A^2\chi_B^2(\nu_A^0 - \nu_B^0)^2(\tau_A + \tau_B) \tag{22}$$

where χ_A is the equilibrium fraction of A and χ_B that of B. It may be noted that χ_A and χ_B can also be found from the n.m.r. spectrum, since the frequency of the collapsed resonance is simply $\chi_A\nu_A^0 + \chi_B\nu_B^0$. The last term in equation (22) represents the additional broadening of the resonance due to exchange and will be significantly large unless τ_A and τ_B are very small, i.e. the reaction is very fast. Similarly, in the slow exchange limit the widths of the two lines centered at ν_A^0 and ν_B^0 will be given by

$$\begin{aligned} \frac{1}{T_{2A}'} &= \frac{1}{T_{2A}} + \frac{1}{\tau_A} \\ \frac{1}{T_{2B}'} &= \frac{1}{T_{2B}} + \frac{1}{\tau_B} \end{aligned} \tag{23}$$

where $1/T_{2A}$ and $1/T_{2B}$ are the widths of the unperturbed lines and τ_A and τ_B are the lifetimes of the A and B species, respectively. This approximation is useful in the region where the exchange is fast enough to cause appreciable broadening but not fast enough to bring about the onset of collapse. Finally, in the intermediate region, a rate can be obtained from the frequency separation of the collapsing lines using the relationship

$$\tau = \frac{1}{\sqrt{2}\pi}\left[(\nu_A^0 - \nu_B^0)^2 - (\nu_A - \nu_B)^2\right]^{-\frac{1}{2}} \tag{24}$$

It should be emphasized that these equations are approximate and apply only to the simplest two-spin cases. An alternative experimental method is to obtain the relaxation times from spin-echo experiments. This method has been developed in a recent series of papers by Allerhand and Gutowsky[24].

IV. APPLICATIONS TO STRUCTURAL PROBLEMS

The literature on the application of n.m.r. to inorganic structural problems is very extensive, and no comprehensive review is feasible in an article of this length. The examples below have been chosen rather with a view towards providing a sampling of the different nuclei which have been used and of the different types of problems encountered. The data given in Table 1 shows that nuclei of the majority of elements are available in principle for the n.m.r. experiment. In practice, all but a relatively small number of experiments have been performed on about half a dozen nuclei.

Table 1. Magnetic properties of nuclei[a].

Isotope (* indicates radioactive)	N.M.R. frequency (in Mc for a 10 kG field)	Natural abundance (%)	Relative sensitivity for equal number of nuclei		Magnetic moment μ (in multiples of the nuclear magneton, $eh/4\pi$ Mc)	Spin I (in multiples of $h/2\pi$)	Electric quadrupole moment Q (in multiples of $e \times 10^{-24}$ cm^2)
			at constant field	at constant frequency			
^{1}H	42.5759	99.9844	1.000	1.000	2.79268	$\frac{1}{2}$	—
^{2}H	6.53566	1.56×10^{-2}	9.65×10^{-3}	0.409	0.857386	1	2.77×10^{-3}
^{6}Li	6.265	7.43	8.50×10^{-3}	0.392	0.82192	1	4.6×10^{-4}
^{7}Li	16.547	92.57	0.294	1.94	3.2560	$\frac{3}{2}$	-0.1
^{9}Be	5.983	100.	1.39×10^{-2}	0.703	-1.1773	$\frac{3}{2}$	2×10^{-2}
^{10}B	4.575	18.83	1.99×10^{-2}	1.72	1.8005	3	7.4×10^{-2}
^{11}B	13.660	81.17	0.165	1.60	2.6880	$\frac{3}{2}$	3.55×10^{-2}
^{13}C	10.705	1.108	1.59×10^{-2}	0.251	0.70220	$\frac{1}{2}$	—
^{14}N	3.076	99.635	1.01×10^{-3}	0.193	0.40358	1	7.1×10^{-2}
^{15}N	4.315	0.365	1.04×10^{-3}	0.101	-0.28304	$\frac{1}{2}$	—
^{17}O	5.772	3.7×10^{-2}	2.91×10^{-2}	1.58	-1.8930	$\frac{5}{2}$	-4×10^{-3}
^{19}F	40.055	100.	0.833	0.941	2.6273	$\frac{1}{2}$	—
^{23}Na	11.262	100.	9.25×10^{-2}	1.32	2.2161	$\frac{3}{2}$	0.1
^{25}Mg	2.606	10.05	2.68×10^{-3}	0.714	-0.85471	$\frac{5}{2}$	—
^{27}Al	11.094	100.	0.206	3.04	3.6385	$\frac{5}{2}$	0.149
^{29}Si	8.458	4.70	7.84×10^{-3}	0.199	-0.55477	$\frac{1}{2}$	—
^{31}P	17.236	100.	6.63×10^{-2}	0.405	1.1305	$\frac{1}{2}$	—
^{33}S	3.266	0.74	2.26×10^{-3}	0.384	0.64274	$\frac{3}{2}$	-0.053
^{35}Cl	4.172	75.4	4.70×10^{-3}	0.490	0.82091	$\frac{3}{2}$	-7.9×10^{-2}
^{37}Cl	3.472	24.6	2.71×10^{-3}	0.408	0.68330	$\frac{3}{2}$	-6.21×10^{-2}
^{39}K	1.987	93.08	5.08×10^{-4}	0.233	0.39094	$\frac{3}{2}$	$+0.07$
^{41}K	1.092	6.91	8.44×10^{-5}	0.128	0.21488	$\frac{3}{2}$	—
^{43}Ca	2.865	0.13	6.40×10^{-2}	1.41	-1.3153	$\frac{7}{2}$	—
^{45}Sc	10.344	100.	0.301	5.10	4.7492	$\frac{7}{2}$	-0.22
^{47}Ti	2.400	7.75	2.09×10^{-3}	0.658	-0.78711	$\frac{5}{2}$	[illegible]

^{49}Ti	2.401	5.51	3.76×10^{-3}	1.18	−1.1022	$\frac{7}{2}$	—
^{51}V	11.193	~100.	0.382	5.52	5.1392	$\frac{7}{2}$	0.2
^{53}Cr	2.406	9.54	9.03×10^{-4}	0.28	−0.47354	$\frac{3}{2}$	—
^{55}Mn	10.553	100.	0.178	2.89	3.4611	$\frac{5}{2}$	0.6
^{57}Fe	1.38	2.245	3.38×10^{-5}	3.2×10^{-2}	0.0903	$\frac{1}{2}$	—
^{59}Co	10.103	100.	0.281	4.98	4.6388	$\frac{7}{2}$	0.5
^{61}Ni	3.79	1.25	3.53×10^{-3}	0.445	0.746	$\frac{3}{2}$	—
^{63}Cu	11.285	69.09	9.31×10^{-2}	1.33	2.2206	$\frac{3}{2}$	−0.16
^{65}Cu	12.090	30.91	1.14	1.42	2.3790	$\frac{3}{2}$	−0.15
^{67}Zn	2.664	4.12	2.86×10^{-3}	0.730	0.87354	$\frac{5}{2}$	0.18
^{69}Ga	10.219	60.2	6.91×10^{-2}	1.200	2.0108	$\frac{3}{2}$	0.178
^{71}Ga	12.984	39.8	0.142	1.525	2.5549	$\frac{3}{2}$	0.112
^{73}Ge	1.485	7.61	1.40×10^{-3}	1.15	−0.87677	$\frac{9}{2}$	−0.2
^{75}As	7.292	100.	2.51×10^{-2}	0.856	1.4349	$\frac{3}{2}$	0.3
^{77}Se	8.131	7.50	6.93×10^{-3}	0.191	0.5325	$\frac{1}{2}$	—
^{79}Br	10.667	50.57	7.86×10^{-2}	1.25	2.0991	$\frac{3}{2}$	0.34
^{81}Br	11.499	49.43	9.85×10^{-2}	1.35	2.2626	$\frac{3}{2}$	0.28
^{85}Rb	4.111	72.8	1.05×10^{-2}	1.13	1.3482	$\frac{5}{2}$	0.28
^{87}Rb	13.932	27.2	0.175	1.64	2.7414	$\frac{3}{2}$	0.14
^{87}Sr	1.845	7.02	2.69×10^{-3}	1.43	−1.0893	$\frac{9}{2}$	—
^{89}Y	2.086	100.	1.18×10^{-4}	4.90×10^{-2}	−0.13682	$\frac{1}{2}$	—
^{91}Zr	3.958	11.23	9.4×10^{-3}	1.08	−1.298	$\frac{5}{2}$	—
^{93}Nb	10.407	100.	0.482	8.07	6.1435	$\frac{9}{2}$	−0.16
^{95}Mo	2.774	15.78	3.23×10^{-3}	0.760	−0.9099	$\frac{5}{2}$	—
^{97}Mo	2.833	9.60	3.44×10^{-3}	0.776	−0.9290	$\frac{5}{2}$	—
^{99}Ru	1.9	12.81	1.07×10^{-3}	0.526	−0.63	$\frac{5}{2}$	—
^{101}Ru	2.1	16.98	1.41×10^{-3}	0.576	−0.69	$\frac{5}{2}$	—
^{103}Rh	1.340	100.	3.12×10^{-5}	3.15×10^{-2}	−0.0879	$\frac{1}{2}$	—

[a] Data from Varian n.m.r. table [reproduced by permission].

(continued)

Table 1 (*continued*)

Isotope (* indicates radioactive)	N.M.R. frequency (in Mc for a 10 kG field)	Natural abundance (%)	Relative sensitivity for equal number of nuclei		Magnetic moment μ (in multiples of the nuclear magneton, $eh/4\pi$ Mc)	Spin I (in multiples of $h/2\pi$)	Electric quadrupole moment Q (in multiples of $e \times 10^{-24}$ cm^2)
			at constant field	at constant frequency			
^{105}Pd	1.74	22.23	7.94×10^{-4}	0.48	−0.57	$\frac{5}{2}$	—
^{107}Ag	1.723	51.35	6.62×10^{-5}	4.05×10^{-2}	−0.1130	$\frac{1}{2}$	—
^{109}Ag	1.981	48.65	1.01×10^{-4}	4.65×10^{-2}	−0.1299	$\frac{1}{2}$	—
^{111}Cd	9.028	12.86	9.54×10^{-3}	0.212	−0.5922	$\frac{1}{2}$	—
^{113}Cd	9.444	12.34	1.09×10^{-2}	0.222	−0.6195	$\frac{1}{2}$	—
^{113}In	9.310	4.16	0.345	7.22	5.4960	$\frac{9}{2}$	0.750
^{115}In*	9.329	95.84	0.347	7.23	5.5073	$\frac{9}{2}$	0.761
^{117}Sn	15.17	7.67	4.52×10^{-2}	0.356	−0.9949	$\frac{1}{2}$	—
^{119}Sn	15.87	8.68	5.18×10^{-2}	0.373	−1.0409	$\frac{1}{2}$	—
^{121}Sb	10.19	57.25	0.160	2.79	3.3417	$\frac{5}{2}$	−0.53
^{123}Sb	5.518	42.75	4.57×10^{-2}	2.72	2.5334	$\frac{7}{2}$	−0.68
^{125}Te	13.45	7.03	3.16×10^{-2}	0.316	−0.8824	$\frac{1}{2}$	—
^{127}I	8.519	100.	9.34×10^{-2}	2.33	2.7937	$\frac{5}{2}$	−0.75
^{129}Xe	11.78	26.24	2.12×10^{-2}	0.277	−0.77255	$\frac{1}{2}$	—
^{131}Xe	3.490	21.24	2.75×10^{-3}	0.410	0.68680	$\frac{3}{2}$	−0.12
^{133}Cs	5.585	100.	4.74×10^{-2}	2.75	2.5642	$\frac{7}{2}$	−0.004
^{135}Ba	4.230	6.59	4.90×10^{-3}	0.497	0.83229	$\frac{3}{2}$	—
^{137}Ba	4.732	11.32	6.86×10^{-3}	0.556	0.93107	$\frac{3}{2}$	—
^{139}La	6.014	99.911	5.92×10^{-2}	2.97	2.7615	$\frac{7}{2}$	0.5
^{141}Pr	11.95	100.	0.258	3.28	3.92	$\frac{5}{2}$	-5.4×10^{-2}
^{143}Nd	2.72	12.20	5.49×10^{-3}	1.34	−1.25	$\frac{7}{2}$	−0.57
^{147}Sm	1.5	15.07	8.8×10^{-4}	0.730	−0.68	$\frac{7}{2}$	0.72
^{151}Eu	10.49	47.77	0.175	2.87	3.441	$\frac{5}{2}$	—
^{153}Eu	4.638	52.23	1.51×10^{-2}	1.27	1.521	$\frac{5}{2}$	—
^{155}Gd	1.2	14.68	1.33×10^{-4}	0.149	−0.25	$\frac{3}{2}$	1.1
^{157}Gd	1.7	15.64	3.34×10^{-4}	0.203	−0.34	$\frac{3}{2}$	1.0

^{161}Dy	1.2	18.73	2.35×10^{-4}	0.317	−0.38	$\frac{5}{2}$	—
^{163}Dy	1.6	24.97	6.38×10^{-4}	0.443	−0.53	$\frac{5}{2}$	—
^{165}Ho	7.22	100.	0.102	3.56	3.31	$\frac{7}{2}$	2
^{167}Er	1.04	22.82	3.11×10^{-4}	0.516	0.48	$\frac{7}{2}$	(10)
^{169}Tm	3.49	100.	5.51×10^{-4}	8.20×10^{-2}	−0.229	$\frac{1}{2}$	—
^{171}Yb	7.51	14.27	5.50×10^{-3}	0.176	0.4926	$\frac{1}{2}$	—
^{173}Yb	2.1	16.08	1.33×10^{-3}	0.566	−0.677	$\frac{5}{2}$	—
^{175}Lu	4.86	97.40	3.12×10^{-2}	2.40	2.230	$\frac{7}{2}$	5.7
^{177}Hf	1.3	18.39	6.38×10^{-4}	0.655	0.61	$\frac{7}{2}$	3
^{179}Hf	0.80	13.78	2.16×10^{-4}	0.617	−0.47	$\frac{9}{2}$	3
^{181}Ta	5.09	100.	3.60×10^{-2}	2.52	2.340	$\frac{7}{2}$	4.0
^{183}W	1.75	14.28	6.98×10^{-5}	4.12×10^{-2}	0.115	$\frac{1}{2}$	—
^{185}Re	9.586	37.07	0.133	2.63	3.1437	$\frac{5}{2}$	2.8
^{187}Re	9.684	62.93	0.137	2.65	3.1760	$\frac{5}{2}$	2.6
^{189}Os	3.307	16.1	2.34×10^{-3}	0.388	0.6507	$\frac{3}{2}$	2.0
^{191}Ir	0.813	38.5	3.5×10^{-5}	9.5×10^{-2}	0.16	$\frac{3}{2}$	1.5
^{193}Ir	0.86	61.5	4.2×10^{-5}	0.101	0.17	$\frac{3}{2}$	1.5
^{195}Pt	9.153	33.7	9.94×10^{-3}	0.215	0.6004	$\frac{1}{2}$	—
^{197}Au	0.731	100.	2.53×10^{-5}	8.59×10^{-2}	0.1439	$\frac{3}{2}$	0.56
^{199}Hg	7.60	16.86	5.67×10^{-3}	0.178	0.4979	$\frac{1}{2}$	—
^{201}Hg	2.80	13.24	1.42×10^{-3}	0.329	−0.5513	$\frac{3}{2}$	0.45
^{203}Tl	24.33	29.52	0.187	0.571	1.5960	$\frac{1}{2}$	—
^{205}Tl	24.57	70.48	0.192	0.577	1.6115	$\frac{1}{2}$	—
^{207}Pb	8.899	21.11	9.13×10^{-3}	0.209	0.5837	$\frac{1}{2}$	—
^{209}Bi	6.842	100.	0.137	5.30	4.0389	$\frac{9}{2}$	−0.4

There are many nuclei which for one reason or another have attracted virtually no attention. Thus ^{25}Mg, for example, may be classified as an unfavorable nucleus since it has a low resonance frequency, probably rather a high quadrupole moment, and only occurs in 10% natural abundance. To the author's knowledge, no examples of ^{25}Mg n.m.r. have been reported in the literature. However, the resonance can be quite easily observed in concentrated aqueous solutions of the Mg^{2+} ion, where the Mg has presumably a very symmetric environment in the form of a hexahydrate. It has not been observed in less symmetric environments such as those provided by Grignard reagents[25]. Thus the utility of ^{25}Mg n.m.r. spectroscopy is likely to remain limited, but the possibility should probably not be neglected for certain types of problem, e.g. questions relating to complexing of Mg^{2+} ions in solution. Perusal of Table 1 reveals numerous other nuclei where the same situation is likely to hold.

A. Transition Metal Hydrides

Nuclear magnetic resonance has proved to be of outstanding utility in studying the chemistry of transition metal hydrides. This area provides a good illustration of both the virtues and the limitations of the method. Resonances of M–H hydrogens (M is a transition metal) occur to extreme high field, and the chemical shift is a reliable diagnostic test for such types of bonds. Typically, they are found in the region from 10 to 30 ppm to high field of TMS and, since the only other resonances known to occur in this region arise from porphyrins and certain paramagnetic complexes, the observation of such a high field line provides excellent evidence for the existence of a hydride. The n.m.r. data for some typical transition metal hydrides are shown in Table 2. A much more complete compilation can be found in a recent review article by Green and Jones[26]. Further evidence regarding the hydride structure can sometimes be obtained from the spin–spin coupling. Thus, if the metal has a magnetic moment, the observation of the M–H coupling provides further evidence of the hydride structure. In the series of compounds $Pt(PEt_3)_2HX$, the magnitude of J_{PtH} varied markedly with X and can be correlated with the frequencies of the first ligand-field transitions of a corresponding series of Co^{III} complexes. This can be understood in terms of equation (12), since the mean triplet excitation energy ΔE will depend on the ligand-field strength of X, and this in turn determines the frequency of the optical absorption. It is assumed that the relative ligand-field strengths of the various X groups are the same for the two series of compounds. The magnitude of J_{MH} may also reflect the *s* character of the M—H bond, and some correlations have been made on this basis. The smaller value of J_{WH} in $[(\pi\text{-}C_5H_5)_2WH_3]^+$ than in $(\pi\text{-}C_5H_5)_2WH_2$ has been explained in this manner: the higher coordination

Table 2. P.M.R. spectra of transition metal hydrides[26,27].

Compound	Shift (ppm from TMS)	J_{MH} (cps)	Other J (cps)
$[FeH(CO_3)(PPh_3)_2]^+$	+7.75	—	J_{HP} = 35.3
trans-RuHCl $[C_2H_4(PEt_2)_2]_2$	+22.5	—	—
trans-OsHCl $[C_2H_4(PEt_2)_2]_2$	+26.7	—	—
$IrH_2(CO)Cl$ $[PPh_3]_2$	+11.2	—	—
trans-$Pt(PEt_3)_2$ HI	+12.8	1369	J_{HP} = 13
trans-$Pt(PEt_3)_2$ HCl	+16.9	1275	J_{HP} = 14
trans-$Pt(PEt_3)_2$ HNO_2	+19.6	1003	J_{HP} = 16.5
trans-$Pt(PEt_3)_2$ HCN	+7.9	778	J_{HP} = 15.5
$[(\pi\text{-}C_5H_5)_2WH_3]^+$	+6.44, +6.08	47.8	J_{HH} = 8.5
$(\pi\text{-}C_5H_5)_2TaH_3$	+3.02, +1.63		J_{HH} = 9.6
$(\pi\text{-}C_5H_5)_2WH_2$	+12.28	73.2	J_{HCp} = 0.75

number leads to a bond with a smaller *s* character as it does in the molecules NH_3 and NH_4^+. In $[(\pi\text{-}C_5H_5)_2WH_3]^+$, two types of hydridic hydrogens with intensity ratio 2:1 have been observed and the coupling between them measured. This indicates that the three hydrogens are not all equivalent. This observation provides an important clue to the geometric structure of the molecule.

It might be expected that the unusually high-field chemical shifts of transition metal hydrides would provide some insight into the very interesting problem posed by the chemical bonding in these compounds. However, as has been pointed out previously, the theoretical interpretation of chemical shift data is still fraught with uncertainty, and in spite of a considerable amount of work in this area no completely satisfactory account of the origin of the shifts has yet been given. The most refined theoretical treatment is due to Buckingham and Stephens[28] who have considered both the diamagnetic term due to the direct shielding effect of the metal *d* electrons and the contribution arising from the temperature-independent paramagnetism. This latter contribution is particularly important in transition metal compounds. This treatment gives a qualitatively satisfactory account of the high-field shifts observed, but detailed explanations of the differences between individual molecules are still lacking.

One of the main limitations on the theoretical discussion of bonding in transition metal hydrides has been the scarcity of reliable data on metal–hydrogen bond lengths. These are not in general accessible from x-ray crystallography. The study of dipole–dipole interactions by broad-line n.m.r. can also make a contribution in this area, as is illustrated by a recent study of $HMn(CO)_5$[29]. This is a particularly favorable case, since all the structural parameters except the Mn—H had been determined so that measurement of the second moment and use of equation (15) suffices to determine the single unknown. In this case, it was found that about 97% of the observed value of S_2 arises from intramolecular Mn—H dipole interaction so that the bond-length determination can be made with some precision. A value of 1.28 ± 0.01 Å was obtained.

B. Proton Magnetic Resonance—Organometallic Compounds

In the structural work on transition metal hydrides, the most useful parameter is generally the chemical shift. In other cases, the spin–spin coupling constants can provide the most significant information. An example is provided by the work of Moy, Emerson, and Oliver[30] on propenylmercury compounds. Such compounds can be prepared by the reactions:

$$C_3H_5Li \xrightarrow{HgBr_2} C_3H_5HgBr \xrightarrow{Na_2SnO_2} (C_3H_5)_2Hg$$

It is a question of some interest to determine whether the configuration of the propenyl groups is retained in the course of these reactions. The n.m.r. spectra of the products obtained by using iso-, *cis*-, and *trans*-propenyllithium in these reactions were analyzed. Each was found to correspond to a single pure product which must be one of the isomers **2**, **3**, or **4**. A consideration of the spin–spin coupling constants sufficed to

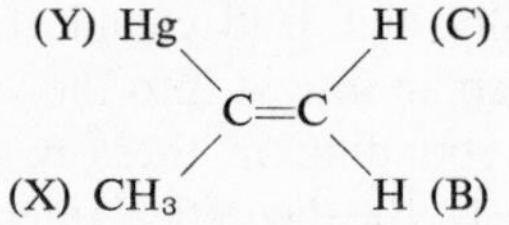

```
(Y) Hg        H (C)          (Y) Hg        CH3 (X)
      \      /                     \      /
       C=C                          C=C
      /      \                     /      \
(X) CH3       H (B)          (A) H         H (B)
      2, iso                       3, cis

              (Y) Hg        H (C)
                    \      /
                     C=C
                    /      \
              (A) H         CH3 (X)
                    4, trans
```

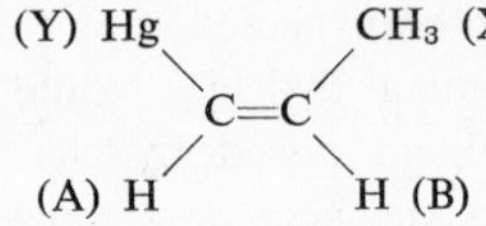

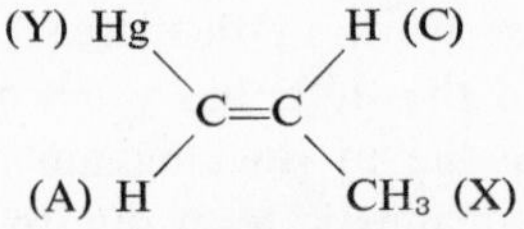

make the appropriate assignments. The data are shown in Table 3. The labeling of the groups corresponds to that of structure **2** to **4**. It has been well established both experimentally and theoretically that proton–proton coupling constants in olefins obey the relationship $J_{trans} > J_{cis} > J_{gem}$

Table 3. Coupling constants of dipropenylmercury isomers[30].

	Diiso-propenylmercury (cps)	Di-*cis*-propenylmercury (cps)	Di-*trans*-propenylmercury (cps)
J_{AB}	—	11.1	—
J_{BC}	4.1	—	—
J_{AC}	—	—	19.2
J_{AX}	—	1.3	1.3
J_{BX}	1.4	6.5	—
J_{CX}	1.4	—	5.0
J_{AY}	—	134	125
J_{BY}	256.5	244	—
J_{CY}	127.8	—	140
J_{XY}	88.2	12	5.0

and that the *trans* coupling constants vary from 14 to 20 cps, the *cis* from 5 to 15 cps, and the *gem* from 0 to 5 cps. There had been some previous evidence to show that ^{199}Hg–^{1}H coupling constants show the same trend. In the present case, the proton–proton coupling constants clearly indicate the correct structural assignments. The mercury–proton coupling constants are in the expected order and provide supporting evidence. The coupling constants involving the methyl groups do not contribute directly to the structural determination but are consistent with values obtained from similar compounds of known structure. It was concluded from this study that the formation of the propenylmercury compounds occurred with complete retention of configuration.

It has been pointed out earlier that the observation of a single resonance peak does not necessarily provide evidence for the structural equivalence of groups if rapid exchange processes are possible. An example of such a situation is provided by the work of Whitesides, Nordlander, and Roberts[31] on γ,γ-dimethylallylmagnesium bromide. At room temperature, a single methyl resonance was observed for this compound. Previously two possibilities have been suggested for the structure of allyl Grignard reagents; namely, either the symmetrical bridge structure **5** or a rapidly equilibrating mixture of the two equivalent classical structures **6**.

H
CH₃ C CH₃
C C
H MgBr H

5

$$\begin{matrix} CH_3 & & H \\ & C{=}C & \\ H & & CH(CH_3)MgBr \end{matrix} \rightleftharpoons \begin{matrix} MgBr(CH_3)CH & & H \\ & C{=}C & \\ H & & CH_3 \end{matrix}$$

6

The temperature dependence of the methyl resonance is shown in Figure 9. At low temperatures, it is observed to split into two equal components. This observation eliminates the suggested bridge structure **5** and is in accord with the equilibrium **6**. The two methyl groups of structures **6** are

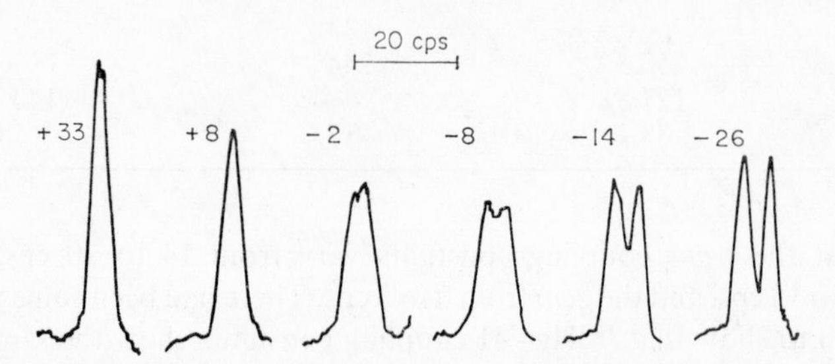

Figure 9. Temperature dependence (°c) of the methyl resonance of γ,γ-dimethylallylmagnesium bromide.

magnetically nonequivalent and under conditions where the rate of exchange is slow give rise to separate resonances. From the n.m.r. data, the activation energy of the exchange process was estimated to be 7 ± 3 kcal and the frequency factor 10^3–10^8 sec^{-1}.

C. Beryllium

The study of the n.m.r. spectra of single crystals containing quadrupolar nuclei is in principle capable of providing information of fundamental interest to inorganic chemists. Relatively little has been done in this area, but an example which illustrates the possibilities is provided by the recent work of Reaves and Gilmer[32] on the ^{9}Be resonance of chrysoberyl. The changes in the n.m.r. spectrum are studied as the single crystal is rotated in the magnetic field. By a suitable choice of several different axes for rotation, it is possible to determine the quadrupole coupling constant, the asymmetry parameter, and the orientation of the principal axes of the electric field gradient with respect to the crystalline axes. This information amounts to a complete description of the electrical environment of the ^{9}Be atom in the crystal. In the present case, it was found that there were two different types of beryllium site but that these were equivalent. A quadrupole coupling constant of 318 ± 2 kc/s and an asymmetry parameter of

0.90 ± 0.1 were obtained. One of the electric field gradient axes corresponds to a crystallographic axis, but the other two are inclined at angles of $85 \pm 1°$ from the corresponding crystal axes.

D. Boron

There are two naturally occurring isotopes of boron, ^{11}B (spin $\frac{3}{2}$, 81% abundant) and ^{10}B (spin 3, 19% abundant). ^{11}B resonance has been used for many structural studies, particularly of boron hydrides. An example, which also serves to illustrate the advantages of high magnetic fields, is provided by the spectrum of decaborane[33], $B_{10}H_{14}$, shown in Figure 10. This molecule contains one set of four equivalent nuclei and three sets of

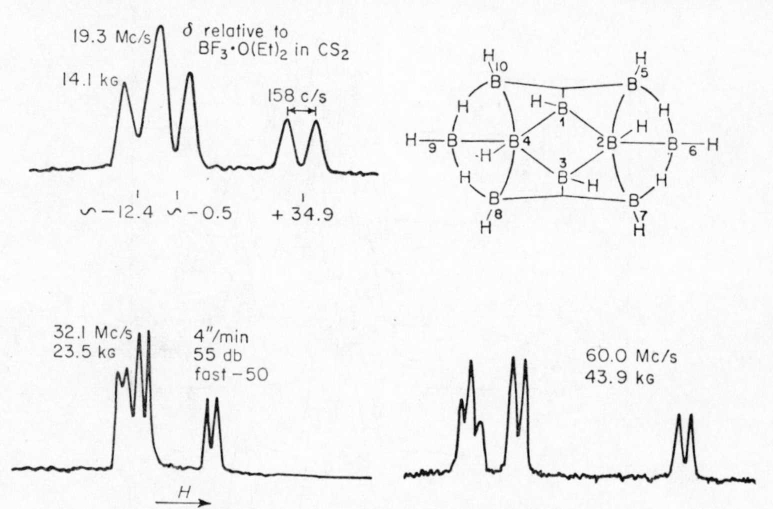

Figure 10. ^{11}B spectrum of decaborane.

two equivalent nuclei. Each boron is split into a doublet by the hydrogen ($I = \frac{1}{2}$) directly bonded to it. The bridging hydrogens do not give rise to resolvable splittings. It is apparent that the spectrum obtained at 14.1 kG is incompletely resolved and could have been structurally misleading. Even at 43.9 kG one pair of doublets is not completely separated. However, combination of the data at two different field strengths shows which spacings are field independent (and therefore correspond to spin–spin splittings) and which are field dependent (and therefore reflect chemical shifts between nonequivalent nuclei) and leads to the correct structural deductions. Comparison of spectra obtained at different fields is always advisable when such ambiguities are possible. Alternatively, similar information can be obtained by replacing the hydrogens with deuterium. This leaves the chemical shifts unchanged but reduces the spin–spin coupling constants by an amount depending on the ratio of the magnetogyric ratios of ^1H and ^2D (a factor of about 6 is involved in this case). This latter technique was used to correctly analyze the decaborane spectrum[34].

Nuclear magnetic resonance can often be used very conveniently to follow the course of relatively slow chemical reactions. This type of application is well illustrated by the work of Dupont and Hawthorne[35] on

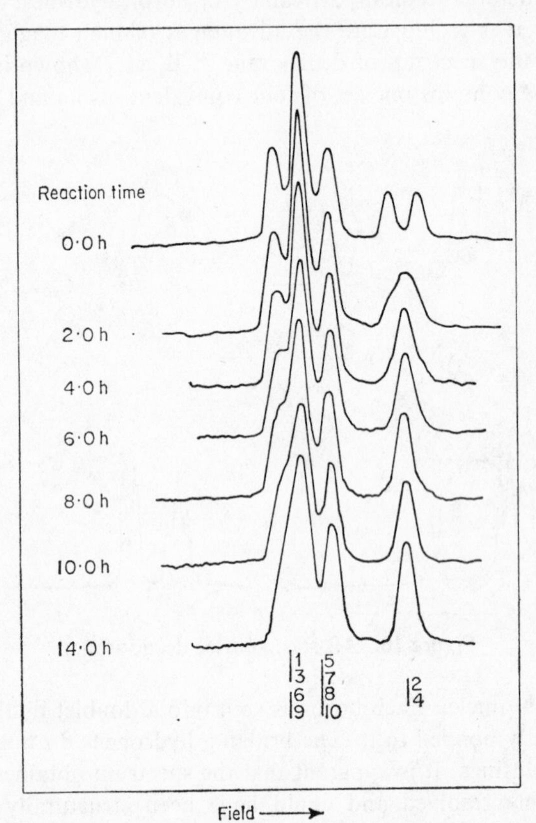

Figure 11. Changes in the ^{11}B spectrum of decaborane on deuteration.

deuterium exchange in decaborane. By reaction with deuterium chloride and aluminum chloride, decaborane readily exchanges four of its fourteen hydrogens. The question arises as to which four hydrogens are labile. In Figure 11, the changes in the ^{11}B spectrum of decaborane on reaction with DCl and $AlCl_3$ are illustrated. Inspection of these spectra shows, for example, that the doublet associated with the 2,4 hydrogens is collapsed to a singlet in the course of the reaction. The elimination of the proton coupling shows that the 2,4 hydrogens are labile. It is also apparent that the high-field peak associated with the 5,7,8,10 protons is unaffected, indicating that these protons are not labile. In a series of further experiments, Dupont and Hawthorne were able to show that the remaining two electrophilic deuteration sites were located at the 1,3-positions. These data on reactivity towards electrophilic substitution were related to theoretical considerations of the charge distribution and boron atom hybridization in decaborane.

It has been pointed out previously that single-crystal n.m.r. studies of quadrupolar nuclei can be used to characterize completely the electrical environment of the nucleus under study. The amount of information available from studies on polycrystalline solids or glasses is more limited but may, nevertheless, be quite useful. This type of experiment can be well illustrated by the work of Silver and Bray[36] on boron-containing glasses. Since in a glassy structure the principal axes of the field gradient tensor for the different atoms are randomly oriented with respect to the magnetic field, the observed resonance consists of the envelope of transitions corresponding to all possible orientations. Analysis of the resulting line shape can yield the quadrupole coupling constant (eqQ) and in favorable cases the asymmetry parameter (η). The relevant theory has been discussed by a number of authors[37]. The magnitude of the quadrupole coupling constant depends markedly on the type of bonding. Thus, for boron tetrahedrally coordinated to four identical atoms, there will be no resultant field gradient at the nucleus and eqQ in theory will be zero. In practice, small distortions of the tetrahedral symmetry or fields from neighboring ions are likely to give rise to a small quadrupole coupling. On the other hand, for boron in a planar BO_3^- ion, a much larger ^{11}B quadrupole coupling constant is expected. Silver and Bray found two different types of ^{11}B resonance attributable to two different boron environments in boron-containing glasses. One type showed a quadrupole coupling constant of 2.76 Mc/s and is associated with boron in a planar environment. This coupling constant, although large, is significantly less than that of a free boron atom (eqQ = 5.39 Mc/s). This type of boron is bonded as in **7a** and the dimination of the coupling constant is considered to arise from contributions due to valence structure **7b**. In structure **7a**, two of the boron $2p$ orbitals

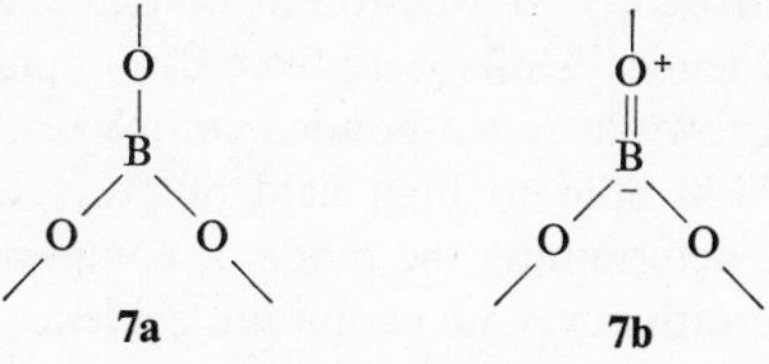

are filled but the third is empty, leading to an imbalance in the charge distribution and hence to a field gradient. The double bonding in **7b** removes this imbalance and reduces the quadrupole coupling constant. The second type of boron site corresponds to $eqQ = 570$ kc/s and is associated with BO_4 tetrahedra. The amount of boron in this type of environment increases with the amount of Na_2O modifier present. Experiments of this type have been used in extensive studies of the structure of glasses.

E. Nitrogen

Nitrogen has two magnetic isotopes, neither of which are particularly favorable for magnetic resonance studies. ^{14}N ($I = 1$) has a small magnetic moment and is often badly broadened by quadrupole effects, especially if present in an unsymmetrical environment. ^{15}N ($I = \frac{1}{2}$) has a very low natural abundance (0.36%). However, in view of the widespread occurrence of nitrogen in inorganic compounds, n.m.r. studies of these nuclei are likely to attract some adherents.

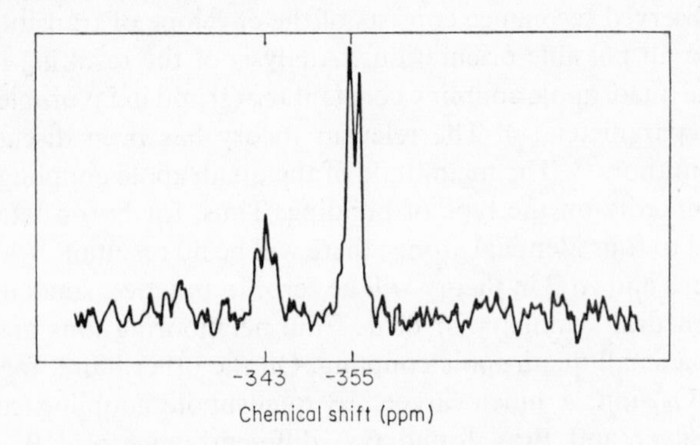

Figure 12. ^{15}N spectrum of the $S_4N_3^+$ ion.

The possibilities of the useful application of ^{14}N chemical shifts are illustrated by the work of Howarth, Richards, and Venanzi[38] on metal thiocyanate complexes. The thiocyanate ion (NCS^-) can complex either through the nitrogen or through the sulfur. In a number of cases, the point of attachment has been reliably determined either by x-ray crystallography or by infrared spectroscopy. It was found that in known sulfur-bonded complexes the ^{14}N chemical shifts were from 0–20 ppm to low field of the NCS^- resonance. Known nitrogen-bonded complexes, on the other hand, appear from 50 to 130 ppm to high field of NCS^-. A rather straightforward method for determining the mode of complexation is thus available. In disodium tetrathiocyanate cadmium, evidence for an equilibrium between nitrogen- and sulfur-bonded species was obtained. The difference between the chemical shifts of the two types of complexes can be rationalized if it is assumed that the effect of temperature-independent paramagnetism is predominant. In this case, the appropriate value of ΔE in an equation of the type (9) is likely to be correlated with the frequency of the n–π^* transition involving nitrogen lone-pair electrons. It is argued that N bonding lowers the energy of the nonbonding orbital, increasing the n–π^* frequency and shifting the ^{14}N resonance to high field, whereas S bonding does not affect the n orbital but lowers the π^* orbital by mixing with the S d orbitals and hence gives rise to a low-field n.m.r. shift.

The widths of the ^{14}N resonances in the metal thiocyanates discussed above varied from 0.5 to 5 G. These broad lines are typical of ^{14}N resonances, and there is not usually therefore much chance of resolving informative spin–spin structure. ^{15}N resonances, on the other hand, are much narrower and the multiplets may be resolved as is illustrated[39] by the spectrum of the $S_4N_3^+$ ion shown in Figure 12. This spectrum is consistent with the structure **8**, the multiplets arising from an ^{15}N–^{15}N coupling constant of 7 cps. The spectrum was obtained with a sample 97.2% enriched in ^{15}N, and the necessity for enrichment is likely to constitute the main barrier to widespread application of ^{15}N n.m.r.

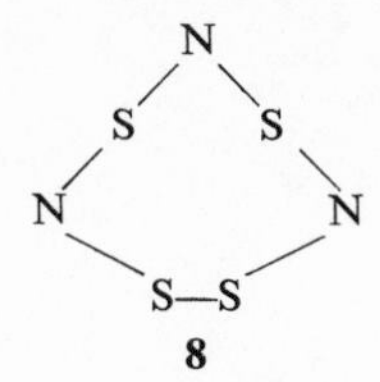

8

F. Oxygen

Oxygen has only a single magnetic isotope, namely ^{17}O which has $I = \frac{5}{2}$ and the very low isotopic abundance of 0.037%. In spite of this latter unfavorable figure, it has still proved possible to obtain spectra either in natural abundance or with only a modest amount of enrichment, although spectrometer sensitivity is always a limiting factor. The quadrupole moment is rather small and in this case probably exerts an advantageous influence. Thus T_1 is sufficiently short so that the lines are not easily saturated (as they are with ^{15}N) but neither are they excessively broad (as they are with ^{14}N). Figgis, Kidd, and Nyholm[40], for example, have examined the ^{17}O n.m.r. spectra of a variety of inorganic compounds and report the linewidths to be about 0.3 G. In a particularly interesting series of measurements, they found a good correlation between the ^{17}O chemical shifts of oxygen bonded to transition metals and the lowest energy electronic transition of these compounds. This is illustrated in Figure 13. They have carried out a theoretical treatment of these data analogous to that developed by Griffith and Orgel[11] to account for ^{59}Co chemical shifts. The success of this treatment shows that the dominant factor is again the paramagnetic contribution arising from terms involving the mixing of the orbitally non-degenerate ground state with excited states possessing orbital angular momentum. The very large chemical shifts involved (more than 1000 ppm in some cases) make the ^{17}O resonances a rather delicate probe for changes

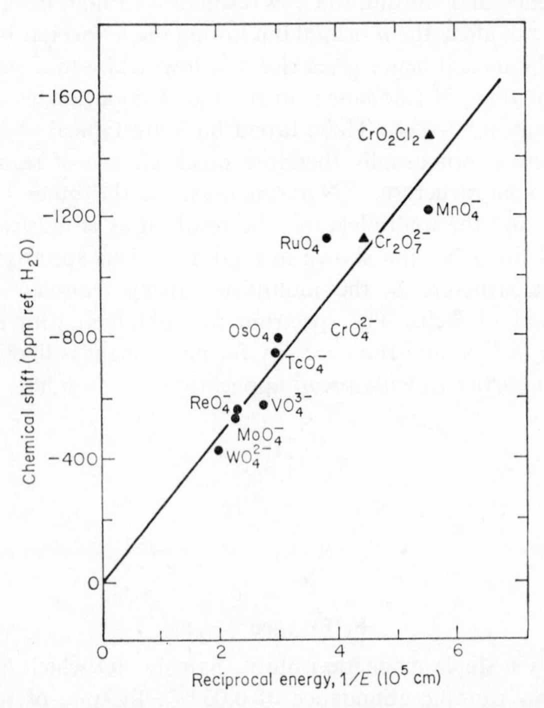

Figure 13. Correlation between ^{17}O chemical shifts and lowest energy electronic transitions of transition metal–oxygen compounds.

in electronic structure. In a later paper[41], the same authors have used ^{17}O n.m.r. to make an elegant study of the rapid reaction which occurs between chromate and dichromate ions, i.e.

$$Cr^*O_4^{2-} + O_3Cr—O—CrO_3^{2-} \rightleftharpoons O_3Cr^*—O—CrO_3^{2-} + CrO_4^{2-} \quad (25)$$

Separate signals can be observed for both the terminal and bridging oxygens of the dichromate ion and for the chromate ion. The relative proportions of chromate and dichromate ions can be varied by changing the basicity of the solution, and reaction rates obtained from an analysis of the resulting linewidth variation. A bimolecular rate coefficient of

$2.3 \pm 0.4 \times 10^3$ l/mole s was obtained for reaction (25). The reaction can be considered a special case of a general type of nucleophilic attack on the dichromate ion, and by comparison of this reaction rate with those of other nucleophiles it was concluded that the polarizability of the nucleophile rather than its basicity is the important rate-determining property.

G. Fluorine

^{19}F ($I = \frac{1}{2}$, 100% abundant) is a particularly favorable nucleus for n.m.r. studies, and its usage has been second only to that of ^1H. There is a wide literature concerned with applications to inorganic compounds. The data on the fluorides of formula MF_n obtained by Muetterties and Phillips[42] and shown in Table 4 serve to illustrate the value of n.m.r. in determining

Table 4. ^{19}F n.m.r. data on binary fluorides[42].

Compound	Chemical shift[a] (ppm)	J_{MF} (cps)	Compound	Chemical shift[a] (ppm)	J_{MF} (cps)
NF_3	−219	160	BrF_5	−349, −219	—
ClF_3	−193, −81	—	IF_5	−138, −95.8	—
BrF_3	−54.3	—	PF_5	−5.2	916
PF_3	−42.3	1441	AsF_5	−11.3	—
AsF_3	−35.0	—	SbF_5	+6.83	—
				+26.2, +52.0	
SbF_3	−23.9	—	MoF_6	−355	44
BF_3	+54.2	10	WF_6	−242	48
SF_4	−19.5, −14.8	—	SeF_6	−128	1400
SeF_4	−14.1	—	SF_6	−127	—
TeF_4	−51.4	—	TeF_6	−20.6	3052
					3688
CF_4	−11.9	—	PF_6^-	−11.6	710
BF_4^-	+71.0	—	SiF_6^{2-}	+49.8	110
SiF_4	+83.3	178	AsF_6^-	−18.1	930
GeF_4	+99.0	—	SbF_6^-	+32.3	1843

[a] Reference trifluoracetic acid.

geometric structures and also some of the possible pitfalls. Thus the fluorides of known pyrimidal (e.g. NF_3), tetrahedral (e.g. BF_4^-), and octahedral (e.g. SF_6) geometries show only a single resonance frequency as expected. Spin–spin structure is often observable in cases where M has a magnetic moment. This is not always the case, though, if M has also a quadrupole moment (i.e. $I > \frac{1}{2}$). Thus the spectrum of NF_3 at different temperatures is illustrated in Figure 14. It is apparent that the spin–spin coupling is lost at low temperatures leaving a sharp singlet. This effect is ascribed to rapid relaxation and corresponding shortening of the lifetimes of the nuclear spin states by quadrupole interaction. The theory has been considered by Pople[43]. The fluorides which show more than one resonance must have lower symmetries, e.g. ClF_3 and IF_5. In some cases, though, the additional structure cannot be observed at high temperatures. Thus SF_4 shows only a single resonance at room temperature but gives a spectrum of two triplets at $-98°c$. This observation demonstrates that the ground state of the molecule has $\mathbf{C}_{2v}$ rather than $\mathbf{T}_d$ symmetry and that

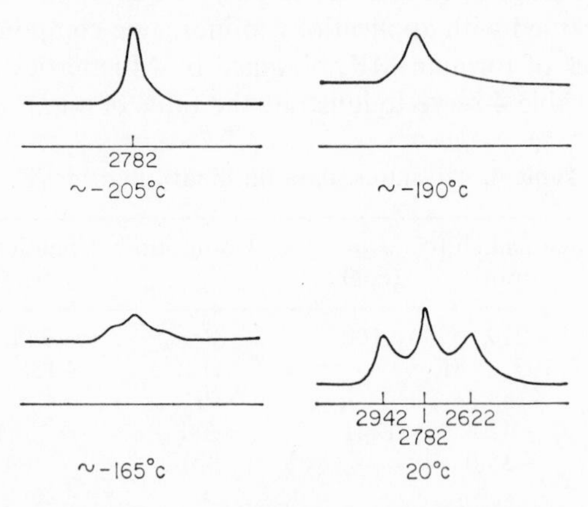

Figure 14. Temperature dependence of the ^{19}F spectrum of NF_3.

there is a process which can lead to rapid exchange of the nonequivalent fluorines at higher temperatures. There are other cases such as PF_5 and AsF_5 where nonequivalent fluorines are expected but not observed. It must be concluded either that the chemical shifts are fortuitously very similar for the different environments or that rapid exchange processes can occur. The latter alternative is the more common and has been demonstrated by Muetterties and coworkers[44] for a variety of five-coordinated compounds. It is suggested by these authors that the ground state of a five-coordinated complex will usually be a trigonal bipyramid but that the energy of a tetragonal pyramid will only be slightly higher and that this will lead to a low barrier for intramolecular exchange processes. When both exchange and quadrupolar relaxation effects are present, the n.m.r. can be quite misleading if care is not taken in the analysis. Thus the ^{19}F resonance of IF_7 appears to be a broad doublet[45]. The spectrum has been interpreted, though, in terms of a single chemical shift (indicating rapid exchange of the fluorines) split by the spin–spin coupling with I ($I = \frac{5}{2}$) and then broadened by the quadrupole interaction to give the apparent doublet. If the n.m.r. spectrum can be measured over a range of temperatures in which the rate of exchange is of the same order as the chemical shift difference (in cps), activation energies for the exchange process can be obtained. Thus Muetterties and Phillips[42] found a value of 4.5 ± 0.8 kcal for the fluorine exchange in SF_4.

H. Sodium and Bromine

Studies of solvation and association of ions in solution comprise another important group of problems which is susceptible to attack by n.m.r. The type of information which can be obtained is illustrated by the work of Richards and Yorke[46] on ^{79}Br, ^{81}Br, and ^{23}Na resonances in solutions of electrolytes. All these nuclei possess large quadrupole moments, and the linewidths are expected to be largely determined by quadrupolar interactions. Theory suggests that the dependence will be of the form (26),

$$\frac{1}{T_1} \propto (eqQ)^2 \frac{\eta}{T} \tag{26}$$

where η is the viscosity of the solution and T is the temperature. The observed linewidth is a measure of $1/T_1$ in these circumstances. It was found that for aqueous solutions of sodium and calcium bromide plots of the bromine linewidths against viscosity gave good straight lines over a wide range of concentrations. The slope of such plots therefore provides a measure of the quadrupole coupling constant. Confirmation of this point is provided by the observation that the ratios of the slopes for the ^{79}Br and ^{81}Br lines were in good agreement with the ratio of the quadrupole moments of these nuclei. The fact that the quadrupole coupling constant does not change with concentration indicates that the environment of the bromide ion is invariant and implies that ion–solvent interactions are dominant. A different result was obtained for aqueous cesium bromide solutions. In this case, it was found that the quadrupole coupling constant, as measured by the linewidth/viscosity ratio, was proportional to the molarity as is shown in Figure 15. This result provides a clear indication of ion–ion interaction in this electrolyte.

Further information on the nature of the ion–solvent interactions was obtained by using mixed solvents of alcohols or acetone and water. The quadrupole coupling arises from the electric field gradient resulting from the polarization induced in the surrounding solvent by the charged ion. Using this model, theory predicts proportionality between the quadrupole coupling constant and the function (27) of the dielectric constant (D) of the mixture. It was found that this relationship is obeyed well for both the

$$\frac{2D + 3}{5D} \tag{27}$$

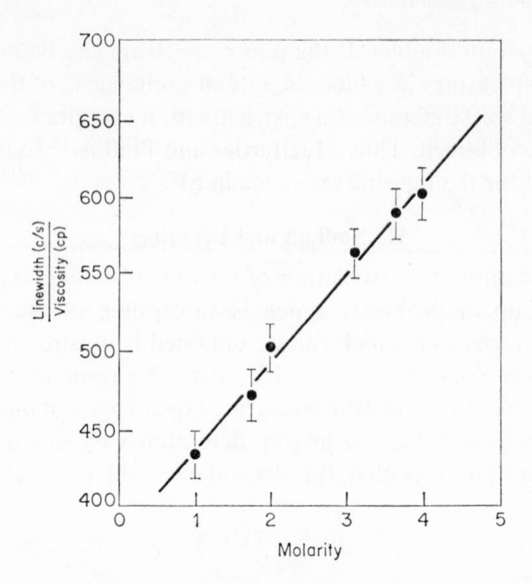

Figure 15. Variation of ^{81}Br linewidths with molarity in aqueous solutions of CsBr.

bromine and sodium resonances, and this lends considerable support to this model of solvent–ion interactions.

I. Phosphorus

^{31}P has probably been the nucleus most widely used for purely inorganic purposes. An extensive compilation of results is contained in the monograph of Fluck[6]. The present examples illustrate no new principles but are typical of a large number of studies. Thus the spectrum in Figure 16 is that of the isohypophosphate ion and was obtained by Callis and co-workers[47]. It shows two chemically different phosphorus atoms, one split into two sets of doublets and the other into a simple doublet. The integrated intensities of the two types are equal. The large coupling constant of 620 cps is in the range expected for hydrogen directly attached to phosphorus. The small coupling constant (17 cps) is the same for both types of P and must be a phosphorus–phosphorus coupling. The small value indicates that the two phosphorus atoms are not directly bonded. All this evidence is consistent with the structure **9**.

$$\left[\begin{array}{ccccc} & O & & O & \\ & | & & | & \\ H- & P & -O- & P & -O \\ & | & & | & \\ & O & & O & \end{array}\right]^{3-}$$

9

The work of Meriwether and Leto[48] on nickel–carbonyl–phosphine complexes shows how far chemical shift data can be used to obtain information in excess of the purely structural type illustrated above. They measured the chemical shifts of a large number of phosphines and compared them with the shifts obtained for the same phosphines coordinated to Ni(o). In most cases, coordination lead to a fairly substantial low-field shift. They consider no less than eight different factors which could contribute to this shift. It is apparent that to make useful comparisons a substantial number of these factors must be kept constant by

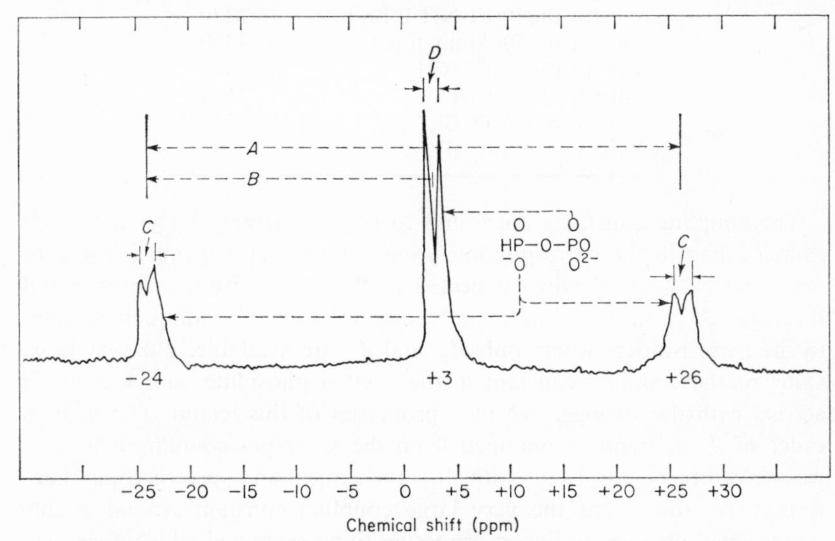

Figure 16. ^{31}P spectrum of the isohypophosphate ($HP_2O_6^{3-}$) ion. Chemical shifts in ppm from 85% H_3PO_4. $A = J_{PH} = 620$ cps. $C = D = J_{PP} = 17$ cps.

choosing a series of closely related compounds. Thus, in the series $Ni(CO)_2(PL_3)_2$, it is considered the major effect on the chemical shift arises from the strong donor σ bond from P to Ni which results in decreased shielding at the phosphorus nucleus. On this assumption, the ^{31}P shifts provide some measure of the strength of the bond. They were unable, however, to obtain unambiguous information on d_π–d_π back donation, on ring current effects, or on inductive effects of phosphine substituents.

It is often found that spin–spin coupling constants can be better correlated with bonding characteristics than can chemical shifts. Thus Pidcock, Richards, and Venanzi[49] have established such a correlation between platinum–phosphorus d_π–d_π bonding and the corresponding spin–spin coupling constants. Some of their experimental results are shown in

Table 5. These data were obtained by a combination of ^{31}P and ^{195}Pt n.m.r. experiments.

Table 5. Platinum–phosphorus spin coupling constants.

Compound	J_{PtP} (cps)
trans-$[(n\text{-}Bu_3P)_2PtBr_2]$	2310
trans-$[(n\text{-}Bu_3P)_2PtCl_2]$	2460
trans-$[(n\text{-}Bu_3P)(Et_2NH)PtBr_2]$	3270
trans-$[(n\text{-}Bu_3P)(\text{Amine})PtCl_2]$	3340
trans-$[(n\text{-}Bu_3P)_2PtHCl]$	3510
cis-$[(n\text{-}Bu_3P)_2PtCl_2]$	3620
trans-$[(n\text{-}Bu_3P)_2Pt_2Cl_4]$	3810
cis-$\{[(EtO)_3P]_2PtCl_2\}$	5700

The coupling constants are found to be considerably larger in the *cis* isomers than in the corresponding *trans* isomers. This is consistent with the stronger d_π–d_π bonding expected in the former isomers because all three d_{xy}, d_{yz}, and d_{xz} orbitals can be used for such π bonding, in contrast to the *trans* isomers where only d_{xz} and d_{yz} are available. The very large value of the coupling constant in the triethyl phosphite complexes is in accord with the stronger acceptor properties of this ligand. The relative order of d_π–d_π bonding obtained from the spin–spin coupling constants was shown to be consistent with thermochemical and spectroscopic data. It may be noted that the very large coupling constants found enable rather small changes in ligand properties to be examined which gives real value to this type of correlation.

J. Selenium

^{77}Se has a nuclear spin of $\frac{1}{2}$ and a natural abundance of 8.3%. The only extensive n.m.r. work on this nucleus is that of Birchall, Gillespie, and Vekris[50]. This study serves to illustrate rather well the possible chemical information obtainable from some of the less common nuclei. The chemical shifts of the twenty-six compounds examined varied from -80 ppm to $+1700$ ppm from $SeOCl_2$ which was chosen as a standard. This is a wide range of shifts, and it can be expected that different selenium species will be readily distinguished by n.m.r. Generally, Se^{VI} compounds were found to give resonances to high field of Se^{IV} compounds. Couplings as large as 1580 cps (in SeO_2F_2) were observed. A large downfield shift of the $SeCl_4$ resonance when dissolved in SO_3 gave evidence for the formation of a complex, probably of the type $[SeCl_3]^+[SO_3Cl]^-$. An equimolar mixture of $SeOF_2$ and $SeOCl_2$ gave three ^{77}Se resonances, indicating exchange to give SeOFCl as well as the starting materials. Mixtures of $SeOF_2$ and $SeOBr_2$, on the other hand, gave only the two initial resonances. A third possibility was found with $SeOBr_2$ and $SeOCl_2$ mixtures which showed a single, slightly broadened peak at an intermediate resonance frequency, i.e. a result indicative of fast exchange. Mixtures of Se_2Cl_2 and Se_2Br_2 showed two peaks each slightly broadened and increased broadening on heating, i.e. evidence for slow exchange of the halogens. The above series of experiments throws some interesting light on the stability of selenium–halogen bonds in different compounds.

Addition of $SbCl_5$ to $SeOCl_2$ gives a high-field shift which increases linearly with the mole ratio $SbCl_5/SeOCl_2$. This observation is consistent with the equilibrium

$$SbCl_5 + SeOCl_2 \rightleftharpoons SbCl_6^- + SeOCl^+$$

Several other chlorides were found to react analogously and could be classified as behaving as strong or weak acids in this reaction according to the magnitude of the chemical shift produced. Several other interesting applications of ^{77}Se n.m.r. can be found in the original paper.

K. Vanadium, Niobium, and Antimony

Many of the transition metal nuclei which possess magnetic moments also have rather large quadrupole moments. The experiment problems involved in n.m.r. studies of these less common nuclei are then rather different from those encountered in the case of Se discussed above. The crux of the problem is that the experimental linewidths depend on the quadrupole coupling constant and may become sufficiently large that the resonances are no longer observable. In such cases, useful n.m.r. studies may be restricted to compounds of rather high symmetry in which the electric field gradient at the nucleus under consideration is small. Nevertheless, much useful information is often available, as is illustrated by the work of Hatton, Saito, and Schneider[51] on ^{51}V, ^{93}Nb, ^{121}Sb, and ^{123}Sb resonances.

^{93}Nb resonances were observed in various solutions of NbF_5. In dry ethanol as solvent, a symmetrical septet with $J = 345$ cps was observed, and this must be assigned to the NbF_6^- ion. The equilibrium

$$2\,NbF_5 \rightleftharpoons NbF_6^- + NbF_4^+$$

is indicated, and this postulate is consistent with the high electrical conductivity of the solution. It was not possible to locate a resonance from NbF_4^+. The cation present must therefore have a structure with rather low electrical symmetry, and a solvated species such as $NbF_4(OEt)_2^+$ was suggested as one possibility. NbF_5 in aqueous HF solution, on the other hand, shows a sharp singlet resonance with no spin coupling apparent. Rapid fluorine exchange by reactions such as

$$NbF_5 + HF \rightleftharpoons NbF_6^- + H^+$$

must be occurring. Solution of NbF_5 in dry ether and in triethylamine show very broad resonances attributable to the complexes $NbF_5 \cdot OEt_2$ and $NbF_5 \cdot NEt_3$, respectively. Analogous solutions of SbF_5 were also examined using both the ^{121}Sb and ^{123}Sb resonances. The aqueous HF solutions in this case showed a septet spectrum from SbF_6^-, indicating the fluorine exchange is relatively slow. In dry ethanol, though, the SbF_6^- resonances were relatively broader than those of NbF_6^-, and the spin–spin structure was lost on either heating or cooling. It was deduced that some additional complexing interactions were occurring.

The ^{51}V spectrum of solutions of V_2O_5 in aqueous HF when cooled comprises a well-resolved quintet. This is assigned to VOF_4^- which is apparently formed in preference to a hexafluoride. The chemistry of vanadium oxyions is known to be complex with the relative stabilities of the various species depending markedly on pH. The possibility of using ^{51}V n.m.r. to disentangle some of these complexities was demonstrated by examining aqueous solutions of NH_4VO_3 and V_2O_5 over a wide pH range. No less than seven different ^{51}V resonances were observed. Five of these were assigned to the species VO_4^{3-}, VO_3^-, $V_2O_7^{4-}$, $V_3O_{10}^{5-}$, and VO_2^+, and the pH ranges over which these species are stable were established.

L. Cobalt

The above examples of transition metal n.m.r. have involved only simple structural interpretations. It is, however, sometimes possible to make some theoretical interpretation of the chemical shift data. This may be illustrated by the work of Freeman, Murray, and Richards[52] on the ^{59}Co spectra of Co^{III} complexes. This work, in fact, represents one of the earliest successful treatments of chemical shift data. ^{59}Co has a nuclear quadrupole moment, and the restriction mentioned above to reasonably symmetric environments applies. The compounds examined were a series of octahedral or pseudooctahedral Co^{III} complexes. The chemical shifts are large and the dominant contribution arises from temperature-independent paramagnetism. By the use of crystal-field theory, it is possible to evaluate this contribution in terms of Δ, the splitting of the Co d orbitals by the octahedral crystal field. An expression of the form (28) is obtained, where

$$\sigma_j = A - \frac{B_0}{\Delta} \tag{28}$$

σ is the shielding parameter and A and B_0 are constants. Δ may be equated to the frequency of the longest wavelength absorption maximum. The success of this treatment is illustrated by Figure 17 which shows a plot of the ^{59}Co resonance frequency against optical absorption wavelength. The theory also predicts a relationship between the temperature dependence of the n.m.r. frequency and that of the optical absorption, but experimentally this is much more difficult to verify.

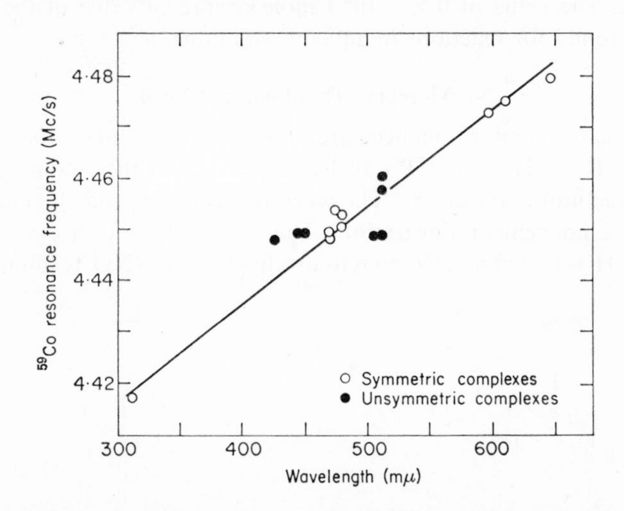

Figure 17. Correlation of ^{59}Co chemical shifts with optical absorption frequencies for Co^{III} complexes.

M. Copper

In the preceding sections, a number of examples in which nuclear-exchange processes have been studied by n.m.r. have been encountered. It should also be pointed out that the method is also applicable to electron-exchange processes such as those occurring in oxidation–reduction reactions. Circumstances are particularly favorable when one of the species involved is paramagnetic. Thus McConnell and Weaver[53] used ^{63}Cu n.m.r. to obtain a rate coefficient for the reaction

$$Cu^{2+} + Cu^{+} \rightleftharpoons Cu^{+} + Cu^{2+}$$

in hydrochloric acid solution. The width of the ^{63}Cu resonance is given by (29), where T_2(a) is the linewidth in a solution containing only cuprous

$$\frac{1}{T_2} = \frac{1}{T_2(\mathrm{a})} + \frac{1}{T_2(\mathrm{b})} \tag{29}$$

chloride. This width is due mostly to quadrupole effects. Addition of a small amount of cupric ion produces an additional broadening described by the term $1/T_2(\mathrm{b})$. It was demonstrated that this additional broadening was due to electron exchange rather than to simply the broadening effect of a paramagnetic ion by adding an equivalent amount of paramagnetic Ni^{II} ions. In this latter case, no significant additional broadening was observed. $T_2(\mathrm{b})$ can be simply related to the rate coefficient for electron exchange. The value of 0.5×10^8 1/mole s represents one of the highest rates yet found for reactions in aqueous solutions.

N. Mercury, Thallium, and Lead

Not all heavy-metal resonances are complicated by quadrupole effects. ^{199}Hg, ^{203}Tl, ^{205}Tl, and ^{207}Pb all have spin $\frac{1}{2}$, and the resonances have found structural applications. The work of Schneider and Buckingham[54] provides a convenient illustration. The chemical shifts are very large, e.g. $Hg(CH_3)_2$ is shifted 2460 ppm to low field of a $Hg(NO_3)_2$ solution used

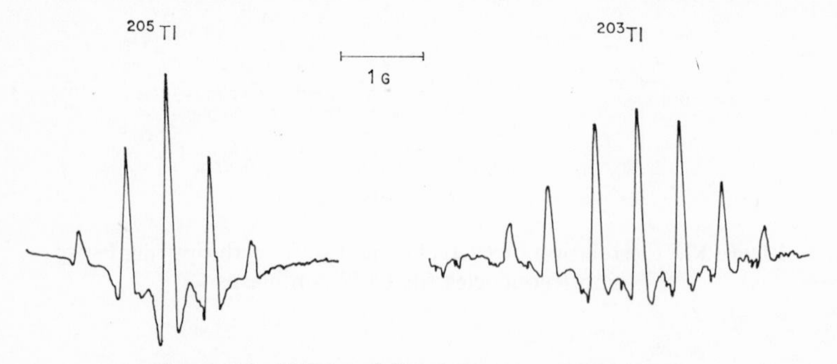

Figure 18. ^{205}Tl and ^{203}Tl Spectra of $[Tl(OEt)]_4$

as reference. The shifts of these nuclei also tend to show large solvent effects. They are therefore rather useful for studying solvation and association effects, and a number of such studies have been reported. The lines are usually sufficiently narrow to resolve spin–spin couplings, e.g. metal–proton couplings of 102 cps, 251 cps, and 61.2 cps were found in $Hg(CH_3)_2$, $Tl(CH_3)_3$, and $Pb(CH_3)_4$, respectively. This feature enhances their utility in structural problems. An example of such a problem is provided by the work on thallous ethoxide. The ^{205}Tl and ^{203}Tl spectra obtained from this compound are shown in Figure 18. These symmetric patterns show the very large spacing of 1280 cps. The fact that the spectra from the two isotopes are quite different can only be explained if this structure arises from ^{203}Tl–^{205}Tl spin–spin coupling. The spectra have been interpreted in terms of a tetramer $(TlOEt)_4$ in which Tl and O atoms occupy alternate corners of a cube. Alternatively, the structure can be viewed as a regular tetrahedron of Tl atoms superimposed on a regular tetrahedron of oxygens. Thus all four thallium atoms are equivalent. There are, however, five different types of molecules to consider; namely, those containing four ^{205}Tl nuclei, those containing three ^{205}Tl and one ^{203}Tl, etc. The relative abundance of these molecules and the simple multiplet patterns expected from each in the ^{205}Tl and ^{203}Tl spectra are shown in Table 6.

Table 6. Multiplets predicted from $(TlOEt)_4$.

Tetramer configuration[a]	Relative abundance	^{205}Tl resonance	^{203}Tl resonance
5–5–5–5	0.2468	singlet	—
5–5–5–3	0.4134	doublet	quartet
5–5–3–3	0.2597	triplet	triplet
5–3–3–3	0.0725	quartet	doublet
3–3–3–3	0.0076	—	singlet

[a] 5 indicates ^{205}Tl, 3 indicates ^{203}Tl.

The observed spectra consists of superpositions of the appropriate multiplets, each with the same coupling constant but weighted by its relative abundance. Satisfactory agreement between experimental and observed intensities was obtained. This analysis leads to the very large spin–spin coupling constant of 2.56 kc/s. This value provides excellent evidence for metal–metal bonding in the tetramer.

O. Conclusion

The examples discussed above by no means include all the nuclei whose n.m.r. has been profitably used to explore inorganic problems. Thus ^{27}Al, ^{29}Si, and ^{119}Sn, although not mentioned, have all been the subject of fruitful studies. Perusal of Table 1 will reveal a number of other nuclei of potential or actual utility which have been similarly omitted. However, the types of experiments possible and the likely limitations should be apparent from the preceding discussion. In summary, the following general points may be noted. Lack of sensitivity is a problem in virtually all n.m.r. experiments involving nuclei other than ^{1}H or ^{19}F. As indicated later in Section VI of this chapter, several new approaches to alleviate this problem are currently being developed. A corresponding growth of interest in less common nuclei can be expected in the near future. Chemical shifts for the heavier nuclei are usually large. This generally brings the advantage of simpler spectra and less ambiguous structural inferences. However, it is also a fair generalization that, for compounds involving heavier atoms, bond strengths are often less than those of the common C—C, C—H, C—O bonds, etc., and exchange processes are correspondingly more common. The expected number of resonances is reduced by such processes, and the utility of n.m.r. becomes increasingly in the area of kinetic rather than purely structural studies. For quadrupolar nuclei, great success can only be expected if the environment of the nucleus is rather symmetric. In other cases, future progress may rest with n.q.r. rather than n.m.r. Finally, it should be pointed out that the theoretical treatment of neither chemical shifts nor spin–spin coupling constants involving heavy nuclei has been carried out to any great degree of sophistication. The present applications to structural problems are largely empirical, but suitable advances in the theory could well greatly enhance the value of experimental n.m.r. data on inorganic compounds.

V. PARAMAGNETIC MOLECULES AND METAL–LIGAND BONDING

A. Contact Interactions and Spin Densities

Paramagnetism is encountered in many transition metal complexes. The presence of unpaired electrons opens the way for electron spin–nuclear spin interactions of the types mentioned in Section II and greatly expands the potential usefulness of n.m.r. investigations. Isotropic hyperfine interactions, when present, usually lead to shifts much larger than the normal chemical shifts (at least for protons) and a detailed interpretation in terms of the electronic and geometric structure of the complex is often possible. The development of this area of research, therefore, offers much of interest to the inorganic chemist.

In order to observe nuclear magnetic resonance in a paramagnetic molecule, it is necessary that one of the conditions

$$T_1^{-1} \gg a_N \quad \text{or} \quad T_e^{-1} \gg a_N$$

is met[55]. In these inequalities, T_1 is the electron-spin relaxation time and T_e an electron-spin exchange time relating to condensed systems. If neither of these conditions is met, T_2 for the nuclear spin will be very short and the resonance line correspondingly very broad and probably unobservable. The physical meaning of these conditions is not easy to define precisely, but it is not excessively misleading (at least in the author's opinion) to consider that slow electron relaxation will produce fluctuating magnetic fields by flipping of the electron spins which have components with frequencies suitable for inducing nuclear relaxation, whereas fast electron relaxation will not. In the latter case, the nuclear spins will sense only the average magnetic field produced by the electron spins. Since in the presence of an applied magnetic field there are always slightly more electron spins oriented with the field (α) than against it (β), this average field is not equal to zero. The excess of α spins will be greater at low temperatures, and this leads to a $1/T$ (Curie law) dependence for isotropic hyperfine shifts. By good fortune, electronic T_1 values for paramagnetic transition metal complexes are usually much shorter than those for organic free radicals, and n.m.r. spectra can often, but not always, be observed. The increased importance of spin–orbit coupling is probably the principal reason for this circumstance. It may be noted that there is a partial exclusion principle involved here, since a short electronic T_1 aids the observation of the n.m.r. spectrum but broadens and makes more difficult the observation of the e.p.r. spectrum. However, this latter difficulty can sometimes be avoided by making the e.p.r. measurements at low temperature.

Much of the theory involved in the interpretation of contact shifts is exactly the same as that previously developed to account for hyperfine splittings in the e.p.r. spectra of free radicals. Thus the hyperfine contact interaction constant a of equations (10) and (11) is just the splitting which would be observed in an e.p.r. spectrum of the molecule. Following the lead of the e.p.r. spectroscopists, it is usual to interpret hyperfine coupling constants in terms of spin densities (ρ). In its simplest form, a spin density can be regarded as the probability of finding an unpaired electron in a specified atomic orbital. Thus for the case of the benzene negative ion ($C_6H_6^-$), for example, the odd electron is in an antibonding π molecular orbital comprised of only carbon $p\pi$ atomic orbitals. By symmetry, there is an equal probability of finding the electron at any of the carbons so that each bears a spin density of $\frac{1}{6}$. However, before proceeding to interpret contact shifts and hyperfine coupling constants in terms of such spin-density distributions, two additional problems must be faced. A special case of the first problem can be again illustrated by reference to the benzene negative ion. As the orbital containing the spin is comprised of carbon $p\pi$ orbitals, since these orbitals have nodes in the plane containing the hydrogen atoms, and as the contact effect demands a spin density at these same hydrogen atoms, there is an immediate problem in accounting for the observed splittings. The general problem is one of relating observed coupling constants, implying spin densities at the nucleus under examination, to the spin densities in the molecular orbitals which theory predicts will contain the unpaired electron. The second problem relates to the experimental observation of negative spin densities. Both of these problems require some consideration of electron correlation.

Electron correlation in molecules arises from Coulomb repulsion and spin-exchange effects. Much recent work in theoretical chemistry has been directed toward the proper incorporation of these effects into molecular wave functions. Qualitatively, the important results for the present discussion can be illustrated by two simple examples. The electron distribution in a C—H fragment of an aromatic radical can be represented by **10**.

C↑↓H

10

The unpaired electron is to be found in the carbon $p\pi$ orbital, but there is a polarization of the electrons in the C—H σ-bonding orbital in the sense shown. As a result, unpaired spin on the carbon induces unpaired spin in the H 1s orbital and leads to a hyperfine interaction. It has been shown that the simple relationship (30), where Q is a constant, holds to a rather

$$a_{\mathrm{H}} = Q\rho_{\mathrm{C}} \tag{30}$$

good approximation[55]. It may be noted that positive spin on the carbon leads to negative spin at the hydrogen. Hence Q is a negative quantity. The second example is provided by the spin distribution in the allyl radical (**11**). There are two equivalent valence-bond structures for this

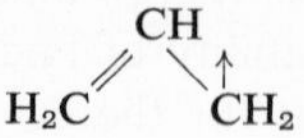

CH
H_2C CH_2

11

radical obtained by placing the spin at either CH_2 group. In the simple molecular-orbital description, the π orbital containing the odd electron has a node at the CH carbon and gives rise to a spin density of $\frac{1}{2}$ at both CH_2 carbons. The effect of spin correlation is to distort the *spin* distribution (not the *charge* distribution) in the filled bonding π orbital in such a way as to pile up positive (α) spin on the CH_2 carbons and leave negative (β) spin at the CH carbon. There are two rather simple types of calculation which take account of this correlation effect. One is the extended Hartree–Fock method developed by McLachlan[56] which uses molecular orbitals with slightly different Coulomb integrals for the α and β spins. The other is the valence-bond method which includes spin correlation from the start, since the basis wave functions are constructed with paired spins. Either of these methods gives qualitatively correct results, but considerably greater sophistication is required if good quantitative agreement is to be expected. However, as will be illustrated later, qualitative agreement is often sufficient to decide the correct assignment of the spin-containing orbital and to derive the chemically significant information. Combination of equations (11) and (30) leads to an important generalization: for aromatic fragments, positive carbon spin densities lead to high-field proton contact shifts, and negative carbon spin densities lead to low-field proton shifts. The sign of the spin density can therefore be obtained directly from the n.m.r. spectrum, whereas at best it can only be obtained indirectly from the e.p.r. spectrum.

The observation of contact shifts is, of course, not restricted to aromatic-type protons. Both other bonding systems, e.g. methyl groups attached to aromatic carbons, and other nuclei, e.g. fluorine, may also be encountered. In general, each situation must be considered separately. Thus, the most important factor in the C—CH_3 fragment is a hyperconjugative effect which leads to direct transfer of positive spin to the methyl hydrogens. An equation of the form (31) can again be written, but in this case Q_{CH_3}

$$a_{CH_3} = Q_{CH_3}\rho_C \tag{31}$$

is positive (since positive spin on the aromatic carbon results in positive spin at the hydrogens) and has been found not to be a true constant. This is to be expected, since the amount of hyperconjugative interaction with the π system will vary from molecule to molecule. For the C—F fragment, both a polarization effect (giving a negative contribution to Q_F) and a conjugative effect (giving a positive contribution to Q_F) must be considered. The situation in aliphatic and σ-bonded systems is less well defined, but it appears that both direct delocalization to the protons and correlation effects will have to be considered.

There is one other complication which is quite frequently encountered and should be mentioned. This is the type of situation which involves rapid exchange between diamagnetic and paramagnetic environments. Such a situation may arise, for example, if there is rapid exchange between a free ligand and a ligand coordinated to a paramagnetic metal ion, i.e. an equilibrium of the type

$$M^+ + L \rightleftharpoons (ML)^+$$

In such cases, the observed contact shifts will be reduced by a factor equal to the fraction of coordinated ligand. The existence of such an equilibrium makes it more difficult to evaluate the contact shifts absolutely but, on the other hand, the study of the equilibria *per se* is often of considerable interest and can easily be accomplished by n.m.r. if large contact shift effects are present. Rapid exchange may also take place between the ground state and an electronically excited state of a transition metal complex. If one such state is a singlet and the other has a higher multiplicity, an exchange between diamagnetic and paramagnetic environments is accomplished in this way. Thus, Ni^{II} complexes have been studied in which the ground state corresponds to a square planar spin-paired molecule (i.e. a singlet electronic state) but in which there is a thermally accessible spin-free tetrahedral excited state (i.e. a triplet). In this case, the contact shift will be given by the more complicated expression (32), in which ΔF

$$\left(\frac{\Delta H}{H_0}\right)_i = a_i \frac{\gamma}{\gamma_N} \frac{g\beta S(S+1)}{3kT[\exp(\Delta F/kT)+1]} \tag{32}$$

is the free-energy difference between the magnetic states or configurations. It may be noted that the diamagnetic and paramagnetic states can differ in both enthalpy and entropy, i.e. ΔF is itself a function of temperature, so that the overall temperature dependence of the contact shifts can become quite complex.

B. Pseudocontact Effects

In addition to the contact effects discussed above, there is also the possibility of pseudocontact effects[57] of the type briefly mentioned in Section II. Pseudocontact shifts involve no delocalization of the electron spin but arise from direct dipole–dipole interactions between the magnetic dipoles of the unpaired electrons and those of the nuclear spins. As was the case for chemical shifts, much structural information may be obtained from the observation of spectra involving isotropic hyperfine shifts without attempting any detailed analysis of the origin of the shifts. This is particularly true since such shifts are usually large and allow easy differentiation of chemically different protons. Some applications along these lines will be discussed below. However, a more complete interpretation of the spectra is very desirable, since it can lead to a wealth of information about the geometric and electronic structure of the complex. The initial problem is always that of deciding whether the shifts arise from contact effects, pseudocontact effects, or a mixture of both.

Formulas appropriate for the description of pseudocontact effects have been derived by McConnell and Robertson[57]. For a molecule possessing tetragonal symmetry for which the g tensor is defined by the components $g_{\parallel}$ and $g_{\perp}$ (parallel and perpendicular, respectively, to the principal axis of the molecule), the resulting shift in the solid is given by (33). In this

$$\frac{\Delta H}{H} = -\frac{(3\cos^2\theta - 1)}{r^3}(g_{\parallel}^2 - g_{\perp}^2)\frac{\beta^2 S(S+1)}{9kT} \tag{33}$$

equation, r is the distance between the paramagnetic electrons (considered to be confined to the transition metal ion) and the resonating nucleus, and θ is the angle between the distance vector and the tetragonal axis of the complex. Equation (33) refers to the solid, and the situation is a little different in solution. This difference arises because quantization of the electron spin in the direction of the applied magnetic field occurs in spite of the g-value anisotropy because of the rapid tumbling of the molecules in solution. Two cases may be distinguished depending on whether (i) $T_{1e} \gg \tau_c$, or (ii) $T_{1e} \ll \tau_c$, where T_{1e} is the electron spin lattice relaxation time and τ_c the correlation time for molecular tumbling. Case (i) leads to formula (34) and case (ii) to formula (35). Comparison of equations

$$\frac{\Delta H}{H} = \frac{(3\cos^2\theta - 1)}{r^3}(g_\parallel + 2g_\perp)(g_\parallel - g_\perp)\frac{\beta^2 S(S+1)}{27kT} \tag{34}$$

$$\frac{\Delta H}{H} = \frac{(3\cos^2\theta - 1)}{r^3}(3g_\parallel^2 + g_\parallel g_\perp - 4g_\perp^2)\frac{\beta^2 S(S+1)}{45kT} \tag{35}$$

(33), (34), and (35) shows that the geometric factor is identical in each case, i.e. passing from one case to another does not change the pattern of shifts but in effect applies a different scaling factor. The difference between solid and solution immediately suggests a method of distinguishing pseudo-contact shifts, since the latter will not depend directly on the state of matter. McConnell and Holm[58] applied this criterion to the shift observed in vanadocene and concluded that it was contact in origin. However, the method has not been widely applied because of the lack of resolution in the solid state.

It is apparent that pseudocontact shifts can be readily calculated from a knowledge of the principal g values (obtainable from e.p.r.) and the molecular geometry (obtainable from x-ray crystallography). In most cases, the bond distances and angles can probably be estimated with sufficient precision to establish the pseudocontact nature of the shifts. Jesson[59] has carried out such a study on Co^{II} trispyrazolylborate and Co^{II} trispyrazolylmethane chelate systems, the structures of which are shown in Figure 19. In these compounds, the cobalt has a very short electron spin relaxation time so that well-resolved n.m.r. spectra are obtained in solution at room temperature. The Co^{II} ion is an odd-electron system so that the ground state must be a Kramers doublet, and it is possible to obtain the e.p.r. spectrum at liquid-helium temperature. Single-crystal and powder e.p.r. experiments show a very large g-value anisotropy, e.g. $g_\parallel = 8.46$, $g_\perp = 0.97$ for the structure of Figure 19 when X = B, Y = H, so that the situation is favorable for pseudocontact shifts. To give a complete account of the shifts, it was necessary to modify equation (35) to allow for the thermal population of the Kramers doublets other than the ground state at room temperature. The resulting formula is (36).

$$\frac{\Delta H}{H} = \frac{(3\cos^2\theta - 1)}{r^3}\frac{\beta^2 S(S+1)}{45kT} \times \frac{\sum_i (3g_\parallel^2 + g_\parallel g_\perp - 4g_\perp^2)_i \exp(-E_i/kT)}{\sum_i \exp(-E_i/kT)} \tag{36}$$

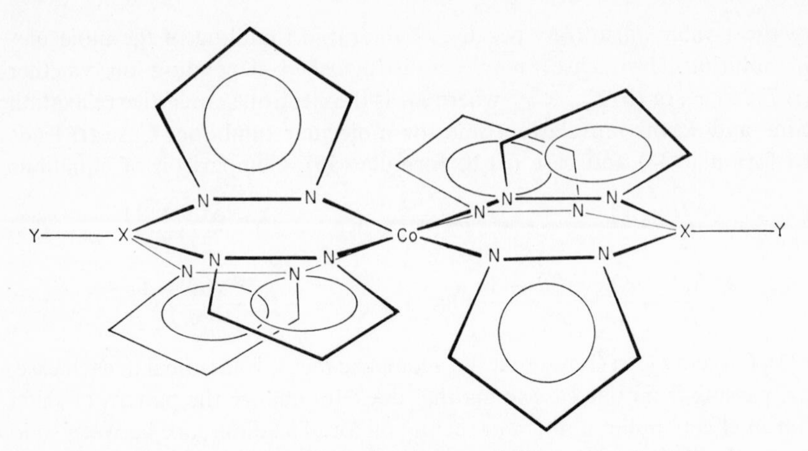

Figure 19. Structures of Co^{II} trispyrazolylborate and Co^{II} trispyrazolylmethane chelates.

The g values for excited states can be calculated using parameters obtained from an analysis of the e.p.r. and optical spectra. However, to establish the pseudocontact nature of the shift with reasonable certainty it is not necessary to carry out the above extensive calculations. Thus the data in Table 7 show some of the n.m.r. shifts obtained in the above compounds in which X = B, Y = H (compound I), and X = B, Y = phenyl (compound II).

Table 7. N.M.R. shifts of Co^{II} trispyrazolylborate chelates.

Compound I			Compound II		
Nucleus	Observed shift[a]	r(Å)[b]	Nucleus	Observed shift[a]	r(Å)
B	−3770 (14.2 Mc/s)	3.40	B	−3700 (14.2 Mc/s)	3.40
H	−6625 (60 Mc/s)	4.61	para-H	−1070 (60 Mc/s)	8.84

[a] Solvent $CDCl_3$, reference corresponding diamagnetic zinc chelate at room temperature.
[b] Estimated using bond lengths from ref. 60.

These particular data have been chosen because the nuclei all lie on an axis perpendicular to the symmetry axis of the molecule, i.e. the angular factor is the same in all cases so that the shifts should be proportional to $1/r^3$. In compound I, the experimental ratio of shifts is 1.7 and the calculated ratio 1.7. In compound II, the corresponding figures are 0.29 and 0.24. With a little more arithmetic, this approach can be readily applied to other nuclei in the molecule, and the broad measure of agreement between observed and calculated ratios constitutes excellent proof that the shifts do indeed arise from pseudocontact effects. It should be noted that the $1/r^3$ dependence makes the shifts very sensitive to bond lengths which becomes an advantage if pseudocontact effects are to be used to obtain geometric data.

The most certain way of establishing the origin of hyperfine shifts is to compare the magnitudes of the shifts calculated from known g values and the patterns calculated from known geometric factors with the experimental results. This was done with some of the transition metal acetylacetonates discussed below and leads to the conclusion that the shifts were predominantly contact in origin. However, the appropriate e.p.r. and x-ray data are often not available and less reliable arguments must be used. Thus, in a number of cases, the decision has been reached on the basis of good agreement with calculated spin-density distributions. This argument is fairly reliable when the spin densities refer to π systems but is much less so for σ systems. Schulman[61] has pointed out that the different dependences of nuclear T_1 and T_2 values on the hyperfine coupling constant provide a way to discriminate contact and pseudocontact interactions. This approach has not as yet been widely used.

C. Structural Studies

The very large proton shifts observed in paramagnetic chelates have proved very useful in purely structural work. Thus McDonald and Phillips[62] have exploited such shifts in a detailed study of the complexes formed between histidine and Co^{II} ions in aqueous solutions. The mode of complexation of this important amino acid may well be of biological significance and detailed information is difficult to obtain by other techniques. Evidence was found for four different types of complexes stable in different pH ranges. The structures of these complexes are shown in Figure 20. The spectra were spread over a range of more than 16,000 cps at 60 Mc/s so that in spite of the relative broadness of the lines resonances due to different types of complex were easily distinguished. Thus complex I of Figure 20 is stable between pH 1 and 3.5. In this pH region, complexed and free histidine exchange rapidly. In the pH range 4.5–10.5, two additional spectra due to the octahedral complexes II and III appear. Formation of these complexes is accompanied by expulsion of water molecules from the Co coordination shell which leads to shifts in the water resonance. The number of molecules of water expelled can be obtained from this shift, and this allows a determination of the stoichiometry of the complex. Different spectra were observed for the diastereoisomers of complex III which result from the use of mixtures of D- and L-histidine. Finally, at pH > 11.5, a fourth spectra due to the tetrahedral complex IV. Optical and magnetic measurements were used to determine the stereochemistry of these complexes. Information regarding the probable binding sites of the histidine molecules in the different complexes can be obtained by noting the positions of largest proton shift. For either a pseudocontact

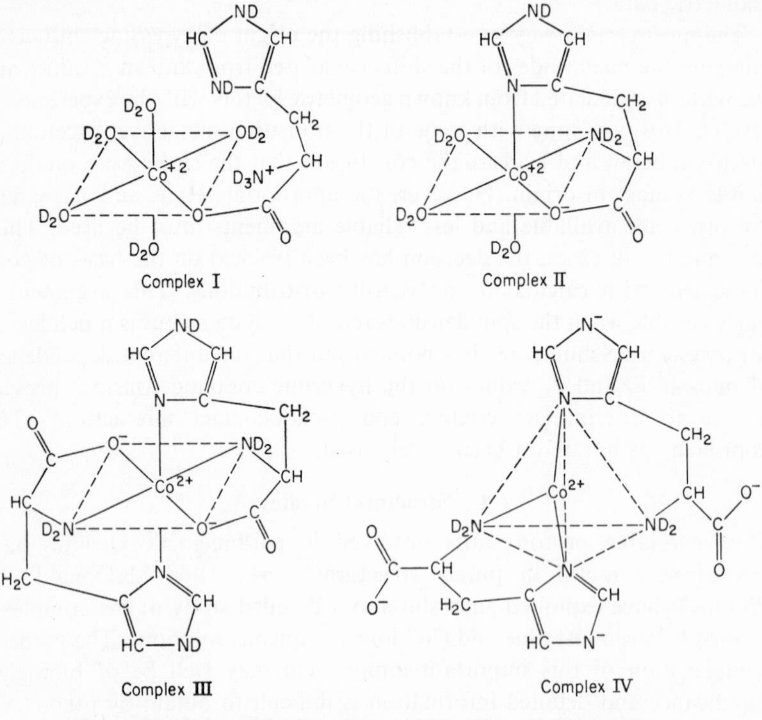

Complex I

Complex II

Complex III

Complex IV

Figure 20. Schematic structures of histidine–Co^{II} complexes.

or σ-contact shift mechanism, such positions are likely to be occupied by the protons closest to the metal ion. This study also yielded useful information regarding rates of ligand exchange in the different complexes.

D. Metal–Ligand Bonding

The general method of approach used to interpret contact shift results on transition metal complexes in terms of bonding interactions can be illustrated by reference to results obtained with some of the first-row transition metal acetylacetones[63]. This is a particularly simple system to study since there are only two chemically distinct types of ligand protons which are present in the ratio of 6:1 so that the chances of incorrect assignments are minimal. The complexes have the structure **12** and the

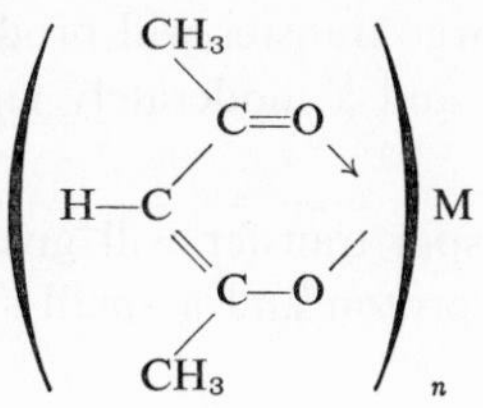

$$\left(\begin{array}{c} CH_3 \\ \diagdown \\ C{=}O \\ H{-}C \qquad\quad \\ C{-}O \\ CH_3 \end{array} \right)_n M$$

12

observed shifts for some of the complexes involving first-row transition metals are shown in Table 8.

Table 8. Contact shifts of transition metal acetylacetonates[a].

	Ti^{III} [b,c]	V^{III}	Cr^{III} [c]	Mn^{III}	Fe^{III}	Co^{II} [d]
CH_3	~ −3500	−2744	−2320	−1505	−1243	−1577
H	?[c]	−2404	?[c]	−1085	+1644	−1404

[a] Shifts at 60 Mc/s from TMS. Solvent $CDCl_3$.
[b] In C_6D_6.
[c] These resonances very broad so that single proton not located.
[d] In C_6D_5N; predominant species in solution $Co(AA)_2py_2$.

A number of arguments were developed to show that the shifts in this series are predominantly contact in origin. These included calculation of pseudocontact contributions where the g values were known and some less direct arguments in other cases. The shifts were interpreted in terms of delocalization of spin in ligand π-molecular orbitals. Either metal-to-ligand charge transfer involving the lowest antibonding π^* orbital or ligand-to-metal charge transfer involving the highest bonding π orbital could occur. In the former case, an α electron will be partly delocalized. In the latter case, if the metal d shell is half-filled or more, a β spin must be donated from the π orbital, leaving excess α spin on the ligand. If, however, the metal d shell is less than half-filled, donation of an α electron, leaving excess β on the ligand, is more likely since this preserves maximum spin multiplicity at the metal ion. The next step in the analysis is to calculate the spin-density distributions in the π and π^* orbitals using in this case McLachlan's[56] MO treatment. The signs of the spin densities will be reversed if a β rather than an α electron is delocalized. Predicted spin densities can then be transformed to predicted contact shifts using equation (11) and the relationship $a_H = Q\rho_C$, where Q is positive for the CH_3 group but negative for the single proton. The qualitative results of this analysis are as follows:

(a) Metal-to-ligand charge transfer will produce a large low-field shift for the CH_3 group and a moderately large low-field shift for the single proton.

(b) Ligand-to-metal α spin transfer will give rise to a large low-field shift for the single proton and a small shift of indeterminate sign for the CH_3.

(c) Ligand-to-metal β spin transfer leads to a large high-field shift for the single proton and a small shift of indeterminate sign for the CH_3.

The experimental shifts can now be compared with these predictions. The CH_3 shift depends predominantly on the metal-to-ligand charge transfer, and the smooth decrease from Ti^{III} to Fe^{III} indicates a corresponding decrease in this interaction. The single proton shift reflects the additional effect of ligand-to-metal charge transfer. This effect becomes increasingly important with higher atomic number. The change in sign when Fe^{III}, which has a half-filled shell, is reached reflects the change from α to β spin transfer. Calculation shows that the fraction of an unpaired electron delocalized to the antibonding orbital drops from about 7×10^{-2} for Ti^{III} to 1×10^{-2} for Fe^{III}. These observations are consistent with a simple energy-level scheme in which the *d* orbitals are relatively high in energy and close to the π^* orbital at the beginning of the series but drop in energy with increasing nuclear charge and eventually approach the π orbital. The results are also consistent with chemical expectations in that ions showing important metal-to-ligand charge transfer, e.g. Ti^{III}, are easily oxidized while those favoring ligand-to-metal charge transfer, e.g. Fe^{III}, are more easily reduced.

The above example was concerned with general principles and rather simple molecules. In favorable cases, much more complex molecules can be treated and more sophisticated conclusions drawn regarding the energy level scheme. An example of such a case is provided by work done on mixed Ni^{II} chelates with aminotroponimine ligands[64] which have the general structure **13**. Solutions containing such mixed chelates can be prepared by mixing a symmetric chelate and an appropriate ligand. Resonances

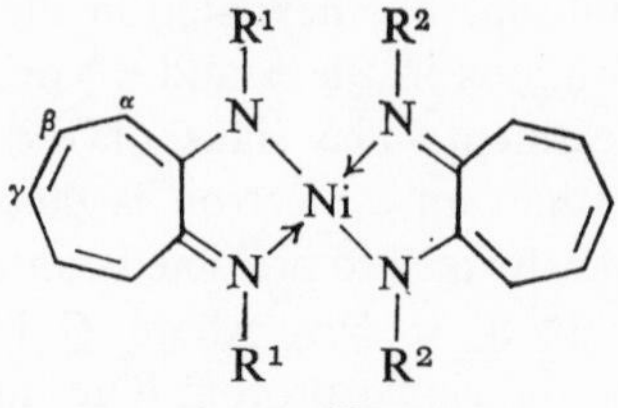

13

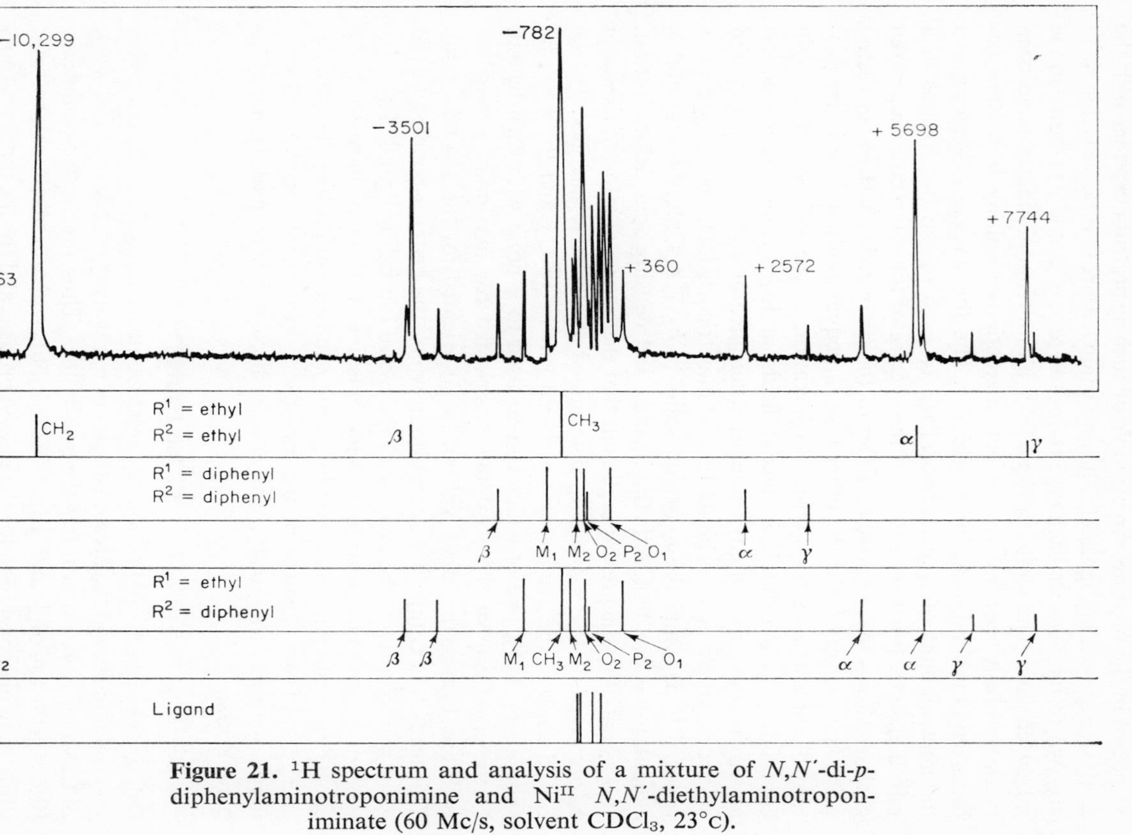

Figure 21. ^{1}H spectrum and analysis of a mixture of *N,N′*-di-*p*-diphenylaminotroponimine and Ni^{II} *N,N′*-diethylaminotroponiminate (60 Mc/s, solvent $CDCl_3$, 23°c).

from both symmetric chelates, both ligands, and the mixed chelate can be identified in the mixture. An example of such a spectrum together with the analysis is shown in Figure 21. The important feature to observe is that there are two resonances for each type of proton (e.g. α, β, and γ protons) in the mixed chelate. The shifts in these Ni^{II} aminotroponiminates have been unambiguously ascribed to contact interactions resulting from spin delocalized in a bonding ligand π orbital[65]. The above result indicates that a different amount of spin is delocalized to each of the two ligands in a mixed chelate. Further analysis of the spectra of the symmetric and mixed chelates shows that relative to either of the symmetric chelates one ligand gains spin density while the other loses an equal amount. The interpretation of these results rests on a consideration of the splitting of the spin containing Ni d orbitals in ligand fields of low symmetry. To a first approximation, the Ni environment is tetrahedral and the two unpaired electrons are in a t_2 orbital. In a symmetric chelate, the symmetry is reduced to $\mathbf{D}_{2d}$ and the t_2 orbital splits to $b_2 + e$. Only the e orbital is capable of π bonding with the ligands and causing spin delocalization. It is essential to the present argument that the b_2 orbital is lower, as would be expected if the π bonding determined the relative energies. In the mixed chelate, the symmetry is reduced further to $\mathbf{C}_{2v}$ and the e orbital splits to $b_1 + b_2$, one of which can be used to π bond with each ligand. However, if one of these orbitals is oriented for maximum π bonding with one ligand, the other will not be so oriented for the second ligand. This leads to a competitive situation in which the better π-bonding ligand gains spin density at the expense of the worse π-bonding ligand. Thus this type of experiment provides a very sensitive method of measuring the effects of substituents on π-bonding ability and also of probing the detailed energy level arrangement of the metal ion. Very good correlations were found between the substituent effects discussed here and Hammett σ parameters.

E. Ion Pairing

One area in which application of pseudocontact interactions should prove very valuable lies in the area of studies of ion pairing in solution. The work of LaMar[66] has demonstrated the possibilities. Thus, for the two complexes $[Bu_4N][(Ph_3P)CoI_3]$ and $[Bu_4N][(Ph_3P)NiI_3]$ dissolved in chloroform, shifts were observed both for the phenyl protons and for the butyl protons. Since the tetrabutylammonium ion is coordinately saturated and incapable of forming more covalent bonds, there can be no contact contribution to the butyl proton shifts. Pseudocontact interaction is, however, feasible if the molecule exists in solution as an ion pair with a well-defined geometry. This was the model adopted to interpret the results. It was found that the positions of the butyl resonances shifted with the addition of excess $[Bu_4N]^+$ ion indicating rapid exchange between ion paired and free $[Bu_4N]^+$ ions. Spectra for the Ni complex are shown in Figure 22. The

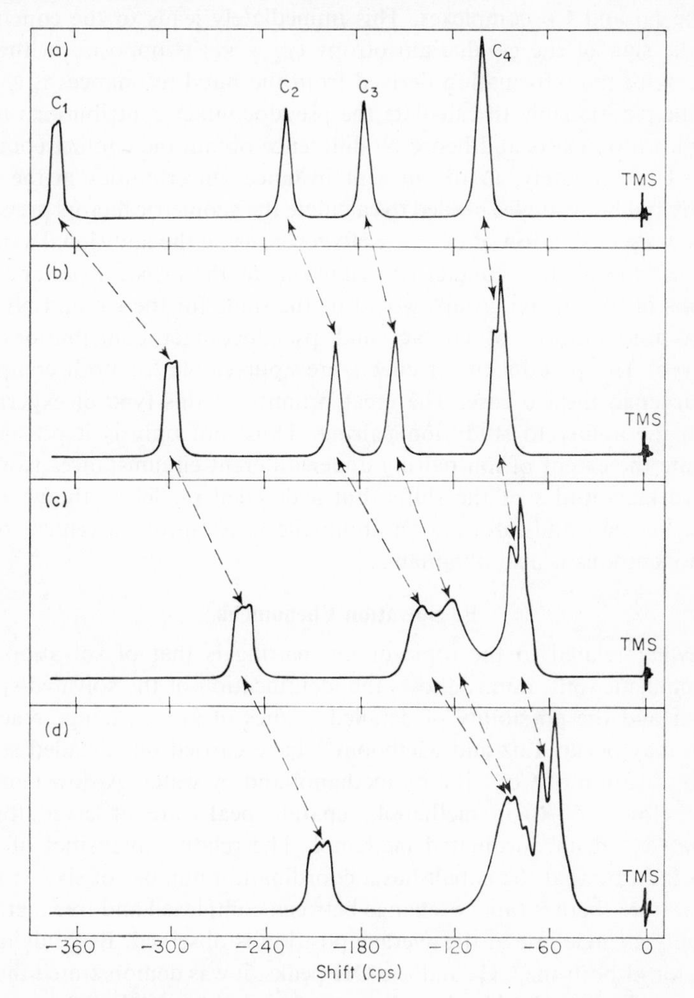

Figure 22. ^1H spectra of butyl protons in $[Bu_4N][(Ph_3P)NiI_3]$. (a) Pure complex, (b) slight excess $[Bu_4N]^+[I]^-$, (c) moderate excess $[Bu_4N]^+[I]^-$, (d) pure $[Bu_4N]^+[I]^-$.

shifts could be fitted to the pseudocontact formula by placing the $[Bu_4N]^+$ ion on the C_3 axis of the $[(Ph_3P)MI_3]^-$ ion and at the opposite side to the bulky triphenylphosphine group. Since only one resonance is observed for the CH_3 group and each of the CH_2 groups of the butyl, the $[Bu_4N]^+$

ion must be tumbling rapidly. Evaluation of the geometric factors for the different protons gave a good fit for the relative shifts based on this model of the ion pair. It was also found that the signs of the shifts were opposite for the Ni and Co complexes. This immediately leads to the conclusion that the sign of the *g*-value anisotropy ($g_{\parallel} - g_{\perp}$) is opposite in the two cases. With the information derived from the butyl resonances it is then, in principle, possible to calculate the pseudocontact contributions to the phenyl proton shifts and hence by difference obtain the contact contributions. Unfortunately, in the present instance, uncertainties in the bond lengths and bond angles needed to calculate the geometric factors prevented an accurate evaluation of $g_{\parallel} - g_{\perp}$. However, using the additional assumption that the relative contact contributions at the *ortho*, *meta*, and *para* protons of the phenyl groups would be the same for the Co and Ni complexes a separation of contact and pseudocontact contributions was achieved. The pseudocontact effects are appreciable for both complexes but larger in the Co case. The great promise of this type of experiment lies in the ability to study ion pairing. Thus, not only is it possible to estimate the extent of ion pairing under different circumstances from the relative magnitudes of the shifts, but a detailed model of the geometry of the ion pair and information about the freedom of movement of the component ions is also obtainable.

F. Solvation Phenomena

Somewhat related to the topic of ion pairing is that of solvation. For paramagnetic ions, n.m.r. allows the identification of the solvated species present and the possibility of detailed studies of any exchange reactions which may occur. Luz and Meiboom[67] have carried out detailed studies of the solvation of Co^{2+} ion by methanol and by water. At low temperature (below $-50°C$) in methanol, separate peaks are observed for coordinated and uncoordinated methanol. The relative intensities of these peaks indicate that the cobalt has a coordination number of six. At room temperature there is rapid exchange between complexed and free methanol and only a single line at the average position is observed. By studying the behavior of both the CH_3 and the OH peaks, it was demonstrated that the exchange process involved exchange of complete methanol molecules rather than of protons. The exchange itself is best studied by relaxation methods[68] and such measurements yielded values of 13.8 kcal/mole and 7.2 e.u. for the activation energy and entropy, respectively. The shifts were considered to be contact in origin and hyperfine coupling constants were evaluated. The metal–ligand bonding must be at least partly covalent. On the addition of a small amount of water to the methanolic solution at low temperatures, additional resonances which also showed large shifts from the usual diamagnetic frequencies appeared. These correspond to solvated species such as **14**, **15**, and **16**. Thus the monohydrate (**14**) shows

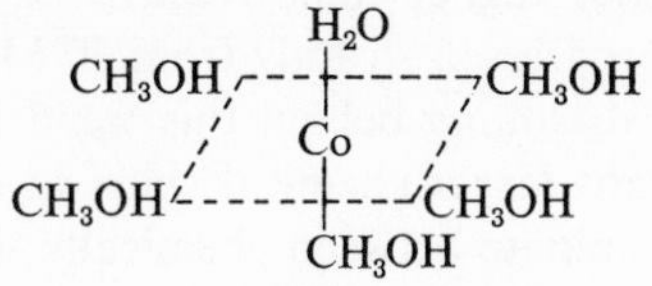

```
             H2O                               H2O
  CH3OH ,---+-----7 CH3OH          CH3OH ,---+-----7 CH3OH
       /    Co    /                     /    Co    /
CH3OH ------+---- CH3OH         CH3OH ------+---- CH3OH
            CH3OH                              H2O
             14                                 15

                       CH3OH
             CH3OH ,---+-----7 H2O
                  /    Co    /
          CH3OH ------+---- H2O
                       CH3OH
                        16
```

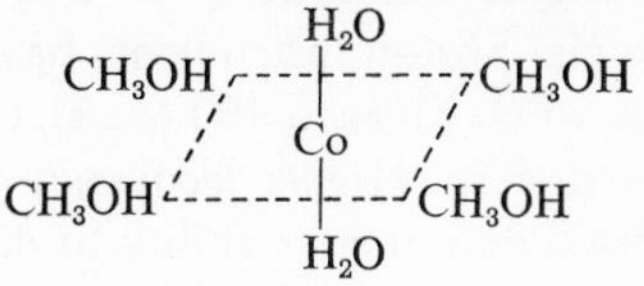

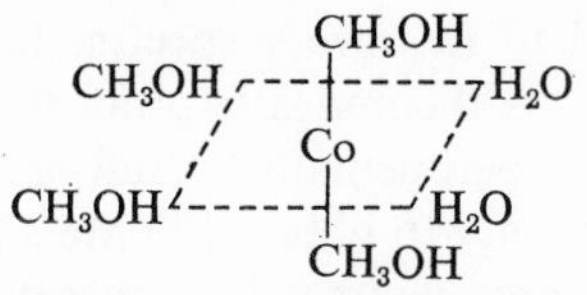

two sets of methanolic resonances with approximate intensity ratios 4:1 corresponding to equatorial and axial ligands. Two sets of resonances attributable to dihydrates were found and these originate from *trans* and *cis* isomers **15** and **16**. By following the intensity of these lines with changes in water concentration, equilibrium constants for the formation of these mixed complexes were obtained. Measurements of this kind are in general only feasible when contact or pseudocontact interactions provide the large shifts in resonance frequency necessary to achieve resolution of the lines of these chemically very similar solvation complexes.

VI. RECENT DEVELOPMENTS IN INSTRUMENTATION AND TECHNIQUES

Commercial instruments suitable for carrying out all straightforward n.m.r. experiments have now been available for several years and are almost universally used. Any extensive description of such conventional instrumentation would not be desirable in the present chapter. There has, however, been continued progress in the instrumental aspects of the subject, and some brief discussion of recent developments and less widely used techniques would seem to be appropriate. Most of the techniques discussed below were not developed to solve inorganic problems and to the present date have not been widely applied in the inorganic field. All of them, though, have potential inorganic applications, and no doubt in due course each will make a contribution to the subject. In the present section, only brief descriptions will be given, together with references to more detailed sources of information.

A. Superconducting Solenoids

In high-resolution n.m.r., there has been a continued trend toward the use of higher magnetic fields and hence higher radiofrequencies. Thus commercial proton instruments have progressed steadily from 30 Mc/s to 40 Mc/s to 60 Mc/s to 100 Mc/s. The reasoning behind this trend is not hard to discern. Higher frequency means larger chemical shifts in cycles per second and greater ability to discriminate between chemically similar nuclei. It also very often means spectra simpler to interpret, since the spin–spin coupling constants are field independent so that increases in chemical shifts often lead to first-order spectra. It also leads to greater sensitivity since sensitivity is theoretically proportional to the 3/2 power of the frequency, although experimentally the full gain is often not achieved. However, the 23.5 kG necessary to obtain 100 Mc/s for protons is close to the practical limit for homogeneous fields generated by electromagnets. Much higher fields can in principle be obtained from superconducting solenoids, and spectrometers incorporating these devices are now in use. A description of a 200 Mc/s (47 kG) proton instrument has been given by Nelson and Weaver[33]. The problem of constructing solenoids to give a sufficiently homogeneous field for high-resolution work (1 part in 2×10^8 or better over the sample volume) is formidable, although the restriction can be eased a little by rapid spinning of the sample. The real advantage of such an instrument lies in the study of very complex molecules such as polymers, and application to this area has been reported[69]. For inorganic chemists, though, the most interesting possibility relates to nuclei with small magnetogyric ratios, the resonances for which occur at low frequencies with electromagnets. The instrument described by Nelson and Weaver suffered from low sensitivity, but a later version developed by these workers, which operates at 54 kG, is much improved in this respect. Considerable utilization of such instruments in the inorganic field can be anticipated.

B. Use of Digital Computers for Signal-to-Noise Enhancement

Lack of sensitivity is a very general problem in n.m.r. and is particularly pronounced for some of the nuclei of most interest to an inorganic chemist. A helpful technique which is now in widespread use involves the use of a CAT (computer of average transients) or similar computing device. The principle involved is very simple. The spectrum to be examined is divided into a number of sections each containing a small frequency range. The spectrometer is then allowed to scan the spectrum, the integrated signal intensity from each segment being stored in one channel of the computer. Two commonly used instruments possess 400 and 1024 channels, respectively. Scanning of the spectrum is repeated as many times as necessary, and the signals accumulated in each computer channel then read out. The readout is of the same form as the initial spectrum but has an enhanced signal-to-noise ratio. Theory indicates that the enhancement will be proportional to the square root of the number of spectral scans. The main limitations of the method are degradation of the resolution due to the limited number of channels or to small drifts in the field during the accumulation of data. The excessive amount of time needed to obtain large enhancements, e.g. a factor of 100 would require 10^4 scans, is also a limiting factor. Other methods of achieving the same result by computer use have also been suggested. Papers by Ernst[70] give a general discussion of the problem. It appears that optimum enhancement could be more rapidly obtained by accumulating a limited number of CAT scans and allowing the computer to fit the best smooth curve to the resulting data. Ernst and Anderson[71] have also suggested a completely different approach in which the r.f. energy is applied in pulses without scanning the spectrum, and the information obtained by Fourier transforming the response of the system. There is an analogy to interferrometric techniques in optical spectroscopy. Data can be accumulated much more rapidly with such a system. Technical advances along these lines are likely to make possible many experiments now deemed impracticable because of lack of sensitivity.

C. Double Resonance and Spin Decoupling

In the usual n.m.r. experiment, the sample is irradiated with a single monochromatic radiofrequency. Experiments involving simultaneous irradiation at two frequencies are referred to as involving double-resonance techniques. A comprehensive review of the application of such techniques to chemical problems has been given by Baldeschwieler and Randall[72]. From the chemical point of view, there are several interesting possibilities. The most widely used has been the so-called spin-decoupling technique which has proved valuable for the analysis of complex spectra. This depends on the experimental observation that if two sets of nuclei A and B are coupled by a spin–spin coupling constant J_{AB}, under the correct conditions, irradiation at the A frequency results in the loss of the spin–spin structure due to the coupling at the B resonance. Similarly, irradiation of B collapses the multiplet structure of the A resonance. The reader is referred to the review of Baldeschwieler and Randall and references therein for the theory underlying these observations, and it will suffice to point out here that there is an analogy, though not a perfect one, to the previously discussed case in which spin multiplets or chemically shifted resonances can be collapsed by shortened relaxation times resulting from chemical exchange. The value of the method is two-fold. Firstly, the complex patterns which result particularly in intermediate coupling cases are simplified and analysis made easier. Secondly, the spectral splitting parameters can be unambiguously assigned to specific coupling constants. A further bonus results from certain types of double-resonance experiments in that they allow determination of the relative signs of spin–spin coupling constants. Such information is never available from first-order spectra and is only infrequently obtained from more complex spectra. Combination of these data on relative signs with the absolute signs of one or two coupling constants obtained by the electric field method of Buckingham and McLachlan[73] offers the possibility of building up an extensive body of data on the absolute signs of spin–spin coupling constants. It has already become apparent that both positive and negative J values are common. Knowledge of the sign is critically important if theoretical interpretation of the coupling constant is to be attempted.

Double-resonance experiments can also be used to obtain chemical shifts which might otherwise be inaccessible. Thus Freeman[74], for example, has described experiments in which observations were made at a frequency corresponding to a ^{13}C satellite line in a proton spectrum. The second irradiating frequency was simultaneously swept through the ^{13}C region. At a frequency corresponding to the appropriate ^{13}C resonance, changes in the proton signal are observed. Thus, effectively, ^{13}C resonances can be detected at the proton frequency. Combination of this method with signal-to-noise enhancement techniques aimed at boosting the weak ^{13}C satellites and eliminating the strong proton resonances from ^{12}C molecules was suggested and has many interesting possibilities. The reverse experiment of irradiating the protons and observing the ^{13}C also offers some attractive features and has been used by Grant[75]. Additional sensitivity for ^{13}C n.m.r. is gained by collapse of the spin multiplets and also by a nuclear Overhauser effect (see below). Wider application of these techniques to inorganic problems will most probably be forthcoming.

D. Endor

Endor is a commonly employed acronym for electron nuclear double resonance. The theory is quite closely related to that of nuclear–nuclear double resonance discussed above. In this case, there is simultaneous irradiation at a microwave frequency corresponding to e.p.r. absorption and at a radiofrequency corresponding to n.m.r. absorption. Endor experiments are, of course, confined to paramagnetic molecules. In the usual experiments, the r.f. field is swept and changes are observed in the microwave absorption. Thus, at best, Endor combines the very high sensitivity of e.p.r. with the very high effective resolution of n.m.r. Most experiments to date have been carried out on solid samples containing transition metal or rate earth ions, and the impetus has been physical rather than chemical.

Baker and Williams[76], for example, have studied Eu^{2+} ions present as impurities in a CaF_2 crystal. Europium has two stable isotopes ^{151}Eu and ^{153}Eu, each of which has $I = \frac{5}{2}$. Nuclear resonance is therefore possible. The ground state of Eu^{2+} which has a half-filled *f* shell is $^8S_{\frac{7}{2}}$. To a first approximation, no hyperfine coupling due to electron–nuclear contact interactions is to be expected for this spherically symmetric ion. However, such interaction is observed and is attributed to admixture of excited configurations in the ground-state wave functions. The high resolution made available by the Endor experiment allowed an accurate measurement of the hyperfine splitting and a theoretical analysis of its origin. The nuclear moments of the two Eu isotopes were also obtained. There are obvious possibilities for the application of Endor to more chemical problems. Thus e.p.r. spectra of transition metal chelates should show hyperfine splittings from magnetic nuclei located on the ligand, but the linewidths are usually too great to allow the resolution of such splittings. Information obtained from Endor experiments on such chelates would be analogous from that obtained from the contact shift measurements described above, but the range of chelates systems which it would be feasible to examine would probably be greater. Recent work by Hyde and Maki[77] and by Hyde[78] has shown that the Endor technique can also be applied to free radicals in solution. It is particularly valuable when the normal e.p.r. spectrum is very complex. Application to inorganic free radicals has some interesting possibilities.

E. Overhauser Effect

The Overhauser effect involves observation of a nuclear resonance with simultaneous irradiation of an electron spin resonance. It differs from Endor in that it is concerned with the effects of microwave irradiation on the nuclear resonance rather than with the effect of radiofrequency radiation on the electron resonance. The effects of importance are attributed to dynamic polarization of the nuclear spins. Thus, in a series of papers by Richards and White[79], the nuclear magnetic resonance of various solvent molecules was examined in the presence of various free radicals as solutes. On irradiating at the e.p.r. frequency of the free radical, the n.m.r. absorption is either enhanced or diminished. The absorption can, in fact, become negative, i.e. emission of r.f. energy rather than absorption is occurring at the resonance frequency. Such results indicate coupling between the electron spin states of the radical and the nuclear spin states of the solvent. Thus saturation of the e.p.r. line gives rise to rapidly fluctuating magnetic fields from the electron spins which exert a polarizing effect on the nuclear spins. Such polarization may arise from two sources; namely, dipole–dipole interactions or scalar (contact) interactions. The two can sometimes be distinguished on the basis of the sign of the nuclear polarization. The increase in the nuclear polarization can be very large leading to greatly improved signal-to-noise ratios. Thus there are possibilities of using the effect to enhance weak resonances such as those of ^{13}C. Overhauser experiments may also make a direct contribution to the study of solvent–solute interactions. If the effects are dipolar in origin, they lead to valuable information regarding molecular collisions. If they are contact in origin, some weak chemical complexation is indicated, since in this case the unpaired electron must have some finite probability of being found in a solvent orbital. Evidence for both types of interaction has been found by Richards and White.

F. Spin-echo Experiments

Most n.m.r. experiments involve continuous irradiation of the sample by the r.f. energy. Information of a rather different kind becomes available if pulses of r.f. energy are applied, and experiments of this latter kind have played an important part in the development of the theory of n.m.r. More recently, it has become clear that such techniques can also make a valuable contribution to the study of exchange processes and fast chemical reactions. This concept has been developed in an important series of papers by Allerhand and Gutowsky[24], although earlier workers had also realized the possibilities. The review by Johnson[23] includes a good account of the area, and the basic theory is covered in the standard texts. The most usual type of experiment is referred to as the spin-echo method. Briefly, application of a pulse giving rise to a strong r.f. field H_1 for a length of time t_W rotates the magnetization due to the nuclear spins through an angle given by

$$\theta = \gamma H_1 t_W$$

Pulses for which the values of H_1 and t_W are chosen so as to make θ equal to 90° or 180° are referred to as 90° and 180° pulses, respectively. Application of a 90° pulse followed by a 180° pulse after an interval t_p results in the detection of a signal by the receiver coil at time $2t_p$ which is referred to as the 'echo'. The Carr–Purcell sequence comprises a 90° pulse followed by 180° pulses at t_p, $3t_p$, $5t_p$, ... with echoes occurring at $2t_p$, $4t_p$, $6t_p$, ... The intensity of this series of echoes decays at a rate proportional to $\exp(-t/T_2)$ which allows the determination of T_2. This is the crux of the application to rate measurements, although there are a number of other intricacies to be taken into account. Allerhand and Gutowsky[24] obtained activation energies and frequency factors for some cases of restricted rotation which were significantly different from the corresponding values obtained by conventional n.m.r. methods. It seems probable that the spin-echo values are to be preferred since they do not involve assumptions regarding the temperature independence of chemical shifts and T_2 values which are implicit in the continuous irradiation methods. It also appears that the spin-echo method will be applicable to much faster reactions. It does, however, have the disadvantage that an essentially low-resolution spectrum is examined, i.e. it is not easy to differentiate the echoes obtained from different kinds of protons in the molecule. Luz and Meiboom have applied spin-echo techniques to investigate the protolysis of trimethylamine[80] and the exchange of methanol coordinated to Co^{2+} ions. Much more extensive application to inorganic systems is to be anticipated.

VII. REFERENCES

1. J. A. Pople, W. G. Schneider, and H. J. Bernstein, *High-Resolution Nuclear Magnetic Resonance*, McGraw-Hill, New York, 1959.
1a. J. W. Emsley, J. Feeney, and L. H. Sutcliffe, *High-Resolution Nuclear Magnetic Resonance Spectroscopy*, Pergamon Press, Oxford, 1965.
2. C. P. Slitcher, *Principles of Magnetic Resonance*, Harper and Row, New York, 1963.
3. A. Abragam, *The Principles of Nuclear Magnetism*, Oxford University Press, London, 1961.
4. J. D. Roberts, *Nuclear Magnetic Resonance*, McGraw-Hill, New York, 1959.
5. L. M. Jackman, *Application of Nuclear Magnetic Resonance in Organic Chemistry*, Pergamon Press, Oxford, 1959.
6. E. Fluck, *Der Kernmagnetische Resonanz und ihre Anwendung im der Anorganischen Chemie*, Springer, Berlin, 1963.
7. E. L. Muetterties and W. D. Phillips, *Advan. Inorg. Nucl. Chem.*, **4**, 231 (1962).
8. P. C. Lauterbur in *Determination of Organic Structures by Physical Methods*, Vol. 2 (Ed. F. C. Nachod and W. D. Phillips), Academic Press, New York, 1962.
9. W. E. Lamb, *Phys. Rev.*, **60**, 817 (1941).
10. A. Saika and C. P. Slichter, *J. Chem. Phys.*, **22**, 26 (1954).
11. J. S. Griffith and L. E. Orgel, *Trans. Faraday Soc.*, **53**, 601 (1957).
12. R. W. Taft, *J. Am. Chem. Soc.*, **79**, 1045 (1957); *J. Phys. Chem.*, **64**, 1805 (1960).
13. R. A. Y. Jones and A. R. Katritzky, *Angew. Chem.* (*Intern. Ed. Engl.*), **1**, 33 (1962).
14. E. L. Muetterties, T. A. Bither, M. W. Farlow, and D. D. Coffman, *J. Inorg. Nucl. Chem.*, **16**, 52 (1960).
15. M. Karplus, D. H. Anderson, T. C. Farrar, and H. S. Gutowsky, *J. Chem. Phys.*, **27**, 597 (1957).
16. M. Karplus, *J. Chem. Phys.*, **30**, 11 (1959).
17. E. R. Andrew, *Nuclear Magnetic Resonance*, Cambridge University Press, Cambridge, 1956.

18. H. S. Gutowsky, G. E. Pake, and R. Bersohn, *J. Chem. Phys.*, **22**, 643 (1954).
19. L. C. Snyder, *J. Chem. Phys.*, **43**, 4041 (1965).
19a. L. C. Snyder and E. W. Anderson, *J. Chem. Phys.*, **42**, 3336 (1965).
20. T. P. Das and E. L. Hahn, *Nuclear Quadrupole Resonance Spectroscopy*, Suppl. 1 to *Solid State Physics*, Academic Press, New York, 1958.
21. M. H. Cohen and F. Reif, *Solid State Phys.*, **5**, 322 (1957).
22. J. C. Rowell, W. D. Phillips, L. R. Melby, and M. Panar, *J. Chem. Phys.*, **43**, 3442 (1965).
23. C. S. Johnson in *Advances in Magnetic Resonance*, Vol. 1 (Ed. J. S. Waugh), Academic Press, New York, 1965, p. 33.
23a. D. R. Eaton and W. A. Sheppard, unpublished results.
24. A. Allerhand and H. S. Gutowsky, *J. Chem. Phys.*, **41**, 2115 (1964); **42**, 1587, 3040, 4203 (1965).
25. D. R. Eaton, unpublished results.
26. M. L. H. Green and D. J. Jones, *Advan. Inorg. Chem. Radiochem.*, **7**, 115 (1965).
27. M. L. H. Green, J. A. McCleverty, L. Pratt, and G. Wilkinson, *J. Chem. Soc.*, **1961**, 4854.
28. A. D. Buckingham and P. J. Stephens, *J. Chem. Soc.*, **1964**, 2747; **1964**, 4583
29. T. Farrar, W. Ryan, A. Davison, and J. W. Faller, *J. Am. Chem. Soc.*, **88**, 184 (1966).
30. D. Moy, M. Emerson, and J. P. Oliver, *Inorg. Chem.*, **2**, 1261 (1963).
31. G. M. Whitsides, J. E. Nordlander, and J. D. Roberts, *J. Am. Chem. Soc.*, **84**, 2010 (1962).
32. H. L. Reaves and T. E. Gilmer, *J. Chem. Phys.*, **42**, 4138 (1965).
33. F. A. Nelson and H. E. Weaver, *Science*, **146**, 223 (1964).
34. R. E. Williams and I. Shapiro, *J. Chem. Phys.*, **29**, 677 (1958).
35. J. A. Dupont and M. F. Hawthorne, *J. Am. Chem. Soc.*, **84**, 1804 (1962).
36. A. H. Silver and P. J. Bray, *J. Chem. Phys.*, **29**, 984 (1958).
37. J. F. Hon and P. J. Bray, *Phys. Rev.*, **110**, 624 (1958).
38. O. W. Howarth, R. E. Richards, and L. M. Venanzi, *J. Chem. Soc.*, **1964**, 3335.
39. M. Logan and W. L. Jolly, *Inorg. Chem.*, **4**, 1508 (1965).
40. B. N. Figgis, R. G. Kidd, and R. S. Nyholm, *Proc. Roy. Soc.* (*London*), *Ser. A*, **269**, 469 (1962).
41. B. N. Figgis, R. G. Kidd, and R. S. Nyholm, *Can. J. Chem.*, **43**, 145 (1965).
42. E. L. Muetterties and W. D. Phillips, *J. Am. Chem. Soc.*, **81**, 1084 (1959).
43. J. A. Pople, *Mol. Phys.*, **1**, 168 (1958).
44. E. L. Muetterties, W. Mahler, K. J. Packer, and R. Schmutzler, *Inorg. Chem.*, **3**, 1298 (1964).
45. E. L. Muetterties and K. J. Packer, *J. Am. Chem. Soc.*, **86**, 293 (1964).
46. R. E. Richards and B. A. Yorke, *Mol. Phys.*, **6**, 289 (1963).
47. C. F. Callis, R. J. Van Wazer, J. N. Shoolery, and W. A. Anderson, *J. Am. Chem. Soc.*, **79**, 2719 (1957).
48. L. S. Meriwether and J. R. Leto, *J. Am. Chem. Soc.*, **83**, 3192 (1961).
49. A. Pidcock, R. E. Richards, and L. M. Venanzi, *Proc. Chem. Soc.*, **1962**, 184.

50. T. Birchall, R. J. Gillespie, and S. L. Vekris, *Can. J. Chem.*, **43**, 1672 (1965).
51. J. V. Hatton, Y. Saito, and W. G. Schneider, *Can. J. Chem.*, **43**, 47 (1965).
52. R. Freeman, G. R. Murray, and R. E. Richards, *Proc. Roy. Soc.* (*London*), *Ser. A*, **242**, 455 (1957).
53. H. M. McConnell and H. E. Weaver, *J. Chem. Phys.*, **25**, 307 (1956).
54. W. G. Schneider and A. D. Buckingham, *Discussions Faraday Soc.*, **34**, 147 (1962).
55. H. M. McConnell and D. B. Chesnut, *J. Chem. Phys.*, **28**, 107 (1958).
56. A. D. McLachlan, *Mol. Phys.*, **3**, 233 (1960).
57. H. M. McConnell and R. E. Robertson, *J. Chem. Phys.*, **29**, 1361 (1958).
58. H. M. McConnell and C. H. Holm, *J. Chem. Phys.*, **28**, 749 (1958).
59. J. P. Jesson, *J. Chem. Phys.*, **47**, 579, 582 (1967).
60. *Tables of Interatomic Distances* (Ed. L. E. Sutton), The Chemical Society, London, 1958.
61. R. G. Schulman, private communication quoted by D. R. Eaton and W. D. Phillips in *Advances in Magnetic Resonance*, Vol. 1 (Ed. J. S. Waugh), Academic Press, New York, 1965, p. 108.
62. C. C. McDonald and W. D. Phillips, *J. Am. Chem. Soc.*, **85**, 3736 (1963).
63. D. R. Eaton, *J. Am. Chem. Soc.*, **87**, 3097 (1965).
64. D. R. Eaton and W. D. Phillips, *J. Chem. Phys.*, **43**, 392 (1965).
65. D. R. Eaton, A. D. Josey, W. D. Phillips, and R. E. Benson, *J. Chem. Phys.*, **37**, 347 (1962).
66. G. N. LaMar, *J. Chem. Phys.*, **41**, 2992 (1964).
67. Z. Luz and S. Meiboom, *J. Chem. Phys.*, **40**, 1058, 1066 (1964).
68. Z. Luz and S. Meiboom, *J. Chem. Phys.*, **40**, 2686 (1964).
69. R. C. Ferguson, *Kautschuk Gummi*, **18**, 723 (1965).
70. R. R. Ernst, *Rev. Sci. Instr.*, **36**, 1689 (1965); **36**, 1696 (1965).
71. R. R. Ernst and W. A. Anderson, *Rev. Sci. Instr.*, **37**, 93 (1966).
72. J. D. Baldeschwieler and E. W. Randall, *Chem. Rev.*, **63**, 81 (1963).
73. A. D. Buckingham and K. A. McLauchlan, *Proc. Chem. Soc.*, **1963**, 144.
74. R. Freeman and W. A. Anderson, *J. Chem. Phys.*, **39**, 806 (1963).
75. D. M. Grant and E. G. Paul, *J. Am. Chem. Soc.*, **86**, 2977, 2984 (1964).
76. J. M. Baker and F. I. B. Williams, *Proc. Roy. Soc.* (*London*). *Ser. A*, **267**, 283 (1962).
77. J. S. Hyde and A. H. Maki, *J. Chem. Phys.*, **40**, 3117 (1964).
78. J. S. Hyde, *J. Chem. Phys.*, **43**, 1806 (1965).
79. R. E. Richards and J. W. White, *Proc. Roy. Soc.* (*London*), *Ser. A*, **269**, 287 (1962); **269**, 301 (1962); **279**, 474 (1964); **279**, 481 (1964); **283**, 458 (1965).
80. Z. Luz and S. Meiboom, *J. Chem. Phys.*, **39**, 366 (1963).

11

Thermochemistry in Inorganic Solution Chemistry*

J. J. CHRISTENSEN† and R. M. IZATT

* Supported in part by United States Atomic Energy Commission Contract AT(04-3)-299.

† Part of this work was carried out while on leave from Brigham Young University as a National Institutes of Health Special Fellow at Oxford University, England. Grant No. 1-F3-GM-24, 361-01.

I. INTRODUCTION

Calorimetry, although used successfully in many fields of chemistry, has not received extensive attention in inorganic solution chemistry. During the past two decades, a very large number of equilibrium constants have been reported for metal–ligand interaction in aqueous solutions. Also, much effort has gone into understanding the mechanisms of inorganic reactions. Unfortunately, few studies have appeared reporting quantitative ΔH and ΔS values for metal–ligand reactions. The lack of calorimetric data in this field has been noted by many workers, and several research groups have recently begun to collect data of interest to the inorganic chemist. The availability of ΔH and ΔS values for carefully selected inorganic systems would be worthwhile for several reasons. Firstly, heats of reaction in aqueous solution can be used to investigate the bonding between metal ions and ligands, and to relate the strength of chemical bonding to the nature of the reacting species. These relationships can then be used to make predictions about the bonding and interactions of substances which have not been experimentally investigated, and to verify and extend theories concerning metal–ligand bonding. Secondly, the need to make approximations, such as assuming ΔH proportional to ΔG in a series of metal ions to obtain stabilization energies, would not be necessary[1]. Thirdly, ΔS data obtained from ΔH and ΔG data should provide insight into the interactions between the species and the solvent.

The ΔH and ΔS data obtainable by calorimetry are also a rich potential source of information about the interactions of metal ions in fields other than inorganic chemistry. For example, metal ions are important as catalysts in organic and biochemical reactions and as the constituent atoms in many vitally important biological compounds, e.g. Mg in chlorophyll, Fe in hemes, Co in vitamin B_{12}, and Cu in hemocyanin. Undoubtedly, the role of the metal atom in each of these situations would be much better understood if data were available for the energetics of the reactions involved.

Enthalpy data have been obtained either from log K versus $1/T$ plots or by direct calorimetry. Of the two methods, direct calorimetry is generally considered to be superior. Log K versus $1/T$ plots appear to be satisfactory in the case of most proton ionization reactions where pK can be determined very accurately, primarily because a single reaction occurs. However, in the case of metal complexation, where there are usually several simultaneous reactions, including one or more involving proton ionization, errors are compounded, resulting in ΔH and ΔS values of doubtful accuracy. Generally, the use of temperature variation studies to determine ΔH in the case of metal complexation should be discouraged except in those few cases where extremely accurate equilibrium constants are known or can be determined for all reactions involved.

Enthalpy data can be obtained by either of two calorimetric procedures: conventional solution calorimetry or thermometric titration calorimetry. Of these, conventional solution calorimetry has been adequately discussed in many standard reference sources and review articles[2-4]. Thermometric titration calorimetry, however, has developed rapidly in recent years as a tool in inorganic research and no complete description of it exists in one place. Because of its particular advantage in studying the simultaneous equilibria common to metal complex formation, a brief description of the method and a summary of published results pertinent to inorganic chemistry are given here. Thermometric titration calorimetry offers several distinct advantages to the worker in inorganic solution chemistry. The method lends itself well to the determination of a much larger number of data in a given period of time than would be possible by conventional solution calorimetry. Also, for many systems, it is possible to obtain confirmation of the number and types of reactions taking place and of the concentrations of the various reactants entering into the reaction. In addition, in many cases there is the possibility of calculating ΔH, ΔG, and ΔS values from the calorimetric data alone.

As it would be impossible to cover all aspects of thermochemistry applied to inorganic solution chemistry in one chapter, we have limited the discussion to the following three topics. (1) A general discussion of the relationships which exist between the heat of reaction in aqueous solution and the bonding in a metal–ligand system. This discussion includes topics such as standard states, internal and environmental parts of the enthalpy change, and various thermodynamic cycles. (2) The analytical and calorimetric applications of thermometric titration calorimetry. (3) A discussion of some recent calorimetric studies of metal–ligand reactions in aqueous solution, including the usefulness and limitations of calorimetric data in understanding class A and B metals, ligand-field stabilization energies, and high-spin and low-spin complexes.

II. THERMODYNAMICS OF METAL–LIGAND INTERACTION

A. General

In this section, metal–ligand interactions in aqueous solution as represented by equation (1) are considered with respect to the various energy terms

$$M(H_2O)_m^{n+}(aq) + rL(H_2O)_t^{s-}(aq) \xrightarrow{\Delta H_c} ML_r(H_2O)_u^{n-rs}(aq) + (m + rt - u)(H_2O)(l) \qquad (1)$$

which contribute to the magnitude of the enthalpy change on complex formation, ΔH_c. The different standard states to which the value of ΔH_c can be referred are examined and their relative merits are discussed. It is shown how, in principle, metal–ligand bonding both in the aqueous and in the gas phase can be investigated from measurements of ΔH_c at room temperature and the use of thermodynamic cycles. Various thermodynamic cycles useful in evaluating energy terms relative to metal–ligand bonding together with the type and source of thermodynamic data required for their use are presented.

B. Standard States for Heats of Metal–Ligand Interaction

At the present time, there is no one accepted standard state to which calorimetric ΔH_c values are referred. Several different standard states have been used in reporting ΔH_c values, and it is pertinent to ask whether any one of these is preferable to the rest.

1. Temperature and pressure

Most ΔH_c measurements refer to a temperature of 25°C and a pressure of 1 atm. Since this is the standard temperature and pressure for other calorimetry measurements, they would seem to also be the most suitable for solution calorimetry.

2. Ionic strength

Values for ΔH_c have been reported for solutions ranging from $\mu = 0$ to as high as $\mu = 4$. Many of the determinations of ΔH_c at high μ values were made because this was the μ value at which the stability constant had been determined. Extensive reviews have appeared in the literature[5–8] which discuss the determination of concentration quotients (equilibrium constants valid at specific μ values greater than zero) in a high concentration of a presumed inert electrolyte (constant ionic medium method). This method has been used by many investigators in studies of solution equilibria as a means of circumventing the problems of determining the activity coefficients necessary for calculation of thermodynamic stability constants (constants valid at $\mu = 0$). By using solutions of fixed ionic strength, concentration quotients can be obtained which are useful for comparison purposes with respect to other constants obtained at the same μ value in the same electrolyte. However, it is very difficult to relate concentration quotients, enthalpy values, or entropy values determined in solutions having high μ values back to a standard state of $\mu = 0$ where they may be combined with other necessary thermodynamic data to investigate bonding and structural features in metal–ligand interaction. The three main disadvantages of determining ΔH_c using the constant ionic medium method are (1) the bulk electrolyte competes with the metal, ligand, and complex for waters of hydration resulting in the system being in a state of partial hydration and giving a ΔH_c value which refers to this state rather than the fully hydrated state, (2) the interaction between the metal and/or the ligands and the bulk electrolyte may result in additional complexes or ion pairs which would contribute to the value of ΔH_c, and (3) the ΔH_c values are valid only for the given electrolyte and for a given concentration of that electrolyte, and cannot be compared (without considerable uncertainty) to ΔH_c values determined in other electrolytes or at other μ values or referred to thermodynamic data valid at $\mu = 0$. It should be noted that in some systems, e.g. Hg^{2+}–, Tl^{3+}–, Fe^{3+}–ligand complexes, it is impossible to avoid extensive metal ion hydrolysis except at high acidities and therefore ΔH_c, ΔG_c, and ΔS_c values must be obtained at high μ values. It is the authors' contention, however, that every effort should be made to determine ΔH_c at low enough concentration of metal, ligand, and supporting electrolyte, if any, to allow extrapolation of ΔH_c to $\mu = 0$, and that this should constitute the standard state for solution calorimetry with respect to ionic strength.

It is possible to evaluate ΔH_c at $\mu = 0$, ΔH_c^0, using equation (2)[10], where

$$\Delta H_c^0 = \Delta H_c + RT^2 \sum \gamma_i \frac{\partial \ln \gamma_i}{\partial T} \tag{2}$$

the temperature variation of the activity coefficients γ_i can be evaluated either from experimental data or from one of the extended forms of the Debye–Hückel equation. However, for metal–ligand reactions involving several complexes, the amount of work required to determine experimentally the temperature variation of γ_i is prohibitive. The extended form of the Debye–Hückel equation such as the one given by Harned and Owen[9] in equation (3) can be used up to concentrations of approximately 0.1.

$$\ln \gamma_i = \frac{-2.303\delta\mu^{\frac{1}{2}}}{1 + A\mu^{\frac{1}{2}}} + 2.303B\mu \tag{3}$$

Harned and Owen[9] have differentiated equation (3) with respect to temperature to obtain the temperature variation of the γ_i values and have reported in table form the numerical values of the resulting parameters for aqueous solution. An approximate relationship for estimating the temperature variation of the γ_i values at μ values greater than 0.1 and room temperature[10] is given in equation (4), where γ_i can be estimated

$$\frac{\partial \log \gamma_i}{\partial T} = 2 \times 10^{-3} \log \gamma_i \tag{4}$$

from (3). Equation (4) should only be used to obtain some idea concerning the magnitudes of the changes of ΔH_c with μ. If sufficient data in the form of ΔH_c versus $\mu^{\frac{1}{2}}$ are available and extend to $\mu < 0.1$, an extrapolation method for evaluating ΔH_c^0 is justified. At lower μ values, $\partial \ln \gamma_i/\partial T$ can be evaluated from the limiting law form of the Debye–Hückel equation[11] and equation (2) becomes

$$\Delta H_c^0 = \Delta H_c + \left(\sum A_i\right)\mu^{\frac{1}{2}} \tag{5}$$

where A_i is a specific constant for each species at a given temperature. A plot of ΔH_c versus $\mu^{\frac{1}{2}}$ should result at low μ values in a straight line which can be extrapolated to $\mu = 0$ to give ΔH_c^0.

Estimations of the effect on the magnitude of ΔH_c of changing ionic strength are shown in Table 1, where ΔH_c as a function of ionic strength is given.

Table 1. Effect of ionic strength on reaction enthalpies[46].

Reaction	$-\Delta H_c$ (kcal/mole)	μ
$Ni^{2+} + 4CN^- \longrightarrow Ni(CN)_4^{2-}$	43.20 ± 0.20	0
	43.35 ± 0.12	0.040
	43.50 ± 0.07	0.082
	43.73 ± 0.05	0.134
	44.30 ± 0.03	0.255

3. Concentration scales

The value of ΔH_c^0 also depends to a small degree upon the choice of concentration scale at infinite dilution. The standard states commonly used in solution chemistry can be classified[13] according to which of the following three concentration scales are used in their definition, mole fraction (x = moles i/total moles), molality (m = moles i/1000 g solvent), and molarity (c = moles i/liter).

Of these three scales, that involving mole fraction has more theoretical significance[13] and has been recommended as a more logical starting point from which to consider specific structural effects[14].

A table giving corrections for converting molar quantities to molal quantities for ΔH^0, ΔG^0, ΔS^0, and ΔC_p^0 between 0 and 60° for aqueous solution has been published[15]. Values from this table are given in Table 2.

Table 2. Corrections to be added to molar values to convert to molal[15].

Temperature (°c)	ΔG^0 (cal/mole)	ΔH^0 (cal/mole)	ΔS^0 (cal/deg mole)	$\Delta C_p{}^0$ (cal/deg mole)
0	0.1	−10.2	−0.04	13.2
10	0.2	13.9	0.05	10.9
20	1.0	35.4	0.12	9.3
25	1.7	45.1	0.16	8.8
30	2.6	55.1	0.19	8.4
40	4.8	74.9	0.26	7.9
50	7.7	95.1	0.32	7.8
60	11.2	115.3	0.38	7.7

Equations (6)–(9) can be used to calculate the same corrections as given in Table 2 for solvents other than water.

$$\Delta G^0(\text{molal}) = \Delta G^0(\text{molar}) + 2.3026RT \log \rho_{\text{solv}} \tag{6}$$

$$\Delta G^0(\text{molal}) = \Delta H^0(\text{molar}) - 2.3026RT^2 \frac{\mathrm{d} \log \rho_{\text{solv}}}{\mathrm{d}T} \tag{7}$$

$$\Delta S^0(\text{molal}) = \Delta S^0(\text{molar}) - 2.3026R\left(\log \rho_{\text{solv}} + T\frac{\mathrm{d} \log \rho_{\text{solv}}}{\mathrm{d}T}\right) \tag{8}$$

$$\Delta C_p^0(\text{molal}) = \Delta C_p^0(\text{molar}) - 2.3026RT\left(2\frac{\mathrm{d} \log \rho_{\text{solv}}}{\mathrm{d}T} + \frac{T\,\mathrm{d}^2 \log \rho_{\text{solv}}}{\mathrm{d}T^2}\right) \tag{9}$$

ΔH_c^0 values are most commonly referred to the molality scale, where the standard state corresponds to a hypothetical ideal solution in which the partial molal enthalpy and heat capacity of the solute have the values of the infinitely dilute solution but the free energy and entropy correspond to unit activity (hypothetical one molal state). ΔH_c^0 values are the same on the molality and mole fraction scale and no corrections are necessary to convert ΔH_c^0 from the one scale to the other.

In summary, the authors feel that the preferred standard state for calorimetric ΔH_c values should be: temperature = 25°c, pressure = 1 atm, ionic strength = 0, and a concentration based on the molarity or mole fraction scale.

C. Unitary and Cratic Parts of Reactions in Aqueous Solution

Those processes occurring in solution which result in changes in ΔH_c, ΔG_c, and ΔS_c can be considered as the sum of a unitary and a communal part[16]. The unitary part of a process is that which is independent of the composition of the solution as a whole while the communal part is that which depends on the amount of a solvent and on the composition of the solution.

The communal part can be further divided into that resulting from interionic forces and a cratic term resulting from the mixing together of a certain number of solute and solvent particles. Extrapolation of these quantities to the hypothetical one molal state should result in the contribution of interionic forces to the communal term becoming negligible.

The cratic terms for the three thermodynamic quantities ΔH^0, ΔG^0, and ΔS^0 are given by equations (10)–(12) and can be thought of as correction

$$\Delta H^0_{cr} = 0 \tag{10}$$

$$\Delta G^0_{cr} = \Delta n_c RT \ln 55.5 = 2.360\,(\Delta n_c)\ \text{kcal/mole at } 25°\text{C}^* \tag{11}$$

$$\Delta S^0_{cr} = -\Delta n_c R \ln 55.5 = -7.9\,(\Delta n_c)\ \text{e.u. at } 25°\text{C} \tag{12}$$

terms that change the standard state from the hypothetical one molal solution to the hypothetical unit mole fraction state. The unitary quantities ΔH^0_u, ΔG^0_u, and ΔS^0_u have as a standard state the unit mole fraction scale and are given by equations (13)–(15). Adamson[14] has pointed out that the

$$\Delta H^0_u = \Delta H^0_c \tag{13}$$

$$\Delta G^0_u = \Delta G^0_c + 2.360\,(\Delta n_c) \tag{14}$$

$$\Delta S^0_u = \Delta S^0_c - 7.9\,(\Delta n_c) \tag{15}$$

hypothetical unit mole fraction state is one of minimum translational entropy for the solute which retains the properties it possesses in dilute solution and suggests that this state corresponding to the quantities ΔH^0_u, ΔG^0_u, and ΔS^0_u be employed as a logical starting point from which to consider specific structural effects on ΔH_c, ΔG_c, and ΔS_c. Adamson[14] and Rossotti[17] have shown the influence of the cratic term on the magnitude of ΔG^0_c and ΔS^0_c and how, for example, the chelate effect is primarily due to the cratic term (chelated complexes are usually more stable than their nonchelated analogs). Although there is no cratic part of ΔH^0_c and hence ΔH^0_c is the same in both the hypothetical one molal state and the hypothetical unit mole fraction state, the authors feel that Adamson's suggestion is pertinent with respect to ΔH^0_c and that only the unitary parts of ΔG^0_c and ΔS^0_c should be combined with ΔH^0_c in deriving other thermodynamic quantities and/or in relating these quantities to structural effects. Also, for purposes of intercomparison and tabulation of ΔH_c, ΔG_c, and ΔS_c data for metal–ligand interaction, the mole fraction unit state should be used as the standard state.

*Δn_c = (moles of products) – (moles of reactants), exclusive of solvent.

D. Internal and Environmental Parts of Enthalpy Changes

The interaction terms that contribute to the magnitude of ΔH^0_c or ΔH^0_u (as has been shown, $\Delta H^0_c = \Delta H^0_u$; we will use ΔH^0_u to designate the change in enthalpy when referring either to the molal or mole fraction scale) can be further considered to consist of an internal part ΔH_I and an environmental part ΔH_E. The internal part consists of those contributions to ΔH^0_u which arise due to the potential, translational, rotational, or vibrational energies of the reacting atom, molecules, or ions and are intramolecular in nature. These effects are independent of intermolecular interaction between the reacting species and the solvent. However, the environmental part is the result of intermolecular interaction between the solvent molecules and the various atoms, molecules, or ions taking part in the reaction. Through proper evaluation of these two terms, much information can be gained concerning the relationship of bonding to structure (internal effects) and solute–solvent interaction (environmental effects). Since the internal part is that related to the bonding in metal–ligand complexes, it will now be examined. ΔH_I can be represented as the change in enthalpy for reaction (16) in the gas phase.

$$M^{n+}(g) + rL^{s-}(g) \longrightarrow ML_r^{n-rs}(g) \qquad \Delta H_I \tag{16}$$

The following cycle illustrates how, for a given complex, the internal part ΔH_I can be evaluated from ΔH^0_u. Also shown are the various thermodynamic terms which are necessary for the calculation.

$$M(H_2O)_m^{n+}(aq) + rL(H_2O)_t^{s-}(aq) \xrightarrow{\Delta H^0_u} ML_r(H_2O)_u^{n-rs}(aq) + (m + rt - u)\,H_2O(l)$$

$$\downarrow -\Delta H_{h,ML_r}$$

$$ML_r^{n-rs}(g) + (m + rt)\,H_2O(l) + aq$$

$$\downarrow -\Delta H_I$$

$$M^{n+}(g) + rL^{s-}(g) + (m + rt)\,H_2O(l) + aq$$

$$\downarrow \Delta H_{h,M}$$

$$M(H_2O)_m^{n+}(aq) + rL^{s-}(g) + rt\,H_2O(l)$$

$$\text{(return to start via } r\,\Delta H_{h,L})$$

where

$\Delta H_{h,ML_r}$ = heat of hydration of complex ML_r^{n-rs}

$\Delta H_{h,M}$ = heat of hydration of metal ion M^{n+}

$\Delta H_{h,L}$ = heat of hydration of ligand L^{s-}

The species in solution are written in a manner which represents the ions complete with their coordinated water molecules, i.e. water molecules in the first coordination sphere. These water molecules are represented separately from the water molecules making up the solvent which are represented by the symbol aq. As pointed out by George and McClure[18], there is extensive evidence that the fully hydrated ions of the transition metals with +2 charge contain six water molecules coordinated octahedrally to the metal ions. For simplicity in writing the equations, they assumed m to be 6 and that each monodentate ligand group replaced one water molecule. Both the method of George and McClure and the present method of representing the reactions led to the same results with respect to the value of ΔH_I, as determined from the above cycle. We choose to represent the reactions in the above manner as it gives a more correct picture of the factors that influence the ΔG and ΔS terms (i.e. cratic terms).

According to Hess's law, ΔH_I can be determined by summing up the changes in enthalpy around the cycle

$$\Delta H_u^0 - \Delta H_{h,ML_r} - \Delta H_I + \Delta H_{h,M} + r\,\Delta H_{h,L} = 0 \qquad (17)$$

and solving for ΔH_I

$$\Delta H_I = \Delta H_u^0 + \Delta H_{h,M} + r\,\Delta H_{h,L} - \Delta H_{h,ML_r} \qquad (18)$$

It can easily be seen that the environmental part of ΔH_u^0 is given by

$$\Delta H_E = \Delta H_u^0 - \Delta H_I = \Delta H_{h,ML_r} - \Delta H_{h,M} - r\,\Delta H_{h,L} \qquad (19)$$

Thus, if adequate information exists for the heats of hydration of the various species taking part in the reaction, both ΔH_I and ΔH_E can be evaluated.

Unfortunately, however, for many ions the heats of hydration are not available. To show why this situation exists it is necessary firstly to illustrate how heats of hydration are measured. The heat of hydration is the heat effect accompanying the introduction of a gaseous ion into water and the resulting separation of the water molecules from each other, together with their subsequent electrostatic attraction to the ion, and is represented for a metal ion by equation (20). It cannot be measured

$$M^{n+}(g) + m\,H_2O(l) + aq \rightarrow M(H_2O)_m^{n+}(aq) \qquad (20)$$

directly but may be determined from the ΔH_R value, obtained by dissolving a metal in an acid solution, together with other thermodynamic quantities through the use of the following cycle.

$$M(s) + m\,H_2O(l) + n\,H^+(aq) \xrightarrow{\Delta H_R} M(H_2O)_m^{n+}(aq) + n/2\,H_2(g)$$

$$\downarrow -\Delta H_{h,M(H_2O)m}$$

$$M(H_2O)_m^{n+}(g) + n/2\,H_2(g) + (aq)$$

$$\downarrow -\Delta H_{L,M}$$

$$M^{n+}(g) + m\,H_2O(g) + n/2\,H_2(g) + (aq)$$

$$\downarrow -\Delta H_{I,M}$$

$$M(g) + m\,H_2O(g) + n/2\,H_2(g) + (aq) - ne^-$$

$$\downarrow -\Delta H_S$$

$$M(s) + m\,H_2O(g) + n/2\,H_2(g) + (aq) - ne^-$$

$$\downarrow -m\Delta H_V$$

$$M(s) + m\,H_2O(l) + n/2\,H_2(g) + (aq) - ne^-$$

$$\downarrow n/2\,\Delta H_D$$

$$M(s) + m\,H_2O(l) + n\,H(g) + (aq) - ne^-$$

$$\downarrow n\,\Delta H_{I,H}$$

$$M(s) + m\,H_2O(l) + n\,H^+(g) + (aq)$$

(the final state returns to the initial state via $n\,\Delta H_{h,H}$)

where

ΔH_R = heat of dissolving metal in acid solution
$\Delta H_{h,M(H_2O)m}$ = heat of hydration of $M(H_2O)_m^{n+}$
$\Delta H_{L,M}$ = heat of ligation of metal ion with water
$\Delta H_{I,M}$ = sum of 1st and 2nd ionization energies of M^{n+}
ΔH_S = heat of sublimation of M^{n+}
ΔH_V = latent heat of vaporization of water
ΔH_D = dissociation energy of hydrogen molecule
$\Delta H_{I,H}$ = ionization energy of hydrogen atom
$\Delta H_{h,H}$ = heat of hydration of hydrogen ion

The heat of hydration of a gaseous metal ion, as represented by equation (20), in terms of the quantities appearing in the above cycle is given by

$$\Delta H_{h,M} = \Delta H_{h,M(H_2O)m} + \Delta H_{L,M} + m\,\Delta H_V \tag{21}$$

By adding together the changes in ΔH around the cycle and solving for $\Delta H_{h,M}$, we obtain

$$\Delta H_{h,M} = \Delta H_R - \Delta H_{I,M} - \Delta H_S + n/2\,\Delta H_D + n\,\Delta H_{I,H} + n\,\Delta H_{h,H} \quad (22)$$

Using the convention that ΔH for the following reaction is set equal to zero

$$\tfrac{1}{2}\,H_2(g) + (aq) \longrightarrow H^+(aq) + e^- \quad (23)$$

then

$$n/2\,\Delta H_D + n\,\Delta H_{I,H} + n\,\Delta H_{h,H} = 0 \quad (24)$$

and the heat of hydration $\Delta H_{h,M}$ becomes equal to the following:

$$\Delta H_{h,M} = \Delta H_R - \Delta H_{I,M} - \Delta H_S \quad (25)$$

Heats of hydration can therefore be determined experimentally for metal ions if values exist for the other energy terms appearing in equation (25). For many metals, there exist standard reference sources containing heats of reaction, ionization energies, and heats of sublimation, making possible extensive compilations of heats of hydration[18].

The heats of hydration of ligands can be determined from similar cycles which differ slightly depending on the nature of the ligand. However, a major problem is the difficulty of obtaining values for electron affinities for the various ligands. Few values exist and these are primarily for the simple inorganic ligands such as Cl^-, Br^-, CN^-, etc.[19]. The lack of heat of hydration data for ligands is not a serious deterrent where relative values of ΔH_I are to be obtained with respect to a given ligand. For example, if the heat of reaction of a given ligand with a series of metal ions is measured, the term for the heat of hydration of the ligand (equation 18) will cancel out when comparing ΔH_I values for different metal ions.

The problem of determining heats of hydration for complex ions is much more serious and at the present time there is no satisfactory way to evaluate this term experimentally. Many complex ions do not even exist outside of aqueous solutions and for most of those that do ionization and dissociation energies are not available.

The heats of hydration of ions can be estimated with the Born charging equation

$$\Delta H_h = (-Z^2e^2/2R)[(1 - 1/D) + (T/D^2)(\partial D/\partial T)_p] \quad (26)$$

where

Ze = product of charge on ion and electron charge
R = radius of central ion plus diameter of water molecule, Å
D = dielectric constant of medium
T = temperature, °K

For aqueous solutions at 25°C, this equation reduces to

$$\Delta H_{\mathrm{h}} = -167\, Z^2/R \text{ (kcal/mole)} \tag{27}$$

Values of ΔH_{h} determined from equation (27) should be considered approximate, as the Born equation is based on an electrostatic model using a continuous dielectric medium in place of the actual molecular structure of the medium. However, values of the heats of hydration of complex ions and ligands calculated from equation (27) are useful in showing trends both in the hydration energy and in the values of ΔH_{I} (equation 18) along a series of metal ions. Thus heats of hydration are extensively available for metal ions, moderately available for ligands, and rarely available for metal complexes.

It is important to remember that all experimentally determined hydration energies are based on a relative scale where the heat of the hydrogen half-reaction is set equal to zero (equation 23) and cannot be used individually to calculate from thermodynamic cycles absolute values of other quantities. However, by taking differences between hydration energies the reference state is eliminated and absolute differences for hydration energies can be obtained. The values of ΔH_{I} and ΔH_{E} calculated from equations (18) and (19), respectively, are absolute values as each calculation involves using the difference in hydration energies.

E. Relative Heats of Reaction

It should be pointed out that even if the heats of hydration of the metal, ligand, and complex were known, the value of ΔH_{I} obtained from equation (18) would be valid only at 298°K. Thus ΔH_{I} would contain, in addition to the change in potential energy (bond energies), contributions from changes in kinetic, rotational, and vibrational energies for the reaction. Although environmental effects have been eliminated by removing the reaction to the gas phase, we still must take into account that the particles have kinetic, rotational, and vibrational energies above the ground state of 0°K. The contribution of these terms to ΔH_{I} at any temperature T above absolute zero is given by expression (28), where $C_{p,j}$ and $C_{p,k}$ are

$$\Delta H^0_{\mathrm{I},298} = \Delta H^0_{\mathrm{I},0} + \left(\sum_i^j n_j \int_0^{298} C_{p,j}\, \mathrm{d}T\right)_{\text{products}} - \left(\sum_1^k n_k \int_0^{298} C_{p,k}\, \mathrm{d}T\right)_{\text{reactants}} \tag{28}$$

heat capacities at constant pressure and n_j and n_k are moles of products and reactants, respectively.

A value of ΔH^0_{I} for reaction (16) in which not only environmental effects but also contributions due to kinetic, rotational, and vibrational energies have been eliminated can be obtained if $\Delta H^0_{\mathrm{I},0}$ can be evaluated from equation (28). Rearranging equation (28) gives $\Delta H^0_{\mathrm{I},0}$ in terms of the heat capacities of the metal, ligand, and complex (equation 29). $\Delta H^0_{\mathrm{I},0}$ is the change in both potential energy and zero-point vibration energies for the

$$\Delta H^0_{\mathrm{I},0} = \Delta H^0_{\mathrm{I},298} + \int_{298}^{0} (C_{p,\mathrm{ML}_r{}^{n-rs}} - C_{p,\mathrm{M}^{n+}} - rC_{p,\mathrm{L}^{s-}})\,\mathrm{d}T$$

$$= \Delta H^0_{\mathrm{I},298} + (H_{298} - H_0)_{\mathrm{M}^{n+}} + r(H_{298} - H_0)_{\mathrm{L}^{s-}} - (H_{298} - H_0)_{\mathrm{ML}_n{}^{n-rs}} \quad (29)$$

reaction and is the most fundamental heat of formation for evaluating bond energies and ligand-field stabilization. If values of heat capacities and hydration energies were available for the metal ion, ligand, and complex ion, it would be possible to calculate $\Delta H^0_{\mathrm{I},0}$ from solution calorimetry measurements and to relate it directly to bond-dissociation and ligand-field stabilization energies.

An example of such a calculation is given in Table 3 for the mercury halide system (Hg—X; X = Cl, Br, or I). Data for these calculations were taken from the literature[20-23]. The changes $(H_{298} - H_0)$ for HgX_2 were not available but were estimated from available data[21] for $MgCl_2$ to be 3 kcal. Values of $(H_{298} - H_0)$ for the ionized atoms were calculated from the corresponding values for the unionized atoms by assuming that the values of $(H_{298} - H_0)$ for the ionized and unionized atoms are the same. The values of ΔH_c valid at $\mu = 0.50$ were used without attempting to correct to zero ionic strength.

Table 3. Values of $\Delta H_{\mathrm{u},298}$, $\Delta H_{\mathrm{I},298}$, $\Delta H_{\mathrm{E},298}$, $\Delta H_{\mathrm{I},0}$, and $\Delta H_{\mathrm{E},0}$ for the reaction of Hg^{2+} with X (X = Cl^-, Br^-, or I^-) to form HgX_2.

X	$\Delta H_{\mathrm{u},298}$ (kcal/mole)	$\Delta H_{\mathrm{I},298}$ (kcal/mole)	$\Delta H_{\mathrm{E},298}$ (kcal/mole)	$\Delta H_{\mathrm{I},0}$ (kcal/mole)	$\Delta H_{\mathrm{E},0}$ (kcal/mole)
Cl^-	−12.75	−608	595	−606	593
Br^-	−21.23	−600	579	−598	577
I^-	−34.15	−596	562	−594	560

From the data in Table 3, it can be seen that of the two corrections applied to $\Delta H_{\mathrm{u},298}$ to obtain $\Delta H_{\mathrm{I},0}$ and $\Delta H_{\mathrm{E},0}$ the one involving the heats of hydration is by far the most significant. These results indicate that not only is it important to have heat of hydration data, i.e. the observed reversal in Table 3 of the order of the HgX_2 series in the gaseous and aqueous phases, but that such data must be very precise. However, for a given system not only are *precise* values unavailable but usually *no* values are available for ligands and complexes for reasons previously discussed, and alternate methods must be devised to relate ΔH_u^0 values from solution calorimetry to bonding and structural effects in complex ions.

One approach is that of matching reactions of the form of equation (1) through which the unknown heats of hydration can be paired-off one against another and, if the matching is successful, made to cancel one another. For example, one may consider the case of two different metals complexing with the same ligand at 0°K. These reactions are represented by equations (30) and (31). By subtracting reaction (31) from reaction (30)

$$M^{n+}(g) + r\,L^{s-}(g) \longrightarrow ML_r^{n-rs}(g) \qquad \Delta H_{I,0}^0 \tag{30}$$

$$M'^{n+}(g) + r\,L^{s-}(g) \longrightarrow M'L_r^{n-rs}(g) \qquad \Delta H_{I,0}'^0 \tag{31}$$

we obtain the difference in bond energies $\Delta H_{I,0}^D$

$$\Delta H_{I,0}^D = \Delta H_{I,0}^0 - \Delta H_{I,0}'^0 \tag{32}$$

where $\Delta H_{I,0}^D$ is for the reaction

$$M^{n+}(g) + M'L_r^{n-rs}(g) \longrightarrow ML_r^{n-rs}(g) + M'^{n+}(g) \tag{33}$$

$\Delta H_{I,0}^D$ can be related by the following equations to the difference in the heat of reaction in aqueous solution at 25°C, ΔH_u^D, where ΔH_u^D is equal to $\Delta H_u^0 - \Delta H_u'^0$.

$$M^{n+}(g) + M'L_r^{n-rs}(g) \longrightarrow ML_r^{n-rs}(g) + M'^{n+}(g) \qquad \Delta H_{I,0}^D$$

$$\Big\downarrow \Delta H_1 = \int_0^{298} (C_{p,ML_r} + C_{p,M'} - C_{p,M} - C_{p,M'L_r})\,dT$$

$$M^{n+}(g) + M'L_r^{n-rs}(g) \longrightarrow ML_r^{n-rs}(g) + M'^{n+}(g) \qquad \Delta H_{I,298}^D$$

$$\Big\downarrow \Delta H_2 = \Delta H_{h,ML_r} + \Delta H_{h,M'} - \Delta H_{h,M} - \Delta H_{h,M'L_r}$$

$$M(H_2O)_m^{n+}(aq) + M'L_r(H_2O)_u^{n-rs}(aq) \longrightarrow$$
$$ML_r(H_2O)_u^{n-rs}(aq) + M'(H_2O)_m^{n+}(aq) \qquad \Delta H_u^D$$

Adding the changes in ΔH gives as the relationship between $\Delta H_{I,0}^D$ and ΔH_u^D

$$\Delta H_{I,0}^D = \Delta H_u^D - \Delta H_1 - \Delta H_2 \tag{34}$$

Substituting in the expressions for ΔH_1 and ΔH_2, we obtain

$$\Delta H_{I,0}^D = \Delta H_u^D - \int_0^{298} (C_{p,ML_r} + C_{p,M'} - C_{p,M} - C_{p,M'L_r})\,dT$$
$$-\Delta H_{h,ML_r} - \Delta H_{h,M'} + \Delta H_{h,M} + \Delta H_{h,M'L_r} \tag{35}$$

By proper matching of reactions an attempt can be made to have $C_{p,ML_r} = C_{p,M'L_r}$, $C_{p,M} = C_{p,M'}$ and $\Delta H_{h,ML_r} = \Delta H_{h,M'L_r}$ which simplifies equation (35) to

$$\Delta H_{I,0}^D = \Delta H_u^D - \Delta H_{h,M'} - \Delta H_{h,M} \tag{36}$$

Since the hydration energies of metal ions are usually known, we have in equation (36) an expression relating the difference in the heat of complexation in the gaseous state at 0°K to the difference in the heat of complexation determined by calorimetry in aqueous solution at 298°K for a pair of reactions. The validity of the relationship depends on how well the reactions are matched but should hold quite well for a series of closely related metal ions, e.g. the trivalent lanthanide or bivalent first transition series metal ions. One must be certain, however, that the same reaction is being considered for the metal ions being compared, i.e. that the same number of ligands are being complexed, the same number of waters of hydration are involved, and the same stereochemistry exists, etc.

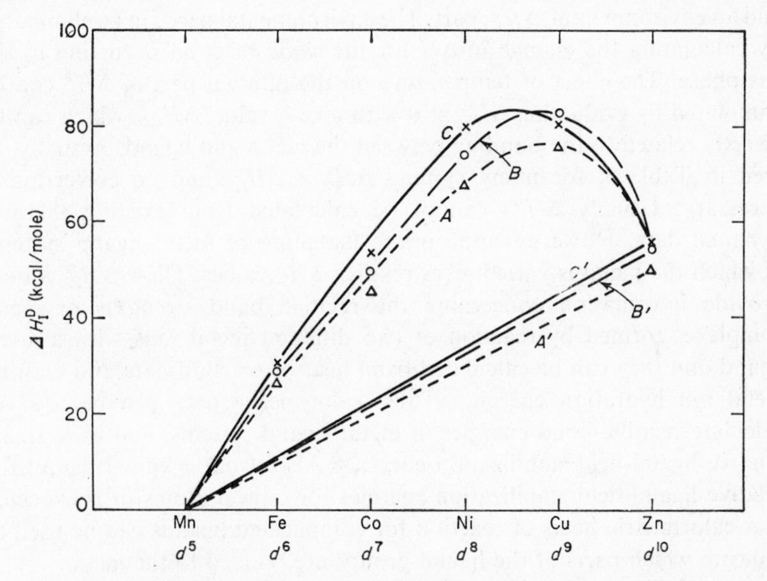

Figure 1. Plot of $\Delta H^{\mathrm{D}}_{\mathrm{I},0}$ versus atomic number (Mn^{2+} to Zn^{2+}).

George and McClure[1] have obtained relative values of $\Delta H^{\mathrm{D}}_{\mathrm{I},0}$ for the first transition series metal ions from Fe^{2+} to Zn^{2+} for 94 complexes of 67 ligands with reference to Mn^{2+} using values of $\Delta H^{\mathrm{D}}_{\mathrm{u}}$ determined from free-energy data. The values of $\Delta H^{\mathrm{D}}_{\mathrm{I},0}$ for mono-, bis-, and trisethylenediamine complexes (*A*, *B*, and *C*, respectively) obtained by them are given in Figure 1. For a given metal and ligand, the difference between the observed $\Delta H^{\mathrm{D}}_{\mathrm{I},0}$ value and that on a straight line (*A*′, *B*′, or *C*′) drawn between Mn and Zn has been interpreted as representing the ligand-field stabilization energy. They pointed out that caution must be used in interpreting the data shown in Figure 1 in terms of the ligand field theory, as the ΔH_u values were not obtained from calorimetry but from the relationship $\Delta H_u = \Delta G_u$. This relationship is based on the assumption that ΔS for the formation of a complex in the series Mn^{2+} to Zn^{2+} is independent of the particular metal ion. There is a need for much more calorimetric work to obtain ΔH_u values which can be used to determine ligand-field stabilization energies. The application of calorimetry to ligand-field theory is covered in a subsequent section.

In summary, it has been shown that to relate calorimetric heats of reaction to bond strengths and ligand-field stabilization energies it is first necessary to obtain thermodynamic values for the heats of reaction in aqueous solution relative to the mole fraction or molality standard state, ΔH_u^0. ΔH_u^0 can then be considered to consist of an internal, ΔH_I^0, and an environmental, ΔH_E^0, part. The environmental part can be eliminated by calculating the change in ΔH for the same reaction occurring in the gas phase. The effect of temperature on the internal part of ΔH_I^0 can be eliminated by evaluating ΔH_I^0 at 0°K to give a value, $\Delta H_{I,0}^0$, which can be directly related to the bonding between the metal and ligand. Actually, as seen in Table 3, for many systems $\Delta H_I^0 \simeq \Delta H_{I,0}^0$ and no correction is necessary. Usually $\Delta H_{I,0}^0$ cannot be calculated from existing thermodynamic data. However, appropriate matching of metal–ligand systems in which the metal is varied gives relative ΔH_I^0 values. These $\Delta H_{I,0}^D$ values provide information concerning the relative bond strengths of metal complexes formed by reaction of two different metal ions with a given ligand and they can be calculated from heat of reaction data and existing metal ion hydration energies. Thus, calorimetric data can be used to calculate relative bond energies in metal–ligand systems, and to evaluate relative ligand-field stabilization energies. Also, from a knowledge of the relative ligand-field stabilization energies for typical atoms, it is expected that calorimetric heats of reaction for complicated ligands can be used to indicate which parts of the ligand groups are bonded to the metal.

III. THERMOMETRIC TITRATION CALORIMETRY

A. General

Thermometric titration calorimetry is a technique in which the temperature of a system is measured as a function of a titrant added. The titrate (material titrated) and titrant may be either a liquid, a gas, or a solid, the temperature change being produced by chemical or physical reaction between the titrate and titrant. The resultant curve of temperature versus volume of titrant can be analyzed analytically with respect to quantity or concentration of titrate and calorimetrically with respect to heat of reaction, heat of solution, etc. Titrations are usually carried out under conditions which are as nearly adiabatic as possible, the exact conditions and equipment depending on the type of information desired.

Thermometric titration calorimetry differs from conventional solution calorimetry in the manner in which the reactants are introduced into the calorimeter. In a conventional solution calorimeter, the total amount of each reactant is introduced into the calorimeter before reaction and the reaction rate is controlled by the kinetics of the reaction, diffusion, and the mechanics of stirring. In contrast, in titration calorimetry one of the reactants is introduced into the calorimeter continuously or incrementally with the degree of reaction being controlled by the rate of addition of the titrant. Although conventional calorimetry is not discussed here, a comprehensive discussion has appeared on the design and operation of reaction calorimeters[24]. The application of conventional calorimetry to the measurement of heats of metal–ligand interaction has been reported by several workers[25-29].

In many respects, thermometric titration calorimetry is ideally suited for measuring heats of metal–ligand interaction in aqueous solutions, since it combines rapidity of measurement with analytical checks on the concentrations of reactants and number of reactions taking place. Both the analytical and calorimetric uses of thermometric titrimetry will be discussed in this section; however, the main emphasis will be on the calorimetric uses with respect to measuring heats of metal complexation.

B. Analytical and Calorimetric Applications

An idealized thermogram is shown in Figure 2 for the titration of a mixture of strong and weak acids ($HCl/CH_2ClCOOH$) with NaOH[30]. The thermogram is idealized in that the stirring effects, heat losses, and heats of dilution have been assumed to be zero. Regions *a* and *d* represent those time periods during which no titrant is being added to the titrate while regions *b* and *c* represent thermal effects due to the addition of the titrant. Region *b* also includes the effects of the chemical reactions. Analysis of this curve can lead to the determination of the concentration of both acids in the solution and will give the heat of reaction of each of the acids with NaOH. The first thermograms of the type shown in Figure 2 were obtained by Bell and Cowell[31] in 1913 by the laborious process of adding the titrant incrementally from a buret and measuring with a Beckmann thermometer the temperature rise after each addition. Although thermograms obtained in this manner can be analyzed either for the concentration of one of the reactants if the stoichiometry is known or the stoichiometry if the concentrations of all the reactants are known, the crudeness of the method and equipment give rise to many difficulties. The more serious of these difficulties are the length of the experiment due to the incremental addition and the erroneous slopes or inflection points due to large response lags of the thermometer and heat transfer between the solution and the surroundings. It was not until the introduction of rapid response electrical sensing devices (thermistors) by Müller[32] in 1941 and Linde and colleagues[33] in 1953 together with an electronic recorder, and an automatic continuous delivery buret by Jordan[34] in

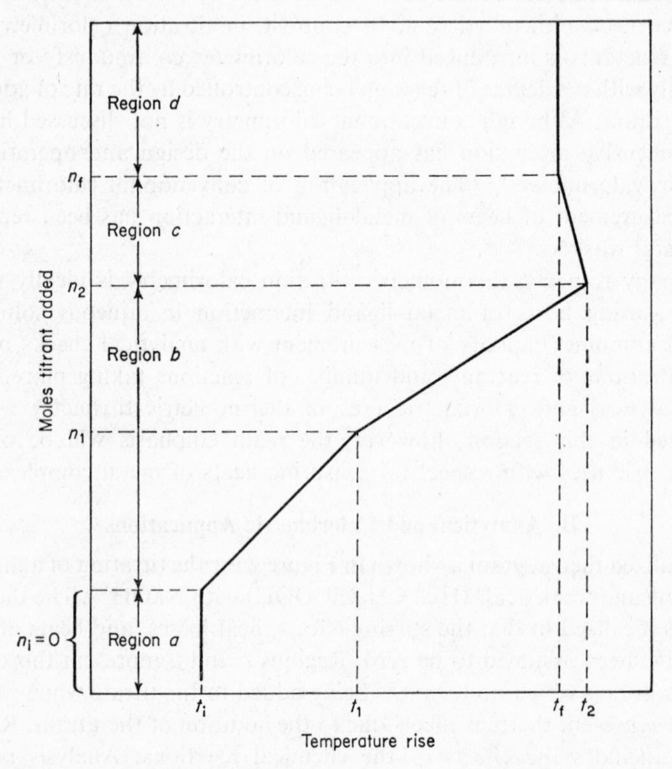

Figure 2. Typical thermogram for the titration of an HCl/$CH_2ClCOOH$ solution with NaOH. Significance of regions and symbols given in text.

1957 that the method could be considered a rapid, effective, analytical procedure. Further developments have led to the use of a differential thermistor circuit with twin titration vessels to eliminate errors due to heat losses and heat of dilution corrections[35] and to circuits for differentiating the thermograms to give 1st and 2nd derivatives for a more accurate estimation of the equivalence point[36,37]. Recent review articles cover in detail the use of thermometric titration as an analytical tool[30,38–41].

The first application of the thermometric titration procedure to the field of calorimetry and the determination of enthalpy changes for reactions in aqueous solution was by Poulsen and Bjerrum[42] in 1955, followed by Keily and Hume[43] and Zenchelsky and coworkers[44] in 1956.

The value of the thermometric titration procedure as applied to calorimetry is three-fold. Firstly, it allows in one experimental run the determination of the energy change as a function of the degree of completion of the reaction(s). This means that each point on the thermometric titration curve corresponds to an experimental run in a conventional reaction calorimeter. One thermogram is thus theoretically equivalent to a large number of determinations by conventional calorimetry. Secondly, the thermometric titration procedure gives a check on the process thought to be occurring in the calorimeter and the analytical procedures used in preparing the reactants. From the slopes of the various parts of the thermogram, and the end point(s), it is often possible to obtain confirmation of the number and types of reactions taking place and of the concentrations of the various reactants. Thirdly, in certain cases, the thermometric titration procedure provides a method for determining the free energy and entropy changes in addition to the enthalpy change for a chemical reaction.

The thermometric titration calorimeters described in the literature can be classified as continuous or incremental depending on whether the titrant is added continuously or incrementally. The continuous method gives one curve or trace on a recording potentiometer, representing the heat produced from the reactions as a function of titrant added or time. The incremental method gives a series of points, each one corresponding to the addition of a portion of the titrant. In this latter procedure, the contents of the calorimeter are usually returned to the initial temperature of the solution after each addition. The continuous method has the advantage of being more rapid but in general is not as accurate as the incremental method. Keily and Hume[43] and Schlyter and Sillén[45] were the first to describe the continuous and incremental thermometric titration calorimetry techniques, respectively.

In much of the following discussion, we will refer to the curves resulting from the continuous addition of titrant and therefore to continuous titration calorimetry, but it should be remembered that similar curves can be constructed from the data obtained from incremental titration calorimetry. The two methods are therefore comparable with respect to the determination of heats of reaction. The curves obtained from the continuous method, however, are more useful than those obtained from the incremental method in determining end points and checking stoichiometry.

In thermometric titration calorimetry, the measured quantity is the heat released upon reaction of the titrant with the titrate. The heat released is a function of the magnitude of the heat of reaction and the amount of reaction, the latter being controlled by the addition of titrant. The usefulness of the titration procedure for determining consecutive heats of reaction can be seen by examining the reaction of Hg^{2+} with Cl^- where it is desired to determine the heats of the following reactions:

$$Hg^{2+} + Cl^- \longrightarrow HgCl^+ \qquad \Delta H_1$$
$$HgCl^+ + Cl^- \longrightarrow HgCl_2(aq) \qquad \Delta H_2$$

A typical titration would be one in which a chloride-free solution containing Hg^{2+} is titrated with a NaCl solution of appropriate concentration.

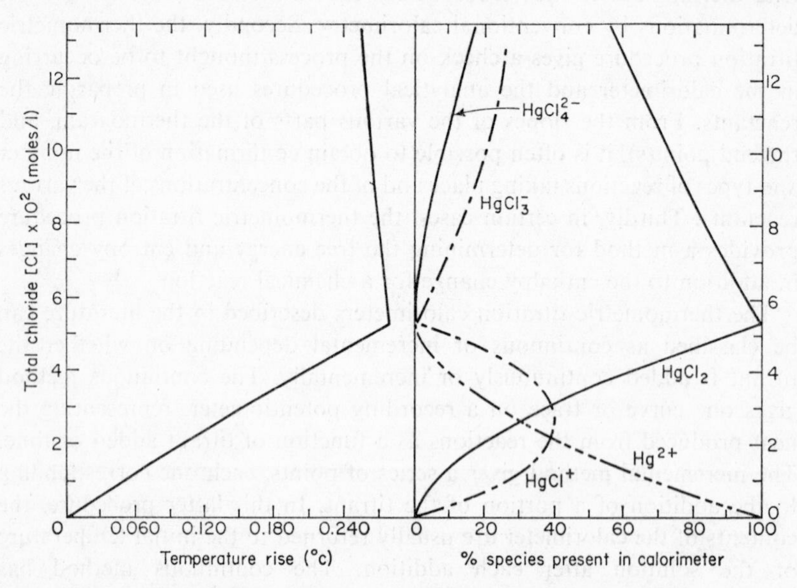

Figure 3. Correlation of temperature rise with percent species present in calorimeter for the titration of Hg^{2+} with NaCl. [Reproduced, by permission, from *Inorg. Chem.* 3, 130 (1964).]

The concentration of the species present in the calorimeter as a function of the amount of titrant added can be obtained from the appropriate equilibrium constants and mass balance expressions. This system has been investigated by Christensen and coworkers[20], and the thermogram and species distribution obtained by them is given in Figure 3. From the thermogram and species distribution, it is possible to obtain the following data:

1. The consecutive heats of reaction, ΔH_1 and ΔH_2, through combination of data from the two plots according to the methods outlined in the subsequent section on data analysis.

2. A check on the stoichiometry of the reactants by comparing the end point for the formation of $HgCl_2$ on the thermogram with the calculation from the species distribution.

3. A check on the rapidity of the reaction by comparing calculated and observed times for the appearance of the end point corresponding to the formation of $HgCl_2$.

The main advantage of thermometric titration calorimetry in this system is that the two enthalpy changes, ΔH_1 and ΔH_2, can be determined rapidly from one titration or set of titrations with all measurements being made on the same solution. In contrast, a minimum of two runs would be necessary in conventional calorimetry and several solutions would be required. It is interesting that, because of the stoichiometric formation of the $HgCl_2$ species and because no other reactions occur at the end point, the overall heat of formation for this reaction can be obtained directly from the thermogram without the use of equilibrium constants.

The Hg^{2+}/Cl^- system is one example of an inorganic system for which the thermometric titration calorimetry procedure can be profitably used to investigate metal–ligand bonding. Before looking at other representative systems which have been investigated by titration calorimetry, let us consider the criteria which determine the shape of the thermograms. The form of the thermogram obtained from a titration is determined by (1) the number of reactions, (2) the relative equilibrium constants for these reactions, (3) the relative heats of reaction of the reacting species, (4) the titrant concentration relative to that of the titrate, (5) the titrant delivery rate, and (6) heat effects due to temperature differences between titrant and solution, heat losses, heats of dilution, etc.

The first of these factors determines the maximum number of end points or break points obtained on the thermogram. In general, less than the maximum number will be observed as was seen in the Hg^{2+}/Cl^- system. The relative magnitudes of the equilibrium constants determine the actual number of break points observed and together with the relative magnitudes of the consecutive heats of reaction determine whether the break points are sharp, rounded, or appear as slight inflections in the thermogram. In the special case where the consecutive heats of reaction are equal in magnitude within the limit of measurement of the equipment, no break point will be observed. For consecutive reactions, sharp break points will usually be observed when the logs of two consecutive equilibrium constants differ by more than approximately 6. The relative magnitude of the heats of reaction determine whether the slope of the thermogram increases or decreases at the break or inflection points. For example, if $-\Delta H_2 > -\Delta H_1$, the slope of the thermogram decreases and, if $-\Delta H_2 < -\Delta H_1$, the slope of the thermogram will increase, assuming these are the only two reaction heats determining the shape of the thermogram.

Factors 4 and 5 influence the form of the thermogram with respect to the temperature rise obtained in a given time interval. Since the titrant is delivered at a constant rate, the thermogram can be considered to be a plot of either the amount of titrant added or the time versus temperature change. Low titrant concentration and slow titrant-delivery rate result

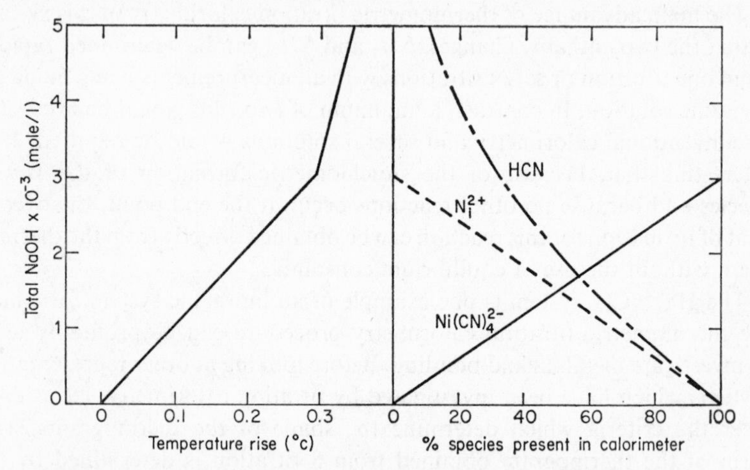

Figure 4. Correlation of temperature rise with percent species present in calorimeter (calculated as total Ni for Ni^{2+} and $Ni(CN)_4^{2-}$, and total CN for HCN) for the titration of an Ni^{2+}/CN^- solution with NaOH.

in less reaction and, therefore, less temperature rise per unit time than would be the case with high concentrations and fast delivery rates. The effect is to extend or compress the time scale with respect to the temperature scale and the species distribution. This freedom to compress or expand the time scale is useful when it is desirable to investigate only one region of the thermogram. A given region can be expanded to increase the accuracy of the measurement or contracted so that the entire reaction region can be investigated.

Factor 6 covers all the effects of the environment on the calorimeter. These effects will be discussed in detail in a subsequent section on instrumentation and practice of thermometric titration calorimetry.

Let us now look at four representative thermograms and species distribution curves for selected metal–ligand systems, at the same time remembering that the shape of the thermograms are primarily determined by items 1–3. It is useful to look at these systems to obtain an idea of the types of reactions for which titration calorimetry can be used to obtain heats of reaction, and to become acquainted with the analytical significance of the thermograms. The reactions taking place and the titrant composition are given on each figure.

Figure 4 shows the thermogram and species distribution for the Ni^{2+}/CN^- system[46] in which a solution initially containing Ni^{2+}, CN^-, and H^+ is titrated with NaOH. The initial straight line is the result of the stoichiometric reactions of OH^- with HCN and CN^- with Ni^{2+} and

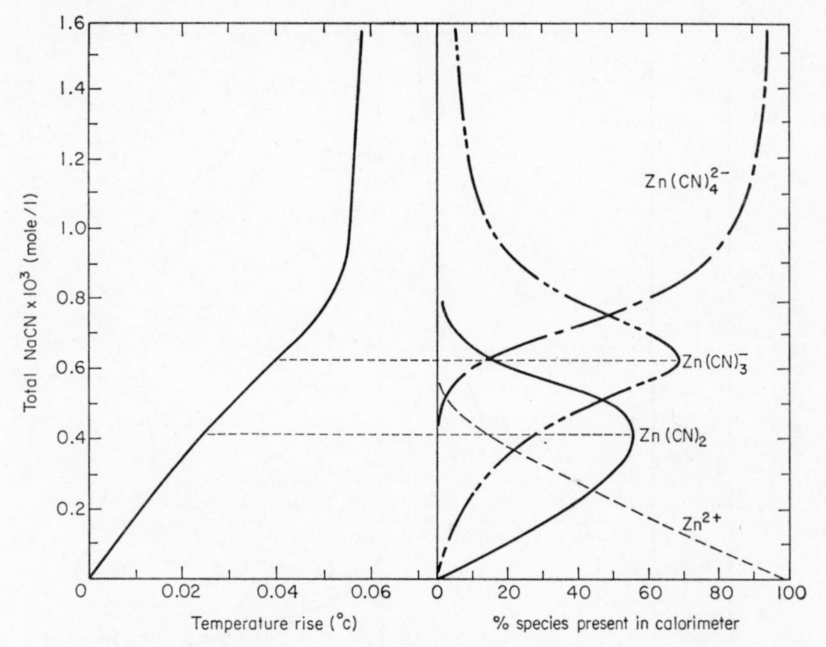

Figure 5. Correlation of temperature rise with percent species present in calorimeter for the titration of Zn^{2+} with NaCN.

confirms that in this system Ni^{2+} is only present as Ni^{2+} and $Ni(CN)_4^{2-}$ [46]. The sharp end point appears at the stoichiometric formation of $Ni(CN)_4^{2-}$, and can be used to check the stoichiometry of the reaction. In the Zn^{2+}/CN^- system[47] (Figure 5), there are no sharp end points as the differences between the logs of the formation constants of $Zn(CN)_4^{2-}$ (19.62) and $Zn(CN)_3^-$ (16.05) and those of $Zn(CN)_3^-$ (16.05) and $Zn(CN)_2$ (11.07) are only 3.57 and 4.98, respectively. The downward curvature of the thermogram results from a progressively greater rate of heat production per increment of NaCN added indicating that the third and fourth stepwise ΔH values are larger than the ΔH value for the formation of $Zn(CN)_2$ per ligand added. It is interesting to note that the thermogram gives evidence confirming equilibrium constant measurements that the species

$ZnCN^+$ was not formed in detectable concentration under the condition of the study. Constant ΔH values could be calculated at different points along the thermogram only if formation constants for the species $Zn(CN)_2$, $Zn(CN)_3^-$, and $Zn(CN)_4^{2-}$ were used. The thermogram for the Hg^{2+}/CN^- system (Figure 6) shows three distinct reaction regions separated by end points[48]. The sharp first end point indicates the stoichiometric formation of $Hg(CN)_2$ and results from a difference of 23.54 between the logs of the

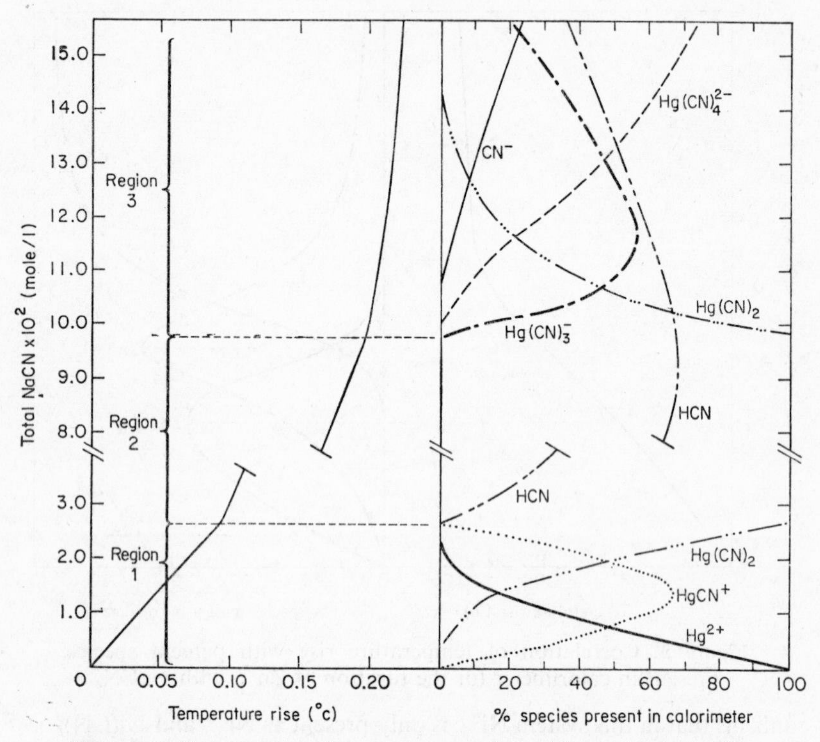

Figure 6. Correlation of temperature rise with percent species present in calorimeter (calculated as total Hg for Hg^{2+}, $HgCN^+$, $Hg(CN)_2$, $Hg(CH)_3^-$, and $Hg(CN)_4^{2-}$, and total CN for HCN and CN^-) for the titration of Hg^{2+} with NaCN.

formation constants for $Hg(CN)_2$ (32.75) and HCN (9.21). The more rounded second end point corresponds to the stoichiometric disappearance of the original excess H^+ and to the formation of HCN in the middle region. This system is interesting in that from a single thermogram the heat of ionization of HCN and the heats of formation of the four metal–ligand species $HgCN^+$, $Hg(CN)_2$, $Hg(CN)_3^-$, and $Hg(CN)_4^{2-}$ can be determined. However, in actual practice, more accurate values of the heats of formation can be obtained from thermograms determined separately for each of the three reaction regions. Heats of formation which can then be determined for the indicated species from their constituent ions are: region 1, $HgCN^+$ and $Hg(CN)_2$; from region 2, HCN; and from region 3, $Hg(CN)_3^-$ and $Hg(CN)_4^{2-}$.

Figure 7 shows the thermogram and species distribution for the Cu^{2+}/glycine (A) system[49]. This thermogram is interesting in that it has both a

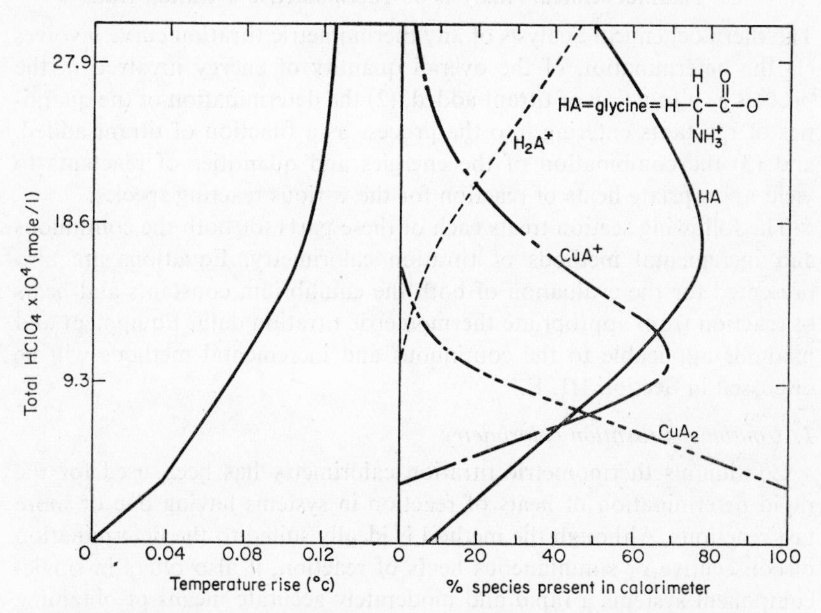

Figure 7. Correlation of temperature rise with percent species present in calorimeter for the titration of Cu(glycinate ion)$_2$ with $HClO_4$. [Reproduced, by permission, from *Inorg. Chem.*, 3, 1565 (1964).]

sharp and a well-rounded end point. The thermogram is the result of the following competing reactions: (1) combination of OH^- and H^+ to form water, (2) the reaction of protons with the carboxyl group of HA, (3) the reaction of protons with the amino group of A^-, (4) the dissociation of CuA_2 and CuA^+ to form Cu^{2+} and A^-. The first break point is due essentially to the difference of 6.9 between the logs of the ion product of H_2O (14.0) and formation constant of CuA^+ (7.10). The rounded region is primarily the result of the difference in the log K values for HA (9.78) and H_2A^+ (2.35) but is complicated by the presence of CuA^+. No break point is seen at the maximum concentration of CuA^+, probably because the difference in the log K values for CuA^+ (7.10) and HA (9.78) is only 2.68.

By examining several thermograms and species distribution curves, we have seen how the thermometric titration procedure in addition to being ideally suited for the determination of heats of complexation can also give much information concerning the stoichiometry of the reactions in the system.

C. Thermochemical Analysis of Thermometric Titration Data

The thermochemical analysis of any thermometric titration curve involves (1) the determination of the overall quantity of energy involved in the process as a function of titrant added, (2) the determination of the quantities of reactants entering into the process as a function of titrant added, and (3) the combination of the energies and quantities of reactants to yield appropriate heats of reaction for the various reacting species.

The following section treats each of these parts for both the continuous and incremental methods of titration calorimetry. Equations are also presented for the evaluation of both the equilibrium constants and heats of reaction from appropriate thermometric titration data. Equipment and methods applicable to the continuous and incremental methods will be discussed in Section III. E.

1. Continuous titration calorimetry

Continuous thermometric titration calorimetry has been used for the rapid determination of heats of reaction in systems having one or more fast reactions. Although the method is ideally suited to the determination of consecutive or simultaneous heats of reaction, it also offers in single-component systems a rapid and moderately accurate means of obtaining heats of solution in addition to the usual analytical information.

a. Single-reaction systems. Jordan[41,50] has discussed the application of continuous titration calorimetry to single-reaction systems and reviews the methods for obtaining heats of reaction from thermograms. In general, the methods used for evaluation are the 'corrected temperature rise' method[34], the 'initial slope' method[43], and the 'point' or 'section' method[51,52]. The first two methods are represented for a single reaction system on the thermogram given in Figure 8. The 'corrected temperature rise' method consists of graphically determining the temperature rise corresponding to the formation of the product by extrapolating the line representing region c back to the start of the titration (line c to point Y). The temperature difference represented by $Y - Z$ is the 'corrected temperature rise' $\Delta\theta_{co}$, and is considered to have been adequately corrected for heats of dilution, heat losses from calorimeter, heat of stirring, and other extraneous effects. The heat of reaction can then be evaluated from (37),

$$\Delta H = -C\,\Delta\theta_{co}/\Delta n \tag{37}$$

where C is the heat capacity in kcal/°c of the calorimeter and reacting system and Δn is the number of moles of product formed. C is determined either by electrical or chemical calibration. Δn is determined either from

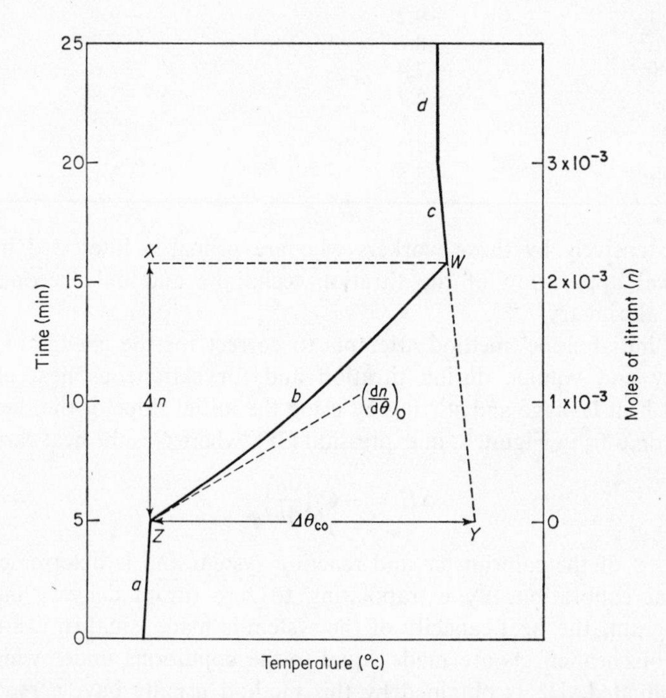

Figure 8. Typical continuous titration calorimetry thermogram illustrating the 'corrected temperature rise' and 'initial slope' methods.

the concentration of the titrant and the distance $X - Z$ on the thermogram or from prior analytical analysis of the solutions. The precision of heats of reaction evaluated by this method for EDTA complexes of several bivalent cations has been reported to be approximately $\pm 3\%$[34]. These values are compared with those for the same reactions determined by classical calorimetric methods[53] in Table 4.

For systems where the reaction is not complete or the heat of reaction is small (less than ± 2 kcal/mole), the error is much larger ($\pm 20\%$ or greater)[54]. This method of analysis for single heats of reaction has been

Table 4. Heats of formation at 25° of EDTA complexes from their constituent ions in aqueous solution. $\mu = 0.07$.

Metal ion	ΔH_c (determined by thermometric titration[34])	ΔH_c (determined by classical method of calorimetry[53])
Pb^{2+}	−12.8	−13.1
Cd^{2+}	−9.2	−9.1
Cu^{2+}	−8.2	−8.2
Ni^{2+}	−7.4	−7.6
Ca^{2+}	−5.7	−5.8
Zn^{2+}	−4.6	−4.5
Co^{2+}	−4.2	−4.1
Mg^{2+}	+5.5	+3.1

used extensively by those workers who are primarily interested in the analytical application of the titration technique and only secondarily in the calorimetry[55–65].

The 'initial slope' method attempts to correct for the change in heat capacity and volume during titration and for extraneous heat effects such as heat leakage and stirring by using the initial slope of the thermogram, $(dn/d\theta)_0$ in Figure 8, in expression (38), where C is the heat capacity

$$\Delta H = -C/\left(\frac{dn}{d\theta}\right)_0 \tag{38}$$

in kcal/°c of the calorimeter and reacting system and is determined by electrical calibration. By extrapolating to zero titrant delivery on the thermogram, the heat capacity of the system is made equal to C and all extraneous heat effects are made equal to the conditions under which C was evaluated. Heats obtained by this method usually have a reported precision of between 2 and 10%[43,66–70]. It is reported[35] that by using a differential or twin thermometric titration circuit the slope of the titration and calibration runs can be compared directly without extrapolation to zero titrant delivery. This method appears to be more accurate than the initial slope method, giving heats of reaction with accuracies of ±1%. The initial slope method of analysis has also been extended to systems having two or more reactions[52] and to the evaluation of heats of dilution and heats of solution[43,52]. Although this method of analysis is rapid and theoretically accurate, it is difficult in practice to obtain accurate initial sections of the thermograms and to evaluate the initial slope with high precision except in the case of very simple systems.

The point or section method of analysis is based on describing the four regions a, b, c, and d of the thermogram given in Figure 8 by energy and mass balance equations[51,52,71–75]. These equations together with the thermogram can then be used to calculate the energy released by the reaction in a given time interval. This method is applicable to systems having incomplete[51,73] as well as complete reactions[46,74]. The amount of reaction can be calculated using appropriate equilibrium constants and material balances or thermogram end points. The method is also applicable to multiple reaction systems and will be discussed in more detail in the following section.

b. Multiple-reaction systems. The thermometric titration calorimetric procedure is ideally suited for determining heats for stepwise complex formation in aqueous solution. Equations and methods have been developed[52,71,73,74] which allow heats to be determined from a single thermogram where several consecutive reactions are involved. These methods are based on describing by heat and material balance relationships the three regions *a*, *b*, and *d* of the thermogram. Region *c* does not appear in multireaction systems characterized by incomplete reaction; however, the resulting equations are valid for thermograms which have a region *c*. The following equations presented by Christensen and colleagues[73] are given as representative of those that have been used for describing the regions of the thermogram. In their derivation, it is assumed that the stirring and temperature sensor effects are constant and that the rate of heat transfer from the calorimeter is proportional to the temperature difference between the calorimeter and its surroundings.

$$(\mathrm{d}H/\mathrm{d}t)_a = C_a(\mathrm{d}\theta/\mathrm{d}t)_a \tag{39}$$

$$(\mathrm{d}H/\mathrm{d}t)_a = [w + k(\theta_s - \theta)] \tag{40}$$

$$(\mathrm{d}H/\mathrm{d}t)_r = [C_a + f(t)_b](\mathrm{d}\theta/\mathrm{d}t)_r \tag{41}$$

$$(\mathrm{d}H/\mathrm{d}t)_r = w + k(\theta_s - \theta) - \sum_p \Delta H_p(\mathrm{d}n_p/\mathrm{d}t)_r - (\theta_i - \theta_T)c_T(\mathrm{d}v/\mathrm{d}t)_r - (\phi_b - \phi_T)(\mathrm{d}v/\mathrm{d}t)_r m \tag{42}$$

$$(\mathrm{d}H/\mathrm{d}t)_d = C_d(\mathrm{d}\theta/\mathrm{d}t)_d \tag{43}$$

$$(\mathrm{d}H/\mathrm{d}t)_d = [w + k(\theta_s - \theta)] \tag{44}$$

Christensen and colleagues[73] have obtained from these equations expressions which when combined with the slopes and temperatures from regions *a* and *d* give the total heat of reaction between any two points on the thermogram (Q_n). Changes in the concentration of the reacting species (Δn) can be obtained from equilibrium constants (see species distributions in Figures 3, 4, 5, 6, and 7) and heats for the stepwise reactions calculated by solving the following equations where Q and Δn have been evaluated at points 1, 2, . . ., r on the thermogram.

$$Q_1 = (\Delta H_1\, \Delta n_1)_1 + (\Delta H_2\, \Delta n_2)_1 + \cdots + (\Delta H_p\, \Delta n_p)_1 \tag{45}$$

$$Q_2 = (\Delta H_1\, \Delta n_1)_2 + (\Delta H_2\, \Delta n_2)_2 + \cdots + (\Delta H_p\, \Delta n_p)_2 \tag{46}$$

. . . .

$$Q_r = (\Delta H_1\, \Delta n_1)_r + (\Delta H_2\, \Delta n_2)_r + \cdots + (\Delta H_p\, \Delta n_p)_r \tag{47}$$

These equations can be solved simultaneously when r is equal to, or by a standard least-squares method when r is greater than, the number of heats to be determined[73].

Examples of heats of metal–ligand interaction evaluated by this method are given in Table 5. It is essential to use electronic computers to reduce the time necessary for calculation of overall heats and concentrations of species.

Table 5. Heats of reaction determined by continuous titration calorimetry.

System studied	$-\Delta H_c$ (kcal/mole)	Ionic strength	Temp. (°C)	Ref.
Hg/CN		0	25	48
$Hg^{2+} + CN^- \longrightarrow HgCN^+$	23.0 ± 0.6			
$HgCN^+ + CN^- \longrightarrow HgCN_2$	25.5 ± 0.5			
$HgCN_2 + CN^- \longrightarrow HgCN_3^-$	7.6 ± 0.2			
$Hg(CN)_3^- + CN^- \longrightarrow HgCN_4^{2-}$	7.2 ± 0.2			
Hg/Cl		0.5		20
$Hg^{2+} + Cl^- \longrightarrow HgCl^+$	6.8 ± 0.2		8	
	5.5 ± 0.2		25	
	5.6 ± 0.2		40	
$HgCl^+ + Cl^- \longrightarrow HgCl_2$	7.3 ± 0.2		8	
	7.2 ± 0.2		25	
	6.9 ± 0.2		40	
Hg/Br		0.5		20
$Hg^{2+} + Br^- \longrightarrow HgBr^+$	11.1 ± 0.2		8	
	10.2 ± 0.2		25	
	10.0 ± 0.2		40	
$HgBr^+ + Br^- \longrightarrow HgBr_2$	11.5 ± 0.2		8	
	11.0 ± 0.2		25	
	11.4 ± 0.2		40	
Hg/I		0.5		20
$Hg^{2+} + I^- \longrightarrow HgI^+$	18.9 ± 0.2		8	
	18.0 ± 0.2		25	
	17.3 ± 0.2		40	
$HgI^+ + I^- \longrightarrow HgI_2$	15.3 ± 0.2		8	
	16.2 ± 0.2		25	
	17.2 ± 0.2		40	

Table 5 (*continued*)

System studied	$-\Delta H_c$ (kcal/mole)	Ionic strength	Temp. (°C)	Ref.
$HgCl_2$/ethylenediamine (L)		0	25	77
$HgCl_2 + L \longrightarrow HgClL^+ + Cl^-$	8.66 ± 0.17			
$HgClL^+ + L \longrightarrow HgL_2^{2+} + Cl^-$	9.03 ± 0.13			
$HgCl_2$/glycinate ion (L)		0	25	77
$HgCl_2 + L^- \longrightarrow HgClL + Cl^-$	6.10 ± 0.07			
$HgClL + L^- \longrightarrow HgL_2 + Cl^-$	2.94 ± 0.06			
$HgCl_2$/methylamine (L)		0	25	77
$HgCl_2 + L \longrightarrow HgClL^+ + Cl^-$	6.8 ± 0.2			
$HgClL^+ + L \longrightarrow HgL_2^{2+} + Cl^-$	1.0 ± 0.3			
Cu/glycine (L)		0	25	49
$Cu^{2+} + L^- \longrightarrow CuL^+$	6.0			
$CuL^+ + L^- \longrightarrow CuL_2$	6.4			
Cu/aminoisobutyric acid (L)		0	25	49
$Cu^{2+} + L^- \longrightarrow CuL^+$	5.4			
$CuL^+ + L^- \longrightarrow CuL_2$	5.7			
Cu/sarcosine (L)		0	25	49
$Cu^{2+} + L^- \longrightarrow CuL^+$	4.6			
$CuL^+ + L^- \longrightarrow CuL_2$	5.4			
Cu/threonine (L)		0	25	49
$Cu^{2+} + L^- \longrightarrow CuL^+$	5.3			
$CuL^+ + L^- \longrightarrow CuL_2$	6.1			
Cu/CN		0	25	78
$Cu^+ + 2\,CN^- \longrightarrow Cu(CN)_2^-$	29.1 ± 0.2			
$Cu(CN)_2^- + CN^- \longrightarrow Cu(CN)_3^{2-}$	11.1 ± 0.08			
$Cu(CN)_3^{2-} + CN^- \longrightarrow Cu(CN)_4^{3-}$	11.2 ± 0.1			
Ag/CN		0	25	78
$Ag^+ + 2\,CN^- \longrightarrow Ag(CN)_2^-$	32.9 ± 0.1			
$Ag(CN)_2^- + CN^- \longrightarrow Ag(CN)_3^{2-}$	0.6 ± 0.2			
Ni/CN		0	25	46
$Ni^{2+} + 4\,CN^- \longrightarrow Ni(CN)_4^{2-}$	43.2 ± 0.2			
Zn/CN		0	25	47
$Zn^{2+} + 2\,CN^- \longrightarrow Zn(CN)_2$	10.8 ± 0.5			
$Zn(CN)_2 + CN^- \longrightarrow Zn(CN)_3^-$	8.4 ± 0.5			
$Zn(CN)_3^- + CN^- \longrightarrow Zn(CN)_4^{2-}$	8.6 ± 0.5			

From equations similar to (39) through (44), Becker and coworkers[52] have also obtained expressions for the evaluation of thermograms by both a sectional and an initial slope method. The sectional method, which is closely related to the corrected temperature rise method for single-reaction systems, allows the total heat of reaction to be obtained for any number of regions on the thermogram by a relatively simple graphical technique[51]. The initial slope method is also very similar to the one for single-reaction systems, but has been extended so that in principle it is also applicable for a multiple-reaction system[52]. These two methods have been applied to the analysis of thermograms for several acid–base systems with one, two, and three reactions[52]. Their only application to metal–ligand systems has been in conjunction with the simultaneous determination of equilibrium constants and heats of reaction (see Section III.D).

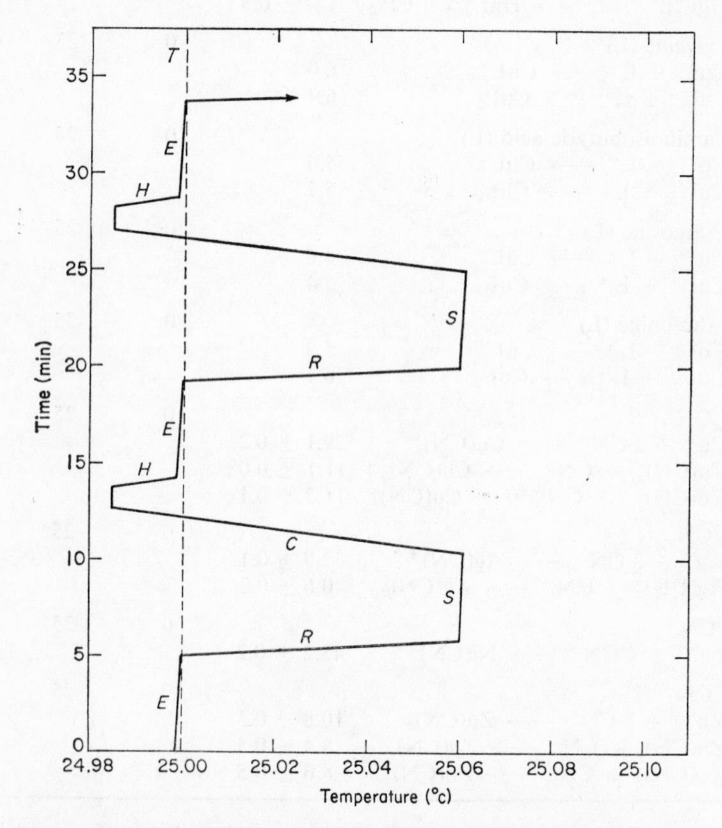

Figure 9. Typical incremental titration calorimetry thermogram with temperature adjustment. T = thermostat temperature, E = equilibration to thermostat temperature, R = reaction heat, S = equilibration period after reaction, C = cooling period, and H = heating period.

2. *Incremental titration calorimetry*

Incremental titration calorimetry, similar to continuous titration calorimetry, has been developed and used for the determination of heats of stepwise complex ion formation because of the large amount of calorimetric data that can be obtained in a given period of time. Typical thermograms resulting from incremental additions of titrant to the calorimeter are shown in Figures 9 and 10. The titration processes represented on these two diagrams differ only in that in one (Figure 10) the temperature of the

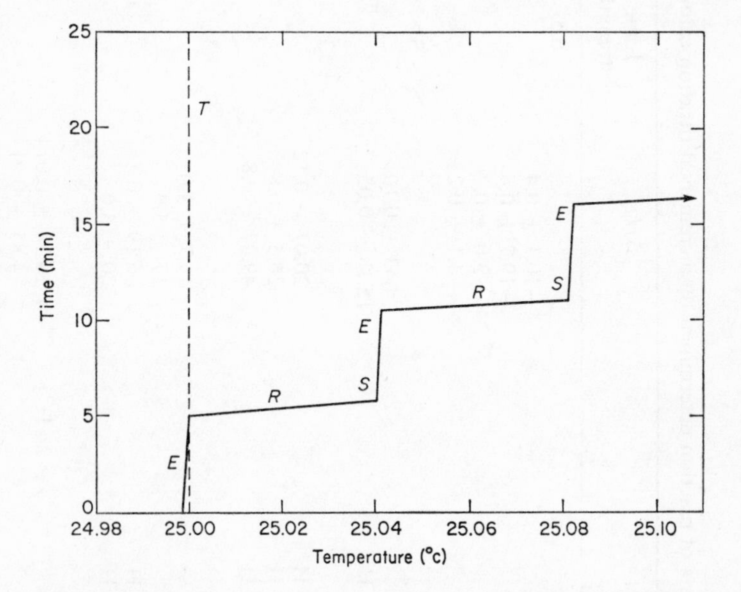

Figure 10. Typical incremental titration calorimetry thermogram without temperature adjustment. Letters are defined in Figure 9.

calorimeter is allowed to continually increase with each addition of titrant[79]. By bringing the temperature back to *T* between runs, the overall operation of the calorimeter is made nearly isothermal and the change in equilibrium with temperature eliminated[45,80,81]. The regions *E, R,* and *S* on the thermograms correspond to the initial drift period, reaction period, and final drift period, respectively, found in conventional solution calorimetry. The heats for the particular reactions occurring in each region can be determined by standard techniques[80,82].

Methods for calculating heats for stepwise reactions from incremental titration calorimetry have been discussed[42,83,84] and have been used to

Table 6. Stepwise heats of reaction determined by incremental titration calorimetry.

Reaction	ΔH_c	Ionic strength	Temp. (°c)	Ref.
Hg/Br		0.5	25	90
$Hg^{2+} + Br^- \longrightarrow HgBr^+$	-10.1 ± 0.4			
$HgBr^+ + Br^- \longrightarrow HgBr_2$	-10.7 ± 0.4			
$HgBr_2 + Br^- \longrightarrow HgBr_3^-$	-3.0 ± 0.2			
$HgBr_3^- + Br^- \longrightarrow HgBr_4^{2-}$	-4.1 ± 0.2			
Be/H_2O		3	25	86
$2\ Be^{2+} + H_2O \longrightarrow Be_2OH^{3+} + H^+$	4.43 ± 0.10			
$3\ Be^{2+} + 3\ H_2O \longrightarrow Be_3(OH)_3^{3+} + 3\ H^+$	15.18 ± 0.05			
Pb/H_2O		3	25	85
$4\ Pb^{2+} + 4\ H_2O \longrightarrow Pb_4(OH)_4^{4+} + 4\ H^+$	20.07 ± 0.12			
$3\ Pb^{2+} + 4\ H_2O \longrightarrow Pb_3(OH)_4^{2+} + 4\ H^+$	26.5 ± 0.8			
$6\ Pb^{2+} + 8\ H_2O \longrightarrow Pb_6(OH)_8^{4+} + 8\ H^+$	49.44 ± 0.8			
Fe/H_2O		3	25	88
$Fe^{3+} + H_2O \longrightarrow Fe(OH)^{2+} + H^+$	19.7 ± 3.0			
$Fe^{3+} + 2\ H_2O \longrightarrow Fe(OH)_2^+ + 2\ H^+$	17 ± 5.0			
$2\ Fe^{3+} + 2\ H_2O \longrightarrow Fe_2(OH)_2^{4+} + 2\ H^+$	6.19 ± 0.3			
$3\ Fe^{3+} + 4\ H_2O \longrightarrow Fe_3(OH)_4^{5+} + 4\ H^+$	20 ± 3.0			
In/H_2O		3	25	89
$(n + 1)\ In^{3+} + 2n\ H_2O \longrightarrow In[In(OH)_2]_n^{(3+n)+} + 2n\ H^+$	$n\ 10.18 \pm 0.014$			
$In^{3+} + H_2O \longrightarrow In(OH)^{2+} + H^+$	4.85 ± 0.91			
$In^{3+} + 2\ H_2O \longrightarrow In(OH)_2^+ + 2\ H^+$	14 ± 9			

$UO_2(R)/H_2O$		3	25	87
$2\,R^{2+} + 2\,H_2O \longrightarrow R_2OH_2^{2+} + 2\,H^+$	9.442 ± 0.109			
$3\,R^{2+} + 4\,H_2O \longrightarrow R_3OH_4^{2+} + 4\,H^+$	18 ± 5			
$3\,R^{2+} + 5\,H_2O \longrightarrow R_3OH_5^{+} + 5\,H^+$	25 ± 2			
$4\,R^{2+} + 6\,H_2O \longrightarrow R_4OH_6^{2+} + 6\,H^+$	24 ± 5			
Mn/bipyridine (L)		0	25	79
$Mn^{2+} + L^- \longrightarrow MnL^+$	−4.30			
$MnL^+ + L^- \longrightarrow MnL_2$	−4.30			
$MnL_2 + L^- \longrightarrow MnL_3^-$	−4.30			
Ni/bipyridine (L)		0	25	79
$Ni^{2+} + L^- \longrightarrow NiL^+$	−8.03			
$NiL^+ + L^- \longrightarrow NiL_2$	−8.03			
$NiL_2 + L^- \longrightarrow NiL_3^-$	−8.03			
Cu/bipyridine (L)		0	25	79
$Cu^{2+} + L^- \longrightarrow CuL^+$	−8.33			
$CuL^+ + L^- \longrightarrow CuL_2$	−8.33			
$CuL_2 + L^- \longrightarrow CuL_3^-$	−8.33			
Zn/bipyridine (L)		0	25	79
$Zn^{2+} + L^- \longrightarrow ZnL^+$	−5.27			
$ZnL^+ + L^- \longrightarrow ZnL_2$	−5.27			
$ZnL_2 + L^- \longrightarrow ZnL_3^-$	−5.27			

19*

Table 7. Stepwise heats of reaction determined by incremental titration calorimetry for the formation of rare earth diglycolates[93], $\mu = 1.0$.

$$\begin{array}{lll} M + L \longrightarrow ML & \Delta H_{c,1} \\ ML + L \longrightarrow ML_2 & \Delta H_{c,2} \\ ML_2 + L \longrightarrow ML_3 & \Delta H_{c,3} \end{array}$$

Metal ion	$-\Delta H_c \pm$ (standard deviation)		
	$-\Delta H_{c,1}$	$-\Delta H_{c,2}$	$-\Delta H_{c,3}$
La	0.070 ± 0.016	0.828 ± 0.021	0.474 ± 0.040
Ce	0.401 ± 0.017	1.270 ± 0.023	1.764 ± 0.032
Pr	0.680 ± 0.016	1.712 ± 0.021	2.500 ± 0.030
Nd	0.848 ± 0.014	2.103 ± 0.019	3.000 ± 0.024
Sm	1.048 ± 0.018	2.878 ± 0.023	4.291 ± 0.028
Eu	0.781 ± 0.017	2.943 ± 0.021	4.507 ± 0.024
Gd	0.360 ± 0.019	2.672 ± 0.022	4.590 ± 0.027
Tb	−0.765 ± 0.021	1.887 ± 0.023	4.464 ± 0.026
Dy	−1.323 ± 0.022	1.081 ± 0.019	4.413 ± 0.022
Ho	−1.591 ± 0.023	0.261 ± 0.025	4.384 ± 0.029
Er	−1.660 ± 0.072	−0.699 ± 0.078	4.202 ± 0.093
Tm	−1.574 ± 0.056	−1.165 ± 0.061	3.819 ± 0.076
Yb	−1.423 ± 0.026	−1.046 ± 0.028	3.857 ± 0.036
Lu	−1.230 ± 0.021	−0.781 ± 0.022	3.819 ± 0.030
Y	−1.732 ± 0.031	−0.489 ± 0.035	3.709 ± 0.040

determine stepwise heats of hydrolysis[85–89] of complexation[42,90], and of proton ionization[91]. These methods apply equally well to the treatment of continuous titration data. Conversely, the methods for calculating heats for consecutive reactions from continuous titration data apply to the treatment of incremental titration data. Computer programs based on the principle of minimizing the error square sum have been used to aid in the calculation of heats for stepwise reactions[92–94]. Other investigators have used incremental titration data to construct continuous titration type thermograms which are subsequently analyzed to obtain overall heat of reaction values[95–97].

Examples of the application of incremental titration calorimetry to the determination of heats for the stepwise interaction of ligands with metal ions are given in Table 6. The data of Grenthe[93,94] on the thermodynamic properties of rare earth complexes, part of which are given in Table 7, illustrate the use of titration calorimetry in evaluating the heats for stepwise interaction of several ligands (diglycolate, dipicolinate, acetate, glycolate, and thioglycolate) with a series of metal ions (rare earths). This study would have required much longer if conventional calorimetric techniques had been employed.

D. Determination of Equilibrium Constants

For certain classes of reactions, $K(\Delta G)$, ΔH, and ΔS values can be determined from calorimetric data alone. The successful application of this method (entropy titration) to a given system depends on (1) the equilibrium constant(s) and the reaction conditions being such that the amount of reaction occurring is measurable, but the reaction is not quantitative, and (2) the ΔH value(s) for the reaction(s) being measurably different from zero[72,73]. Consider the simple incomplete reaction of reactant A with B to form product AB (i.e. $A + B \longrightarrow AB$) which can be described by equations (48), (49), (50), and (51). These equations can

$$Q = \Delta H[AB]V \tag{48}$$

$$K = [AB]/[A][B] \tag{49}$$

$$[A_{total}] = [A] + [AB] \tag{50}$$

$$[B_{total}] = [B] + [AB] \tag{51}$$

be combined to give one equation with two unknowns, ΔH and K

$$\Delta H/K = V[B_{total}][A_{total}](\Delta H)^2/Q - [B_{total} + A_{total}]\Delta H + Q/V \tag{52}$$

(equation 52). By carrying out two calorimetric measurements of Q at different concentrations of A and B, two equations of the form of equation (52) are obtained from which the values of K and ΔH can be calculated. These equations illustrate for a system having only one reaction the method by which both the equilibrium constant and the heat of reaction can be evaluated from calorimetric data.

Similar equations can be developed for multiple-component systems such as the stepwise formation of the complex ML_r from metal ion M and ligand L. For such a system the reactions can be represented by

$$M + L \longrightarrow ML$$
$$M + 2\,L \longrightarrow ML_2$$
$$\cdot \quad \cdot \quad \cdot \quad \cdot \quad \cdot \quad \cdot$$
$$M + r\,L \longrightarrow ML_r$$

Equations similar to (48), (49), (50), and (51) can be written:

$$Q = V\sum_{1}^{r}[ML_r][\Delta H_r] \tag{53}$$

$$\beta_r = \frac{[ML_r]}{[M][L]^r} \tag{54}$$

$$[L_{total}] = [L] + [ML] + 2[ML_2] + \cdots + r[ML_r] \tag{55}$$

$$[M_{total}] = [M] + [ML] + [ML_2] + \cdots + [ML_r] \tag{56}$$

These equations can be combined to give one equation which will now contain $2r$ unknowns, $r\beta$ and $r\,\Delta H$ values. This has been done for the case where $r = 2$[72,73,74]. In general, however, the equations are solved by approximation methods[55,72-74,90,98,99,146]. Values of β_r are assumed and the corresponding concentrations of the complexes formed are calculated from equations (54), (55), and (56) for each of $2r$ calorimetric measurements. Values of ΔH are then calculated from (53), new β_r values are obtained and the process repeated until the best set of constants $\beta_1, \beta_2, \ldots, \beta_r$ are obtained. Sillen[100-102] and Christensen, Izatt, and coworkers[146] have developed computer programs for performing these calculations based on minimizing the error square sum:

$$U = \sum_{n=1}^{s} \left(Q_n - V \sum_{m=1}^{r} [\mathrm{ML}_m]\,\Delta H_m \right)^2 \tag{57}$$

Values of the equilibrium constants are varied in a systematic manner until the minimum value of U is found which corresponds to the best set of values for the constants.

Graphical methods have also been developed for systems having only one reaction for the determination of equilibrium constants or solubility constants from titration thermograms[56,61,62,93,103-106]. In general, these methods give only crude estimations of the constants and are of only limited value to the inorganic chemist.

The calorimetric data used for equilibrium constant determinations can be obtained by means of conventional calorimetry[98,107-109], incremental titration calorimetry[90,110], or continuous titration calorimetry[51,52,72-75, 99,111,112,146]. The titration techniques are particularly well suited to the evaluation of K, ΔH, and ΔS from calorimetry data because of the rapidity with which the required heats can be determined. Continuous titration calorimetry is especially suitable since in a single run one obtains the equivalent of a large number of determinations by either incremental or conventional calorimetry. The investigations of Christensen, Izatt, and coworkers[72,73,111,146] show this method to be capable of determining constants to the same accuracy as conventional methods such as pH titration and spectrophotometry. Examples of systems where K has been determined from calorimetry are given in Table 8.

The technique of determining equilibrium constants and heats of reactions from calorimetric data is of general usefulness in studying weak complexes (i.e. Cl^-, Br^-, I^-, and SO_4^{2-} complexes of type 'A' metal ions), acids with pK values greater than 10 or less than 4, and reactions in nonaqueous or mixed solvents. Izatt, Christensen, and Eatough have recently measured ΔG^0, ΔH^0, and ΔS^0 by this technique for the interaction of SO_4^{2-} with 31 uni-, di-, and trivalent metal ions[146] and the reaction of

Table 8. Equilibrium constants and enthalpies of reaction determined by calorimetry.

System	Log K	ΔH_c	Ionic strength	Temp. (°c)	Method	Ref.
Cu/CN						
$Cu(CN)_2^- + CN^- \longrightarrow Cu(CN)_3^{2-}$	5.00		0.6	25	Incremental	110
$Cu(CN)_3^{2-} + CN^- \longrightarrow Cu(CN)_4^{3-}$	2.64					
H/SO_4						
$HSO_4^- \longrightarrow H^+ + SO_4^{2-}$	2.00 ± 0.04	−4.86 ± 0.15	0	25	Continuous	146
H/PO_4						
$OH^- + HPO_4^{2-} \longrightarrow H_2O + PO_4^{3-}$	1.61 ± 0.03	−9.1 ± 0.4	0	25	Continuous	73
	1.61	−8.9	0.1	25	Continuous	75
Phenol (M)/dimethylformamide(A) in isooctane						
M + A ⟶ MA	2.06	−9.37		25	Continuous	51
Phenol (M)/dimethylformamide(A) in toluene						
M + A ⟶ MA	1.56	−5.52		25	Continuous	51
$MA + A \longrightarrow MA_2$	0.67	−2.07				

(continued)

Table 8 (*continued*)

System	Log K	ΔH_c	Ionic strength	Temp. (°C)	Method	Ref.
Ag/pyridine(py)						
$Ag^+ + py \longrightarrow Ag\,py^+$	2.24	−4.77		25	Continuous	52
	2.00 ± 0.04	−4.83 ± 0.05		36	Conventional	98
$Ag\,py^+ + py \longrightarrow Ag\,py_2^+$	1.95	−6.76		25	Continuous	52
	2.11 ± 0.08	−6.51 ± 0.06		25	Conventional	98
Hg/Br						
$HgBr_2 + Br^- \longrightarrow HgBr_3^-$	2.26	−2.08		25	Continuous	52
	2.1 ± 0.1	−3.0 ± 0.2	0	25	Incremental	90
$HgBr_3^- + Br^- \longrightarrow HgBr_4^{2-}$	1.38	−3.46		25	Continuous	52
	1.7 ± 0.1	−4.1 ± 0.2	0	25	Incremental	90
Cu/pyridine(py)						
$Cu^{2+} + py \longrightarrow Cu\,py^{2+}$	2.50 ± 0.02	−4.02 ± 0.08	0	25	Continuous	150
$Cu\,py^{2+} + py \longrightarrow Cu\,py_2^{2+}$	1.80 ± 0.05	−4.84 ± 0.10	0	25	Continuous	150
$Cu\,py_2^{2+} + py \longrightarrow Cu\,py_3^{2+}$	0.86 ± 0.06	−7.2 ± 0.6	0	25	Continuous	150
$Cu\,py_3^{2+} + py \longrightarrow Cu\,py_4^{2+}$	0.88 ± 0.10	−5.4 ± 1.5	0	25	Continuous	150

thiourea with $HgCN_2$ in ethanol/water mixtures[151]. Christensen, Izatt, and colleagues have measured pK, ΔH^0, and ΔS^0 values for proton ionization from several sugars, ribonucleosides, purines, and pyrimidines in regions of high pH[111,147,148].

It is also possible to extend the technique to metal–ligand systems which form strong complexes and to acids with pK values between 4 and 10. One approach is to match the reaction which is to be studied with another reaction for which the K and ΔH values are known. For example, the proton ionization of HPO_4^{2-} was determined by studying the ionization reaction in conjunction with the one for the combination of H^+ and OH^- to form water[72–74] (see Table 8). Since the equilibrium constant of the

$$HPO_4^{2-} \longrightarrow H^+ + PO_4^{3-} \qquad \Delta H_1 = 4.2 \quad \log K_1 = -12.386 \tag{58}$$

$$H^+ + OH^- \longrightarrow H_2O \qquad \Delta H_2 = -13.3 \quad \log K_2 = 13.996 \tag{59}$$

$$HPO_4^{2-} + OH^- \longrightarrow H_2O + PO_4^{3-} \qquad \Delta H_3 = -9.1 \quad \log K_3 = 1.610 \tag{60}$$

reaction given by equation (60) is small, the reaction does not go to completion and both K_3 and ΔH_3 can be evaluated from the calorimetric data. Combining K_3 and ΔH_3 with the known values of K_2 and ΔH_2 give K_1 and ΔH_1. This method of matching can in principle be extended to the determination of K and ΔH for any reaction by matching that reaction with another appropriate reaction for which the K and ΔH values are known. An evaluation of this procedure has been made by Christensen, Wrathall and Izatt,[149]. They determined the pK, ΔH^0, and ΔS^0 values for proton ionization from metanilic acid (pK = 3.75), pyridine (pK = 5.17), imidazole (pK = 6.99), THAM (pK = 8.08), and glycine (pK = 9.78) using acetic acid (pK = 4.756) as the titrant. Excellent correlation with literature pK values was obtained (± 0.01 pK) when the pK for the overall reaction was less than 4. The difficulty in matching reactions comes in finding a second reaction which meets all the necessary criteria. This field of determining equilibrium constants, and especially stepwise constants from calorimetric data, should be a very productive research area in the future; however, additional experimentation must be done to learn the limitations and precision of the method.

E. Instrumentation and Practice

1. General

Thermometric titration calorimetry is still in the developmental stages with respect to equipment design, accuracy of heat measurements, and data analysis. This is especially true for continuous titration calorimetry with its stringent demands for instantaneous measuring and recording of temperature changes. The present state of the art with respect to calorimeter design, titrant delivery methods, environment control, and data recording will be described for both incremental and continuous titration calorimetry.

The main components of a titration calorimetry apparatus are indicated in the block diagram shown in Figure 11. The titrant containing one of the

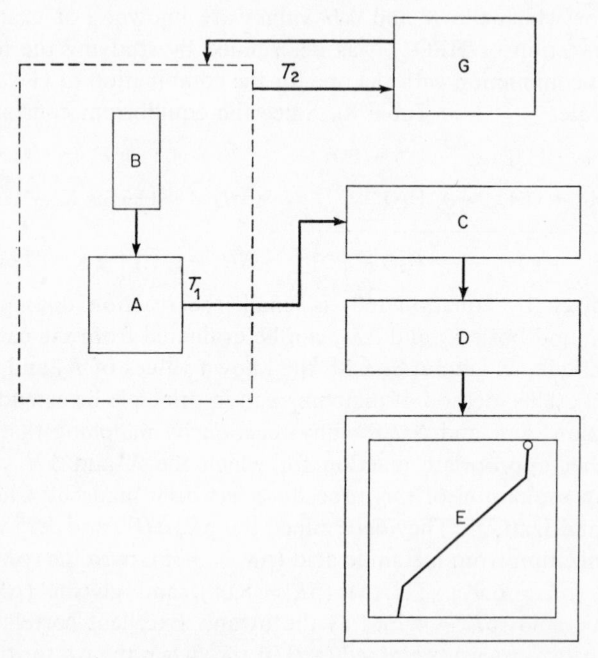

Figure 11. Schematic diagram of a continuous titration calorimeter apparatus. A = titration calorimeter with temperature sensor, B = constant delivery motor buret, C = temperature-measuring circuit and buret controls, D = amplifier, E = strip chart recorder, and G = environmental control circuit.

reactants is introduced either continuously or incrementally from the buret B into the calorimeter A. The resulting temperature change of reaction is sensed by the temperature sensor T_1 and converted to a corresponding voltage in a Wheatstone bridge circuit C. This voltage is amplified in the amplifying circuit D and recorded on a strip chart recorder E. The temperatures of the calorimeter A and the buret B are measured by the sensor T_2, and controlled by temperature controller G. The temperatures of the titrant and titrate must be equal or their difference known very precisely. Several different continuous[51,71,76] and incremental [45,79,80,81,113–115] titration calorimeters have been described in the literature.

2. Continuous titration calorimeters

The calorimeter and equipment developed by Christensen, Izatt, and coworkers[76] can be used to illustrate the design and operation of continuous titration calorimeters. The essentials of the design of the calorimeter are shown in Figure 12. This calorimeter was reported to have a low leakage modulus, a short equilibration time, and a rapid response time, all of which are necessary for accurately determining the temperature change as a function of titrant added or time.

In operation, the calorimeter was totally immersed along with titrant reservoir and titrant lines in a constant temperature water-bath. The buret was contained in a separate dry compartment of the same bath. This bath was in turn placed in a larger water-bath. The outer bath was controlled to ±0.005°C and the inner bath was controlled to ±0.0003°C.

A temperature adjustment circuit was installed to allow the temperature of the calorimeter to be rapidly and accurately brought to within 0.007°C of the inner water-bath temperature and, therefore, increase the number of determinations per given time.

The temperature measuring circuit was a d.c. Wheatstone bridge arrangement using a thermistor as a temperature-sensing element. The bridge output was amplified by a microvolt-indicating amplifier and recorded by an extended range recorder. The calorimeter was tested by determining the heat of ionization of water at 25°C. The value obtained, 13.34 ± 0.02 kcal/mole, was in excellent agreement with the accepted value of 13.34 kcal/mole[116,117]. Temperature changes of 0.01°C and heats of 1 cal have been measured with this equipment to an accuracy of better than 0.1%.

Jordan[34] has described a titration apparatus which, although used mainly for analytical applications, can be used for calorimetric determinations. The calorimeter was a simple Dewar flask fitted with a stirrer, titrator tip, heater, and thermistor temperature sensor. The thermistor was incorporated into a Wheatstone bridge circuit. The output voltage from the bridge was fed directly to the recorder. The calorimeter and buret were not temperature controlled. The apparatus has been reported to measure heats of reaction with moderate precision and accuracy (±1%).

Tyson and coworkers[35,118] developed a differential titration calorimeter technique for determining heats of reactions using two calorimeters and bridge circuits similar to those of Jordan. The titrant was titrated simultaneously into the two calorimeters, both of which were filled with solvent but only one with solute. By this procedure, the heats of dilution and other

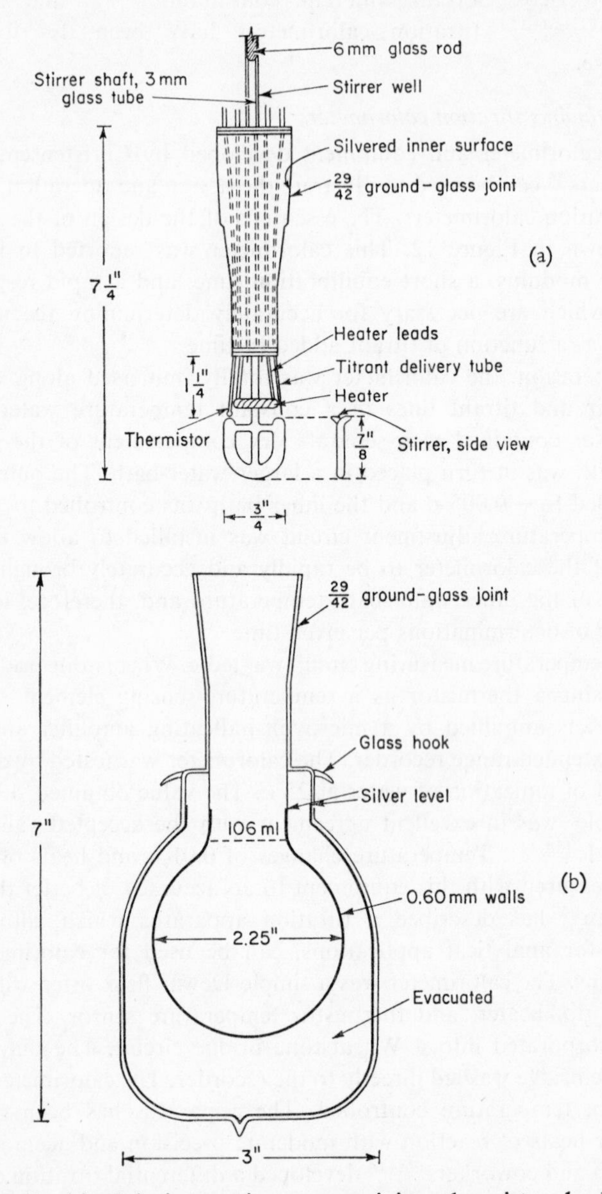

Figure 12. (a) Calorimeter insert containing thermistor leads, stirrer well, stirrer, titrator tip, and heater leads. (b) Continuous thermometric titration calorimeter. [Reproduced, by permission, from *Rev. Sci. Instr.*, **36**, 779 (1965).]

extraneous heats were supposedly canceled by comparing the temperature changes in the two calorimeters. The temperature difference was recorded on a strip chart recorder and analyzed for the appropriate heat of reaction.

3. Incremental titration calorimeters

Incremental titration calorimeters are essentially conventional reaction calorimeters in which have been incorporated methods for repeatedly introducing one of the reactants (titrant) in an unbroken series of measurements and, in many cases, for cooling the calorimeter down to the standard reference temperature between experiments. The calorimeter described by Johanasson[80] (shown in Figure 13) is representative of many incremental

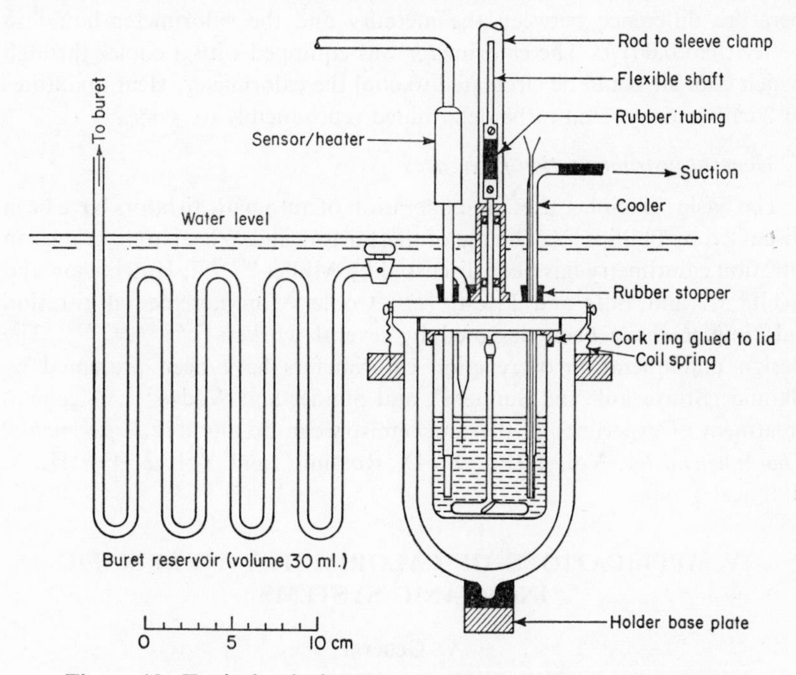

Figure 13. Typical calorimeter assembly for incremental thermometric titration calorimetry. [Reproduced, by permission, from *Arkiv Kemi*, **24**, 196 (1965).]

titration calorimeters. This calorimeter and auxiliary devices have been modified from those described by Schlyter[45]. The precision of the calorimeter was reported to be $\pm 0.5\%$ or better.

Gerding and coworkers[115] have reported a calorimeter which is similar to that shown in Figure 13 except that the titrant is introduced by means of a built-in pipet that can be repeatedly emptied and filled. The temperature was measured by means of a thermistor and the calorimeter was brought back to the thermostat temperature by blowing a precooled gas through a built-in cooler. The calorimeter was reported capable of giving results accurate to $\pm 0.2\%$ and $\pm 0.5\%$ when the total amounts of heat evolved per titrant addition are about 3.5 and 1.8 cal, respectively.

An isothermal jacket titration calorimeter which was a modification of their ordinary reaction calorimeter but with a buret added has been reported by Danielsson and colleagues[113,114]. The titrant flowed from the buret through a glass spiral, freely suspended in the thermostat water, then through a second spiral which was immersed in mercury. The mercury served as a heat buffer to level out any small temperature variations of the thermostat. A five-junction thermocouple was used to control the temperature difference between the mercury and the calorimeter liquid to better than 0.001°c. The calorimeter was equipped with a cooler through which cool air could be circulated to cool the calorimeter. Heat quantities of 2 cal were reported to be determined reproducibly to $\pm 1\%$.

4. General instrumentation references

The basic principles, uses, and operation of automatic titrators have been discussed by Phillips[119]. The use of thermistors and Wheatstone bridges in titration calorimetry has been discussed by Müller[120–122], Hutchinson and White[123], and Pitts and Priestley[124]. Coolers for incremental titration calorimeters have been designed by several workers[45,80,81,113,114]. The design and operation of reaction calorimeters have been presented by Skinner, Sturtevant, and Sunner[24], and Sunner and Wädso[125]. A general treatment of experimental thermochemistry can be found in *Experimental Thermochemistry*, Vol. 1 (Ed. F. D. Rossini)[2] and Vol. 2 (Ed. H. A. Skinner)[2].

IV. APPLICATIONS OF CALORIMETRY TO SPECIFIC INORGANIC SYSTEMS

A. General

Calorimetry provides a means for the direct measurement of the enthalpy changes involved in chemical reactions and of the absolute entropies for chemical species. These quantities give information concerning bond energies and solvent interactions, respectively, which provide the basis for understanding metal–ligand reactions. Heat of reaction data for metal–ligand interactions are not numerous and many of the reported values were obtained from K versus $1/T$ plots making them of questionable validity. Many workers who have attempted correlations of thermodynamic data have assumed ΔH_c to be proportional to ΔG_c and have used ΔG_c (or K) in their correlations. This can in certain cases, not always predictable, lead to serious errors, since entropy changes are not necessarily constant in any series of metal ion–ligand reactions.

Calorimetric data coupled with free energy data for reactions in aqueous or nonaqueous solutions provide useful information concerning the relation between the magnitude of the entropy or enthalpy change and the magnitude of the equilibrium constant. This information is often helpful in understanding stability trends in a series of complexes. However, it is not generally appreciated that one can usually say very little about actual causes as distinguished from trends of observed stabilities from solution ΔH_c and ΔS_c data alone, since these data contain terms which are often difficult to separate out and evaluate. For example, heats of hydration are of the order of 500 kcal/mole for many bivalent metal ions (see Table 3). An unknown fraction of this heat of hydration is involved each time that a ligand replaces a water molecule in its reaction with the metal ion, and this fraction may be different for different metal ions. Therefore, since the ΔH_c values which are observed in the formation of aqueous metal ion complexes are of the order of 10 to 100 times smaller than the heat of hydration values, their magnitude is usually much less than the uncertainties of the heat of hydration values. Although the trends in such cases may be significant, and may lead to interesting correlations in the particular solvent employed and for the particular metal ions studied, certainly the overall values have little significance with respect to understanding either the absolute or relative metal–ligand bond strengths. Without the $\Delta H^0_{I,0}$ or $\Delta H^D_{I,0}$ values for the thermochemical cycle outlined in Section II, it would seem pointless to speculate on the possible existence of, for example, π bonding, resonance, steric factors, etc. Very few quantitative data aimed at evaluating the energy terms outlined in Section II are presently available. This undoubtedly will be an important area of research in the future, since a knowledge of these energy terms is required for a complete understanding of the bond energies involved in metal–ligand reactions. However, their measurement will be much more difficult and require more sophisticated experiments than the kind which involve the simple replacement of water molecules from stable hydrated metal ions.

Jones[126] has summarized and discussed much of the calorimetric work done with inorganic systems prior to about 1961. The resurgence of interest about 1952 in crystal-field effects in inorganic chemistry created a need for ΔH_c data to test both crystal-field and ligand-field theories. Unfortunately, these data were nearly nonexistent at that time and early attempts to correlate experimental data with theory were made[127] using available heat of hydration, dihalide lattice energy, and metal sublimation energy data. The first transition series metal ions present an especially interesting series for study because of the expected interplay of ligand-field stabilization and spin-pairing energies in determining the magnitude of the ΔH_c values and the possibility of correlating the experimental ΔH_c data with ligand-field theory. Despite this theoretical interest, few studies seem to have been designed to obtain calorimetric data for other than high-spin complexes of the bivalent transition series metal ions.

The following discussion is limited to a few specific examples which illustrate several kinds of problems that have been studied using calorimetry, the type of information which has been obtained, and some of the difficulties involved in the interpretation of the data. It is hoped that this approach will provide the incentive for interested and qualified investigators to obtain the calorimetric data necessary to understand better the bonding forces in chemical systems. Most studies to date, although providing interesting and useful data, have not resulted in extensive insight into the actual bonding forces in the complexes.

The discussion will be divided into the following sections: (1) alkaline earth complexes with EDTA-type ligands, (2) rare earth complexes with ligands containing oxygen donor atoms, (3) transition metal complexes where the metal ion is of the d^n (n = 1–9) type, and (4) transition metal complexes where the metal ion is of the d^{10} type. The original references should be consulted for additional details in each case.

B. Studies Involving Alkaline Earth Complexes

Extensive thermodynamic studies have been made during the past two decades of the interaction of polyaminocarboxylate-type ligands with alkaline earth metal ions. The early work involved the determination of formation constants. Later studies, summarized by Wright, Holloway, and Reilley[128] and Boyd, Bryson, Nancollas, and Torrance[26], have reported calorimetric results. ΔG_c^0, ΔH_c^0, and $\Delta S_c^0 + S^0$ (M^{2+}) values for the interaction of Mg^{2+}, Ca^{2+}, Sr^{2+}, and Ba^{2+} with representative ligands of the polyaminocarboxylate type are plotted in Figure 14 in a manner similar to that used by Boyd and colleagues[26]. A prominent feature of the data in Figure 14 is the decrease in $-\Delta G_c^0$, and hence of the formation constant value in each case between Ca^{2+} and Mg^{2+}. An inverse proportionality between the formation constants of the alkaline earth complexes and the ionic radius of the metal ion would be predicted from electrostatic considerations alone; however, the data show clearly that this prediction is not realized. Attempts to explain the anomalous decrease in the formation constant between Ca^{2+} and Mg^{2+} led to the collection of the calorimetric data.

It is now of interest to evaluate the extent to which the calorimetric data have provided an explanation for the observed trend in the ΔG_c^0

values. The data in Figure 14 show that in these systems both the ΔH_c^0 and the ΔS_c^0 values are important in determining the magnitude of the ΔG_c^0 value and in approximately equal proportion in most cases. However,

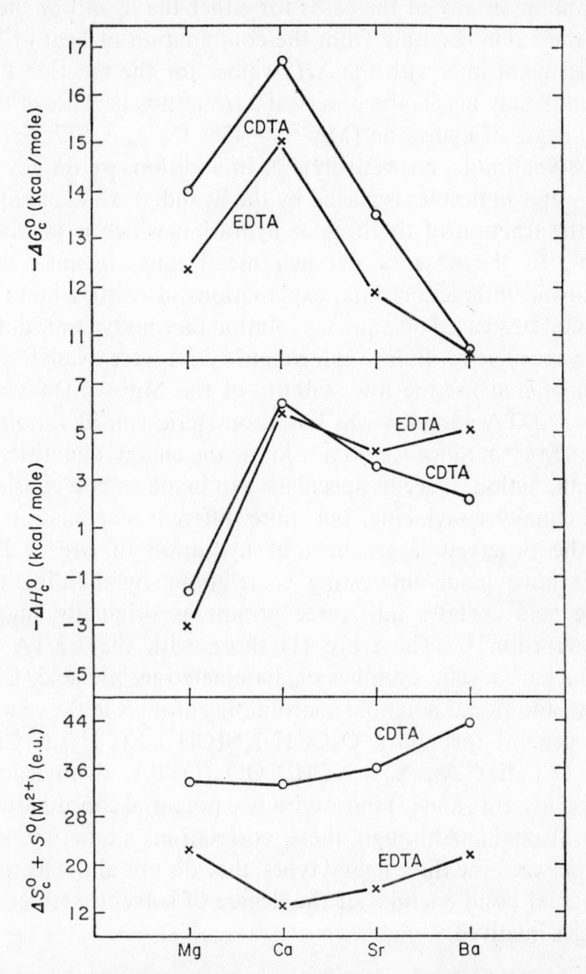

Figure 14. Plots of ΔG_c^0, ΔH_c^0, and $\Delta S_c^0 + S^0$ (M^{2+}) values for the interaction of Mg^{2+}, Ca^{2+}, Sr^{2+}, and Ba^{2+} with ethylenediaminetetraacetic acid (EDTA) and carbocyclic cyclohexanediaminetetraacetic acid (CDTA).

the calorimetric data do not provide the answer to the more fundamental question concerning the relative importance of the solvent interaction and bond-energy terms discussed in Section II which together determine the magnitudes of ΔH_c^0 and ΔS_c^0. This question would be answered only if the additional data for the thermochemical cycles mentioned in Section II were available. The use of S^0 (M^{2+}) by Boyd and colleagues[26] is an attempt to make a partial correction. However, corresponding S^0 data are not available in any of the cases for either the ligand or the complex ion. Also, the value resulting from the combination of heat of hydration data for the metal ions with the ΔH_c^0 values for the reaction in aqueous solution completely masks the observed ΔH_c^0 trends because of the magnitude of the heats of hydration (Mg^{2+}, -456; Ca^{2+}, -377; Sr^{2+}, -342; Ba^{2+}, -308 kcal/mole, respectively)[129]. In addition, we do not know the number of water molecules replaced by the ligand(s), and consequently do not know the fraction of the heat of hydration which is involved in the replacement. In the absence of such metal ion–, ligand–, and metal complex–solvent interaction data, explanations of relative bond strengths in metal complexes based on aqueous solution thermodynamic data must of necessity be speculative. Before calorimetric data were available, the usual explanation offered for the low stability of the Mg^{2+}–EDTA compared to the Ca^{2+}–EDTA complex was based on steric considerations (i.e. the small size of Mg^{2+}). Since we do not know the energy quantities involved in solvent interaction, it seems speculative to invoke steric considerations. In fact, an equally convincing, but quite different, explanation could be based on the observed larger heat of hydration of Mg^{2+}. Boyd and colleagues[26] have made interesting correlations by dividing the polyaminoacetic acid chelates into three groups as originally suggested by Kroll and Gordon[130]. These are (1) those with the EDTA skeleton, including the carbocyclic cyclohexanediaminetetraacetic acid, CDTA, (2) those with an additional potential coordinating atom X in the central chain, having the general formula $(^-O_2CCH_2)_2N(CH_2)_2X(CH_2)_2N(CH_2CO_2^-)_2$ where X = O (EEDTA), X = NCH_2COO (DTPA, diethylenetriaminepentaacetic acid), etc., and (3) those with two potential coordinating centers in the central chain. Although these correlations point out significant differences between the three ligand types, they do not allow us to estimate either the actual bond energies or the degree of solvent interaction by the several species involved.

C. Studies Involving Rare Earth Complexes

The previous discussion is particularly applicable to studies involving these complexes, since the hydration energies of the trivalent metal ions are very large (La^{3+} = 780 kcal/mole) and the heats for metal ion–ligand interaction in aqueous solution are invariably very small ($\sim$0.5 kcal/mole). The conclusion to be drawn from these facts is that the metal–ligand bond energy is of the order of that involving the attachment of the water molecules to the rare earth ion. Some interesting conclusions have been reached by

Grenthe[93,94,131] based on the interaction of diglycolate and dipicolinate with the metal ions La^{3+} to Lu^{3+}. These results are shown in Figure 15 where $T\Delta S_c^0$ and ΔH_c^0 are plotted against atomic number for La^{3+} to Lu^{3+}. The main feature of these curves is their S-shape character. Grenthe inter-

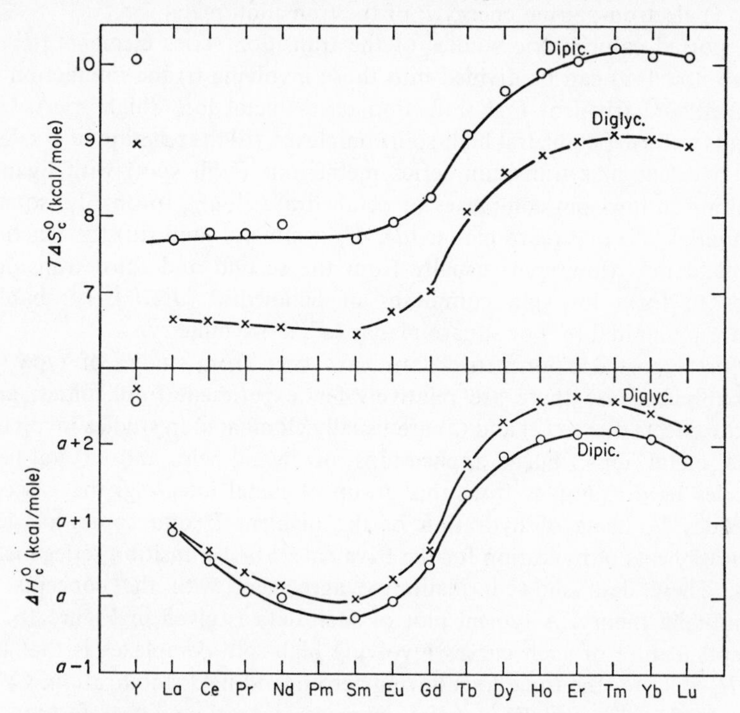

Figure 15. ΔH_c^0 and $T\Delta S_c^0$ values for interaction of trivalent rare earth metal ions with diglycolate and dipicolinate ions. $a = -1$ and -4 for diglycolate and dipicolinate ion, respectively. [Reproduced, by permission, from *Acta Chem. Scand.*, **18**, 283 (1964).]

prets these data as indicating a structural change, possibly involving a change in metal ion coordination number in the middle of the series. However, since no data are available for the ligands, rare earth metal ions, or metal complexes of the type mentioned in Section II, we cannot discount the possibility that the peculiar shapes of these curves have other explanations.

D. Studies Involving Transition Series Elements

The energy quantities which collectively or in part constitute ΔH_c^0 when n ligands interact with a transition metal ion in aqueous solution can be summarized as follows: (1) electrostatic and covalent σ-type interaction between the positively charged metal ion and the negatively charged anion or dipole molecule, (2) π bonding involving donation of electrons between metal d orbitals and ligand d or antibonding orbitals, (3) energy involved in release of water molecules from ligand and metal ion during complexation, (4) ligand-field stabilization energy Δ of the transition metal ion, and (5) electron-pairing energy P of the transition metal ion.

Reported calorimetric studies of the transition series elements (d^{10-n} where n = 1–9) can be divided into those involving (i) the interaction of divalent and trivalent first transition series metal ions (high spin) with ligands to form octahedral high-spin complexes, (ii) the reaction of divalent and trivalent first transition series metal ions (high spin) with ligands resulting in low-spin complexes of octahedral (d^1–d^6), (probably) square pyramidal (d^7) or square planar (d^8, d^9) symmetry, and (iii) the reaction of metal ions (low spin) usually from the second and third transition series to form low-spin complexes of octahedral (d^1–d^6), (probably) square pyramidal (d^7) or square planar (d^8, d^9) symmetry.

Most reported calorimetric data have been from studies of type (i), probably because there are relatively few experimental difficulties, and energy effects due to (2) and (5) are usually eliminated in studies involving these metal ions. Early applications of ligand-field and crystal-field theories used examples from this group of metal ions, e.g. the gaseous dihalides[132], heats of hydration of the divalent[133] and somewhat less accurate heats of hydration for the trivalent[134] first transition series metal ions. These data showed qualitative agreement with the concepts of ligand-field theory. A typical plot of such data is given in Figure 16. A general feature of such curves involving high-spin complexes is that the $-\Delta H_h$ data for the three ions having zero ligand-field stabilization, Ca^{2+} (d^0), Mn^{2+} (d^5), and Zn^{2+} (d^{10}), increase in nearly a linear fashion as would be predicted by electrostatics if ΔS_c^0 is constant in the series. It has been shown[135] that the deviation of the remaining M^{2+} from the nearly straight line relationship shown by Ca^{2+}, Mn^{2+}, and Zn^{2+} is qualitatively accounted for by ligand-field stabilization. Although the general features of these curves are understood, adequate explanations have not been given for the facts that Mn^{2+} is usually slightly above the line drawn between Ca^{2+} and Zn^{2+}, and that the corrected heats for the other M^{2+} do not fall exactly on the line between Ca^{2+} and Zn^{2+}. However, since the heats of hydration of these metal ions are uncertain, it does not seem profitable to attempt further explanation at this time. Subsequent work has shown that this same qualitative agreement with theory can be obtained for many metal–ligand complexes. For example, complexes of the first transition series divalent metal ions with ethylenediamine[27], tetraethylenepentamine[136], bipyridine[25,79,137,138], and *o*-phenanthroline[137] give curves similar to that in Figure 16 although with the last two ligands Fe^{2+} lies considerably above the curve. Reasons for this will be presented later. Generally, calorimetric data are available for the series Mn^{2+} to Zn^{2+} but not for V^{2+} and Cr^{2+} or for most members of the first transition series trivalent metal ions. The lack of data in these last two cases is undoubtedly a result of the experimental difficulties involved in working with these metal ions.

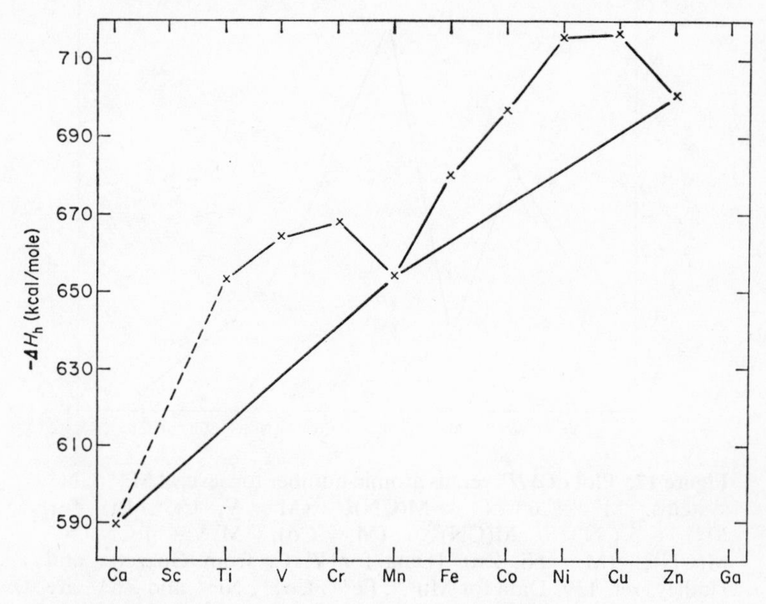

Figure 16. Plot of ΔH_h versus atomic number (Ca^{2+} to Zn^{2+}). [After George and McClure, ref. 1, p. 416.]

Considerably more complex systems are found among those of type (ii). An example of such a system is found in the interaction of cyanide ion with the first transition series divalent metal ions. In this case CN^-, a strong field ligand, forms complexes of varying geometry which are always low spin in contrast to the high-spin metal ions from which they are derived. This series of complexes has been studied and the results reported by Guzzetta and Hadley[139], and Izatt, Christensen, and coworkers[29, 46–48,140,141]. The following presentation will be limited primarily to a discussion of the latter results, since these are considered to be the more reliable. The values reported by Guzzetta and Hadley for divalent vanadium and chromium can be considered only tentative, since the experimental conditions used by these authors could easily have resulted in oxidation of these metal ions and/or metal complexes with resulting erroneous results. Also, as has been shown recently[140], exceptionally careful experimentation is required in order to prevent oxidation of the Co^{II} in the Co^{2+}/CN^- system. There is no indication that such care was exercised by Guzzetta and Hadley and for this reason the ΔH_c^0 value for the Co^{II}/CN^- system reported by Watt[140] is preferred.

Figure 17 gives a plot of ΔH_c^0 against atomic number for the interaction

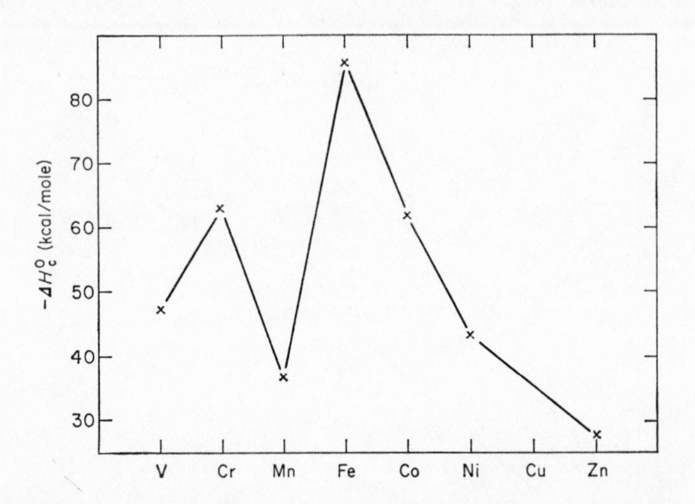

Figure 17. Plot of ΔH_c^0 versus atomic number for several M^{2+}/CN^- systems. $M^{2+} + 6\ CN^- \rightarrow M(CN)_6^{4-}$ (M = V, Cr, Mn, Fe). $M^{2+} + 5\ CN^- \rightarrow M(CN)_5^{3-}$ (M = Co). $M^{2+} + 4\ CN^- \rightarrow M(CN)_4^{2-}$ (M = Ni, Zn). [Data for V, Cr from Guzzetta and Hadley, ref. 139. Data for Mn^{2+}, Fe^{2+}, Co^{2+}, Ni^{2+} and Zn^{2+} are from ref. 140, ref. 29, ref. 140, ref. 46, and ref. 47, respectively.]

of CN^- with bivalent V^{2+}, Cr^{2+}, Mn^{2+}, and Fe^{2+} to form octahedral $M(CN)_6^{4-}$, Co^{2+} to form $M(CN)_5^{3-}$, Ni^{2+} to form planar $M(CN)_4^{2-}$, and Zn^{2+} to form tetrahedral $Zn(CN)_4^{2-}$. In each case, except that of Zn^{2+}, P is exceeded by Δ, and spin-paired complexes are observed. Since Zn^{2+} is a d^{10} metal ion, these arguments do not pertain to it. The value of Δ varies from ligand to ligand and that of P varies from metal ion to metal ion. It is interesting that all first transition series cyanide complexes are spin paired whereas all bipyridine complexes are high spin except that of Fe^{2+} [25,79,138]. It can now be seen that the much larger (compared to Mn^{2+}, Co^{2+})ΔH_c^0 value in the case of the bipyridine–Fe^{2+} and *o*-phenanthroline–Fe^{2+} complexes can be accounted for by the increased stabilization of the low-spin states[138] of Fe^{2+} as reflected in Δ being much larger than P in these cases.

In Figure 17, a very large $-\Delta H_c^0$ increase from Mn^{II} to Fe^{II} is followed by nearly constant successive decreases to Co^{II} and Ni^{II}. The increase from Mn^{II} to Fe^{II} parallels the much greater pairing energy and lower ligand-field stabilization energy of Mn^{II} compared to those of Fe^{II}. The decrease from Fe^{II} to Co^{II} to Ni^{II} parallels nicely the observed decrease from 6 to 5 to 4 in the number of CN^- ligands coordinated. The pairing energies for Fe^{II}, Co^{II}, and Ni^{II} are also comparable[140]. The species $Zn(CN)_4^{2-}$ is tetrahedral and direct comparison with the planar $Ni(CN)_4^{2-}$ is, therefore, not possible.

Again, the calorimetric data provide interesting trends relative to water as a solvent. However, we know very little about actual bond energies, since the solvation terms for metal ion, ligand, and metal–complex are not known.

The only example of type (iii) is provided by the recent work of Izatt and coworkers[141] on the Pd^{2+}/CN^- system. The data are given in Table 9

Table 9. ΔH_c^0 value for a reaction involving Pd^{2+} at 25°c.

Reaction	ΔH_c^0 (kcal/mole)	μ
$Pd^{2+} + 4\,CN^- \longrightarrow Pd(CN)_4^{2-}$	-92.3 ± 0.9	0
$Ni^{2+} + 4\,CN^- \longrightarrow Ni(CN)_4^{2-}$	-43.2 ± 0.2	0

together with corresponding Ni^{2+}/CN^- data[46]. A comparison of the ΔH_c^0 values for the reaction $M^{2+} + 4\,CN^- \rightarrow M(CN)_4^{2-}$ where M = Pd and Ni is informative. The ΔH_c^0 value in the case of Ni^{II} is 50 kcal/mole less negative than that seen in Table 9 for Pd^{II}. This difference, resulting from the fact that the ΔH_c^0 value reflects spin pairing of M^{II} in the case of nickel, but not palladium, is nearly the expected one on the basis of the estimated spin-pairing energy of the Ni^{II} alone[142]. The large difference in ΔH_c^0 values for the Pd^{II}/CN and Ni^{II}/CN systems is especially remarkable when compared to the corresponding data for the d^{10} ions, Zn^{2+}[47] and Cd^{2+}[143], where ΔH_c^0 is identical, 28 kcal/mole, for both systems.

The increasing stability of complexes formed between a given type 'b' metal ion and Cl^-, Br^-, I^-, respectively, has been ascribed to increasing availability of d orbitals for π bonding in this series[144,145]. However, it must be realized that attempts to explain the bonding must take into account the fact that these reactions usually occur in the aqueous phase. It has been observed[144,145] that all reactions are type 'a' in the gaseous phase indicating that electrostatic terms are dominant. This fact is shown by the data for the Hg^{2+}/X^-(X = Cl, Br, I) system[20]. This system shows typical type 'b' behavior in aqueous solution but, as is seen in Table 3, this behavior is reversed in the gaseous phase. The reversal results primarily from the relative magnitudes of the heats of hydration of the halide ions which become more positive in the order Cl^-, Br^-, I^-. Results for divalent Zn, Cd, and Hg with the halides show similar results, Zn being the more stable in the gaseous phase primarily as a result of its larger heat of hydration.

The Hg^{II}/halide system shows that interpretations based on aqueous data alone can lead to false conclusions concerning, in this case, relative metal–ligand bond strengths. However, the gaseous data show that the bond energies follow expected electrostatic trends. It is, therefore, not necessary to invoke π bonding or similarly vague terms to explain the order of stability, since they can be understood using available ligand hydration energy terms. In this system, the data mentioned in Section II are available (except for heat capacity data over a wide temperature range), and the interpretation of the results is straightforward. It is probable that other systems will also appear much simpler when data for their proper analysis are available.

V. REFERENCES

1. P. George and D. S. McClure in *Progress in Inorganic Chemistry*, Vol. 1 (Ed. F. A. Cotton), Interscience, New York, 1959, p. 428.
2. *Experimental Thermochemistry*, Vol. 1 (Ed. F. D. Rossini), Interscience, New York, 1956; and Vol. 2 (Ed. H. A. Skinner), Interscience, New York, 1962.
3. J. M. Sturtevant in *Physical Methods in Organic Chemistry*, Vol. 1, Part 1, 3rd ed. (Ed. A. Weissberger), Interscience, New York, 1959, Chap. 10.
4. W. P. White, *The Modern Calorimeter*, Chemical Catalog Co., New York, 1928.
5. G. Biedermann and L. G. Sillén, *Arkiv Kemi*, **5**, 425 (1953).
6. F. J. C. Rossotti and H. Rossotti, *The Determination of Stability Constants*, McGraw-Hill, New York, 1961, p. 19.
7. F. J. C. Rossotti in *Modern Coordination Chemistry* (Ed. J. Lewis and R. G. Wilkins), Interscience, New York, 1960, pp. 58–59.
8. J. E. Prue, *Ionic Equilibria*, Pergamon, New York, 1966, p. 44.
9. H. S. Harned and B. B. Owen, *The Physical Chemistry of Electrolytic Solutions*, Am. Chem. Soc. Monograph Series, No. 95, Reinhold, New York, 1943, p. 48.
10. D. B. Scaife and H. J. V. Tyrrell, *J. Chem. Soc.*, **1958**, 392.
11. G. N. Lewis and M. Randall, *Thermodynamics*, 2nd ed. (revised by K. S. Pitzer and L. Brewer), McGraw-Hill, New York, 1961, p. 339.
12. G. H. Nancollas, *J. Chem. Soc.*, **1955**, 1458.
13. K. Denbigh, *The Principles of Chemical Equilibrium*, Cambridge University Press, Cambridge, 1961, p. 273.
14. A. W. Adamson, *J. Am. Chem. Soc.*, **76**, 1578 (1954).

15. J. H. Ashby, E. M. Crook, and S. P. Datta, *Biochem. J.*, **56**, 190 (1954).
16. R. W. Gurney, *Ionic Processes in Solution*, Dover, New York, 1953, p. 90.
17. Ref. 7, p. 59.
18. Ref. 1, p. 416.
19. W. M. Latimer, *The Oxidation States of the Elements and Their Potentials in Aqueous Solution*, 2nd ed., Prentice-Hall, Englewood Cliffs, N.J., 1952.
20. J. J. Christensen, R. M. Izatt, L. D. Hansen, and J. D. Hale, *Inorg. Chem.*, **3**, 130 (1964).
21. Ref. 11, pp. 672, 683.
22. *Selected Values of Chemical Thermodynamic Properties*, Circular 500, U.S. Department of Commerce, Washington, D.C., 1952.
23. *Selected Values of Chemical Thermodynamic Properties*, Technical Notes 270–1 and 270–2, U.S. Department of Commerce, Washington, D.C., 1965.
24. H. A. Skinner, J. M. Sturtevant, and S. Sunner in *Experimental Thermochemistry* (Ed. H. A. Skinner), Interscience, New York, 1962, Chap. 9.
25. G. Anderegg, *Helv. Chim. Acta*, **46**, 2813 (1963).
26. S. Boyd, A. Bryson, G. H. Nancollas, and K. Torrance, *J. Chem. Soc.*, **1965**, 7353.
27. L. Sacconi, P. Paoletti, and M. Ciampolini, *J. Chem. Soc.*, **1964**, 5046.
28. L. A. K. Staveley and T. Randall, *Discussions Faraday Soc.*, **26**, 157 (1958).
29. G. D. Watt, J. J. Christensen, and R. M. Izatt, *Inorg. Chem.*, **4**, 220 (1965).
30. J. Jordan, *J. Chem. Educ.*, **40**, A5 (1963).
31. J. M. Bell and C. F. Cowell, *J. Am. Chem. Soc.*, **35**, 49 (1913).
32. R. H. Müller, *Ind. Eng. Chem., Anal. Ed.*, **13**, 667 (1941).
33. H. W. Linde, L. B. Rogers, and D. N. Hume, *Anal. Chem.*, **25**, 404 (1953).
34. J. Jordan and T. G. Alleman, *Anal. Chem.*, **29**, 9 (1957).
35. B. C. Tyson, W. H. McCurdy, and C. E. Bricker, *Anal. Chem.*, **33**, 1640 (1961).
36. P. T. Priestley, *Analyst*, **88**, 194 (1963).
37. S. T. Zenchelsky and P. R. Segatto, *Anal. Chem.*, **29**, 1856 (1957).
38. G. W. Ewing, *Instrumental Methods of Chemical Analysis*, 2nd ed., McGraw-Hill, New York, 1960, Chap. 20, pp. 347–349.
39. H. H. Willard, L. L. Merritt, and J. A. Dean, *Instrumental Methods of Analysis*, 2nd ed., Van Nostrand, Princeton, N.J., 1951, pp. 321–323.
40. W. W. Wendlandt, *Thermal Methods of Analysis*, Interscience, New York, 1964, Chap. VIII.
41. J. Jordan in *Treatise on Analytical Chemistry*, Part 1 (Ed. I. M. Kolthoff and P. J. Elving), Interscience, New York, Chap. 86.
42. I. Poulsen and J. Bjerrum, *Acta. Chem. Scand.*, **9**, 1407 (1955).
43. H. J. Keily and D. N. Hume, *Anal. Chem.*, **28**, 1294 (1956).
44. S. T. Zenchelsky, J. Periale, and J. C. Cobb, *Anal. Chem.*, **28**, 67 (1956).
45. K. Schlyter and L. G. Sillén, *Acta Chem. Scand.*, **13**, 385 (1959).
46. J. J. Christensen, R. M. Izatt, J. D. Hale, R. T. Pack, and G. D. Watt, *Inorg. Chem.*, **2**, 337 (1963).
47. R. M. Izatt, J. J. Christensen, J. W. Hansen, and G. D. Watt, *Inorg. Chem.*, **4**, 718 (1965).

48. J. J. Christensen, R. M. Izatt, and D. Eatough, *Inorg. Chem.*, **4**, 1278 (1965).
49. R. M. Izatt, J. J. Christensen, and V. Kothari, *Inorg. Chem.*, **3**, 1565 (1964).
50. J. Jordan, *Chimia* (*Aarau*), **17**, 101 (1963).
51. F. Becker, J. Barthel, N. G. Schmahl, G. Lange, and H. M. Lüschow, *Z. Phys. Chem.* (*Frankfurt*), **37**, 33 (1963).
52. F. Becker, J. Barthel, N. G. Schmahl, and H. M. Lüschow, *Z. Phys. Chem.* (*Frankfurt*), **37**, 52 (1963).
53. R. G. Charles, *J. Am. Chem. Soc.*, **76**, 5854 (1954).
54. J. Jordan and W. H. Dumbaugh, Jr., *Bull. Chem. Thermodynam.* (*IUPAC*), *Sect. A*, **2**, 9–11 (1959).
55. T. G. Alleman, *Abstr. Papers*, 132nd Meeting Am. Chem. Soc., New York, September, 1957, p. 11B.
56. F. J. Cioffi and S. T. Zenchelsky, *J. Phys. Chem.*, **67**, 357 (1963).
57. J. Jordan and E. J. Billingham, Jr., *Chem. Thermodynam. Props. High Temp. IUPAC, 18th, Montreal, Canada, 1961*, pp. 144–148.
58. J. Jordan and G. J. Ewing in *Handbook of Analytical Chemistry* (Ed. L. Meites), McGraw-Hill, New York, 1963, Section 8, pp. 3–7.
59. J. Jordan and J. Pendergrast, *Proc. Intern. Conf. Coordination Chem., 7th, Uppsala, Sweden, 1962*, p. 102.
60. J. Pendergrast, *Ph.D. Thesis*, Pennsylvania State University, 1962.
61. J. Jordan, J. Meier, E. J. Billingham, Jr., and J. Pendergrast, *Anal. Chem.*, **31**, 1439 (1959).
62. J. Jordan, J. Meier, E. J. Billingham, Jr., and J. Pendergrast, *Anal. Chem.*, **32**, 651 (1960).
63. J. Jordan, J. Meier, E. J. Billingham, Jr., and J. Pendergrast, *Nature*, **187**, 318 (1960).
64. J. Jordan, P. T. Pei, and R. A. Javick, *Anal. Chem.*, **35**, 1533 (1963).
65. E. J. Billingham, Jr., *Dissertation Abstr.*, **22**, 44 (1961).
66. W. L. Everson, *Anal. Chem.*, **36**, 854 (1964).
67. E. J. Forman and D. N. Hume, *Talanta*, **11**, 129 (1964).
68. D. N. Hume and H. J. Keily, *Abstr. Papers*, 132nd Meeting Am. Chem. Soc., New York, September, 1957, p. 6B.
69. H. J. Keily and D. N. Hume, *Anal. Chem.*, **36**, 543 (1964).
70. T. E. Mead, *J. Phys. Chem.*, **66**, 2149 (1962).
71. J. J. Christensen and R. M. Izatt, *J. Phys. Chem.*, **66**, 1030 (1962).
72. L. D. Hansen, J. J. Christensen, and R. M. Izatt, *Chem. Commun.*, No. 3, 36 (1965).
73. J. J. Christensen, R. M. Izatt, L. D. Hansen, and J. A. Partridge, *J. Phys. Chem.*, **70**, 2003 (1966).
74. L. D. Hansen, *Dissertation Abstr.*, **26**, 5000 (1966).
75. P. Papoff, G. Torsi, and P. G. Zambonin, *Gazz. Chim. Ital.*, **95**, 1031 (1965).
76. J. J. Christensen, R. M. Izatt, and L. D. Hansen, *Rev. Sci. Instr.*, **36**, 779 (1965).
77. J. A. Partridge, J. J. Christensen, and R. M. Izatt, *J. Am. Chem. Soc.*, **88**, 1649 (1966).
78. R. M. Izatt, H. D. Johnston, G. D. Watt, and J. J. Christensen, *Inorg. Chem.*, **6**, 132 (1967).

79. G. Atkinson and J. E. Bauman, *Inorg. Chem.*, **1**, 900 (1962).
80. S. Johansson, *Arkiv Kemi*, **24**, 189 (1965).
81. K. Schlyter, *Trans. Roy. Inst. Technol. Stockholm*, **1959**, 132.
82. Ref. 2, Vol. 1, Chap. 3.
83. K. Schlyter, *Dissertation*, K.T.H., 1–15 (1962), Roy. Inst. Technol., Stockholm.
84. K. Schlyter, *Trans. Royal Inst. Technol. Stockholm*, **1960**, 152.
85. B. Carell and A. Olin, *Acta Chem. Scand.*, **16**, 2350 (1962).
86. B. Carell and A. Olin, *Acta Chem. Scand.*, **16**, 2357 (1962).
87. K. Schlyter, *Trans. Roy. Inst. Technol. Stockholm*, **1962**, 195.
88. K. Schlyter, *Trans. Roy. Inst. Technol. Stockholm*, **1962**, 196.
89. K. Schlyter, *Trans. Roy. Inst. Technol. Stockholm*, **1961**, 182.
90. M. Björkman and L. G. Sillén, *Trans. Roy. Inst. Technol. Stockholm*, **1963**, 199.
91. K. Schlyter and D. L. Martin, *Trans. Royal Inst. Technol. Stockholm*, **1961**, 175.
92. R. Arnek, unpublished correspondence, Dept. Inorg. Chem., Roy. Inst. of Technol., Stockholm, Sweden, 1965.
93. I. Grenthe, *Acta Chem. Scand.*, **17**, 2487 (1963).
94. I. Grenthe, *Acta Chem. Scand.*, **18**, 283 (1964).
95. I. P. Gol'dshtein, E. N. Gur'yenova, and I. R. Karpovich, *Russ. J. Phys. Chem.* (*English Transl.*), **39**, 491 (1965).
96. E. Popper, L. Roman, and P. Marcu, *Talanta*, **11**, 515 (1964).
97. E. Popper, L. Roman, and P. Marcu, *Talanta*, **12**, 249 (1965).
98. P. Paoletti, A. Vacca, and D. Arenare, *J. Phys. Chem.*, **70**, 193 (1966).
99. C. A. Seitz, *M.S. Thesis*, Washington State University, 1964.
100. N. Ingri and L. G. Sillén, *Arkiv Kemi*, **23**, 97, 1964.
101. L. G. Sillén, *Acta Chem. Scand.*, **16**, 159 (1962).
102. L. G. Sillén, *Acta Chem. Scand.*, **18**, 1085 (1964).
103. S. K. Siddhanta, *J. Indian Chem. Soc.*, **25**, 579 (1948).
104. S. K. Siddhanta, *J. Indian Chem. Soc.*, **25**, 584 (1948).
105. S. K. Siddhanta and M. P. Guha, *J. Indian Chem. Soc.*, **32**, 355 (1955).
106. S. T. Zenchelsky and P. R. Segatto, *J. Am. Chem. Soc.*, **80**, 4796 (1958).
107. T. H. Benzinger and R. Hems, *Proc. Natl. Acad. Sci. U.S.*, **42**, 896 (1956).
108. T. F. Bolles and R. S. Drago, *J. Am. Chem. Soc.*, **87**, 5015 (1965).
109. M. H. Dilke and D. D. Eley, *J. Chem. Soc.*, **1949**, 2601.
110. A. Brenner, *J. Electrochem. Soc.*, **112**, 611 (1965).
111. R. M. Izatt, L. D. Hansen, J. H. Rytting, and J. J. Christensen, *J. Am. Chem. Soc.*, **87**, 2760 (1965).
112. P. Papoff and P. G. Zambonin, *Ric. Sci. Riv.*, **5**, 93 (1965).
113. I. Danielsson, B. Nelander, S. Sunner, and I. Wädso, *Acta Chem. Scand.*, **18**, 995 (1964).
114. I. Danielsson, *Acta Chem. Fennic*, **B38**, 43 (1965).
115. P. Gerding, I. Leden, and S. Sunner, *Acta Chem. Scand.*, **17**, 2190 (1963).
116. J. D. Hale, R. M. Izatt, and J. J. Christensen, *J. Phys. Chem.*, **67**, 2605 (1963).
117. C. E. Vanderzee and J. A. Swanson, *J. Phys. Chem.*, **67**, 2608 (1963).

118. B. Tyson, Jr., and W. H. McCurdy, Jr., *Abstr. Papers*, 140th Meeting, Am. Chem. Soc., Chicago, Illinois, 1961, p. 2B.
119. J. P. Phillips, *Automatic Titrators*, Academic, New York, 1959, pp. 110–116.
120. D. C. Müller, *Abstr. Papers*, 132nd Meeting, Am. Chem. Soc., New York, September, 1957, p. 6B.
121. R. H. Müller, *Abstr. Papers*, 132nd Meeting, Am. Chem. Soc., New York, September, 1957, p. 5B.
122. R. H. Müller and H. J. Stolten, *Anal. Chem.*, **25**, 1103 (1953).
123. W. P. Hutchinson and A. G. White, *J. Sci. Instr.*, **32**, 309 (1955).
124. E. Pitts and P. T. Priestley, *J. Sci. Instr.*, **39**, 75 (1962).
125. S. Sunner and I. Wädso, *Acta Chem. Scand.*, **13**, 97 (1959).
126. M. M. Jones, *Elementary Coordination Chemistry*, Prentice-Hall, Englewood Cliffs, N.J., 1964, Chap. 12.
127. Ref. 1, pp. 381–463.
128. D. L. Wright, J. H. Holloway, and C. N. Reilley, *Anal. Chem.*, **37**, 884 (1965).
129. J. P. Hunt, *Metal Ions in Aqueous Solution*, Benjamin, New York, 1963, p. 16.
130. H. Kroll and M. Gordon, *Ann. N.Y. Acad. Sci.*, **88**, 341, 1960.
131. I. Grenthe, *Acta Chem. Scand.*, **18**, 283 (1964).
132. L. E. Orgel, *An Introduction to Transition Metal Chemistry: Ligand-Field Theory*, Wiley, New York, 1960.
133. Ref. 1, p. 421.
134. B. N. Figgis, *Introduction to Ligand Fields*, Interscience, New York, 1966.
135. Ref. 1, p. 418.
136. P. Paoletti and A. Vacca, *J. Chem. Soc.*, **1964**, 5051.
137. G. Anderegg, *Helv. Chim. Acta*, **46**, 2397 (1963).
138. R. L. Davies and K. W. Dunning, *J. Chem. Soc.*, **1965**, 4168.
139. F. H. Guzzetta and W. B. Hadley, *Inorg. Chem.*, **3**, 259 (1964).
140. G. D. Watt, *Ph.D. Thesis*, Brigham Young University, 1966.
141. G. D. Watt, D. Eatough, R. M. Izatt, and J. J. Christensen, *J. Chem. Soc., A*, **1967**, 1304.
142. D. P. Graddon, *An Introduction to Coordination Chemistry*, Pergamon, New York, 1961, p. 29.
143. J. J. Christensen and R. M. Izatt, unpublished data.
144. F. Basolo and R. G. Pearson, *Mechanisms of Inorganic Reactions*, Wiley, New York, 1963, p. 179.
145. R. G. Pearson, *J. Am. Chem. Soc.*, **85**, 3533 (1963).
146. D. Eatough, *Ph.D. Dissertation*, Brigham Young University, 1967.
147. R. M. Izatt, J. H. Rytting, L. D. Hansen, and J. J. Christensen, *J. Am. Chem. Soc.*, **88**, 2641 (1966).
148. J. J. Christensen, J. H. Rytting, and R. M. Izatt, *J. Phys. Chem.*, **71**, 2700, (1967).
149. J. J. Christensen, D. P. Wrathall, and R. M. Izatt, *Anal. Chem.*, in press.
150. R. M. Izatt, D. Eatough, J. J. Christensen and R. L. Snow, *J. Phys. Chem.*, in press.
151. R. M. Izatt, D. Eatough, and J. J. Christensen, *J. Phys. Chem.*, in press.

12

General Conclusions

P. DAY and H. A. O. HILL

I. INTRODUCTION

The preceding chapters have each given an account of one physical method and the way in which it can be applied to obtain information about inorganic molecules. In this chapter, we wish to offer some more general observations about the type of information furnished by the various methods, the difficulties inherent in them, and the relationships between them. In this manner, we hope that potential users of the techniques described in this book will be able to make rational choices of the approach best suited to the problems they wish to solve. The generalizations which we shall make are supported by a detailed consideration of specific examples from past history. These show how the interests of practitioners in different techniques have from time to time converged on certain areas, to the mutual enhancement of all and with a deepening in our understanding of the molecules in question. What we call 'understanding' comes ultimately, of course, from our success in fitting the results of experimental measurement into the general framework of some theory and checking the theoretical predictions against further measurements. The difficulty which must be emphasized in a book of this kind is that the types of approximation used in theories appropriate to each type of experiment are very different—we obtain what might be called a 'gross' or a 'fine' view of the electronic or nuclear situation, according to the method of examination.

As a preliminary, we may attempt a classification of the methods according to whether they give information primarily about molecular geometry and size, i.e. about the spatial distribution of the nuclei, or about the bonding forces, i.e. the spatial and energy distribution of the electrons. Of course, in many cases we can use a method in either way, but to discuss the electronic situation in a molecule often requires a prior theoretical framework, such as that provided by MO theory, while locating the equilibrium positions of the nuclei is often a matter of symmetry arguments and selection rules which do not depend on a specific model. Diffraction, for example, has only rarely been used to investigate electron distribution, though in principle it is possible because, except for neutron diffraction, it is the electrons which scatter the incident beam. Similarly, with spectroscopic techniques such as infrared and n.m.r., the numbers of lines observed often provide sufficient information, via selection rules, to deduce molecular shapes with great precision. Information about electron densities, however, can only be obtained through the medium of a theory which defines the shapes of the shells of electron distribution more precisely than a simple expression of their symmetries. At the other extreme, we should frequently be unwilling to accept the evidence of ultraviolet spectroscopy alone for the size and symmetry of a molecule in the absence of confirming crystallographic or vibrational evidence. When such confirmation is available though, ultraviolet spectroscopy comes into its own as a major tool for laying bare the electronic structures of molecules. A distinguishing feature of the methods in most frequent use for determining the shapes and sizes of molecules then, is their reliance on the operation of selection rules so that, to begin with at least, we do not have to concern ourselves with intensity measurements but only use the crudest arguments about whether a transition is present or not. For example, the chapter on vibrational spectroscopy in this book makes it clear that both measurement and interpretation of the intensities of infrared and Raman spectra is by no means easy. Nevertheless, simply to count the number of bands appearing in the spectrum is often an unambiguous way of assigning molecular geometry, as we shall illustrate below in the case of the xenon fluorides.

Though some of them may be more frequently used in this cruder 'selection rule' fashion, the raw data from all the methods described in this book, when subjected to suitable theoretical analysis, will give information on the spatial and energy distribution of the electrons. The question is whether, if we applied all the methods to a given molecule and analysed the data according to the theory appropriate to each, we would arrive at superposable views of the nuclear and electronic structure. Unfortunately, at the present time the answer is not an obvious 'yes' for two reasons, one inherent in nature, the other a confession of our own inadequacy. The natural obstacle to obtaining comparable views of molecules by different physical methods is the uncertainty principle. High-energy spectroscopy gives a 'view' of a molecule in 10^{-12} s or so, while n.m.r. requires 10^{-8} s and Mössbauer spectroscopy as long as 10^{-7} s. A well-known example is PF_5, which is shown by electron diffraction to be a trigonal bipyramid but whose n.m.r. spectrum shows that all the fluorine atoms at equivalent[1]. The relaxation times of n.m.r. and Mössbauer spectroscopy fall in the range of the fastest chemical reactions and so both could in principle be used to study rates. The use of n.m.r. line broadening for this purpose is well established but there do not seem to have been any reports of similar experiments using the Mössbauer effect. If the nuclear arrangement of a molecule is time dependent, as a result either of vibrations or of chemical reactions in which groups of atoms enter or leave, the different methods will thus give different 'views' of the molecule.

The theoretical obstacle to obtaining strictly comparable views of the energy and spatial distribution of the electrons stems from the extremely wide range of energy differences covered by the various methods. Table 1

Table 1. Energy characteristics of spectroscopic techniques.

	Customary energy units	Approximate energy difference detected (ev)
X-ray	electronvolt	10^3 upwards
Photoelectron	electronvolt	0–20
Electronic	cm^{-1}	1–10
Vibrational	cm^{-1}	10^{-2}–1
Electron spin resonance	kMc/s	10^{-4}
Nuclear spin resonance	Mc/s	10^{-7}
Mössbauer	mm/s	10^{-8}

shows that from Mössbauer to x-ray spectroscopy spans eleven orders of magnitude of electronvolts. As we have remarked, in order to translate spectroscopic observations into conclusions about electron distributions, theoretical arguments must be applied to the raw data; while the ultimate ideal would be a single theory that would rationalize quantitatively all the physical properties, this is at present far from realization. In the past ten years, LCAO–MO theory has emerged as the common ground of many of the methods, almost completely displacing the older valence-bond approach, though it is interesting that the latter continues to find application in the calculation of nuclear spin–spin coupling constants[2]. Also, bond lengths calculated by superposing valence-bond canonical structures are no less accurate than those obtained from molecular-orbital bond order–bond length correlations[3]. The breadth of current applications of MO theory can be gauged by referring to the accounts in this book of two of the methods at opposite ends of the energy spectrum, namely nuclear quadrupole resonance and electronic spectroscopy.

The different approaches to a self-consistent molecular-orbital formulation required to rationalize ground states and excited states are well illustrated by the way in which approximations to Roothaan's SCF–LCAO–MO equations[4] were made by Pople[5] and by Pariser and Parr[6]. The elements of the secular determinant are related to the charges and bond orders, which can only be obtained by solving the determinant, so an iterative method must be employed to obtain fully self-consistent solutions, between which there are no offdiagonal elements. Pople, being interested primarily in ground-state properties such as charge distributions, required wave functions which were as close to self consistency as possible and therefore used the iterative procedure. On the other hand, Pariser and Parr wished to rationalize the energies of excited states. They were content to employ very approximate wave functions (actually solutions of the Hückel equations) and use empirical electron-repulsion integrals to calculate the off-diagonal elements. Configuration interaction via the latter then made up for deficiencies in the quality of the wave functions. Current electronic theories of complex molecules almost always take a semiempirical form in which the diagonal elements of the secular determinant are estimated from experimental atomic ionization data, while overlap integrals are obtained from computed atomic wave functions. A lack of reliable atomic spectral data as well as wave functions for heavier atoms has inhibited the precise interpretation of the results from numerous physical examinations, for example the many spin–spin coupling constants now known (see Section IV).

II. EXPERIMENTAL PROBLEMS

It will be noted that throughout this book there is little emphasis on the practical details of each physical method and, consequently, only limited discussion of the problems associated with the design of energy sources, detection systems, and the attainment of high accuracy and precision. This is not intended to suggest that such questions are of lesser importance. Indeed, most chemists using any of these methods seek the best obtainable sensitivity, resolution, accuracy, reliability, or versatility. We collect here some general comments about the different kinds of experimental problems associated with the various methods.

As noted earlier, there is an enormous difference between the transition energies associated with each of the spectroscopic methods. It is not surprising, therefore, to find problems peculiar to particular energy ranges. Low-energy methods, such as nuclear magnetic resonance or nuclear quadruple resonance, are insensitive in that they require a high sample concentration. This is a direct consequence of the small energy difference between the ground and excited state or states. With a normal distribution between the energy levels, the excess population of the ground state will be small. Since the energy absorbed by the sample is proportional to this excess population, poor signal-to-noise ratios usually result. In order to preserve the excess population of the ground state and thereby observe a *resonance* effect, the excited state must be depopulated. Spontaneous emission in transitions between different nuclear magnetic energy levels is highly improbable and so the excess population of the ground state can be maintained only if electromagnetic radiation induces emission, i.e. brings about relaxation. The requirement of an excess population in the ground state also limits the intensity of the radiofrequency field; too intense a field tends to equalize the populations of the energy states and no absorption is observed. However, if the relaxation process is too efficient, we may encounter another problem. The uncertainty in the energy of the transition and the lifetime of the excited state are related by the Heisenberg uncertainty principle and consequently a very fast relaxation time may lead to a large linewidth. Because of these rather stringent conditions, we are usually content to observe 'allowed' transitions, e.g. $\Delta M = \pm 1$ in nuclear magnetic resonance. When we consider methods which use electromagnetic radiation of higher energy, e.g. electron spin resonance spectroscopy, sensitivity problems are not so critical but those caused by relaxation phenomena become correspondingly more important. Relaxation times are critically dependent on the environment of the paramagnetic species and the temperature. Again, fast relaxation leads to loss of resolution due to line broadening, or even to the complete absence of a signal.

In infrared spectroscopy, the energy separation between the ground and excited states is sufficiently great that there is now no sensitivity problem for transitions of high probability. Though rotational fine structure cannot usually be resolved when measurements are made on condensed phases, the problem of coexcitation is not insuperable. However, we are now interested in transitions of low probability. This is more obvious in electronic absorption spectroscopy where we often seek to derive information from transitions having 10^{-7} the probability of a fully allowed one. The usual limitation is one of getting enough power into the sample, an important matter, since we may be interested in extinction coefficients, in contrast to the low-energy methods in which, for technical reasons, most of the incident radiation is discarded before detection. The principal problem in the higher energy regions is one of resolution, since the co-excitation of vibrational and rotational transition often leads to very broad bands. In some situations, this can be overcome by lowering the temperature of the sample.

The high-energy methods, such as Mössbauer and x-ray spectroscopy, involve problems of handling and supplying large amounts of energy. As pointed out in Chapter 1, the choice of a diffraction method is often governed by the physical state of the molecule under investigation, or even by considerations of cost. The major problem in all the diffraction methods is the extraction of information from the collected data.

X-ray spectroscopy has never been widely used because it is afflicted by all the problems expected from high-energy radiation, i.e. coexcitation and overlapping bands leading to poor resolution and even decomposition of compounds.

Mössbauer spectroscopy occupies a unique position among the spectroscopic methods, using a very high-energy source to detect very small energy changes. This feat is possible only because the linewidth is exceptionally narrow, owing to the relatively long lifetime of the excited nuclear energy level. Problems of resolution still occur, of course, but the principal problem is the low photon density at the source, and consequently at the detector. To overcome this requires a statistical treatment of the results. Thus, among the high-energy methods, resolution is a general problem and the loss of information which results is critically damaging both to assignments of the spectra and, in consequence, to their interpretation in terms of electronic and molecular parameters.

III. RATIONALE OF OBSERVABLES

The simplest use of any of the physical methods described in this book is that of 'finger-printing', in which the absorption, emission, or diffraction pattern ascribed to a compound is used as a means of identification or as a criterion of purity. Some techniques are more valuable than others in this respect but, in general, the more observables which result from the experiment, the higher the definition or resolution, and the more the technique involves the interaction of radiation with the whole molecule, the more valuable it will be, providing it is economical and experimentally simple. Thus infrared spectroscopy is usually a reliable finger-printing method, since vibrations involving each part of a molecule may be detected, whereas electronic absorption spectra may result from transitions confined to a small part of a molecule and thus remain insensitive to changes in other parts of the molecule. Of the methods which are available for qualitative analysis, some may also be used quantitatively but only those methods which present results in the form of extinction coefficients are useful, and electronic absorption spectroscopy is by far the most reliable. The reason for this lies in its superior sensitivity and the ease of preparing suitable solutions for examination.

However, the major concern of previous chapters has been with the use of the various methods for determining molecular and electronic structure. Among the spectroscopic methods, the first problem encountered is usually one of assignment, a situation not often emphasized in textbooks. Thus, research students frequently expect instruments to present not only spectra but also detailed interpretations. For anything other than the most general considerations, the assignment must be based on some theoretical model. For example, in n.m.r. we choose to discuss the spectra using the concepts of chemical shifts and spin–spin coupling constants; in infrared spectroscopy, we might be satisfied in certain applications with interpretations based on group frequencies, whilst spectra of transition metal complexes are often interpreted in terms of the ligand-field model. We cannot be sure that parameters extracted from the experimental data by these models (for example, $10Dq$ in the crystal-field model) are physically meaningful. By this we mean that, if it were possible to perform completely *a priori* calculations, such a parameter would appear in an internally consistent manner or, less stringently, that the parameter is transferable outside the original frame of reference. Nevertheless, the calculation of such parameters as shielding constants or hyperfine coupling constants in e.s.r. occupies the attention of many inorganic chemists. It may sometimes happen that no model can be found which gives a convincing assignment and, even when an assignment is possible, there may not be enough observables to determine unambiguously the parameters used by the model. If the molecular structure is known, symmetry arguments frequently allow a prediction of the number of observable transitions. If this is not possible, because the selection rules are not understood or the molecular symmetry is too low, the inorganic chemist often resorts to empirical correlations to aid the assignment, e.g. the use of group frequencies in infrared, chemical shifts in n.m.r., and even the change in energy of a *d–d* transition when one ligand is replaced by another. Sometimes valuable information is lost, which could greatly aid assignment; for example, in electronic absorption spectroscopy, where the direction of polarization of a transition cannot be derived from a solution spectrum but, in favourable circumstances, is obtainable from measurements on single crystals. Similarly, a knowledge of the magnetic dipole component of an electronic transition would increase the number of observables.

These considerations underline an important aspect of the interpretation of spectra: the relationship between the number of observables and the number of parameters that define them. Thus in n.m.r., in the absence of accidental degeneracy or exchange, the energy of a transition is defined by two parameters only, $\nu_i\delta$, the chemical shift, and $|J_{ij}|$, the coupling constant. Usually there are more observables than parameters, which can therefore be determined unambiguously. However, it is noteworthy that when further information, such as the sign of a coupling constant or indeed the magnitude of the coupling constant between equivalent nuclei, is required, then other kinds of experiments must be performed. A further example of the need to enhance the information content of an experiment comes from nuclear quadrupole resonance. In a non-symmetric field gradient, nuclei with $I = \frac{3}{2}$ exhibit only one transition, whose energy is governed by two unknowns, eqQ and η, the asymmetry parameter. However, if the sample is placed in a weak magnetic field and the degeneracy of the ΔM levels is removed, four transitions result, and the two parameters may thus be determined. The crystal-field model requires three parameters to define the *d–d* spectrum of a d^n ion in a cubic field and in the absence of appreciable spin–orbit coupling: the splitting parameter 10 Dq and two electron-repulsion parameters. Nevertheless, it is often difficult, because of poor resolution, to assign even three bands with confidence; in lower symmetries, the situation is very much worse. Sometimes, e.g. in *f–f* spectra, there are enough observable transitions to permit statistical fitting of observables to theoretical parameters, a situation rarely encountered (with the exception of x-ray crystallography) in the other methods described in this book. Such considerations as these determine the acceptable level of agreement between theory and observation in each of the methods. Thus, small low-symmetry crystal-field components in f^n spectra may be readily identified, but there is no theory of charge-transfer spectra which achieves agreement as good as 10%.

IV. EXAMPLES

To illustrate some of the generalizations that we have been making, specific examples of applications of physical methods will now be considered. The two classes of compound which we shall examine are the xenon fluorides and ferrocene, since both were unexpected when discovered and of completely unknown structures. By a combination of many techniques, we now have sound theoretical rationalizations of their structures and stability.

Though it cannot be said that the successful preparation of inert gas compounds in 1962 was entirely unexpected, since their existence had been predicted by Pauling in 1938 and unsuccessful attempts had been reported before, yet their constitution was quite unknown when they were discovered and the elucidation of their molecular structures by physical means is an object lesson in the application of many of the methods described in this book. In this respect, they provide an interesting contrast with ferrocene, discovered almost exactly a decade before, when many now familiar techniques were in their infancy (e.g. n.m.r. and e.s.r.) or had not yet been discovered (Mössbauer spectroscopy). Also, the greater molecular complexity and lower symmetry of ferrocene led to ambiguities in interpreting the physical measurements in contrast to the xenon fluorides, where high symmetries permitted unequivocal assignments of geometry and size within months of their first discovery.

Apart from verifying the formula by measuring the combining weights of xenon and fluorine, the first report on the preparation of XeF_4[7] contained infrared data which enabled the authors to state that the molecular symmetry could only be square planar or tetrahedral. A further report one month later[8], in which Raman data were added to the infrared, concluded that the balance of probability was in favour of square planar symmetry, though it was difficult to rule out the possibility that some peaks may have been due to impurities. Mass spectra were also reported and peaks due to XeF and XeF_2 were identified, showing that the latter were more volatile than XeF_4. It is worth noting that at this point in the investigation the ultraviolet spectrum of XeF_4 in HF solution was most unhelpful, since the absorption merely increased uniformly from an extinction coefficient of 30 at 300 mμ to 100 at 285 mμ and no peaks were observed. Since XeF_4 sublimes easily into crystals, diffraction studies were at once started, and five months after it was first prepared no less than three independent reports had appeared[9–11]. We see, therefore, that in the determination of molecular structure those methods are of greatest use which not only yield a high number of observables but also demand a relatively large number of parameters for their rationalization.

The infrared spectrum of XeF_4, measured as a gas at low pressure[8], contained three bands, of which one had a Q branch, while the Raman spectrum measured in the solid state had two major and one very weak band, though we should bear in mind that the low-frequency spectra of molecular solids may contain bands due to lattice as well as molecular vibrations. In the infrared spectrum, only one of the bands was found in the band-stretching region from 500–700 cm^{-1}, which suggested that the molecule was either tetrahedral or square planar. However, the former has two and the latter only one infrared active bending mode, so that the correct symmetry must be square planar. Only the out-of-plane bending mode A_{2u} of a square planar molecule is expected to have a Q branch, so that the other two are at once assigned. The only remaining doubtful feature was that the doubly degenerate bending mode was expected to have a triplet structure, whereas only a doublet was resolved, a discrepancy which the authors put down to Coriolis coupling between the degenerate vibration and rotations. It even proved possible to investigate whether the fluorine atoms were puckered out of a plane through the xenon by observing whether the B_{2g} Raman band became infrared allowed. Thus it was concluded that the upper limit for the deviation of the Xe—F bonds from this plane was about 0.5° or 0.02 Å.

In the case of xenon difluoride, vibrational spectroscopy provided a picture even more complete than that of XeF_4, since enough frequencies are observed to define completely the force constants[12]. The asymmetric stretch ν_3 at 559 cm^{-1} has no *Q* branch and the molecule is therefore straight. The symmetric stretch ν_1, of which there is no sign in the infrared spectrum, appears strongly in the Raman spectrum of the solid at 496 cm^{-1}, and the infrared band at 1070 cm^{-1} is therefore $\nu_1 + \nu_3$; ν_2 has a *Q* branch as expected and the bond length could be calculated from the *P–R* separation of ν_3. Comparisons of the force constants with those of ICl_2^- [13] are of interest in that the ionic characters of the bonds have been estimated in the latter by nuclear quadrupole resonance and in XeF_2 from the ^{19}F nuclear magnetic resonance. The principal stretching constant of the interhalogen is one-third that of the xenon compound, while the constant for interaction between the stretching modes is more than twice as large. A similar comparison can be made between XeF_4 and ICl_4^-.

The various diffraction studies of xenon tetrafluoride provide material for some interesting comparisons. Very high absorption of Cu *Kα* radiation by the inert gas combined with the relatively weak scattering power of the halogen made x-ray work difficult, but despite this handicap three x-ray investigations[9–11] were quickly reported, followed by neutron-[14] and electron-diffraction work[15]. Of the two detailed x-ray studies, one used visual intensity measurement and the other a counter. 268 reflections were measured visually, of which 54 were given zero weight because of experimental factors, while the remainder were subjected to a least-squares analysis, leading to a final reliability index of 0.097 when anisotropic thermal parameters had been included. However, we should note that with isotropic thermal parameters this value was insignificantly different (0.100) so that the physical meaningfulness of anisotropic thermal motion in this crystal is open to question. 286 reflections were measured with the counter, of which 96 with values other than zero were said to result from the fluorines alone. The *R* value was lower than for the visually estimated intensities (0.059) but the difference between the Xe—F distances in these independent determinations, 1.961 ± 0.026[10] and 1.921 ± 0.021[5], were not thought to be significant. The space group requires that the molecule is planar but not necessarily square; both studies, however, agree that it clearly is.

One extreme disadvantage of neutron diffraction exemplified by the work on XeF_4[14] is the long exposure times necessitated by low neutron fluxes. With a crystal of about 25 mg, data collection took a month, and during the exposure the crystal grew at the expense of others sealed in the tube, so that varying absorption corrections were required according to the day on which each set of data was obtained. However, more reflections were observed than in either x-ray study (599) and 26 structural parameters were introduced into the least-squares analysis. These were the neutron-scattering factors and coordinates of Xe and the two inequivalent fluorines, and six thermal parameters. The reliability factor fell to 0.067 for a Xe—F distance of 1.953 ± 0.002 Å, so that all three studies agree to within less than twice the standard deviation of each.

By comparison, electron diffraction is a much less precise structural tool, but apart from infrared spectroscopy, is the only one which provides evidence about the molecular structure in the gas phase. Theoretical scattering curves computed for various assumed geometries[15] did not differ sufficiently to choose between a square planar or staggered square arrangement for XeF_4. As in x-ray work, scattering by the xenon dominates the results.

Once information on the nuclear configurations of the xenon fluorides had been secured by vibrational spectroscopy and diffraction, other techniques were quickly brought in to probe the electronic configurations in space and energy. The compounds are almost ideally suited to study by nuclear magnetic resonance[16], since ^{19}F ($S = \frac{1}{2}$) is 100% abundant, ^{129}Xe ($S = \frac{1}{2}$) is 25%, and ^{131}Xe ($S = \frac{3}{2}$) is 25% abundant. Magnetic coupling alone can occur between ^{129}Xe and ^{19}F but between ^{131}Xe and ^{19}F coupling is also possible between the quadrupole moment and any electric field gradient that may exist at the xenon nucleus. Fully resolved ^{19}F nuclear magnetic resonance spectra of XeF_4 contain two lines due to coupling with ^{129}Xe. If the molecule were tetrahedral, a four-line spectrum would be expected to arise from coupling between ^{131}Xe and ^{19}F but quadrupole relaxation collapses these to a single line. The simplicity of the spectrum shows at once that all the fluorines are equivalent, but of course the time scale of the n.m.r. experiment is important here, as we have already remarked. From the linewidths it is also possible to say that the average lifetime of a fluorine atom attached to xenon in a hydrofluoric acid solution is greater than one second. In principle, three parameters are available from the experiment, the chemical shifts of F and Xe and the Xe–F coupling constant, but the xenon resonance turns out to be difficult to measure. Interpretation of the data in terms of electronic parameters is difficult and only qualitative comparisons have been attempted. The ^{19}F chemical shift should be related to the ionicity of the Xe—F bond and it is possible to construct a scale of shifts ranging from the fluorine molecule (zero ionicity) to the fluoride ion (ionicity of one). The fluorine atoms in the xenon fluorides then appear to have comparable ionicities to that in HF but it is important to note that changes in hybridization could have the same effect on the measured shift as changes in ionicity. We have once again the problem of insufficient observables to match the number of parameters in the electronic model. Neither is the Xe–F coupling constant interpretable at present, because wave functions of sufficient accuracy are not available for xenon.

A similar problem arises in the interpretation of the Mössbauer spectra of the xenon fluorides[17]. In order to analyse quadrupole splitting, the quadrupole moment of the excited state of ^{129}Xe is required. Usually, such quantities are calculated from the quadrupole coupling e^2qQ (see Chapter 8 for a case where q can be calculated) but the problem here is the opposite one. If Q is taken from optical measurements on ^{131}Xe, which has the same low-lying excited states as ^{129}Xe, the experimental value of the quadrupole coupling yields an unreasonably large value for q. It was assumed by Perlow and colleagues that the field gradient in XeF_4 could be attributed solely to the doubly occupied p_z, the p_x, p_y orbitals contributing nothing, and thus they were able to arrive at a value of $\langle r^{-3} \rangle$ empirically. The next step was to calculate $\langle r^{-3} \rangle$ for a singly ionized $5p$ orbital of xenon using approximate wave functions but the calculated quadrupole splitting which resulted was too small by a factor of 2.7. It was said, however, that if a correction was introduced to take account of the decreased shielding of the xenon nucleus resulting from its high ionic charge, the agreement became very much better. Basically, the problem is that the Mössbauer experiment yields only two major observables, the isomer shift and quadrupole splitting (Chapter 8), and it is extremely difficult to arrive at an electronic model which at the same time is realistic in its assumptions and yet contains sufficiently few freely chosen parameters that unambiguous conclusions are possible.

Paramagnetic resonance has an advantage in permitting a distinction between the isotropic part of the nuclear hyperfine interaction tensor (Chapter 7) which contains information only about the s character of the unpaired electron and the anisotropic parts which contain information about p and d contributions. The method cannot be applied to the xenon fluorides as such, since they are diamagnetic but, by γ-irradiating crystals of XeF_4, radicals are produced which remain trapped at fixed orientations. These appear to be XeF molecules[18]. To obtain the orbital character of the unpaired electron, the experimental g values and hyperfine coupling tensors must be compared with those calculated using atomic wave functions. In this way it was found that the unpaired electron had 3% fluorine $2s$, 5% xenon $5s$, 47% fluorine $2p$, and 36% xenon $5p$ character.

The contribution of xenon 6*d* orbitals could not be estimated owing to the lack of suitable wave functions.

We see, therefore, that within months of their discovery, the shapes and sizes of this group of unusual molecules had been fully elucidated and their electronic properties comprehensively investigated. Material was thus available for intensive theoretical analysis. If to date this analysis has not been wholly successful, the reason is in the inadequacy of the methods now available for solving the Schrödinger equation of such complicated electronic systems, and not in the lack of experimental data against which to test the calculations.

In contrast, the investigation[19,20] of the chemistry of ferrocene during the last 15 years makes a fascinating story in which the early, if limited, success in interpreting the molecular and electronic structure has been followed by years of difficulty in attempts to understand the details. In 1952, the well-known staggered and eclipsed forms were both proposed[21] and the main structural problem since then has been to understand the structure in various phases and at different temperatures in terms of these two conformations. However, before we see how far these attempts have been successful, it is of interest to examine the evidence on which the now-accepted structures were proposed. The most suggestive pieces of evidence were, at least when viewed retrospectively, the single sharp band in the infrared spectrum at 3.25 μ and the absence of a dipole moment. The former was taken as an indication of a single type of hydrogen, whilst the latter showed that the molecule had a centre of symmetry or an improper axis of rotation or belonged to one of the point groups $\mathbf{C}_{nh}$ or $\mathbf{D}_{nh}$. The diamagnetism was evoked as evidence for the proposed structure and an interesting analogy was drawn between ferrocene, the ferricinium cation, and the ferro- and ferricyanide anions. The electronic structure was, understandably, not understood, and the authors reveal the thought-forms of the time by stating, 'details of hybridization will *determine* the precise geometry' (our italics). However, the maximum use was made of the limited amount of information then available.

A number of features of the proposed structures were substantiated when other methods were applied or became available. For example, the nuclear magnetic resonance of ferrocene in solution showed[22] the equivalence of the hydrogens. Obviously, it was expected that the diffraction methods would confirm the structure and determine the details, mainly whether the rings were eclipsed ($\mathbf{D}_{5h}$) or staggered ($\mathbf{D}_{5d}$). X-ray diffraction measurements[23] were best interpreted by assuming that the staggered form predominated in the crystal at room temperature, but the electron-density map, which showed a very high electron density between the carbons, led the authors to suggest that there was extensive torsional vibration of the rings about the five-fold axis. This shows one of the difficulties of the x-ray method, in that the data are collected over a long period of time and so it is difficult to differentiate between 'static' and 'dynamic' disorder in the crystal. Electron-diffraction studies had previously shown[24] that the eclipsed form predominates in the vapour phase, though the energy barrier to rotation about the five-fold axis was estimated to be about 1–2 kcal/mole. If the energy barrier to rotation is so small, then rotation must be considered when discussing phenomena in the solid state, unless specific intermolecular effects intervene. In solution, each ring in any ferrocene molecule must be free to rotate with respect to the other ring since a study of the dipole moments of mono- and disubstituted ferrocenes gave[25] results which were only consistent with free rotation. Attention was drawn to the importance of intermolecular effects in the solid state by the study of the temperature dependence of thermodynamic properties[26] and of the proton magnetic resonance spectra[27,28] of crystalline ferrocene. Thus, there exists a striking transition point in the heat capacity curve at 163.9°K and over approximately the same temperature range an anomalous decrease in the proton magnetic resonance absorption linewidth.

It now seems likely that these effects, and the anomalous electron density, have the same cause: that at room temperature the staggered form does indeed predominate but there is rotational disorder, either of one staggered ferrocene molecule with respect to another in the same plane, or with a significant amount of the eclipsed form. The anomalous electron-density distribution and the experimental entropy of transition are thereby explained. Such a disordered state would include a proportion of intermolecular configurations where the hydrogen–hydrogen distance would diminish on cooling to a value which must result in considerable repulsion. Other arrangements possible in more ordered structures would allow the molecules to 'fit' giving a system of 'interlocking gears'. Presumably, the λ point represents the transition to the more ordered state which, in view of the similarity between the x-ray diffraction data obtained at 77° K and at room temperature, must have all the molecules in the staggered form. This is consistent with the fact that below the λ point the reorientation of the ferrocene molecules can be described by a single correlation time in the nuclear magnetic resonance experiment. There is a sharp discontinuity in the temperature dependence of the correlation time at the λ point, and at higher temperatures the data are more consistent with the existence of at least two different values of the spin lattice relaxation time. It is possible that the different relaxation times are related to the four possible arrangements of the staggered and eclipsed forms, and the relative amounts present at any temperature above the λ point. So we see that the application of more than one technique leads to a more satisfactory understanding of the molecular geometry.

Physical methods have been useful in the characterization of substituted ferrocenes. The large number of transitions amongst vibrational levels may make a complete analysis[29] of the infrared spectrum difficult but it is the ideal technique for characterization or finger-printing. Empirical rules have been formulated in an attempt to relate the infrared absorption spectra to particular substitution patterns. For example the '9–10 rule' states[30] that ferrocene derivatives which have one ring unsubstituted absorb near 9–10 μ while those having both rings substituted lack absorption. Altering the number of substituents in ferrocene changes[31] both the energy and number of transitions in the infrared absorption spectra. This is also true of the nuclear magnetic resonance spectra[32] where electron-withdrawing substituents lead to chemical shift differences of approximately 0.3 ppm between the 2- and 3-hydrogens. With alkyl- and acetylferrocenes, it is possible to distinguish between 1,2- and 1,3-substitution. However, not all physical methods are sensitive to substituents. N.M.R. and i.r. spectra are intimately connected with the cyclopentadiene rings, whereas the ultraviolet absorption spectrum and the Mössbauer spectrum depend strongly on the properties of the iron atom. Therefore, although there is[33] a slight bathochromic shift in the absorption bands at 440 and 325 mμ when electron-attracting substituents are introduced into the ring, it is not valuable for the characterization of derivatives, giving few transitions which, anyway, are insensitive to small perturbations in the molecule. Similarly, in the Mössbauer spectra, the isomer shift and quadrupole splitting are[34] only slightly dependent on the substituent.

The remarkable structure of ferrocene has made it a favourite subject of theoretical treatments, most of which, as we shall see, seek to account for the data derived from the application of physical methods. However, the early treatments were more concerned with an explanation of the 'stability' of the molecule and its general chemistry. In this connexion, there was considerable interest in the 'aromatic character' of ferrocene. The chemical evidence showed that ferrocene is indeed highly reactive in electrophilic substitution reactions, e.g. it is 10^6 times more reactive in the Friedel–Crafts acylation than benzene[35]. The physical evidence now accepted as evidence of aromaticity is the ability to sustain a 'ring current'. The proton chemical shift is[36] at higher field than that of benzene and this is due supposedly to charge delocalization on to the ring and to the proximity of the metal ion, though no exact description of these factors has been presented. The ^{13}C chemical shift is[37] different to that of benzene and is close to that calculated for cyclopentadienide anion, but the ^{13}C–^{1}H spin-coupling constant is similar to that in benzene. Again, force constants in the valence force field approximation were[29] shown to be quite comparable to those calculated for benzene. Nuclear magnetic resonance and vibrational spectroscopy thus yield results which, though invaluable for empirical comparisons of different molecules, are not yet amenable to a more fundamental theoretical treatment.

Knowledge of the physical properties of a metal complex should present not only a challenge but also a guide to theoretical treatments. Though the problem of an *a priori* calculation of the electronic energies in ferrocene is at present beyond our theoretical powers, a number of calculations involving many approximations have been made and the latest one[38] is reasonably consistent with a number of the facts. It is interesting that Cotton and Wilkinson in their review suggested[19] that extensive measurements of the electronic spectra and *a fortiori* electron spin resonance experiments would furnish the most useful data. Some progress has been made on the former but not on the latter. It is surprising that the e.s.r. of the ferricinium cation has not been measured at liquid-helium temperatures. The fact that no e.s.r. is observed at liquid-nitrogen temperatures has been invoked[39] as evidence of a doubly degenerate ground state, $^2E_{2g}$, giving rise to a fast relaxation time. This is consistent with the large orbital contribution to the magnetic moment, which for the iodide is[40] 2.34 ± 0.12 B.M. compared to the spin-only value of 1.73 B.M. Indeed, those measurements which give some indication of the nature of the highest filled levels are particularly valuable. The ability to calculate the ionization potentials of the highest filled orbitals is a most important criterion of success, particularly now that accurate data are available from photoelectron spectroscopy (see page 102)[41]. The most demanding criterion, however, is a successful interpretation of the electronic spectrum, since this involves many eigenstates of the molecule. The spectrum must first be assigned. This has been achieved[42] for the visible absorption spectrum but not for the ultraviolet nor for the x-ray absorption spectrum[43], and this has prevented their use in the interpretation of energy levels in the molecule. The assignment of the absorption spectrum depended on a gaussian analysis of one observed absorption band into four components, one of which had an intensity 0.002 times that of the component of maximum intensity. Such a procedure is not without risk. However, having obtained five observables, these could be fitted using a theory containing four parameters, Ds, Dt, F_2, and F_4, the first two being one-electron splitting parameters, the latter the effective Slater–Condon electron-repulsion parameters. Employing these parameters involved an assumption of the order of the one-electron energy levels. Most calculations had assigned the ground-state configuration as $(a_{1g})^2$ $(e_{2g})^4$ $^1A_{1g}$, consistent with the photoelectron spectrum, giving ferricinium as $(a_{1g})^2$ $(e_{2g})^3$ $^2E_{2g}$ in agreement with susceptibility and e.s.r. measurements. From the values of Ds and Dt, the order of the d levels was found to be $e_{1g}^* > a_{1g} > e_{2g}$ with $e_{1g}^* - a_{1g} = 25.5$ kK and $a_{1g} - e_{2g} = 4.59$ kK. These values agree with a calculation which used[38] the Wolfsberg–Helmholz approximation but not with two earlier calculations[44,45] which were formally more complex and differed from each other in the type of wave function used. The Wolfsberg–Helmholz calculation gave an energy for the highest filled molecular orbital in agreement with the observed ionization potential and, like the other two calculations, agreed with the observed sign of the electric field gradient as determined[46] by the magnetic perturbation of the quadrupole splitting in the Mössbauer spectrum. However, the same note of caution sounded by Cotton and Wilkinson must be mentioned: that a theory with serious defects could give a meritriciously good account of the spectral data. It is interesting to note those techniques which have proved, in this instance at least, of value in guiding theoretical treatments or provided data amenable to calculation: electronic absorption spectroscopy, Mössbauer spectroscopy, the ionization potential, and magnetic measurements. All are associated with the properties of the metal, presumably since the electrons in the orbital of highest energy are associated mainly with the metal. E.S.R. measurements would provide the same sort of information, and details of the energies of the ring electrons must await analysis of the ultraviolet spectrum. Information of great potential value is furnished by the photoelectron spectrum.

V. REFERENCES

1. R. S. Berry, *J. Chem. Phys.*, **32**, 933 (1960).
2. M. Karplus, D. H. Anderson, T. C. Farrar, and H. S. Gutowsky, *J. Chem. Phys.*, **27**, 597 (1957).
3. O. Chalvet and R. Daudel, *J. Phys. Chem.*, **56**, 365 (1952).
4. C. C. J. Roothaan, *Rev. Mod. Phys.*, **23**, 69 (1951).
5. J. A. Pople, *Trans. Faraday Soc.*, **49**, 1375 (1953).
6. R. Pariser and R. G. Parr, *J. Chem. Phys.*, **21**, 466, 767 (1953).
7. H. H. Claassen, H. Selig, and J. G. Malm, *J. Am. Chem. Soc.*, **84**, 3593 (1962).
8. C. L. Chernick, H. H. Claassen, P. R. Fields, H. H. Hyman, J. G. Malm, W. M. Manning, M. S. Matheson, L. A. Quaterman, F. Schreiner, H. H. Selig, I. Sheft, S. Siegel, E. N. Sloth, L. Stein, M. H. Studier, J. L. Weeds, and M. H. Zirin, *Science*, **138**, 136 (1962).
9. J. A. Ibers and W. C. Hamilton, *Science*, **139**, 106 (1963).
10. S. Siegel and E. Gebert, *J. Am. Chem. Soc.*, **85**, 240 (1963).
11. D. H. Templeton, A. Zalkin, J. D. Forrester, and S. M. Williamson, *J. Am. Chem. Soc.*, **85**, 242 (1963).
12. D. F. Smith, *J. Chem. Phys.*, **38**, 276 (1963).
13. W. B. Person, G. R. Anderson, J. N. Fordemwalt, H. Stammreich, and R. Forneris, *J. Chem. Phys.*, **35**, 908 (1961).
14. H. A. Levy and P. A. Agron, *J. Am. Chem. Soc.*, **85**, 241 (1963); J. H. Burns, P. A. Agron, and H. A. Levy, *Science*, **139**, 1209 (1963).
15. R. K. Bohn, K. Katada, J. V. Martinez, and S. H. Bauer, *Noble Gas Compounds*, University of Chicago Press, Chicago, 1963, p. 238.

16. T. H. Brown, E. B. Whipple, and P. H. Verdier, *J. Chem. Phys.*, **38**, 3209 (1963); *Science*, **140**, 178 (1963); J. C. Hindeman and A. Svirmickas, *Noble Gas Compounds*, University of Chicago Press, Chicago, 1963, p. 251.
17. C. L. Chernick, C. E. Johnson, J. G. Malm, G. J. Perlow, and M. R. Perlow, *Phys. Letters*, **5**, 103 (1963); G. J. Perlow and M. R. Perlow, *Rev. Mod. Phys.*, **36**, 353 (1964).
18. W. E. Falconer and J. R. Morton, *Proc. Chem. Soc.*, **1963**, 95.
19. G. Wilkinson and F. A. Cotton, *Progr. Inorg. Chem.*, **1959**, 1.
20. M. Rosenblum, *Chemistry of the Iron Group Metallocenes*, Interscience, New York, 1965.
21. G. Wilkinson, M. Rosenblum, M. C. Whiting, and R. B. Woodward, *J. Am. Chem. Soc.*, **74**, 2125 (1952).
22. T. S. Piper and G. Wilkinson, *J. Inorg. Nucl. Chem.*, **3**, 104 (1956).
23. J. D. Dunitz, L. E. Orgel, and A. Rich, *Acta Cryst.*, **9**, 373 (1956).
24. E. A. Siebold and L. E. Sutton, *J. Chem. Phys.*, **23**, 1967 (1955).
25. H. H. Richmond and H. Freiser, *J. Am. Chem. Soc.*, **77**, 2022 (1954).
26. J. W. Edwards, G. L. Kington and R. Mason, *Trans. Faraday Soc.*, **56**, 660 (1960).
27. C. H. Holm and J. A. Ibers, *J. Chem. Phys.*, **30**, 885 (1959).
28. L. N. Mulay and A. Attalla, *J. Am. Chem. Soc.*, **85**, 702 (1963).
29. E. R. Lippincott and R. D. Nelson, *Spectrochim. Acta*, **10**, 307 (1958).
30. M. Rosenblum, *Chem. Ind. (London)*, **1958**, 953.
31. M. Rosenblum and W. G. Howells, *J. Am. Chem. Soc.*, **84**, 1167 (1962).
32. Y. Nagai, J. Hooz, and R. A. Benkeser, *Bull. Soc. Chem. Japan*, **37**, 53 (1964).
33. D. R. Scott and R. S. Becker, *J. Chem. Phys.*, **35**, 516, 2246 (1962).
34. G. K. Wertheim and R. H. Herber, *J. Chem. Phys.*, **38**, 2106 (1963).
35. M. Rosenblum, J. O. Santer, and W. G. Howells, *J. Am. Chem. Soc.*, **85**, 1450 (1963).
36. J. R. Leto, F. A. Cotton, and J. S. Waugh, *Nature*, **180**, 978 (1957); G. Fraenkel, R. Carter, A. McLachlan, and J. H. Richards, *J. Am. Chem. Soc.*, **82**, 5846 (1960).
37. P. C. Lauterbur, *J. Am. Chem. Soc.*, **83**, 1838 (1961).
38. R. D. Fischer, *Theoret. Chim. Acta*, **1**, 418 (1963).
39. D. A. Levy and L. E. Orgel, *Mol. Phys.*, **4**, 93 (1961); R. M. Golding and L. E. Orgel, *J. Chem. Soc.*, **1962**, 363.
40. F. Englemann, *Z. Naturforsch*, **8b**, 775 (1953); L. Friedman, A. P. Irsa, and G. Wilkinson, *J. Am. Chem. Soc.*, **77**, 3689 (1955).
41. J. C. D. Brand and W. Snedden, *Trans. Faraday Soc.*, **53**, 894 (1957).
42. D. R. Scott and R. S. Becker, *J. Organometal. Chem.*, **4**, 409 (1965).
43. See F. A. Cotton and H. P. Hemson, *J. Chem. Phys.*, **26**, 1758 (1956).
44. E. M. Shustorovich and M. E. Dyatkina, *Dokl. Akad. Nauk SSSR*, **128**, 1234 (1959); *Zh. Strubr. Khim.*, **1**, 109 (1960).
45. J. P. Dahl and C. J. Ballhausen, *Kgl. Danske Videnskab. Selskab, Mat.-Fys. Medd.*, **33**, 5 (1961).
46. R. L. Collins, *J. Chem. Phys.*, **42**, 1072 (1965).

Subject Index